SECOND EDITION

INTRODUCTION TO

Nonimaging Optics

SECOND EDITION

INTRODUCTION TO Nonimaging Optics

JULIO CHAVES

LIGHT PRESCRIPTIONS INNOVATORS
MADRID, SPAIN

CRC Press
Taylor & Francis Group
Boca Raton London New York

CRC Press is an imprint of the
Taylor & Francis Group, an **informa** business

CRC Press
Taylor & Francis Group
6000 Broken Sound Parkway NW, Suite 300
Boca Raton, FL 33487-2742

First issued in paperback 2017

ISBN-13: 978-1-4822-0673-9 (hbk)
ISBN-13: 978-1-138-74790-6 (pbk)

Library of Congress Cataloging-in-Publication Data

Chaves, Julio.
Introduction to nonimaging optics / Julio Chaves. -- Second edition.
pages cm
"A CRC title."
Includes bibliographical references and index.
ISBN 978-1-4822-0673-9 (alk. paper)
1. Light--Measurement. 2. Solar collectors. 3. Optical measurements. 4. Optics. 5. Reflectors, Lighting. I. Title. II. Title: Nonimaging optics.

TJ812.C427 2016
621.36--dc23 2015003356

Visit the Taylor & Francis Web site at
http://www.taylorandfrancis.com

and the CRC Press Web site at
http://www.crcpress.com

Contents

Preface

Over the past few years some significant nonimaging optical devices have been developed. The second edition of *Introduction to Nonimaging Optics* reflects these developments. Both solar energy concentration and LED illumination have benefited from these late developments. In particular, Köhler illumination combined with nonimaging optics methods led to new optics that address some of the challenges in these fields.

This new edition includes about 45% new material in four new chapters and additions to existing chapters. As with the first edition, it assumes no previous knowledge of nonimaging optics and now covers a wider range of subjects. Some chapters contain intuitive descriptions of the design methods or reasons to use nonimaging devices, while others delve deeper into the theoretical fundamentals or descriptions of more advanced optics.

New Chapter 1 contains an intuitive description of the advantages of nonimaging optics. It assumes no previous knowledge of the field and lays out some of its basic concepts and justifications for its use.

Chapter 9 was extended to include 3-D free-form optics. These more complex designs have more degrees of freedom, which allows them to be used in more challenging situations. For that reason, 3-D free-form optics are a new important trend in optical design.

New Chapter 10 describes some methods for generating output wavefronts used in the design of optics for prescribed output (intensity or irradiance), a very common problem in optical design. These wavefronts are then combined with nonimaging configurations to obtain optical devices. Some illustrative examples are given, but the same methods may be applied to other situations.

New Chapter 11 describes the limit case in which the étendue of the radiation crossing an optic goes to zero (becomes infinitesimal). Although more limited when compared to SMS optics (described in Chapter 9), these infinitesimal étendue optics are much easier to design. These types of optics have found application mainly in the high concentration of solar energy since the angular aperture of direct sunlight is very small.

New Chapter 12 describes Köhler optics combined with nonimaging methods. This is a powerful combination with many uses. A main application with increasing importance is LED color mixing. New lamps coming to market try to combine a tunable emission spectrum with a prescribed output pattern in compact and efficient devices. Köhler optics by themselves or used in combination with other nonimaging optics devices are a serious contender with good potential in this new trend.

When combined, the SMS prescribed intensity wavefronts and Köhler configurations constitute powerful design tools that can address challenging illumination design problems.

Part II of the book includes new material on integral invariants and their applications to nonimaging optics. Chapter 14 describes some of the theoretical aspects, which are common to other fields such as classical mechanics or analytical dynamics. Chapter 18 applies these concepts to optics, in particular to étendue 2-D, one of the invariants used in nonimaging optics. Section 18.4 also derives the expressions for étendue 2-D, but from a geometrical point of view, without having to rely on the Hamiltonian theory. The remainder of Chapter 18 provides examples of applications.

Julio Chaves

Preface to the First Edition

This book is an introduction to nonimaging optics or anidolic optics. The term *nonimaging* comes from the fact that these optics do not form an image of an object, they are nonimaging. The word *anidolic* comes from the Greek (an+eidolon) and has the same meaning. The words *anidolico/anidolica* are mostly used in the Latin languages, such as Spanish, Portuguese, or French, whereas nonimaging is more commonly used in English.

Many optical systems are designed to form the image of an object. In these systems, we have three main components: the object, the optic, and the image formed. The object is considered as being a set of light-emitting points. The optic collects that light (or part of it) and redirects it to an image. The goal of this image is that the rays of light coming out of one point on the object are again concentrated onto a point. Therefore, it is desirable that there be a one-to-one correspondence between the points on the object and those of the image. Only a few "academic" optical systems achieve this perfectly.

Instead, in nonimaging optical systems, in place of an object there is a light source, and the optic is differently designed; and in place of an image there is a receiver. The optic simply transfers the radiation from the source to the receiver, producing a prescribed radiation distribution thereupon.

Although there has been some pioneering work in nonimaging physical optics, nonimaging optics has been developed mostly under the aegis of geometrical optics. Its applications are also based on geometrical optics. Accordingly, this book deals only with nonimaging geometrical optics.

This branch of optics is relatively recent. Its development started in the mid-1960s at three different locations—by V. K. Baranov (Union of Soviet Socialist Republics), Martin Ploke (Germany), and Roland Winston (United States)—and led to the independent origin of the first anidolic concentrators. Among these three earliest works, the one most developed was the American one, resulting in what nonimaging optics is today.

The applications of this field are varied, ranging from astrophysics to particle physics, in solar energy, and in illumination systems. Solar energy was the first substantial big application of nonimaging optics, but recently, illumination has become the major application driving development. These two applications are of prime importance today, as lighting's cost of energy increases, and awareness of its environmental consequences mounts. Nonimaging optics is the ideal tool for designing optimized solar energy collectors and concentrators, which are becoming increasingly important, as we search for alternative and cleaner ways to produce the energy we need. It is also the best tool for designing optimized illumination optics, which engenders more efficient designs and, therefore, lower energy consumption. In addition, with the advent of solid-state lighting, nonimaging optics is clearly the best tool to design the optics to control the light that these devices produce. With the considerable growth that these markets are likely to have in the near future, nonimaging optics will, certainly, become a very important tool.

This book is an introduction to this young branch of optics. It is divided into two sections: The first one deals with nonimaging optics—its main concepts and methods. The second section is a summary of the general concepts of geometrical optics and some other topics. Although the first section is meant to be complete by itself, many general concepts have a different usage in nonimaging optics than in other branches of optics. That is why the second part may be very useful in explaining those concepts from the perspective of nonimaging optics. It is, therefore, a part of the book that the reader can refer to while reading the first section, should some concepts seem obscure or used differently from what he or she is used to.

Julio Chaves

Acknowledgments

This book is the result of many years of studying and designing nonimaging optical devices. Throughout the whole effort my wife Ana stood by my side. This work would have never been possible without her love and dedication through the years, even as the writing took so much of my time away from her.

My parents Julio and Rosa Maria and my brother Alexandre Chaves always tried to provide the best for me and have always been supportive.

I am fortunate to work with extraordinary and talented people who share a passion for nonimaging optics. They have been a source of inspiration and I have discussed with them, through the years, many of the topics covered in the book: My colleagues who have joined Light Prescriptions Innovators, LLC (LPI) over the years, and especially those who, like me, work in optical design: Waqidi Falicoff, Rubén Mohedano, Maikel Hernández, José Blen, and Aleksandra Cvetković. People at the Universidad Politécnica de Madrid (Technical University of Madrid), in particular those with whom I have worked more closely: Pablo Benítez and Juan Carlos Miñano and also people on their team, Pablo Zamora, Dejan Grabovičkić, and Marina Buljan. Manuel Collares-Pereira and Diogo Canavarro, first at the Instituto Superior Técnico (Higher Technical Institute) in Lisbon, and then at the University of Évora in Portugal.

Acknowledgments

Author

Julio Chaves was born in Monção, Portugal. He completed his undergraduate studies in physics engineering at the Instituto Superior Técnico (Higher Technical Institute), Universidade Técnica de Lisboa (Technical University of Lisbon), Lisbon, Portugal in 1995. He received his PhD in physics from the same Institute. Dr. Chaves did postgraduate work in Spain during 2002 at the Solar Energy Institute, Universidad Politécnica de Madrid (Technical University of Madrid). In 2003, he moved to California, and joined Light Prescriptions Innovators, LLC (or LPI). In 2006, he moved back to Madrid, Spain, and since then has been working with LPI.

Dr. Chaves developed the new concepts of stepped flow-line optics and ideal light confinement by caustics (caustics as flow lines). He is the coinventor of several patents, and the coauthor of many papers in the field of nonimaging optics. He participated in the early development of the simultaneous multiple surface design method in three-dimensional geometry.

List of Symbols

$\dot{x}$	Total derivative of $x(t)$ where t is time: $\dot{x}(t) = dx/dt$
x'	Total derivative of $x(y)$ where y is a geometrical quantity: $x'(y) = dx/dy$
∇	Gradient of a scalar function: $\nabla F(x_1, x_2, x_3) = (\partial F/\partial x_1, \partial F/\partial x_2, \partial F/\partial x_3)$
$\nabla \times$	Rotational operator (curl)
$[\mathbf{A}, \mathbf{B}]$	Distance between points **A** and **B**
$[[\mathbf{A}, \mathbf{B}]]$	Optical path length between points **A** and **B**
$\lvert a \rvert$	Absolute value of a
$\lVert \mathbf{v} \rVert$	Magnitude of vector **v**
$\langle Z \rangle$	Average value of Z
(x_1, x_2)	Two-dimensional vector or point with coordinates x_1 and x_2
A	Area A in a three-dimensional system
a	Length a in a two-dimensional system
$\mathbf{c}(\sigma)$	Curve with parameter σ
$F(x, y)$	Function F of parameters x and y
F_{A1-A2}	Shape factor from area A_1 to area A_2
H	Hamiltonian (when light paths are parameterized by coordinate x_3)
i_1, i_2, i_3	Generalized coordinates
$\mathbf{J}$	Vector flux
L	Radiance Lagrangian (in the context of Lagrangian optics)
L_V	Luminance
L^*	Basic radiance: $L^* = L/n^2$
L_V^*	Basic luminance: $L_V^* = L_V/n^2$
$\mathbf{n}$	Unit vector normal to a surface or curve. It has components $\mathbf{n} = (m_1, m_2, m_3)$.
n	Refractive index
$\mathbf{p}$	Optical momentum: $\mathbf{p} = n\mathbf{t}$ where n is the refractive index and $\mathbf{t}$ is a unit vector tangent to the light ray. It has components $\mathbf{p} = (p_1, p_2, p_3)$.
$\mathbf{s}(\sigma)$	Path of a light ray with parameter σ
S	Optical path length
u_1, u_2, u_3	Generalized momenta
U	Étendue
Φ	Flux (energy per unit time)
Ω	Solid angle
P	Hamiltonian (when light paths are parameterized by a generic parameter σ)

$\mathbf{P}$	Point $\mathbf{P}$. It has coordinates $\mathbf{P} = (P_1, P_2, P_3)$ along the axes x_1, x_2, x_3 axes
$\mathbf{P}_1$	Point $\mathbf{P}_1$. It has coordinates $\mathbf{P}_1 = (P_{11}, P_{12}, P_{13})$ along the axes x_1, x_2, x_3
$\mathbf{v}$	Vector $\mathbf{v} = (v_1, v_2, v_3)$
x_1, x_2, x_3	Spacial coordinates
//	Parallel to (in some figures)

List of Abbreviations

2-D	Two-dimensional geometry
3-D	Three-dimensional geometry
CAP	Concentration Acceptance Product
CPC	Compound Parabolic Concentrator
CEC	Compound Elliptical Concentrator
DTIRC	Dielectric Total Internal Reflection Concentrator
I	Total Internal Reflection in the SMS design method
R	Refraction in the SMS design method
RR	SMS lens made of two refractive surfaces
RX	SMS optic made of a refractive and a reflective surface
RXI	SMS optic in which light undergoes a refraction, then a reflection, and then a Total Internal Reflection (TIR)
SMS	Simultaneous Multiple Surfaces (design method)
TERC	Tailored Edge Ray Concentrator
TIR	Total Internal Reflection
X	Reflection in the SMS design method
XR	SMS optic made of a reflective and a refractive surface
XX	SMS optic made of two reflective surfaces

Angle Transformer	Device that accepts radiation with a given angle θ_1 and puts out radiation with another angle θ_2
Angle Rotator	Device that rotates light, maintains the area and angle of the light
Trumpet	Concentrator made of two hyperbolic mirrors with the same foci. All the radiation headed toward the line between the foci and intersected by the mirrors is concentrated onto the line between the vertex of the hyperbolas

Functions defined in Chapter 21:
nrm(...), ang(...), angp(...), angpn(...), angh(...), $R(\alpha)$, islp(...), isl(...), par(...), eli(...), hyp(...), winv(...), uinv(...), wmp(...), ump(...), wme(...), ume(...), cop(...), cco(...), dco(...), coptpt(...), ccoptpt(...), dcoptpt(...), coptsl(...), rfx(...), rfr(...), rfrnrm(...), rfxnrm(...).

Section I

Nonimaging Optics

1

Why Use Nonimaging Optics

1.1 Area and Angle

Typically, optical systems have an emitter (light source) E which emits light, an optic O which deflects it, and a receiver (target) R which is illuminated by it, as shown in Figure 1.1.

As light travels in space, or through an optic, it spreads over some area, and also over some angle. Area and angle are related and that is the basis for illumination optics. Figure 1.2 shows some flashlights emitting light downwards. In Figure 1.2a, the flashlights emit light toward a large area a_1. The angular aperture θ_1 of the light crossing a_1 is small. In Figure 1.2b, the same flashlights now emit light toward a small area a_2. The angular aperture θ_2 of the light crossing a_2 is now large. The reason for this is that we cannot place all flashlights on top of one another; they must be placed side by side, and this increases the angle of the light going through a_2.

If light crosses a large area contained within a small angular aperture (as in Figure 1.2a), and we now try to "squeeze" that light through a small area (as in Figure 1.2b), its angular aperture increases. So, for the same amount of light crossing an aperture, if the area is large, the angle is small; and if the area is small, the angle is large. This characteristic of light is called conservation of étendue.

As referred above, in an illumination system, we have an emitter (source), an optic, and a receiver (target). The emitter and receiver may be of different sizes. If the emitter E is small and the receiver R is large and distant, the optic is called a collimator, as in Figure 1.3a. If the emitter E is large and distant and the receiver R is small, the optic is called a concentrator, as in Figure 1.3c. When emitter E and receiver R are of similar size, the optic is a condenser, as in Figure 1.3b.

An example of a collimator is a flashlight emitting light rays r coming from a small source inside it and illuminating, for example, a large wall R, as in Figure 1.4a. An example of a concentrator is a magnifying glass pointed at the sun (the emitter), collecting sun rays r and concentrating sunlight onto a small receiver R, as in Figure 1.4b.

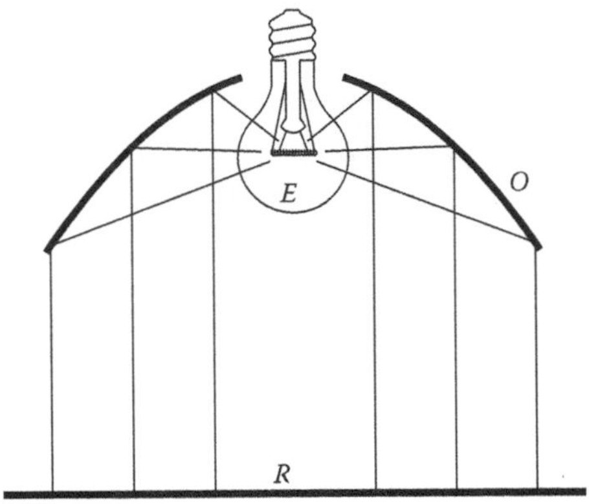

FIGURE 1.1
Optical system composed of an emitter E, an optic O, and a receiver R.

Figure 1.5a shows an optic (lens in this example) with emitter E and receiver R. Figure 1.5b shows a diagrammatic representation of the same. For the sake of simplicity, this diagrammatic representation will be used below.

Two different optics, O_1 and O_2, may capture the same amount of light from an emitter E, as shown diagrammatically in Figure 1.6. These two optics capture light from the same size emitter E within the same emission angle α.

Different size optics for the same purpose may have an important impact on the system in which these optics are included.

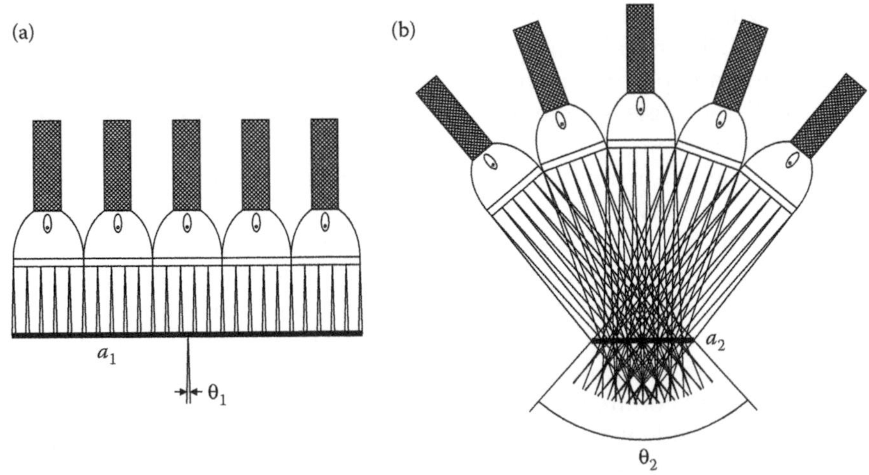

FIGURE 1.2
The same amount of light emitted by some flashlights either (a) crosses a large area a_1 with a small angle θ_1 or (b) crosses a small area a_2 with a large angle θ_2.

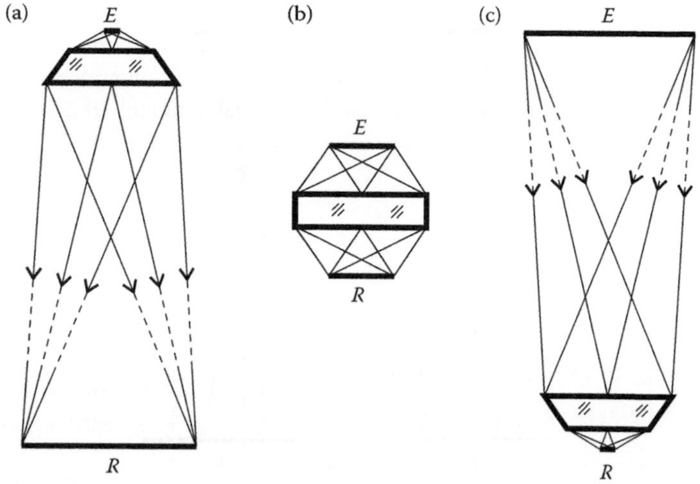

FIGURE 1.3
(a) If the emitter *E* is small and the receiver *R* is large, the optic is a collimator. (b) If emitter *E* and receiver *R* are of similar size, the optic is a condenser. (c) If the emitter *E* is large and the receiver *R* is small, the optic is a concentrator.

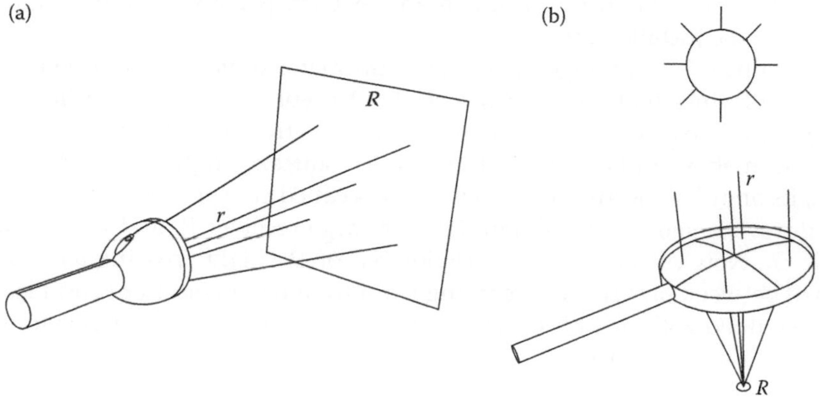

FIGURE 1.4
(a) Flashlight, example of a collimator. (b) Magnifying glass pointing at the sun, example of a concentrator.

1.2 Collimators: Illumination of a Large Receiver

Consider the situation shown in Figure 1.7a with a small emitter *E* emitting within angle α and a large receiver *R*. The light from *E* can be transferred to *R* by a collimator optic O_1, as shown in Figure 1.7b. However, it may also be that a smaller collimator optic O_2 can also transfer the same light from *E* to *R*.

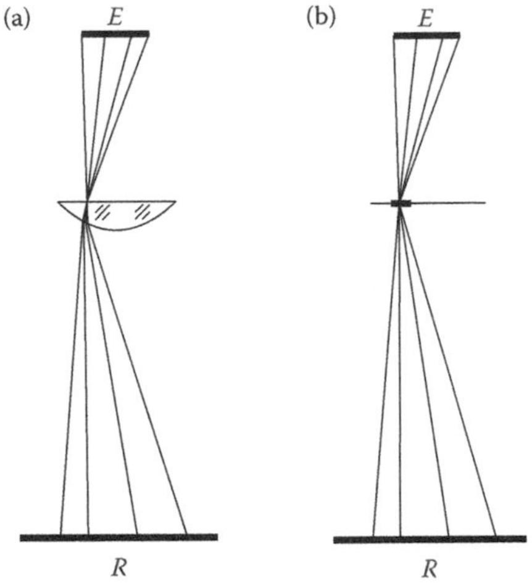

FIGURE 1.5
(a) Optic (lens) with emitter *E* and receiver *R*. (b) Diagrammatic representation of the same.

From a practical point of view, however, a smaller collimator optic may be preferable, since it may be cheaper to make, package, or ship, and needs less volume to be installed into.

Typically, it is advantageous to have collimator optics which are efficient and small, since that minimizes cost. By efficient, we mean an optic which transfers to the receiver all the light it intersects from the emitter.

As seen above in Figure 1.2, if we try to "squeeze" light through a small area, its angular aperture increases. This is called conservation of étendue.

Figure 1.8a shows a small emitter *E*, a large receiver *R*, and a collimator optic O_1. Now consider a small section da_1 of O_1. Light crossing da_1 is contained between rays r_1 and r_2, coming from the edges of emitter *E*, and leaves da_1 contained within angle θ_1. This optic is efficient, since it redirects to *R* all the light it intersects from *E*.

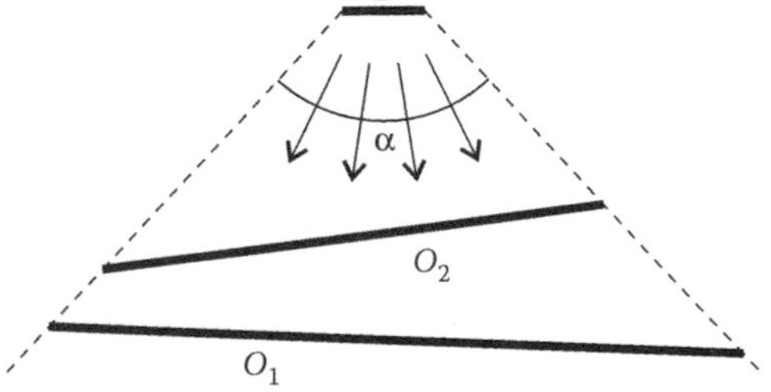

FIGURE 1.6
Optics O_1 and O_2 "see" the same size emitter *E* and same emission angle α and, therefore, capture the same amount of light from *E*.

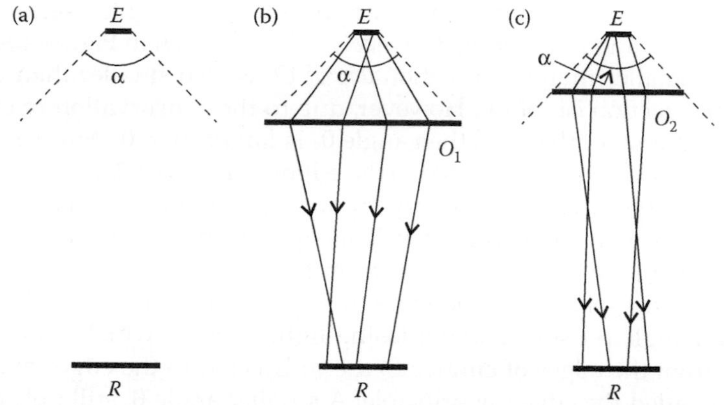

FIGURE 1.7
(a) Small emitter E emitting light within an angle α and a large receiver R. (b) Large collimator optic O_1 captures the light from E and transfers it to receiver R. (c) Small collimator optic O_2 captures the light from E and transfers it to R.

Now consider a smaller optic O_2 capturing the same amount of light from E and redirecting it to receiver R, as shown in Figure 1.8b. Since now O_2 is smaller than O_1, a section da_2 of O_2 is also smaller than the corresponding section da_1 of O_1. The amount of light from E intersected by da_2, is the same as the amount of light from E intersected by da_1. However, due to the conservation of étendue, if area da_2 is smaller than da_1, angle θ_2 is larger than θ_1. In this case, rays r_1 and r_2, coming from the edges of emitter E, are redirected toward the edges of receiver R. This optic is efficient, since it redirects to R all the light it intersects from E.

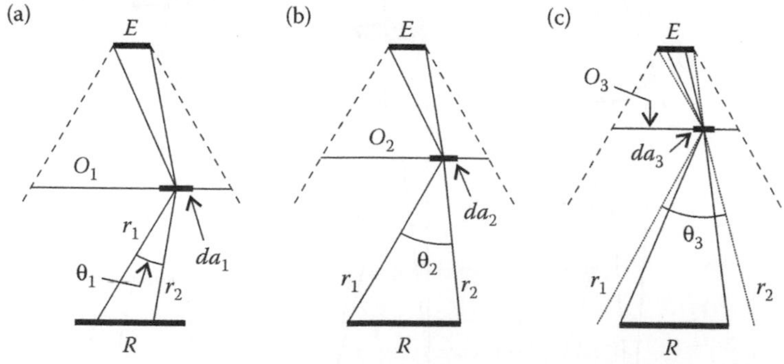

FIGURE 1.8
(a) Light from E crossing a large section da_1 of large optic O_1 falls inside receiver R. (b) Optimum situation in which the rays from the edges of E are redirected toward the edges of R. (c) Inefficient optic since some light from E crossing a small section da_3 of small optic O_3 fails receiver R.

Finally, consider yet another smaller optic, O_3, capturing the same amount of light from E, and redirecting it to receiver R, as shown in Figure 1.8c. Since now O_3 is smaller than O_2, a section da_3 of O_3 is also smaller than the corresponding section da_2 of O_2. However, due to the conservation of étendue, if area da_3 is smaller than da_2, then angle θ_3 is larger than θ_2. Since, in Figure 1.8b, rays r_1 and r_2 were redirected to the edges of receiver R, and θ_3 in Figure 1.8c is now larger than θ_2, these rays r_1 and r_2 must be redirected by O_3 in wider directions that fail receiver R. This optic is inefficient, since it fails to redirect to R all the light it intersects from E.

The optimum solution is, therefore, that given in Figure 1.8b. At da_2, the maximum angle possible without losing light, is θ_2, in which rays r_1 and r_2 (coming from the edges of emitter E) are redirected to the edges of receiver R. This is called the edge ray principle. A smaller angle θ_1 still puts all light on R, and results in an efficient optic, but da_1 (and optic O_1) could be smaller. On the other hand, a wider angle θ_3 results in a small da_3 (and small optic O_3), but some light fails R, resulting in a low efficiency optic.

It may then be concluded that, for collimators, efficiency and compactness combined result in the edge ray principle.

Figure 1.9 shows the same optical configuration as Figure 1.8, only now with real lenses, instead of a diagrammatic representation.

Section da_2 of the optic in Figure 1.8b redirects to the edges of the receiver R the rays coming from the edges of the emitter E. Since da_2 may be at any position across the optic, an optic with these characteristics is as shown in Figure 1.10a. This optic obeys the edge ray principle: rays coming from edge $\mathbf{E}_1$ of emitter E and crossing any position $\mathbf{P}$ on the optic are redirected to edge

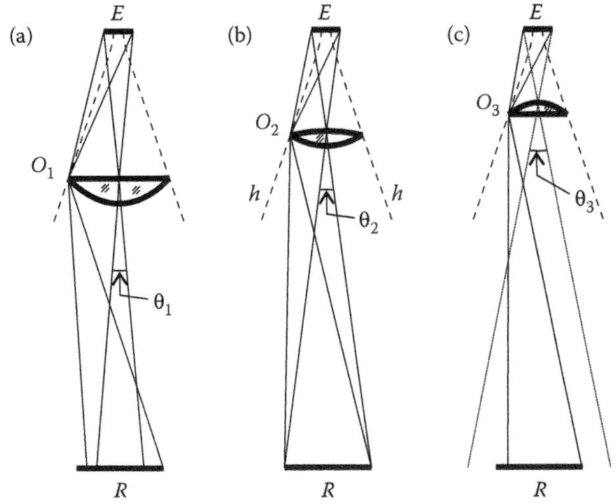

FIGURE 1.9
Same as Figure 1.8, only now with real lenses instead of a diagrammatic representation. (a) Efficient but large and optic. (b) Optimum optic. (c) Small but inefficient optic.

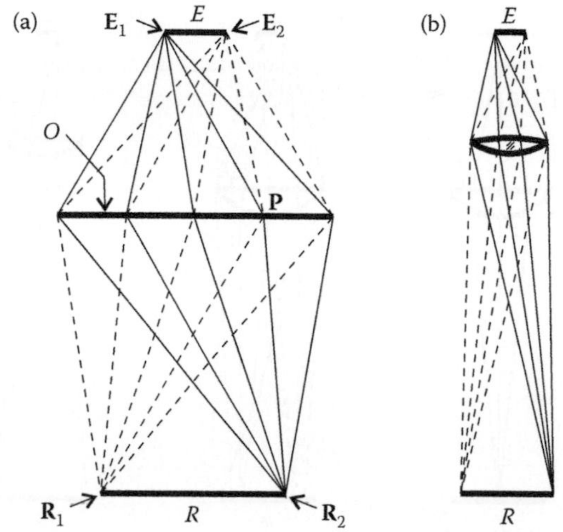

FIGURE 1.10
(a) Diagrammatic representation of a collimator optic *O* that redirects the rays coming from the edges of the emitter *E* to the edges of the receiver *R*. (b) Real lens instead of a diagrammatic representation.

$\mathbf{R}_2$ of the receiver *R*, and, accordingly, rays coming from edge $\mathbf{E}_2$ of emitter *E* and crossing any position **P** on the optic, are redirected to edge $\mathbf{R}_1$ of the receiver *R*. Figure 1.10b shows the same, but with a real lens instead of a diagrammatic representation.

One option to transfer light from an emitter to a receiver is to focus the center of the emitter *E* onto the center of the receiver *R*, like optic O_1 in Figure 1.11a. In general, however, this optic O_1 will not satisfy the edge ray principle, and the rays coming from the edges of the emitter *E* are not focused at the edges of the receiver *R*, as shown in Figure 1.11b, in which the rays coming from the edges of *E* spread over *R*. This optic is large and must be placed far away from the emitter. For that reason, emitter *E* and optic O_1 use a large volume V_1. Optic O_1 is designed for a point source (emitter) at the center of the extended emitter *E*. Also, O_1 is large when compared to emitter *E*. For those reasons, O_1 is called a point source optic.

However, it is possible to design a different optic O_2, which verifies the edge ray principle as shown in Figure 1.11c. Optic O_2 captures the same amount of light from *E* as O_1 and transfers it to receiver *R* (as does O_1). However, optic O_2 and emitter *E* now use a smaller volume V_2. Also, O_2 is designed for the edges of source (emitter) *E* and, therefore, its full extent. For that reason, O_2 is called an extended source optic.

The receiver *R* of a collimator O_1 may have a very large size, and be at a very large distance from O_1, as shown in Figure 1.12. In the limit case we get an infinite receiver *R* placed at an infinite distance. An extended source collimator optic (Figure 1.12a) must emit light confined to an angle α to fully

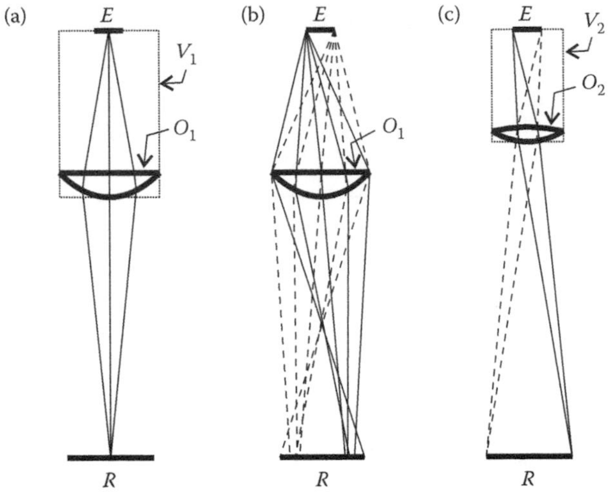

FIGURE 1.11
(a) Optic O_1 is designed to focus the center of E onto the center of R; it needs a large volume V_1. (b) Optic O_1 does not obey the edge ray principle and rays from the edges of E do not converge to the edges of R. (c) Optic O_2 designed with the edge ray principle needs a small volume V_2.

illuminate the receiver. Receiver R subtends an angle α when "seen" from optic O_1. Light emitted from O_1 at a wider angle than α will fail receiver R, resulting in low efficiency.

A point source collimator O_2 (Figure 1.12b) focuses the center of the source (emitter) E onto the center of the target (receiver) R. In the limit case in

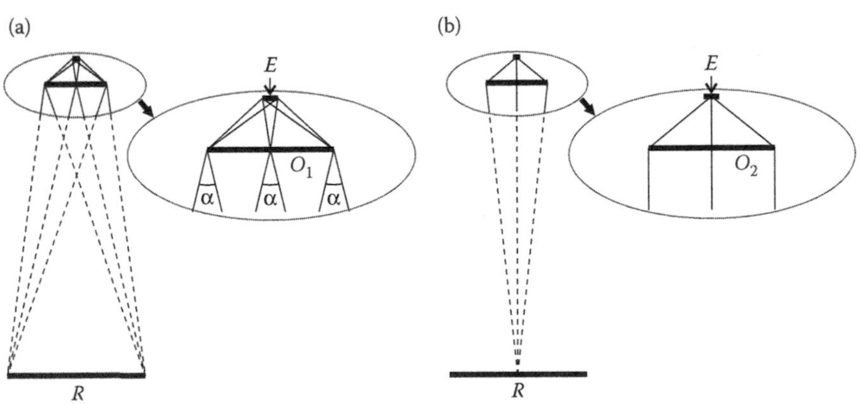

FIGURE 1.12
Receiver R is very large and very far away from optics O_1 and O_2. In the limited case, we have an infinite size receiver R placed at an infinite distance. (a) Extended source optic O_1 must emit light within an angle α to illuminate R. (b) Point source optic O_2 takes the rays from the center of E and emits them parallel to each other toward the center of R.

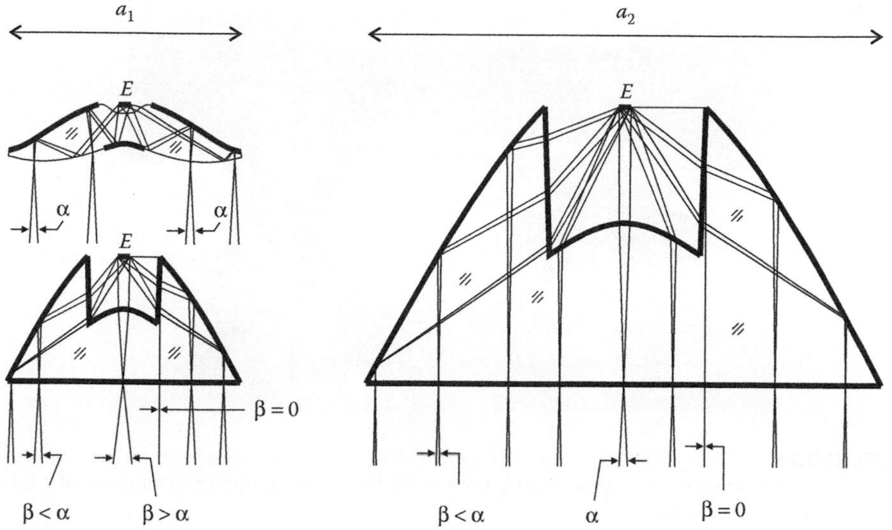

FIGURE 1.13
Left, top: optic designed with the edge ray principle, compact and efficient. Left, bottom: point source optic, compact but inefficient. Right: point source optic, efficient but very large.

which R is at an infinite distance from O_2, the rays coming from the center of E are emitted parallel to each other.

As an example of different kinds of collimator optics, Figure 1.13 shows three optics for the same emitter E and a large and distant (infinite) receiver R that subtends an angle α when "seen" from these optics.

The optic on the left, top, is an extended source optic (as in Figure 1.12a) designed with the edge ray principle. Its exit aperture is a_1, and its angular emission angle is α across its whole aperture. This optic, called an RXI[1] (see Chapter 9), is compact and efficient. The optic on the left, bottom, is a point source optic (as in Figure 1.12b) with the same emitter E and same exit aperture a_1. At some positions, it emits light within a cone $\beta < \alpha$, and therefore, all that light will fall on the receiver. However, at other positions, it emits light within a cone $\beta > \alpha$, and therefore, some of that light will fail the receiver and be lost. This optic is therefore compact but inefficient. The optic on the right is again a point source optic (as in Figure 1.12b) with the same emitter E but with a much larger exit aperture a_2. Its emission angle is $\beta \leq \alpha$ across the whole aperture a_2. All the light will then hit the receiver. This optic is therefore efficient but it is not compact.

Figure 1.14 shows an RXI (left) and a point source optic (right), both with the same exit aperture a_1. It illustrates the comparison in Figure 1.13 on the left.

Note that in Figure 1.14 the optics point up, while in Figure 1.13 the optics point down.

FIGURE 1.14
RXI (left) and point source optic (right), both with the same diameter. (Courtesy of Light Prescriptions Innovators.)

1.3 Concentrators: Illumination of a Small Receiver

Small collimator optics are desirable, but that is not the case with concentrator optics. Actually, the opposite is true. In a concentrator, the larger its aperture, the more light it captures from the source (emitter). Figure 1.15a shows diagrammatically two concentrator optics, O_1 and O_2, for large emitter E and small receiver R. Both optics "see" the same size receiver R at the same angle α and, therefore, have the same "ability" to put light onto R. Optic O_2, however, is larger and, therefore, captures more light from the emitter E.

Figure 1.15b shows the same, but for solar concentrators O_1 and O_2. By being larger, optic O_2 captures more sunlight and redirects it to receiver R.

Typically, it is advantageous to have concentrator optics that are efficient and large, since that maximizes the amount of light captured from the emitter. By "efficient," we mean an optic which transfers to the receiver all the light it intersects from the emitter.

Figure 1.16a shows a large emitter E, a small receiver R, and a concentrator optic O_1. Now consider a small section da_1 of O_1. Light crossing da_1 contained between rays r_1 and r_2 would fall on receiver R. Since O_1 is small, da_1 is also small, and therefore, angle θ_1 is large (conservation of étendue). Rays r_1 and r_2 would fall on R, but the emitter E does not extend that far and, therefore, no light is coming from those directions. However, this optic is efficient, since it redirects to R all the light it intersects from E.

Now consider a larger optic O_2, as shown in Figure 1.16b. Since now O_2 is larger than O_1, a section da_2 of O_2 is also larger than the corresponding section da_1 of O_1. However, due to the conservation of étendue, if area da_2 is larger than da_1, angle θ_2 is smaller than θ_1. In this case, rays r_1 and r_2 coming

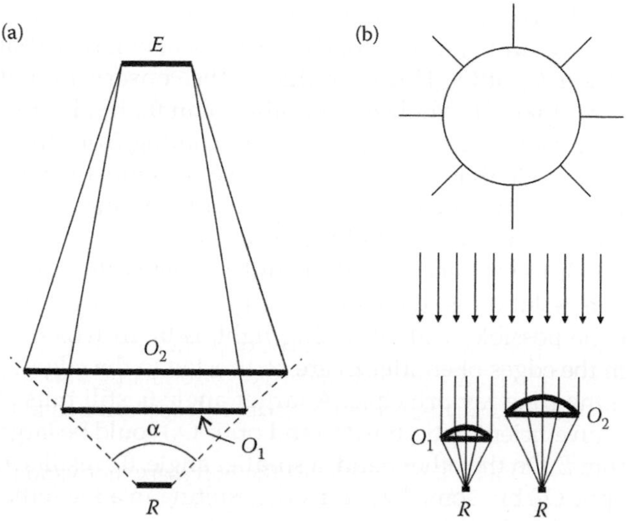

FIGURE 1.15
(a) A larger concentrator O_2 captures and redirects to receiver R more light from the emitter E than a smaller optic O_1. (b) Situation in which optics O_1 and O_2 are solar concentrators and the emitter is the sun.

from the edges of emitter E are redirected toward the edges of receiver R. This optic is efficient, since it redirects to R all the light it intersects from E.

Here, the "ability" to transfer light from da_1 to R is the same as that from da_2 to R. The étendue of the light transferable between da_1 and R is the same as the étendue of the light transferable between da_2 and R. However, in the case of Figure 1.16a, that "ability" is not fully used, since no light is coming from directions r_1 or r_2.

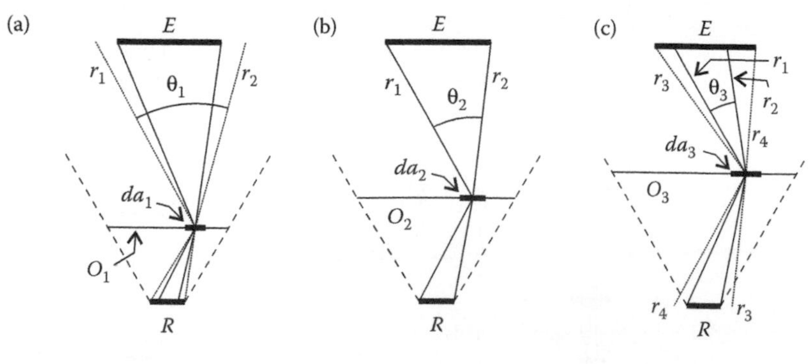

FIGURE 1.16
(a) A small section da_1 of small optic O_1 could accept light from directions r_1 and r_2 beyond emitter E and where there is no light. (b) Optimum situation in which the rays from the edges of E are redirected toward the edges of R. (c) Inefficient optic since some light from E crossing a large section da_3 of large optic O_3 fails receiver R.

Finally, consider yet another larger optic O_3 as shown in Figure 1.16c. Since now O_3 is larger than O_2, a section, da_3, of O_3 is also larger than the corresponding section da_2 of O_2. However, due to the conservation of étendue, if area da_3 is larger than da_2, angle θ_3 is smaller than θ_2. In Figure 1.16b, rays r_1 and r_2 reaching the edges of receiver *R* were coming from the edges of the emitter *E*. Since angle θ_3 is now smaller, these rays must come from points inside the emitter *E*. However, now rays r_3 and r_4 coming from the edges of emitter *E* and crossing da_3 will fail the receiver *R*. This optic is then inefficient, since it fails to redirect to *R* all the light it intersects from *E*.

The optimum solution is, therefore, that given in Figure 1.16b. At da_2, the minimum angle possible, without losing light, is θ_2, in which rays r_1 and r_2 (coming from the edges of emitter *E*) are redirected to the edges of receiver *R*. This is again the edge ray principle. A larger angle θ_1 still puts all light on *R* and results in an efficient optic, but da_1 (and optic O_1) could be larger, capturing more light from *E*. On the other hand, a smaller angle, θ_3, results in a larger da_3 (and larger optic O_3) but some light fails *R*, resulting in a low-efficiency optic.

It may then be concluded that, for concentrators, efficiency and maximum size, combined, result in the edge ray principle.

Figure 1.17 shows the same as Figure 1.16, only now with real lenses instead of a diagrammatic representation.

Section da_2 of the optic in Figure 1.16b redirects to the edges of the receiver *R* the rays coming from the edges of the emitter *E*. Since da_2 may be at any position across the optic, an optic with these characteristics is as shown in Figure 1.18a. This optic obeys the edge ray principle: rays coming from edge $\mathbf{E}_1$ of emitter *E* and crossing any position **P** on the optic are redirected to edge

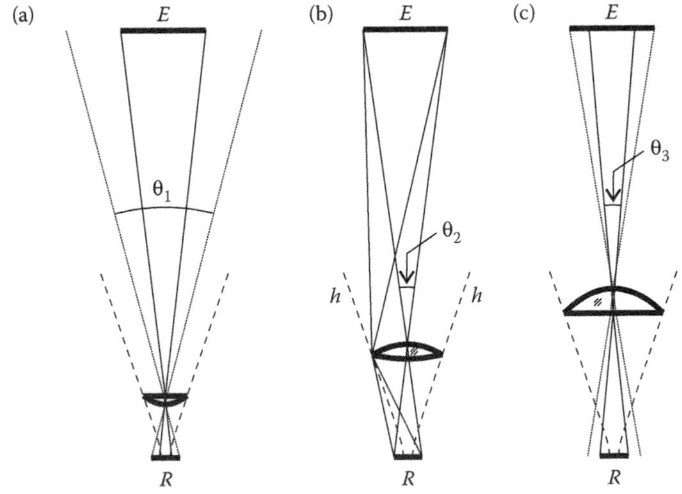

FIGURE 1.17
Same as Figure 1.16 only now with real lenses instead of a diagrammatic representation. (a) Efficient but small optic. (b) Optimum optic. (c) Large but inefficient optic.

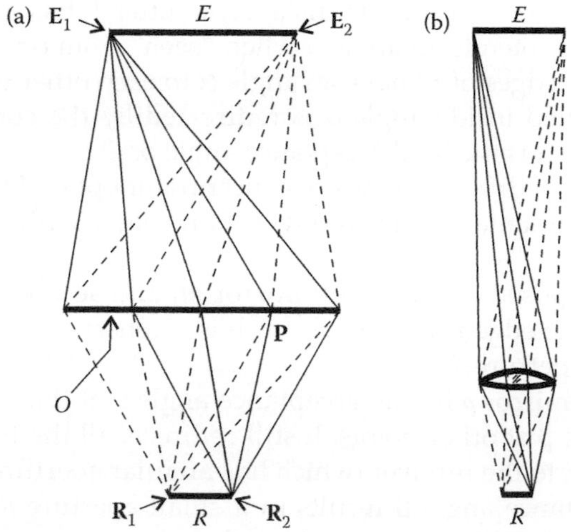

FIGURE 1.18
(a) Diagrammatic representation of a concentrator optic O that redirects the rays coming from the edges of the emitter E to the edges of the receiver R. (b) Real lens instead of a diagrammatic representation.

$\mathbf{R}_2$ of the receiver R, and accordingly rays coming from edge $\mathbf{E}_2$ of emitter E and crossing any position $\mathbf{P}$ on the optic are redirected to edge $\mathbf{R}_1$ of the receiver R. Figure 1.18b shows the same optical configuration, but with a real lens, instead of a diagrammatic representation.

The emitter E of a concentrator, O_1, may have a very large size, and be at a very large distance from O_1, as shown in Figure 1.19. In the limit case of an

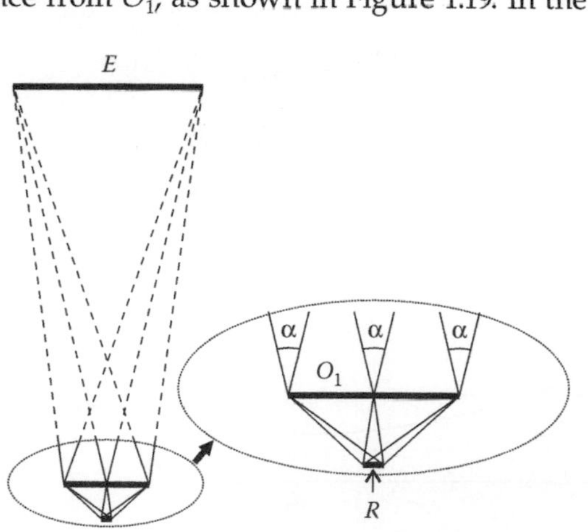

FIGURE 1.19
Emitter E is very large and very far away from concentrator optic O_1. In the limited case of an infinite size emitter at an infinite distance, the incoming rays make an angle α to each other across the whole aperture of O_1. Angle α is called total acceptance angle.

infinite size emitter at an infinite distance, emitter E has an angular aperture α, that is, it subtends an angle α when "seen" from optic O_1. Light rays coming from the edges of E make an angle α to each other when they reach O_1. Light contained inside angle α is redirected by the concentrator to its receiver R. Angle α is the total acceptance angle of O_1.

As an example of different kinds of concentrator optics, Figure 1.20 shows two optics for the same circular receiver R and same emitter angular aperture α.

The optic composed of a primary mirror m_1 and secondary mirror m_2 is designed with the edge ray principle.[2] It has a constant acceptance angle α across its large aperture a_1.

The parabolic mirror p has an acceptance angle α at the edge, but a wider acceptance angle β at other points. It still redirects all the light, it intersects from the emitter, to the receiver (which has angular aperture $\alpha < \beta$), but this increased acceptance angle β results in a small aperture a_2. Edge rays for circular receivers R will be discussed in other chapters.

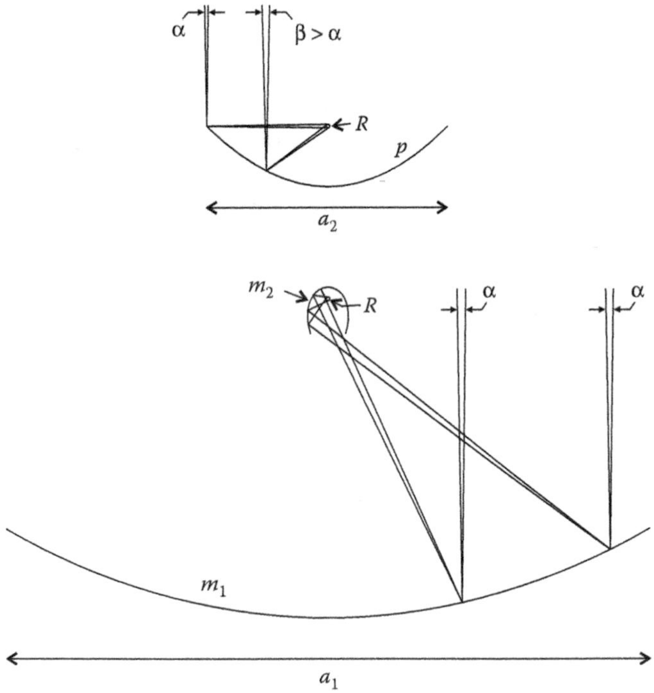

FIGURE 1.20
Large optic made of a primary mirror m_1 and secondary mirror m_2 has a large aperture a_1 and the same acceptance angle α across the whole aperture. Small parabolic mirror p has the same size receiver R but a smaller aperture a_2.

1.4 Collimators and Concentrators Summary

The results obtained above for collimators and concentrators may now be summarized in Figures 1.21 and 1.22. In the case of collimators, as shown in Figure 1.21, the optimum optical solution is the one in the middle, designed with the edge ray method. To the left of it, we have optics which are efficient but large, and to the right of it we have optics which are small but inefficient. All these optics capture the same amount of light from emitter *E*.

In the case of concentrators, as shown in Figure 1.22, the optimum optical solution is again the one in the middle, designed with the edge ray method. To the left of it, we have optics which are large but inefficient, and to the right of it we have optics which are efficient but small. All these optics have the same "ability" to put light onto receiver *R*.

An optic designed with the edge ray principle focuses the light from edge $\mathbf{E}_1$ of the source onto the edge $\mathbf{R}_2$ of the receiver, and the light from edge $\mathbf{E}_2$ of

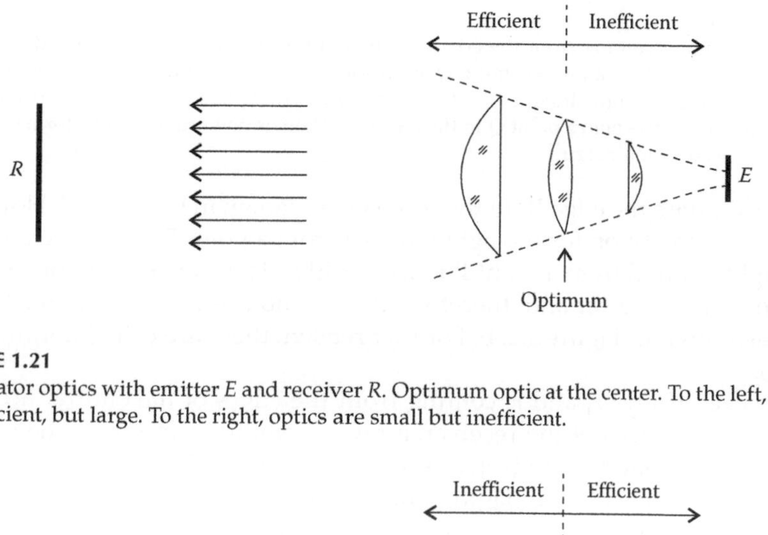

FIGURE 1.21
Collimator optics with emitter *E* and receiver *R*. Optimum optic at the center. To the left, optics are efficient, but large. To the right, optics are small but inefficient.

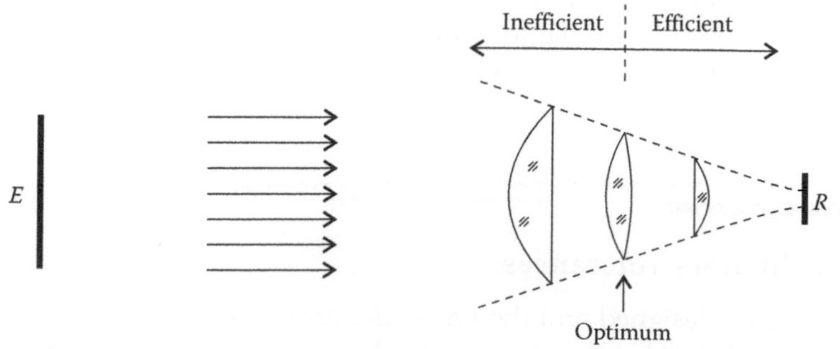

FIGURE 1.22
Concentrator optics with emitter *E* and receiver *R*. Optimum optic at the center. To the left, optics are large but inefficient. To the right, optics are efficient but small.

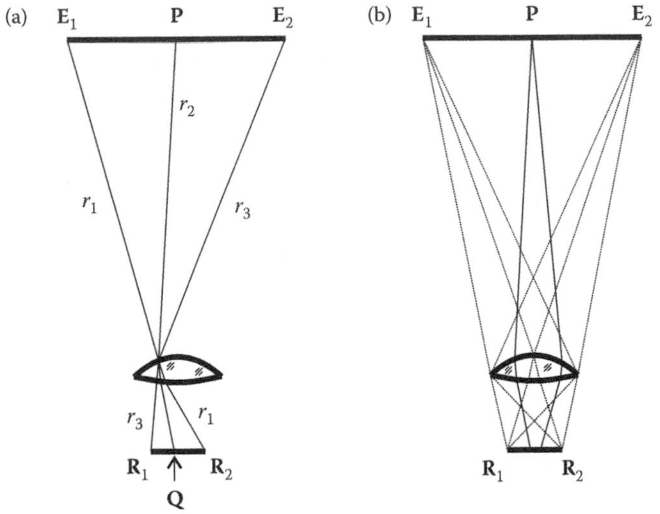

FIGURE 1.23
(a) Rays r_1 and r_3 coming from the edges of the emitter are redirected to the edges of the receiver and, therefore, a ray r_2 coming from inside the emitter is redirected to a point inside the receiver. (b) An optic designed with the edge ray principle does not necessarily focus a point **P** in the emitter onto a point **Q** in the receiver. There is no need for image formation and the optic is called nonimaging.

the source onto the edge $\mathbf{R}_1$ of the receiver, as shown in Figure 1.23. However, there is no condition for the light emitted from a point **P** inside the receiver. The light emitted from a point **P** in the emitter does not need to converge on a point at the receiver and, therefore, there is no condition for image formation, as shown in Figure 1.23b. For that reason, these are called nonimaging optics.

However, if rays r_1 and r_3 coming from the edges of the emitter are redirected to the edges of the receiver, a ray r_2 coming from inside the emitter will end up inside the receiver, as shown in Figure 1.23a. For that reason, the edge ray principle of nonimaging optics ensures that all light from the emitter that is intersected by the optic is transferred to the receiver.

1.5 Collimators Tolerances

Optics are first designed and then manufactured. Manufacturing, however, introduces different kinds of errors, which include imperfections in the optical surfaces, or imperfect assembly of components. Usage also introduces further errors, such as relative movements of the components or deterioration over time. There are, therefore, many sources of errors when making and using optics. It is, therefore, highly desirable to design the optics which are as tolerant as possible to all these errors.[3]

High tolerance to errors leads to relaxed manufacture, assembly and usage, which in turn leads to lower cost.

Figure 1.24a shows a perfect optic which focuses a point in the emitter *E* onto a point on the receiver *R*. If this optic is made, the manufactured optic may have errors—for example, oscillations on optical surface *s*, as shown in Figure 1.24b. Light is no longer focused at a point on *R*, but, instead, spreads out over some area. Errors are quite often random variations, and light is deviated randomly (diffused) relative to its ideal path.

Now consider a nonimaging collimator with emitter *E*, receiver *R*, and fixed aperture size **AB**, as shown in Figure 1.25a. This optic fully illuminates the receiver *R* since the rays coming from the edges of *E* are redirected to the edges of *R*. Suppose now that we wanted to decrease the spot size on *R*, that is, we wanted to put all the light in a smaller spot inside of *R*. One could then increase the area of a small section of the collimator from a_1 to a_3. From conservation of étendue, angle α_1 would decrease to α_3, and a_3 would illuminate a smaller spot inside *R*, as desired (see Figure 1.25b). However, since the total aperture **AB** is fixed, somewhere else in the optic, another area a_2 must decrease to an area a_4. Again from conservation of étendue, angle α_2 must increase to α_4, and this will make the light to spread over a larger area, increasing the spot to a larger size *s*. It can then be concluded that the spot created by a nonimaging optic

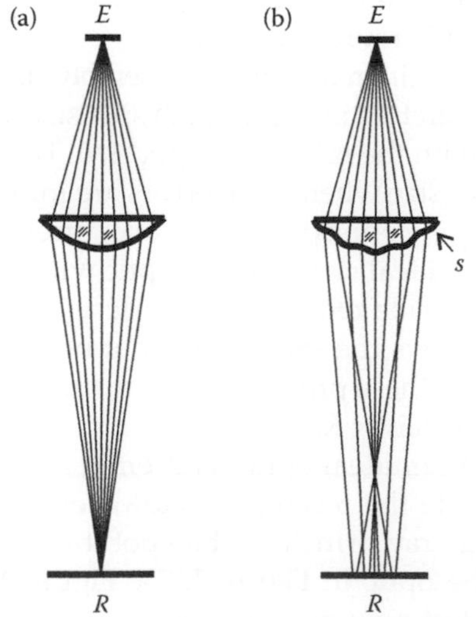

FIGURE 1.24
(a) Perfect optic focuses a point of emitter *E* onto a point of receiver *R*. (b) Imperfectly manufactured optic: oscillations in one of its surfaces *s* scatter the light as it passes the optic, which no longer focuses the light onto a point.

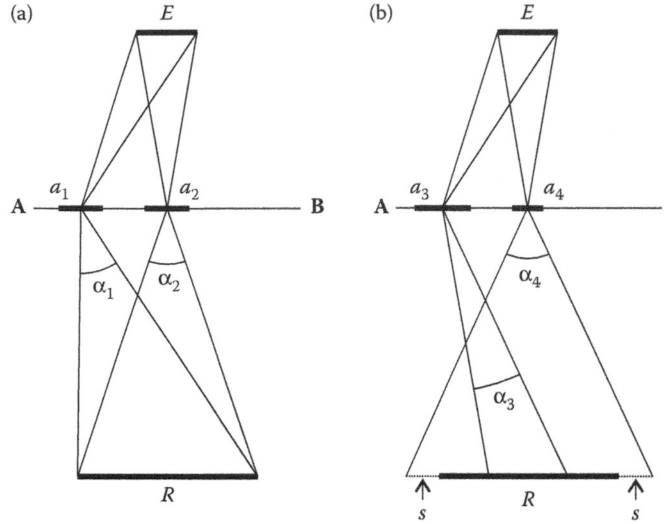

FIGURE 1.25
(a) Nonimaging optic collimator of fixed aperture size **AB**. (b) By increasing a_1 to a_3 then α_1 decreases to α_3 illuminating a smaller spot in *R*. However a_2 decreases to a_4 and α_2 increases to α_4 leading to larger spot size *s*.

cannot be decreased. The nonimaging optic in Figure 1.25a then minimizes the angular aperture α across its whole aperture.

Figure 1.26 shows the same as Figure 1.25, only now with real lenses, instead of a diagrammatic representation.

We have seen that errors (in manufacture, assembly, or usage) diffuse the light as it crosses the optic, therefore increasing the spot size. However, with nonimaging optics, we start with the minimum spot size. Therefore, we also end up with the minimum spot size when these errors are introduced. It may then be concluded that nonimaging optics maximize tolerances to errors in collimators.

Figure 1.27a shows an ideal nonimaging optic with emitter *E* and aperture **AB** that creates a spot inside receiver *R*. It is designed for a smaller receiver than *R*. Figure 1.27b shows the same optic, but now manufactured, assembled, and used. Errors in these processes now diffuse the light that now fully illuminates the whole receiver *R*.

Figure 1.27c shows a nonideal optic with emitter *E* and aperture **AB** that also creates a spot inside the receiver *R*. However, since this optic was not designed with the edge ray principle, the spot it creates is now larger than the spot created by the optic in Figure 1.27a. Figure 1.27d shows the same optic in Figure 1.27c, but now as a real optic. Errors now diffuse the light that spreads over an area larger than the receiver *R*, as shown by rays *r*. Since some light now fails to reach the receiver, efficiency decreases.

It may then be concluded that the nonimaging optic in Figure 1.27a is more tolerant to errors, and maintains efficiency. Even with errors, all light falls on

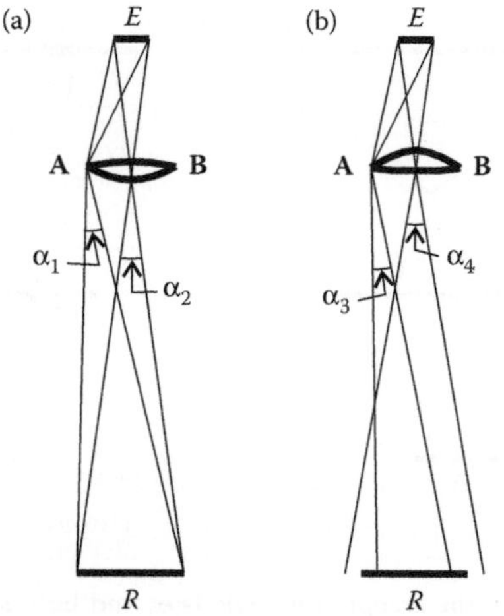

FIGURE 1.26
Same as Figure 1.25, only now with real lenses instead of a diagrammatic representation. (a) Nonimaging optic. (b) Optic with reduced angle α_3 at some position results in an increased angle α_4 somewhere else if the aperture **AB** is fixed.

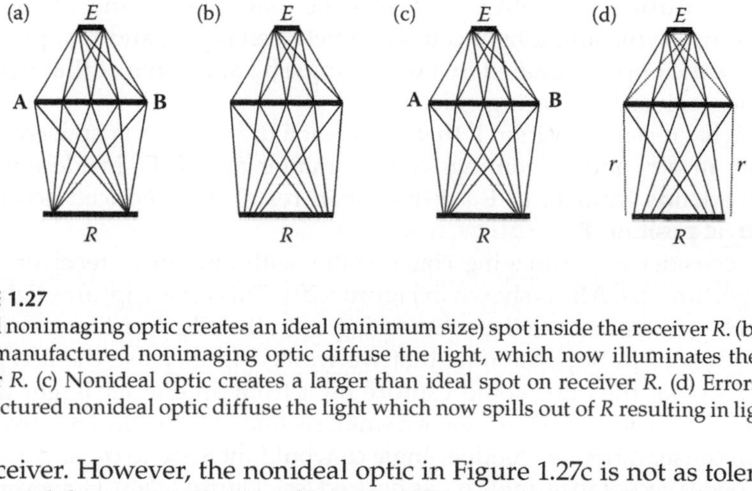

FIGURE 1.27
(a) Ideal nonimaging optic creates an ideal (minimum size) spot inside the receiver *R*. (b) Errors in the manufactured nonimaging optic diffuse the light, which now illuminates the whole receiver *R*. (c) Nonideal optic creates a larger than ideal spot on receiver *R*. (d) Errors in the manufactured nonideal optic diffuse the light which now spills out of *R* resulting in light loss.

the receiver. However, the nonideal optic in Figure 1.27c is not as tolerant to errors, and does not maintain efficiency. With errors, some light will miss the receiver, lowering efficiency.

1.6 Concentrators Tolerances

Concentrators, like collimators, will have imperfections when they are made, assembled, and used. Also in the case of concentrators, errors will diffuse

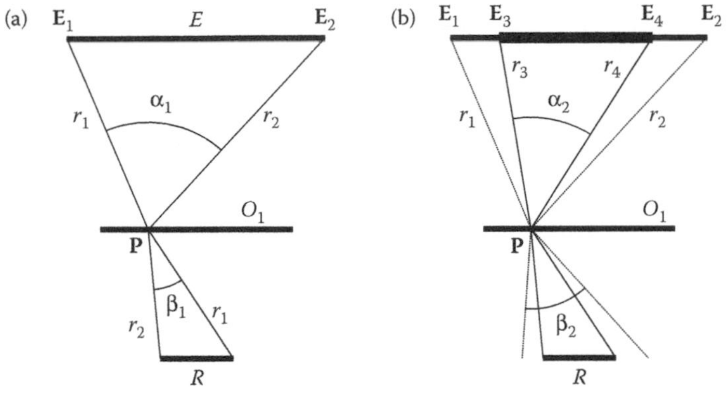

FIGURE 1.28
(a) At position **P** of optic O_1 the acceptance angle is α_1 and light leaves **P** within angle β_1. (b) Errors diffuse the light, which now leaves **P** within a wider angle β_2. Acceptance angle is reduced to α_2.

the light as it passes the optic. Figure 1.28a shows a nonimaging concentrator O_1. Incoming rays inside acceptance angle α_1 at position **P** leave inside angle β_1, fully illuminating the receiver R. When this optic is made and used, errors will diffuse the light as it passes the real optic, as shown in Figure 1.28b. Now, the incoming light contained between rays r_1 and r_2 at position **P** will leave the optic contained in a wider angle β_2, and some of that light will miss the receiver R, lowering efficiency.

The light crossing position **P** that does reach the receiver is now contained between rays r_3 and r_4, coming from a smaller area $\mathbf{E}_3\mathbf{E}_4$ inside emitter E (which extends from $\mathbf{E}_1$ to $\mathbf{E}_2$). Errors then result in a reduced acceptance angle α_2 at position **P**.

Now consider a nonimaging concentrator with emitter E, receiver R, and fixed aperture size **AB**, as shown in Figure 1.29a. This optic captures light from the whole emitter E, since the rays coming from the edges of E are redirected to the edges of R. Suppose now that we wanted to increase the acceptance angle α so that the optic could capture light from an emitter larger than E. One could then decrease the area of a small section of the collimator from a_1 to a_3. From conservation of étendue, angle α_1 would increase to α_3, and a_3 would have a wider acceptance angle α_3, as desired (see Figure 1.29b). However, since the total aperture **AB** is fixed, somewhere else in the optic, another area a_2 must increase to an area a_4. Again from conservation of étendue, angle α_2 must decrease to α_4, and this will reduce the acceptance angle at a_4. Now some rays, including r_1 and r_2, coming from the outer portions of emitter E will miss the receiver, reducing efficiency. It can then be concluded that the acceptance angle of a nonimaging optic cannot be increased. The nonimaging optic in Figure 1.29a then maximizes the acceptance angle α across its whole aperture.

Figure 1.30 shows the same optical configuration as Figure 1.29, only now with real lenses instead of a diagrammatic representation.

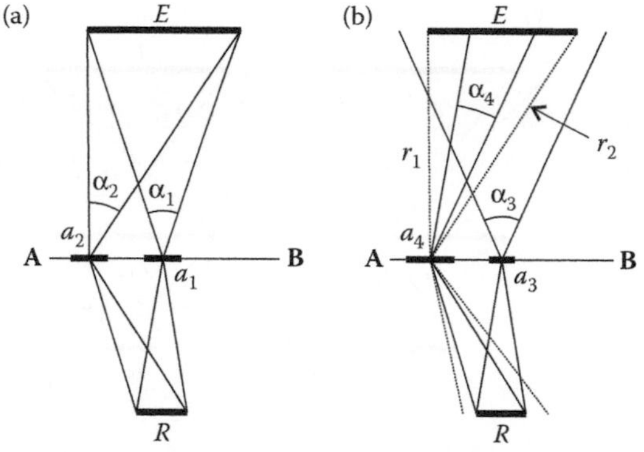

FIGURE 1.29
(a) Nonimaging optic concentrator of fixed aperture size **AB**. (b) By decreasing a_1 to a_3 then α_1 increases to α_3 increasing the acceptance angle. However a_2 increases to a_4 and α_2 decreases to α_4 leading to light loss of rays including r_1 and r_2.

We have seen that errors (in manufacture, assembly, or usage) diffuse the light as it crosses the optic, thereby reducing the acceptance angle. However, with nonimaging optics, we start with the maximum acceptance angle. Therefore, we also end up with the maximum acceptance angle when these errors are introduced. It may then be concluded that nonimaging optics maximize tolerances to errors in concentrators.

Figure 1.31a shows a nonimaging concentrator optic with aperture **AB** and acceptance angle α_1 at position **P**, wider than the angular aperture of emitter E. When this optic is manufactured and used, errors diffuse the light and

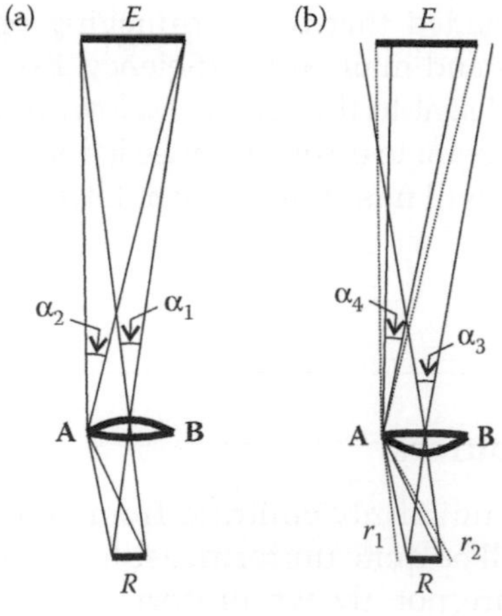

FIGURE 1.30
Same as Figure 1.29, only now with real lenses instead of a diagrammatic representation. (a) Nonimaging optic. (b) Optic with increased angle α_3 at some position results in a decreased angle α_4 somewhere else if the aperture **AB** is fixed.

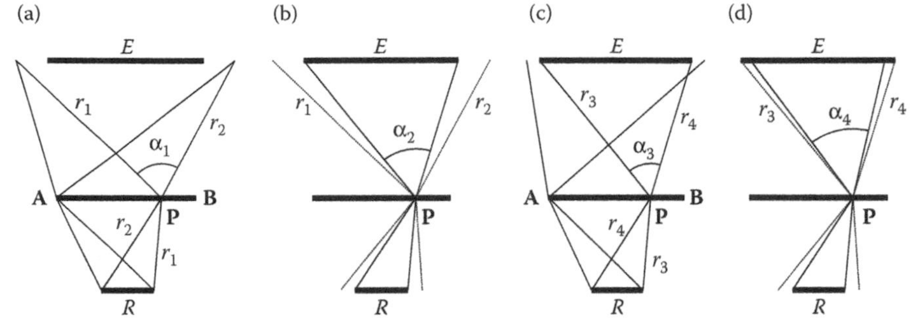

FIGURE 1.31
(a) Ideal nonimaging optic has a wide acceptance angle α_1 wider than *E*. (b) Errors in manufactured nonimaging optic diffuse the light, reducing the acceptance angle to α_2. (c) Nonideal optic has a small acceptance angle α_3. (d) Errors in manufactured nonideal optic further reduce the acceptance angle to α_4 resulting in light loss.

part of the light contained between rays r_1 and r_2 will now miss the receiver, as shown in Figure 1.31b. The acceptance angle is then reduced to α_2, which is still wide enough to capture the light emitted by *E*.

Figure 1.31c shows a nonideal concentrator with aperture **AB**. Since this optic was not designed with the edge ray principle, it has a smaller acceptance angle α_3 at position **P**. When manufactured, the light contained between rays r_3 and r_4 will be diffused as it crosses the optic, resulting in a reduced acceptance angle α_4, as shown in Figure 1.31d. Now the optic is no longer able to capture all the light from emitter *E* and some rays, including r_3 and r_4, will miss the receiver *R* reducing efficiency.

It may then be concluded that the nonimaging optic in Figure 1.31a is more tolerant to errors and maintains efficiency. Even with errors, all light from the emitter is still sent to the receiver. However, the nonideal optic in Figure 1.31c is not as tolerant to errors because it does not maintain efficiency. With errors, some light will miss the receiver, lowering efficiency.

1.7 Nonuniform Sources

When the emitter *E* is uniformly emitting from all its points, the illumination on a receiver *R* will be quite uniform, as shown in Figure 1.32.

However, emitters are not always uniform. They may, for example, be made of several individual small sources. Or there may be tolerances in assembling the light sources, which makes the position of the source vary. Figure 1.33 shows that possibility in which an emitter *E*, when assembled in a real device, may be placed anywhere inside a position tolerance *t*. Depending

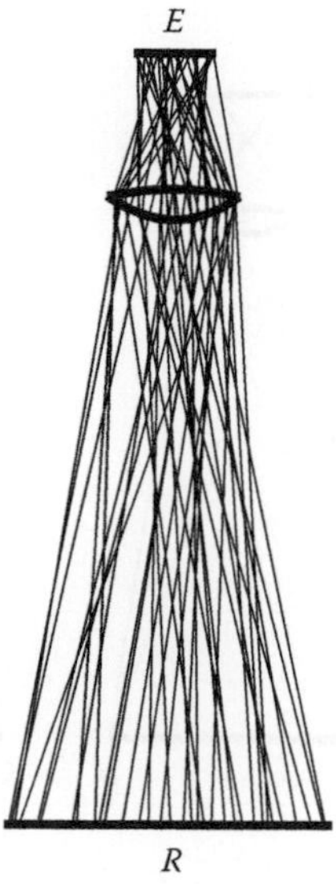

FIGURE 1.32
If the emitter E is fully lit and uniform, the optic creates a quite uniform illumination on receiver R.

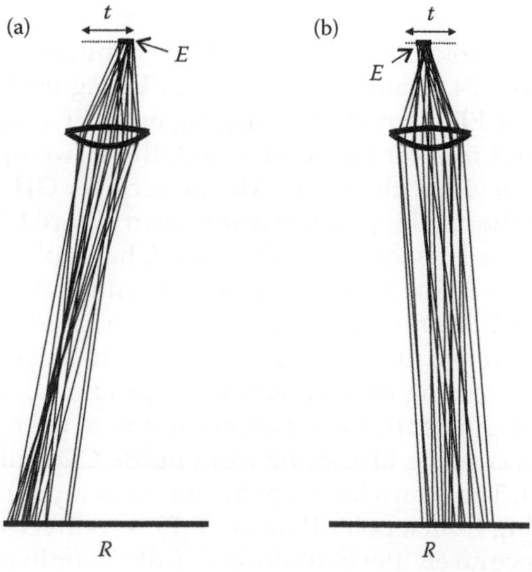

FIGURE 1.33
Due to the manufacturing tolerances, emitter E may be positioned anywhere inside position tolerance t. However, this uncontrolled positioning of emitter E results in an uncontrolled non-uniform illumination of receiver R. (a) and (b) show light crossing the optic for two possible positions of emitter E inside a position tolerance t.

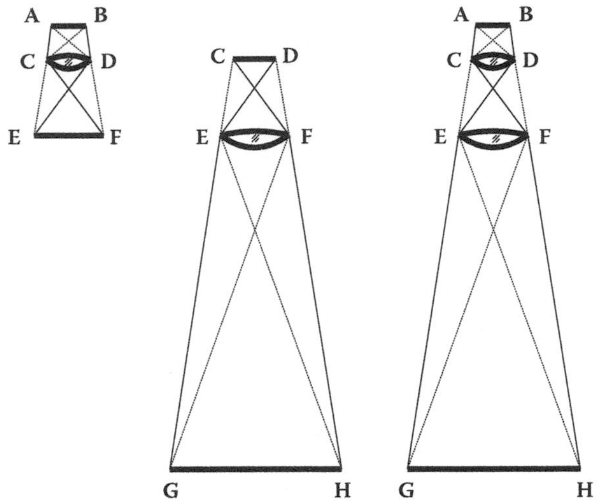

FIGURE 1.34
Left: nonimaging optic **CD** with emitter **AB** and receiver **EF**. Middle: nonimaging optic **EF** with emitter **CD** and receiver **GH**. Right: the two previous optics combined into a device with emitter **AB** and receiver **GH**.

on the position of E inside tolerance t, the illuminance on the receiver may vary considerably. This may be an undesirable characteristic of the collimator optic.

A possible way to avoid this is to combine two nonimaging optics in series, as shown in Figure 1.34.[4] A nonimaging optic **CD** (Figure 1.34 left) has emitter **AB** and receiver **EF**. Another nonimaging optic **EF** (Figure 1.34 middle) has emitter **CD** and receiver **GH**. Combining these two optics (Figure 1.34 right), results in a device with emitter **AB** and receiver **GH**.

The behavior of the resulting device is shown in Figure 1.35. Due to assembly tolerances of a real device, a small emitter E has a placement tolerance t extending from **A** to **B**. Ray r_1 coming from an emitter at position E_1 crosses optic **CD** at point **C**. Nonimaging optic **EF** is designed for a source with edges **CD** and, therefore, it redirects this ray to the edge **H** of receiver R. Accordingly, ray r_2 coming from an emitter at position E_1 crosses optic **CD** at point **D**. Nonimaging optic **EF** redirects this ray to the edge **G** of receiver R. Also, any ray r_3 crossing **EF** coming from inside **CD** will be redirected to a point inside **GH**. These rays have a path similar to ray r_3 in Figure 1.23a, in which a ray coming from a point **P** inside $\mathbf{E_1E_2}$ is redirected toward a point **Q** inside $\mathbf{R_1R_2}$. Since an emitter at position E_1 fully illuminates **CD**, there will be rays crossing all points between **C** and **D**. These rays will be spread out by optic **EF** inside **GH**, fully illuminating it.

Even if the emitter moves to other positions, E_2 or E_3, inside position tolerance t, the illumination pattern on R will be approximately maintained, as shown in Figure 1.35.

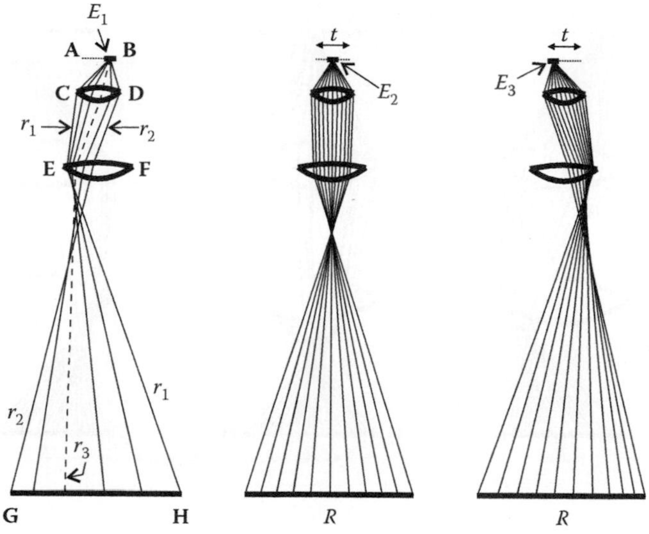

FIGURE 1.35
The emitter may be at different positions E_1, E_2, or E_3 inside position tolerance t and still the optic will fully illuminate the receiver R, extending from **G** to **H**.

Another option for this configuration is color mixing. Figure 1.36 shows a combination of an optic **CD** designed for an emitter **AB** and receiver **EF**, and another optic **EF** with emitter **CD** and receiver **GH**. This system is now used with an emitter composed of three different light sources. These may be, for example, a red source E_R, a green source E_G, and a blue source E_B, forming an RGB (Red, Green, Blue) emitter. When only one of these sources is lit, the receiver R is illuminated by the light of that color. However, when the three are lit simultaneously, these three colors superimpose on the receiver, which will be illuminated by white light. Also, sources E_R, E_G, and E_B may be moved inside position tolerance t (stretching from **A** to **B**), and the optic will still maintain the illumination on receiver R quite constant.

The concepts presented above can also be applied to concentrators. Figure 1.40 shows that for sunlight incidence, angles inside the acceptance angle of the concentrator light is captured and reaches the receiver. However, the irradiance on the receiver may be very nonuniform, with all the light concentrated onto a small spot inside the receiver. This may be an undesirable characteristic of the concentrator optic.

A possible way to avoid this is to combine two nonimaging optics in series, as shown in Figure 1.37. A nonimaging optic **CD** (Figure 1.37 left) has acceptance angle 2θ and receiver **EF**. Another nonimaging optic **EF** (Figure 1.37 middle) has emitter **CD** and receiver **GH**. Combining these two optics (Figure 1.37 right) results in a device with acceptance angle 2θ and receiver **GH**.

The behavior of the resulting device is shown in Figure 1.38. Even for different incidence angles of sunlight inside the acceptance angle, the whole

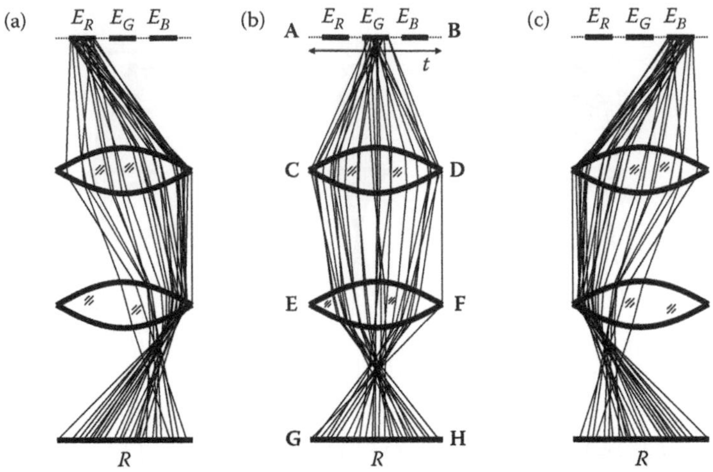

FIGURE 1.36
Suppose that E_R, E_G, and E_B represent light sources of different colors (e.g., red, green, and blue). If only one source is lit, the receiver will be illuminated by the light of that color. If all sources are lit simultaneously, the receiver will be illuminated by white light resulting from mixing all colors. Sources E_R, E_G, and E_B may move inside position tolerance t extending from **A** to **B**. (a), (b), and (c) show, respectively, the situations in which the red, green, or blue light sources are on.

receiver R is always illuminated and the irradiance is quite uniform. The way this optic works is similar to the one in Figure 1.35.

An incoming ray r_1 inside the acceptance angle of optic **CD** will be redirected toward a point inside its receiver **EF**, as shown in Figure 1.38b. Now, if ray r_1 crosses edge **C** of **CD** (the emitter for optic **EF**), it will be redirected by optic **EF**

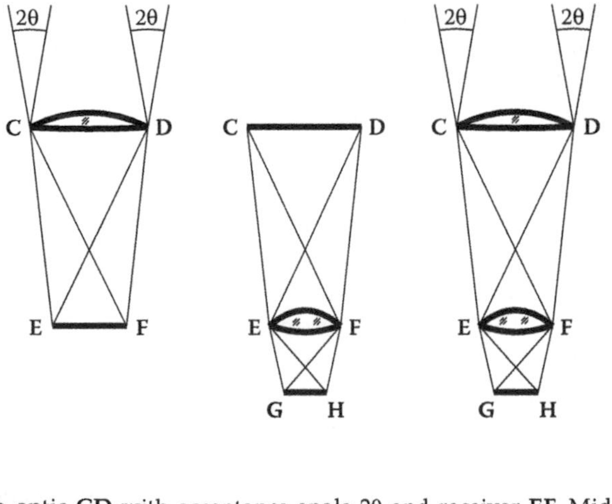

FIGURE 1.37
Left: nonimaging optic **CD** with acceptance angle 2θ and receiver **EF**. Middle: nonimaging optic **EF** with emitter **CD** and receiver **GH**. Right: the two previous optics combined into a device with acceptance angle 2θ and receiver **GH**.

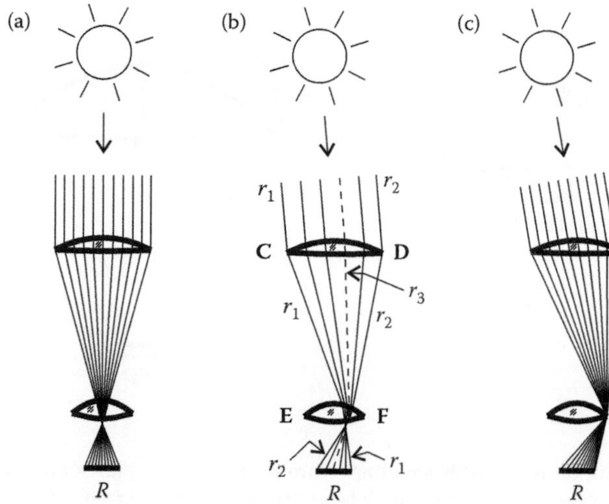

FIGURE 1.38
The receiver R is fully illuminated for different incidence angles of sunlight inside the acceptance angle. (a), (b), and (c) show the paths of light through the system for three different directions of incoming sunlight.

to the edge of receiver R (the receiver of optic **EF**). Something similar happens with another ray r_2. However, since this ray now crosses optic **CD** at point **D**, it is redirected by optic **EF** to the other edge of R. Finally, another ray r_3 inside the acceptance angle of optic **CD** is also redirected by this optic to a point inside its receiver **EF**. Now, for optic **EF**, ray r_3 is coming from a point inside its emitter **CD** and, therefore, it is redirected by optic **EF** to a point inside its receiver R.

1.8 Solar Concentrators

An important application of nonimaging optics is as solar energy concentrators. Figure 1.39a shows a concentrator with entrance aperture a_A and exit aperture (receiver) a_2. The acceptance angle is α_A and the half-acceptance angle is $\theta_A = \alpha_A/2$. This concentrator accepts radiation coming within an angle $\pm\theta_A$ relative to the vertical. Its geometrical concentration is defined as $C_A = a_A/a_2$. Figure 1.39b shows another concentrator with a larger entrance aperture a_B and the same exit aperture (receiver) a_2. Due to the conservation of étendue, since the area of the entrance aperture increased from a_A to a_B, the acceptance angle decreased from α_A to $\alpha_B = 2\theta_B$. This concentrator accepts radiation coming within an angle $\pm\theta_B$ relative to the vertical. Its geometrical concentration is also defined as $C_B = a_B/a_2$, but is now larger than C_A. Therefore, the optic in Figure 1.39a has low concentrations with a wide

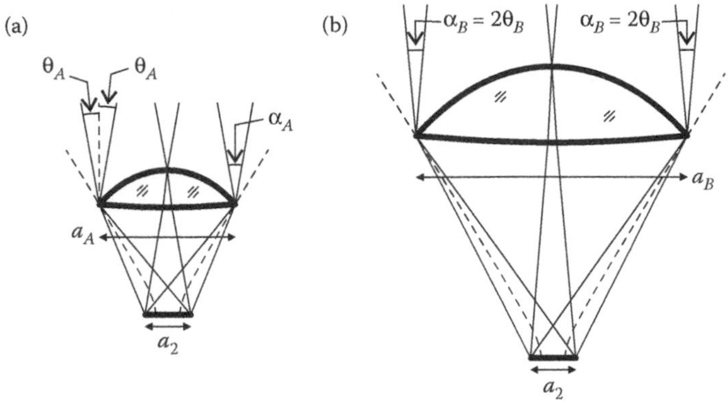

FIGURE 1.39
(a) Low concentration optic with small aperture a_A receiver a_2 and wide acceptance angle $\alpha_A = 2\theta_A$. (b) High concentration optic with large aperture a_B, same receiver a_2 but small acceptance angle $\alpha_B = 2\theta_B$.

acceptance angle, while the optic in Figure 1.39b has high concentration with a small acceptance angle.

We now look at the behavior of an optic with a wide half-acceptance angle θ (low concentration) when exposed to sunlight. In this example, the concentrator is static. As the sun moves across the sky, it first makes an angle $|\beta| > \theta$ to the axis of the concentrator, as shown in Figure 1.40 left. Sunlight enters the optic but fails the receiver, and no light is captured. Eventually, sunlight will make an angle θ to the axis of the concentrator, and light finally reaches the receiver, being captured. As time goes on, sunlight makes an angle $|\beta| < \theta$ to the axis of the concentrator (Figure 1.40 center), and light continues to reach the receiver, being captured. Eventually, sunlight will again make an angle θ to the axis of the concentrator, and then again a larger angle $\beta > \theta$, in which case sunlight again fails to reach the receiver (Figure 1.40 right).

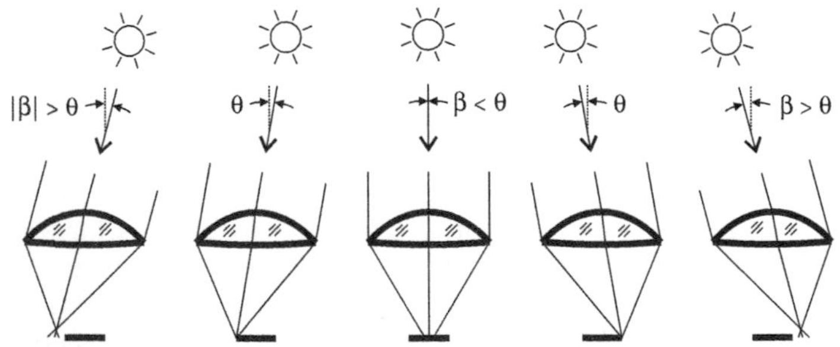

FIGURE 1.40
For a static solar concentrator, as the sun moves across the sky, sunlight is captured if its incidence angle β to the axis of the optic is less than the half-acceptance angle θ of the concentrator.

Nonimaging optics maximize the half-acceptance angle θ and, therefore, also maximize the time a solar concentrator can remain static, while capturing sunlight.

We now look at the behavior of an optic with a small half-acceptance angle θ (high concentration) when exposed to sunlight. Since now the acceptance angle is small, the concentrator can only remain static for a short period of time while capturing sunlight. For that reason, the concentrator must track the sun, and must be constantly orientated to point at the sun, as shown in Figure 1.41. In real systems, however, there will be errors when pointing at the sun, due to imprecision of the tracking system. For a given direction *s* of sunlight, if the axis *b* of the concentrator makes an angle to *s* contained between ±θ, sunlight will still be captured. Therefore, the concentrator has a tolerance of ±θ when pointing at the sun.

Since nonimaging optics maximize the half-acceptance angle θ, they also maximize the tolerance of a concentrator to deviations (errors) when pointing at the sun.

In some applications, several solar concentrators are assembled in modules, as shown in Figure 1.42. In this example, the module is made of three concentrators. The module points in direction **u** but, due to assembly errors, individual concentrators point in slightly different directions $\mathbf{v}_1$, $\mathbf{v}_2$, $\mathbf{v}_3$.

The acceptance angle of each individual concentrator is α but the module assembly has a smaller acceptance angle α_M. It results from the intersections of the acceptance angles of the individual concentrators. Only within a narrow angle α_A, all the three concentrators are able to simultaneously capture the radiation coming from direction **u**.

As illustrated in Figure 1.40, an ideal concentrator will capture all light if the incidence angle β is inside its half-acceptance angle θ, that is $|\beta| < \theta$. Also, light will miss the receiver if the incidence angle β is outside the acceptance angle $|\beta| > \theta$. The transmission curve is then as shown in Figure 1.43. It is

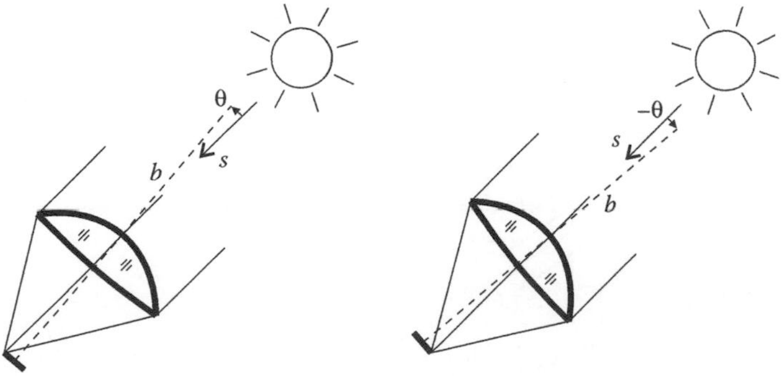

FIGURE 1.41
When pointing at the sun, a solar concentrator has a tolerance in the pointing direction of ±θ where θ is its half-acceptance angle.

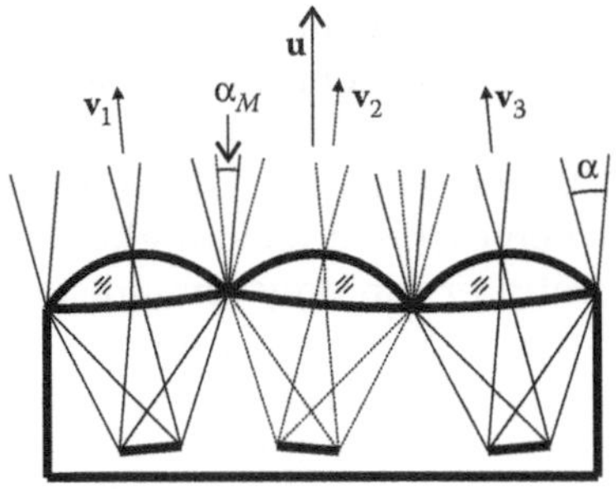

FIGURE 1.42
Module made of three individual concentrators, each with acceptance angle α. Due to assembly errors, individual concentrators point in different directions $\mathbf{v}_1$, $\mathbf{v}_2$, and $\mathbf{v}_3$, reducing the acceptance angle of the module to α_M.

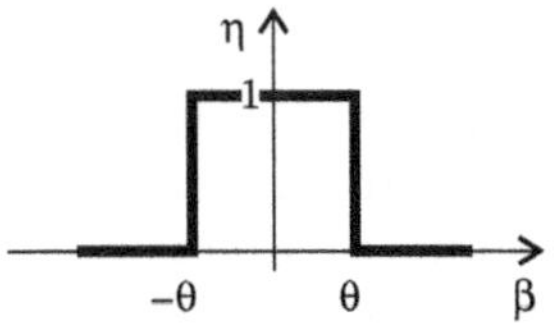

FIGURE 1.43
An ideal concentrator accepts light for incidence angles $|\beta| < \theta$ and rejects light for incidence angles $|\beta| > \theta$ resulting in a stepped transmission (efficiency) curve $\eta(\beta)$.

zero (no light reaches the receiver) for $|\beta| > \theta$, and unity (all light reaches the receiver) for $|\beta| < \theta$.

This result, however, assumes that the incident light is made of parallel rays *r*, as shown in Figure 1.44a. Sunlight, however, has a small angular aperture α_S as shown in Figure 1.44b.

When the finite angular aperture of real sunlight is taken into consideration, the transmission curve of an ideal concentrator changes. Figure 1.45 shows the behavior of a concentrator under real sunlight, with angular aperture α_S.

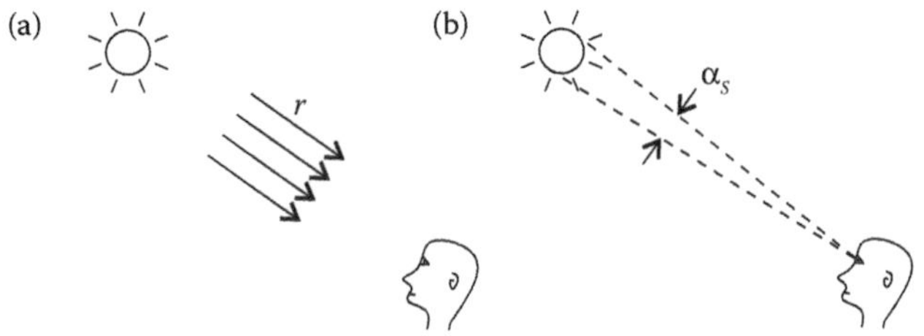

FIGURE 1.44
(a) In some situations, sunlight may be approximated as parallel rays. (b) Real sunlight has a small angular aperture α_S.

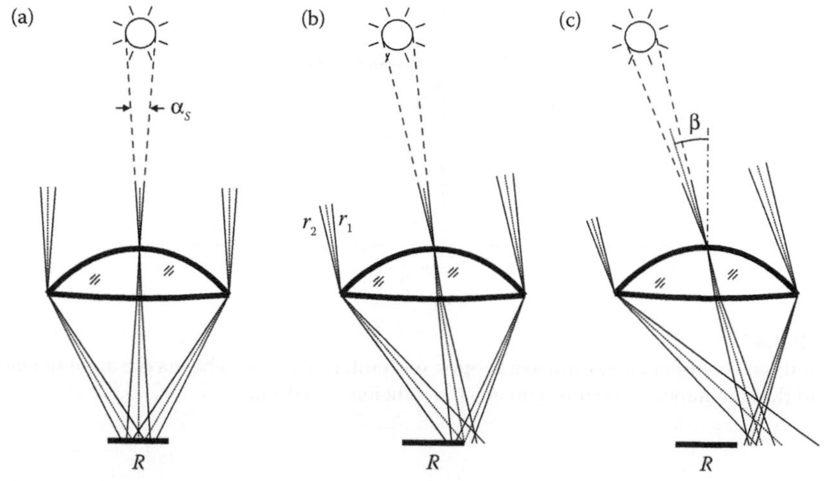

FIGURE 1.45
Sunlight has a finite (although small) angular aperture α_S. (a) All sunlight is captured. (b) Some sunlight is captured and some is not resulting in a smooth transition from capturing to not capturing sunlight. (c) Sunlight is not captured.

When sunlight reaches the concentrator along its axis (as in Figure 1.45a), all light is captured by receiver R. As the incidence angle increases, a situation is reached (Figure 1.45b) in which some light reaches the inside of R (rays parallel to r_1), and some light misses the receiver (rays parallel to r_2). As the incidence angle β increases still further, all light will miss the receiver, as in Figure 1.45c. This results in a transmission curve with a smooth transition from all light being captured to all light being lost. Figure 1.46 shows the transmission (efficiency) curve $\eta(\beta)$ for different incidence angles β of

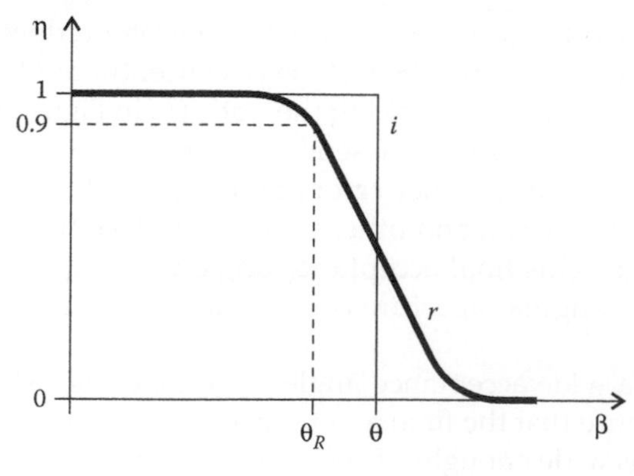

FIGURE 1.46
Ideal transmission curve *i* for different directions of incoming parallel rays and real transmission curve *r* for different directions of incoming sunlight which has a finite angular aperture.

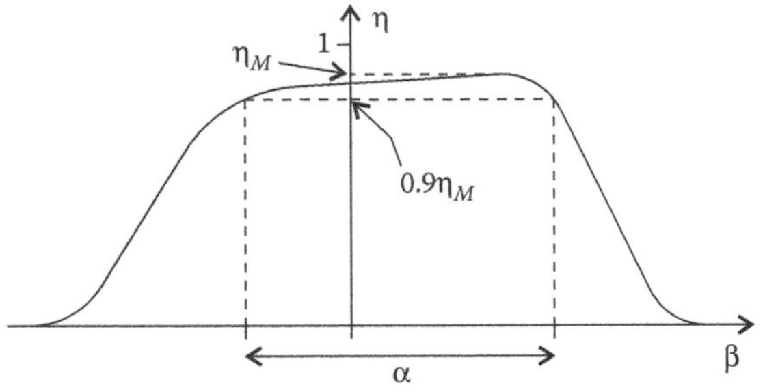

FIGURE 1.47
Transmission curve of an asymmetrical optic with internal losses. The maximum efficiency is η_M and the acceptance angle α is defined at an efficiency of 0.9 η_M.

parallel radiation (ideal curve *i*), and for different incidence angles β of sunlight, which has a finite angular aperture (real curve *r*).

Under parallel radiation, the concentrator has a half-acceptance angle θ. Under sunlight with finite angular aperture, the concentrator has a (reduced) real half-acceptance angle θ_R commonly defined as the angle for which the transmission (efficiency) of the concentrator drops to 90% of its maximum.[3,5]

A more general situation occurs when the optic is asymmetric, and has internal losses. The transmission curve η(β) is, in this case, also asymmetric, and the maximum efficiency is now a value $\eta_M < 1$, as shown in Figure 1.47. The acceptance angle α is defined at an efficiency of $0.9\eta_M$. In general, the transmission will be a 3-D function of the incidence direction. In that case, the acceptance angle is defined by the circular cone whose acceptance is $0.9\eta_M$.[5] A concentrator is characterized by its maximum efficiency η_M, and its acceptance angle α.

Different imperfections and errors in a concentrator reduce the acceptance angle, as illustrated in Figure 1.48. It shows a concentrator O_1 designed for an acceptance angle α_1. Errors in making the optic (as in Figure 1.24b) reduce its acceptance angle to α_2; errors in assembly (as in Figure 1.42) further reduce the acceptance angle to α_3; other errors and imperfections (tracking errors, wind, dust, overtime wear, and others) further reduce the acceptance angle to a final value α_4. This final acceptance angle value α_4 must still be wider than the sunlight angular aperture α_S, so that all the light coming from the sun is captured.

Starting with a wide acceptance angle α_1 at the design stage is then very important to ensure that the final acceptance angle α_4, of a real device working in the field, is wide enough for the system to have high efficiency.

The acceptance angle may then be seen as a tolerance budget[3] that is spent in different kinds of errors: in the optical surfaces, assembly, installation, tracker structure, sun-tracking, or the finite angular aperture of sunlight.

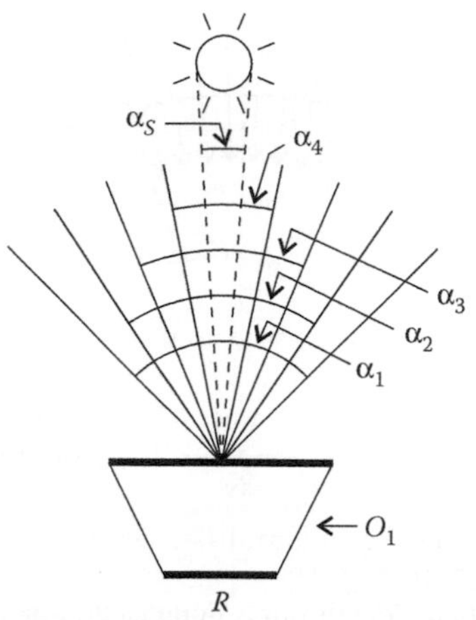

FIGURE 1.48
Solar concentrator O_1 is designed with acceptance angle α_1. Errors in making the optic reduce the acceptance angle to α_2, errors in assembly reduce it to α_3 and other errors and imperfections reduce it to α_4 which must still be wide enough to capture sunlight with angular aperture α_S.

1.9 Light Flux

The amount of light flowing through an aperture depends on how bright the light source is, but also on how much area and angle is available for light to flow through. Figure 1.49 shows light incident on an aperture dx (solid lines). If now we place another aperture dx next to it, doubling the total aperture, the amount of light going through will also double (dotted lines). This means that the amount of light crossing an aperture is proportional to its area: $d\Phi \propto dx$.

Something similar happens relative to angle. Figure 1.50 shows light incident on dx inside an angle $d\theta$ (solid lines). If now we place another cone of light with angle $d\theta$ next to it, doubling the total angle, the amount of light going through will also double (dotted lines). This means that the amount of light crossing an aperture is also proportional to its angle: $d\Phi \propto dx\, d\theta$.

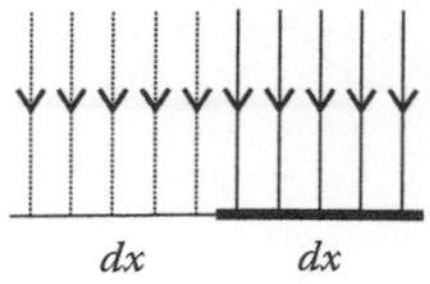

FIGURE 1.49
Doubling the aperture dx, for light to flow through, doubles the amount of light.

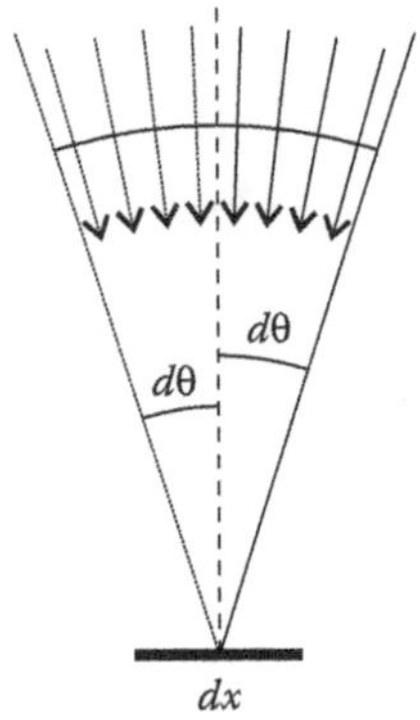

FIGURE 1.50
Doubling the angle $d\theta$ for light to flow through, doubles the amount of light.

Now, if light comes at dx tilted by an angle θ relative to the normal to dx, the projected area of dx in the direction of the incident light is $dx\cos\theta$, as shown in Figure 1.51. The amount of light going through dx is then reduced by a cos θ factor, and is, therefore, $d\Phi \propto dx\cos\theta\, d\theta$.

Finally, light can be brighter or dimmer and the light flux is given by

$$d\Phi = L dx \cos\theta \, d\theta \tag{1.1}$$

where L is the brightness (luminance) of the light.

In three-dimensional geometry, the situation is similar, but now the light flux, instead of being proportional to the plane angle, is proportional to the solid angle. Figure 1.52a shows light incident on dA inside a solid angle $d\Omega = dA_S/r^2$ defined by area dA_S. If now we place another area dA_S next to it, doubling the total solid angle, the amount of light going through will also

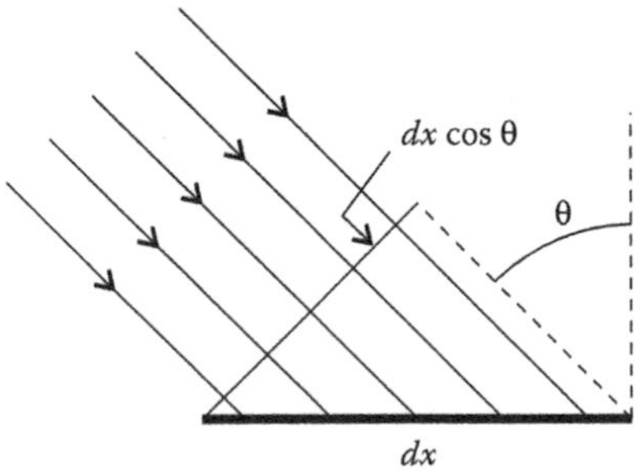

FIGURE 1.51
If the direction of light makes an angle θ to the aperture normal, the projected area in the direction of the incident light is dx cos θ and, therefore, the amount of light crossing dx is proportional to dx cos θ.

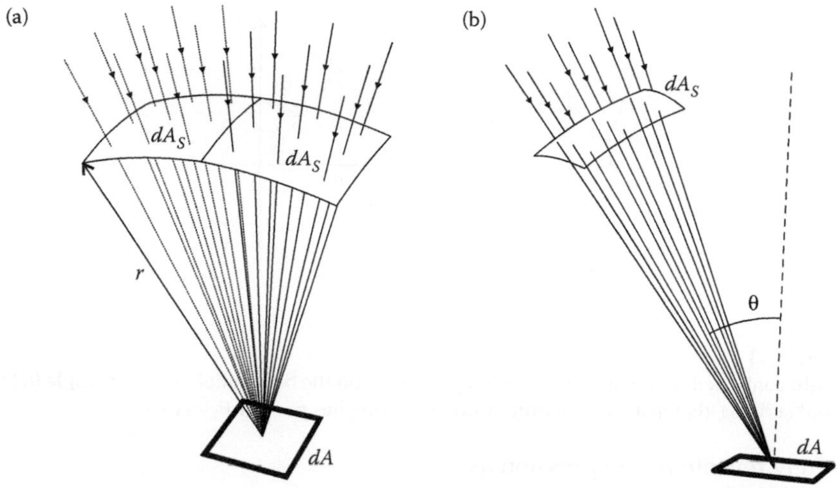

FIGURE 1.52
The solid angle of the light is defined by $d\Omega = dA_S/r^2$. (a) Doubling the area dA_S for light to flow through doubles the solid angle and also doubles the amount of light. (b) The direction of light makes an angle θ to the normal to dA decreasing the amount of light through dA by a factor cos θ.

double (dotted lines). This means that the amount of light crossing an aperture, dA, is proportional to its solid angle in three-dimensional geometry:

$$d\Phi = L dA \cos\theta \, d\Omega$$

where θ is the angle between the direction of light, and the normal to the surface dA, as shown in Figure 1.52b.

Going back to two-dimensional geometry, the flux of light in expression (1.1) may be written as:

$$d\Phi = L dU \tag{1.2}$$

where

$$dU = dx \cos\theta \, d\theta \tag{1.3}$$

is called the étendue of the light (in air or vacuum, where the refractive index is $n = 1$).

If area dx is on the x_1 axis, as in Figure 1.53, this expression becomes:

$$dU = dx_1 \cos\theta_2 \, d\theta_2 \tag{1.4}$$

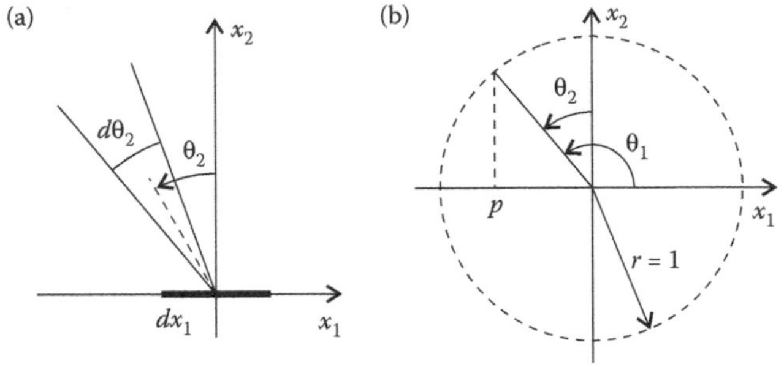

FIGURE 1.53
(a) Light contained within angle $d\theta_2$ crossing area dx_1 on the horizontal axis at an angle θ_2 to its normal (axis x_2). (b) On a circle of unit radius $r = 1$, one has $p = -\sin\theta_2 = \cos\theta_1$.

We may rewrite this expression as

$$dU = dx_1(\cos\theta_2\, d\theta_2) = dx_1\, d(\sin\theta_2) = -dx_1\, dp$$

where $p = -\sin\theta_2 = \cos\theta_1$, as shown in Figure 1.53b (note that in the case in this figure $p < 0$ and $\sin\theta_2 > 0$).

The configuration shown in Figure 1.54 is similar to that in Figure 1.53a, with light crossing dx_1 on the horizontal axis x_1, confined within angle $d\theta_2$ bound by directions $\mathbf{v}_A$ and $\mathbf{v}_B$.

The étendue is in this case given by $dU = -dx_1\, dp = -dx_1\,(p_B - p_A) = dx_1\,(p_A - p_B)$. This expression may now be applied to the étendue of the light exchanged between an emitter **CD** and a receiver **BA** at a height h, as shown in Figure 1.55.

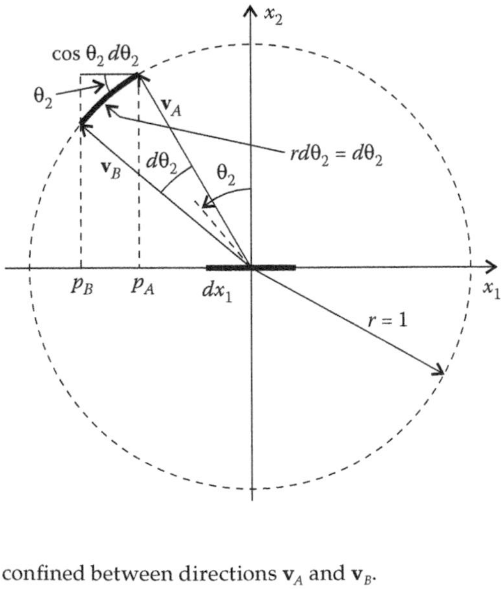

FIGURE 1.54
Light crossing dx_1 is confined between directions $\mathbf{v}_A$ and $\mathbf{v}_B$.

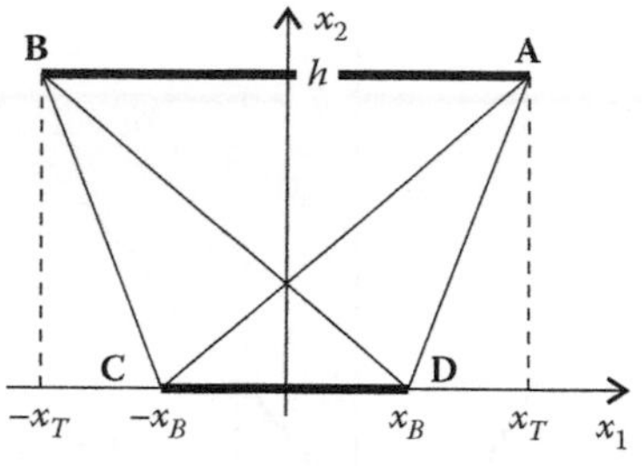

FIGURE 1.55
Emitter **CD** extending from $-x_B$ to x_B and receiver **BA** extending from $-x_T$ to x_T at a height h.

At a point $(x, 0)$ along emitter **CD**, the light reaching receiver **BA** is confined to the cone defined by unit vectors $\mathbf{v}_A$ and $\mathbf{v}_B$ given by

$$\begin{aligned}\mathbf{v}_A &= \frac{\mathbf{A}-(x,0)}{\|\mathbf{A}-(x,0)\|} = \left(\frac{x_T - x}{\sqrt{(x_T-x)^2+h^2}}, \frac{h}{\sqrt{(x_T-x)^2+h^2}}\right)\\ \mathbf{v}_B &= \frac{\mathbf{B}-(x,0)}{\|\mathbf{B}-(x,0)\|} = \left(\frac{-x_T - x}{\sqrt{(-x_T-x)^2+h^2}}, \frac{h}{\sqrt{(-x_T-x)^2+h^2}}\right)\end{aligned} \tag{1.5}$$

The horizontal x_1 coordinates of vectors $\mathbf{v}_A$ and $\mathbf{v}_B$ are given by

$$\begin{aligned}p_A(x) &= \frac{x_T - x}{\sqrt{(x_T-x)^2+h^2}}\\ p_B(x) &= \frac{-x_T - x}{\sqrt{(-x_T-x)^2+h^2}}\end{aligned} \tag{1.6}$$

as shown in Figure 1.56.

The étendue of the light emitted by **CD** and captured by **BA** is then given by

$$U = \int_{-x_B}^{x_B} p_A(x) - p_B(x)dx = \int_{-x_B}^{x_B} p_A(x)dx - \int_{-x_B}^{x_B} p_B(x)dx \tag{1.7}$$

Replacing the values for $p_A(x)$ and $p_B(x)$ from expressions (1.6) we get

$$\begin{aligned}U &= \int_{-x_B}^{x_B} \frac{x_T - x}{\sqrt{(x_T-x)^2+h^2}}dx + \int_{-x_B}^{x_B} \frac{x_T + x}{\sqrt{(x_T+x)^2+h^2}}dx\\ &= \left[-\sqrt{(x_T-x)^2+h^2}\right]_{-x_B}^{x_B} + \left[\sqrt{(x_T+x)^2+h^2}\right]_{-x_B}^{x_B}\\ &= 2\sqrt{(x_T+x_B)^2+h^2} - 2\sqrt{(x_T-x_B)^2+h^2}\end{aligned} \tag{1.8}$$

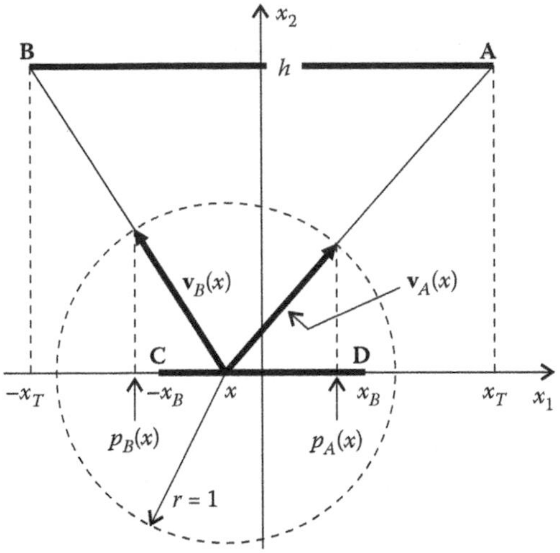

FIGURE 1.56
The étendue (and therefore the light flux) emitted by **CD** that is captured by **BA** is given by integrating the étendue of the light emitted from each point x of **CD** and captured by **BA**.

which can also be written as:

$$U = 2([\mathbf{C},\mathbf{A}] - [\mathbf{D},\mathbf{A}]) \tag{1.9}$$

where [**X**, **Y**] is the distance between points **X** and **Y**.

Referring now to Figure 1.57, points $\mathbf{A}_1$ and $\mathbf{A}_2$ are on hyperbola h with foci **C** and **D**. By being on the hyperbola, points $\mathbf{A}_1$ and $\mathbf{A}_2$ verify $[\mathbf{C}, \mathbf{A}_1] - [\mathbf{D}, \mathbf{A}_1] = [\mathbf{C}, \mathbf{A}_2] - [\mathbf{D}, \mathbf{A}_2]$ and, therefore, the étendue of the light emitted by **CD** and captured by $\mathbf{B}_1\mathbf{A}_1$ is the same as the étendue of the light emitted by **CD** and captured by $\mathbf{B}_2\mathbf{A}_2$.

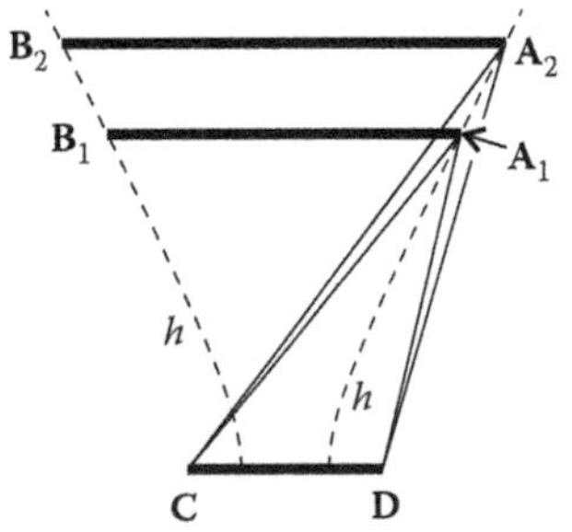

FIGURE 1.57
Points $\mathbf{A}_1$ and $\mathbf{A}_2$ are on a hyperbola h with foci **C** and **D** and, therefore, verify $[\mathbf{C}, \mathbf{A}_1] - [\mathbf{D}, \mathbf{A}_1] = [\mathbf{C}, \mathbf{A}_2] - [\mathbf{D}, \mathbf{A}_1]$. The amount of light exchanged between **CD** and $\mathbf{B}_1\mathbf{A}_1$ is the same as between **CD** and $\mathbf{B}_2\mathbf{A}_2$.

Since the light flux is proportional to the étendue, an optic with aperture $\mathbf{B}_1\mathbf{A}_1$ then captures the same amount of light from emitter **CD** as another optic with aperture $\mathbf{B}_2\mathbf{A}_2$. These hyperbolas are called flow-lines, and are also shown in Figures 1.9 and 1.17 or in Figures 1.21 and 1.22.

1.10 Wavefronts and the SMS

As light travels through an optical system, it may encounter different optical surfaces, where it is refracted or reflected. The optical path length of a ray section between two optical surfaces is defined as the product of the refractive index and the distance the ray travels between those surfaces. The total optical path length of a ray is the sum of the optical path lengths for all ray sections. Figure 1.58 shows a light ray travelling from **P** to **Q** while crossing optical surfaces $\mathbf{c}_1$ and $\mathbf{c}_2$. The refractive index of the material is n_1 between **P** and $\mathbf{c}_1$, n_2 between $\mathbf{c}_1$ and $\mathbf{c}_2$, and n_3 between $\mathbf{c}_2$ and **Q**. Its optical path length from **P** to **Q** is $S = n_1d_1 + n_2d_2 + n_3d_3$.

Figure 1.59 shows a set of light rays, $r_1, r_2, r_3, \ldots$ perpendicular to wavefront w_1. After being deflected at an optical surface $\mathbf{c}(\sigma)$, these rays are now perpendicular to wavefront w_2. If $\mathbf{c}(\sigma)$ is a refractive surface, a ray incident with an angle α_1 to the surface normal emerges at the other side, making an angle α_2 with the surface normal. These angles are related by the law of refraction, which states that $n_1 \sin \alpha_1 = n_2 \sin \alpha_2$, where n_1 is the refractive index before $\mathbf{c}(\sigma)$, and n_2 the refractive index after it.

We now look at the optical path length for two light rays crossing $\mathbf{c}(\sigma)$ at two neighboring points, $\mathbf{C}_1 = \mathbf{c}(\sigma)$ and $\mathbf{C}_2 = \mathbf{c}(\sigma + d\sigma)$, a distance, dc, apart from each other, as shown in Figure 1.60.

The optical path length of the ray through $\mathbf{C}_1$ is $S_1 = n_1d_7 + n_2(d_{10} + d_8)$ or

$$S_1 = n_1d_7 + n_2dc \sin \alpha_2 + n_2d_8 \tag{1.10}$$

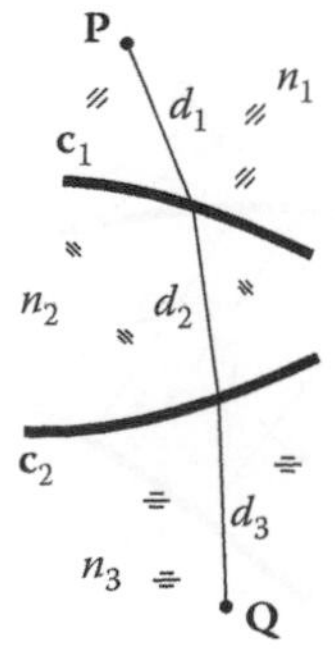

FIGURE 1.58
The optical path length between two points **P** and **Q** is given by $S = n_1d_1 + n_2d_2 + n_3d_3$.

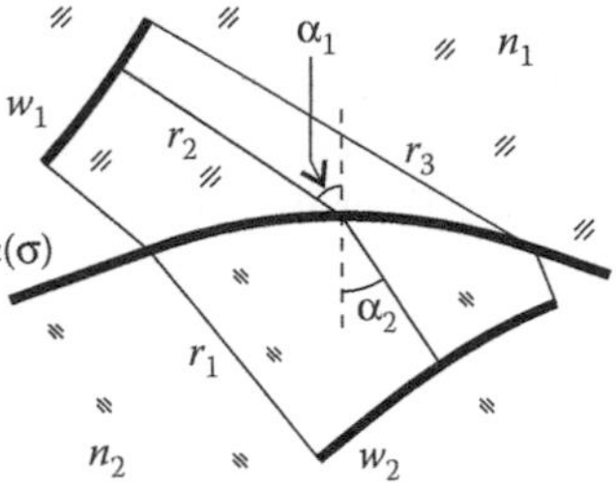

FIGURE 1.59
Light rays r are perpendicular to wavefront w_1, refract at surface **c**(σ), and emerge at a side which is perpendicular to wavefront w_2.

The optical path length of the ray through $\mathbf{C}_2$ is $S_2 = n_1(d_5 + d_9) + n_2 d_6$ or

$$S_2 = n_1 d_5 + n_1 dc \sin\alpha_1 + n_2 d_6 \tag{1.11}$$

Now, from Figure 1.60, $d_5 = d_7$ and $d_6 = d_8$. Also, from the law of refraction, $n_1 \sin\alpha_1 = n_2 \sin\alpha_2$ and, therefore, $dS = S_2 - S_1 = 0$. Since neighboring rays have the same optical path length between wavefronts w_1 and w_2, all rays between these two wavefronts will also have the same optical path length.

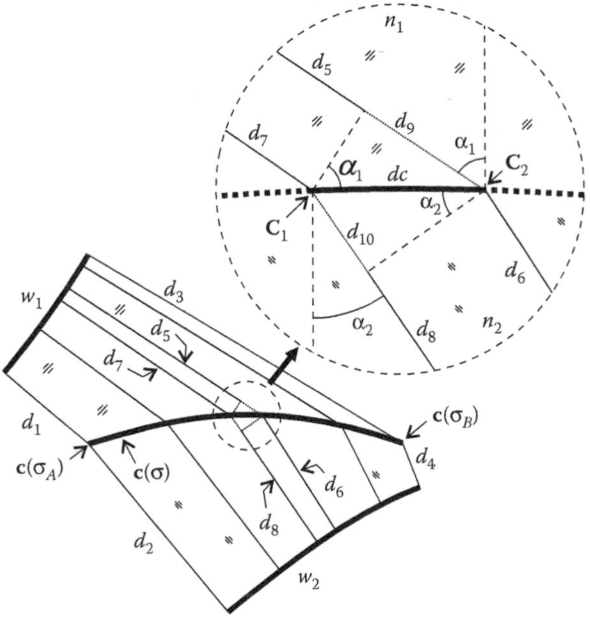

FIGURE 1.60
The optical path length between wavefronts w_1 and w_2 is the same for all light rays between these two wavefronts.

The difference in optical path length for the rays going through $\mathbf{c}(\sigma_A)$ and $\mathbf{c}(\sigma_B)$ is given by

$$S_B - S_A = \int_A^B dS = \int_{\sigma_A}^{\sigma_B} \frac{dS}{d\sigma} d\sigma = 0 \tag{1.12}$$

which means that the optical path length is the same for the rays going through $\mathbf{c}(\sigma_A)$ and $\mathbf{c}(\sigma_B)$. In general, the optical path length is the same for all rays between wavefronts w_1 and w_2.[6,7]

Also, a set of rays perpendicular to a wavefront will remain perpendicular to a wavefront, as the rays travel through an optical system, with refractions and reflections (theorem of Malus and Dupin).[6]

The same conclusion may be obtained for reflection, in which case $n_1 = n_2$. Figure 1.61 shows two neighboring rays reflected on a small portion dc of a mirror. These rays are perpendicular to incoming wavefront w_1 and outgoing wavefront w_2. Here $d_5 + d_{6M} = d_7 + d_{8M}$ and, since $d_8 = d_{8M}$ and $d_6 = d_{6M}$, we get $d_5 + d_6 = d_7 + d_8$ and the optical path length is the same for the two rays reflected at the edges of dc. Note that the distances d_{6M} and d_{8M} and wavefront w_{2M} do not exist physically, they are just a geometrical construction. Just as in the case of refraction, this result may be extrapolated and the optical path length between the wavefronts w_1 and w_2 is the same for all rays between them.

As light travels through multiple optical surfaces, the optical path length is also the same for all rays. Figure 1.62 shows two surfaces, $\mathbf{c}_1$ and $\mathbf{c}_2$, separating three media of refractive indices n_1, n_2, and n_3. Optical path length is the same for all rays between the wavefronts w_1 and w_2, and it is also the same for all rays between wavefronts w_2 and w_3. It must, therefore, also be the same for all rays between the wavefronts w_1 and w_3.

Now, a ray r may be launched from a point $\mathbf{W}_1$ in wavefront w_1 in a direction $\mathbf{v}$, as shown in Figure 1.63. Given the optical path length between the wavefronts w_1 and w_2; this defines the position of point $\mathbf{P}$ along direction $\mathbf{v}$ (perpendicular to w_1), for which, refraction occurs on surface $\mathbf{c}(\sigma)$ separating

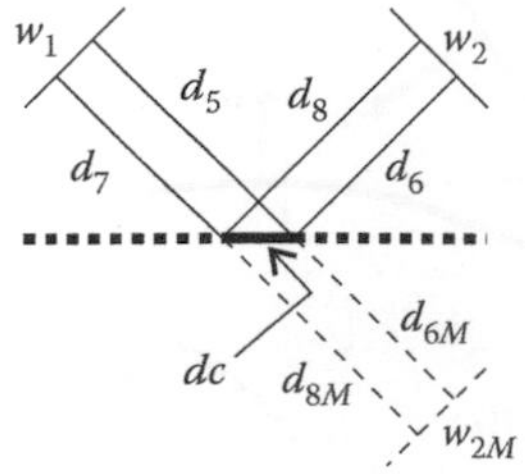

FIGURE 1.61
The optical path length between wavefronts w_1 and w_2 is the same for two neighboring rays reflected on mirror with length dc.

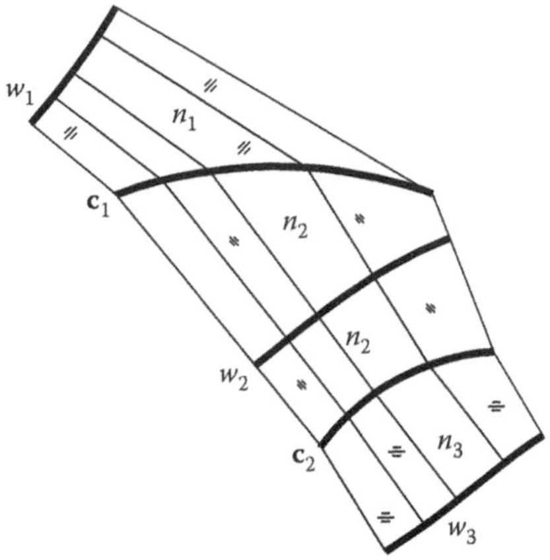

FIGURE 1.62
Optical path length is the same for all rays between the wavefronts w_1 and w_2. It is also the same for all rays between w_2 and w_3 and, therefore, is the same for all rays between w_1 and w_3.

materials of refractive indices n_1 and n_2. That is, there is only one point **P** along **v** for which $S = n_1[\mathbf{W}_1, \mathbf{P}] + n_2[\mathbf{P}, \mathbf{W}_2]$, where S is the optical path length between w_1 and w_2, and $[\mathbf{X}, \mathbf{Y}]$ the distance between points **X** and **Y**. Since we have the direction of the incident and refracted rays at point **P**, we can also calculate the normal $\mathbf{n}_P$ to the surface $\mathbf{c}(\sigma)$ at point **P** (see Chapter 16). Moving point $\mathbf{W}_1$ along wavefront w_1 (rays $r_1, r_2, r_3, \ldots$ in Figure 1.59), with a constant optical path length between w_1 and w_2, allows us to calculate the complete shape of surface $\mathbf{c}(\sigma)$, which is called a Cartesian Oval. The same holds true for reflection, in which case $n_1 = n_2$.

These results may be used for the design of nonimaging optics. Figure 1.64 shows an emitter $\mathbf{E}_1\mathbf{E}_2$ and receiver $\mathbf{R}_1\mathbf{R}_2$, and in between an optic (lens) made of optical surfaces $\mathbf{c}_1$ and $\mathbf{c}_2$. Wavefronts w_1, w_2, w_3, and w_4 are circles centered at $\mathbf{E}_1$, $\mathbf{E}_2$, $\mathbf{R}_1$, and $\mathbf{R}_2$, respectively.

Now, we choose a value for the optical path length S_{14} between wavefronts w_1 and w_4. We also choose an initial point $\mathbf{P}_0$ and its normal $\mathbf{n}_0$ on the top

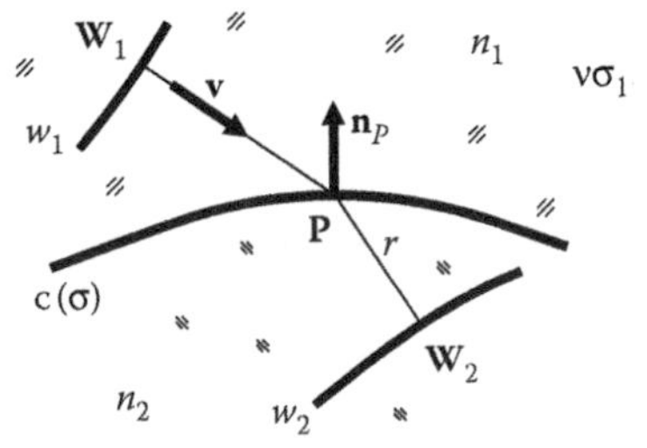

FIGURE 1.63
Given the optical path length between wavefronts w_1 and w_2, this defines the position of point **P** where refraction occurs. The same is valid in the case of reflection (where $n_1 = n_2$).

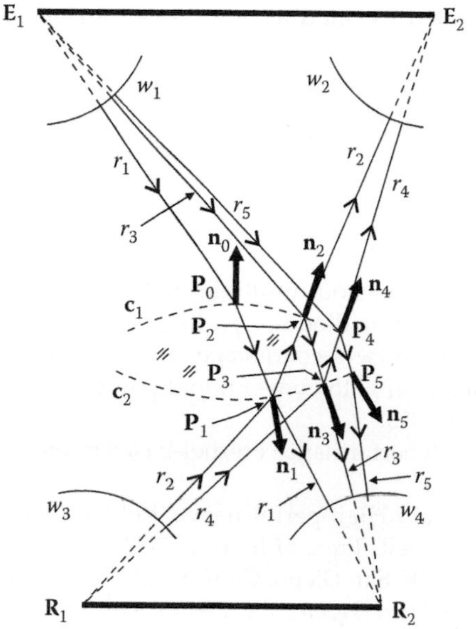

FIGURE 1.64
Constant optical path length between wavefronts w_1 and w_4 and also between w_3 and w_2 allows us to calculate the set of points on a lens focusing $\mathbf{E}_1$ onto $\mathbf{R}_2$ and $\mathbf{E}_2$ onto $\mathbf{R}_1$.

surface $\mathbf{c}_1$ of the lens. Now take the ray r_1 perpendicular to w_1 through $\mathbf{P}_0$. Since we know the normal $\mathbf{n}_0$ at $\mathbf{P}_0$ we may refract r_1 into the lens, calculating the direction of the ray inside the lens (see Chapter 16). Also, since we know the distance ray r_1 travelled between w_1 and $\mathbf{P}_0$, we may calculate the optical path length between $\mathbf{P}_0$ and w_4. With it, it is possible to determine the position of point $\mathbf{P}_1$ and its normal $\mathbf{n}_1$ on the bottom surface $\mathbf{c}_2$ of the lens. Now, from symmetry of the system, the optical path length between w_3 and w_2 is also S_{14}, as between w_1 and w_4. We may now repeat the same process as before. Take the ray r_2, perpendicular to w_3 through $\mathbf{P}_1$. Since we know the normal $\mathbf{n}_1$ at $\mathbf{P}_1$, we may refract r_2 into the lens, calculating the direction of the ray inside the lens. Also, since we know the distance ray r_2 travelled between w_3 and $\mathbf{P}_1$, we may calculate the optical path length between $\mathbf{P}_1$ and w_2. With it, it is possible to determine the position of point $\mathbf{P}_2$ and its normal $\mathbf{n}_2$ on the top surface $\mathbf{c}_1$ of the lens. The process is now repeated with ray r_3 using the same S_{14} optical path length between w_1 and w_4 and a new point $\mathbf{P}_3$ and normal $\mathbf{n}_3$ is obtained on the bottom surface $\mathbf{c}_2$ of the lens. The process is again repeated with ray r_4 using the same S_{14} optical path length between w_3 and w_2, and a new point $\mathbf{P}_4$ and normal $\mathbf{n}_4$ is obtained on the top surface $\mathbf{c}_1$ of the lens. Another ray r_5 allows us to calculate a new point $\mathbf{P}_5$ and corresponding normal $\mathbf{n}_5$ on the bottom surface $\mathbf{c}_2$ of the lens. This process is further repeated, calculating a set of points on both the top and bottom surfaces of the lens simultaneously. This lens will focus point $\mathbf{E}_1$ onto $\mathbf{R}_2$, and $\mathbf{E}_2$ onto $\mathbf{R}_1$.

This is the method used to calculate the lenses in previous figures, such as Figures 1.9b and 1.17b.

References

1. Muñoz, F. et al., Simultaneous multiple surface design of compact air-gap collimators for light-emitting diodes, *Opt. Eng.*, 43, 1522, 2004.
2. Canavarro, C. et al., New second-stage concentrators (XX SMS) for parabolic primaries; Comparison with conventional parabolic trough concentrators, *Sol. Energy*, 92, 98–105, 2013.
3. Pablo, B. et al., High performance Fresnel-based photovoltaic concentrator, *Opt. Express*, 18(S1), 2010.
4. Hernandez, M. et al., High-performance Köhler concentrators with uniform irradiance on solar cell, *Proc. SPIE Vol. 7059, Nonimaging Optics and Efficient Illumination Systems V*, San Diego, California, USA, September 2, 2008.
5. Koshel, R. J., *Illumination Engineering: Design with Nonimaging Optics*, Wiley-IEEE Press, 2013.
6. Born, M. and Wolf, E., *Principles of Optics*, Pergamon Press, Oxford, 1980.
7. Welford, W.T., *Useful Optics*, The University of Chicago Press, Chicago, USA, 1991.

2

Fundamental Concepts

2.1 Introduction

Imaging optical systems have three main components—the object, the optic, and the image it forms. The object is considered as a set of points that emit light in all directions. The light (or part of it) from each point on the object is captured by the optical system and concentrated onto a point in the image. The distances between points on the image may be scaled relative to those on the object, resulting in magnification.

Nonimaging optical systems, instead of an object, have a light source, and instead of an image, have a receiver. Instead of an image of the source, the optic produces a prescribed illuminance (or irradiance) pattern on the receiver.

The first application of nonimaging optics was in the design of concentrators that could perform at the maximum theoretical (thermodynamic) limit. The compound parabolic concentrator (CPC) was the first two-dimensional (2-D) concentrator ever designed, and the success of the device gave birth to nonimaging optics.

This chapter introduces some of the differences between imaging and nonimaging optics, presents the CPC as a concentrator, and shows that it is ideal in two dimensions.

2.2 Imaging and Nonimaging Optics

Figure 2.1 shows a schematic representation of an imaging setup. On the left we have an object **EF**, at the center an optic **CD**, and on the right an image **AB**.

Light coming from edge point **F** on the object must be concentrated onto edge point **A** of the image. Accordingly, light coming from point **E** must be concentrated onto point **B**. This condition would still be valid for any point **P** on the object. Light leaving point **P** is concentrated onto a point **Q** in the

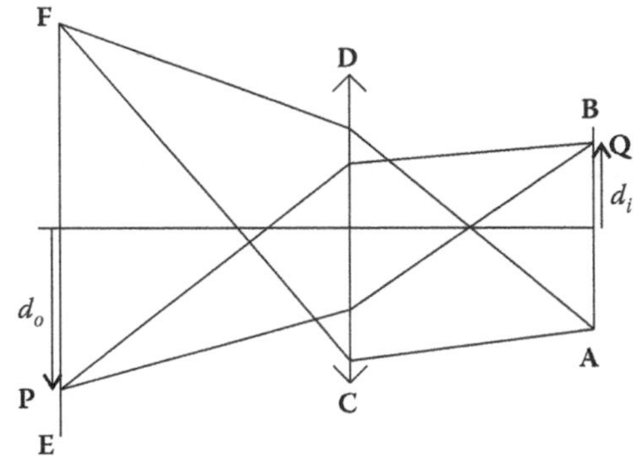

FIGURE 2.1
In an imaging optical system, light coming from any point **P** in the object is concentrated onto a point **Q** in the image, in such a way that $d_i = Md_o$, with d_o and d_i being the distances between **P** and the optical axis, and **Q** and the optical axis, respectively. In particular, light coming from the edge points **E** and **F** of the object, is concentrated onto edge points **B** and **A** of the image, respectively.

image. The distances to the optical axis d_o and d_i from points in the object, and the image, respectively, are related by the following:

$$d_i = Md_o \tag{2.1}$$

where M is the magnification of the system.[1–4] This condition requires that the relative dimensions of several parts of the object are maintained in the image.

Let us now see how to design such a system, using lenses. We can start by concentrating light coming from a point in the object onto the corresponding point in the image. To solve this problem, a Cartesian oval can be used.[1,5] We have, in this case, a set of rays to be focused, and a surface to be defined, as shown in Figure 2.2.

The optical path length along a straight line between **P** and **Q** is given as $S = D + nD_1$. The optical path length of a light ray passing from **P** to **Q** through a point **R** on the surface must also be given by S, so we must have $S = d + nd_1$. This condition enables us to obtain all the points of the Cartesian oval.

If we now want, nevertheless, to focus two points of the object onto two points of the image **AB**, a surface is no longer sufficient. We then need at least two surfaces. Let us then suppose that, in fact, two surfaces are sufficient. We now have two sets of edge rays that are to be focused, those coming from **E** and **F** (that must be focused to **B** and **A** respectively), and we have two surfaces to be defined. Let us then suppose that a lens similar to the one presented in Figure 2.3 can be designed so that it focuses the two sets of edge rays of the object onto the two sets of edge rays of the image (later in this chapter, a way to design such a lens is presented).

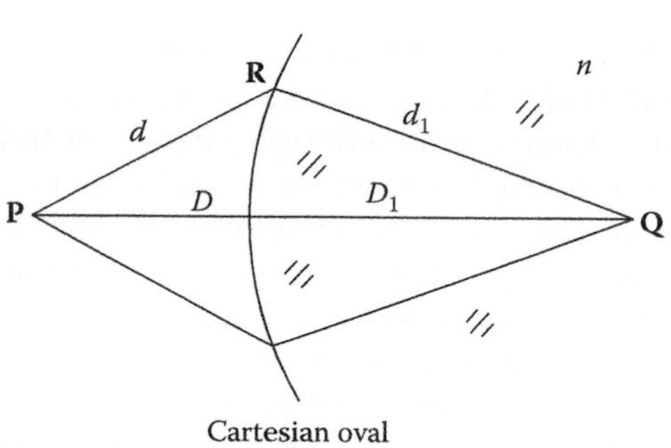

FIGURE 2.2
To solve the problem of forming an image through an optical system, we can start by trying to focus light coming from a point on the image onto a point on the object. A way to achieve this is by using a Cartesian oval. In this case, each point on the surface is crossed by just one ray of light coming from the object. It is then possible to choose the slope of the surface so that convergence is guaranteed.

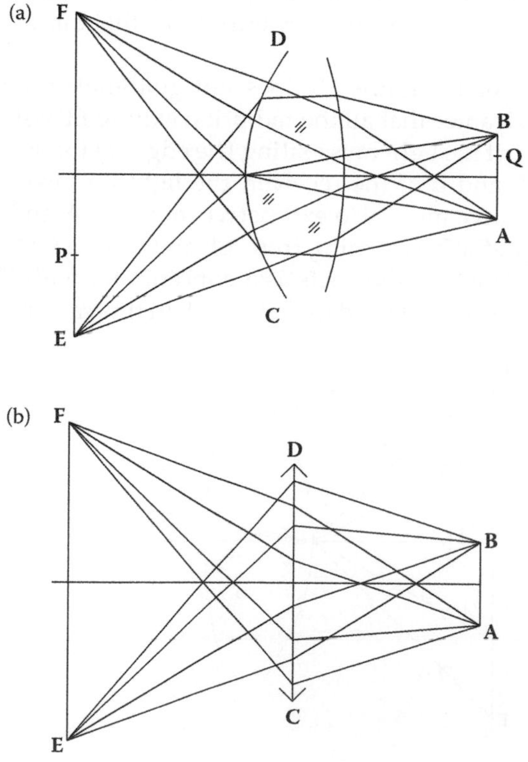

FIGURE 2.3
(a) A lens that focuses onto **A** and **B** the light coming from **F** and **E**, respectively. Note that **E** and **F** are edges of the "object" and that **A** and **B** are edges of the "image," and (b) the same optical system, but in a schematic way.

However, this new lens does not guarantee that light coming from an intermediate point **P** in the object is concentrated onto the corresponding point **Q** in the image, because there are not enough degrees of freedom to do so. To add new degrees of freedom, however, more surfaces must be added. Since a lens can have only two surfaces, more lenses must be added. To guarantee that the light coming from more points in the object is concentrated onto the corresponding points in the image, the systems become more complex. Eventually, this would lead us to systems having an infinite number of lenses.[6,7]

If we do not intend to increase the number of lenses, a new degree of freedom must be found, that allows the focusing of several points of the object onto the corresponding points in the image. One way is to consider a lens whose refractive index varies from point-to-point in its interior.[3,6,7] This kind of solution is, nonetheless, hard to implement, because it is difficult to build a material with a refractive index varying in accordance with the results of the calculations.

Owing to these and other difficulties in designing an ideal imaging device, the optical devices available do not produce perfect images, but images with aberrations. These arguments do not prove that it is impossible to make (build) a perfect imaging system, they only show that this task does not seem to be easy.

Although the lens of Figure 2.3 does not guarantee the formation of an image, it does guarantee that all the radiation exiting **EF** will eventually pass across **AB**. In fact, if the light rays exiting the edges of the source **E** and **F** pass through edges **A** and **B** of the receiver, the light rays exiting intermediate points **P** of the source must also exit between points **A** and **B**. Therefore, in this case, all the radiations coming from **EF** and hitting **CD** will end up concentrated at **AB**. This lens then acts as a concentrator with **EF** as source and **AB** as receiver. This is illustrated in Figure 2.4. In this case, ray r_1 coming from

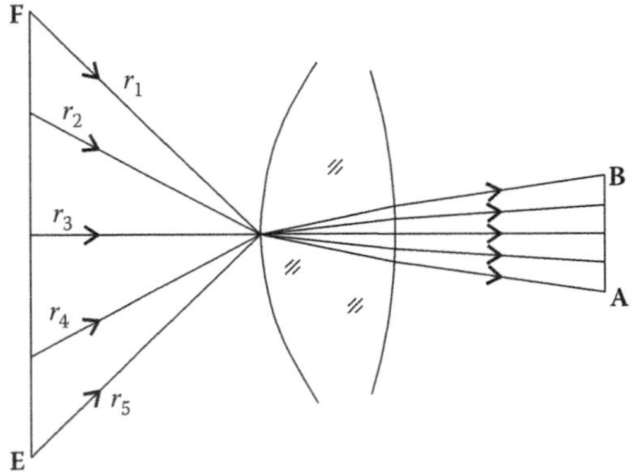

FIGURE 2.4
If ray r_1 coming from the edge **F** of the source is deflected to edge point **A** on the receiver, and ray r_5 coming from the edge **E** of the source is deflected to edge **B** of the receiver, all other rays—r_2, r_3, r_4—coming from intermediate points in source **EF** will end between points **A** and **B** on the receiver.

edge point **F** of the source is deflected toward edge point **A** of the receiver and ray r_5 coming from edge **E** of the source is deflected to edge point **B** of the receiver. Therefore, rays r_2, r_3, and r_4, coming from intermediate points in the source, are deflected to intermediate points on the receiver.

Generally, nevertheless, the light rays coming from a point **P** in the object, as shown in Figure 2.3, will not converge onto a point **Q**, so that no image will be formed at **AB**.

As seen, many degrees of freedom are required for the design of an imaging system, because the formation of an image imposes a large number of conditions that must be fulfilled simultaneously. From these results the difficulty of designing a perfect imaging device, since the number of available degrees of freedom for the design of an optical system is usually not sufficient. If the objective is, nonetheless, just to transfer the energy from a source to a receiver, image formation is unnecessary. Instead, it suffices to require that the light rays coming from the edges of the source are transformed into rays going to the edges of the receiver, as shown in Figure 2.4. Now there are far fewer requirements, and only a small number of degrees of freedom will result in an ideal device.

If the light source is displaced to infinity, becoming infinitely large, the situation presented in Figure 2.3 becomes that of Figure 2.5.

In this case, the incoming radiation can be characterized by the angular aperture θ. This lens now works as a device concentrating onto **AB** all the radiation with half-angular aperture θ falling on **CD**. This device must be designed such that the parallel rays d_1 are concentrated onto **A** and the

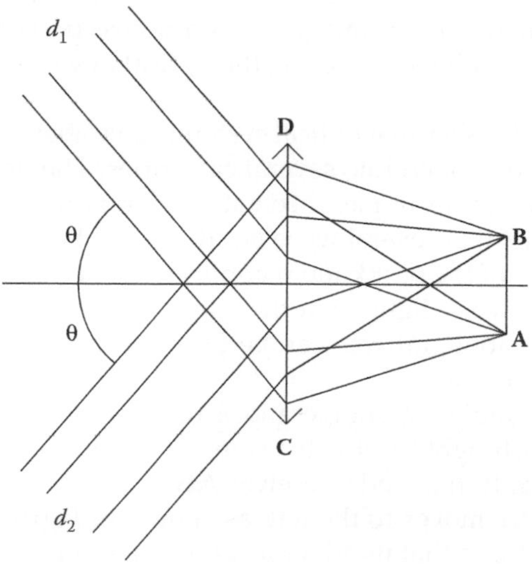

FIGURE 2.5
The limit case of Figure 2.3b, in which the edge points **E** and **F** are displaced to infinity. Now the radiation arriving to the optical system **CD** has an angular aperture θ for each side. Edge rays d_1 are concentrated onto point **A** and edge rays d_2 are concentrated onto point **B**.

parallel rays d_2 are concentrated onto **B**. In this manner, all the radiation falling on the device, making an angle to the optical axis smaller than θ, must pass between **A** and **B**.

We can also compare the optical devices presented in Figure 2.1 and 2.3. In both the cases, the condition is such that the light coming from **EF** must pass through **AB**. In the case of the device presented in Figure 2.1, it is also required that light coming from **F** must be concentrated onto **A**, and that the light coming from **E** must be concentrated onto **B**. Besides, light coming from any other point **P** must be concentrated onto a point **Q** on the image, being the distances d_0 and d_i of **P** and **Q** to the optical axis related by Equation 2.1.

In the case of the device presented in Figure 2.3 the only requirement is that the light coming from **F** must be concentrated onto **A**, and that the light coming from **E** must be concentrated onto **B**. The light coming from a generic point **P** of the object will not be necessarily concentrated onto any point along **AB**, so generally no image will be formed.

The device presented in Figure 2.1 is imaging, and the one presented in Figure 2.3 is nonimaging. Note that both perform the same when used as radiation collectors.

2.3 The Compound Parabolic Concentrator

As described earlier, nonimaging devices can be used as concentrators. In this case, the formation of an image is not a necessary condition. The only condition is that the radiation entering the optical device ends up being concentrated at its exit.

It was mentioned earlier that optical systems have aberrations. As a matter of fact, these can be divided into several categories. The device presented in Figure 2.3 can have, for example, chromatic aberrations.[1,2,8] This nonideality results from the fact that several wavelengths of light are refracted in different directions. One of the best known applications of this effect is the use of prisms to separate white light from the sun, into its several spectral colors. To avoid this aberration, mirrors can be used, because all wavelengths are reflected in the same way.

We start with a radiation source and a receiver onto which we want to concentrate as much light as possible coming out of the source. Figure 2.6a shows a source (emitter) E_1 and a receiver **AB**.

If now this source moves to the left, as shown in Figure 2.6b, and grows in size from E_1 to E_2, so that its edges always touch the rays r_1 and r_2, which make an angle 2θ between each other, the radiation field at **AB** will tend to be the one in Figure 2.7, in which the receiver **AB** is shown in a horizontal orientation. At each point, the receiver **AB** "sees" the incoming radiation contained between two edge rays that make an angle 2θ between each

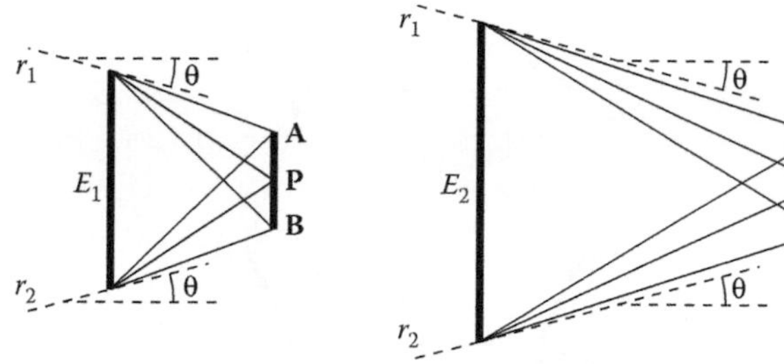

FIGURE 2.6
As the source E moves to the left, and grows so that its edges always touch the rays r_1 and r_2, its size will be E_1, E_2, ... The radiation received at **AB** tends to be confined at every point to an angle 2θ.

other. These edge rays are coming from the infinite source E at an infinite distance.

Our goal is to concentrate this radiation to the maximum possible extent, that is, to send the maximum power through the aperture **AB**. Our approach is to let **AB** be the exit aperture of the device, and then generate mirror profiles upward from points **A** and **B**. We may start with simple flat mirrors, placing one on point **A**, and another on point **B**. Owing to the symmetry of the problem about the vertical line through mid point **P**, these mirrors are also symmetrical. This situation is presented in Figure 2.8.

To deflect onto **AB** the maximum possible radiation, angle β must be as small as possible, so that the entrance aperture $\mathbf{C_1D_1}$ can be as large as possible. But there is a limit to the minimum value of β, which is reached when the ray of light r_1, reflected at $\mathbf{D_1}$, is redirected to point **A**. If β is smaller, there will be rays reflected by $\mathbf{BD_1}$ onto $\mathbf{AC_1}$, and from there, away from **AB**. After placing the first mirror, a second one can be added above it. Figure 2.9 presents this possibility.

Also in this case, the slope of the mirrors is chosen so as to maximize the width of the entrance aperture, which is now $\mathbf{C_2D_2}$. Again this means that this mirror must redirect the edge rays coming from the left, so that the ray r_2 is reflected at $\mathbf{D_2}$ toward point **A**. We can now add more and more mirrors

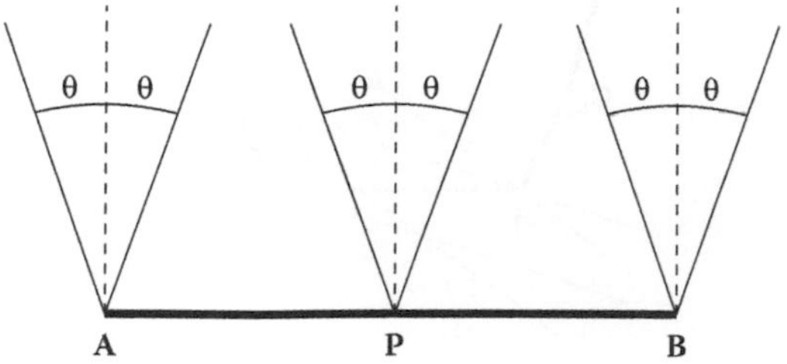

FIGURE 2.7
Uniform radiation of angular aperture θ for each side, and falling on a surface **AB**.

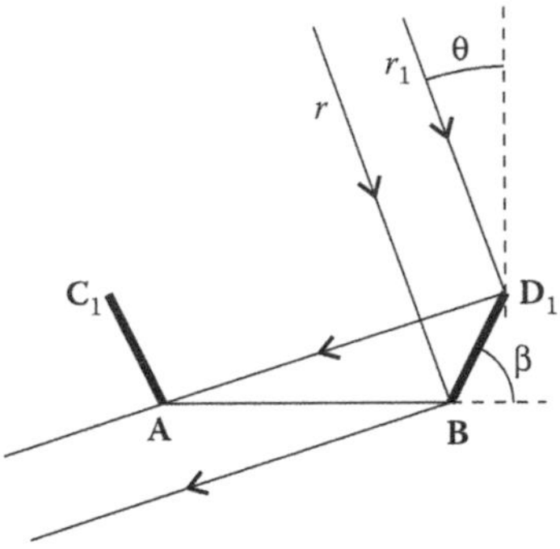

FIGURE 2.8
To concentrate radiation onto **AB**, we can place mirrors at **A** and **B**. To capture the maximum amount of radiation, entrance aperture $\mathbf{C_1D_1}$ must also be a maximum. Therefore, the angle β that these mirrors make with the horizontal must be a minimum. This minimum value of β is obtained when the edge ray r_1 coming from the left, and falling on $\mathbf{D_1}$, is reflected toward **A**. If β decreases, this light ray would be reflected at $\mathbf{D_1}$, then at mirror $\mathbf{AC_1}$, and from there would be reflected away from **AB**. Mirror $\mathbf{AC_1}$ is symmetrical to $\mathbf{BD_1}$.

atop one another. These mirrors have a finite size, but they can be made as small as desired. As this happens, more and more smaller mirrors can be added. The mirrors together tend to adjust to a curve. This situation is presented in Figure 2.10. Angle β, which was minimized previously for each small mirror, is now the slope of the curve and must also be minimized at each point.

Considering the way this curve is defined, it must deflect onto a point **A** the edge rays *r* coming from the left. We then have a curve that deflects a set of parallel rays onto a point. The geometrical curve having this characteristic is a parabola, so that the curve is a parabola with its axis parallel to the edge rays *r* coming from the left, and having its focus at point **A**. It can also be

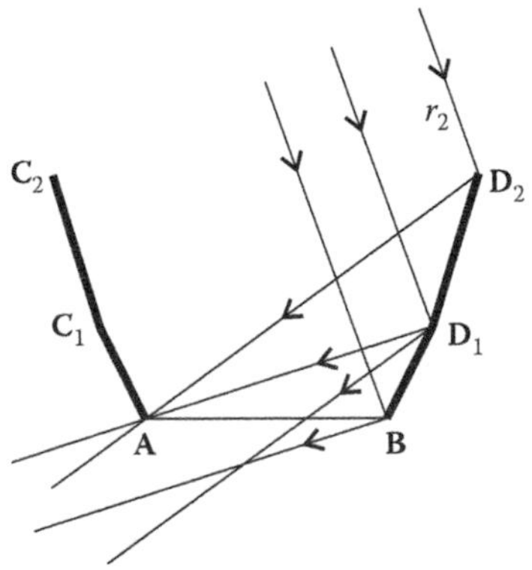

FIGURE 2.9
Using the same method presented in Figure 2.8, it is now possible to add new mirrors at points $\mathbf{C_1}$ and $\mathbf{D_1}$, enlarging even more the dimension of the entrance aperture that now becomes $\mathbf{C_2D_2}$.

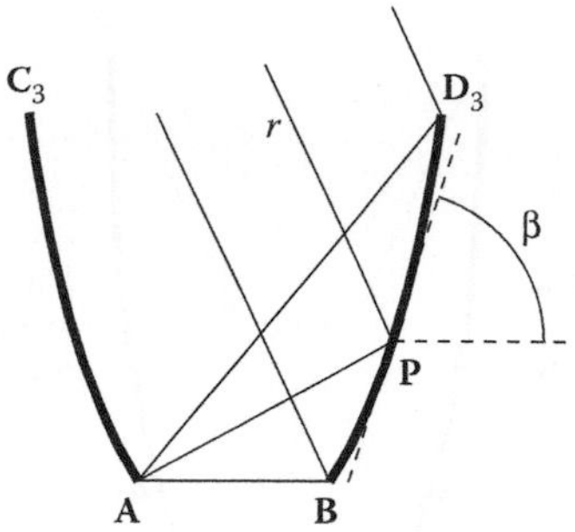

FIGURE 2.10
The procedure presented in Figure 2.9 can now be extended by adding more mirrors and diminishing their size.

noted that this curve is the one that, at each point **P**, produces the smallest value for β, that is, the one that leads to a maximum entrance aperture $\mathbf{C_3D_3}$.

As can be seen, from Figure 2.11, if the parabola is extended upward, there comes a point where it starts tilting inside, reducing the size of the entrance aperture.

When this happens, the top of the right mirror starts to shadow the bottom of the left, and vice versa. Since we are interested in obtaining the maximum possible entrance aperture, the parabolas must be cut at line **CD** where the distance between them is maximum. The final concentrator must then look like Figure 2.12.

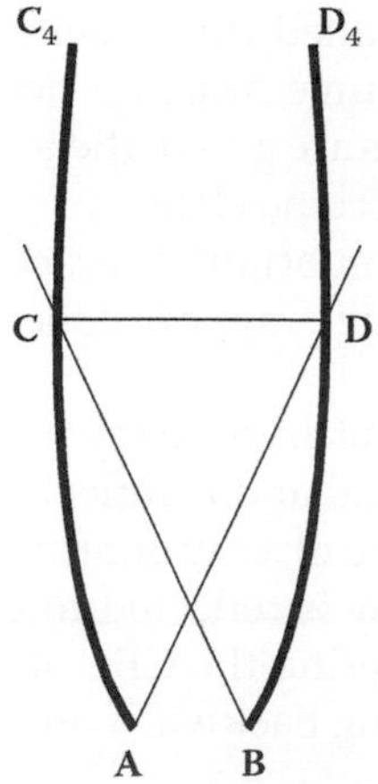

FIGURE 2.11
As the parabolas are extended upward, the distance between the mirrors increases until a maximum **CD** is reached, and then starts to decrease. Also, portions $\mathbf{DD_4}$ and $\mathbf{CC_4}$ of the mirror shadow the other portions of mirror **AC** and **BD**, respectively. Since the goal is to maximize the size of the entrance aperture, the parabolas must be cut at **CD**.

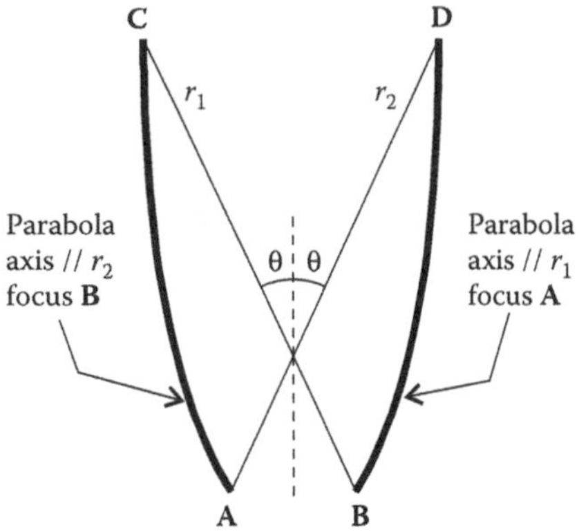

FIGURE 2.12
A CPC is a concentrator with entrance aperture **CD**, that accepts radiation, making a maximum angle of ±θ with the vertical, and concentrates it into **AB**.

The profile of this device consists of two parabolic arcs, **AC** and **BD**. The arc **BD** is part of a parabola having its axis parallel to direction **BC** (i.e., tilted θ to the left) and focus **A**. Arc **AC** is symmetrical to **BD**.[9–14] It is called the CPC because of these two parabolic arcs.

Note that the initial goal was a concentrator having the largest possible entrance aperture. The design at which we arrived, is a combination of two curves deflecting the rays coming from the edges of the source of radiation onto the receiver's edges. This is the basic principle in the design of nonimaging concentrators, and is called the edge-ray principle—light rays coming from the edges of the source must be deflected onto the edges of the receiver.[15–17] As more examples are given, the terms "edges of the source" and "edges of the receiver" will become clear.

We can now analyze an important characteristic of this device. Figure 2.13 shows how the parallel edge rays are concentrated onto the edges of the receiver.

Figure 2.14a shows the path of an edge ray inside a CPC. This ray enters the CPC at an angle θ to the vertical and is reflected toward the receiver's edge.

Figure 2.14b presents the case of a ray entering the CPC making an angle to the vertical $\theta_1 < \theta$. Now the ray is reflected toward the receiver. Figure 2.14c presents the case of a ray entering the CPC at an angle $\theta_2 > \theta$. The ray, after some reflections, ends up going backward and exiting through the entrance aperture.

This behavior of the rays inside a CPC is general, in the sense that all the rays entering the CPC with an angle $\theta_1 < \theta$ hit the receiver, and all the rays entering the CPC with an angle $\theta_2 > \theta$ reflect on its walls until they exit the CPC through the entrance aperture. A ray ending on the receiver is said to be accepted, and a ray that goes back again is said to be rejected. The ratio

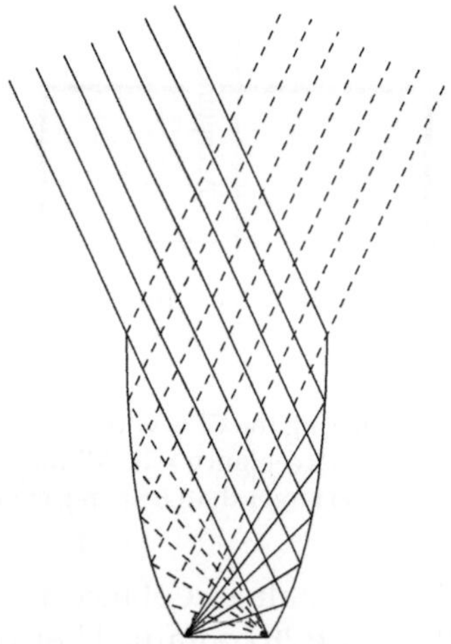

FIGURE 2.13
Trajectories of the edge rays inside a CPC.

between the number of accepted rays, and the number of rays entering the CPC, is called the acceptance:

$$\text{Acceptance} = \frac{\text{Number of rays hitting the receiver}}{\text{Number of rays entering the CPC}} \tag{2.2}$$

Therefore, for $\theta_1 < \theta$ and $\theta_1 > -\theta$, the acceptance is 1 (all the rays entering the CPC hit the receiver) and for $\theta_2 > \theta$ or $\theta_2 < -\theta$, the acceptance is 0 (all the rays entering the CPC are rejected, ending with exit through the entrance aperture). Therefore, the acceptance of a CPC has the shape presented in

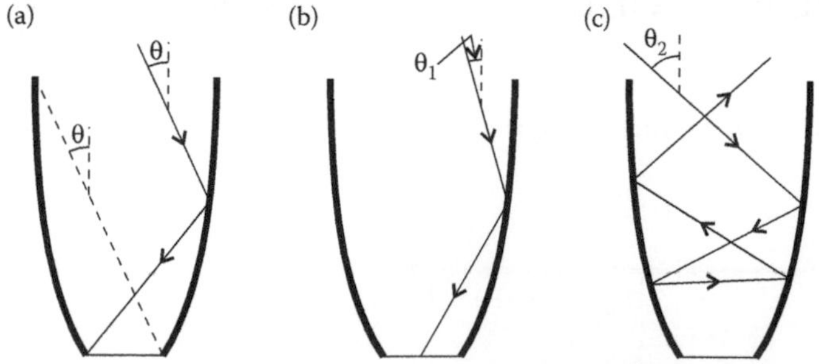

FIGURE 2.14
Trajectories of three kinds of rays inside a CPC. (a) A ray entering the CPC at an angle to the vertical of half-acceptance angle θ. This ray is reflected to the edge of the receiver. (b) A ray entering the CPC at an angle to the vertical smaller than θ is accepted (hits the receiver). (c) A ray entering the CPC at an angle larger than θ is rejected by retroreflection (ends up exiting through the entrance aperture).

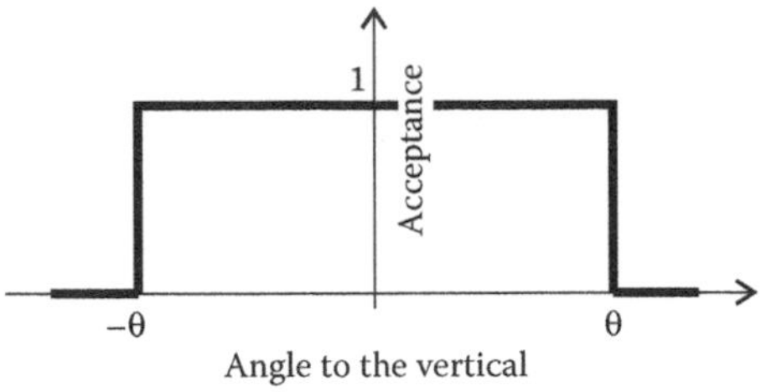

FIGURE 2.15
Acceptance of a CPC. All the rays entering the CPC with an angle to the vertical (axis of symmetry) smaller than θ hit the receiver (acceptance = 1). All the rays entering the CPC with an angle to the vertical (axis of symmetry) larger than θ are rejected (acceptance = 0).

Figure 2.15. Angle θ is called the half-acceptance angle since the CPC accepts all the radiation within the angle 2θ contained between −θ and +θ.

The concentrator obtained earlier must be one capable of delivering the maximum possible concentration because it was designed so as to maximize the size of the entrance aperture without losses of radiation.

We can now calculate the concentration that such a device attains. To do this, we need to remember a property of the parabola presented in Figure 2.16. If a line passing through **A** and **B** is perpendicular to the optical axis, we have [**A**, **C**] + [**C**, **F**] = [**B**, **D**] + [**D**, **F**], where **F** is the focus and **AC** and **BD** are rays parallel to the optical axis. Here, [**X**, **Y**] represents the distance between two arbitrary points **X** and **Y**.

In Figure 2.17, we have a CPC with entrance aperture a_1 and exit aperture a_2. The half-acceptance angle is θ. Parabola **BD** has focus **A** and its axis parallel to **BC**. From the property of the parabola mentioned earlier, we can write[7,18]:

$$[\mathbf{C},\mathbf{B}] + a_2 = [\mathbf{E},\mathbf{D}] + [\mathbf{D},\mathbf{A}] \Leftrightarrow a_2 = a_1 \sin\theta \Leftrightarrow \frac{a_1}{a_2} = \frac{1}{\sin\theta} \tag{2.3}$$

since [**C**, **B**] = [**D**, **A**] and [**E**, **D**] = a_1 sinθ.

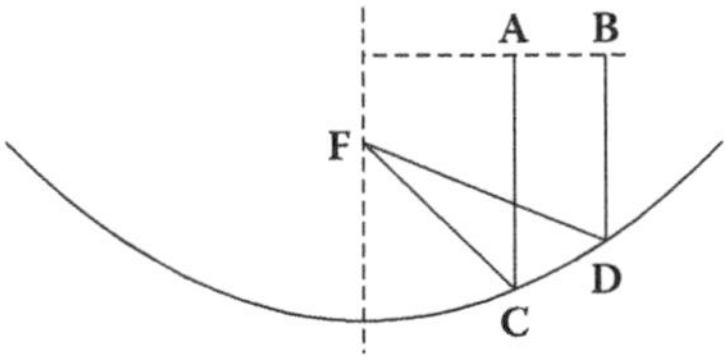

FIGURE 2.16
In a parabola, the path length of two light rays **ACF** and **BDF** is the same, as long as **A** and **B** are placed on a line perpendicular to the optical axis, and **AC** and **BD** are parallel to the optical axis.

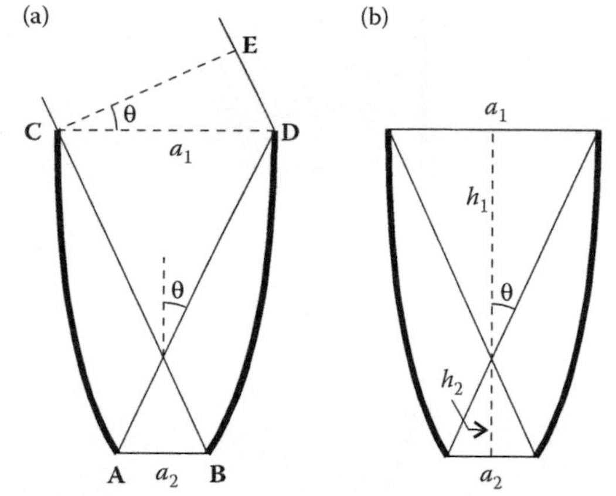

FIGURE 2.17
(a) The maximum concentration of a CPC, (b) its height is $h = h_1 + h_2$.

We now have a relationship between the sizes of the entrance aperture and exit aperture for the concentrator we derived.

Line **CE** is perpendicular to the edge rays coming from the left. The optical path length between the wavefront **CE** and the focus **A** is the same for all the edge rays perpendicular to **CE**.

It is also possible to obtain the height h of the CPC. From Figure 2.17b we obtain:

$$h = h_1 + h_2 = \frac{a_1/2}{\tan\theta} + \frac{a_2/2}{\tan\theta} = a_1 \frac{1 + \sin\theta}{2\tan\theta} \tag{2.4}$$

Note that, when $\theta \to 0$, $h \to \infty$ so, for small acceptance angles, the CPC becomes very tall.[7,19]

The CPC, although ideal in two dimensions, is not ideal when made into a three-dimensional (3-D) device. Figure 2.18 shows a 3-D CPC with circular symmetry obtained by rotating the profile of a 2-D CPC around its axis of symmetry.

The CPC aperture has normal **n**. If now we consider a set of parallel rays at an angle α to **n**, we can trace those rays through the CPC and see how much of that light gets to the small exit aperture at the bottom. Figure 2.19 shows the result of such a calculation for CPCs designed for $\theta = 10°$, 20°, 40°, and 60° acceptance angles. For each one of these design angles, we have a transmission curve as a function of incidence angle α. As can be seen, the transmission (acceptance) is not a perfect step function; that is, it does not fall on a vertical straight line as in Figure 2.15 for the 2-D case. Instead, it falls off on a sharp curve. Therefore, the circular 3-D CPC is close to ideal, but not ideal. Some skew rays inside the design angle are rejected by the CPC. They keep bouncing around until they end up coming out through the entrance

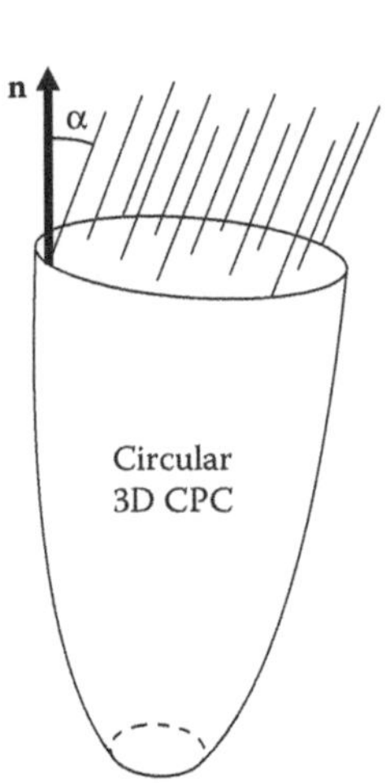

FIGURE 2.18
Circular 3-D CPC and a set of parallel rays at an angle α to the vertical (normal to the entrance aperture).

aperture again. Also, some rays outside the design angle end up hitting the small aperture.

If we look at the flux transmission inside the design angle θ, we see that all is not transmitted. These results are shown in Figure 2.20.

For the points of the large (entrance) aperture of the CPC, we consider all the light contained inside a vertical cone of angle θ, and see how much of that flux ends on the small exit aperture of the CPC. As we can see, the light transmitted inside the design angle is not 100%. This is due to the fact that the transmission is not ideal, either inside the design angle, as we can see from Figure 2.19, or some skew rays are rejected by the 3-D CPC. As the design angle θ increases, the transmission inside θ also increases. Note that as angle θ increases, the mirrors of the CPC get smaller, and more light hits the small aperture directly. Figure 2.21 shows a square CPC made of plastic.

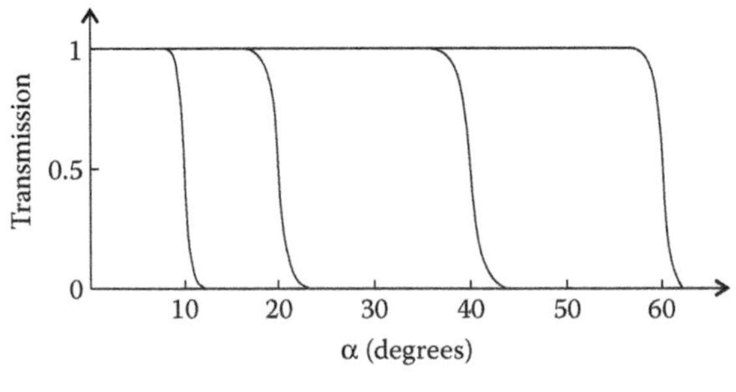

FIGURE 2.19
Transmission curves for circular CPCs designed for acceptance angles of θ = 10°, 20°, 40°, and 60°.

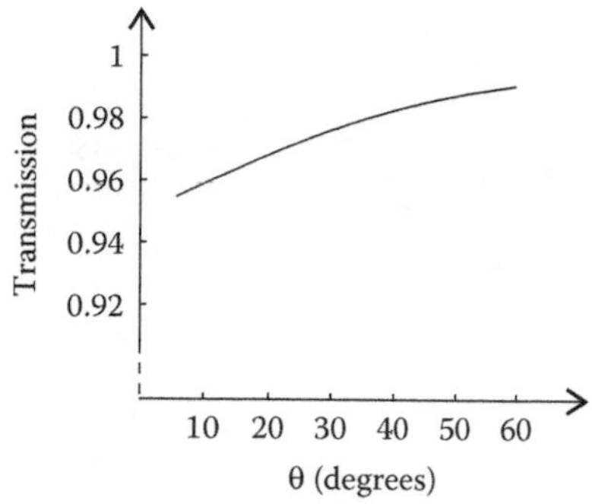

FIGURE 2.20
Total flux transmission inside the design angle θ for circular CPCs.

FIGURE 2.21
Square CPC made of plastic. (Courtesy of Light Prescriptions Innovators. With permission.)

2.4 Maximum Concentration

The CPC is a 2-D concentrator that was designed for maximum concentration. To verify that its concentration is, in fact, maximum, we use the second principle of thermodynamics.

We consider a trough optical system as in Figure 2.22. It extends to infinity in both directions, and consists of a cylindrical black body S_R of radius r

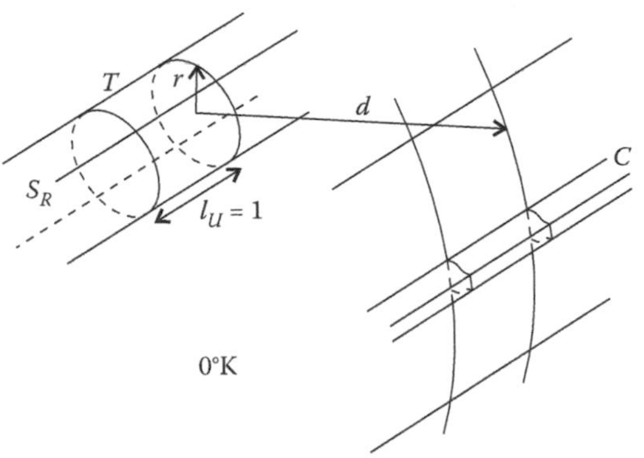

FIGURE 2.22
Linear system.

(on the left) at a temperature T, and emits light into space at a temperature of 0 K. As radiation travels through space, it eventually reaches an imaginary cylinder of radius d. On the face of this imaginary cylinder, there is a linear concentrator C.

A blackbody emitter of area dA at a temperature T emits Lambertian radiation, and the total flux (Watts) emitted into a hemisphere is given by[20,21]

$$d\Phi_{\text{hem}} = \sigma T^4 dA \tag{2.5}$$

where σ is the Stephan–Boltzmann constant. A length l_U of the cylindrical black body then emits a radiation flux given by

$$\Phi_U = 2\pi r \sigma T^4 l_U \tag{2.6}$$

In the case where $l_u = 1$ (i.e., when we consider a unit length), we obtain the flux emitted per unit length, which is given by

$$\Phi = 2\pi r \sigma T^4 \tag{2.7}$$

The optical system of Figure 2.22 is shown again in Figures 2.23 and 1.22 (top view). The concentrator C has an entrance aperture of width a_1 and exit aperture of width a_2. Entrance aperture a_1 can only exchange radiation with the radiation source S_R or with the rest of the universe, which is at 0 K. The amount of radiation that a_1 receives per unit length is given by

$$\Phi = \sigma T^4 \frac{2\pi r}{2\pi d} a_1 \tag{2.8}$$

This power can now be concentrated without losses onto area a_2 by concentrator C.

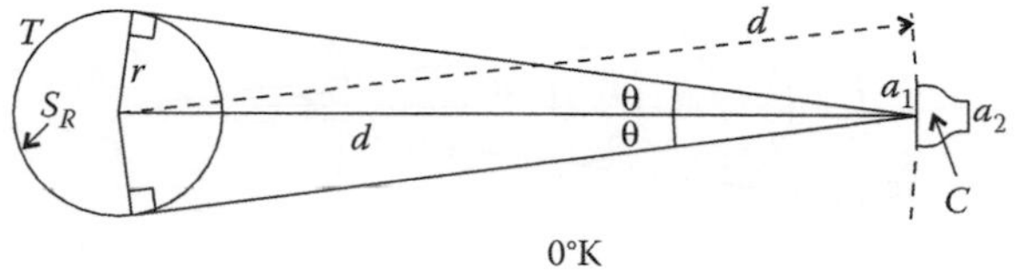

FIGURE 2.23
Top view of linear system.

Let us suppose that on its exit aperture a_2, concentrator C has a black body that absorbs radiation, and, therefore, gets heated up. The second principle of thermodynamics states that temperature T_{a2} of a_2 can never be higher than the temperature T of the radiation source S_R, that is, $T_{a2} \leq T$. If we had $T_{a2} > T$, we could place a heat engine working between a_2 and the S_R, and we would have perpetual motion engine, which is impossible. Let us then suppose that a_2 heats up to the maximum possible temperature, that is, the temperature T of S_R, where it stabilizes. In this case, it will emit a power per unit length given by

$$\Phi_2 = \sigma a_2 T^4 \tag{2.9}$$

To maintain a stable temperature, it is necessary that a_2 is in thermal equilibrium, that is, the radiation that it receives from S_R equals the radiation emitted to the exterior. In this case, we must have

$$\Phi = \Phi_2 \Leftrightarrow a_2 = a_1 r/d \Leftrightarrow a_2 = a_1 \sin\theta \tag{2.10}$$

Note that a_2 exchanges radiation with S_R through the entrance a_1 of the concentrator. The radiation exiting a_1 and coming from a_2 can only be headed to S_R. In fact, if a_2 could send radiation to space, it could also receive radiation from space, which is at 0 K and, in this case, it could not attain the temperature of S_R. The acceptance angle of the device having entrance aperture a_1 and exit aperture a_2, cannot be higher than angle θ represented in Figure 2.23. This means that concentrator C cannot accept any radiation that could come from a direction outside the angle 2θ. Accordingly, the radiation emitted by a_2 and exiting through a_1 must be confined to the same angle 2θ.

In Figure 2.23, the entrance aperture a_1 is curved with radius d. However, we can make the cylindrical source S_R larger, and push it further to the left, so that r/d = constant and, therefore, angle θ is also constant, as shown in Figure 2.24.

As the radius d of the entrance aperture a_1 of the concentrator C increases, it tends to a flat surface (or straight line in two dimensions). In this limit case, the maximum concentration is also given by expression (2.10) that is,

$$\frac{a_1}{a_2} = \frac{1}{\sin\theta} \tag{2.11}$$

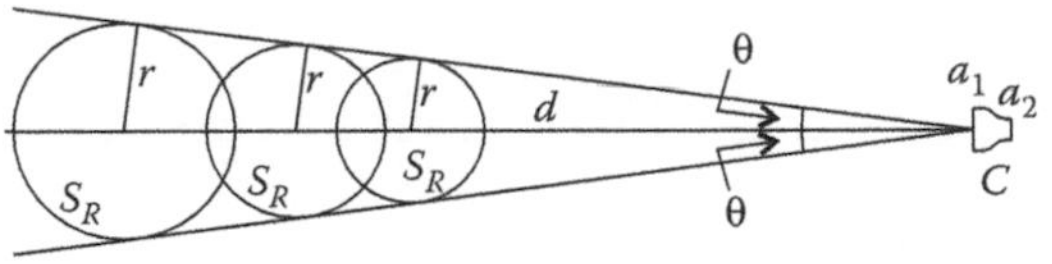

FIGURE 2.24
The cylindrical radiant source S_R gets larger while maintaining the ratio r/d and, therefore, the angle θ is at the entrance aperture a_1 of concentrator **C**.

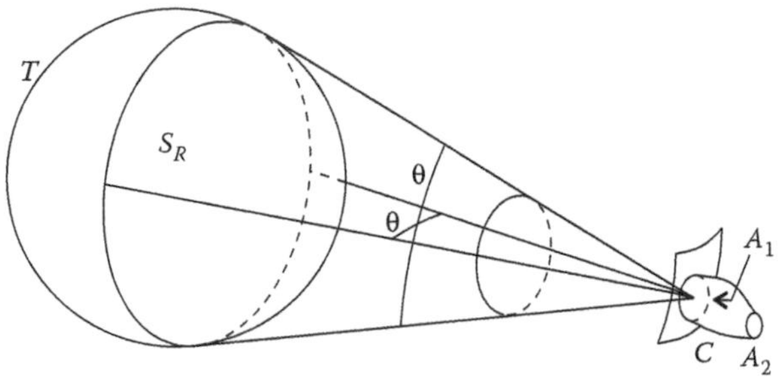

FIGURE 2.25
A concentrator C collects and concentrates radiation emitted by a spherical source S_R.

which is the same value we obtained for the concentration of the CPC. Thus, it can be concluded that the CPC is, in fact, an ideal concentrator.

A similar reasoning can be used to calculate the maximum possible concentration for 3-D concentrators. Now, instead of the source S_R being an infinite cylinder, it is a sphere, as shown in Figure 2.25. The concentrator C has an entrance aperture of area A_1 and an exit aperture of area A_2, and the source defines at A_1 a circular cone of half-angle θ. An example of one of these optical systems is when source S_R is the sun, and concentrator C is on earth collecting and concentrating the sun's energy.

Figure 2.26 shows a vertical cut of this setup, where the source S_R has a radius r, temperature T, and emits radiation into space, which is at a temperature of 0 K. As the emitted radiation travels through space, it will eventually illuminate an imaginary spherical surface of radius d. On this spherical

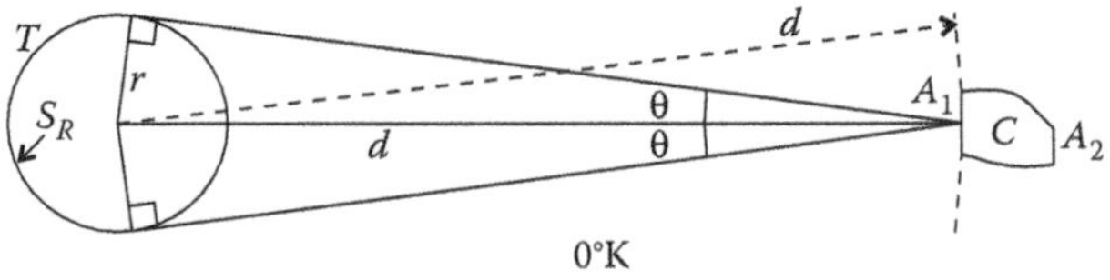

FIGURE 2.26
A vertical cut of the geometry of Figure 2.25 showing a spherical source S_R at a temperature T, emitting radiation to space, and a concentrator C whose entrance aperture A_1 is on a sphere of radius d.

surface, we have the entrance aperture A_1 of the concentrator C, which concentrates the radiation that falls on A_1 onto exit aperture A_2.

The flux emitted by the spherical source S_R is given by

$$\Phi = 4\pi r^2 \sigma T^4 \tag{2.12}$$

and the radiation that A_1 captures, is given by

$$\Phi_{A1} = \frac{4\pi r^2 \sigma T^4}{4\pi d^2} A_1 \tag{2.13}$$

This radiation will be concentrated onto a black body placed at A_2, which will heat up to a maximum temperature T, equal to that of the source S_R. The radiation that A_2 emits must equal to the one it receives to maintain thermal equilibrium. We have, in this case,

$$\frac{4\pi r^2 \sigma T^4}{4\pi d^2} A_1 = A_2 \sigma T^4 \Leftrightarrow \frac{A_1}{A_2} = \frac{d^2}{r^2} \Leftrightarrow \frac{A_1}{A_2} = \frac{1}{\sin^2\theta} \tag{2.14}$$

Although, in this construction, the entrance aperture A_1 is on a sphere of radius d, as d goes to infinity and the source grows in the way shown in Figure 2.24, in which the angle θ is kept constant, the entrance aperture will tend to a flat surface.

A further generalization of this result is obtained when the concentrator C is made of a material of refractive index n. In this case, the black body at A_2 is immersed in this medium of refractive index n and, for its radiation emission, we must use the value of the Stephan–Boltzmann constant σ in a material of refractive index n, which is given by[22]

$$\sigma = n^2 \frac{2\pi}{15} \frac{k^4}{c_0 h^3} = n^2 \sigma_V \tag{2.15}$$

where $\sigma_V = 5.670 \times 10^{-8}$ Wm^{-2}K^{-4} is the value it has in vacuum ($n = 1$), k the Boltzmann constant, h the Planck's constant, and c_0 the speed of light in vacuum. Source S_R continues to be in vacuum and, therefore, we continue to use $n = 1$. Expression (2.14) now becomes

$$\frac{4\pi r^2 \sigma_V T^4}{4\pi d^2} A_1 = A_2 n^2 \sigma_V T^4 \Leftrightarrow \frac{A_1}{A_2} = n^2 \frac{d^2}{r^2} \Leftrightarrow \frac{A_1}{A_2} = \frac{n^2}{\sin^2\theta} \tag{2.16}$$

Because A_2 now emits n^2 times more light, the light concentration may then be n^2 times higher.

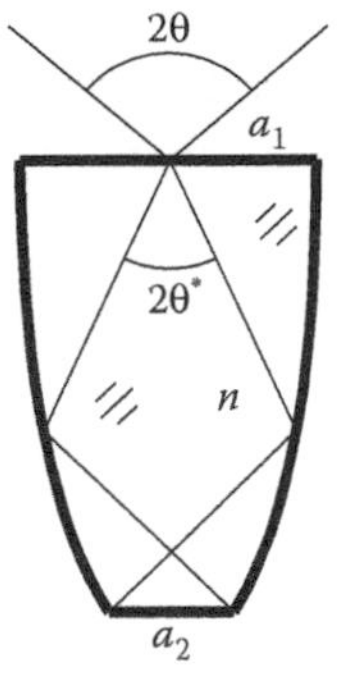

FIGURE 2.27
The light entering a dielectric CPC refracts, and its angular aperture diminishes from 2θ to 2θ*. The acceptance of the CPC is still 2θ, but since it is dielectric, it must be designed for the light angular aperture 2θ* after refraction.

In 2-D geometry, this expression becomes

$$\frac{a_1}{a_2} = \frac{n}{\sin\theta} \tag{2.17}$$

It may be seen that the CPC attains this ideal concentration by considering the case in which the CPC is made of a material of refractive index n as shown in Figure 2.27.

In this case, when the light enters the CPC, it refracts, and its angular aperture diminishes from 2θ to 2θ*, where $\sin\theta = n \sin\theta^*$. For the dielectric CPC, we have $a_1 \sin\theta^* = a_2$ and expression (2.17) follows.

The maximum concentration that a concentrator can provide is $C_{max} = n/\sin\theta$, as given by expression (2.17) and, in the case where $n = 1$, (a concentrator filled with air), the maximum concentration becomes $C_{max} = 1/\sin\theta$. Nonimaging concentrators may reach (or get close to) this maximum limit, and this makes them very important in solar energy concentration (see Section 4.12, after Equation 4.77).

2.5 Examples

The following examples use expressions for the curves and functions that are derived in Chapter 21.

EXAMPLE 2.1

Design a CPC for an acceptance angle of 30° and a receiver of unit length.

We start by calculating the general expression for the mirrors of a CPC and then apply them to the particular case in which the acceptance angle is 30°. A general CPC for an acceptance angle θ is shown in Figure 2.28.

It consists of two symmetrical parabolic arcs. The parabola on the right has focus **F**, passes through point **P**, and its axis r is tilted by an angle $\alpha = \pi/2 + \theta$ to the horizontal.

A parabola with focus $\mathbf{F} = (F_1, F_2)$, tilted by an angle α to the horizontal, and passing through a point **P**, can be parameterized as

$$\frac{\sqrt{(\mathbf{P}-\mathbf{F})\cdot(\mathbf{P}-\mathbf{F})} - (\mathbf{P}-\mathbf{F})\cdot(\cos\alpha, \sin\alpha)}{1-\cos\phi}(\cos(\phi+\alpha), \sin(\phi+\alpha)) + (F_1, F_2) \tag{2.18}$$

where the parameter ϕ is the angle to the axis of the parabola as shown in Figure 2.29.

In the particular case of the right-hand side parabola of the CPC in Figure 2.28, we can make $\mathbf{F} = (-a, 0)$ and $\mathbf{P} = (a, 0)$ with $a > 0$. Replacing these values in the expression for the parabola, we get

$$\left(a\frac{1-\cos(\phi+2\theta)+2\sin(\phi+\theta)}{\cos\phi - 1}, a\frac{\cos(\phi+\theta)}{\sin^2(\phi/2)}(1+\sin\theta) \right) \tag{2.19}$$

with $3\pi/2 - \theta \le \phi \le 2\pi - 2\theta$. The left-hand side of the CPC is obtained by symmetry about the x_2-axis (by changing the sign of the first component).

Now, we may apply this result to the particular case of a CPC, with an acceptance angle of 30°. We assume that the small aperture **FP** has a unit length so that $\mathbf{F} = (-0.5, 0)$ and $\mathbf{P} = (0, 0.5)$. We also have $\theta = 30\pi/180$ rad. Replacing these values in the above expression for the right-hand side parabola, we obtain

$$\left(\frac{0.5(1-\cos(\pi/3+\phi)+2\sin(\pi/6+\phi))}{\cos\phi - 1}, 0.75\frac{\cos(\pi/6+\phi)}{\sin^2(\phi/2)} \right) \tag{2.20}$$

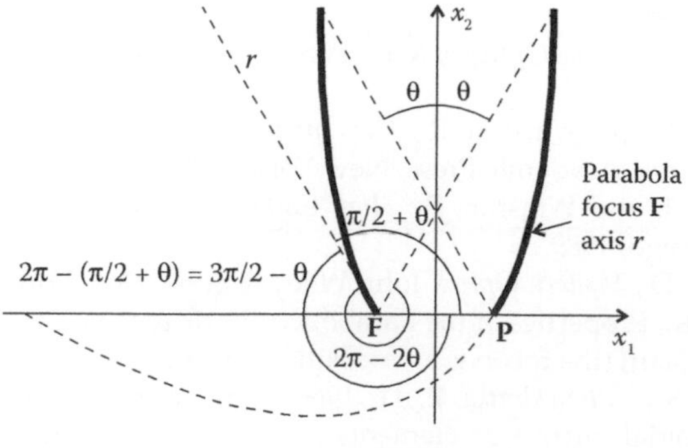

FIGURE 2.28
A CPC is composed of two parabolic arcs tilted by an angle θ to the vertical. The right-hand side arc is tilted counterclockwise and the left-hand side one is its symmetrical.

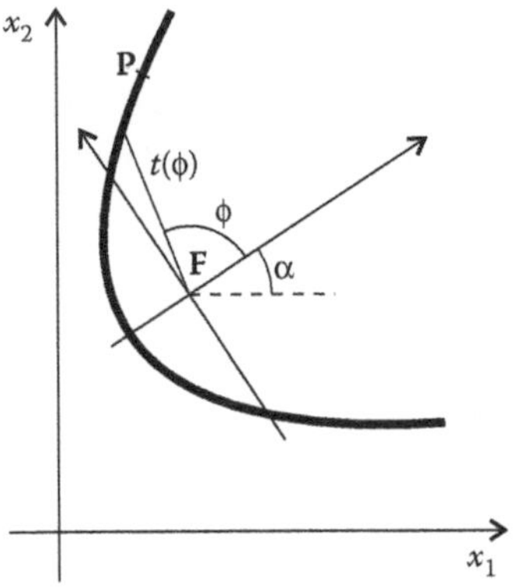

FIGURE 2.29
Parabola with focus **F**, tilted by an angle α to the horizontal and passing through a point **P**.

for $4\pi/3 \leq \phi \leq 5\pi/3$. The left-hand side parabola is obtained by symmetry around the vertical axis, that is, by changing the sign of the first component of the parameterization.

References

1. Hecht, E., *Optics*, 3rd ed., Addison-Wesley Longman, Inc., Reading, Massachusetts, 1998.
2. Jenkins, F. A. and White, H. E., *Fundamentals of Optics*, 3rd ed., McGraw-Hill Book Company, New York, 1957.
3. Luneburg, R. K., *Mathematical Theory of Optics*, University of California Press, Berkeley and Los Angeles, 1964.
4. Meyer-Arendt, J. R., *Introduction to Classical and Modern Optics*, Prentice-Hall, New Jersey, 1989.
5. Stavroudis, O. N., *The Optics of Rays, Wave Fronts, and Caustics*, Academic Press, New York, 1972.
6. Welford, W. T. and Winston, R., *The Optics of Nonimaging Concentrators—Light and Solar Energy*, Academic Press, New York, 1978.
7. Welford, W. T. and Winston, R., *High Collection Nonimaging Optics*, Academic Press, San Diego, 1989.
8. Guenther, R. D., *Modern Optics*, John Wiley & Sons, New York, 1990.
9. Baranov, V. K., Properties of the Parabolico-thoric focons, *Opt.-Mekh. Prom.*, 6, 1, 1965 (in Russian) (the focon is a "focusing cone").
10. Baranov, V. K., *Geliotekhnika*, 2, 11, 1966. (English translation: Baranov, V. K., Parabolotoroidal mirrors as elements of solar energy concentrators, *Appl. Sol. Energy*, 2, 9, 1966.)

11. Ploke, M., Lichtführungseinrichtungen mit starker Konzentrationswirkung, *Optik*, 25, 31, 1967. (English translation of title: A light guiding device with strong concentration action.)
12. Hinterberger, H. and Winston, R., Efficient light coupler for threshold Čerenkov counters, *Review of Scientific Instruments*, 37, 1094, 1966.
13. Winston, R., Radiant energy collection, United States Patent 4.003.638, 1977.
14. Winston, R., Radiant energy collection, United States Patent 3.923.381, 1975.
15. Ries, H. and Rabl, A., Edge-ray principle of nonimaging optics, *J. Opt. Soc. Am. A*, 11, 2627, 1994.
16. Davies, P. A., Edge-ray principle of non-imaging optics, *J. Opt. Soc. Am. A*, 11, 1256, 1994.
17. Benitez, P. and Miñano, J. C., Offence against the Edge Ray Theorem?, *Nonimaging Opt.ics and Efficient Illumination Systems*, SPIE Vol. 5529, 108, 2004.
18. Winston, R., Light collection within the framework of geometrical optics, *J. Opt. Soc. Am.*, 60, 245, 1970.
19. Rabl, A., *Active Solar Collectors and Their Applications*, Oxford University Press, New York, Oxford.
20. Siegel, R. and Howell, J. R., *Thermal Radiation Heat Transfer*, McGraw-Hill Book Company, New York, 1972.
21. Sparrow, E. M. and Cess, R. D., *Radiation Heat Transfer—Augmented edition*, Hemisphere Publishing Corporation, Washington, London; McGraw-Hill, New York, 1978.
22. Rabl, A., Comparison of solar concentrators, *Sol. Energy*, 18, 93, 1976.

3

Design of Two-Dimensional Concentrators

3.1 Introduction

The compound parabolic concentrator (CPC) is a 2-D concentrator designed for capturing and concentrating a radiation field with a given angular aperture, onto a flat receiver. This radiation field can be thought of as being created by an infinitely large source at an infinite distance. The edge rays of the incoming radiation come from the edges of the (infinite) source, and are concentrated onto the edges of the receiver. This basic principle can be used to generate many other nonimaging devices. Its generalization is called the "edge-ray principle" and is the basis of nonimaging optics.

This chapter explores generalizations of the CPC design. These include, for example, different sources and receiver shapes, different light entrance and exit angular apertures, and nonparallel entrance and exit apertures.

3.2 Concentrators for Sources at a Finite Distance

The CPC was designed for an infinitely large source at an infinite distance. It is, however, possible to generalize the CPC design to other sources and receiver shapes. Figure 3.1 shows the case in which the radiation source has a finite size and is at a finite distance. Here we have a source **EF** and a receiver **AB**. We may now design an optic to concentrate as much radiation as possible coming out of **EF** onto **AB**. We will use the edge-ray principle, which tells us that the light rays coming from the edges of the source must be deflected to the edges of the receiver. In this case, the edges of the source are, naturally, points **E** and **F**, and the edges of the receiver are **A** and **B**. The edge-ray principle then states that we must concentrate the rays of light coming from **E** and **F** onto **A** and **B**. Similar to what has been done in the case of the CPC, here too, we will use mirrors. As seen in Figure 3.1, the upper mirror of this new concentrator must have a slope at each point **P**, such that it deflects the rays coming from edge point **E** of the source of radiation onto edge point

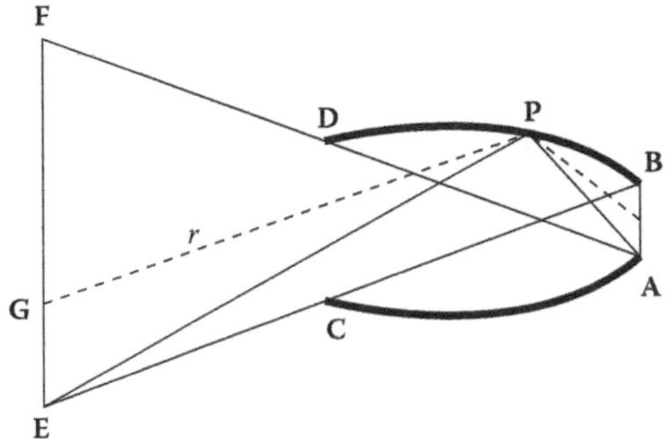

FIGURE 3.1
Optical device concentrating onto **AB** the radiation coming from **EF**. Each point **P** on the mirror **BD** must have a slope such that it reflects to edge point **A** of the receiver the ray of light coming from the edge **E** of the source. Mirror **DB** must then reflect to **A** the light rays coming from **E** and, therefore, it must be an elliptical arc having foci **E** and **A**. Mirror **CA** is symmetrical to **DB**. Because this receiver is composed of two elliptical arcs, it is called the CEC.

A of the receiver. This curve is then an ellipse with foci **E** and **A**, and passing through **B**. This construction principle ensures that any light ray coming from a point **G** on the source, and reflected at any point **P** on the reflector, will hit the receiver. The concentrator thus obtained is called the compound elliptical concentrator (CEC) because it is composed of two arcs of ellipses.[1]

To design one of these systems, we can start, for instance, by defining the source **EF** and receiver **AB**. Then the elliptical arc having foci **E** and **A**, and passing through **B**, can be drawn. This elliptical arc extends from point **B** until it finds line **AF** at point **D**. Elliptical arc **CA** can be obtained by symmetry. Note that the CPC obtained earlier is a particular case of this new configuration. If points **E** and **F** are displaced to infinity along lines **CB** and **AD**, respectively, the elliptical arcs tend to become two parabolic arcs, and the CEC turns into a CPC. As was the case with the CPC, the CEC is an ideal device.

The CEC can now be compared to an imaging system such as a lens.

Figure 3.2 compares a CEC with an ideal imaging lens. As can be seen, in the case of the lens, light coming from each point **P** in the source of light **EF** is concentrated onto a point **Q** in the image **AB**. Thus, an observer to the right of the image **AB** will not see light coming from the set of points **P** that forms **EF**, but instead sees light coming from points **Q**, forming **AB**. Therefore, instead of seeing **EF**, the observer sees **AB**, which is an image produced by the lens.

This does not happen with the CEC, where only light coming from edge points **E** and **F** is concentrated onto points **A** and **B**. For a generic point **P** of **EF**, there is no convergence to a point of **AB**; thus, no image is formed. For this reason, they are called nonimaging or anidolic devices.

Note that there is a similarity between the lens in Figure 2.3 and the CEC. The lens ideally guarantees the convergence onto point **A** of the rays of light coming from **F**, as well as the convergence to **B** of the rays of light coming

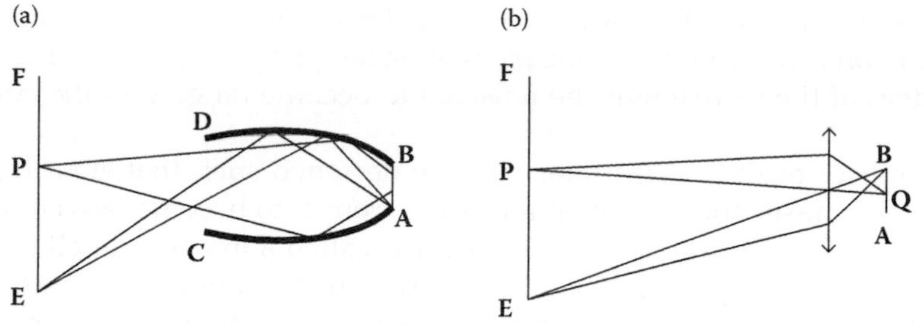

FIGURE 3.2
(a) The CEC is a nonimaging (anidolic) device. Thus, the rays of light exiting point **P** in the source of energy **EF** will not, in general, converge to a point on the receiver **AB**, so that no image is formed. (b) A quite different imaging system, where the rays of light exiting **P** meet at **Q** on **AB**.

from **E**, but it does not guarantee the convergence of the rays coming from **P** to **Q**. With the CEC, something similar happens: rays coming from **E** converge at **A**, and those coming from **F** converge at **B**, but there is no guarantee of convergence for the rays coming from **P**. These devices concentrate radiation, and transmit it in an ideal way, but lose image pattern, that is, information that the image might contain.

3.3 Concentrators for Tubular Receivers

Until now, only solutions for linear receivers were presented. Now consider receivers having convex shapes, such as circular. This is presented in Figure 3.3, where the edge rays for the source are still those coming from **E**

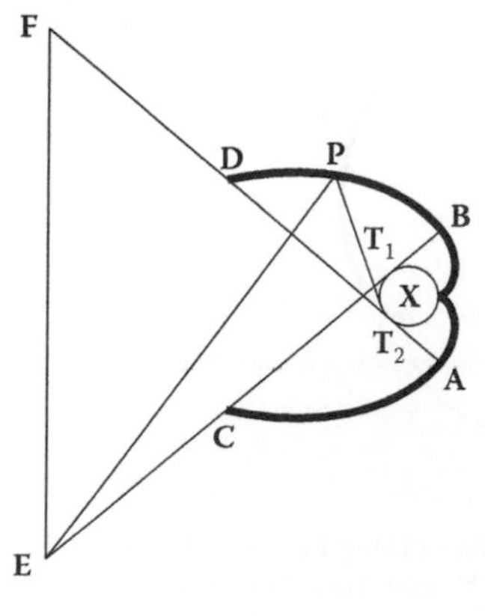

FIGURE 3.3
A nonimaging device concentrating onto the circular receiver, the radiation coming from a source **EF**.

and **F**, but now the edge rays for the circular receiver are those tangent to it. Therefore, the edge-ray principle now states that the rays coming from the edges of the source must be reflected to become tangent to the circular receiver.

The mirror of the concentrator must be designed such that at each point **P**, its slope causes the rays of light coming from **E** to become tangent to the circular receiver. The shape of this mirror is called a macrofocal ellipse (see Chapter 21), having focus **E**, and as macrofocus the circular receiver. This design method enables us to obtain the portion of the mirror from point **D** to point **B**, where it meets line $\mathbf{ET_1}$ passing through **E** and tangent to the receiver at point $\mathbf{T_1}$. From this point forward, the mirror takes the shape of an involute extending from point **B** to point **X**, which is on the axis of symmetry.

To justify the introduction of the involute, it is necessary to examine its optical properties. An involute can be obtained by unrolling a string of constant length around a circle, as presented in Figure 3.4a. Its optical behavior is presented in Figure 3.4b. A ray of light tangent to the receiver and coming from a point **T′** is reflected by the mirror at a point **B′** back to **T′**. Therefore, any ray *r* coming from the space between the receiver and the mirror will be reflected toward the receiver, as desired. Note also that this curve obeys the edge-ray principle: an edge ray leaving tangentially from **T′** is reflected at **B′** again headed to **T′**, that is, tangentially to the tube, and so the reflected ray at **B′** is also an edge ray.

As mentioned earlier regarding the CEC, in the case of the concentrator in Figure 3.3 also, start by designating the source **EF** and the circular receiver, so that the mirrors can be calculated. As mentioned earlier, the mirror must touch a point **X** on the receiver. To design the concentrator, it may be simpler to start by calculating the involute to the receiver, starting at point **X** and

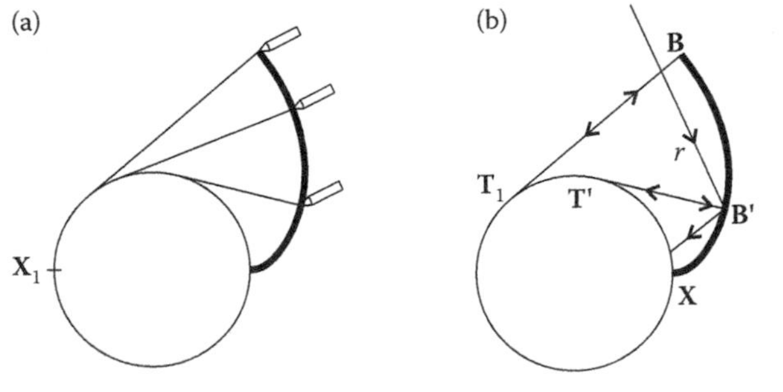

FIGURE 3.4
(a) An involute to a circle can be obtained by unrolling a string of constant length. One tip of the string is attached to point $\mathbf{X_1}$ and the other tip describes an involute. This design method generates a curve perpendicular at every point of the tangent to the circle. (b) In optical terms, this means that a ray of light **T′B′**, leaving the circle at point **T′** tangentially, will be reflected back at the involute at **B′** returning toward **T′**. Therefore, any ray of light *r* passing through the space between the involute and the circle, will be reflected at **B′** toward the circle.

extending it until it touches (at point **B**) the line passing through **E** and $\mathbf{T}_1$. After point **B** is determined, the remaining part of the mirror is designed according to the method presented earlier. This will extend from point **B** until it touches (at point **D**) line $\mathbf{FT}_2$. Points $\mathbf{T}_1$ and $\mathbf{T}_2$ are tangency points of lines **EB** and **AF** with the receiver.

3.4 Angle Transformers

The CPC presented in Figure 3.5a has an acceptance angle θ_1 for each side, and concentrates the incoming radiation to the maximum possible extent, therefore making the exit angle $\pi/2$. The device in Figure 3.5b is an angle transformer.[1]

In this case, the concentration of the device is the maximum for the given entrance and exit angles, θ_1 and θ_2, respectively. Each mirror in this device is composed of two portions. For the right-hand-side mirror, we have parabola **DQ** and flat mirror **QB**. The parabola concentrates to edge **A** of the receiver, the incoming edge rays between r_2 and r_3. Point **Q** is such that ray r_2 is reflected at **Q**, and exits the device making an angle θ_2 to the vertical. Portion **QB** of the mirror reflects incoming edge rays between r_1 and r_2 in a direction making an angle θ_2 to the vertical.

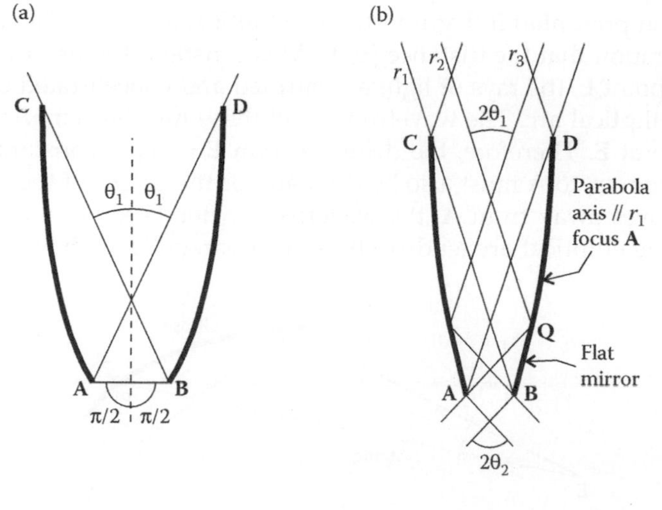

FIGURE 3.5
(a) A CPC with a half-acceptance angle θ_1 and a half-exit angle $\pi/2$. (b) An angle transformer, composed of two parabolic arcs and two flat mirrors. The half-acceptance angle is θ_1 and the half-exit angle is θ_2 which is now smaller than $\pi/2$. Note that as $\theta_2 \rightarrow \pi/2$, the flat mirrors tend to disappear and the angle transformer turns into a CPC.

3.5 The String Method

A simple way to obtain the shape of the mirrors consists in using "the string method" or "the gardener's method."[2]

The optical device (CEC), presented in Figure 3.2, to concentrate onto a receiver **AB** the radiation coming from a source **EF**, is composed of two elliptical arcs. These curves were defined point by point, so as to reflect the rays of light coming from the edges of the source to the edge points of the receiver.

Another way to obtain the elliptical arc is by using the gardener's method. It has this name because it enables us to design an ellipse easily on the ground, using a string and two sticks, just as gardeners do. Let us return to Figure 3.1 and presume that we have a string having length [**E**, **B**, **A**], and whose extremities are fixed at points **E** and **A**. If we stretch it with a marker, and move the marker (along points **P**) so as to maintain the string stretched, we obtain an ellipse. This is because, on an ellipse, the length [**E**, **P**, **A**] (string length) is constant for all its points **P**. This method of designing an ellipse is presented in Figure 3.6.

In the case of the CEC presented in Figure 3.1, the source of radiation, **EF**, is placed at a finite distance. Suppose that the edge points **F** and **E** of the source go to infinity along lines **AD** and **BC**, respectively. The CEC will become a CPC in this case. Lines **EP** and **EB** will now be parallel and the elliptical arc **DB** will become a parabolic arc, with the same thing happening to the elliptical arc **CA**. The string method that defined the ellipses of the CEC can now be adapted to define the parabolas of the CPC. Let us then consider the construction presented in Figure 3.6. The elliptical arc was defined keeping in consideration that the distance [**E**, **P**, **A**] is constant. Let us now consider that from point **E**, the rays of light are emitted and concentrated onto point **A** by the elliptical arc. The wavefront w_1 of these rays is a circular arc centered at point **E**. Therefore, the distance from **E** to w_1 is constant, and the distance from w_1 to **A** must also be constant for the points of the ellipse. As point **E** moves away from **A**, the wavefront w_1 tends to become a straight line, and the elliptical arc tends to become a parabolic arc having focus **A**.

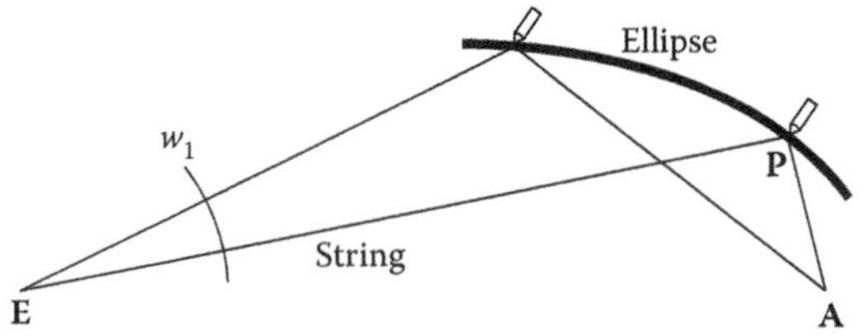

FIGURE 3.6
String method to design an ellipse. Fixing the string extremities at points **E** and **A**, and stretching it with a marker, and moving the marker so as to maintain the string stretched, draws an ellipse.

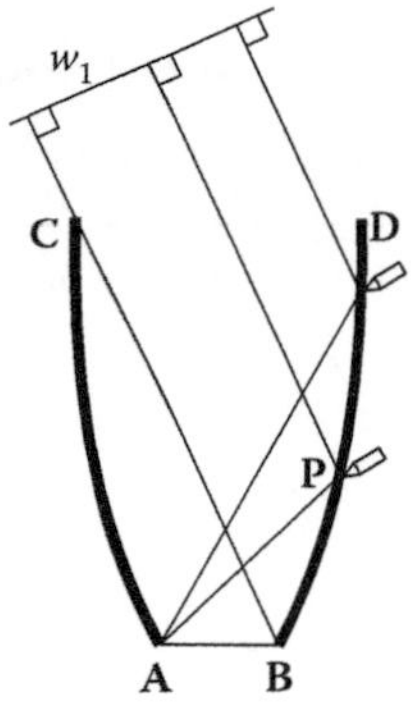

FIGURE 3.7
String method for the design of a parabolic arc. A string having a constant length is fixed at point **A** and kept perpendicular to line (wavefront) w_1 (its tip slides on w_1). Keeping the string stretched with a marker and moving it, draws a parabolic arc. Note that this possibility results from the property of the parabola shown in Figure 2.16.

The string method can, in this case, be adapted to the design of the parabolic arcs, considering the string "fixed" in the wavefront w_1 and at point **A**. Figure 3.7 presents the application of the string method to the generation of the parabolic arcs of a CPC. Consider that the length [**A**, **P**, w_1] is constant, and the string is kept perpendicular to the wavefront w_1. If the string is kept stretched by a marker, moving the marker (along points **P**) draws a parabolic arc. The tip of the string on w_1 slides on it, as point **P** moves on the curve **BD**.

This method can now be applied to the design of the mirrors of the concentrator presented in Figure 3.3. Let us consider that the mirror presented in Figure 3.8 transforms the wavefront w_1 into the wavefront w_2. The optical path length between the two wavefronts is constant (see Chapter 15). In this case, $n = 1$, so that the optical path equals the distance. The distance from w_1 to w_2 must then be constant. If w_1 is a circular arc having center **E**, then the distance between **E** and w_2 is constant. Now consider that the wavefront w_2 has the shape of an involute to the circular receiver. The lines perpendicular to the involute are tangent to the receiver. Because the rays of light are perpendicular to the wavefront, it can be concluded that, in this case, the rays of light are tangent to the receiver. An involute is, by construction, a curve, w_2, such that the distance from w_2 to $\mathbf{X}_1$ is constant for a string attached to a point $\mathbf{X}_1$, and rolled around the receiver as shown in Figures 3.4a and 3.8. Since we have already concluded that the distance from **E** to w_2 is constant, it can now be seen that the distance from **E** to $\mathbf{X}_1$ is constant. Therefore, we can fix a string at **E** and $\mathbf{X}_1$ and generate the mirror by the string method.

The string method enables the generation of generic concentrator profiles for any shape of radiation source or receiver. It can then be applied to the concentrator presented in Figure 3.3 for a source, **EF**, and a circular receiver, as presented in Figure 3.9.[3,4] This method enables us to generate the whole

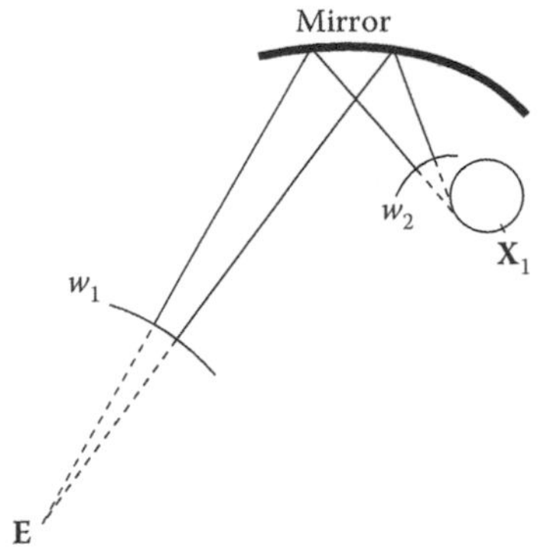

FIGURE 3.8
The distance between **E** and w_1 is the same for all the rays of light, the same happening between the wavefronts w_1 and w_2. Considering that, in an involute w_2, the distance between w_2 and a point $\mathbf{X}_1$ is constant for a string attached to $\mathbf{X}_1$ and rolled on the receiver, then the distance from **E** to $\mathbf{X}_1$ is constant, and a string can be fixed at points **E** and $\mathbf{X}_1$, and the mirror drawn by the string method.

mirror. In this case, a tip of the string must be fixed at point **E** of the source as before, but the other tip must be fixed at point **X** of the receiver. The length of the string must be such that the resulting mirror touches point **X** of the receiver. The string stretches from **X**, around the receiver, and then straight to point **P**, and from there to point **E**. The string method draws a macrofocal ellipse having focus **E**, and as macrofocus the circular receiver (points **P**, **P′**, …) and then the involute to the circular receiver (points **P″**).

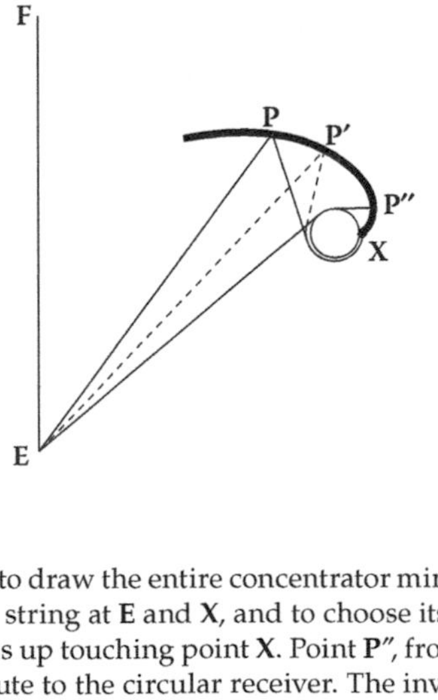

FIGURE 3.9
The string method enables us to draw the entire concentrator mirror of Figure 3.3. For this, it is necessary to fix the tips of the string at **E** and **X**, and to choose its length so that, as the mirror is designed from **P** to **P′**, it ends up touching point **X**. Point **P″**, from where edge **E** of the source cannot be seen, is on an involute to the circular receiver. The involute portion of the mirror is also generated by the same string, as the rest of the mirror.

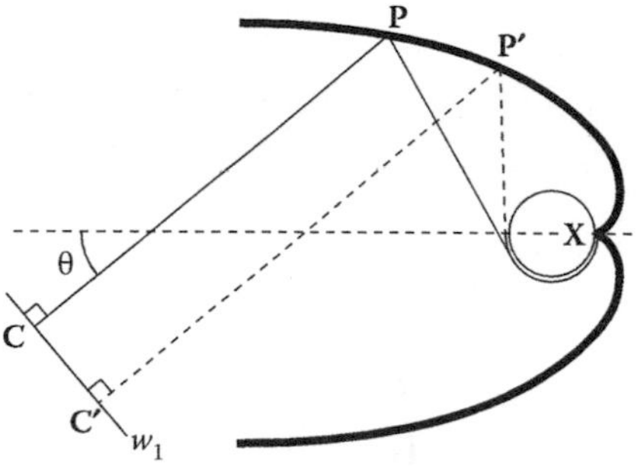

FIGURE 3.10
The string method presented in Figure 3.9, for the design of concentrators for circular receivers, can be extended to the case in which the source is placed at infinity, where the string must be kept perpendicular to the wavefront w_1 for the entire extent of the mirror being drawn.

As seen earlier, the string method can be adapted to the case where the source of radiation is placed at infinity, as shown in Figure 3.10. In this case, the string of constant length is fixed at point **X**, and kept perpendicular to the wavefront w_1.

It is also convenient to note that this method enables us to find the shape of the mirrors of ideal optical devices concentrating the light from arbitrary sources to arbitrary receivers. Consider, for example, a linear receiver where the radiation is concentrated not only from above, but also from below. Figure 3.11a presents such a device, for which the string must initially pass through points **C-B-X-B-A**. The string stretches from **C** to **B**, then goes underneath the receiver to point **X**, then back to **B**, and then over the receiver to point **A**. The design of the concentrator then starts at point **X**. Moving the string, the points (such as $\mathbf{P}_1$) of a circular arc, having center **B**, are obtained. Now the string stretches through points **C-B-**$\mathbf{P}_1$**-B-A**. This arc ends at point $\mathbf{Q}_1$, where the light coming from the source starts to be visible, that is, the edge rays *r* start to illuminate the mirror. For the mirror points (such as $\mathbf{P}_2$) between $\mathbf{Q}_1$ and $\mathbf{Q}_2$, the string unrolls around point **B**, while being kept perpendicular to the wavefront w_1. Therefore, this part of the mirror is a parabola, having its axis perpendicular to w_1 and its focus at **B**. Here the string stretches through points $\mathbf{C}_2$-$\mathbf{P}_2$**-B-A**. From point $\mathbf{Q}_2$ forward, the string unrolls around point **A**. Therefore, between $\mathbf{Q}_2$ and **D**, the mirror is a parabola, having an axis perpendicular to the wavefront w_1, and focus at **A**. Now the string stretches through points $\mathbf{C}_3$-$\mathbf{P}_3$**-A**.

This method can be extended to other shapes of the receiver. Figure 3.11b presents a concentrator for a triangular receiver **A-J-B-A**. In this case, between **X** and $\mathbf{Q}_1$, the mirror is shaped as a circular arc; between $\mathbf{Q}_1$ and $\mathbf{Q}_2$, it is shaped as a parabolic arc having focus **B**; between $\mathbf{Q}_2$ and $\mathbf{Q}_3$, it is shaped as a parabolic arc having focus **J**; and between $\mathbf{Q}_3$ and **D**, it is shaped as a parabolic arc having focus **A**. All the parabolic arcs have axes perpendicular to the wavefront w_1.

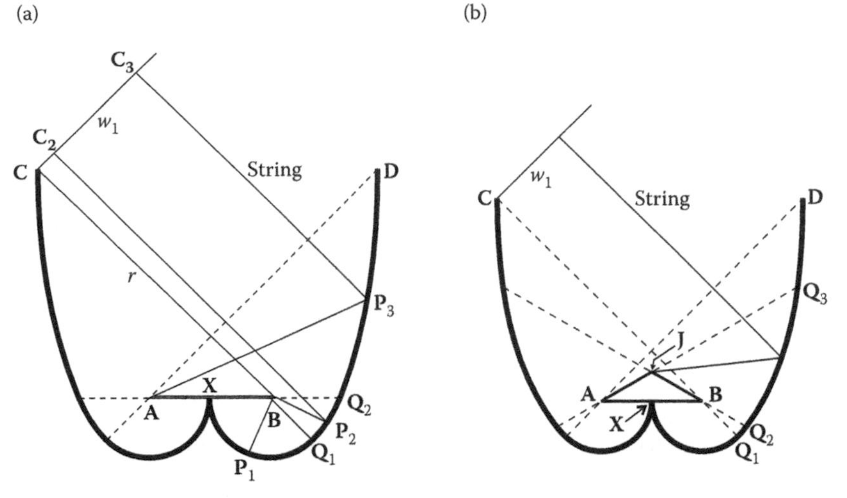

FIGURE 3.11
Concentrators having a linear receiver receiving radiation from above and from below (a) and having a triangular receiver (b). Both devices can be obtained using the string method. In the case of the device presented in (a), the string initially stretches through points **C-B-**$\mathbf{P_1}$**-B-A**. Point $\mathbf{P_1}$ is part of a circumference having center **B**. For points $\mathbf{P_2}$, the string passes through $\mathbf{C_2}$**-**$\mathbf{P_2}$**-B-A**, and describes a parabola having focus **B**. For points $\mathbf{P_3}$, the string passes through $\mathbf{C_3}$**-**$\mathbf{P_3}$**-A**, and describes a parabola having focus **A**. The parabolic arcs have axes perpendicular to the wavefronts w_1. The mirrors for concentrator (b) are obtained in a similar manner.

Figure 3.12 shows a CPC solar thermal collector for water-heating, based on a design similar to that in Figure 3.11b. Part of the side mirrors was truncated at the top where they are almost vertical and, for that reason, contribute largely to the total mirror area, but little to increasing the concentration. The receiver has an inverted-V-shape **A-J-B**, and the bottom **A-B** was removed.[5] This removes the contact between mirror and receiver at point X, avoiding thermal contact that would dissipate the heat from the receiver through the mirror. Removing the bottom **A-B** does not affect efficiency, and the inverted-V receiver **A-J-B** still captures all light reflected toward it by the CPC mirrors.

Figure 3.12b shows a cut through the concentrator. The water flows through small tubes at the vertex **J** of the receiver, shaped as an inverted-V (see Figure 3.11b).

The string method can also be applied in cases where the source has a given shape and is placed at a finite distance. Figure 3.13 presents a device transmitting all the light exiting a circular source onto a receiver having the same shape and size.

In this case, the string is wrapped around the source and the receiver.[6]

The mirror of the device presented in Figure 3.13 is a generalization of an ellipse. An ellipse ideally transmits all the light from a source $\mathbf{E_1F_1}$ onto a receiver $\mathbf{E_2F_2}$, as presented in Figure 3.14, if $\mathbf{F_1}$ and $\mathbf{F_2}$ are its foci.

In this case, the light exiting the source $\mathbf{E_1F_1}$ is transferred to receiver $\mathbf{F_2E_2}$.

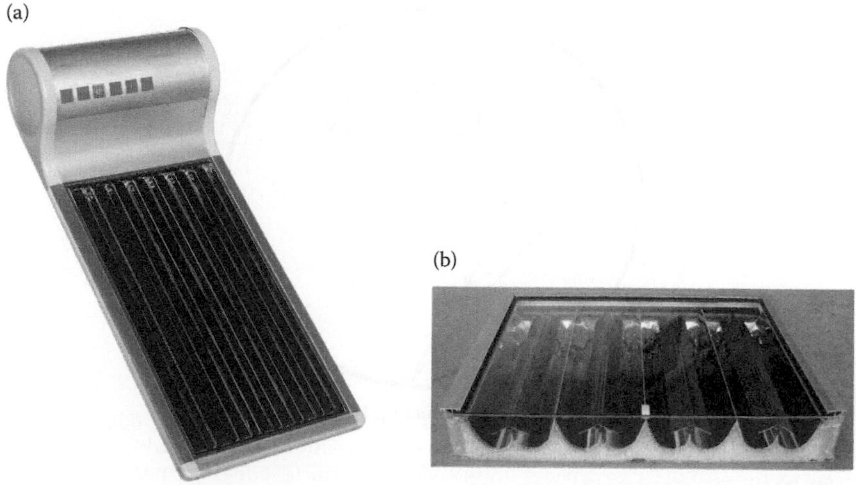

FIGURE 3.12
(a) CPC solar thermal collector. (b) Cut showing its interior structure. (Courtesy of Manuel Collares-Pereira and João Oliveira.)

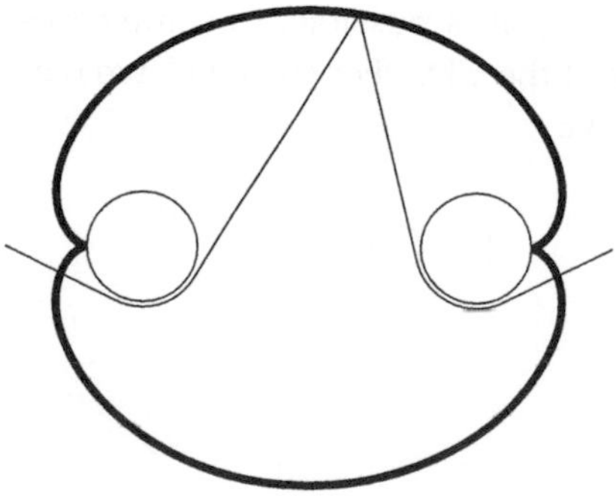

FIGURE 3.13
The string method can also be extended to the case where the source has any shape. This figure presents the case in which the source and receiver are circles having the same size.

3.6 Optics with Dielectrics

The concentrators so far had mirrors with interior air ($n = 1$). Now consider devices made of a material having a refractive index n. Figure 3.15 depicts a CPC made of a dielectric material of refractive index n.

The design of a CPC made of a dielectric material is, in every way, similar to that presented earlier for the case with mirrors and interior air. In this case, a CPC designed for a half-acceptance angle θ_1^* will have a half-acceptance

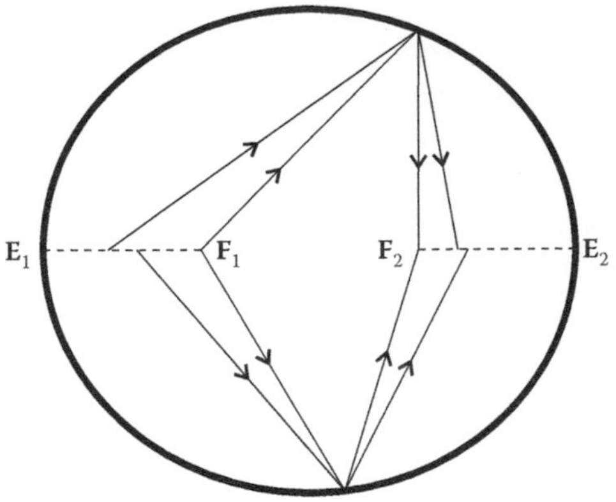

FIGURE 3.14
An ellipse ideally transfers the radiation from a source $\mathbf{E}_1\mathbf{F}_1$ onto a receiver $\mathbf{F}_2\mathbf{E}_2$, where $\mathbf{F}_1$ and $\mathbf{F}_2$ are its foci.

angle θ_1, due to the refraction at the entrance of the CPC. Angles θ_1 and θ_1^* are related by $\sin\theta_1 = n\sin\theta_1^*$. It is possible, in some cases, to use total internal reflection in the walls of the CPC. For the CPC, we have $[\mathbf{A},\mathbf{B}] = [\mathbf{C},\mathbf{D}]\sin\theta_1^*$. Replacing $\sin\theta_1^*$ we have

$$[\mathbf{A},\mathbf{B}] = \frac{[\mathbf{C},\mathbf{D}]\sin\theta_1}{n} \tag{3.1}$$

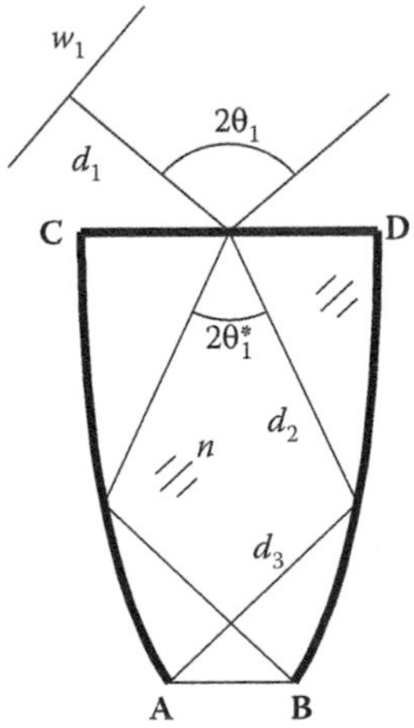

FIGURE 3.15
A CPC of a dielectric material having index of refraction n can be built. This opens up the possibility of using total internal reflection at the walls of the CPC.

Points on the mirror **BD** of the CPC obey $d_1 + nd_2 + nd_3 = C$, where C is the constant optical path length.

The CPC made of dielectric material has a useful feature: its entrance aperture **CD** does not necessarily have to be flat. It is, therefore, possible to design optics with curved entrance aperture, and receiver immersed in dielectric.[7] When the receiver **AB** is flat, these are usually called dielectric total internal reflection concentrators (DTIRCs).[8,9] One such optic is presented in Figure 3.16. The advantage of this possibility is that it enables the design of more compact devices.

For example, the entrance aperture can be shaped as a circular arc. Once the shape of the entrance is defined, the shape of the lateral wall **DB** (and its symmetric **AC**) can be calculated. Rounding the entrance enables us to design more compact devices.

The optical path length between w_1 and **A** is constant. This result was used earlier to define the string method. This method can now be adjusted to this new situation. Let us then suppose that the device is made of a material having a refractive index n. In this case, we have

$$[\mathbf{Q}^*, \mathbf{Q}] + n[\mathbf{Q}, \mathbf{P}] + n[\mathbf{P}, \mathbf{A}] = \text{Constant} \tag{3.2}$$

Given the shape of the entrance aperture, it is then possible to calculate the shape of the lateral walls. Note that the presented concentrator has an entrance whose dimension is equivalent to the distance from **C** to **D**, and not to the arc **CD**. Its concentration is then [**C**, **D**]/[**A**, **B**], where [**C**, **D**] is the distance from **C** to **D**. Also, expression (3.1) still applies to this concentrator.

In the case of the device presented in Figure 3.16, the receiver must be immersed in a medium of refractive index n. If this does not happen, the

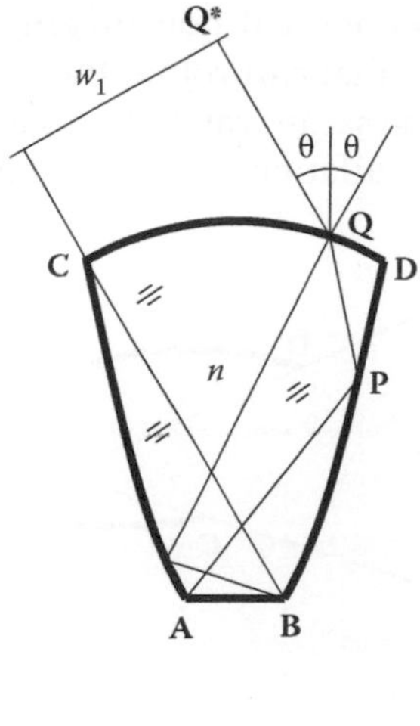

FIGURE 3.16
If a concentrator is made of dielectric, its entrance aperture no longer needs to be flat. In this case, a curved entrance aperture **CD** enables the design of more compact devices.

device must be designed for an exit angle equal to the critical angle, so that there is no total internal reflection at the exit **AB**, and the radiation leaves the device to the air between $\pm\pi/2$.

3.7 Asymmetrical Optics

A CEC is a device allowing us to concentrate radiation coming from a source at a finite distance. In the cases presented earlier, the source and receiver were arranged in a symmetrical configuration. This arrangement can, nonetheless, be generalized. Figure 3.17 presents a CEC concentrator designed for a generic set source-receiver, in which the relative positions and orientations of source and receiver are asymmetrical. In this case, the CEC is designed the same way as the earlier ones. The elliptical arc **BD** has foci **E** and **A**, and the elliptical arc **AC** has foci **F** and **B**.

When the source **EF** tends to infinity, the asymmetrical CEC tends to an asymmetrical CPC. Figure 3.18 presents one such CPC.[10,11]

The asymmetrical CPCs were proposed for stationary collectors of solar energy, which would have different acceptance areas for winter and summer.[1,12,13] In summer, the sun is higher in the sky, and in winter, it is lower. Therefore, in the case presented in Figure 3.18, the direction r_1 could correspond to the direction of the sun in summer and the direction r_2 could coincide with the direction of the sun in winter. In this case, the CPC would accept more radiation, and have a higher concentration in winter, than in summer. This situation could, nonetheless, be inverted if the CPC was used, for example, in a heating system or in an airconditioning system.

Another example of asymmetrical nonimaging optical systems is angle rotators.[14] These are devices that can rotate the radiation without changing their angular aperture, in the same way that angle transformers can modify the angular aperture of the radiation without changing its direction.

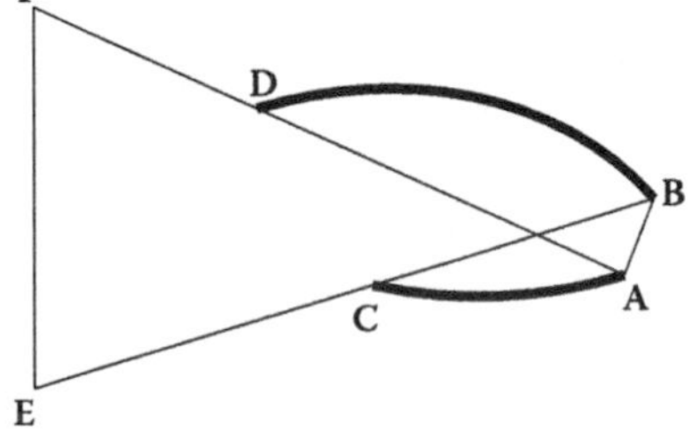

FIGURE 3.17
CEC for a source and receiver in asymmetrical positions. The elliptical arc **BD** has foci **A** and **E**, and the elliptical arc **AC** has foci **F** and **B**.

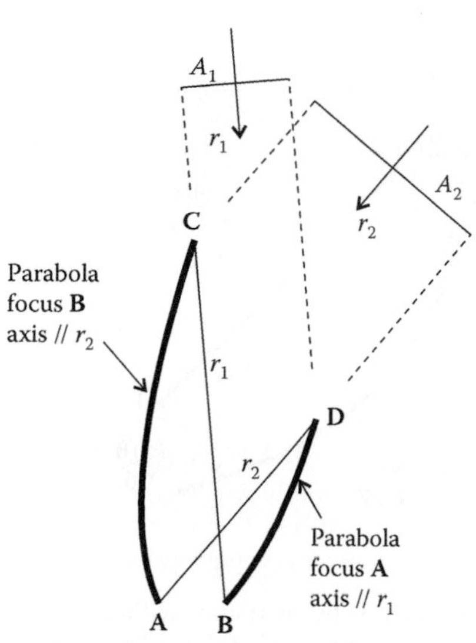

FIGURE 3.18
Asymmetrical CPC. Similar to the CPC, the asymmetrical CPC is also composed of two parabolic arcs. Arc **BD** has focus **A** and axis parallel to r_1. Arc **AC** has focus **B** and axis parallel to r_2. When the radiation comes from direction r_2, the area A_2 intercepted by the concentrator is larger than area A_1 intercepted when the radiation comes from direction r_1. Therefore, this device accepts different amounts of energy and has different concentrations for different acceptance angles.

Figure 3.19 shows an example of an angle rotator. It is composed of flat mirrors and an elliptical arc. The acceptance and exit angles are 2θ. The radiation is rotated by an angle ϕ.

The flat mirror $\mathbf{F}_1\mathbf{F}_2$ is perpendicular neither to the entrance aperture at point $\mathbf{F}_1$ nor to the exit aperture at point $\mathbf{F}_2$. This is because the elliptical arc has a focus at the edge $\mathbf{F}_1$ of the entrance, and another at point $\mathbf{F}_2$ of the exit.

A particular case of this angle rotator occurs when points $\mathbf{F}_1$ and $\mathbf{F}_2$ coincide, and the flat mirrors disappear. In this case, the ellipse tends to a circular arc. The acceptance and the exit angles must, in this particular case, be $\pi/2$, and we get the device presented in Figure 3.20, which is a circular arc with center **C**.[15]

Figure 3.21 shows another example of an angle rotator.[16]

It is composed of a central circular light guide made of two circular arcs with center **C** bound by sections s_1 and s_2, and a compound macrofocal parabolic optic at each end. Each of these optics is made of an exterior macrofocal parabola, with an axis parallel to edge rays that are parallel to the ray r_2 and the macrofocus c_M (see Chapter 21). It reflects these edge rays in directions tangent to c_M. These rays reach the inner mirror of the central circular portion of the light guide with angle α to the normal. From there, they are reflected, and reach the outer mirror of the circular portion, reaching it with

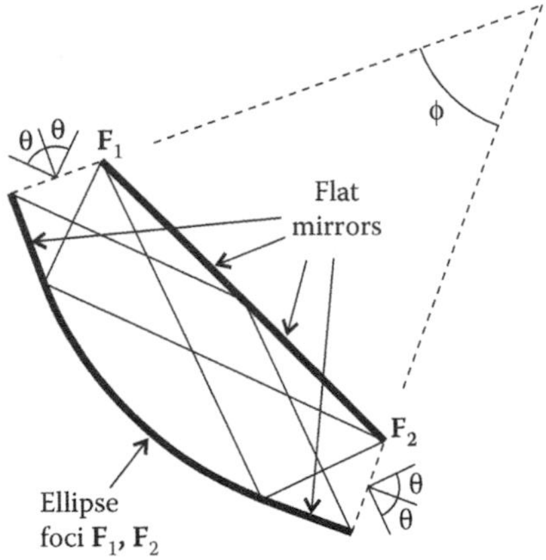

FIGURE 3.19
An angle rotator rotates the radiation by an angle ϕ without changing its angular aperture θ. The angle rotator presented here is composed of three flat mirrors and an elliptical arc.

an angle β to the normal. The inner macrofocal parabola has the axis parallel to the edge ray r_1 and also the macrofocus c_M. It reflects light rays parallel to r_1 in a direction such that they appear to come from the tangent to the macrofocus. These rays reach the outer mirror of the central circular portion of the light guide with an angle β to the normal.

Inside this circular portion, the edge rays keep bouncing back and forth between the two circular mirrors, hitting the inner mirror with angle α, and the outer mirror with angle β. At the other end of the light guide, a symmetrical compound macrofocal parabolic optic "undoes" what the first did, and we recover the radiation confined between angles $\pm\theta$. Angle ϕ for the central portion of the angle rotator can be chosen freely.

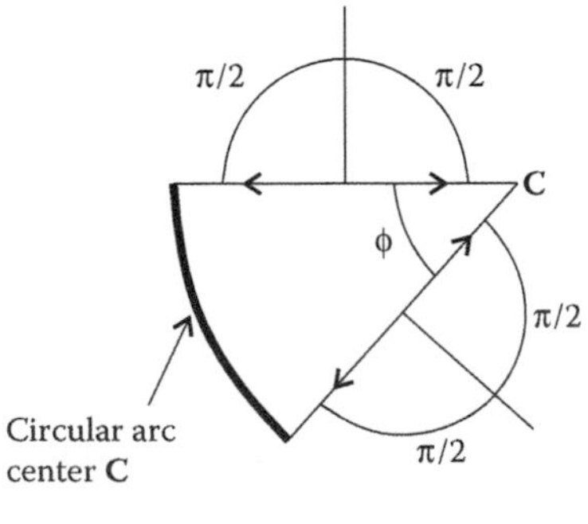

FIGURE 3.20
An arc of circumference. (This device accepts radiation having a half-angle $\pi/2$ and rotates it by an angle ϕ.)

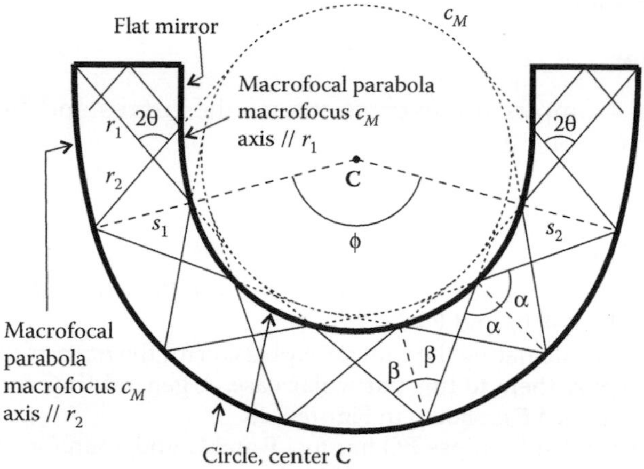

FIGURE 3.21
Angle rotator composed of a central circular light guide with a compound macrofocal parabolic optic at each end.

A limit case of this kind of optic is shown in Figure 3.22. Now, angle α was chosen to be 90° and, therefore, the inner circular mirror is no longer needed, as edge rays would now be tangent to it.

In this optic, light is confined in the space between sections s_1 and s_2, between the circular outer mirror with center **C** and the caustic c_M of these edge rays, which is also a circle with center **C**, but has a smaller radius r. The inner surface of the optic between s_1 and s_2 can now be chosen with any arbitrary shape, as it is a nonoptical surface. It can be used for mechanical applications, such as holding the optic in place without introducing light losses. As seen earlier, angle ϕ can be chosen freely.

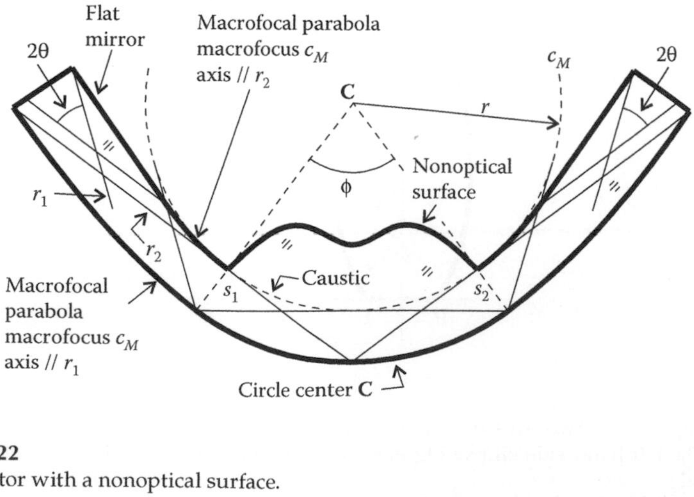

FIGURE 3.22
Angle rotator with a nonoptical surface.

3.8 Examples

The following examples use expressions for the curves and functions that are derived in Chapter 21.

EXAMPLE 3.1

Design a CEC for a source **RG** and a receiver **PF**, where **R** = (–3, 10), **G** = (3, 10), **P** = (–1, 0), and **F** = (1, 0).

We start by calculating the general expression for the mirrors of a CEC, and then apply them to this particular case. A general CEC for source **RG** and receiver **PF** is shown in Figure 3.23.

The left-hand-side ellipse **PQ** has foci **F** and **G** and, therefore, is tilted by an angle α to the horizontal. This ellipse must pass at point **P**, and this defines it.

Consider now the general case of an ellipse with given foci **F** and **G**, and that passes through a point **P** as shown in Figure 3.24.

From the positions of **F**, **G**, and **P**, we can calculate

$$\begin{aligned} K &= t_P + d_P = [\mathbf{F},\mathbf{P}] + [\mathbf{P},\mathbf{G}] \\ f &= [\mathbf{F},\mathbf{G}] \\ \alpha &= \text{angh}(\mathbf{v}) \text{ with } \mathbf{v} = (v_1, v_2) = \mathbf{G} - \mathbf{F} \end{aligned} \tag{3.3}$$

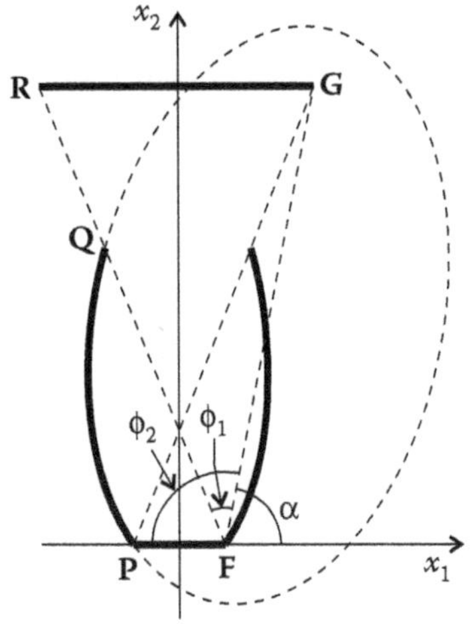

FIGURE 3.23
A CEC composed of two arcs of ellipses with foci at the edges of the source and the receiver. (The left-hand-side ellipse **PQ** has foci **F** and **G**. The right-hand-side ellipse is its symmetrical.)

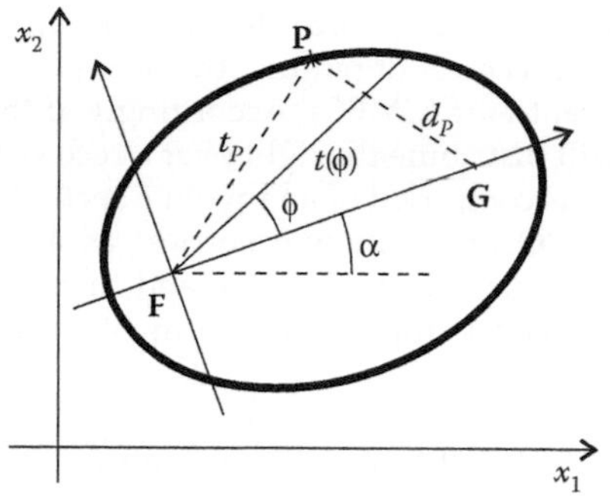

FIGURE 3.24
General ellipse with foci **F** and **G** that passes through a given point **P**.

where angh is the function that gives us the angle of a vector to the horizontal. The ellipse is then given by

$$\frac{K^2 - f^2}{2K - 2f\cos\phi}(\cos(\phi + \alpha), \sin(\phi + \alpha)) + \mathbf{F} \tag{3.4}$$

This same expression can be used to describe the ellipse **PQ** in Figure 3.23. The parameter range in this case is $\phi_1 \le \phi \le \phi_2$ with

$$\begin{aligned} \phi_1 &= \text{ang}(\mathbf{R} - \mathbf{F}, \mathbf{G} - \mathbf{F}) \\ \phi_2 &= \text{ang}(\mathbf{P} - \mathbf{F}, \mathbf{G} - \mathbf{F}) \end{aligned} \tag{3.5}$$

where ang is the function that gives the angle between two vectors.

Now we may apply these results to the particular case in which **R** = (–3, 10), **G** = (3, 10), **P** = (–1, 0), and **F** = (1, 0). Replacing these values in the earlier expressions, for the elliptical arc **PQ** we get

$$\frac{\left(2 + 2\sqrt{29}\right)^2 - 104}{2\left(2 + 2\sqrt{29}\right) - 4\sqrt{26}\cos\phi}\left(\cos\left(\phi + \arccos\left(\frac{1}{\sqrt{26}}\right)\right),\right.$$
$$\left.\sin\left(\phi + \arccos\left(\frac{1}{\sqrt{26}}\right)\right)\right) + (1,0) \tag{3.6}$$

for $\arccos\left(23/\sqrt{754}\right) \le \phi \le \arccos\left(-1/\sqrt{26}\right)$. The right-hand-side elliptical arc is symmetrical to this one, and can be obtained by changing the sign of the first component of the parameterization.

EXAMPLE 3.2

Design a concentrator for a circular receiver where the source is a line **RG** with **R** = (–5, 10) and **G** = (5, 10), and the receiver is centered at the origin and has radius $r = 1$.

The light emitted by a linear source **RG** can be captured and concentrated onto a circular receiver of radius r by means of a compound macrofocal ellipse concentrator (CMEC). Accordingly, if the circle is a light source, the optic will distribute the light over a receiver **RG**.

This concentrator is composed of an involute section **VP** and a macrofocal ellipse section **PQ**, and their symmetrical, as shown in Figure 3.25.

We start by calculating the involute **VP** and then use its end point **P** to calculate the macofocal ellipse **PQ**. The involute section **VP** has the equation

$$r(\cos(\phi+\alpha_I),\sin(\phi+\alpha_I))+r\phi(\cos(\phi-\pi/2+\alpha_I),\sin(\phi-\pi/2+\alpha_I)) \tag{3.7}$$

with $\alpha_I = -\pi/2$ because it touches the circle at point **V** that makes an angle $-\pi/2$ to the horizontal axis x_1. The parameterization of the involute then becomes

$$r(-\phi\cos\phi+\sin\phi,-\cos\phi-\phi\sin\phi) \tag{3.8}$$

Point **T** on the receiver and edge point **G** on the source define a tangent line to the receiver. Point **T** can be obtained from

$$\begin{aligned}\beta &= \arccos\left(r/\|\mathbf{G}\|\right)\\ \mathbf{T} &= rR(\beta)\cdot\mathbf{G}/\|\mathbf{G}\|\end{aligned} \tag{3.9}$$

where $R(\beta)$ is a rotation matrix of an angle β. Angle γ can now be calculated as

$$\gamma = \text{ang}((0,1),\mathbf{G}-\mathbf{T}) \tag{3.10}$$

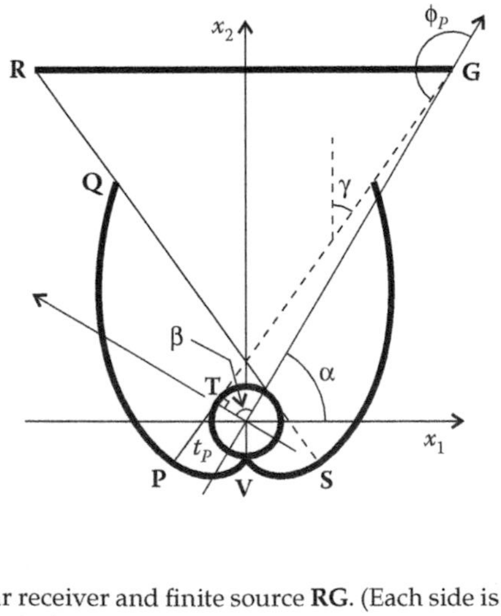

FIGURE 3.25
Concentrator for a tubular receiver and finite source **RG**. (Each side is composed of an involute arc and a macrofocal elliptical arc.)

where ang is a function that gives the angle that the first argument vector makes relative to the second vector argument. The involute **VP** then has the parameter range $-(\pi/2+\gamma) \leq \phi \leq 0$. Point **P**, where the involute ends and the macrofocal parabola starts, is obtained from the involute parametric equation at the parameter value $-(\pi/2+\gamma)$. Angle α that the major axis of the winding macrofocal ellipse **PQ** makes to the horizontal axis x_1 is given by

$$\alpha = \text{angh}(\mathbf{G}) \tag{3.11}$$

where the function angh gives the angle a vector makes to the horizontal. Angle ϕ_P for point **P** is given by

$$\phi_P = \text{ang}(\mathbf{P}-\mathbf{G},\mathbf{G}) \tag{3.12}$$

and distance t_P from point **T** to **P** is given by

$$t_P = \sqrt{\mathbf{P}\cdot\mathbf{P}-r^2} = r\left(\frac{\pi}{2}+\gamma\right) \tag{3.13}$$

We can now calculate

$$f = \sqrt{\mathbf{G}\cdot\mathbf{G}} \tag{3.14}$$

where f is the distance between the center of the macrofocus (circular receiver) and point **G**. We can now calculate

$$K = t_P + r\phi_P + \sqrt{f^2+r^2+t_P^2-2f\left(t_P\cos\phi_P + r\sin\phi_P\right)} \tag{3.15}$$

and the winding macrofocal ellipse **PQ** is parameterized by

$$r(\sin(\phi+\alpha),-\cos(\phi+\alpha)) + \frac{(K-r\phi)^2+2fr\sin\phi-f^2-r^2}{2(K-r\phi-f\cos\phi)}(\cos(\phi+\alpha),\sin(\phi+\alpha)) \tag{3.16}$$

The parameter ranges between the values

$$\begin{aligned}\phi_1 &= \text{ang}(\mathbf{R}-\mathbf{S},\mathbf{G})\\ \phi_2 &= \text{ang}(\mathbf{P}-\mathbf{G},\mathbf{G})\end{aligned} \tag{3.17}$$

where **S** is symmetrical to **P** about the vertical axis x_2. We then have $\phi_1 \leq \phi \leq \phi_2$.

In the particular case in which $r = 1$, the parameterization for the involute becomes

$$(-\phi\cos\phi + \sin\phi, -\cos\phi - \phi\sin\phi) \tag{3.18}$$

and from the positions of $\mathbf{R} = (-5, 10)$ and $\mathbf{G} = (5, 10)$, we also have $\gamma = 0.55321$ rad. Point $\mathbf{P} = (-1.96684, -1.28177)$ and the parameterization for the macrofocal ellipse is

$$(F(\phi) + \sin(1.10715 + \phi), -\cos(1.10715 + \phi) + F(\phi)\sin(1.10715 + \phi)) \tag{3.19}$$

with

$$F(\phi) = -\frac{0.5\cos(1.10715 + \phi)[126 - (18.4356 - \phi)^2 - 22.3607\sin\phi]}{18.4356 - \phi - 11.1803\cos\phi} \tag{3.20}$$

The parameter range is $\phi_1 \le \phi \le \phi_2$, which in this case is $1.01686 \le \phi \le 3.05203$.

EXAMPLE 3.3

Design an angle transformer for an exit (smaller) aperture of dimension 1 (unit length), acceptance angle $\theta_1 = 30° = \pi/6$ rad and exit angle $\theta_2 = 70° = 70\pi/180$ rad.

We start by placing the edge points of the exit aperture in positions $\mathbf{E} = (0.5, 0)$ and $\mathbf{F} = (-0.5, 0)$, as shown in Figure 3.26.

If a ray r is traced backward in a direction $\mathbf{u} = (\cos\beta, \sin\beta)$ with $\beta = \pi/2 - \theta_2$, it reflects at point $\mathbf{E}$ in a direction that makes an angle θ_1 to the vertical in direction $\mathbf{v} = (\cos\alpha, \sin\alpha)$ with $\alpha = \pi/2 + \theta_1$. The direction $\mathbf{t}$ tangent to the mirror is given by

$$\mathbf{t} = \frac{\mathbf{u} + \mathbf{v}}{\|\mathbf{u} + \mathbf{v}\|} \tag{3.21}$$

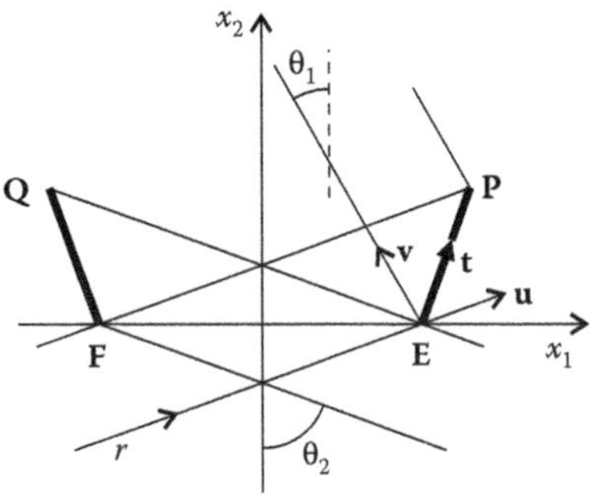

FIGURE 3.26
The edge **P** of the flat mirror **EP** can be obtained from the directions of the incident and reflected rays at point **E** and also the position of point **F**.

Intersecting a straight line that goes through point **F** with direction **u**, with another straight line passing through point **E** with direction **t**, we can find the position of point **P** as

$$\mathbf{P} = \text{isl}(\mathbf{F},\mathbf{u},\mathbf{E},\mathbf{t}) = (0.652704, 0.41955) \tag{3.22}$$

Point **Q** can be obtained by symmetry about the vertical axis. Mirrors **EP** and **FQ** are flat.

We can now calculate the parabola that completes the device. It has focus at point $\mathbf{F} = (F_1, F_2)$, and an axis tilted by an angle $\alpha = \pi/2 + \theta_1$ to the horizontal, and passes through point **P** as shown in Figure 3.27.

The parabola can be parameterized as

$$\begin{aligned}&\frac{\sqrt{(\mathbf{P}-\mathbf{F})\cdot(\mathbf{P}-\mathbf{F})} - (\mathbf{P}-\mathbf{F})\cdot(\cos\alpha, \sin\alpha)}{1-\cos\phi}(\cos(\phi+\alpha), \sin(\phi+\alpha)) + (F_1, F_2)\\ &= \frac{1.43969}{1-\cos\phi}(\cos(2.0944+\phi) - 0.5, \sin(2.0944+\phi))\end{aligned} \tag{3.23}$$

If $\mathbf{w} = (\cos(\pi/2 - \theta_1), \sin(\pi/2 - \theta_1))$, the limits for the parameter ϕ can be obtained as

$$\begin{aligned}\phi_1 &= \text{angp}(\mathbf{P}-\mathbf{F}, \mathbf{v}) = 260 \text{ deg}\\ \phi_2 &= \text{angp}(\mathbf{w}, \mathbf{v}) = 300 \text{ deg}\end{aligned} \tag{3.24}$$

The left-hand-side parabola is symmetrical to the right-hand-side one with respect to the vertical axis.

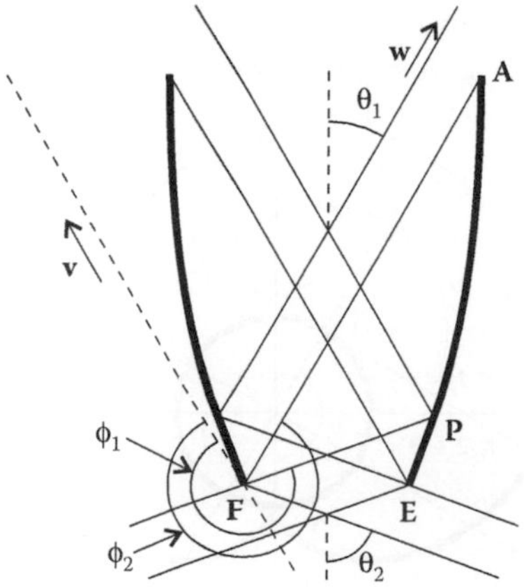

FIGURE 3.27
The parabola **PA** is tilted by an angle $\alpha = \pi/2 + \theta_1$ to the horizontal, has focus **F**, and passes through point **P**.

EXAMPLE 3.4

Design a concentrator with a half-acceptance angle $\theta = 40°$ and a circular receiver centered at the origin, and with radius $r = 1$.

Light with an angular spread of 2θ can be captured and concentrated onto a circular receiver of radius r by means of a compound macrofocal parabola concentrator (CMPC). Accordingly, if the circle is a light source, the optic will distribute the light over a total angle of 2θ.

This concentrator is composed of an involute section and a macrofocal parabola section, as shown in Figure 3.28.

We start by calculating the involute **VP**, and then use its end point **P** to calculate the macofocal parabola **PQ**. The involute section **VP** has the equation

$$r(\cos(\phi + \alpha_I), \sin(\phi + \alpha_I)) + r\phi(\cos(\phi - \pi/2 + \alpha_I), \sin(\phi - \pi/2 + \alpha_I)) \tag{3.25}$$

with $\alpha_I = -\pi/2$ because the involute touches the circle at point **V** that makes an angle $-\pi/2$ to the horizontal axis x_1. Its equation then becomes

$$r(-\phi\cos\phi + \sin\phi, -\cos\phi - \phi\sin\phi) \tag{3.26}$$

with $-(\pi/2 + \theta) \leq \phi \leq 0$. Mirror **PQ** is a winding macrofocal parabola tilted by an angle $\alpha = \pi/2 - \theta$ to the horizontal. For point **P** we may get the values of

$$\begin{aligned} t_P &= r\left(\frac{\pi}{2} + \theta\right) \\ \phi_P &= \pi \end{aligned} \tag{3.27}$$

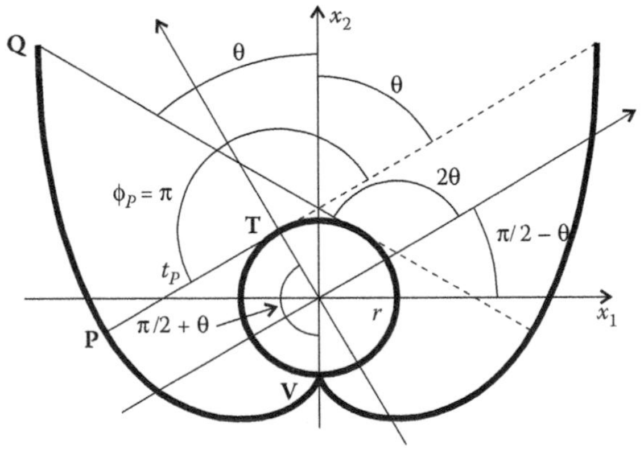

FIGURE 3.28
Concentrator for a tubular receiver and acceptance angle 2θ. (Each side is composed of an involute arc and a macrofocal parabola arc.)

From the values of t_P and ϕ_P, we can calculate constant K by

$$K = t_P - t_P \cos\phi_P + r + r\phi_P - r\pi/2 - r\sin\phi_P \tag{3.28}$$

or

$$K = r\left(1 + \frac{3\pi}{2} + 2\theta\right) \tag{3.29}$$

And now the macrofocal parabola can be calculated from

$$r(\sin(\phi+\alpha), -\cos(\phi+\alpha)) + \frac{K - r(\phi-\pi/2) - r(1-\sin\phi)}{1-\cos\phi}(\cos(\phi+\alpha), \sin(\phi+\alpha)) \tag{3.30}$$

as

$$\frac{r}{\cos\phi - 1}\Big(\cos\theta - \cos(\phi-\theta) + (2\pi - \phi + 2\phi)\sin(\phi-\theta), (-2\pi + \phi - 2\theta)\cos(\phi-\theta) - \sin(\phi-\theta) - \sin\theta\Big) \tag{3.31}$$

with $2\theta \le \phi \le \pi$.

In the particular case where the radius of the receiver is $r = 1$, the parameterization of the involute becomes

$$(-\phi\cos\phi + \sin\phi, -\cos\phi - \phi\sin\phi) \tag{3.32}$$

for the parameter range $-13\pi/18 \le \phi \le 0$. Because the acceptance angle is $\theta = 40\pi/180$ rad, we obtain for the parameterization of the macrofocal parabola

$$\left(\frac{\cos(2\pi/9) - \cos(2\pi/9 - \phi) - (22\pi/9 - \phi)\sin(2\pi/9 - \phi)}{\cos\phi - 1}, \frac{(\phi - 22\pi/9)\cos(2\pi/9 - \phi) - \sin(2\pi/9) + \sin(2\pi/9 - \phi)}{\cos(\phi) - 1}\right) \tag{3.33}$$

for $4\pi/9 \le \phi \le \pi$.

EXAMPLE 3.5

Design an angle rotator for a half-acceptance angle of $\theta = 45°$ and a rotation angle of $\beta = 50°$. The dimension of the entrance and exit apertures is $d = 1$.

Figure 3.29 shows the geometry of an angle rotator and the parameters that define it.

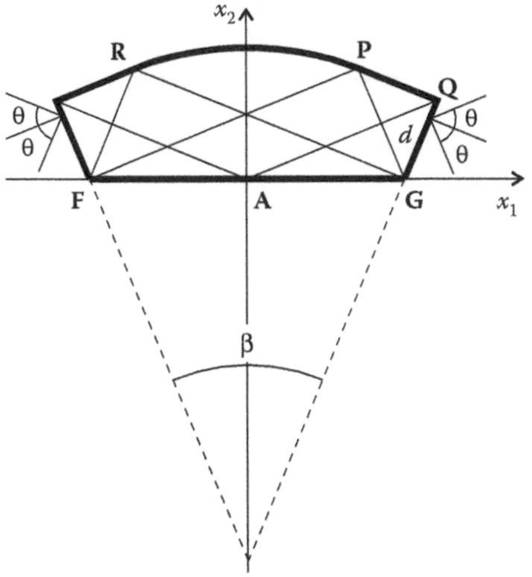

FIGURE 3.29
The geometry of an angle rotator can be defined by the dimension d of the entrance and exit apertures, the half-acceptance angle θ and the rotation angle β for the radiation.

Figure 3.30 shows in more detail the geometry of the entrance aperture of the angle rotator.

The distance $d = [\mathbf{G}, \mathbf{Q}]$ is given and also the angles θ and β. From Figure 3.30b, we obtain

$$\begin{aligned} b\sin(\theta - \beta/2) &= d\sin(\pi/2 - \beta/2) \\ c\cos(\beta/2) &= b\cos\theta \end{aligned} \tag{3.34}$$

and these expressions allow us to determine b and c as $b = 2.64987$ and $c = 2.06744$. Now, if point $\mathbf{A} = (0, 0)$, we can calculate the position of $\mathbf{G}$ as $\mathbf{G} = (c, 0)$ and then $\mathbf{Q} = \mathbf{G} + d(\cos(\pi/2 - \beta/2), \sin(\pi/2 - \beta/2)) = (2.49006,$

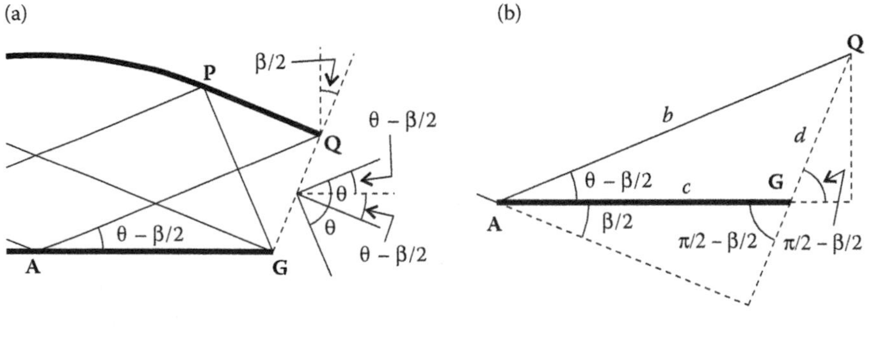

FIGURE 3.30
(a) Geometry of the entrance aperture of the angle rotator. (b) Geometry for calculating the positions of points **G** and **Q**.

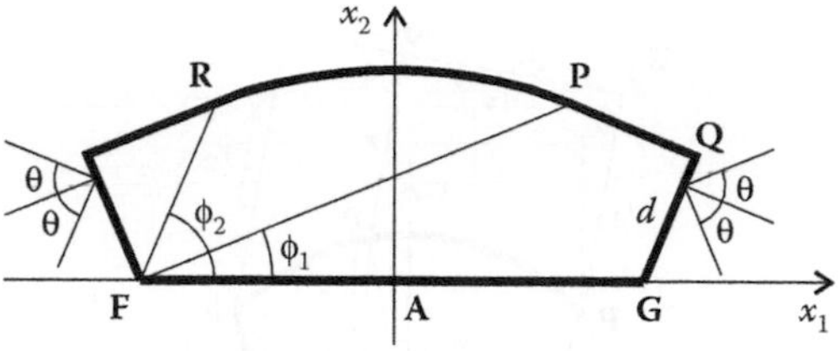

FIGURE 3.31
Arc **PR** is elliptical with foci **F** and **G**. It is parameterized by parameter ϕ, which is limited by the values ϕ_1 and ϕ_2.

0.906308). Point **P** can now be calculated by intersecting the straight line that passes through **G** and makes an angle $-\theta - \beta/2$ to the horizontal and the straight line that passes through point **Q** and makes an angle $-\beta/2$ to the horizontal. Point **P** can then be obtained by

$$\mathbf{P} = \text{isl}(\mathbf{G}, \mathbf{v}, \mathbf{Q}, \mathbf{u}) \tag{3.35}$$

where $\mathbf{v} = (\cos(-\theta - \beta/2), \sin(-\theta - \beta/2))$ and $\mathbf{u} = (\cos(-\beta/2), \sin(-\beta/2))$. We then get $\mathbf{P} = (1.58375, 1.32893)$. The ellipse arc **PR** can now be calculated as a portion of an ellipse that has focus **F** and **G** and passes through point **P**, where **F** is symmetrical to **G** with respect to the vertical axis as shown in Figure 3.31.

From the positions of **F**, **G**, and **P** we get $K = [\mathbf{F}, \mathbf{P}] + [\mathbf{G}, \mathbf{P}]$ and $f = 2c$ and the ellipse is given by

$$\begin{aligned} &\frac{K^2 - f^2}{2K - 2f\cos\phi}(\cos\phi, \sin\phi) + \mathbf{F} \\ &= \frac{10.9899}{10.5995 - 8.26977\cos\phi}(\cos\phi - 2.06744, \sin\phi) \end{aligned} \tag{3.36}$$

for $\phi_1 \le \phi \le \phi_2$, where $\phi_1 = \text{angh}(\mathbf{P} - \mathbf{F}) = 0.349066$ rad and $\phi_2 = \text{angh}(\mathbf{R} - \mathbf{F}) =$ 1.22173 rad.

EXAMPLE 3.6

Design a DTIRC for an acceptance angle $\theta = \pm 10°$ and a circular entrance aperture. The refractive index of the optic is $n = 1.5$.

We start by defining the circular entrance aperture (refractive surface) between points $\mathbf{P}_1$ and $\mathbf{P}_2$ and center $\mathbf{C} = (0, 0)$, as shown in Figure 3.32. We also consider a unit radius $[\mathbf{C}, \mathbf{P}_1]$ for the entrance aperture, as this is just a scale factor. The entrance aperture is then parameterized by

$$\mathbf{P}(\phi) = \left(\cos\left(\frac{\pi}{2} + \phi\right), \sin\left(\frac{\pi}{2} + \phi\right)\right) \tag{3.37}$$

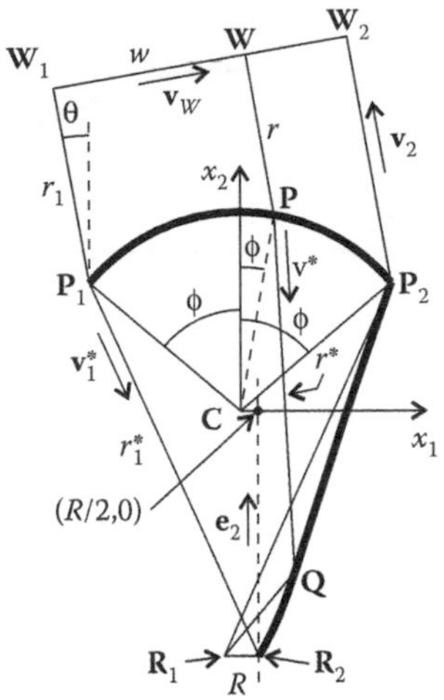

FIGURE 3.32
DTIRC for an acceptance angle ±θ.

with $-\phi \le \phi \le \phi$. We choose angle $\phi = 50°$ and get the positions of points $\mathbf{P}_1$ and $\mathbf{P}_2$ as (Figure 3.32)

$$\begin{aligned}\mathbf{P}_1 &= \left(\cos\left(\frac{7\pi}{9}\right), \sin\left(\frac{7\pi}{9}\right)\right)\\ \mathbf{P}_2 &= \left(\cos\left(\frac{2\pi}{9}\right), \sin\left(\frac{2\pi}{9}\right)\right)\end{aligned} \tag{3.38}$$

We can now define wavefront w perpendicular to the edge rays coming from the left and points $\mathbf{W}_1$ and $\mathbf{W}_2$ on it. Point $\mathbf{W}_1$ is on the ray r_1 through $\mathbf{P}_1$ making an angle θ to the vertical. We choose a unit distance from $\mathbf{P}_1$ to $\mathbf{W}_1$ and get

$$\mathbf{W}_1 = (-0.939693, 1.6276) \tag{3.39}$$

Position of point $\mathbf{W}_2$ can now be obtained as

$$\mathbf{W}_2 = \text{isl}(\mathbf{W}_1, \mathbf{v}_W, \mathbf{P}_2, \mathbf{v}_2) = (0.546198, 1.8896) \tag{3.40}$$

Normal to the receiver entrance aperture at point $\mathbf{P}_1$ is $\mathbf{n}_{P1} = \text{nrm}(\mathbf{C} - \mathbf{P}_1) = -\mathbf{P}_1$, and we can refract ray r_1 at point $\mathbf{P}_1$ and get the refracted ray r_1^* in direction $\mathbf{v}_1^*$ as

$$\mathbf{v}_1^* = \text{rfr}(-\mathbf{v}_2, -\mathbf{P}_1, 1, n) = (0.416693, -0.909047) \tag{3.41}$$

Now, the exit aperture size R of the optic is related to the entrance aperture dimension $[\mathbf{P}_1, \mathbf{P}_2]$ as

$$R = [\mathbf{P}_1, \mathbf{P}_2]\sin\theta/n = 0.177363 \tag{3.42}$$

Edge point $\mathbf{R}_2$ of the exit aperture can now be obtained as

$$\begin{aligned} \mathbf{R}_2 &= \text{isl}(\mathbf{P}_1, \mathbf{v}_1^*, (R/2, 0), \mathbf{e}_2) = (0.0886815, -1.22186) \\ \mathbf{R}_1 &= (-0.0886815, -1.22186) \end{aligned} \tag{3.43}$$

where $\mathbf{e}_2 = (0,1)$ and $\mathbf{R}_1$ is symmetrical to $\mathbf{R}_2$. Edge rays parallel to r_1 must now be concentrated on to point $\mathbf{R}_1$ after refraction on the upper surface and reflection of the side mirror. The optical path length between wavefront w and point $\mathbf{R}_1$ can be obtained from the path of ray r_1 that we already know as the following:

$$S = [\mathbf{W}_1, \mathbf{P}_1] + n[\mathbf{P}_1, \mathbf{R}_2] + nR = 4.34286 \tag{3.44}$$

We now take a value for ϕ of, for example, $\phi = -10°$ and get point $\mathbf{P}$ on the upper refractive surface as $\mathbf{P} = (0.173648, 0.984808)$. The corresponding point $\mathbf{W}$ on the wavefront is given by

$$\mathbf{W} = \text{isl}(\mathbf{W}_1, \mathbf{v}_W, \mathbf{P}, \mathbf{v}_2) = (0.0301537, 1.79861) \tag{3.45}$$

The refracted ray r^* at $\mathbf{P}$ has direction $\mathbf{v}^*$ given by

$$\mathbf{v}^* = \text{rfr}(-\mathbf{v}_2, -\mathbf{P}, 1, n) = (0.0554755, -0.99846) \tag{3.46}$$

Finally, point $\mathbf{Q}$ on the sidewall can be obtained as

$$\mathbf{Q} = \text{coptpt}(\mathbf{P}, \mathbf{v}^*, \mathbf{R}_1, n, S - [\mathbf{P}, \mathbf{W}]) = (0.273056, -0.804356) \tag{3.47}$$

By giving different values to angle ϕ we can calculate different points on the sidewall of the concentrator.

References

1. Rabl, A. and Winston, R., Ideal concentrators for finite sources and restricted exit angles, *Applied Optics*, 15, 2880, 1976.
2. Welford, W. T. and Winston, R., *High Collection Nonimaging Optics*, Academic Press, San Diego, 1989.

3. Winston, R. and Hinterberger, H., Principles of cylindrical concentrators for solar energy, *Solar Energy*, 17, 255, 1975.
4. Rabl, A., Solar concentrators with maximal concentration for cylindrical absorbers, *Applied Optics*, 15, 1871, 1976.
5. Collares-Pereira, M. and Oliveira, J., Solar energy collector of the non-evacuated compound parabolic concentrator type, *European Patent Specification EP0678714B1*, 2000.
6. Kuppenheimer, J. D., Design of multilamp nonimaging laser pump cavities, *Optical Engineering*, 27, 1067, 1988.
7. Miñano, J. C., Ruiz, J. M., and Luque, A., Design of optimal and ideal 2-D concentrators with the collector immersed in a dielectric tube, *Applied Optics*, 22, 3960, 1983.
8. Ning, X., Winston, R., and O'Gallagher, J., Dielectric totally internally reflecting concentrators, *Applied Optics*, 26, 300, 1987.
9. Friedman, R. P. and Gordon, J. M., Optical designs for ultrahigh-flux infrared and solar energy collection: Monolithic dielectric tailored edge-ray concentrators, *Applied Optics*, 35, 6684, 1996.
10. Rabl, A., Comparison of solar concentrators, *Solar Energy*, 18, 93, 1976.
11. Kreider, J. F. and Kreith, F., *Solar Energy Handbook*, McGraw-Hill Book Company, New York, 1981.
12. Rabl, A., *Active Solar Collectors and their Applications*, Oxford University Press, New York, Oxford, 1985.
13. Welford, W. T. and Winston, R., *The Optics of Nonimaging Concentrators—Light and Solar Energy*, Academic Press, New York, 1978.
14. Chaves, J. and Collares-Pereira, M., Ideal concentrators with gaps, *Applied Optics*, 41, 1267, 2002.
15. Collares-Pereira, M., Mendes, J. F., Rabl, A., and Ries, H. Redirecting concentrated radiation, *Nonimaging Optics: Maximum Efficiency Light Transfer III*, SPIE Vol. 2538, 131, 1995.
16. Chaves, J. et al., Combination of light sources and light distribution using manifold optics, *Nonimaging Optics and Efficient Illumination Systems III*, SPIE Vol. 6338, 63380M, 2006.

4

Étendue and the Winston–Welford Design Method

4.1 Introduction

As light travels through an optical system, it requires area and angular space. Figure 4.1 shows a spherical light source S_R (e.g., the sun) of radius r emitting light into space. As the emitted light expands, it will eventually illuminate the inner face of a spherical surface A_1 of radius d_1. When it reaches the surface, the angular spread of the light is confined to angle θ_1, defined by the tangents to S_R on the points of A_1. This angle θ_1 can be obtained from $r/d_1 = \sin\theta_1$. The area of the spherical surface A_1 is given by $A_1 = 4\pi d_1^2$, or by using the expression obtained for $\sin\theta_1$, we get $A_1 \sin^2\theta_1 = 4\pi r^2 = A_S$, where A_S is the area of the source S_R.

We may now compare what happens to the light as it continues to expand, and illuminates a sphere A_2 of a larger radius d_2, as shown in Figure 4.2. Similarly to what we did above for A_1, we have $A_2 \sin^2\theta_2 = A_S$, and, thus, $A_1 \sin^2\theta_1 = A_2 \sin^2\theta_2$. As light travels through space further away from the source, the area it uses increases, but the angle it uses decreases. This happens in a way that quantity $A \sin^2\theta$ is conserved.

Now if area A_2 separates two media of different refractive indices n_1 and n_2, as shown in Figure 4.3, light will refract as it crosses A_2. Its angular aperture will now change from $2\theta_2$ to $2\theta_2^*$, where θ_2 and θ_2^* are related by $n_1 \sin\theta_2 = n_2 \sin\theta_2^*$, and the light will appear to come from a virtual source S_V as it travels in the new medium of refractive index n_2.

We may therefore write $A_1 \sin^2\theta_1 = A_2 \sin^2\theta_2 = A_2(n_2^2/n_1^2)\sin^2\theta_2^*$ or $n_1^2 A_1 \sin^2\theta_1 = n_2^2 A_2 \sin^2\theta_2^*$ and the quantity $n^2 A \sin^2\theta$ is conserved as light travels through space. The quantity $U = \pi n^2 A \sin^2\theta$ is called the étendue of the radiation crossing area A within a cone of angle $\pm\theta$ and is conserved in the geometry presented earlier.

If the geometry of the system was 2-D, the source S_R would be a circle, and the perimeter of another circle of radius d_1 would be $a_1 = 2\pi d_1$ or by using the expression obtained earlier for $\sin\theta_1$ we get $a_1 \sin\theta_1 = 2\pi r = a_S$ where a_S is the perimeter of the source S_R. Similarly, we have $a_2 \sin\theta_2 = a_S$ and, thus,

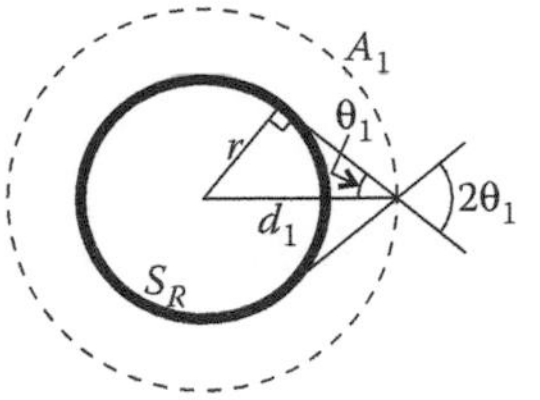

FIGURE 4.1
Angle θ_1 and distance d_1 are related by $r/d_1 = \sin\theta_1$.

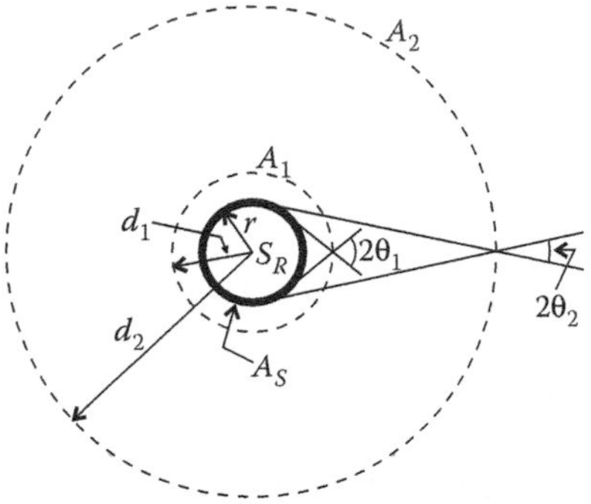

FIGURE 4.2
As the light emitted by a spherical source S_R travels through space, the area it illuminates increases, but the angular spread of the light diminishes.

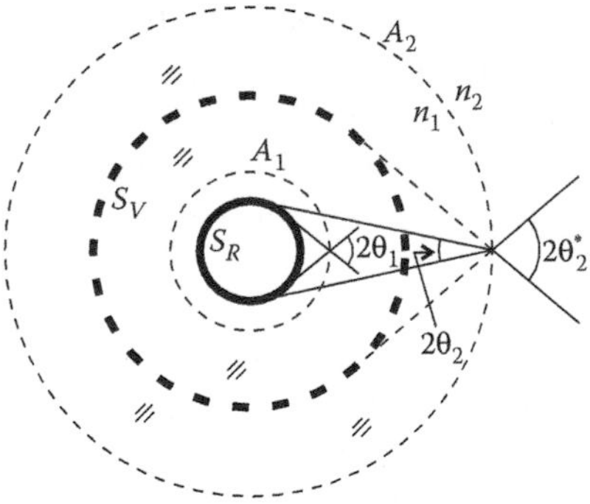

FIGURE 4.3
Light emitted by a spherical source S_R travels through space and hits surface A_2 that separates two media of refractive indices n_1 and n_2. Refraction changes the angular aperture of the light that now appears to come from a spherical virtual source S_V.

$a_1 \sin\theta_1 = a_2 \sin\theta_2$. Therefore, as light travels in the plane, the quantity $a \sin\theta$ is conserved. If the light travels through materials of different refractive indices, the conserved quantity is $na \sin\theta$. The quantity $U_{2D} = 2na \sin\theta$ is called the 2-D étendue of the radiation crossing the length a within an angle $\pm\theta$, and is conserved in the geometry presented earlier.

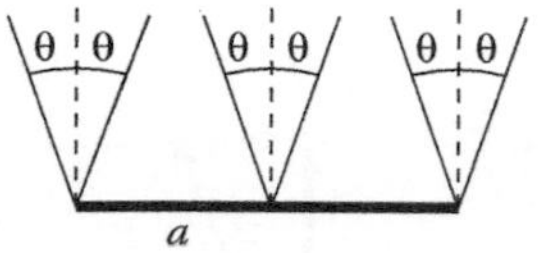

FIGURE 4.4
Length a illuminated by uniform light confined between $\pm\theta$ relative to the normal to a.

In the differential form, the 2-D geometry can be written as $dU = nda \cos\theta\, d\theta$ so that, for the case of a length a illuminated by uniform light confined between $\pm\theta$ relative to the vertical (normal to a), we have (as shown in Figure 4.4)

$$U = na \int_{-\theta}^{\theta} \cos\theta\, d\theta = 2\, na \sin\theta \tag{4.1}$$

In 3-D geometry, étendue is defined as $dU = n^2 dA \cos\theta d\Omega$ where $d\Omega$ is an element of a solid angle. This expression is derived in the following section.

4.2 Conservation of Étendue

A typical application of nonimaging optics is to transfer radiation from a source to a receiver by conserving étendue. From this, we can see that étendue is a central concept in this field. Conservation of étendue can be derived from optical principles (Chapter 18). However, it is also an important concept in other fields such as classical (statistical) mechanics (Liouville's theorem, Chapter 18), radiometry and photometry (geometrical extent, Chapter 20), or radiation heat transfer (reciprocity relation, Chapter 20).

Here, we present the conservation of étendue from the point of view of thermodynamics. This approach has already been used in Chapter 2 when calculating the maximum concentration an optic can provide. This proof of the conservation of étendue is not rigorous, but it is rather intuitive, and therefore we use it here. A more rigorous proof can be given in the context of Hamiltonian optics (Chapter 18).

We first introduce the concept of radiance that is also presented and discussed in detail in Chapter 20 (together with luminance for photometric quantities). If an area dA emits (or is crossed) by radiation of a flux $d\Phi$ (energy per unit time) at an angle θ to its normal, and this flux is contained inside a solid angle $d\Omega$, we may define a quantity L called radiance (as shown in Figure 4.5a)[1–3] by

$$L = \frac{d\Phi}{dA \cos\theta\, d\Omega} \tag{4.2}$$

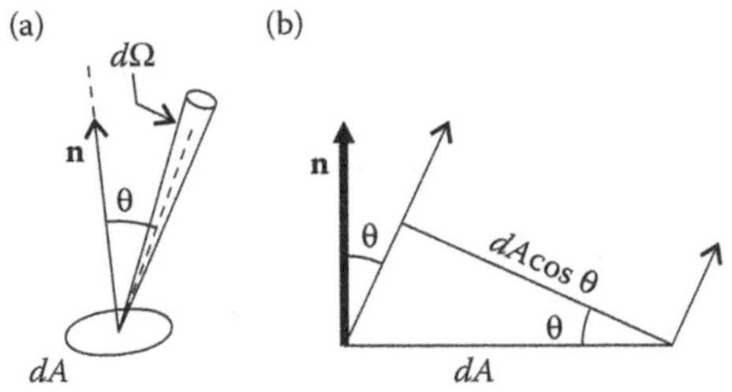

FIGURE 4.5
Definition of radiance in 3-D geometry (a) and 2-D geometry (b).

Note that $d\Phi$ is a second-order differential because it is proportional to the product of dA and $d\Omega$.

If area dA is in a medium of refractive index n, expression (4.2) for the radiance can be rewritten as

$$d\Phi = \frac{L}{n^2} n^2 dA \cos\theta \, d\Omega = L^* dU \tag{4.3}$$

where $L^* = L/n^2$ is called the basic radiance and

$$dU = n^2 dA \cos\theta \, d\Omega \tag{4.4}$$

is the étendue.

The emitted flux per solid angle is called intensity and is given by

$$I = \frac{d\Phi}{d\Omega} = L dA \cos\theta \tag{4.5}$$

where dA is the area of the emitting surface and angle $d\Omega$ is taken in a direction making an angle θ to the normal **n** to the surface.

Generally, L may depend on the direction of the light being emitted, but an important case is obtained when L is a constant. The intensity is proportional to cos θ, that is, to the projected area in the direction θ, as shown in Figure 4.5b. A surface that emits light with this kind of angular distribution is called a Lambertian emitter.

We may now calculate the total flux emission $d\Phi_{hem}$ of an area dA immersed in a medium of refractive index n over a whole hemisphere by integrating expression (4.3) over the solid angle defined by that hemisphere. An area dA^* on the surface of a sphere of radius r defines a solid angle $d\Omega$ given by

$$d\Omega = dA^*/r^2 = \sin\theta \, d\theta \, d\varphi \tag{4.6}$$

as shown in Figure 4.6 in spherical coordinates.

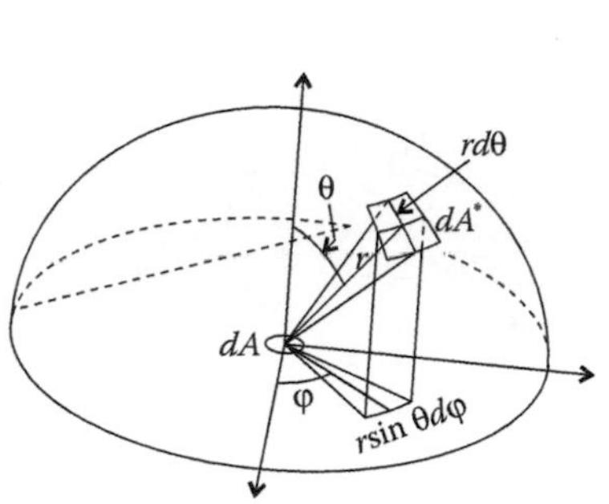

FIGURE 4.6
Solid angle in spherical coordinates.

The light emission over a whole hemisphere for area dA is then given by

$$d\Phi_{hem} = L^* n^2 dA \int_0^{2\pi} \int_0^{\pi/2} \cos\theta \sin\theta \, d\theta d\varphi = \pi n^2 L^* dA \tag{4.7}$$

If area dA is a blackbody emitter at a temperature T, its emission will be Lambertian, and the total flux (in Watts) emitted into the hemisphere is given by[4,5]

$$d\Phi_{hem} = \sigma T^4 dA \tag{4.8}$$

The value of the Stephan–Boltzmann constant σ in a material of refractive index n is given by[5,6]

$$\sigma = n^2 \frac{2\pi}{15} \frac{k^4}{c_0 h^3} = n^2 \sigma_V \tag{4.9}$$

where $\sigma_V = 5.670 \times 10^{-8}\ \mathrm{Wm^{-2}K^{-4}}$ is the value it has in vacuum ($n = 1$), k the Boltzmann constant, h the Planck's constant, and c_0 the speed of light in vacuum. From expressions (4.7) through (4.9) we have

$$L^* = \frac{\sigma_V T^4}{\pi} \tag{4.10}$$

for the basic radiance of a blackbody emitter at a temperature T. With the definition of basic radiance L^* and its value as a function of temperature, we consider a few situations.

Figure 4.7 shows the first of these situations. We have two surfaces, dA_3 and dA_4, that are separated by distance r. The angles that their normals, $\mathbf{n}_3$ and $\mathbf{n}_4$, make to the direction r are θ_3 and θ_4. The medium between these surfaces has a refractive index n_3.

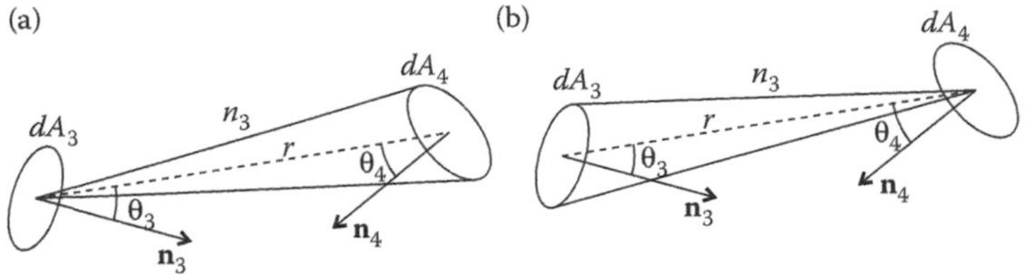

FIGURE 4.7
First situation: The étendue of the light emitted by dA_3 toward dA_4 (a) equals that of light emitted by dA_4 toward dA_3 (b).

If dA_3 emits light toward dA_4, the étendue of this light is (as shown in Figure 4.7a)

$$dU_{34} = n_3^2 dA_3 \cos\theta_3 d\Omega_{34} = n_3^2 dA_3 \cos\theta_3 \frac{dA_4 \cos\theta_4}{r^2} \tag{4.11}$$

If dA_4 emits light toward dA_3, the étendue of this light is (as shown in Figure 4.7b)

$$dU_{43} = n_3^2 dA_4 \cos\theta_4 d\Omega_{43} = n_3^2 dA_4 \cos\theta_4 \frac{dA_3 \cos\theta_3}{r^2} \tag{4.12}$$

From expressions we can conclude that

$$dU_{34} = dU_{43} \tag{4.13}$$

We now consider a second situation as shown in Figure 4.8. Now we consider that the system is in equilibrium, so that the radiation flux $d\Phi_{34}$ that dA_3 emits toward dA_4 equals flux $d\Phi_{43}$ that dA_4 emits toward dA_3.

From $d\Phi_{34} = d\Phi_{43}$ and expressions (4.3) and (4.13) we have

$$L_3^* = L_4^* \tag{4.14}$$

where L_3^* is the basic radiance at dA_3 of the light emitted from dA_3 toward dA_4 and L_4^* the basic radiance at dA_4 of the light emitted from dA_4 toward dA_3.

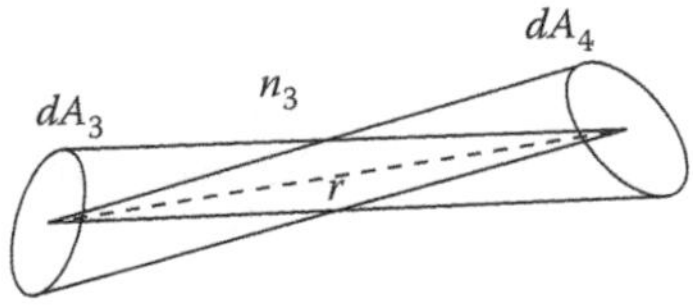

FIGURE 4.8
Second situation: At equilibrium, the basic radiance of the light emitted by dA_3 toward dA_4 equals that of the light emitted by dA_4 toward dA_3.

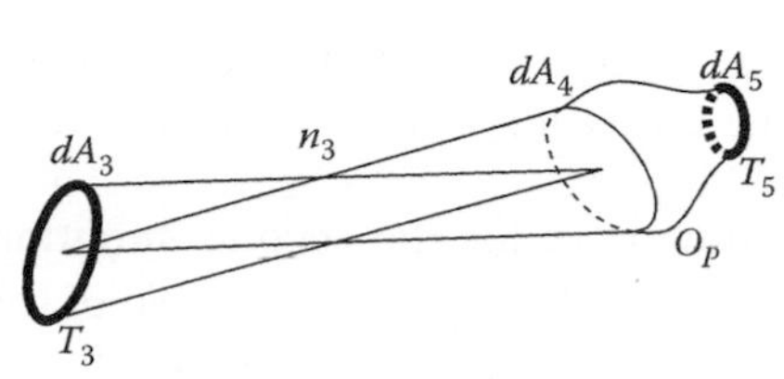

FIGURE 4.9
Third situation: A blackbody dA_3 at a temperature T_3 emits light toward the entrance aperture dA_4 of an optic O_P. The optic redirects this light to another blackbody dA_5. The maximum temperature dA_5 can attain is T_3.

We now consider a third situation as shown in Figure 4.9. Now dA_3 is a blackbody at a temperature T_3 emitting light into a medium of refractive index n_3.

Between areas dA_4 and dA_5 we have an optic O_P that redirects the light it receives from dA_3 toward dA_5. We consider dA_5 as another blackbody at a temperature T_5. The temperature T_5 of dA_5 depends on the radiation exchange with source dA_3. The blackbody dA_5 also emits light due to its temperature, and we consider that the optic between dA_5 and dA_4 redirects this light toward dA_3.

The second principle of thermodynamics states that a process whose only result is to transfer heat from one body to another at a higher temperature is not possible (postulate of Clausius).[7–9] The second principle of thermodynamics then sets a maximum for the temperature of dA_5 equal to that of dA_3, that is, $T_{5max} = T_3$. Therefore, it also sets a maximum for the basic radiance L_5^* at dA_5 since temperature and basic radiance are related by (4.10). In the limit case where we have a system in equilibrium at $T_5 = T_3$, we also have $L_5^* = L_3^*$ and from expression (4.14) we get

$$L_3^* = L_4^* = L_5^* \tag{4.15}$$

We finally consider a fourth situation as shown in Figure 4.10.[10] Now, we have a blackbody dA_1 at a temperature T_1 in a medium of refractive index n_2. It emits light that travels in the medium of refractive index n_2 until it is captured by an area dA_2 at the entrance aperture of an optic between dA_2

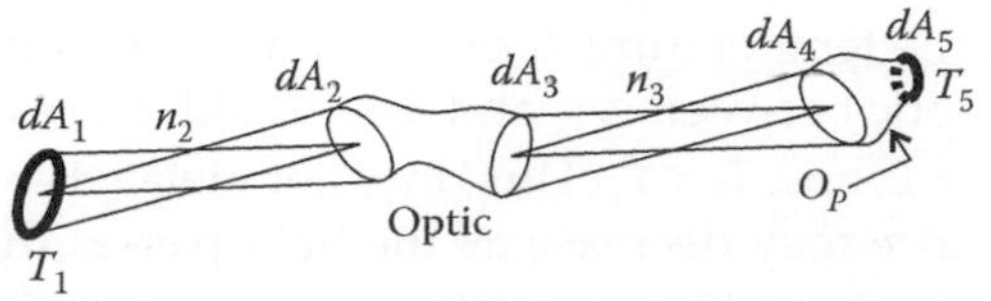

FIGURE 4.10
Fourth situation: A blackbody dA_1 at a temperature T_1 emits light that enters an optic through an area dA_2. This same light exits the optic through dA_3 toward dA_4, and is redirected toward dA_5 by an optic O_P. At thermal equilibrium, $T_5 = T_1$.

and dA_3. After crossing the optic, this light travels between dA_3 and dA_4 in a medium of refractive index n_3, and is redirected to a blackbody dA_5 by another optic O_P.

Using the same argument as for the second situation, we can conclude that $L_1^* = L_2^*$. Also in this case, the second principle of thermodynamics sets a maximum temperature $T_5 = T_1$ for the blackbody dA_5. We then have $L_1^* = L_5^*$. From Equation 4.15 we then get

$$L_1^* = L_2^* = L_3^* = L_4^* = L_5^* \tag{4.16}$$

and the basic radiance is conserved through the system.

In equilibrium, the flux $d\Phi_{12}$ that dA_1 emits toward dA_2 is the same as $d\Phi_{21}$ that dA_2 emits toward dA_1, that is $d\Phi_{12} = d\Phi_{21}$. The optic between dA_2 and dA_3 receives at dA_2 a flux $d\Phi_{12}$ from dA_1 given by $d\Phi_{12} = L_2^* dU_{21}$. Also, the flux that exits the optic through dA_3 toward dA_4 is given by $d\Phi_{34} = L_3^* dU_{34}$. If the flux is conserved, that is, $d\Phi_{12} = d\Phi_{34}$, and since the basic radiance is also conserved through the optic ($L_2^* = L_3^*$), the étendue is also conserved through the optic, and we have

$$dU_{21} = dU_{34} \tag{4.17}$$

which states that the étendue of the light entering the optic at dA_2 equals that of the light exiting the optic at dA_3.

4.3 Nonideal Optical Systems

We have seen that étendue and basic radiance are conserved in optical systems. This, however, is only true in "perfect" optical systems. We now give a few examples of "imperfect" optics in which étendue may be lost or increased, or basic radiance may decrease.

Referring to the system in Figure 4.10, the second principle of thermodynamics states that the temperature T_5 of dA_5 cannot be higher than T_1 of the source dA_1. If the optic between dA_2 and dA_3 would decrease the basic radiance, then $L_3^* < L_2^* = L_1^*$ and $T_5 < T_1$. This does not violate the second principle and the basic radiance may decrease as the light passes through an optic.[11] The basic radiance then is either conserved, or decreases. We may then conclude that the second principle of thermodynamics implies that an optic that increases basic radiance is not possible.

From $d\Phi = L^* dU$, we can see that if the system conserves the flux $d\Phi$ but reduces the basic radiance L^*, then the étendue must increase. Note that

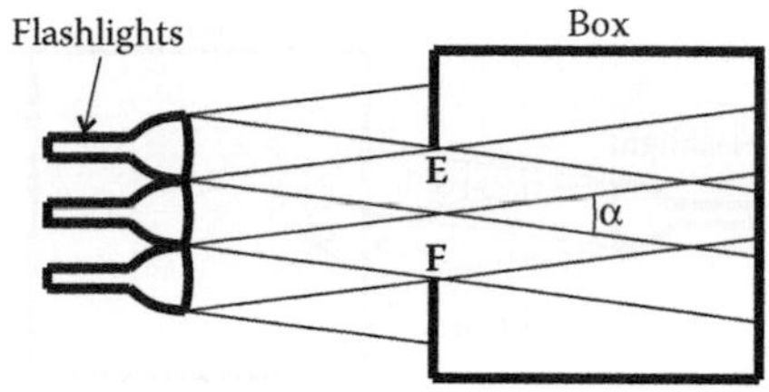

FIGURE 4.11
When there is a loss of light in a system that conserves radiance, part of the étendue is also lost.

étendue cannot decrease without losing flux since, for a constant flux, a lower étendue would mean a higher basic radiance, and this is not possible.

We now consider a few situations of "nonideal" optical systems. In an optical system where the radiance L is conserved (no variations in refractive index), we may lose étendue if we lose flux. From the expression $\Phi = LU$ (the case in which $n = 1$), if a part of the flux Φ is lost, this means a part of the étendue U is also lost. One such possibility is shown in Figure 4.11, where we have a set of flashlights emitting light with angular aperture α toward a box with a hole **EF** on its side. When entering the box, part of the light is shaded by the walls, and only a part of it passes through the hole. The étendue of the light entering the box is reduced because there is loss of light.

There are also situations in which the basic radiance may decrease. An example is when absorption of light takes place as it travels through a material, as shown in Figure 4.12.

In this example, we have light with an angular aperture 2α entering a space between two parallel mirrors, M_1 and M_2. At the other side, the area and angular aperture of light are still the same and, therefore, étendue is also the same. If, however, the material is absorptive, the light flux decreases and, from $\Phi = L^*U$ we can see that the basic radiance also decreases.

There are situations in which the étendue of the light in an optical system may increase. An example is when light hits a diffuser. The angular aperture increases and, therefore, the area-angle (étendue) also increases. Figure 4.13 shows the effect of placing a diffuser at the entrance aperture of a box whose interior is to be illuminated using a flashlight. The diffuser increases the angular spread of the light from α to γ without changing the area **EF** and, therefore, increases the étendue. We cannot, however, "undiffuse" the light

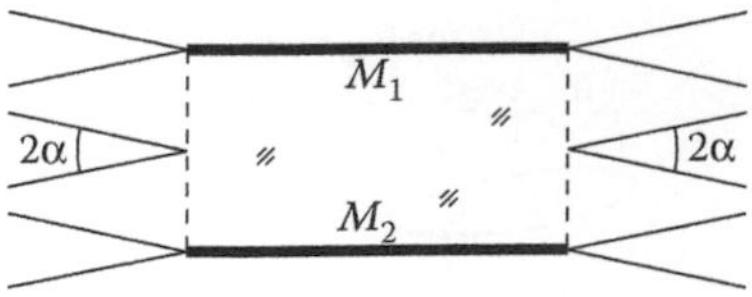

FIGURE 4.12
Light traveling in an absorptive optical system loses flux, and the radiance decreases.

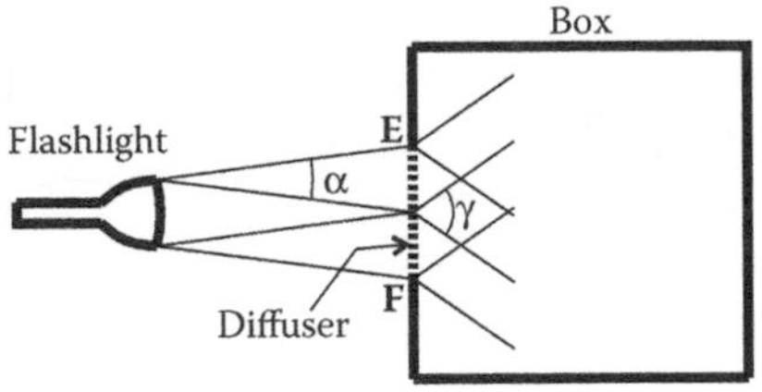

FIGURE 4.13
The étendue may be increased by diffusing the light, but once it has increased, it cannot be decreased. There is no "undiffuser" to undo what the diffuser has done.

once it has been diffused. This means that the étendue of light can be lost or increased, but not decreased, as we have seen earlier.

The diffuser in Figure 4.13 also decreases the basic radiance, because the light flux is assumed to be conserved, and the étendue is increased.

We have now seen that an optic can conserve the basic radiance (in an ideal system), or decrease it (e.g., if the optical system is absorptive), but cannot increase it. The opposite happens with étendue: an optic can conserve it or increase it, but not decrease it. Étendue can, however, be lost by losing light.

4.4 Étendue as a Geometrical Quantity

We can now give a further insight into the physical meaning of étendue. We shall first consider the case in which $n = 1$. Étendue is given by $dU = dA \cos \theta d\Omega$, and it is purely a geometrical quantity as we can see from Figure 4.14.

When light passes through an area dA, it requires "room." This space has two components: "spatial room" measured by the area, and "angular room" measured by the solid angle. However, if light crosses the area dA in a direction θ to its normal, then it "sees" only the projected area $dA \cos \theta$ as the available area for it to pass through. Therefore, the étendue is the product of the available spatial room $dA \cos \theta$ and the angular room $d\Omega$ defined by the solid angle.

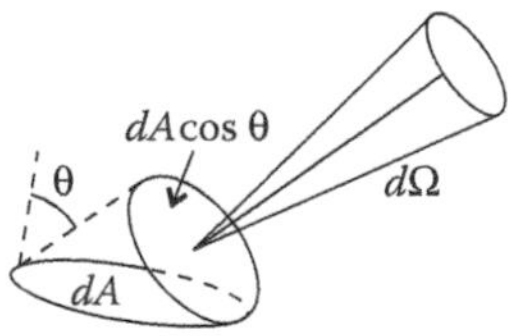

FIGURE 4.14
The étendue is a geometrical quantity that measures the amount of "room" available for the light to pass through.

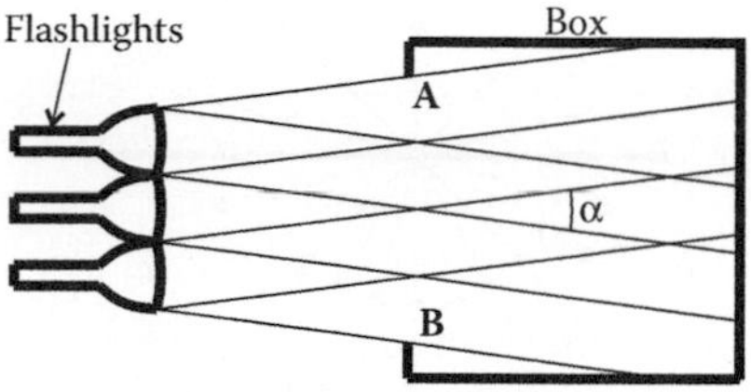

FIGURE 4.15
The interior of a box is illuminated using three parallel flashlights. The hole **AB** on the side of the box is large, but the light has a small angular spread α.

Conservation of étendue then tells us that the product of projected area and solid angle is constant. This means that if the area available for the light is increased, the solid angle decreases. But if the area decreases, the solid angle must increase, so that étendue remains constant. For example, imagine we need to illuminate the interior of a box with the light of three flashlights. These flashlights emit a beam of light with an angular aperture α. We may open a large hole **AB** on the side of the box and send the light through it, as in the case of Figure 4.15. In this case, the light the box receives has a small angular aperture α but is spread out over a large area **AB**.

An alternative way of illuminating the interior of the box is to open a smaller hole **CD**, tilt some of the flashlights, and make the light pass through this smaller aperture, as shown in Figure 4.16. In this case, however, the angular aperture of the light entering the box is larger, as indicated by the angle β.

To illuminate the interior of the box, we have two options: (a) either the area of the hole is large and the angle of the light is small, or (b) the area is small, but the angle of the light is large. It is as if the light needed "space" to move through. Either we give it some area (physical space) for it to pass through or, if we diminish the area, we must give it angular space. This area-angle conservancy is the conservation of étendue. If the area diminishes, the angle increases, and if the area increases, the angle diminishes.

We now consider the case in which $n \neq 1$. In a medium of refractive index n, we can "fit" n^2 times more light than in air ($n = 1$) (see expression

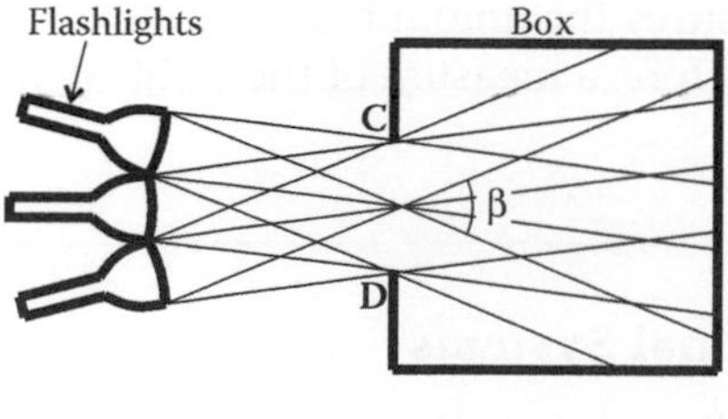

FIGURE 4.16
The interior of a box is illuminated using three converging flashlights. The hole **CD** on the side of the box is small, but the light has a large angular spread β.

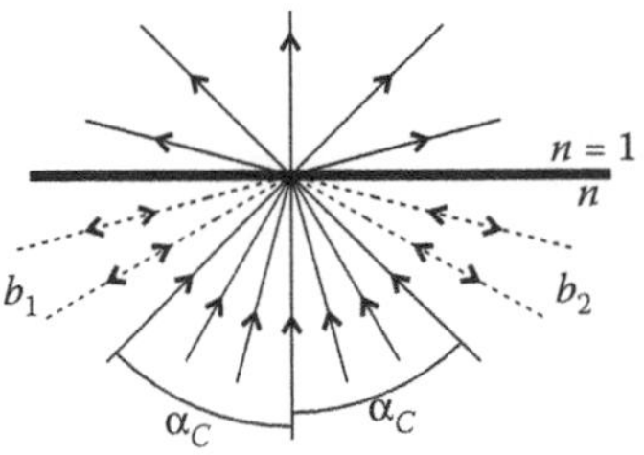

FIGURE 4.17
In a medium of refractive index n, we can have more light than in a medium of refractive index $n = 1$.

(4.9)) and that also "adds" to the room available for the light, which is now $dU = n^2 dA \cos\theta d\Omega$.

This is shown in Figure 4.17. Diffuse light traveling inside a medium of refractive index n refracts into a medium of refractive index $n = 1$ (air). The light contained inside the critical angle $\pm\alpha_C$ refracts into air, and occupies all the angular space available, spreading out to an angle of $\pm\pi/2$. Bundle b_1 outside the critical angle suffers total internal reflection, and continues in medium n as bundle b_2. The same happens to another bundle b_2 coming in the reverse direction and hitting the interface. It suffers total internal reflection, and continues inside the material as bundle b_1. This means that the room for the light in air is smaller and, therefore, some light traveling in the medium of index n does not "fit" and is "rejected" (suffers total internal reflection at the interface).

We now reverse the direction of light, and consider that the diffuse light is coming from air ($n = 1$) into the medium of refractive index n. We can see that, as it refracts into the medium, this light is confined to the critical angle and, therefore, does not use all the angular space available.

As light travels through an optical system, if the amount of light does not change (the flux is constant) the "room" it needs to progress (étendue) is also constant. This is the basis of radiation concentration. We can reduce the area that the light passes through, as long as we increase the solid angle, so that the available room for the light to pass through remains constant.

Also, the basic radiance given by expression (4.3) can be written as $L^* = d\Phi/U$ and it measures the amount of light flux per unit available room for the light. It is, therefore, a measure of the light "density."

4.5 Two-Dimensional Systems

Suppose we have a 2-D system with a Lambertian light source S_R illuminating a curve $\mathbf{c}(\sigma)$, as shown in Figure 4.18. At each point $\mathbf{P}$ of the curve, the light is confined between edge rays r_A and r_B tangent to the source.

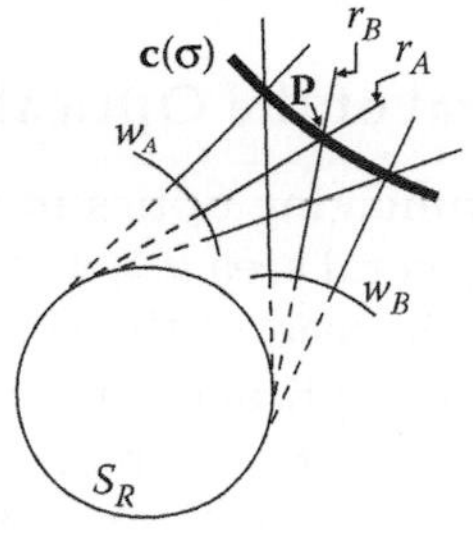

FIGURE 4.18
Light emitted by a source S_R crossing a curve **c**(σ).

Edge rays r_A and r_B are perpendicular to wavefronts w_A and w_B. We may then think of a curve **c**(σ) as being illuminated by light, whose edge rays are perpendicular to the two given wavefronts w_A and w_B, as shown in Figure 4.19. This is a more general situation than that of Figure 4.18, as now the radiation field at **c**(σ) does not necessarily come from a Lambertian light source.

Yet another way to look at this situation is as shown in Figure 4.20. Here we only consider the curve **c**(σ) and the direction of the edge rays r_A and r_B at each point **P**.

In all the cases, we have a given angular distribution of light on the curve **c**(σ), and this light has some étendue.

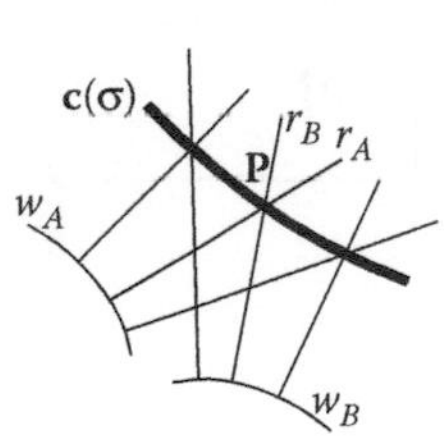

FIGURE 4.19
Light, whose edge rays are perpendicular to wavefronts w_A and w_B, crossing a curve **c**(σ).

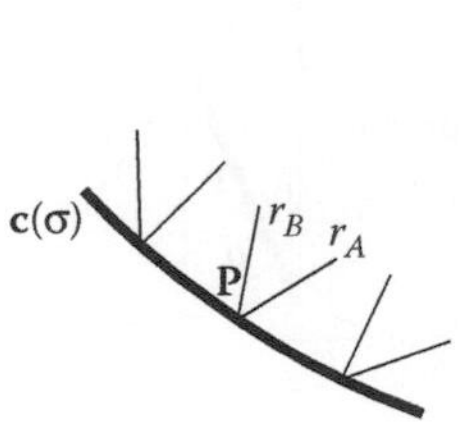

FIGURE 4.20
Light confined between edge rays r_A and r_B crossing a curve **c**(σ).

4.6 Étendue as an Integral of the Optical Momentum

Étendue is often used in nonimaging optics in a different form. Instead of being defined in terms of a general area dA, it is defined in terms of an area, dx_1dx_2, in the x_1x_2 plane, and instead of the solid angle $d\Omega$, it is defined in terms of another quantity called optical momentum. The optical momentum is a vector defined at each point on the path of a ray. It has, as its magnitude, the refractive index of the medium at that point, and the same direction as the light ray at that point. It is tangential to the light ray at each point. This quantity is defined and presented in detail in Chapter 15. Figure 4.21 shows a light ray traveling in a medium of constant refractive index (straight path), and then entering a medium of varying refractive index (curved path), and the optical momentum vector **p** for two points of this light ray.

Vector **p** can be obtained in terms of its components from the geometry shown in Figure 4.22, and can therefore be written as

$$\begin{aligned}\mathbf{p} = (p_1, p_2, p_3) &= (n\cos\theta_1, n\cos\theta_2, n\cos\theta_3) \\ &= n(\sin\theta_3\cos\varphi, \sin\theta_3\sin\varphi, \cos\theta_3)\end{aligned} \quad (4.18)$$

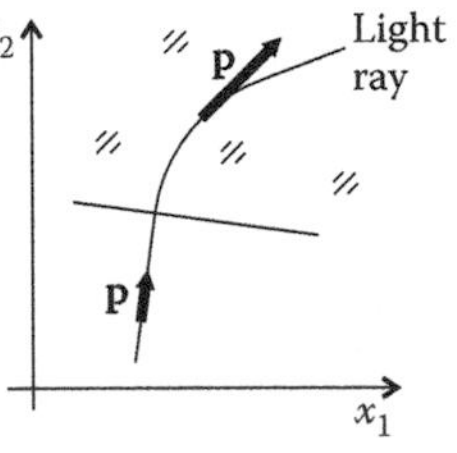

FIGURE 4.21
Optical momentum.

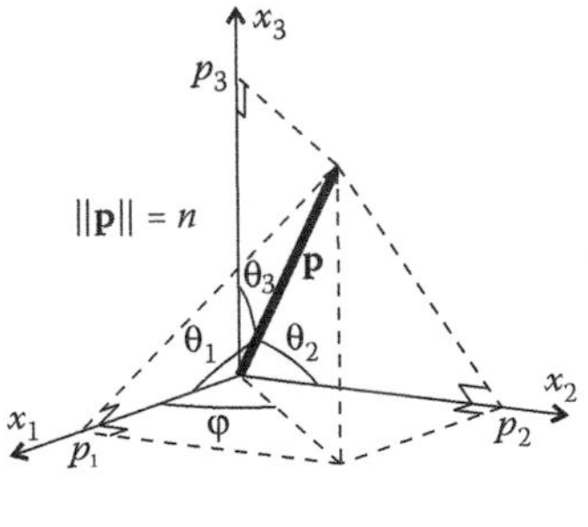

FIGURE 4.22
The momentum of a ray can be obtained in spherical coordinates.

where θ_1, θ_2, and θ_3 are the angles that $\mathbf{p}$ makes to the x_1, x_2, and x_3 axes, respectively, and $\|\mathbf{p}\| = n$. Angle φ is the angle the projection of $\mathbf{p}$ onto the x_1x_2 plane makes with the x_1 axis.

As defined earlier, p_1, p_2, and p_3 are the x_1, x_2, and x_3 components of vector $\mathbf{p}$ given by expression (4.18), and, therefore, dp_1dp_2 can be written as

$$\begin{aligned} dp_1dp_2 &= \frac{\partial(p_1,p_2)}{\partial(\theta_3,\varphi)}d\theta_3 d\varphi = \left(\frac{\partial p_1}{\partial\theta_3}\frac{\partial p_2}{\partial\varphi} - \frac{\partial p_1}{\partial\varphi}\frac{\partial p_2}{\partial\theta_3}\right)d\theta_3 d\varphi \\ &= n^2\cos\theta_3\sin\theta_3 d\theta_3 d\varphi = n^2\cos\theta_3 d\Omega \end{aligned} \tag{4.19}$$

where $d\Omega$ is the solid angle, as seen from Figure 4.6, in the particular case in which the normal to an area dA is in the direction of axis x_3. We can then write:

$$dU = n^2 dx_1dx_2\cos\theta_3 d\Omega = dx_1dx_2dp_1dp_2 \tag{4.20}$$

for the case in which the area dA is in the plane x_1x_2 and is given by $dA = dx_1dx_2$, as in Figure 4.23.

We now analyze the 2-D case. Expression (4.4) simplifies in this case to

$$dU_{2\text{-}D} = n\,da\cos\theta d\theta \tag{4.21}$$

where da is an infinitesimal length, and θ the angle to the normal to da. This situation in 2-D geometry is shown in Figure 4.24a for the particular case in which $da = dx_1$ is on the x_1 axis and, therefore, the normal to da is along the x_2 axis. The expression (4.21) for the étendue in this case is given by

$$dU_{2\text{-}D} = n\,dx_1\cos\theta_2 d\theta_2 \tag{4.22}$$

In this particular case, the small length dx_1 is immersed in a medium of refractive index n and crossed by radiation confined to a small angle $d\theta_2$, making an angle θ_2 to the normal to dx_1.

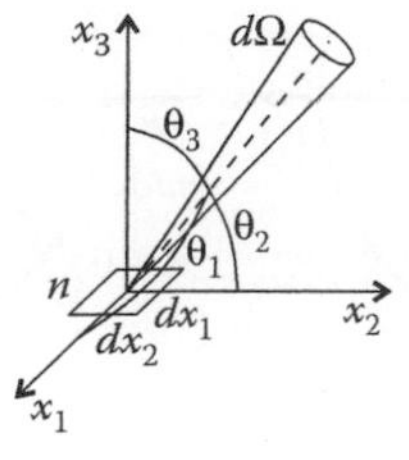

FIGURE 4.23
Étendue for the case in which $dA = dx_1dx_2$.

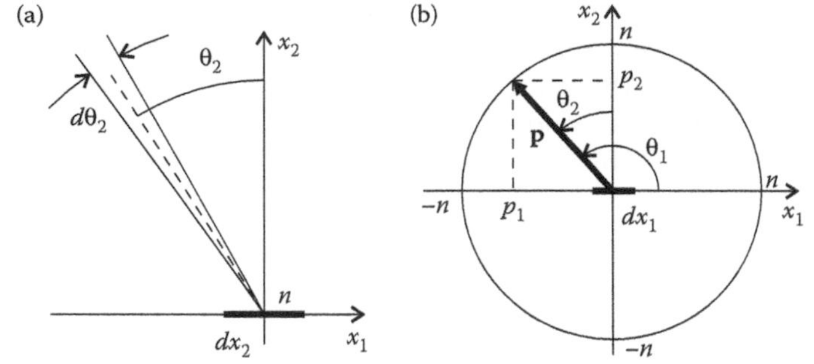

FIGURE 4.24
(a) Étendue in two dimensions. (b) Optical momentum $\mathbf{p}$ of a ray crossing dx_1.

The optical momentum $\mathbf{p}$ in two dimensions is a vector with two components that is given by

$$\mathbf{p} = (p_1, p_2) = n(\cos\theta_1, \cos\theta_2) \tag{4.23}$$

as shown in Figure 4.24b, where it is shown touching a circle of radius equal to the refractive index n at dx_1.

Expression (4.22) for the étendue can now be written as a function of the 2-D optical momentum. Referring to Figure 4.25a for the case in which $p_2 > 0$ (since $\cos\theta_2 > 0$), we see that $(nd\theta_2)\cos\theta_2 = -dp_1$ because p_1 decreases as θ_2 increases. The result $n\cos\theta_2 d\theta_2 = -dp_1$ also holds true for the case in Figure 4.25b, in which $p_2 < 0$ (since $\cos\theta_2 < 0$).

We can then write

$$dU_{2\text{-}D} = ndx_1\cos\theta_2 d\theta_2 = -dx_1dp_1 \tag{4.24}$$

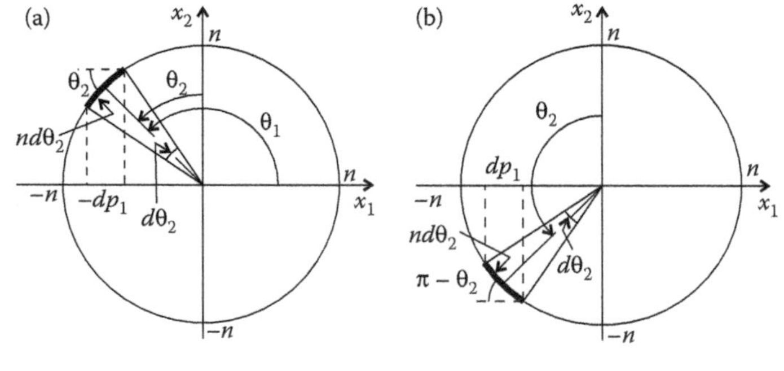

FIGURE 4.25
Étendue in two dimensions as a function of the optical momentum component p_1 when $p_2 > 0$ (a) and when $p_2 < 0$ (b).

If at point $(x,0)$, radiation is confined between $\mathbf{p}_A$ and $\mathbf{p}_B$, the infinitesimal étendue through dx_1 is

$$dU = -dx_1 \int_{\mathbf{p}_A}^{\mathbf{p}_B} dp_1 = -dx_1(p_{B1} - p_{A1}) = (\mathbf{p}_A - \mathbf{p}_B) \cdot d\mathbf{x}_1 \tag{4.25}$$

where $d\mathbf{x}_1$ points in the direction of the positive x_1 axis, and $\|d\mathbf{x}_1\| = dx_1$ as shown in Figure 4.26a.

In the more general case in which light crosses a curve c parameterized by $\mathbf{c}(\sigma)$ for $\sigma_1 < \sigma < \sigma_2$, we have the situation shown in Figure 4.26b. This situation is the same as that presented in Figure 4.20. Now the étendue is

$$U = \int_{\sigma_1}^{\sigma_2} (\mathbf{p}_A - \mathbf{p}_B) \cdot d\mathbf{c} = \int_{\sigma_1}^{\sigma_2} (\mathbf{p}_A - \mathbf{p}_B) \cdot \frac{d\mathbf{c}}{d\sigma} d\sigma \tag{4.26}$$

calculated along the curve.

Another way to calculate the étendue is by rewriting expression (4.26) as

$$U = \int_C n(\mathbf{t}_A - \mathbf{t}_B) \cdot d\mathbf{c} = \int_C n\mathbf{t}_A \cdot d\mathbf{c} - \int_C n\mathbf{t}_B \cdot d\mathbf{c} \tag{4.27}$$

where $\|\mathbf{t}_A\| = \|\mathbf{t}_B\| = 1$. This now enables us to calculate the étendue as a function of the angles θ_A and θ_B that the edge rays make to the normal $\mathbf{n}$ to the curve c at each point, as shown in Figure 4.27a.

We have

$$\begin{aligned} \mathbf{t}_A \cdot d\mathbf{c} &= -\sin\theta_A dc \\ \mathbf{t}_B \cdot d\mathbf{c} &= -\sin\theta_B dc \end{aligned} \tag{4.28}$$

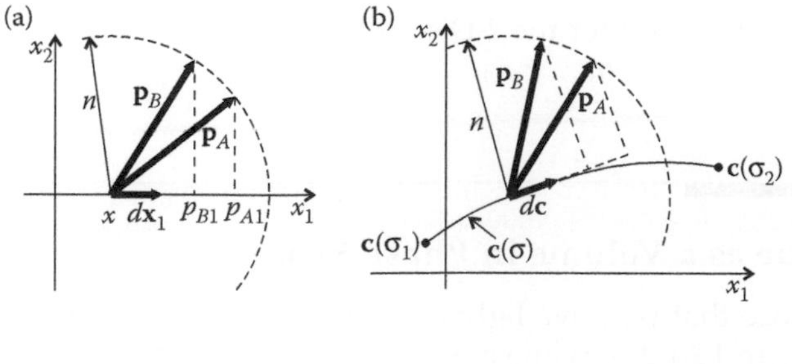

FIGURE 4.26
Étendue of the light (a) at an infinitesimal length dx_1 on the x_1 axis, and (b) through an infinitesimal portion of a curve parameterized by $\mathbf{c}(\sigma)$.

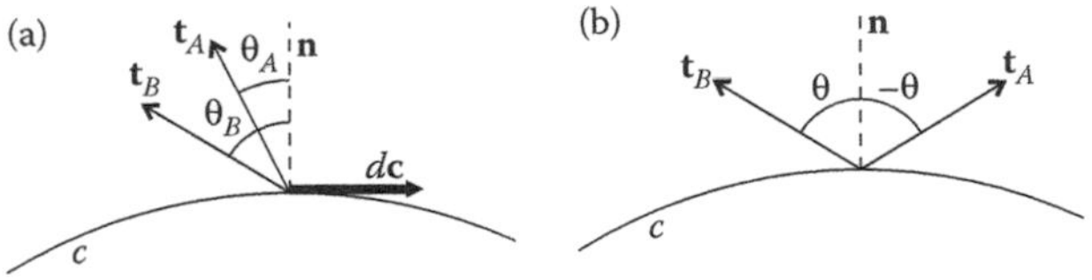

FIGURE 4.27
(a) Edge rays of the radiation crossing curve c make angles θ_A and θ_B to the normal $\mathbf{n}$ to the curve. (b) Shows the particular case in which $\theta_A = -\theta$ and $\theta_B = \theta$.

where $dc = \|d\mathbf{c}\|$. From expression (4.27) we then have

$$U = \int_C n\sin\theta_B\, dc - \int_C n\sin\theta_A\, dc$$

$$= \frac{\int_C n\sin\theta_B\, dc}{\int_C dc}\int_C dc - \frac{\int_C n\sin\theta_A\, dc}{\int_C dc}\int_C dc = \langle n\sin\theta_B\rangle a - \langle n\sin\theta_A\rangle a \quad (4.29)$$

where < > denotes the average and a is the length of the curve c. If the refractive index n does not vary along the curve, we have

$$U = n[\langle\sin\theta_B\rangle - \langle\sin\theta_A\rangle]a \quad (4.30)$$

If θ_A = constant and θ_B = constant on the curve c, then

$$U = na(\sin\theta_B - \sin\theta_A) \quad (4.31)$$

Finally, in the particular case shown in Figure 4.27b, in which $\theta_B = \theta$ and $\theta_A = -\theta$ for all the points of the curve c, we get

$$U = 2na\sin\theta \quad (4.32)$$

which is the expression for the 2-D étendue given earlier.

4.7 Étendue as a Volume in Phase Space

Let us suppose that we have light crossing the x_1 axis in the direction of x_2 positive. Figure 4.28 shows one of those light rays on the left. It crosses the x_1 axis at position $(x,0)$ and has a direction defined by an optical momentum $\mathbf{p}$. The refractive index on the x_1 axis is n.

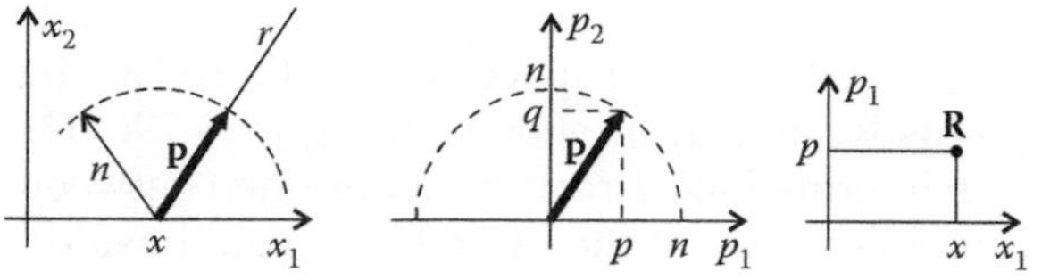

FIGURE 4.28
A light ray crossing the x_1 axis toward x_2 positive can be defined by a point $\mathbf{R}$ in phase space x_1p_1.

The optical momentum has components $\mathbf{p} = (p, q)$ in space (p_1, p_2) as shown in Figure 4.28 (center). Since we know that the light is propagating toward positive x_2 and $||\mathbf{p}|| = n$, we have $p^2 + q^2 = n^2$ with $q > 0$. Coordinate q of the optical momentum can then be obtained as a function of p as

$$q = \sqrt{n^2 - p^2} \tag{4.33}$$

Note that from Figure 4.28 (center) we can also see that $\mathbf{p}$ is a vector from the origin to the semicircle of radius n centered at the origin and above the x_1 axis. Specifying the value of p fully defines (p, q) and, therefore, the vector $\mathbf{p}$, and the direction of the light ray.

Therefore, the p_1 coordinate of the vector $\mathbf{p}$, given by p, is enough to define the direction of propagation of the light ray r at point $(x, 0)$. This light ray at position $(x, 0)$ can then be defined by a point $\mathbf{R} = (x, p)$ in space (x_1, p_1) as shown on the right of Figure 4.28. This space (x_1, p_1) is called phase space, and a point $\mathbf{R}$ on it defines the position and direction of a light ray in a medium of refractive index n.

Let us now suppose that we have radiation contained between edge rays r_A and r_B at a point $(x, 0)$ immersed in a medium of refractive index n at that point, as shown in Figure 4.29a. These rays have optical momentums

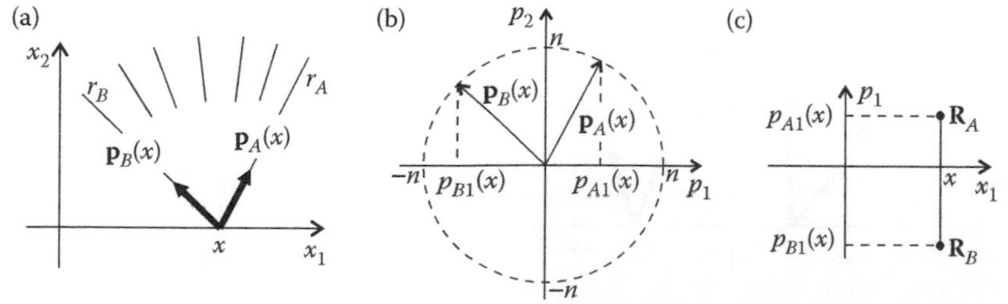

FIGURE 4.29
(a) Radiation contained between edge rays r_A and r_B crossing at point $(x,0)$. (b) The two edge rays r_A and r_B have x_1 components of the momentum vector given by p_{A1} and p_{B1}, and are represented in phase space as two points $\mathbf{R}_A$ and $\mathbf{R}_B$. (c) All the rays crossing at point $(x,0)$ contained between r_A and r_B are represented by a vertical line between $\mathbf{R}_A$ and $\mathbf{R}_B$.

$\mathbf{p}_A(x)$ and $\mathbf{p}_B(x)$, whose x_1 components are $p_{A1}(x)$ and $p_{B1}(x)$, as shown in Figure 4.29b. When represented in phase space (x_1, p_1), these two light rays correspond to two points $\mathbf{R}_A$ and $\mathbf{R}_B$ as shown in Figure 4.29c. The rays passing through $(x, 0)$ with intermediate directions (contained between r_A and r_B) will be represented in phase space at the same horizontal position x, but with p_1 values ranging from $p_{B1}(x)$ to $p_{A1}(x)$ as shown in Figure 4.29c. These rays are presented as a vertical line in phase space from point $\mathbf{R}_B$ to point $\mathbf{R}_A$.

Let us now consider a line extending on the x_1 axis from x_m to x_M crossed by radiation, with variable extreme directions from point to point. We have a situation similar to the one presented in Figure 4.30. For each value of x_1 (e.g., $x_1 = x$), the radiation is contained between the rays, with momentum vectors $\mathbf{p}_A(x_1)$ and $\mathbf{p}_B(x_1)$, which have p_1 components $p_{A1}(x_1)$ and $p_{B1}(x_1)$, as shown in Figure 4.30b. The set of all the edge rays is, therefore, represented in phase space by a line ∂R passing through the points $(x_1, p_{A1}(x_1))$ and $(x_1, p_{B1}(x_1))$. The zone R enclosed by this line represents all the light rays crossing the line $x_m x_M$.

The étendue of the light crossing $x_m x_M$ is given by

$$U = -\iint dp_1 dx_1 = -\int_{x_m}^{x_M}\left(\int_{r_A}^{r_B} dp_1\right)dx_1 = \int_{x_m}^{x_M}[p_{A1}(x_1) - p_{B1}(x_1)]dx_1 \tag{4.34}$$

where

$$dU = [p_{A1}(x_1) - p_{B1}(x_1)]dx_1 \tag{4.35}$$

is the area of a vertical stripe of thickness dx_1 from $p_{B1}(x_1)$ to $p_{A1}(x_1)$, as shown in Figure 4.30b for the particular case of $x_1 = x$. Therefore, $U = -\iint dp_1 dx_1$ gives us the area of zone R in phase space. It can, therefore, be concluded that the étendue of the radiation crossing $x_m x_M$ is given by the area of zone R in phase space enclosed by ∂R. Note that expression (4.35) is the same as expression (4.25).

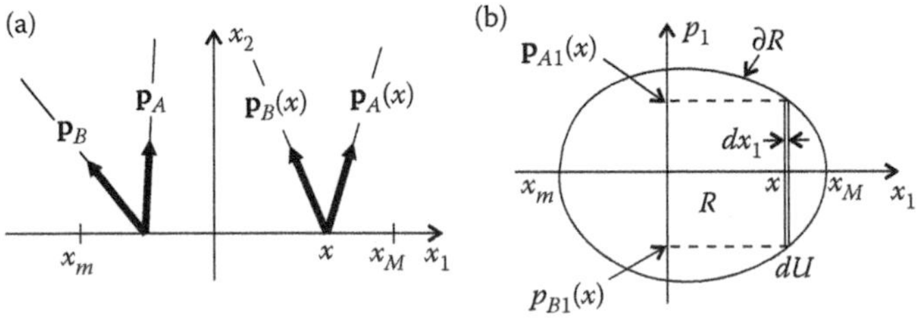

FIGURE 4.30
(a) Radiation crossing line $x_m x_M$ with variable extreme directions. For each value of x_1, the optical momentum $\mathbf{p}$ has x_1 components ranging from $p_{B1}(x_1)$ to $p_{A1}(x_1)$. (b) The zone R represents all the light rays crossing $x_m x_M$, and ∂R represents the edge rays of that radiation. The area inside ∂R corresponds to the étendue of that radiation.

4.8 Étendue as a Difference in Optical Path Length

We start with a general situation that we later apply to the calculation of étendue. We have one set of rays and the corresponding perpendicular wavefronts as in Figure 4.31. We also suppose that if we integrate the optical path length along any of the light rays from a reference wavefront w to wavefront w_1, we get $S = s_1$ and $S = s_2$ if we integrate from w to w_2.

For the particular case, for example, of ray r, this means that

$$\int_{\mathbf{W}}^{\mathbf{W}_1} n\, ds = s_1 \quad \text{and} \quad \int_{\mathbf{W}}^{\mathbf{W}_2} n\, ds = s_2 \tag{4.36}$$

where the integral is taken along r from point $\mathbf{W}$ to points $\mathbf{W}_1$ and $\mathbf{W}_2$.

We now consider the following integral along a curve c:

$$\int_{c} n\mathbf{t} \cdot d\mathbf{c} \tag{4.37}$$

where this curve c extends from point $\mathbf{P}_1$ to point $\mathbf{P}_2$, as shown in Figure 4.32a. Unit vector $\mathbf{t}$ has $\|\mathbf{t}\| = 1$ and is tangent to the light ray and perpendicular to the wavefronts, as shown in Figure 4.32b.

From Figure 4.32 we can see that $n\mathbf{t}{\cdot}d\mathbf{c} = dS$, where dS is an increment in the optical path length, and therefore (as seen in Chapter 15)

$$\int_{c} n\mathbf{t} \cdot d\mathbf{c} = \int_{\mathbf{P}_1}^{\mathbf{P}_2} n\mathbf{t} \cdot d\mathbf{c} = s_2 - s_1 \tag{4.38}$$

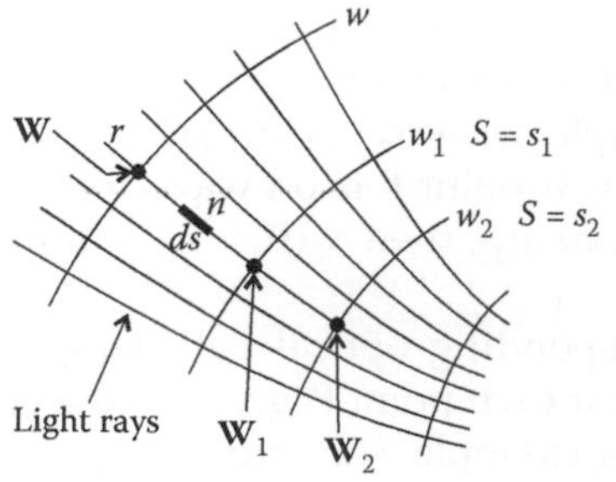

FIGURE 4.31
Light rays perpendicular to the corresponding wavefronts.

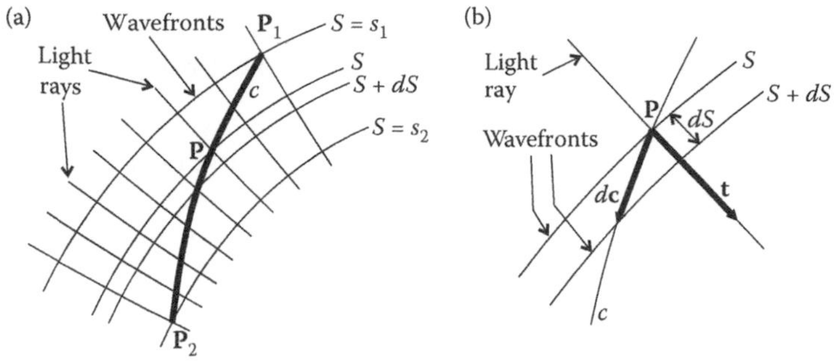

FIGURE 4.32
Integrating $n\mathbf{t} \cdot d\mathbf{c}$ along a curve c from $\mathbf{P}_1$ to $\mathbf{P}_2$, we get the difference in the optical path length between $\mathbf{P}_1$ and $\mathbf{P}_2$, which is $s_2 - s_1$. At each point $\mathbf{P}$ along the curve c, $\mathbf{t} \cdot d\mathbf{c}$ is the projection of $d\mathbf{c}$ in the direction of $\mathbf{t}$ and, therefore, $n\mathbf{t} \cdot d\mathbf{c} = dS$, where dS is an infinitesimal increment in the optical path length. (b) shows a detaile of figure (a) at point **P**.

We now apply this general result to the calculation of étendue. We consider the case in which the edge rays of the radiation are perpendicular to the two wavefronts, as shown in Figure 4.19. Here we consider the general case of a variable refractive index, in which the light rays are curved.

We have seen in expression (4.27) that the étendue of the light crossing a curve c parameterized by $\mathbf{c}(\sigma)$, whose edge rays have extreme directions given by unit vectors $\mathbf{t}_A$ and $\mathbf{t}_B$ at each point on the curve, can be calculated by

$$U = \int_C n\mathbf{t}_A \cdot d\mathbf{c} - \int_C n\mathbf{t}_B \cdot d\mathbf{c} \tag{4.39}$$

The étendue is then obtained by subtracting two integrals, such as the one in expression (4.38). The first integral is relative to one set of edge rays, and the second to the other set. Figure 4.33 shows a curve c illuminated by radiation confined between edge rays r_A and r_B, similar to that in Figure 4.19.

We now define an optical path length function $S_A(\mathbf{P})$ for the bundle of rays r_A that, for each point **P**, gives us the optical path length from a wavefront w_A to point **P**. For example, wavefronts w_{A1} and w_{A2} are those through points $\mathbf{P}_1$ and $\mathbf{P}_2$ of the curve c. If point **P** is on wavefront w_{A1}, then $S_A(\mathbf{P}) = s_{A1}$ and if point **P** is on wavefront w_{A2}, then $S_A(\mathbf{P}) = s_{A2}$. For example, $S_A(\mathbf{P}_1) = s_{A1}$ and $S_A(\mathbf{P}_2) = s_{A2}$.

We define the corresponding optical path length function $S_B(\mathbf{P})$ for the bundle of rays r_B that, for each point **P**, gives us the optical path length from a wavefront w_B to **P**. For example, wavefronts w_{B1} and w_{B2} are those through points $\mathbf{P}_1$ and $\mathbf{P}_2$ of the curve c. If point **P** is on wavefront w_{B1}, then $S_B(\mathbf{P}) = s_{B1}$, and if point **P** is on wavefront w_{B2}, then $S_B(\mathbf{P}) = s_{B2}$. For example, $S_B(\mathbf{P}_1) = s_{B1}$ and $S_B(\mathbf{P}_2) = s_{B2}$.

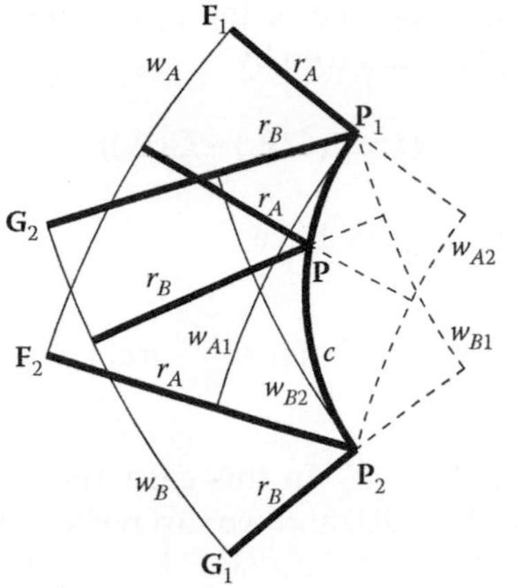

FIGURE 4.33
Radiation confined between two bundles of edge rays r_A and r_B crossing a curve c.

From expressions (4.39) and (4.38) we obtain

$$U = \int_C n\mathbf{t}_A \cdot d\mathbf{c} - \int_C n\mathbf{t}_B \cdot d\mathbf{c} = \int_{\mathbf{P}_1}^{\mathbf{P}_2} n\mathbf{t}_A \cdot d\mathbf{c} - \int_{\mathbf{P}_1}^{\mathbf{P}_2} n\mathbf{t}_B \cdot d\mathbf{c}$$
$$= \int_{\mathbf{P}_1}^{\mathbf{P}_2} n\mathbf{t}_A \cdot d\mathbf{c} + \int_{\mathbf{P}_2}^{\mathbf{P}_1} n\mathbf{t}_B \cdot d\mathbf{c} = (s_{A2} - s_{A1}) + (s_{B1} - s_{B2}) \tag{4.40}$$

Taking the two rays r_A passing through $\mathbf{F}_1\mathbf{P}_1$ and $\mathbf{F}_2\mathbf{P}_2$, and the two rays r_B passing through $\mathbf{G}_2\mathbf{P}_1$ and $\mathbf{G}_1\mathbf{P}_2$, we can write:[12,13]

$$U = ([[\mathbf{F}_2,\mathbf{P}_2]] - [[\mathbf{F}_1,\mathbf{P}_1]]) + ([[\mathbf{G}_2,\mathbf{P}_1]] - [[\mathbf{G}_1,\mathbf{P}_2]])$$
$$= [[\mathbf{F}_2,\mathbf{P}_2]] + [[\mathbf{G}_2,\mathbf{P}_1]] - [[\mathbf{F}_1,\mathbf{P}_1]] - [[\mathbf{G}_1,\mathbf{P}_2]] \tag{4.41}$$

where [[**A**,**B**]] is the optical path length between **A** and **B**. We can also write:

$$U = ([[\mathbf{F}_2,\mathbf{P}_2]] - [[\mathbf{G}_1,\mathbf{P}_2]]) - ([[\mathbf{F}_1,\mathbf{P}_1]] - [[\mathbf{G}_2,\mathbf{P}_1]])$$
$$= [S_A(\mathbf{P}_2) - S_B(\mathbf{P}_2)] - [S_A(\mathbf{P}_1) - S_B(\mathbf{P}_1)] \tag{4.42}$$

Defining

$$G = (S_A - S_B)/2 \tag{4.43}$$

We can conclude that the étendue of the light "passing" between the two points $\mathbf{P}_1$ and $\mathbf{P}_2$ of the plane is given by

$$U = 2(G(\mathbf{P}_2) - G(\mathbf{P}_1)) \tag{4.44}$$

Note that if $\mathbf{P}_2 = \mathbf{P}_1 + (dx_1, dx_2)$, we have

$$G(\mathbf{P}_2) = G(\mathbf{P}_1) + \left(\frac{\partial G}{\partial x_1} dx_1 + \frac{\partial G}{\partial x_2} dx_2 \right) = G(\mathbf{P}_1) + dG \tag{4.45}$$

and, therefore, $G(\mathbf{P}_2) - G(\mathbf{P}_1) = dG$. In this case, the étendue of the radiation passing between $\mathbf{P}_1$ and $\mathbf{P}_2$ is dU, and we can write

$$dU = 2dG \tag{4.46}$$

which gives us the étendue as a function of difference in the optical path lengths.

We can now apply the general results obtained to some particular cases. Let us then suppose, for example, that the rays of light perpendicular to wavefront w_A come from a point **F**, and that the rays of light perpendicular to w_B come from a point **G**. This situation is presented in Figure 4.34a.

In this case, the étendue arriving at $\mathbf{P}_1\mathbf{P}_2$ coming from **GF** is given by expression (4.41) obtained earlier. We can write

$$\begin{aligned} [[\mathbf{G}_2,\mathbf{P}_1]] - [[\mathbf{G}_1,\mathbf{P}_2]] &= [[\mathbf{G},\mathbf{G}_2]] + [[\mathbf{G}_2,\mathbf{P}_1]] - ([[\mathbf{G},\mathbf{G}_1]] + [[\mathbf{G}_1,\mathbf{P}_2]]) \\ &= [[\mathbf{G},\mathbf{P}_1]] - [[\mathbf{G},\mathbf{P}_2]] \end{aligned} \tag{4.47}$$

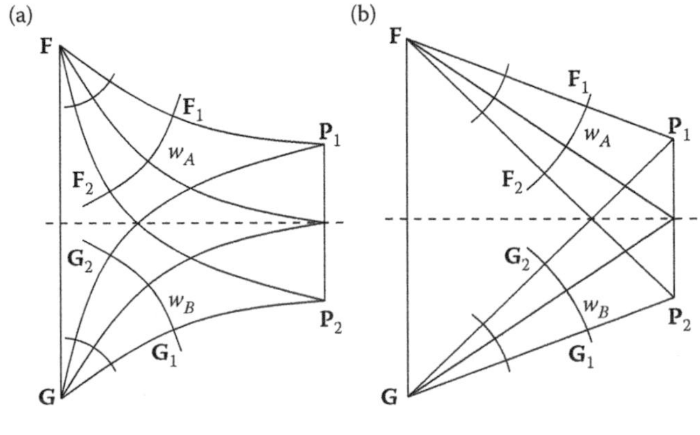

FIGURE 4.34
Calculation of the étendue from a line **FG** to another one $\mathbf{P}_1\mathbf{P}_2$ in the case of a medium, (a) having a variable refractive index from point to point, and (b) having a constant refractive index.

since $[[\mathbf{G}, \mathbf{G}_2]] = [[\mathbf{G}, \mathbf{G}_1]]$. Acting accordingly for $[[\mathbf{F}_2, \mathbf{P}_2]] - [[\mathbf{F}_1, \mathbf{P}_1]]$, we get

$$U = [[\mathbf{F},\mathbf{P}_2]] + [[\mathbf{G},\mathbf{P}_1]] - [[\mathbf{F},\mathbf{P}_1]] - [[\mathbf{G},\mathbf{P}_2]] \tag{4.48}$$

If the horizontal dashed line is an axis of symmetry for the system, expression simplifies to

$$U = 2([[\mathbf{F},\mathbf{P}_2]] - [[\mathbf{F},\mathbf{P}_1]]) = 2(S_A(\mathbf{P}_2) - S_A(\mathbf{P}_1)) \tag{4.49}$$

Given the symmetry of the optical system, we can write $S_A(\mathbf{P}_1) = S_B(\mathbf{P}_2)$ and, therefore,

$$U = 2(S_A(\mathbf{P}_2) - S_B(\mathbf{P}_2)) = 4G(\mathbf{P}_2) \tag{4.50}$$

The étendue is then given by a difference in the optical path lengths.[13] In the particular case of having a medium of refractive index $n = 1$, the optical path lengths coincide with the distances between the points. Therefore, for the case presented in Figure 4.34b, the étendue arriving at $\mathbf{P}_1\mathbf{P}_2$ coming from the source **FG** is given by the preceding expressions, but where the optical path length $[[\mathbf{X}, \mathbf{Y}]]$ between points **X** and **Y** can be replaced by the distance $[\mathbf{X}, \mathbf{Y}]$ between those same points.

4.9 Flow-Lines

Let us consider that we have two wavefronts, w_A and w_B, propagating through an optical system. After propagation, w_{A1} is transformed to w_{A2}, and w_{B1} to w_{B2}, as shown in Figure 4.35.

The optical path length between w_{A1} and w_{A2} is constant and equals, for example, S_{A1A2}. Accordingly, the optical path length between w_{B1} and w_{B2} is constant, and has a value of S_{B1B2}. Now, if we consider a point **P** between the wavefronts, and if S_{A1}, S_{B1}, S_{A2}, and S_{B2} are the optical path lengths between **P** and w_{A1}, w_{B1}, w_{A2}, and w_{B2}, respectively, we can write

$$\begin{aligned} S_{A1} + S_{A2} &= S_{A1A2} \\ S_{B1} + S_{B2} &= S_{B1B2} \end{aligned} \tag{4.51}$$

Now consider a mirror m (mirrored on both sides) placed between the wavefronts and shaped so that it reflects the light rays coming from w_{A1} to w_{B2}, and the rays coming from w_{B1} to w_{A2}. This can be accomplished if the mirror bisects at each point the rays coming from w_{A1} and w_{B1}. Now w_{A1} is reflected

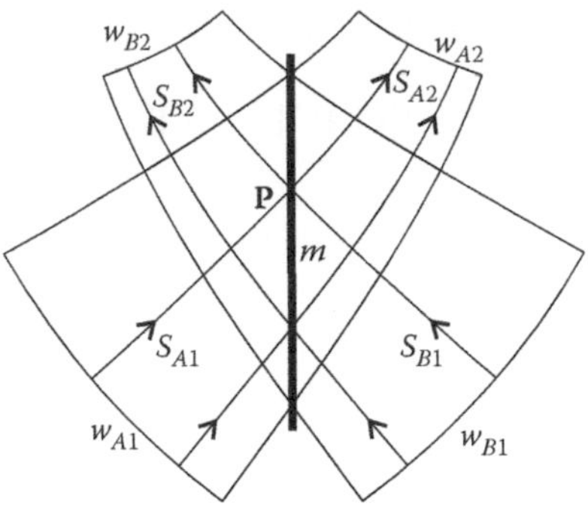

FIGURE 4.35
The wavefront w_{A1} propagates through an optical system becoming w_{A2}, and w_{B1} also propagates to become w_{B2}. It is possible to place a mirror m (mirrored on both sides) reflecting w_{A1} to w_{B2}, and w_{B1} to w_{A2}, when this mirror bisects at each point the rays coming from w_{A1} and w_{B1}. The points of this mirror belong to a line of constant G.

to w_{B2} and w_{B1} to w_{A2}. Therefore, after placing the mirror m, the optical path length between w_{A1} and w_{B2} is now constant, and equals, for example, S_{A1B2}, and is also constant between w_{B1} and w_{A2} having a value of S_{B1A2}. We can then write for the rays reflected at mirror m as follows:

$$\begin{aligned} S_{A1} + S_{B2} &= S_{A1B2} \\ S_{B1} + S_{A2} &= S_{B1A2} \end{aligned} \tag{4.52}$$

These expressions can be used to obtain the shape of the mirror m. It is calculated by imposing constant optical path length between w_{A1} and w_{B2}. Alternatively, it can also be calculated by imposing constant optical path length between w_{B1} and w_{A2}.

Combining the preceding equations, we also get for the points on mirror m

$$S_{A1} - S_{B1} = S_{A1} + S_{A2} - (S_{B1} + S_{A2}) = S_{A1A2} - S_{B1A2} = S_{A1B1} \tag{4.53}$$

where S_{A1B1} is a constant. Accordingly, and also for the points on mirror m

$$S_{A2} - S_{B2} = S_{A2} + S_{B1} - (S_{B2} + S_{B1}) = S_{B1A2} - S_{B1B2} = S_{A2B2} \tag{4.54}$$

where S_{A2B2} is also a constant. It can then be concluded that the points of mirror m are those for which $S_{A1} - S_{B1} = \text{constant}$ (or $S_{A2} - S_{B2} = \text{constant}$). Therefore, this mirror is a line of constant G. The lines $G = \text{constant}$ then bisect, at each point, the rays coming from w_{A1} and w_{B1}. They also bisect the rays going to w_{A2} and w_{B2}, since these rays are the same rays coming from w_{A1} and w_{B1}. Straight or not, such lines of constant G are generally known as flow-lines.

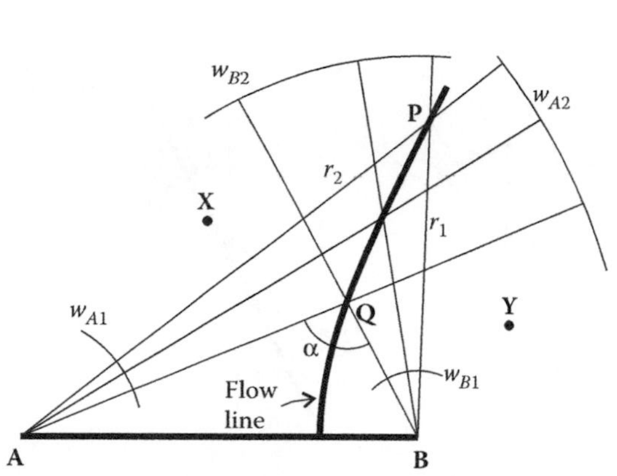

FIGURE 4.36
The flow-lines of the radiation emitted by a flat Lambertian source **AB** are hyperbolas with foci at the edges **A** and **B** of the source.

Now consider the particular case in which the two sets of edge rays come from the edges of a flat source, as shown in Figure 4.36. At a given point **Q**, light is confined between two edge rays making an angle α between them. As point **Q** moves closer to **AB**, angle α increases. It has a maximum of $\alpha = \pi$ for the points of **AB**. At the line **AB**, radiation is then fully Lambertian and confined between directions that make angles $\pm\pi/2$ to its normal. The lines of constant G are now hyperbolas. Figure 4.36 shows a Lambertian source **AB**, and a flow-line (line of constant G). Since, for each point **P** on a hyperbola with foci **A** and **B**, we have **[P, A]** – **[P, B]** = constant, we can conclude that this hyperbola is a flow-line. Wavefronts w_A and w_B are in this case circles centered at **A** and **B**, respectively. Figure 4.36 shows wavefront w_A at positions w_{A1} and w_{A2}, and wavefront w_B at positions w_{B1} and w_{B2}. Since the radiation is fully Lambertian at **AB**, making angles $\pm\pi/2$ to its normal, both wavefronts w_{A1} and w_{B1} and the flow-lines that touch **AB** are perpendicular to **AB** when they reach it.

We now place a mirror (mirrored on both sides) along the flow-line through a point **P**. Before introducing this mirror, light at point **P** is confined between edge rays r_1 and r_2, as shown in Figure 4.37a. After placing the mirror, we have the situation in Figure 4.37b in which the bundle of rays is now split into two.

The incoming bundle is divided into b_1 and b_3. If the mirror was not there, b_1 would come out as bundle b_4, and bundle b_3 would come out as bundle b_2. With the mirror, however, b_1 is reflected as b_2 and b_3 is reflected as b_4. Since the mirror bisects the edge rays r_1 and r_2, bundles b_1 and b_3 are symmetrical. Bundles b_2 and b_4 are also symmetrical. For the points to the left of the mirror, it shades b_3 but at the same time reflects b_1, so that these two effects cancel out. The same is true for the points to the right of the mirror. The consequence of this is that the radiation field is unaffected by introducing the

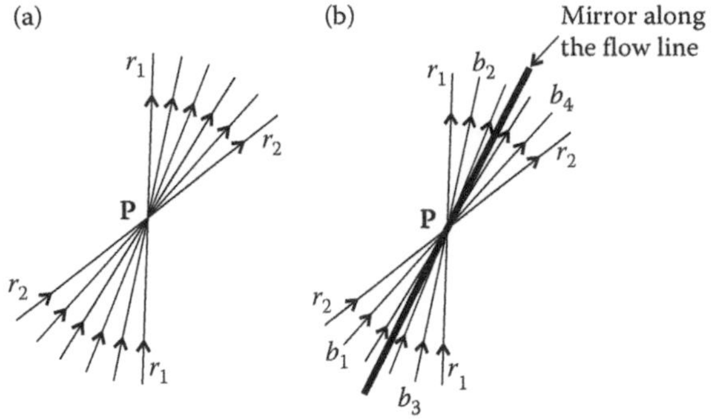

FIGURE 4.37
(a) A bundle of rays crosses a point **P**. (b) Introducing a mirror along the flow-line has no effect on the radiation field.

mirror. This means that, with or without the mirror along the flow-line, points **X** and **Y** in Figure 4.36 see the same radiation field.

We now consider two flow-lines with $G = G_1$ and $G = G_2$, where G_1 and G_2 are constants, as shown in Figure 4.38.

A curve c_Q has one of the end points $\mathbf{Q}_1$ on the first flow-line, and the other end point $\mathbf{Q}_2$ on the second flow-line. The étendue through c_Q is given by

$$U_Q = 2(G(\mathbf{Q}_1) - G(\mathbf{Q}_2)) = 2(G_1 - G_2) \tag{4.55}$$

On the other hand, we have another curve c_P that also has one of the end points $\mathbf{P}_1$ on the first flow-line, and the other end point $\mathbf{P}_2$ on the second flow-line. The étendue through c_P is given by

$$U_P = 2(G(\mathbf{P}_1) - G(\mathbf{P}_2)) = 2(G_1 - G_2) \tag{4.56}$$

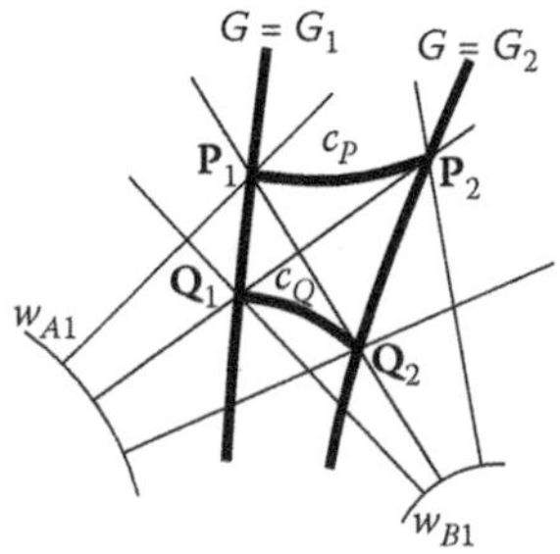

FIGURE 4.38
The étendue of the light crossing a curve c_Q between points $\mathbf{Q}_1$ and $\mathbf{Q}_2$ is the same as that of the radiation crossing a curve c_P between points $\mathbf{P}_1$ and $\mathbf{P}_2$, because edges $\mathbf{P}_1$ and $\mathbf{Q}_1$ on one side, and $\mathbf{P}_2$ and $\mathbf{Q}_2$ on the other, are on the same flow-lines.

The étendue of the radiation crossing the two curves is then the same, showing that the étendue is conserved between the flow-lines.

4.10 The Winston–Welford Design Method

The fact that étendue is conserved between the flow-lines can be used to design nonimaging optical devices.[13,14] The Winston–Welford design method involves placing two mirrors along two flow-lines so that the light is guided between them, while conserving étendue. For this reason, it is also called the flow-line design method. We will now consider some examples of optics designed according to this principle.

As a first example, refer to the situation presented in Figure 4.36, and consider two symmetrical flow-lines $\mathbf{Q}_1\mathbf{P}_1$ and $\mathbf{Q}_2\mathbf{P}_2$ generated by a Lambertian source **AB**, as shown in Figure 4.39.

Étendue is conserved between the flow-lines and, therefore, the étendue of the radiation at line c_P between $\mathbf{P}_1$ and $\mathbf{P}_2$ is the same as the étendue at line c_Q between $\mathbf{Q}_1$ and $\mathbf{Q}_2$. At line c_Q, radiation is fully Lambertian, confined between $\pm\pi/2$. However, at a point **V** on line c_P, radiation is confined between directions r_1 and r_2 pointing to **B** and **A**.

If we make lines $\mathbf{Q}_1\mathbf{P}_1$ and $\mathbf{Q}_2\mathbf{P}_2$ the mirrors along the flow-lines, these will not change the radiation pattern. This means that point **V** will still see the radiation pattern the same as that obtained without the mirrors. This radiation, however, now comes from $\mathbf{Q}_1\mathbf{Q}_2$ guided by the mirrors, instead of coming from the whole source **AB**, as earlier. The light that crosses c_P still appears to come from the source **AB**.

If we invert the direction of light, we can also think of c_P as a light source emitting toward **AB**. Mirrors $\mathbf{P}_1\mathbf{Q}_1$ and $\mathbf{P}_2\mathbf{Q}_2$ then concentrate this radiation onto $\mathbf{Q}_1\mathbf{Q}_2$ where it becomes fully Lambertian. Figure 4.40 shows this

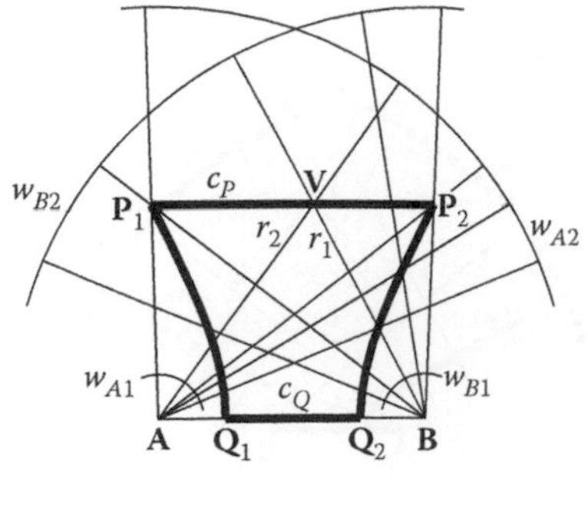

FIGURE 4.39
The étendue is conserved between the flow-lines that connect c_Q and c_P.

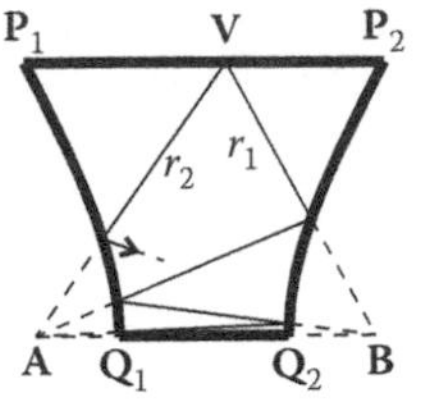

FIGURE 4.40
A concentrator for a source $\mathbf{P_1P_2}$ emitting toward **AB**. The radiation is concentrated onto receiver $\mathbf{Q_1Q_2}$ where it becomes fully Lambertian.

possibility, in which light emitted from $\mathbf{P_1P_2}$ toward **AB** bounces back and forth between the mirrors $\mathbf{P_1Q_1}$ and $\mathbf{P_2Q_2}$, and ends up on $\mathbf{Q_1Q_2}$.

For point **V**, for example, ray r_1 emitted toward point **B** is reflected by the right-hand-side mirror toward point **A**. The left-hand-side mirror then reflects it toward **B** again. This process goes on until (after an infinite number of reflections) this ray reaches $\mathbf{Q_1Q_2}$. The same happens to a ray r_2 emitted from **V** toward **A**. Intermediate rays between r_1 and r_2 either bounce off the mirrors and reach $\mathbf{Q_1Q_2}$, or reach it directly without any reflections. This concentrator is called trumpet,[15] and maximally concentrates onto receiver $\mathbf{Q_1Q_2}$ all radiation entering its aperture $\mathbf{P_1P_2}$ headed toward **AB**. Since the étendue from $\mathbf{P_1P_2}$ toward **AB** is the same as that of a Lambertian source $\mathbf{Q_1Q_2}$, this concentrator is ideal.[16]

As a second example, consider another situation, as shown in Figure 4.41, in which the wavefronts w_{A2} and w_{B2} have different shapes. Wavefront w_{A2} is made of a circle with center **A** up to ray r_1. From that ray to the left it is flat. Wavefront w_{A1} has the same geometry, and results from the propagation of w_{A2} to the left. Wavefront w_{B2} is symmetrical to w_{A2}, and w_{B1} is symmetrical to w_{A1}. Since the sections of the wavefronts that touch **AB** are perpendicular to **AB** when they reach it, the radiation there is again fully Lambertian.

We now look at the flow-lines defined by the rays perpendicular to these wavefronts. We have four different zones, shown as 1, 2, 3, and 4, in

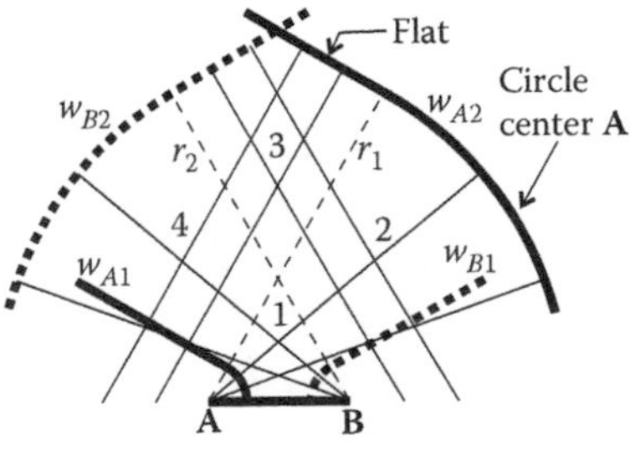

FIGURE 4.41
Wavefronts made of two different sections: circular and flat.

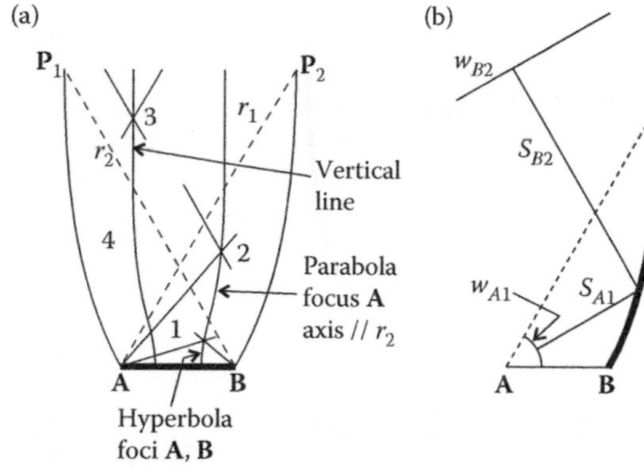

FIGURE 4.42
(a) Flow lines inside a CPC. (b) The right-hand-side mirror of a CPC concentrator is seen as a flow line.

Figure 4.42. In zone 1, the edge rays come from points **A** and **B** and we have the same situation as shown in Figure 4.36. They are hyperbolas with foci **A** and **B**, as shown in Figure 4.42a.

In zone 3, the edge rays are straight and parallel to symmetrical rays r_1 and r_2, so that the flow-lines are vertical straight lines. In zone 2, one of the edge rays comes from point **A** and the other is parallel to r_2. The flow-lines are then parabolas that reflect the rays parallel to r_2 toward point **A**. Zone 4 is symmetrical to zone 2, and the flow-lines are also parabolas, but with focus **B** and axis parallel to r_1, and they reflect the rays parallel to r_1 toward point **B**.

Note that in the case of Figure 4.39, the flow-lines were generated by a Lambertian source **AB**. That is not the case here because only those rays that are perpendicular to the circular sections of the wavefronts are coming from the edges **A** and **B** of **AB**. Therefore, only the portion of the flow-lines in zone 1 can be considered as being generated by a Lambertian source **AB**.

For a particular case of the flow-lines that touch the edges **A** and **B**, the hyperbolas of zone 1 vanish, and we are left with the parabolas of zone 2. The parabolas $\mathbf{AP}_1$ and $\mathbf{BP}_2$ form a compound parabolic concentrator (CPC).

When we want to design a CPC, however, we consider only the portions of the wavefronts that generate the flow-line we want for the optic. Figure 4.42b shows that the possibility in which the parabola $\mathbf{BP}_2$ is generated by the first of expressions (4.52), that is, $S_{A1} + S_{B2} = \text{constant}$. This is also the string method, whereby we attach a string of constant length to wavefronts w_{A1} and w_{B2}, keep it tight with a pencil, allow the string to slide on w_{A1} and w_{A2} so that it is always perpendicular to w_{A1} and w_{A2}, and then move the pencil; we then obtain the parabolic CPC profile, as shown in Figure 4.43. The other mirror of the CPC is generated in the same way.

As a third example, consider a different situation in which we extend the receiver **AB** in Figure 4.41 downward with, say, two vertical lines, and also

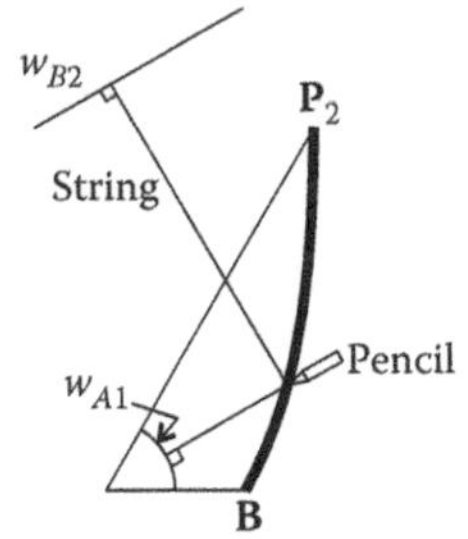

FIGURE 4.43
Drawing the profile of a CPC using the string method.

define wavefronts w_A and w_B below **AB**, as shown in Figure 4.44. In Figure 4.44, wavefront w_B is shown at positions w_{B0}, w_{B1}, and w_{B2}, and wavefront w_A is shown at positions w_{A0}, w_{A1}, and w_{A2}.

Now, besides zones 1–4, we also have zone 5, between the rays r_3 and r_4, and zone 6 between the ray r_4 and the receiver. In zone 5, one of the edge rays comes from point **B**, and the other is parallel to r_2. The flow-lines are then parabolas that reflect the rays parallel to r_2 toward point **B**. As we move closer to ray r_4, both edge rays approach this ray r_4, and the flow-lines become perpendicular to r_4. In zone 6 below r_4, both edge rays come from point **B**, and the flow-lines are circles centered at **B**, perpendicular to those rays at each point (just as what happened at ray r_4).

Figure 4.45 shows a concentrator for a receiver **CABD** designed using the wavefronts w_A and w_B in Figure 4.44.[17]

This concentrator is composed of a parabola with focus **A** and axis parallel to r_2 in zone 2, then it has a parabola with focus **B**, and axis parallel to r_2 in

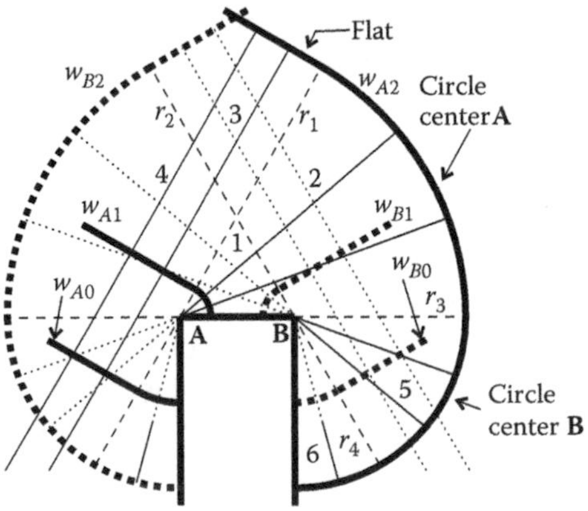

FIGURE 4.44
Wavefronts for a receiver with a horizontal section **AB** and two vertical sections starting at points **A** and **B**.

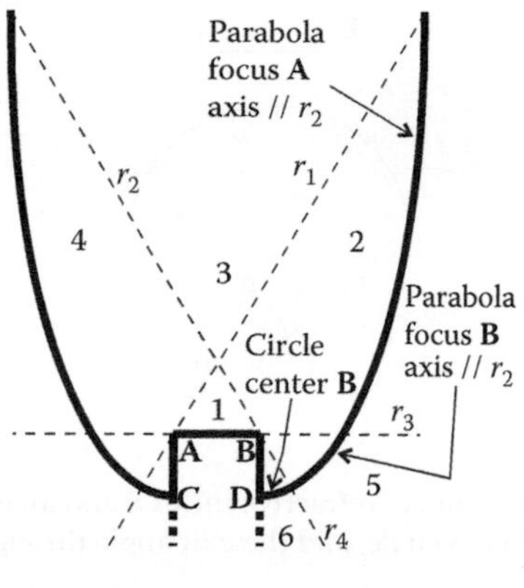

FIGURE 4.45
Concentrator for a receiver **CABD**.

zone 5, and finally, it also has a circle with center **B** in zone 6. Choosing other flow-lines would result in a concentrator for a receiver with different positions of **C** and **D**, as shown in Figure 4.46.

The geometry of this optic is similar to that of the concentrator in Figure 4.45, except for the reflector on the right that extends with a vertical flat mirror along the corresponding flow-line in zone 3.

As a fourth example, now consider another optic designed by the flow-line method. We have a receiver **AB** immersed in a medium of refractive index n, and an emitter $\mathbf{E_1E_2}$ in air. These two media are separated by a curve c, which, in this example, is a circle. The exit angle through **AB** is 2θ, as shown in Figure 4.47.

Now we calculate point **P** and its symmetrical **Q** on the curve c, such that the étendue from $\mathbf{E_1E_2}$ to **PQ** matches that of the receiver, that is,

$$2([\mathbf{E}_2,\mathbf{P}]-[\mathbf{E}_2,\mathbf{Q}]) = 2n[\mathbf{A},\mathbf{B}]\sin\theta \tag{4.57}$$

We now calculate the path of ray r_P starting at $\mathbf{E}_1$ and refracted at **P**, as shown in Figure 4.48a.

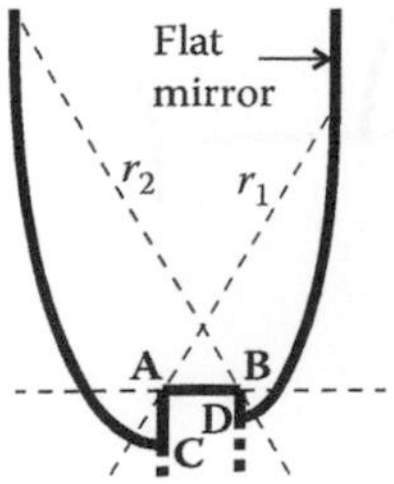

FIGURE 4.46
Concentrator for an asymmetrical receiver **CABD**.

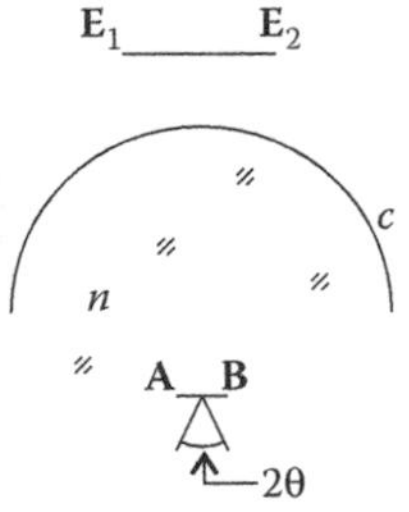

FIGURE 4.47
Receiver **AB** immersed in a medium of refractive index n and an emitter $\mathbf{E_1E_2}$ in air. The curve c that separates the two media is a circle, and the exit angle through **AB** is 2θ.

We define wavefront w_B above the curve c and to the right of ray r_P, as a circle $\mathbf{W_2W_4}$ with center $\mathbf{E_1}$. Below the curve c, the portion $\mathbf{W_1W_3}$ is the propagation of $\mathbf{W_2W_4}$ through the curve c. Also, we define wavefront w_B below the curve c and to the left of ray r_P as a piecewise curve starting at $\mathbf{W_3}$ and defined as flat and perpendicular to r_P, then as a circle with center **B** and as flat and perpendicular to edge ray r_2, ending at $\mathbf{W_5}$. Above the curve c, the portion $\mathbf{W_4W_6}$ of wavefront w_{B2} is the propagation of $\mathbf{W_5W_3}$ through the curve c.

This procedure defines wavefront w_B in position w_{B1} as a curve between $\mathbf{W_1}$ and $\mathbf{W_5}$, and in position w_{B2} as another curve between $\mathbf{W_2}$ and $\mathbf{W_6}$. Wavefront w_A is symmetrical to w_B about axis of symmetry b, which is the perpendicular bisector of emitter $\mathbf{E_1E_2}$ and of receiver **AB**, as shown in Figure 4.48b.

With these wavefronts, we can calculate the flow-line starting at the edge **B** of the receiver. It is also a piecewise curve with several pieces, as shown in Figure 4.49, where ray r_1 is symmetrical to r_2 and ray r_Q is symmetrical to r_P.

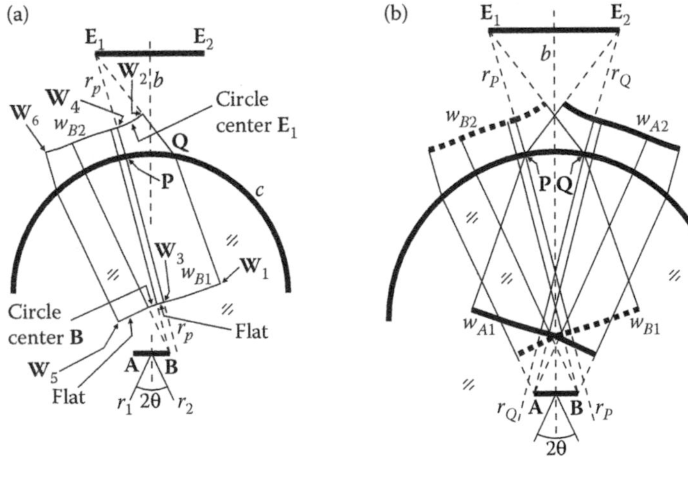

FIGURE 4.48
Wavefronts for defining the concentrator.

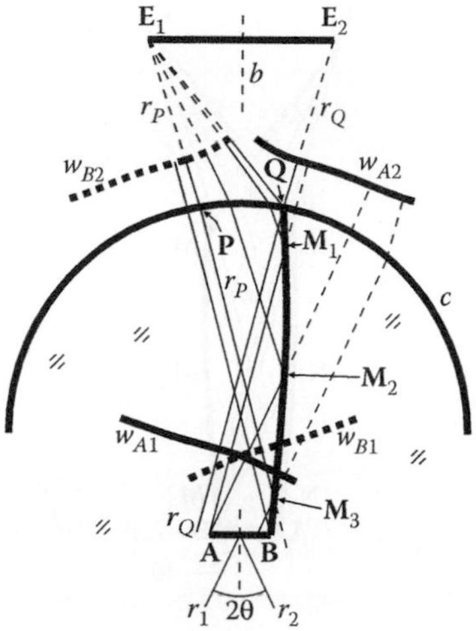

FIGURE 4.49
Flow-line starting at point **B**. It defines the shape of the sidewall for the concentrator.

Between points **B** and $\mathbf{M}_3$, the flow-line is flat, bisecting the rays parallel to r_P and those parallel to r_1. Between points $\mathbf{M}_3$ and $\mathbf{M}_2$, the flow-line bisects the edge rays coming from $\mathbf{E}_1$ and refracted at the curve *c* and those parallel to r_1. Between points $\mathbf{M}_2$ and $\mathbf{M}_1$, the flow-line bisects the edge rays coming from $\mathbf{E}_1$ and refracted at the curve *c* and those converging to point **A**.

Between points $\mathbf{M}_1$ and **Q**, the flow-line bisects the edge rays coming from $\mathbf{E}_1$ and refracted at the curve *c* and those parallel to ray r_Q. This same flow-line would then continue (not shown) above the curve *c* as a hyperbola through point **Q**, and with foci $\mathbf{E}_1$ and $\mathbf{E}_2$.

Figure 4.50 shows the complete optic, in which the side $\mathbf{PN_1N_2N_3A}$ is symmetrical to $\mathbf{QM_1M_2M_3B}$.

Light reflected by the portion $\mathbf{QM}_1$ of the right-hand-side wall is reflected toward portion $\mathbf{N_3A}$ of the left-hand-side wall, and from there toward **AB**.

In the example just presented, wavefront w_{B2} results from propagating w_{B1} through the curve *c*. Also, wavefront w_{A2} results from propagating w_{A1} through the curve *c*.

Now, we consider a possible way of calculating the propagation of wavefronts through surfaces. Consider the curve *c* separating two media of refractive indices n_1 and n_2, as shown in Figure 4.51. Wavefront w propagates through this curve *c*. We start with the wavefront at position w_1 and want to calculate its shape at position w_2. We define the optical path length between w_1 and w_2 as having a value *S*. We may, for example, take a point **P** on *c*, and calculate the corresponding point $\mathbf{W}_1$ on w_1. Point $\mathbf{W}_1$ is such that the perpendicular to w_1 through $\mathbf{W}_1$ crosses *c* at point **P**. Now we can calculate the optical path length between $\mathbf{W}_1$ and **P** as $S_1 = n_1[\mathbf{W}_1, \mathbf{P}]$. We now have the optical path length between **P** and the point $\mathbf{W}_2$ on the propagated wavefront

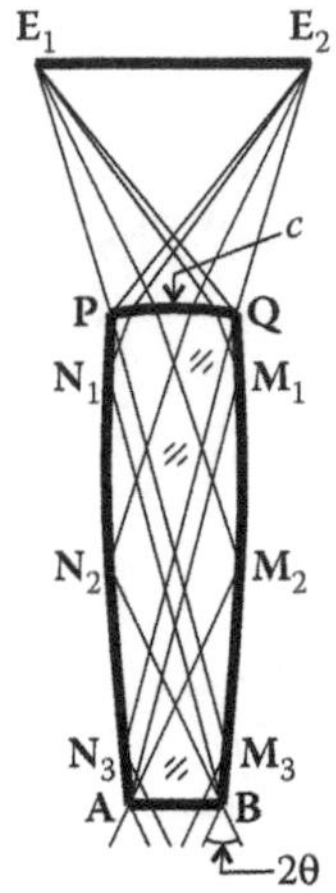

FIGURE 4.50
Concentrator for an emitter $\mathbf{E_1E_2}$ and a receiver **AB** immersed in a medium of refractive index n. The exit angle through **AB** is 2θ, and the entrance aperture of the optic is a circle c.

w_2, which is given by $S_2 = S - S_1$. The distance between **P** and $\mathbf{W}_2$ is $d_2 = S_2/n_2$. Since we know the position of $\mathbf{W}_1$ and **P**, we have the direction $\mathbf{t}_1$ of the ray incident at **P**. We now refract this ray, and obtain the direction $\mathbf{t}_2$ (unit vector) of the refracted ray. Point $\mathbf{W}_2$ is given by $\mathbf{W}_2 = \mathbf{P} + d_2\mathbf{t}_2$. Repeating the process for other points on c, we obtain a set of points on w_2. We can now interpolate between those points using, for example, a spline method.

Optics such as that shown in Figure 4.50 can be designed for other parameter values. Also, we do not have to calculate the shape of the wavefronts to design one of these devices. Another example is shown in Figure 4.52, where the emitter $\mathbf{E_1E_2}$ has the same size E as the entrance aperture size A_E of the optic. Vertical line b is the perpendicular bisector of emitter $\mathbf{E_1E_2}$ and receiver **AB**. The system is symmetrical relative to b.

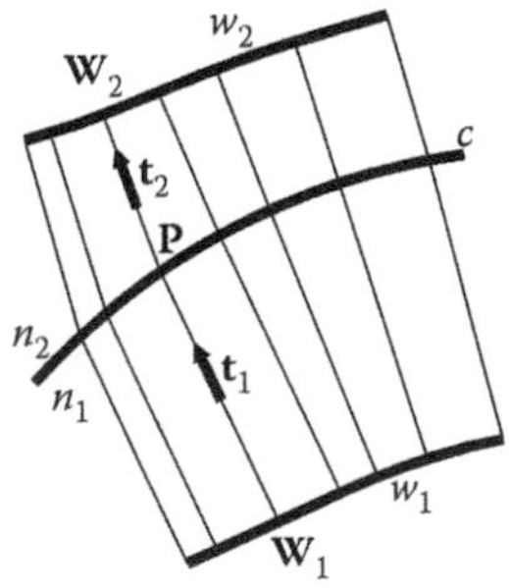

FIGURE 4.51
Propagation of a wavefront through a curve c that separates two media of refractive indices, n_1 and n_2.

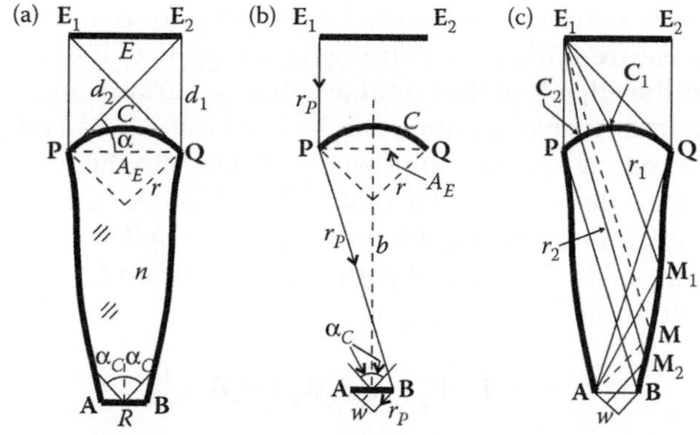

FIGURE 4.52
(a) An optic made of dielectric material for an emitter E and a receiver R. Light reaches the receiver confined to an angle $\pm\alpha_C$. (b) The entrance aperture is a circle C with radius r. The position of the exit aperture **AB** is defined by the path of ray r_P. (c) Portion **Q-M** of the side wall is obtained by constant optical path length from $\mathbf{E}_1$ to **A**. Portion **M-B** of the side wall of the optic is obtained by constant optical path length between $\mathbf{E}_1$ and w.

We start the design by determining the relative positions of the entrance aperture A_E and emitter E as presented in Figure 4.52a. Consider, for example, that the light at the receiver R is confined to an angle $\pm\alpha_C$. The étendue at the receiver R is then

$$U_R = 2nR\sin\alpha_C \tag{4.58}$$

This étendue is the same as that exchanged between A_E and E. Considering that $A_E = E$, from Figure 4.52a it can be seen that

$$\begin{aligned} \alpha &= \arctan(d_1/A_E) \\ d_2 &= d_1/\sin\alpha \end{aligned} \tag{4.59}$$

The étendue from A_E to E is then

$$U = 2(d_2 - d_1) = 2d_1\left(\frac{1}{\sin\alpha} - 1\right) = 2d_1\left(\frac{1}{\sin(\arctan(d_1/A_E))} - 1\right) \tag{4.60}$$

If étendue is to be conserved, we must have

$$nR\sin\alpha_C = d_1\left(\frac{1}{\sin(\arctan(d_1/A_E))} - 1\right) \tag{4.61}$$

To define the device, some parameters must be established: for example, the angle α_C, the value of $R = [\mathbf{A}, \mathbf{B}]$, and the value of $E = [\mathbf{E}_1, \mathbf{E}_2]$ (which equals A_E). Then, using expression (4.61), we can calculate the distance d_1.

The shape of the entrance surface can be chosen as a circular arc with radius r. As shown in Figure 4.52b, ray r_P coming from edge $\mathbf{E}_1$ of the emitter can now be refracted at the edge **P** of the circular entrance surface, and since the size of R is known, the receiver position can be determined. Its end point **B** will be on the refracted ray r_P at the point whose distance to the bisector b is $R/2$.

We now define a flat wavefront w that makes an angle α_C to **AB** and, therefore, is perpendicular to the right-hand edge ray at **AB**, as shown in Figure 4.52b and Figure 4.52c. The optical path length between $\mathbf{E}_1$ and wavefront w can now be determined from ray r_P as

$$S = [\mathbf{E}_1, \mathbf{P}] + n[\mathbf{P}, \mathbf{B}] + n[\mathbf{B}, w] \tag{4.62}$$

where $[\mathbf{B},w]$ is the distance between point **B** and wavefront w. This optical path length now enables us to calculate the shape of the lateral mirror **QB**. Mirror **PA** is symmetrical to **QB**. Points $\mathbf{M}_1$ between **Q** and **M** are calculated for rays r_1 from edge $\mathbf{E}_1$ of the emitter to edge **A** of the receiver, as shown in Figure 4.52c. These rays fulfill

$$S = [\mathbf{E}_1, \mathbf{C}_1] + n[\mathbf{C}_1, \mathbf{M}_1] + n[\mathbf{M}_1, \mathbf{A}] \tag{4.63}$$

Points $\mathbf{M}_2$ between **M** and **B** are calculated for rays r_2 from edge $\mathbf{E}_1$ of the emitter to wavefront w at the receiver. These rays fulfill

$$S = [\mathbf{E}_1, \mathbf{C}_2] + n[\mathbf{C}_2, \mathbf{M}_2] + n[\mathbf{M}_2, w] \tag{4.64}$$

where $[\mathbf{M}_2,w]$ is the distance between point $\mathbf{M}_2$ and wavefront w.

Now consider a particular case of this design, in which α_C is the critical angle for the material of refractive index n. Combine two of these optics, as shown in Figure 4.53. Light enters the optic on the left through its small aperture coming from a Lambertian source R_A separated from the optic by a small air gap. When the light refracts into the material of the optic, it will be confined to the critical angle $\pm\alpha_C$.

Light exits the optic on the left through its aperture $\mathbf{P}_A\mathbf{Q}_A$, and enters the optic on the right through its aperture $\mathbf{P}_B\mathbf{Q}_B$. It will then be concentrated onto its small aperture, where it is confined to the critical angle. When it exits

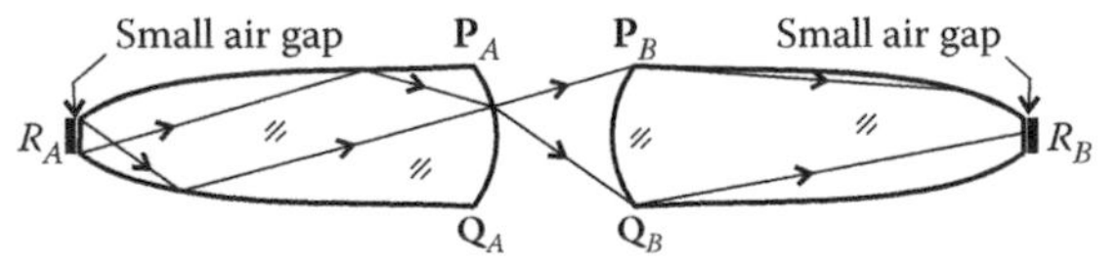

FIGURE 4.53
If a light source is placed at R_A, its light will exit the left-hand-side optic, enter the right-hand-side optic, and be concentrated onto R_B with no losses.

the optic, the angular aperture widens to a fully Lambertian illumination of receiver R_B.

Source R_A and receiver R_B are of the same size, and radiation is fully Lambertian at both places. Also, light travels through air between $\mathbf{P}_A\mathbf{Q}_A$ and $\mathbf{P}_B\mathbf{Q}_B$ without any guiding mirrors.

4.11 Caustics as Flow-Lines

Figure 4.54 shows an angle rotator for an angle 2θ, where Figure 4.54b shows a detail of Figure 4.54a between sections s_1 and s_2. Between these two sections, s_1 and s_2, light is confined between circular mirror m with center **C** and circular caustic c, also centered at **C**. A flow-line g between these two sections is also a circle with radius R centered at **C**. The edge rays make an angle α to flow-line g.

As the radius R of g decreases, angle α also decreases. In the limit case, where this angle becomes zero, the flow-line tends to a caustic c to both sets of edge rays, and R tends to [**C**,**B**]. A mirror placed along the caustic would no longer reflect the edge rays, and light can be confined between mirror m and the caustic c. Mirror m is also a flow-line obtained in the case in which R = [**C**,**A**].

Étendue is conserved between the flow-lines. Since caustic c is a limit case of flow-lines g, étendue is also conserved between flow-line m and caustic c.

Figure 4.55 shows another example of light confinement by caustics. Caustics c_L on the left and c_R on the right confine light as it travels through air between optics O_A and O_B, both made of material with refractive index n. In this example, emitter R_A and receiver R_B touch the medium with refractive

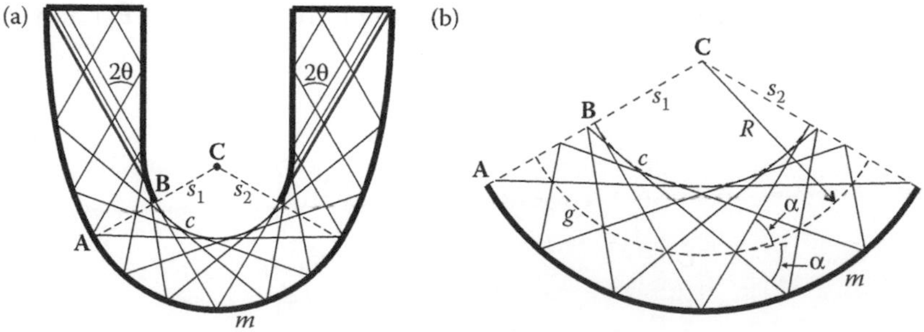

FIGURE 4.54
(a) An angle rotator for an angle 2θ. (b) Caustic c is a limit case of a flow-line g, in which the angle, α, that the edge rays make to the flow-line, becomes zero. This happens as the radius R of flow-line g tends to the distance between **C** and **B**.

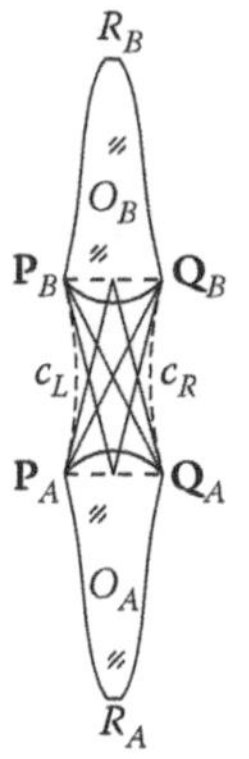

FIGURE 4.55
Optics O_A and O_B exchange light through air confined between caustics c_L on the left, and c_R on the right.

index n. There is no air gap between R_A and optic O_A, or between R_B and optic O_B.

Figure 4.56 shows the geometry of the caustics and the edge rays between the optics in detail.

Caustic c_L on the left is a circle with center $\mathbf{C}_L$. Edge rays between r_1 and r_2 all intersect at point $\mathbf{P}_B$ and edge rays between r_1 and r_3 are tangent to caustic c_L. The system has top-down symmetry about horizontal line l_H and left–right symmetry about vertical line l_V.

Figure 4.57 shows the design method for optic O_A. The curved top surface between $\mathbf{P}_A$ and $\mathbf{Q}_A$ is a circle with center $\mathbf{C}$. The points of the side mirrors may be defined by a string of constant optical path length. Ray r_M is tangent to caustic c_L at point $\mathbf{P}_B$.

For the points of the mirror above point $\mathbf{M}$, the optical path length $[\mathbf{P}_B,\mathbf{E}] + n[\mathbf{E},\mathbf{M}] + n[\mathbf{M},\mathbf{A}]$ is constant, where $\mathbf{E}$ is a point on the circular top surface of the optic. For the points of the mirror below point $\mathbf{M}$, the optical path length is still the same, but the string (still starting at $\mathbf{P}_B$) now rolls around the caustic c_L.

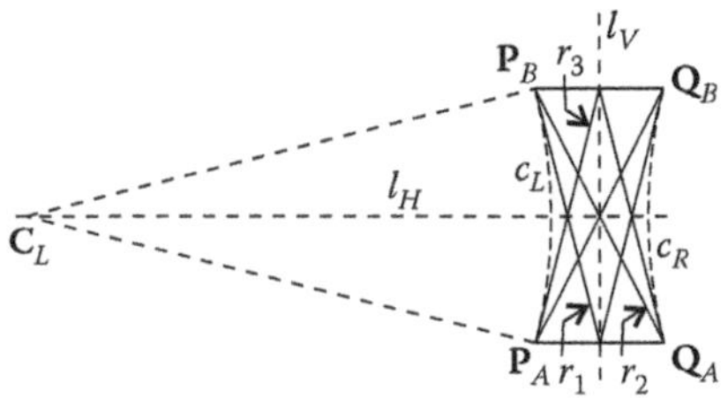

FIGURE 4.56
The geometry of the caustics and the edge rays for the radiation exchanged between optics O_A and O_B.

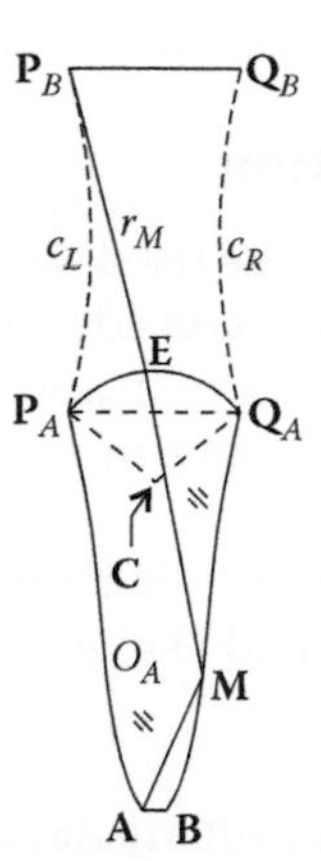

FIGURE 4.57
Design for optic O_A.

The size of emitter **AB** must be such that the étendue of the light it emits matches that exchanged between optics O_A and O_B. This étendue can be obtained from the geometry in Figure 4.58.

The étendue for the light emitted between $\mathbf{P}_A$ and $\mathbf{Q}_A$ is, by symmetry, twice that emitted between $\mathbf{P}_A$ and **X**, where **X** is the midpoint of $\mathbf{P}_A\mathbf{Q}_A$. For each point **P** between $\mathbf{P}_A$ and **X**, light is confined between edge rays r_A crossing point $\mathbf{Q}_B$ and r_B tangent to caustic c_L at point **T**. Caustic c_L is circular with center $\mathbf{C}_L$ and radius R. The étendue is then given by

$$U = -2\int_{\mathbf{P}_A}^{\mathbf{X}} (\cos\theta_B - \cos\theta_A)dx \tag{4.65}$$

where, using function angh(…) defined in Chapter 21, we have $\theta_A = (\mathbf{Q}_B - \mathbf{P})$ and $\theta_B = \theta - \alpha$, where $\theta = \text{angh}\,(\mathbf{C}_L - \mathbf{P})$ and $\alpha = \arcsin\,(R/[\mathbf{C}_L, \mathbf{P}])$.

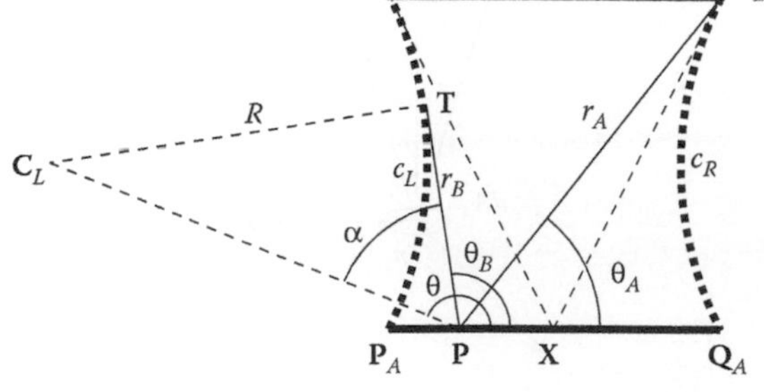

FIGURE 4.58
The geometry for calculating the étendue of the light exchanged between the optics.

4.12 Maximum Concentration

Conservation of étendue can also be used to derive the maximum concentration an optic can provide. We now consider a 3-D situation and calculate the étendue of radiation with half-angular aperture θ_1 and crossing (or being emitted by) an area dA_1 immersed in a medium with a refractive index of n_1, as presented in Figure 4.59.

The étendue of the radiation crossing (or being emitted by) dA_1 can be obtained from expressions (4.4) and (4.6) as

$$U_{dA1} = n_1^2 dA_1 \int_0^{2\pi} \int_0^{\theta_1} \cos\theta \sin\theta \, d\theta \, d\varphi = \pi n_1^2 dA_1 \sin^2\theta_1 \tag{4.66}$$

If dA_1 is a part of an area A_1, where the radiation falling on it is uniform, the total étendue of the radiation falling on A_1 is given by

$$U_1 = \pi n_1^2 \sin^2\theta_1 \int_{A_1} dA_1 = \pi n_1^2 A_1 \sin^2\theta_1 \tag{4.67}$$

We now apply this result to an optical system with entrance aperture A_1 and exit aperture A_2, θ_1 being the half-angular aperture for the radiation at the entrance aperture, and θ_2 the half-angular aperture at the exit aperture, as presented in Figure 4.60.

If the refractive index at the exit aperture is n_2, the étendue of the radiation exiting the device is given as

$$U_2 = \pi n_2^2 A_2 \sin^2\theta_2 \tag{4.68}$$

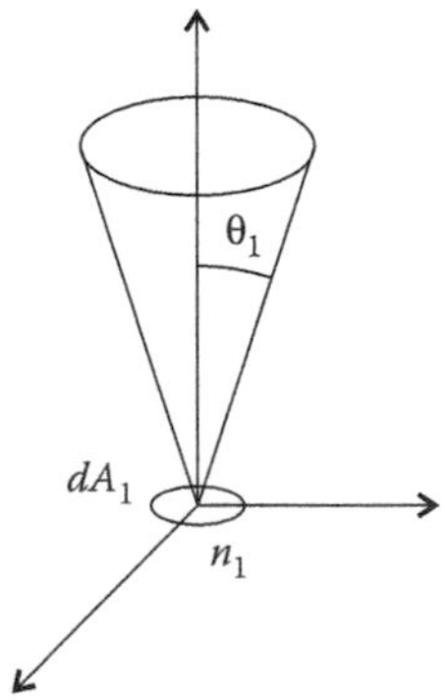

FIGURE 4.59
Surface dA_1 immersed in a medium with index of refraction n_1 receiving radiation with half-angular aperture θ_1.

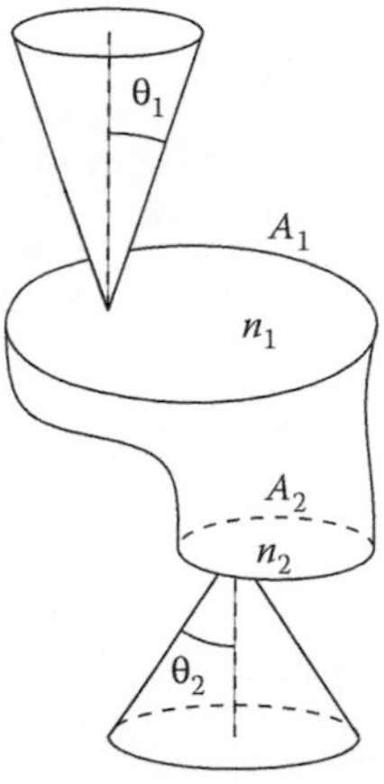

FIGURE 4.60
Optical device with an entrance aperture A_1 and exit aperture A_2. At the entrance aperture, the refractive index is n_1, and the half-angular aperture of the radiation is θ_1. At the exit aperture, the refractive index is n_2 and the half-angular aperture of the radiation is θ_2.

Since the étendue is conserved in the passage through an optical system, the étendue at the entrance aperture must be equal to the one at the exit, that is, $U_1 = U_2$, therefore,

$$\frac{A_1}{A_2} = \frac{n_2^2 \sin^2 \theta_2}{n_1^2 \sin^2 \theta_1} \tag{4.69}$$

the angle θ_2 at the exit aperture cannot be higher than $\pi/2$; therefore, the minimum exit area $A_{2\min}$ can be obtained for $\theta_2 = \pi/2$. This area corresponds to the maximum possible concentration:

$$C_{\max} = \frac{A_1}{A_{2\min}} = \frac{n_2^2}{n_1^2} \frac{1}{\sin^2 \theta_1} \tag{4.70}$$

or, in the particular case in which the refractive index at the entrance of the device is $n_1 = 1$:

$$C_{\max} = \frac{n_2^2}{\sin^2 \theta_1} \tag{4.71}$$

Let us now consider a 2-D optical system with entrance aperture a_1 and exit aperture a_2, and having refractive indices n_1 and n_2 at the entrance and exit apertures, respectively, as shown in Figure 4.61a.

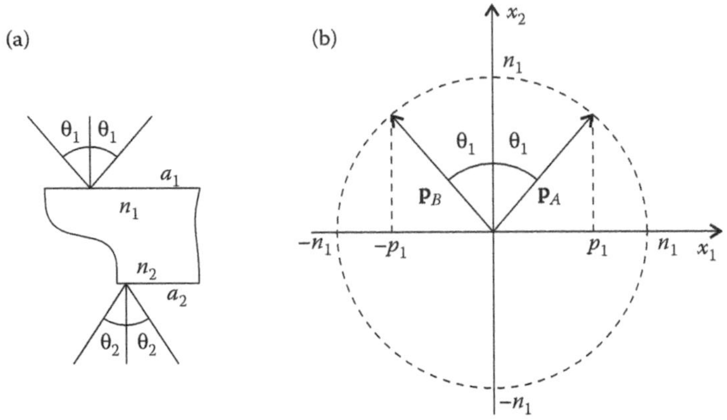

FIGURE 4.61
(a) A 2-D optical device with an entrance aperture a_1 and exit aperture a_2. At the entrance aperture, the refractive index is n_1 and the half-angular aperture of the radiation is θ_1. At the exit aperture, the refractive index is n_2 and the half-angular aperture of the radiation is θ_2. (b) Calculation of the étendue at the entrance aperture of the optic.

Let us further suppose that the radiation is uniform at the entrance and exit apertures. The étendue at the entrance aperture can be obtained from Figure 4.61b as

$$U_1 = -a_1 \int_A^B dp_1 = -a_1(-p_1 - p_1) = 2a_1p_1 = 2n_1a_1 \sin\theta_1 \tag{4.72}$$

where the integration is taken from direction $\mathbf{p}_A$ to $\mathbf{p}_B$, that is, the positive direction of the angles in Figure 4.24. Similarly, the étendue at the exit aperture is given as

$$U_2 = 2n_2a_2 \sin\theta_2 \tag{4.73}$$

Since the étendue is conserved in the passage through the optical system, we must have $U_1 = U_2$, therefore,

$$\frac{a_1}{a_2} = \frac{n_2}{n_1}\frac{\sin\theta_2}{\sin\theta_1} \tag{4.74}$$

As in the 3-D case analyzed earlier, the exit angle cannot be higher than $\pi/2$ and, therefore, the minimum length $a_{2\text{min}}$ of the exit aperture can be obtained for $\theta_2 = \pi/2$. It corresponds to the maximum possible concentration

$$C_{\text{max}} = \frac{a_1}{a_{2\text{min}}} = \frac{n_2}{n_1}\frac{1}{\sin\theta_1} \tag{4.75}$$

or, in the particular case in which the refractive index at the entrance of the device is $n_1 = 1$,

$$C_{max} = \frac{n_2}{\sin\theta_1} \tag{4.76}$$

The preceding expressions allow us to find the maximum possible concentration C_{max} as a function of the half-angular aperture θ_1 of the incoming radiation. An example of this uniform constant-angle radiation is the one coming from the sun. When observed from the earth, the sun has a very small angular aperture. Nonimaging concentrators are capable of providing high concentration of light and, therefore, are well suited for high concentration of solar energy. However, these devices are also useful for smaller concentrations.

Since the relations presented earlier give the maximum concentration, the real concentration must be smaller than that. Let us then consider, to simplify, that we have a concentrator whose interior is filled with air ($n = 1$). The concentration that it can attain must be smaller than C_{max}, that is,

$$C \leq \frac{1}{\sin\theta} \tag{4.77}$$

where θ is the half-angular aperture of the radiation. Expression (4.77) can now be rewritten in a different form, as follows:

$$\theta \leq \arcsin(1/C) \tag{4.78}$$

and it can then be concluded that, for a given concentration C of a device, the half-acceptance angle cannot be higher than arcsin $(1/C)$. The maximum value for the half-acceptance angle is, therefore, given by

$$\theta_{max} = \arcsin(1/C) \tag{4.79}$$

Similar expression can be obtained for 3-D systems and containing materials having an index of refraction $n \neq 1$. It can then be concluded that the nonimaging optical systems have the maximum acceptance angle 2θ for a given concentration C. This characteristic makes them very important in the concentration of solar energy.[18,19]

Let us consider a solar concentrator having concentration C. Let us further consider that the concentration is low so that the acceptance angle is large. Since nonimaging concentrators have the maximum acceptance angle θ for a given concentration C, they allow us to keep the concentrator stationary for the longest time possible, while the sun is moving in the sky, as presented in Figure 4.62.

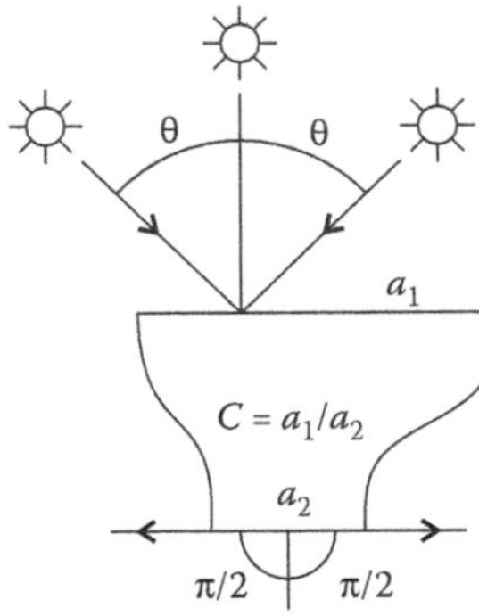

FIGURE 4.62
Solar concentrator having a concentration C. An ideal anidolic device having a concentration C will have the maximum acceptance angle for the incident radiation. In low-concentration solar systems, this characteristic can reduce the need to follow the sun in the sky, in addition to allowing the maximum acceptance of diffuse radiation. For high-concentration solar systems, where the tracking is mandatory, it alleviates the requirement for precise tracking.

As long as the sun moves inside the acceptance angle 2θ, its light is captured and transferred to the absorber (exit aperture of the device). This is an important characteristic, since it alleviates the need to track the angular motion of the sun in the sky. Besides the radiation arriving at the concentrator directly from the sun, there is also some diffuse radiation from the scattering of light in the atmosphere. This radiation arrives from all directions, but with greatest strength near the sun. Since the acceptance angle is maximum, the concentrator will capture the maximum diffuse radiation as well.

Let us now consider that the concentration is high, and the acceptance angle is necessarily small, so that it is mandatory to track the sun. The tracking systems are, nonetheless, as complex and expensive in proportion to their precision. But since the acceptance angle is maximum, the need for precise tracking can be relaxed.

The use of anidolic optics in solar energy systems is, therefore, advantageous, as long as there is some need for concentration.

Note that, for a high concentration, the acceptance angle is small, and the acceptance of diffuse radiation is minor. In particular, when the acceptance angle equals the angular aperture of the sun, there is no longer a collection of diffuse radiation, but only of direct radiation. This is usually not done because of unavoidable tracking errors.

When comparing solar concentrators, a quantity called Concentration Acceptance Product (CAP) is sometimes used.[20,21] It is defined as $CAP = \sqrt{C} \sin\theta$ where C is the concentration and θ the acceptance angle of the solar concentrator. From Equation 4.71 we can see that $CAP < n$ where n is the refractive index in which the receiver is immersed. The goal of a solar concentrator is to have a CAP as high as possible (high concentration C and high acceptance angle θ).

Another method used in comparing solar concentrators is to design for the same acceptance angle. This means designing different concentrators for the same overall tolerances (same tolerance budget–see Chapter 1). Optics with the same tolerances may be made with similar manufacturing and operated in similar conditions and, therefore, at a similar cost. In that case, the optics with a higher concentration are typically preferable (and have a higher *CAP*).

4.13 Étendue and the Shape Factor

One way to calculate the étendue in a homogeneous medium is by making use of the concept of shape factors, from the field of radiative heat transfer. The relation between étendue and shape factor is discussed in detail in Chapter 20, but the results are summarized here for convenience. The étendue of the light emitted by an infinitesimal area dA_1 immersed in air toward another infinitesimal area dA_2 also in air, is

$$\begin{aligned} dU &= dA_1 \cos\theta_1 d\Omega = dA_1 \cos\theta_1 \frac{dA_2 \cos\theta_2}{r^2} \\ &= \pi dA_1 \frac{dA_2 \cos\theta_1 \cos\theta_2}{\pi r^2} = \pi dA_1 dF_{dA1-dA2} \end{aligned} \tag{4.80}$$

as seen from Figure 4.63a. For 3-D systems, the étendue can, therefore, be related to the shape factor $F_{dA1-dA2}$ of an area dA_1 to another area dA_2 by

$$dU = \pi dA_1 dF_{dA1-dA2} \tag{4.81}$$

where

$$dF_{dA1-dA2} = \frac{dA_2 \cos\theta_1 \cos\theta_2}{\pi r^2} \tag{4.82}$$

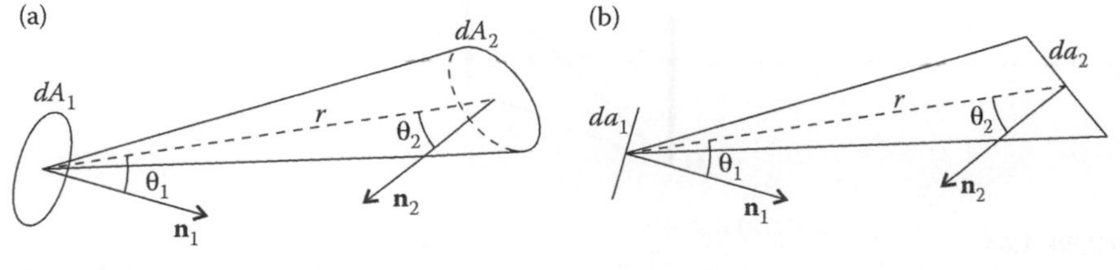

FIGURE 4.63
(a) The étendue of the light emitted from an infinitesimal area dA_1 toward another infinitesimal area dA_2. (b) The étendue for 2-D geometry.

For 2-D systems, the situation is similar, and is shown in Figure 4.63b.

We can now write

$$dU = da_1 \cos\theta_1 d\theta_1 = da_1 \cos\theta_1 \frac{da_2 \cos\theta_2}{r}$$
$$= 2da_1 \frac{da_2 \cos\theta_1 \cos\theta_2}{2r} = 2da_1 F_{da1-da2} \quad (4.83)$$

For 2-D systems, the étendue can, therefore, be related to the shape factor $F_{da1-da2}$ of a length da_1 to another length da_2 by

$$dU = 2da_1\, dF_{da1-da2} \quad (4.84)$$

where

$$dF_{da1-da2} = \frac{da_2 \cos\theta_1 \cos\theta_2}{2r} \quad (4.85)$$

We can then use the known methods for the calculation of shape factors to help us calculate the étendue. One of these methods is Hottel's crossed-string method.[4] An example of this method can be applied to a system similar to that presented in Figure 4.34b. Then consider a Lambertian source of radiation **FG** with dimension a_1, and a line $\mathbf{P}_1\mathbf{P}_2$ with dimension a_2, as presented in Figure 4.64.

The shape factor from a_1 to a_2 is given by

$$F_{a1-a2} = \frac{[[\mathbf{F},\mathbf{P}_2]] + [[\mathbf{G},\mathbf{P}_1]] - [[\mathbf{F},\mathbf{P}_1]] - [[\mathbf{G},\mathbf{P}_2]]}{2a_1} \quad (4.86)$$

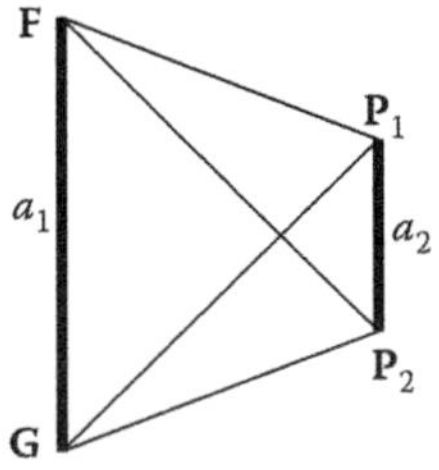

FIGURE 4.64
The shape factor from a_1 to a_2 can be calculated by Hottel's crossed-string method. The étendue can then be calculated using the relation between the two quantities: étendue and shape factor.

Integrating expression (4.84) we obtain

$$U = 2a_1F_{a1-a2} \tag{4.87}$$

Replacing $F_{a1} - a_2$ in expression (4.87) for U, we again obtain expression (4.48).

The relation between the shape factor and the étendue in the 3-D case enables us to calculate the étendue from a circular source A_1 having radius d, to a circular surface A_2 having radius ρ. Let us consider that these two surfaces are separated by a distance h, as presented in Figure 4.65.

The shape factor from A_1 to A_2 is given by[5]

$$F_{A1-A2} = \frac{1}{2}\left(Z - \sqrt{Z^2 - 4X^2Y^2}\right) \tag{4.88}$$

with $X = \rho/h$, $Y = h/d$ and $Z = 1 + (1 + X^2)Y^2$. Integrating expression (4.81) for 3-D systems, we get

$$U = \pi A_1F_{A1-A2} \tag{4.89}$$

Surface A_1 has an area πd^2 therefore, the étendue from A_1 to A_2 is given as

$$U = \frac{1}{2}\left(Z - \sqrt{Z^2 - 4X^2Y^2}\right)\pi^2 d^2 \tag{4.90}$$

Expression (4.90) can also be written in the following form:[13]

$$U = \frac{\pi^2}{4}\left(\sqrt{(\rho - d)^2 + h^2} - \sqrt{(\rho + d)^2 + h^2}\right)^2 = \frac{\pi^2}{4}([\mathbf{F},\mathbf{P}_2] - [\mathbf{F},\mathbf{P}_1])^2 \tag{4.91}$$

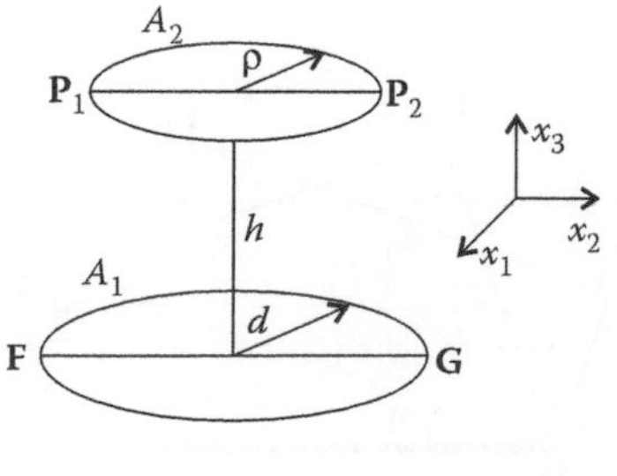

FIGURE 4.65
The étendue, from a circular source having an area A_1 to a circular surface having an area A_2, can be calculated from the shape factor from A_1 to A_2.

4.14 Examples

The examples presented in this section use expressions for the curves and functions that are derived in Chapter 21.

EXAMPLE 4.1

Calculate the étendue of radiation with half-angle $\theta = 20°$ illuminating a straight line $\mathbf{P}_1\mathbf{P}_2$ with length 3.

Figure 4.66 shows the straight line $\mathbf{P}_1\mathbf{P}_2$ and the wavefronts w_1 and w_2 perpendicular to the edge rays making a total angle of 2θ to one another.

The étendue of the incoming radiation is given by

$$U = [\mathbf{F}_2,\mathbf{P}_2] + [\mathbf{G}_2,\mathbf{P}_1] - [\mathbf{G}_1,\mathbf{P}_2] - [\mathbf{F}_1,\mathbf{P}_1] = 2[\mathbf{P}_1,\mathbf{P}_2]\sin\theta \quad (4.92)$$

In this case $[\mathbf{P}_1, \mathbf{P}_2] = 3$ and $\theta = 20°$ and we get $U = 2.05212$.

EXAMPLE 4.2

Calculate the étendue of a source a emitting light with an angular aperture 2θ tilted by an angle γ to the vertical. The length of a is equal to 3, $\gamma = 60°$, and $\theta = 10°$.

We present several possibilities for calculating this étendue.

Possibility 1. We may consider that the light emitted by a comes from a source a^* tilted by an angle γ to the horizontal, and whose light is confined by a flat mirror perpendicular to a^* until it reaches a, as shown in Figure 4.67.

Since the mirror is parallel to the direction of propagation of the light, it does not alter the angular aperture 2θ of the light.

The étendue of the light emitted by a must then be equal to that emitted by a^*. We have

$$U = 2a^* \sin\theta = 2a\cos\gamma\sin\theta \quad (4.93)$$

Now by replacing the values for a, γ, and θ, we get $U = 0.520945$.

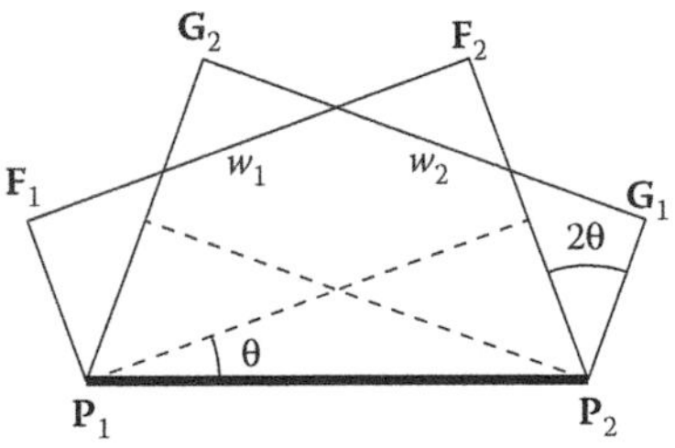

FIGURE 4.66
Radiation with a half-angle θ illuminating a straight line $\mathbf{P}_1\mathbf{P}_2$.

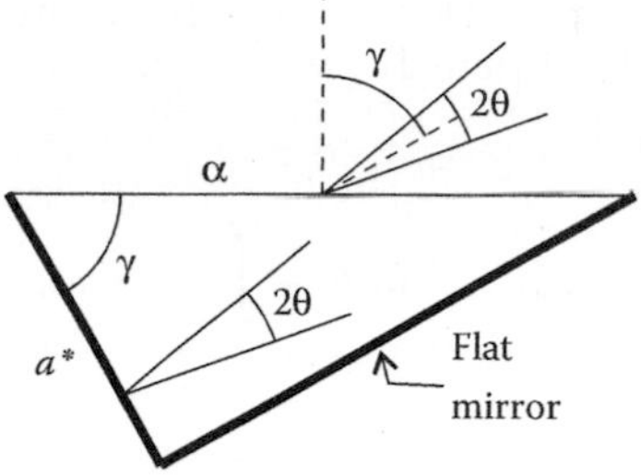

FIGURE 4.67
Source a emitting light with an angular aperture 2θ tilted by an angle γ to the vertical. Light appears to come from another source a^* whose size is the projection of a in the direction of light emission.

Possibility 2. Now consider the same situation, but from a different point of view. We now assume that the same source a is on the x_1 axis between x_m and x_M, that is, $a = x_M - x_m$, as shown in Figure 4.68a.

The étendue of the radiation is given by

$$U = \int_{x_m}^{x_M} (\mathbf{p}_A - \mathbf{p}_B) \cdot d\mathbf{c} \tag{4.94}$$

where $\mathbf{p}_A = (\cos(\varphi - \theta),\ \sin(\varphi - \theta))$, $\mathbf{p}_B = (\cos(\varphi + \theta),\ \sin(\varphi + \theta))$, and $d\mathbf{c} = dc(1, 0)$, since a is on the x_1 axis. Note that $\varphi = \pi/2 - \gamma$.

In this case, dc is an element of length da on a, and we obtain

$$\begin{aligned} U &= \int_{x_m}^{x_M} (\cos(\varphi - \theta) - \cos(\varphi + \theta))\, da = a(\cos(\varphi - \theta) - \cos(\varphi + \theta)) \\ &= 2a \sin\theta \sin\varphi \end{aligned} \tag{4.95}$$

Replacing a, $\varphi = \pi/2 - \gamma$, and θ we get $U = 0.520945$.

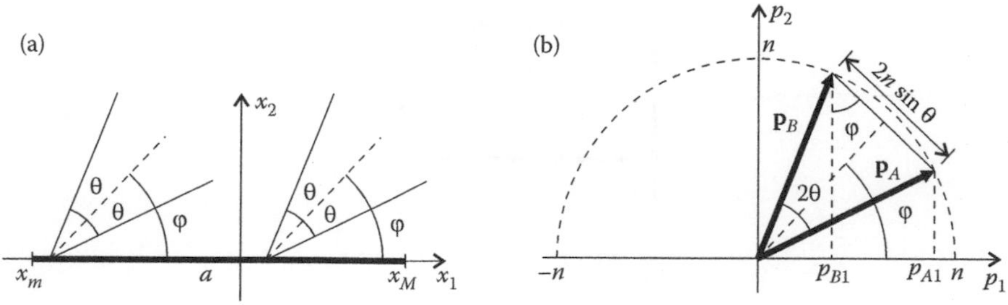

FIGURE 4.68
(a) Uniform radiation for all points of $x_m x_M$. (b) Its étendue can be obtained as $U = 2n\, a \sin\theta \sin\varphi$.

Possibility 3. Another possible way to calculate the same étendue is by using expression (4.94) differently, where we now make $\mathbf{p}_A = (p_{A1}, p_{A2})$ and $\mathbf{p}_B = (p_{B1}, p_{B2})$ to get

$$U = \int_{x_m}^{x_M} ((p_{A1}, p_{A2}) - (p_{B1}, p_{B2})) \cdot (1,0)\, da = \int_{x_m}^{x_M} (p_{A1} - p_{B1})\, da = a(p_{A1} - p_{B1}) \tag{4.96}$$

which can also be written as

$$U = 2a \sin\theta \sin\varphi \tag{4.97}$$

as can be seen from Figure 4.68b with $n = 1$.

Possibility 4. Yet another possibility is to represent the radiation emitted by a in phase space and calculate the area it uses. Figure 4.69 shows this possibility.

In physical space, source a extends from x_m to x_M, and in angular space, the radiation is confined between directions that have p_1 component of the optical momentum between p_{A1} and p_{B1}. The étendue equals the area in phase space occupied by the radiation, and is given by

$$U = (x_M - x_M)(p_{A1} - p_{B1}) = a(p_{A1} - p_{B1}) \tag{4.98}$$

which is the same as expression (4.96).

EXAMPLE 4.3

Calculate the étendue of uniform radiation with a half-angle 5° captured by a parabola of rim angle 30°, when it reaches the parabola and after reflection.

A parabola with focus $\mathbf{F} = (0, 0)$ and a unit distance between focus and vertex $[\mathbf{F}, \mathbf{V}] = 1$, is parameterized by

$$\mathbf{c}(\varphi) = \frac{2}{1 - \cos\varphi}(\cos\varphi, \sin\varphi) \tag{4.99}$$

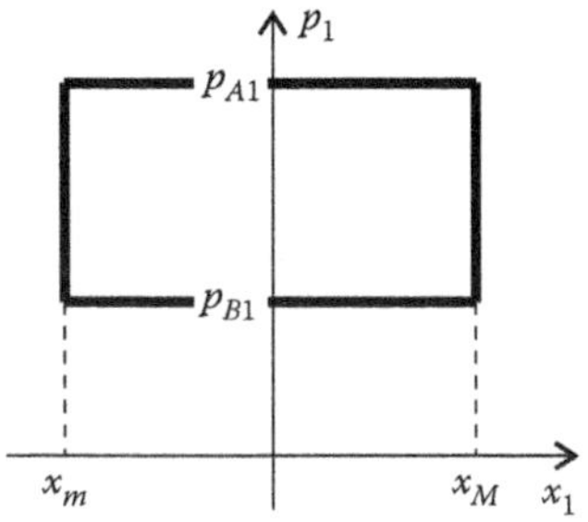

FIGURE 4.69
Area in phase space of radiation with p_1 components for the edge rays p_{A1} and p_{A2}.

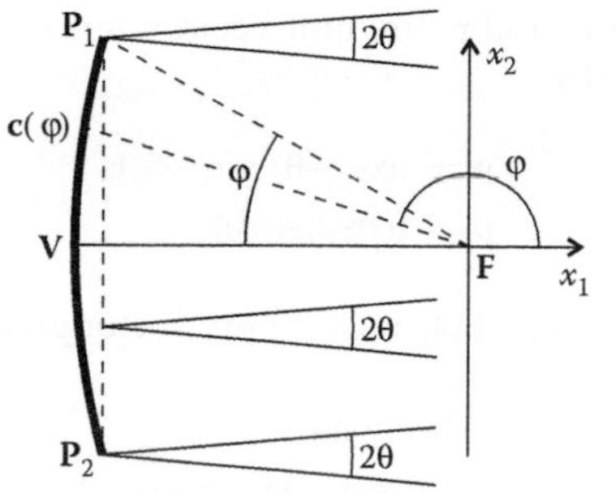

FIGURE 4.70
A parabola with focus **F** collects light with a half-angle θ.

If the rim angle is $\varphi = 30°$, the parameter range for the parabola is $\pi - \varphi \le \varphi \le \pi + \varphi$. Edge points $\mathbf{P}_1$ and $\mathbf{P}_2$ of the parabola are obtained at the edges of the parameter range, and are given by $\mathbf{P}_1 = \mathbf{c}(\pi - \varphi) = \left(6 - 4\sqrt{3}, 4 - 2\sqrt{3}\right)$ and $\mathbf{P}_2 = \mathbf{c}(\pi + \varphi) = \left(6 - 4\sqrt{3}, -4 + 2\sqrt{3}\right)$ as shown in Figure 4.70. The half-angle of the light collected by the parabola is $\theta = 5°$.

The étendue of the radiation captured by the parabola is (as shown in Figure 4.70)

$$U = 2[\mathbf{P}_1, \mathbf{P}_2]\sin\theta = 0.186826 \tag{4.100}$$

We may now calculate the étendue of the same radiation when it hits the parabola. Figure 4.71 shows this situation.

The étendue of the light reaching the parabola is given by

$$U = \int_{\pi-\varphi}^{\pi+\varphi} (\mathbf{p}_1 - \mathbf{p}_2) \cdot d\mathbf{c} = \int_{\pi-\varphi}^{\pi+\varphi} (\mathbf{p}_1 - \mathbf{p}_2) \cdot \frac{d\mathbf{c}}{d\phi} d\phi \tag{4.101}$$

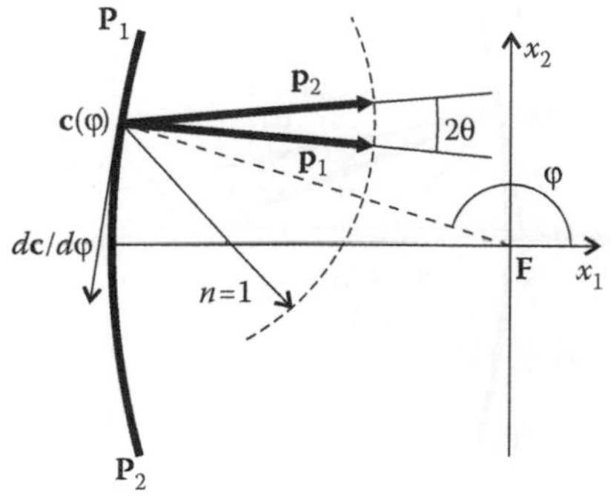

FIGURE 4.71
Radiation with a half-angle θ hits a parabolic mirror.

where momentum $\mathbf{p}_1$ and $\mathbf{p}_2$ are unit vectors, since the refractive index $n = 1$ and are given by

$$\begin{aligned}\mathbf{p}_1 &= (\cos(-\theta), \sin(-\theta)) \\ \mathbf{p}_2 &= (\cos\theta, \sin\theta)\end{aligned} \tag{4.102}$$

The derivative of the parabola is a tangent vector given by

$$\frac{d\mathbf{c}}{d\varphi} = \left(\frac{-2\sin\varphi}{(\cos\varphi - 1)^2}, \frac{2}{\cos\varphi - 1} \right) \tag{4.103}$$

Inserting expressions (4.102) and (4.103) into expression (4.101), the étendue is given by

$$U = \int_{\pi-\varphi}^{\pi+\varphi} \frac{-4\sin(\pi/36)}{\cos\varphi - 1} d\varphi = 0.186826 \tag{4.104}$$

which is the same as calculated earlier.

After reflection by the mirror, $\mathbf{p}_1$ and $\mathbf{p}_2$ are now given by

$$\begin{aligned}\mathbf{p}_1 &= R(-\theta) \cdot \operatorname{nrm}(\mathbf{F} - \mathbf{c}) = \frac{2\sin^2(\varphi/2)}{\cos\varphi - 1}(\cos(\pi/36 - \varphi), \sin(\pi/36 - \varphi)) \\ \mathbf{p}_2 &= R(\theta) \cdot \operatorname{nrm}(\mathbf{F} - \mathbf{c}) = \frac{2\sin^2(\varphi/2)}{\cos\varphi - 1}(\cos(\pi/36 + \phi), \sin(\pi/36 + \varphi))\end{aligned} \tag{4.105}$$

as shown in Figure 4.72.

The étendue of the light reflected by the mirror is also given by expression (4.101), now with the new values for $\mathbf{p}_1$ and $\mathbf{p}_2$, so that

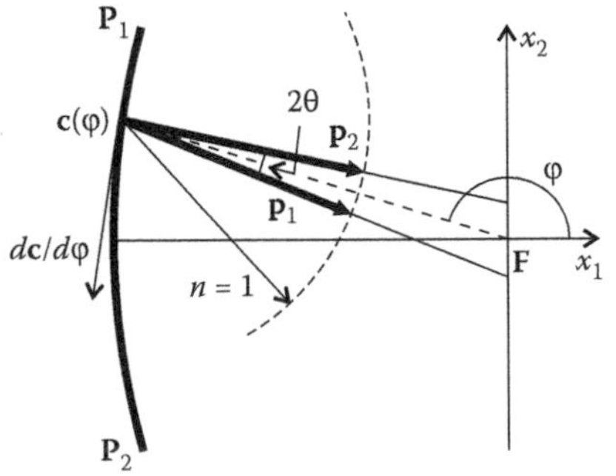

FIGURE 4.72
Radiation with a half-angle θ after reflection of a parabolic mirror.

$$U = 0.348623 \int_{\pi-\varphi}^{\pi+\varphi} \frac{1}{1-\cos\varphi}\, d\varphi = 0.186826 \tag{4.106}$$

The values of the étendue are the same, before light reaches the parabola, at the parabola, and after reflection.

References

1. Nicodemus, F. E., Radiance, *Am. J. Phys.*, 31, 368, 1963.
2. Boyd, R. W., *Radiometry and the Detection of Optical Radiation*, John Wiley & Sons, New York, 1983.
3. McCluney, W. R., *Introduction to Radiometry and Photometry*, Artech House, Boston, 1994.
4. Siegel, R., Howell, J. R., *Thermal Radiation Heat Transfer*, McGraw-Hill Book Company, New York, 1972.
5. Sparrow, E. M. and Cess, R. D., *Radiation Heat Transfer—Augmented edition*, Hemisphere Publishing Corporation, Washington, London; McGraw-Hill Book Company, New York, 1978.
6. Rabl, A., Comparison of solar concentrators, *Sol. Energy*, 18, 93, 1976.
7. Çengel, Y. A. and Boles, M. A., *Thermodynamics—An Engineering Approach*, McGraw-Hill Book Company, New York, 1989.
8. Fermi, E., *Thermodynamics*, Dover Publications, Inc, New York, 1936.
9. Ichimura, H., Usui, T., and Hashitsume, N., *Thermodynamics, An Advanced Course with Problems and Solutions*, North-Holland Publishing Company, Amsterdam, John Wiley and Sons, Inc, New York, 1968.
10. Ries, H., Thermodynamic limitations of the concentration of electromagnetic radiation, *J. Opt. Soc. Am.*, 72, 380, 1982.
11. Smestad, G. et al., The thermodynamic limits of light concentrators, *Sol. Energy Materials*, 21, 99, 1990.
12. Welford, W. T. and Winston, R., *The Optics of Nonimaging Concentrators—Light and Solar Energy*, Academic Press, New York, 1978.
13. Welford, W. T. and Winston, R., *High Collection Nonimaging Optics*, Academic Press, San Diego, 1989.
14. Winston, R. and Welford, W. T., Two-dimensional concentrators for inhomogeneous media, *J. Opt. Soc. Am. A*, 68, 289, 1978.
15. O'Gallagher, J., Winston, R., and Welford, W. T., Axially symmetrical nonimaging flux concentrators with the maximum theoretical concentration ratio, *J. Opt. Soc. Am.*, 4, 66, 1987.
16. Winston, R. and Welford, W. T., Geometrical vector flux and some new nonimaging concentrators, *J. Opt. Soc. Am. A*, 69, 532, 1979.
17. Feuermann, D. and Gordon, J. M., Optical performance of axisymmetric edge-ray concentrators and illuminators, *Appl. Opt.*, 37, 1905, 1998.

18. Winston, R., Principles of solar concentrators of a novel design, *Sol. Energy*, 16, 89, 1974.
19. Gordon, J., *Solar Energy—The State of the Art, ISES Position Papers*, James & James Science Publishers Ltd, London, 2001.
20. Pablo, B. et al., High performance Fresnel-based photovoltaic concentrator, *Opt. Express*, 18, S1, 2010.
21. Koshel, R. J., *Illumination Engineering: Design with Nonimaging Optics*, Wiley-IEEE Press, Piscataway, New Jersey, 2013.

5

Vector Flux

5.1 Introduction

We have seen that if we take the flow-lines generated by a Lambertian source, we obtain an ideal trumpet concentrator. Other shapes of Lambertian sources generate other shapes of flow-lines and those can be taken as concentrators with different geometries. We now consider some more simple examples of nonimaging optics obtained by taking flow-lines generated by Lambertian sources.

An example of uniform illumination is that of sunlight. The sun emits light in all directions, and when it reaches the earth, this light is confined to a small angle $\pm\alpha$ and, therefore, it appears to arrive within a cone of angle 2α as shown in Figure 5.1.

We take a plate and orient it perpendicularly to the direction of the sun, so that all its points will be illuminated by a cone of angle 2α. For point $\mathbf{P}_1$ on the plate, for example, all radiation is confined between edge ray r_U and edge ray r_L, as shown in Figure 5.2.

Imagine now that we put a thin mirror, M, mirrored on both sides, in a direction perpendicular to plate P as shown in Figure 5.3. The mirror shades the light between rays r_{L1} and r_{L2} that would hit point $\mathbf{P}_1$ if the mirror had not been there.

It, however, also mirrors toward $\mathbf{P}_1$ the light confined between r_{U1} and r_{U2} that, without the mirror, would hit point $\mathbf{P}_2$. Therefore, this mirror does not alter the radiation that reaches each of these points, because what it shades on one side, it reflects on the other. The same argument could be used for any other points on the plate.

If we take two of these mirrors, we get the geometry in Figure 5.4. It is a nonimaging device that accepts radiation with half-angle α and emits radiation with the same angle, maintaining the area. This is a light guide, defined by mirrors M_1 and M_2.

As another example, we now consider the round (2-D) Lambertian radiation source S_R shown in Figure 5.5. Two edge rays that are tangent to the source bound the radiation that crosses any point $\mathbf{P}$. A mirror M that is perpendicular to the source, will then bisect the edge rays.

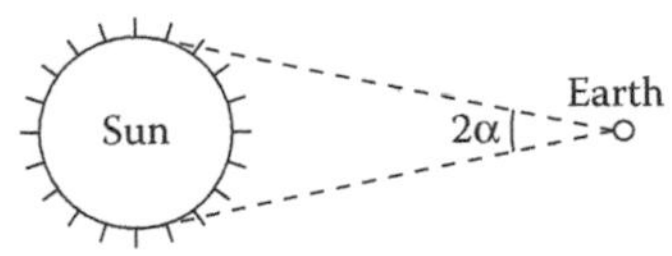

FIGURE 5.1
When it reaches the earth, the sunlight is confined angularly to ±α and, therefore, its total angular aperture is 2α.

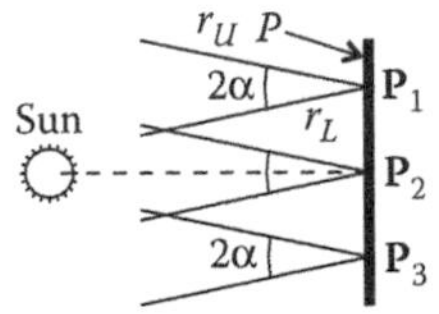

FIGURE 5.2
A plate P perpendicular to the direction of the sun "sees" incoming light confined to an angle 2α. At point $\mathbf{P}_1$, the light is confined between the edge rays r_U and r_L. Something similar happens to the light hitting the other points of plate P.

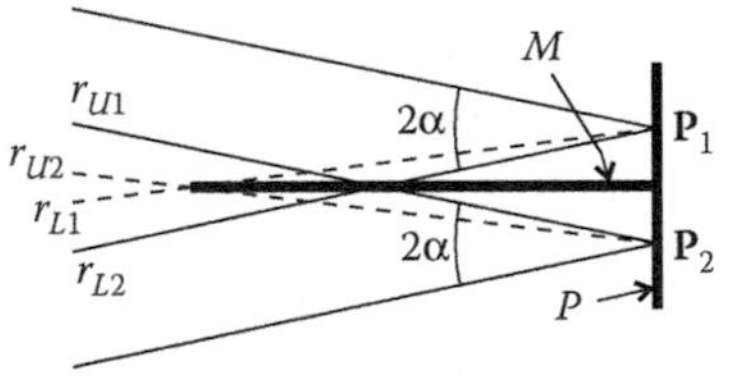

FIGURE 5.3
A mirror M perpendicular to the plate P bisects the edge rays of the incoming light.

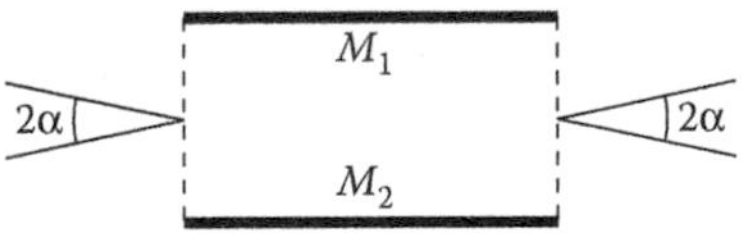

FIGURE 5.4
Two parallel mirrors can be used as a nonimaging optical device that accepts radiation with half-angle α and emits radiation with the same characteristics.

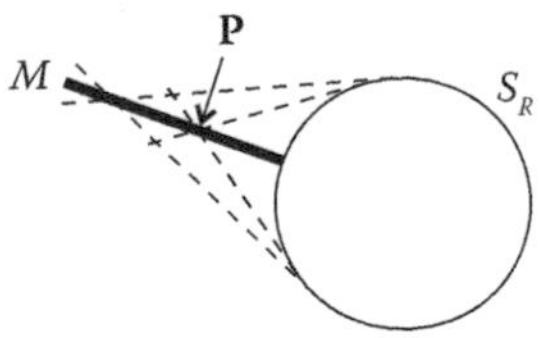

FIGURE 5.5
A flat mirror perpendicular to a circular source bisects the edge rays at each point in space.

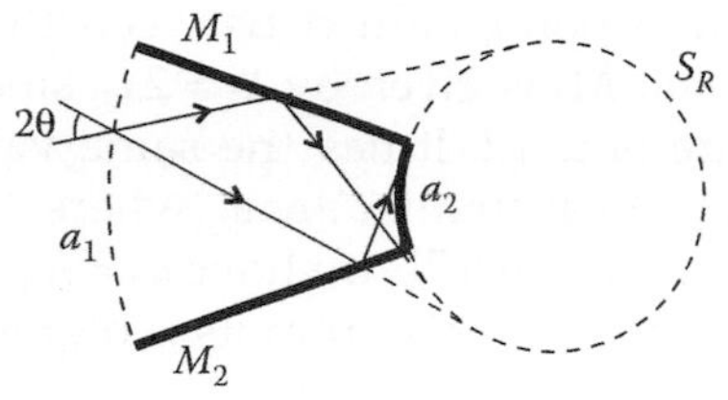

FIGURE 5.6
Light confined to an angle 2θ with bisector perpendicular to a_1 and headed toward S_R is concentrated by mirrors M_1 and M_2 to a_2.

For the reasons presented above, this mirror will not alter the radiation field created by the source S_R. We may then put two radial flat mirrors M_1 and M_2 on the source, as shown in Figure 5.6. These mirrors will not alter the radiation field created by the source. At arc a_1, the source S_R creates a radiation field with an angular aperture 2θ. The presence of the mirrors will not alter that. However, the radiation that a_1 now receives, comes from portion a_2 of the source, since the rest of it is shadowed by the mirrors. We may then remove that portion of the source outside the mirrors, and leave only the portion a_2 that they bound. Circular arc source a_2 will then create at a_1, with the help of M_1 and M_2, a uniform radiation field with angular aperture 2θ.

Inverting now the direction of the radiation, we may imagine that M_1 and M_2 form a concentrator with round entrance aperture a_1 and acceptance angle 2θ, and round receiver a_2. Since the radiation at a_2 is Lambertian (radiation angular aperture of $\pm\pi/2$), this concentrator is ideal, and provides maximum concentration.

Flow-lines bisect the edge rays at every point and the concentrator of Figure 5.6 was constructed by taking two flow-lines of the radiation field created by the source S_R. Choosing other flow-lines (radial lines coming out of S_R), we could get different size concentrators. Also, choosing different heights for these lines would lead to concentrators with different acceptance angles.

Étendue is conserved in the space between these flow-lines, as shown in Figure 5.7a.

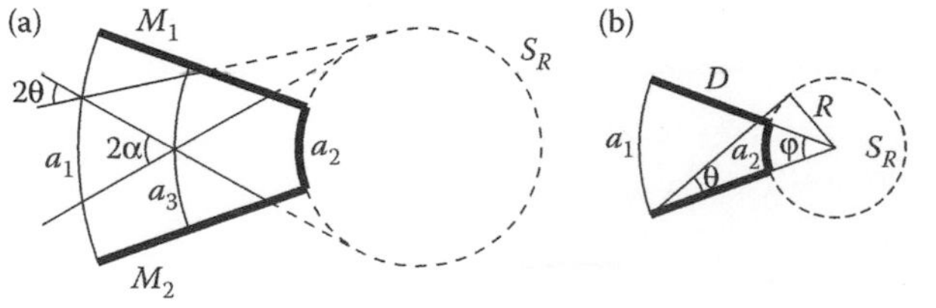

FIGURE 5.7
(a) The étendue U is conserved between two flow-lines M_1 and M_2. (b) This can can be seen from the fact that $U = 2a_2 = 2a_3 \sin\alpha = 2a_1 \sin\theta$.

The étendue of the radiation confined between the mirrors (which follow the flow-lines) M_1 and M_2 is given by $U = 2a_2$, since the radiation there has an angular aperture of $\pm\pi/2$. It has the same value at circular arc a_3, where $U = 2a_3\sin\alpha = 2a_2$ and at circular arc a_1, where it has the value $U = 2a_1 \sin\theta = 2a_2$. Note that from Figure 5.7b we have $a_2 = R\varphi$ and $a_1 = D\varphi$, where R is the radius of S_R, and D the distance from its center to a_1. We then have $a_2/a_1 = R/D = \sin\theta$.

If the whole source S_R is present, the étendue of the radiation crossing between a point $\mathbf{P}_1$ on flow-line M_1 and a point $\mathbf{P}_2$ on flow-line M_2 is constant, as we move $\mathbf{P}_1$ on M_1, and $\mathbf{P}_2$ on M_2 (no mirrors). If M_1 and M_2 are mirrors, the étendue of the radiation is conserved as light travels confined by them.

Note that if we took the portion of the mirrors M_1 and M_2 between a_1 and a_3, we would get an angle transformer with round entrance aperture a_1, and acceptance angle 2θ, with concentric exit aperture a_3 (also round), and exit angle 2α. This optic would fulfill the conservation of étendue: $2a_1 \sin\theta = 2a_3 \sin\alpha$.

The direction of the bisector to the edge rays and the angle between them define a direction and a magnitude and, therefore, a vector. It is called the vector flux $\mathbf{J}$ that, at each point on the plane, points in the direction of the bisector to the edge rays (the same as mirrors M_1 and M_2 mentioned earlier), with scalar magnitude $\|\mathbf{J}\| = 2n\sin\theta$, when the edge rays make a mutual angle 2θ (such as at the aperture a_1 in Figure 5.7), and the refractive index at the point is n. Figure 5.8 shows a source S emitting light that crosses point $\mathbf{P}$ bound by edge rays r_1 and r_2 making an angle 2θ to each other. The vector flux $\mathbf{J}$ at $\mathbf{P}$ points in the direction of the bisector of r_1, and r_2 has magnitude $2n \sin\theta$, where n is the refractive index at $\mathbf{P}$.

The vector flux points in the same direction as the flow-lines. If the flow-lines are straight, the direction of the vector flux is the same as that of these lines. If the flow-lines are curved, then the vector flux is tangent to the flow-lines.

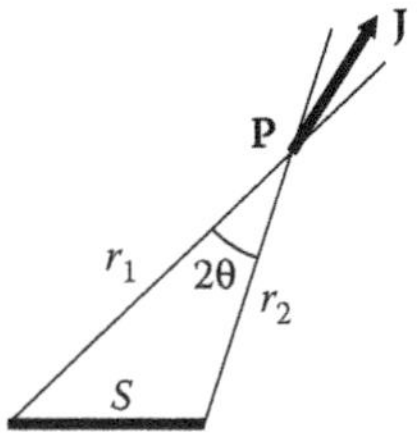

FIGURE 5.8
The vector flux points in the direction of the bisector of the edge rays at each point. Its magnitude is a function of the refractive index and the angle of the edge rays. At point **P**, the edge rays r_1 and r_2 of the radiation emitted by a source S define a vector flux **J**.

5.2 Definition of Vector Flux

Let $d\Phi$ be the energy flux (energy per unit time) crossing a surface dA immersed in a medium with refractive index n, through an element of solid angle $d\Omega$. We can then write

$$d\Phi = L^* dU = L^* n^2 dA \cos\theta \, d\Omega \tag{5.1}$$

where dU is the étendue of the radiation, and $L^* = L/n^2$ the basic radiance (or basic luminance L_V^* if photometric quantities are used). Angle θ is the angle the normal $\mathbf{n}$ to dA makes with the direction $\mathbf{t}$ defined by solid angle $d\Omega$, as shown in Figure 5.9.

If both $\mathbf{n}$ and $\mathbf{t}$ are unit vectors, then $\|\mathbf{t}\| = \|\mathbf{n}\| = 1$ and the dot product of $\mathbf{n}$ and $\mathbf{t}$ is given by

$$\mathbf{t} \cdot \mathbf{n} = \|\mathbf{t}\| \; \|\mathbf{n}\| \cos\theta = \cos\theta \tag{5.2}$$

Therefore, expression (5.1) can be written as

$$d\Phi = \mathbf{t} \cdot \mathbf{n} \, L^* n^2 dA \, d\Omega \tag{5.3}$$

We now consider another situation in Figure 5.10, in which we again have an area dA that has normal $\mathbf{n}$. However, now light crosses dA in two directions, $\mathbf{t}_X$ and $\mathbf{t}_Y$. In the case of the light crossing dA within solid angle $d\Omega_X$ in direction $\mathbf{t}_X$, we have $\mathbf{t}_X \cdot \mathbf{n} > 0$, which means that $d\Phi > 0$. On the other hand, in the case of the light crossing dA within solid angle $d\Omega_Y$ in direction $\mathbf{t}_Y$, we have $\mathbf{t}_Y \cdot \mathbf{n} < 0$ and therefore $d\Phi < 0$.

We now consider the total energy crossing an area dA per unit time. It can be calculated by integrating (5.1) over the solid angle as

$$d\Phi = dA \int L^* n^2 \cos\theta \, d\Omega \tag{5.4}$$

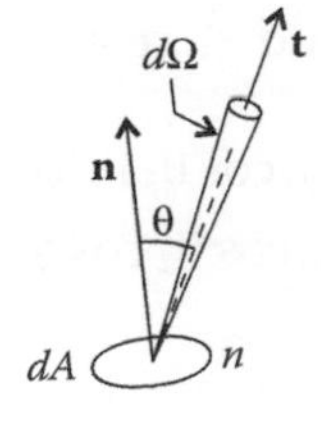

FIGURE 5.9
Radiation flows through area dA, perpendicular to vector $\mathbf{n}$, and through a solid angle $d\Omega$ in the direction of vector $\mathbf{t}$.

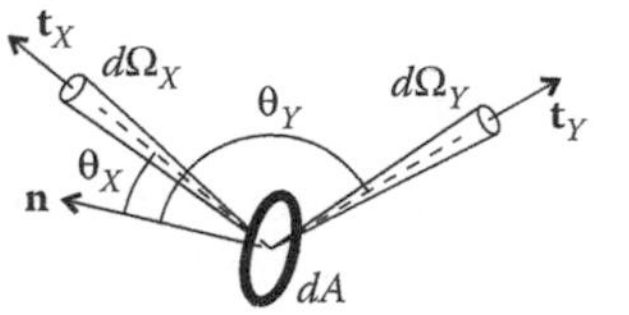

FIGURE 5.10
The flux of the light crossing dA in direction $\mathbf{t}_X$ is positive because $\mathbf{t}_X \cdot \mathbf{n} > 0$, but the flux of the light crossing dA in direction $\mathbf{t}_Y$ is negative because $\mathbf{t}_Y \cdot \mathbf{n} < 0$.

Note that in this expression $d\Phi$ is a first-order differential, because it is proportional to dA, while in expression (5.1) $d\Phi$ is a second-order differential, because it is proportional to the product of dA and $d\Omega$.

If the radiation distribution on dA is Lambertian (isotropic, diffuse) L, and therefore L^*, will not depend on the direction,[1,2] and it can be taken out of the integral and we get

$$d\Phi = L^* dA \int n^2 \cos\theta \, d\Omega \tag{5.5}$$

where this integral is calculated over all directions in which there is light. Defining now

$$J_N = \frac{d\Phi}{L^* dA} \tag{5.6}$$

it can be seen that the radiation (energy) crossing dA per unit time and per unit basic radiance is proportional to the integral

$$J_N = \int n^2 \cos\theta \, d\Omega = \int n^2 \mathbf{t} \cdot \mathbf{n} \, d\Omega \tag{5.7}$$

Let us now consider $\mathbf{n} = (\cos\gamma_1, \cos\gamma_2, \cos\gamma_3)$ where γ_1, γ_2, and γ_3 are the angles that vector $\mathbf{n}$ makes with the axes x_1, x_2, and x_3, respectively. The same way, we can write $\mathbf{t} = (\cos\theta_1, \cos\theta_2, \cos\theta_3)$ where θ_1, θ_2, and θ_3 are the angles that vector $\mathbf{t}$ makes with axis x_1, x_2, and x_3, respectively. We can then write

$$\begin{aligned} \mathbf{t} \cdot \mathbf{n} &= (\cos\theta_1, \cos\theta_2, \cos\theta_3) \cdot (\cos\gamma_1, \cos\gamma_2, \cos\gamma_3) \\ &= (\cos\theta_1 \cos\gamma_1, \cos\theta_2 \cos\gamma_2, \cos\theta_3 \cos\gamma_3) \end{aligned} \tag{5.8}$$

The integral of expression (5.7) can then be written in the form[3]

$$J_N = \int n^2 (\cos\theta_1 \cos\gamma_1 + \cos\theta_2 \cos\gamma_2 + \cos\theta_3 \cos\gamma_3) \, d\Omega \tag{5.9}$$

This integral can now be written in the form

$$J_N = \cos\gamma_1 \int n^2 \cos\theta_1 \, d\Omega + \cos\gamma_2 \int n^2 \cos\theta_2 \, d\Omega + \cos\gamma_3 \int n^2 \cos\theta_3 \, d\Omega \tag{5.10}$$

That is,

$$J_N = \left(\int n^2 \cos\theta_1 \, d\Omega, \int n^2 \cos\theta_2 \, d\Omega, \int n^2 \cos\theta_3 \, d\Omega \right) \cdot (\cos\gamma_1, \cos\gamma_2, \cos\gamma_3) \tag{5.11}$$

or

$$J_N = \mathbf{J} \cdot \mathbf{n} \tag{5.12}$$

where the vector

$$\mathbf{J} = \left(\int n^2 \cos\theta_1 \, d\Omega, \int n^2 \cos\theta_2 \, d\Omega, \int n^2 \cos\theta_3 \, d\Omega \right) \tag{5.13}$$

is called the vector flux or the light vector.[4] From Equation 5.6 we get

$$\frac{d\Phi}{dA} = L^* \, \mathbf{J} \cdot \mathbf{n} \tag{5.14}$$

for the flux per unit area through an area dA with normal $\mathbf{n}$. We can then see that $\mathbf{J}$ points in the direction of maximum flux per unit area. The vector flux can also be related to the étendue. From expression (5.14) and $d\Phi = L^* dU$ we get

$$\frac{dU}{dA} = \mathbf{J} \cdot \mathbf{n} \tag{5.15}$$

$\mathbf{J}$ being a measure of the étendue per unit area at each point, if the radiation in the optical system comes originally from Lambertian sources, since we have considered that L^* does not depend on direction.[5]

It can then be seen from Equation 5.12 that J_N is just the magnitude of the projection of vector $\mathbf{J}$ in the direction of vector $\mathbf{n} = (\cos\gamma_1, \cos\gamma_2, \cos\gamma_3)$ normal to surface dA.

We have seen before in Equation 4.19 that $n^2 \cos\theta_3 d\Omega = dp_1 dp_2$. The same way, $n^2 \cos\theta_1 d\Omega = dp_2 dp_3$ and $n^2 \cos\theta_2 d\Omega = dp_1 dp_3$. We can then write vector $\mathbf{J}$ of expression (5.13) in the form:

$$\mathbf{J} = \left(\int dp_2 dp_3, \int dp_1 dp_3, \int dp_1 dp_2 \right) \tag{5.16}$$

For 2-D systems, the flux, basic radiance, and étendue, are related by the 2-D version of expression (5.1), which is

$$d\Phi = L^* dU = L^* n da \cos\theta \, d\theta \tag{5.17}$$

where $L^* = L/n$. An expression similar to Equation 5.13 can then be written for the 2-D case as

$$\mathbf{J} = \left(\int n \cos\theta_1 \, d\theta_1, \int n \cos\theta_2 \, d\theta_2 \right) \tag{5.18}$$

The angles to axes x_1 and x_2 are defined as presented in Figure 5.11.

We can see from this figure that $\theta_1 = \theta_2 + \pi/2$ and therefore $\sin\theta_1 = \cos\theta_2$ and $\sin\theta_2 = -\cos\theta_1$. We can then write Equation 5.18 as

$$\mathbf{J} = \left(\int n\, d(\sin\theta_1), \int n\, d(\sin\theta_2) \right) = \left(\int n\, d(\cos\theta_2), -\int n\, d(\cos\theta_1) \right) \tag{5.19}$$

and therefore

$$\mathbf{J} = \left(\int dp_2, -\int dp_1 \right) \tag{5.20}$$

For each point on the plane, we can define a vector flux **J**. This defines a vector field on the plane, as shown in Figure 5.12.

Now consider the lines that are tangent to **J** at each point, as in Figure 5.12a. We consider a point **P** on one of these lines, as shown in Figure 5.12b. The net flux through an element of length *da* on one of these lines is given by the 2-D version of expression (5.14) as

$$d\Phi = da \, L^* \, \mathbf{J} \cdot \mathbf{n} \tag{5.21}$$

But since the line is tangent to **J**, the normal **n** to *da* is perpendicular to **J** and, therefore, $\mathbf{J} \cdot \mathbf{n} = 0$. This means that the net flux $d\Phi$ through *da* is zero. In this

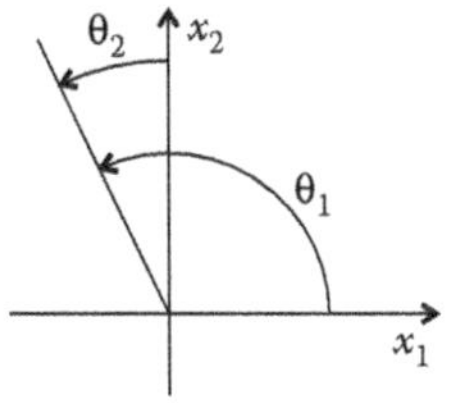

FIGURE 5.11
Angles θ_1 and θ_2 of a light ray to axes x_1 and x_2, respectively.

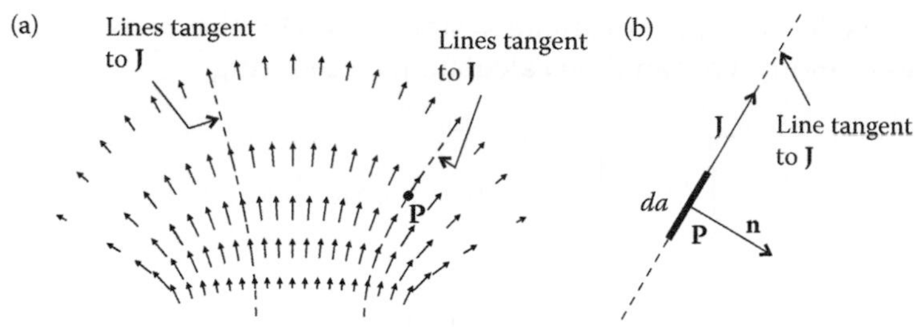

FIGURE 5.12
(a) Vector flux **J** field on the plane. (b) The net flux crossing a line tangent to **J** at each point is zero.

case, the flux crossing *da* from left to right cancels the flux crossing *da* from right to left, making the net flux crossing *da* zero. The flux is then conserved between two of these lines, as shown in Figure 5.12a. Since basic radiance is also conserved in an optical system, we can conclude from expression (5.1) that the étendue is also conserved between two of these lines. We have already seen that the étendue is conserved between flow-lines and, therefore, the lines tangent to **J** are the flow-lines.

In 3-D geometry, these lines become a surface, and the flux is conserved inside that surface. These surfaces are called tubes of flux.[4]

5.3 Vector Flux as a Bisector of the Edge Rays

We now calculate the direction and magnitude of the vector flux **J** at a given point **P** as a function of the directions of the edge rays crossing **P**. Let us then consider a point **P** on the plane, and that all the radiation passing through **P** is contained between rays r_A and r_B, as presented in Figure 5.13.

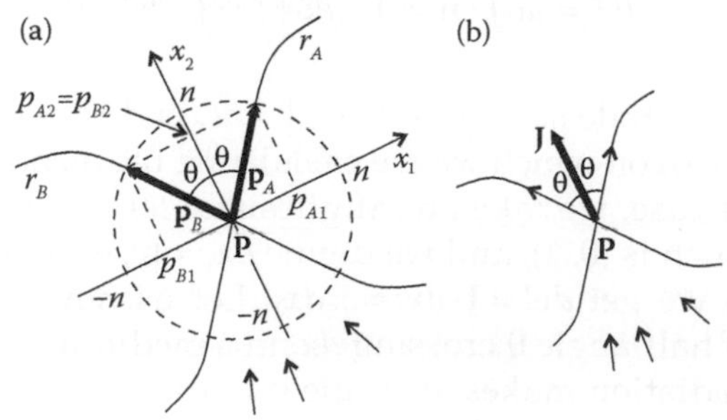

FIGURE 5.13
(a) When all the radiation passing through a point **P** is contained between two edge rays, r_A and r_B, the vector flux **J** points in the direction of the bisector to r_A and r_B at **P**. (b) **J** has magnitude $\|\mathbf{J}\| = 2n \sin\theta$, where 2θ is the angle that the edge rays make to each other.

Now consider a local coordinate system whereby the x_2 axis bisects the edge rays r_A and r_B. We can then calculate **J**, considering

$$\mathbf{J} = \left(\int_A^B dp_2, -\int_A^B dp_1 \right) = (p_{B2} - p_{A2}, -(p_{B1} - p_{A1})) \tag{5.22}$$

But since $p_{A1} = n \sin \theta$ and $p_{B1} = -p_{A1}$, we get

$$\mathbf{J} = (0, -n(-\sin\theta - \sin\theta)) = (0, 2n\sin\theta) \tag{5.23}$$

It can then be concluded that **J** points in the direction of the bisector of the edge rays, and has magnitude $\|\mathbf{J}\| = 2n \sin \theta$. This result is in accordance with the fact that the lines tangent to **J** are flow-lines, since these lines also bisect the edge rays of the radiation field.

It can also be seen that, in a medium of given refractive index $n(x_1, x_2)$, the paths of the edge rays are specified by the vector flux **J**. In fact, the magnitude of **J** gives us the angle between the edge rays at each point, and its direction gives the orientation of these edge rays relative to the coordinate system.

5.4 Vector Flux and Étendue

We now use the result obtained earlier, that the vector flux bisects the edge rays, and consider the relation between the vector flux and the étendue.

For 2-D systems, in expression (5.15) instead of an area dA, we have a length dc along a curve $\mathbf{c}(\sigma)$ on the plane, so that expression (5.15) can be written as

$$dU = dc\,\mathbf{J} \cdot \mathbf{n} = \mathbf{J} \cdot (dc\,\mathbf{n}) = \mathbf{J} \cdot d\mathbf{c}_N \tag{5.24}$$

where $d\mathbf{c}_N$ has magnitude dc, that is $\|d\mathbf{c}_N\| = dc$, and is perpendicular (normal) to the curve $\mathbf{c}(\sigma)$ on which we are calculating the étendue.

As a particular case, we take an infinitesimal length dx_1 on the x_1 axis, the normal of which is (0, 1), and we define $d\mathbf{x}_1 = (0, dx_1)$. If $\mathbf{J} = (J_1, J_2)$, from expression (5.24) we get $dU = \mathbf{J} \cdot d\mathbf{x}_1 = J_2 dx_1$. Let us now suppose that we have radiation of half-angle θ crossing dx_1 in a medium of refractive index n and that this radiation makes an angle φ to the horizontal, as shown in Figure 5.14.

We have $J_2 = \|\mathbf{J}\| \sin\varphi$ and therefore $J_2 = 2n \sin\theta \sin\varphi$ and:

$$dU = 2n \sin\theta \sin\varphi \, dx_1 \tag{5.25}$$

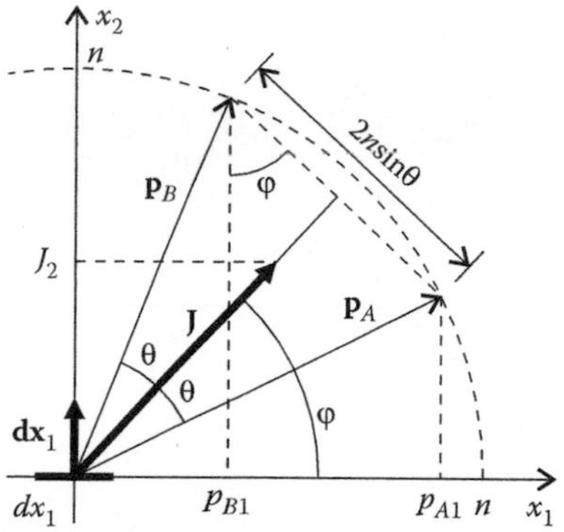

FIGURE 5.14
The étendue of radiation with half-angle θ tilted at an angle φ to the horizontal crossing an area dx_1 can be obtained as $dU = J_2 dx_1$ where J_2 is the x_2 component of the vector flux.

Since from Equation 5.20 we have

$$J_2 = -\int_A^B dp_1 = p_{A1} - p_{B1} \tag{5.26}$$

the expression for the étendue can also be written as $dU = (p_{A1} - p_{B1})dx_1$. The geometrical construction in Figure 5.14 shows that this expression is equivalent to Equation 5.25.

We now consider the general case of light crossing a curve c parameterized by $\mathbf{c}(\sigma)$ on the plane. From Equation 5.24 we then have

$$U = \int_C \mathbf{J} \cdot d\mathbf{c}_N \tag{5.27}$$

If this curve starts at point $\mathbf{P}_1$ and ends at point $\mathbf{P}_2$, we have seen in Chapter 4 that the étendue is given by $U = 2(G(\mathbf{P}_2) - G(\mathbf{P}_1))$.

Now consider the étendue of the radiation passing between two points $\mathbf{P}_1$ and $\mathbf{P}_2$ such that $\mathbf{P}_2 = \mathbf{P}_1 + (dx_1, dx_2)$ as presented in Figure 5.15.

Since from equation $dU = 2dG$, we have

$$dU = \mathbf{J} \cdot d\mathbf{c}_N = \mathbf{J} \cdot (-dx_2, dx_1) = 2dG \tag{5.28}$$

where $(-dx_2, dx_1)$ is a perpendicular vector to (dx_1, dx_2). We can then write

$$-J_1 dx_2 + J_2 dx_1 = 2\left(\frac{\partial G}{\partial x_1} dx_1 + \frac{\partial G}{\partial x_2} dx_2\right) \tag{5.29}$$

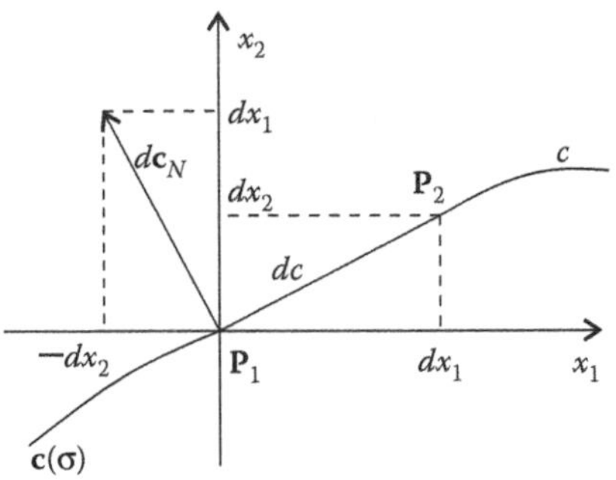

FIGURE 5.15
Line dc connecting points $\mathbf{P}_1$ and $\mathbf{P}_2$ and corresponding normal $d\mathbf{c}_N$.

and therefore we get

$$\left(2\frac{\partial G}{\partial x_2} + J_1\right)dx_2 + \left(2\frac{\partial G}{\partial x_1} - J_2\right)dx_1 = 0 \tag{5.30}$$

Since this equation must hold for any dx_1 and dx_2, we must have

$$J_1 = -2\frac{\partial G}{\partial x_2} \tag{5.31}$$

and

$$J_2 = 2\frac{\partial G}{\partial x_1} \tag{5.32}$$

or[6]

$$\mathbf{J} = 2\left\{-\frac{\partial G}{\partial x_2}, \frac{\partial G}{\partial x_1}\right\} \tag{5.33}$$

The lines that are tangent to the vector flux $\mathbf{J}$ at each point are called lines of the vector flux $\mathbf{J}$. From the expression (5.33), we have $\pm(-\partial G/\partial x_2, \partial G/\partial x_1) \cdot (\partial G/\partial x_1, \partial G/\partial x_2) = 0$ and therefore $\mathbf{J} \cdot \nabla G = 0$. It can then be concluded that the vector flux $\mathbf{J}$ is tangent to the lines $G =$ constant and, therefore, the lines of the vector flux $\mathbf{J}$ coincide with the lines $G =$ constant, which are the flow-lines.

The vector flux lines cannot cross. If they did cross, this would mean that, at a given point, we would have two flux vectors pointing in different directions, and this is impossible, since we can only have a value of the vector flux vector at each point given by expression (5.13) in the 3-D case or expression (5.18) in the 2-D case.

The divergence of vector flux **J** is given by

$$\nabla \cdot \mathbf{J} = \frac{\partial J_1}{\partial x_1} + \frac{\partial J_2}{\partial x_2} = 2\left(-\frac{\partial G}{\partial x_1 \partial x_2} + \frac{\partial G}{\partial x_2 \partial x_1} \right) = 0 \tag{5.34}$$

in a zone free from sources or attenuators.[4]

5.5 Vector Flux for Disk-Shaped Lambertian Sources

As examples of how to calculate the vector flux for a given Lambertian source, we consider the cases of a linear source in 2-D geometry, and a disk-shaped source in 3-D geometry.

We start with a 2-D linear Lambertian source extending between points $\mathbf{F}_1$ and $\mathbf{F}_2$. At a point **P**, the radiation coming from this source is contained between two edge rays, r_A and r_B, as presented in Figure 5.16a. This system is symmetrical with respect to axis x_2.

Since the radiation at point **P** is limited by edge rays r_A and r_B, vector **J** can be obtained from

$$\mathbf{J} = \left(\int_{\mathbf{p}_A}^{\mathbf{p}_B} dp_2, -\int_{\mathbf{p}_A}^{\mathbf{p}_B} dp_1 \right) = (\Delta p_2, -\Delta p_1) = (p_{B2} - p_{A2}, -(p_{B1} - p_{A1})) \tag{5.35}$$

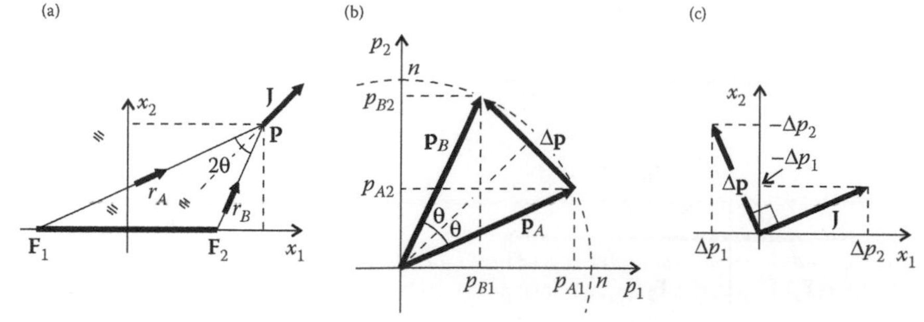

FIGURE 5.16
(a) The vector flux **J** produced by a linear Lambertian source between points $\mathbf{F}_1$ and $\mathbf{F}_2$ at point **P** has a magnitude $\|\mathbf{J}\| = 2n\text{Sin}\theta$, and points in the direction of the bisector of angle 2θ defined by point **P** and the edges $\mathbf{F}_1$ and $\mathbf{F}_2$ of the source. (b) Vector $\Delta\mathbf{p} = \mathbf{p}_B - \mathbf{p}_A$. (c) Vector flux **J** has the same magnitude, but is perpendicular to vector $\Delta\mathbf{p}$.

as can be seen from Figure 5.16b. This vector has the same magnitude and is perpendicular to vector

$$\Delta\mathbf{p} = (\Delta p_1, \Delta p_2) = (p_{B1} - p_{A1}, p_{B2} - p_{A2}) = \mathbf{p}_B - \mathbf{p}_A \tag{5.36}$$

as can be seen from Figure 5.16c. Vector $\Delta\mathbf{p}$ can also be written as

$$\Delta\mathbf{p} = \left(\int_{\mathbf{p}_A}^{\mathbf{p}_B} dp_1, \int_{\mathbf{p}_A}^{\mathbf{p}_B} dp_2 \right) \tag{5.37}$$

From Figure 5.16b, it can be seen that

$$\|\Delta\mathbf{p}\| = 2n \sin\theta \tag{5.38}$$

and therefore,

$$\|\mathbf{J}\| = 2n \sin\theta \tag{5.39}$$

where vector $\mathbf{J}$ points in the direction of the bisector of $\mathbf{p}_A$ and $\mathbf{p}_B$, since it is perpendicular to $\Delta\mathbf{p}$.

The vector flux $\mathbf{J}$ at point $\mathbf{P}$ then points in the direction of the bisector of the edge rays r_A and r_B of the Lambertian source $\mathrm{F_1F_2}$, as presented in Figure 5.16a. Therefore, the lines of flow of the geometrical vector flux $\mathbf{J}$ (these lines are tangent at each point to the direction of $\mathbf{J}$) are hyperbolas with foci $\mathrm{F_1}$ and $\mathrm{F_2}$.[5,7,8] These lines, shown in Figure 5.17b, bisect at each point of the plane the edge rays of the source $\mathrm{F_1F_2}$; that is, they bisect at each point in the plane the rays coming from the edges $\mathrm{F_1}$ and $\mathrm{F_2}$ of the source.

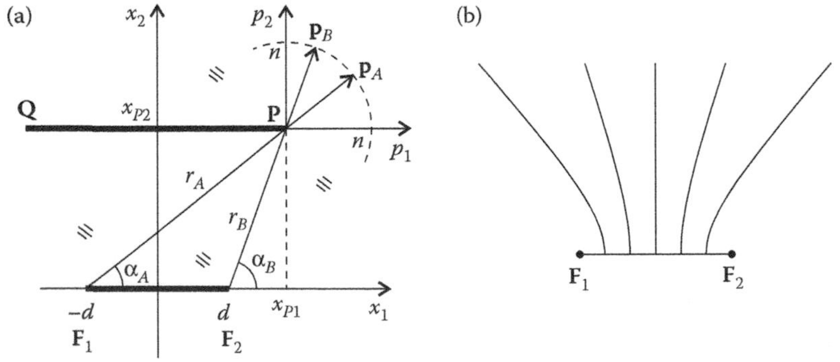

FIGURE 5.17
(a) A Lambertian source $\mathrm{F_1F_2}$ produces at a point $\mathbf{P}$ a vector flux $\mathbf{J}$ pointing in the direction of the bisector of the edge rays r_A and r_B of the source. These edge rays cross point $\mathbf{P}$ with momenta $\mathbf{p}_A$ and $\mathbf{p}_B$. (b) The lines of flow of the geometrical vector flux $\mathbf{J}$ are, therefore, shaped as hyperbolas having foci $\mathrm{F_1}$ and $\mathrm{F_2}$.

Now consider the étendue from source $\mathbf{F}_1\mathbf{F}_2$ to a line defined by point $\mathbf{P}$ and its symmetrical $\mathbf{Q}$, as shown in Figure 5.17a. If $\mathbf{P}_1 = (x_{P1}, x_{P2})$ then $\mathbf{Q} = (-x_{P1}, x_{P2})$, and the étendue from $\mathbf{F}_1\mathbf{F}_2$ to $\mathbf{QP}$ is $U(\mathbf{Q}, \mathbf{P})$. In this case, we have $U = 4G$ from Equation 4.50, and from Equation 5.33 we get

$$\mathbf{J} = \frac{1}{2}\left\{-\frac{\partial U}{\partial x_2}, \frac{\partial U}{\partial x_1}\right\} \tag{5.40}$$

which gives us the vector flux $\mathbf{J}$ at point $\mathbf{P}$.

The étendue from a source $\mathbf{F}_1\mathbf{F}_2$ to a line $\mathbf{PQ}$ (where $\mathbf{Q}$ is symmetrical to $\mathbf{P}$ respect to the x_2 axis) is given as

$$U = 2n([\mathbf{P},\mathbf{F}_1] - [\mathbf{P},\mathbf{F}_2]) \tag{5.41}$$

where $[\mathbf{X}, \mathbf{Y}]$ is the distance between points $\mathbf{X}$ and $\mathbf{Y}$. If $\mathbf{P} = (x_{P1}, x_{P2})$ and $\mathbf{F}_1 = (-d, 0)$, and $\mathbf{F}_2 = (d, 0)$, we have $\mathbf{P} - \mathbf{F}_1 = (x_{P1} + d, x_{P2})$ and $\mathbf{P} - \mathbf{F}_2 = (x_{P1} - d, x_{P2})$. We have then

$$U = 2n\left(\sqrt{(x_{P1} + d)^2 + x_{P2}^2} - \sqrt{(x_{P1} - d)^2 + x_{P2}^2}\right) \tag{5.42}$$

Since this optical system is symmetrical with respect to axis x_2, using expression (5.40) and calculating the derivatives of U given by expression (5.42), we obtain the components of $\mathbf{J}$ given by

$$\begin{aligned} J_1 &= n\left(\frac{x_{P2}}{\sqrt{(x_{P1} - d)^2 + x_{P2}^2}} - \frac{x_{P2}}{\sqrt{(x_{P1} + d)^2 + x_{P2}^2}}\right) = n\left(\frac{x_{P2}}{[\mathbf{P},\mathbf{F}_2]} - \frac{x_{P2}}{[\mathbf{P},\mathbf{F}_1]}\right) \\ J_2 &= n\left(\frac{x_{P1} + d}{\sqrt{(x_{P1} + d)^2 + x_{P2}^2}} - \frac{x_{P1} - d}{\sqrt{(x_{P1} - d)^2 + x_{P2}^2}}\right) = n\left(\frac{x_{P1} + d}{[\mathbf{P},\mathbf{F}_1]} - \frac{x_{P1} - d}{[\mathbf{P},\mathbf{F}_2]}\right) \end{aligned} \tag{5.43}$$

or

$$\begin{aligned} J_1 &= n \sin \alpha_B - n \sin \alpha_A = p_{B2} - p_{A2} \\ J_2 &= n \cos \alpha_A - n \cos \alpha_B = p_{A1} - p_{B1} \end{aligned} \tag{5.44}$$

which is the same as expression (5.35). From expression (5.41) for the étendue, it can also be seen that

$$U = \text{contant} \Rightarrow [\mathbf{P},\mathbf{F}_1] - [\mathbf{P},\mathbf{F}_2] = \text{constant} \tag{5.45}$$

This condition defines hyperbolas having foci $\mathbf{F}_1$ and $\mathbf{F}_2$. Considering a constant étendue, points **P** and **Q** must lie along the hyperbolas corresponding to the lines of flow of the vector flux.

Let us now suppose that the system considered earlier is 3-D, with rotational symmetry around the axis x_3. Let us consider the étendue from a circular source A_1 to a circular surface with radius ρ placed at a distance h, as in Figure 5.18.

As can be observed from Figure 5.18a, a variation $d\rho$ in ρ coordinate corresponds to a circular strip of radius ρ and width $d\rho$ having, therefore, an area $dA = 2\pi\rho d\rho$. From expression (5.15), we get

$$dU = 2\pi\rho\, d\rho\, J_3 \tag{5.46}$$

and therefore,

$$J_3 = \frac{1}{2\pi\rho}\frac{\partial U}{\partial \rho} \tag{5.47}$$

A variation dh in the distance between the source A_1 and the surface of radius ρ, as shown in Figure 5.18b, leads to a variation in the étendue given by

$$dU = -2\pi\rho\, dh\, J_\rho \tag{5.48}$$

and therefore,

$$J_\rho = -\frac{1}{2\pi\rho}\frac{\partial U}{\partial h} \tag{5.49}$$

Note that now the variation dU in the étendue is negative, since the new surface of radius ρ obtained by variation dh is now further away from the

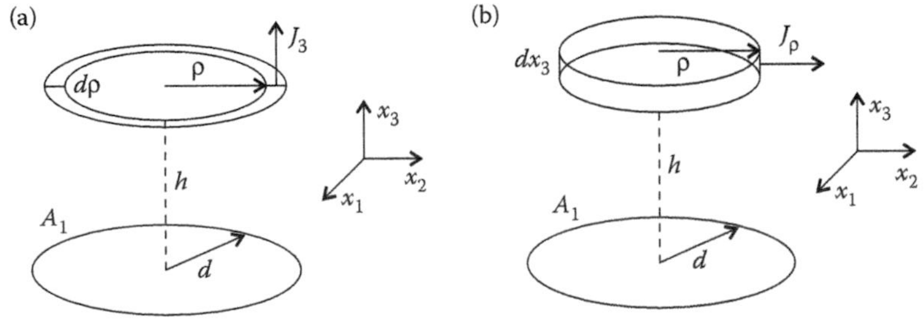

FIGURE 5.18
The étendue from a circular source A_1 to a circular surface with radius ρ and placed at a distance h. Variations in the coordinates ρ and h enable us to relate the components J_3 and J_ρ of vector flux **J** with the corresponding variations of étendue. (a) Variation in ρ. (b) Variation in h.

source. The étendue passing through it is then smaller than before. This can also be seen as a flux decrease, since we have $d\Phi = L^* dU$.

The étendue from a source A_1 and radius d to a surface of radius ρ placed at a distance h has already been calculated, and is given by Equation 4.91 as

$$U = \frac{\pi^2}{4}\left(\sqrt{(\rho - d)^2 + h^2} - \sqrt{(\rho + d)^2 + h^2}\right)^2 \tag{5.50}$$

The components of **J** can now be obtained by[8]

$$J_3 = \frac{1}{2\pi\rho}\frac{\partial U}{\partial \rho} = \frac{\pi}{2}\left(\frac{d^2 - \rho^2 - h^2}{\sqrt{(d^2 + \rho^2 + h^2)^2 - 4d^2\rho^2}} + 1\right) \tag{5.51}$$

and

$$J_\rho = -\frac{1}{2\pi\rho}\frac{\partial U}{\partial h} = \frac{\pi}{2}\frac{h}{\rho}\left(\frac{d^2 + \rho^2 + h^2}{\sqrt{(d^2 + \rho^2 + h^2)^2 - 4d^2\rho^2}} - 1\right) \tag{5.52}$$

5.6 Design of Concentrators Using the Vector Flux

The vector flux can be used as a tool to obtain ideal concentrators. The design method involves placing mirrors along the lines of flow of vector flux **J**,[7,8] which, as we have seen, correspond to the flow-lines. These mirrors do not change the radiation pattern and, therefore, do not change the vector field of **J**.

This result can now be applied to the design of concentrators. We have seen already that a linear flat source generates a trumpet concentrator, whose mirrors are hyperbolic. With hyperbolas, nonetheless, it is not possible to obtain a compound parabolic concentrator (CPC) as a shape that does not disturb the field **J**, since the CPC consists of two parabolic arcs, and we only have hyperbolic lines of flow of **J** available. A parabola can be obtained in the limiting case of a hyperbola when one of its foci moves to infinity. Let us then consider Figure 5.17. Lines of flow of **J** shaped as parabolas can be obtained by keeping, for example, $\mathbf{F}_2$ fixed and allowing $\mathbf{F}_1$ to go to infinity along line $\mathbf{F}_1\mathbf{F}_2$. The Lambertian source then tends to become a straight line starting at $\mathbf{F}_2$ and extending horizontally to infinity. The corresponding lines of flow of **J** are parabolas with focus $\mathbf{F}_2$ and horizontal axis, as presented in Figure 5.19. Again, the vector flux **J** points in the direction of the bisector of the edge rays r_A and r_B of the source.

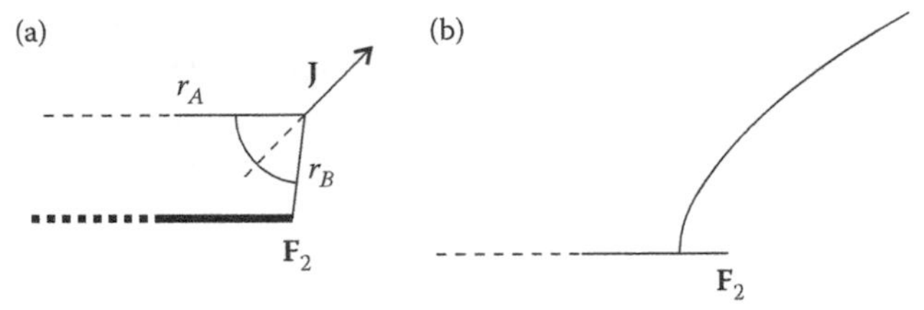

FIGURE 5.19
A source shaped as a straight line starting at $\mathbf{F}_2$ and extending to infinity. (a) The vector flux **J** points in the direction of the bisector of edge rays r_A and r_B of the source. (b) The lines of flow of **J** are shaped as parabolas with focus $\mathbf{F}_2$ and axis coincident with the straight line.

It is now possible to combine two parabolas to form a CPC. Two parabolas can be obtained from two semi-infinite straight lines. Since, in a CPC, the two parabolic arcs make an angle to the vertical, also the two straight lines must make an angle to the vertical. Figure 5.20 presents one such possibility. Here, a Lambertian source $\mathbf{F}_2\mathbf{F}_1\mathbf{F}_3$ is used, where points $\mathbf{F}_1$ and $\mathbf{F}_3$ are considered to be at an infinite distance.

The visible shape of the source is different for the points in zones 1 and 2, and, therefore, vector **J** is calculated in a different manner in these two zones. For points **P** in the right-zone 1, only the source $\mathbf{F}_1\mathbf{F}_3$ is visible, that is, the visible edges of the source are $\mathbf{F}_1$ and $\mathbf{F}_3$ (note that $\mathbf{F}_3$ is a point placed at an infinite distance). Therefore, the edge rays of the source in a point **P** of zone 1 on the right side are r_1 and r_2. Vector flux **J** points in the direction of the bisector to r_1 and r_2, and the lines of **J** are parabolas with focus $\mathbf{F}_1$ and

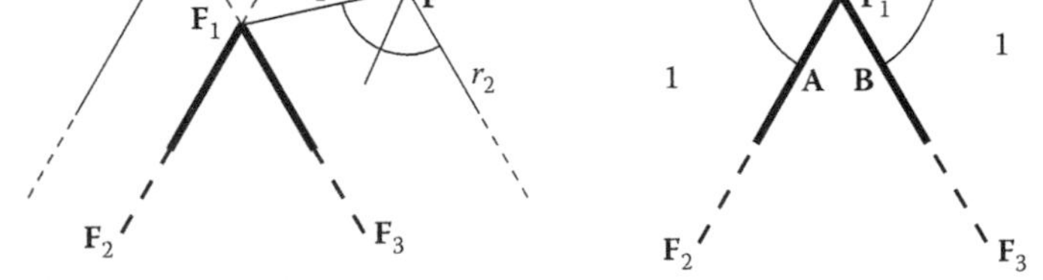

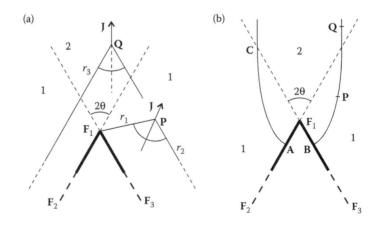

FIGURE 5.20
(a) Direction of vector flux **J** for different regions of a combination of two Lambertian sources shaped as semi-infinite straight lines starting at $\mathbf{F}_1$ and extending toward $\mathbf{F}_2$ and $\mathbf{F}_3$ (at an infinite distance). (b) Flow lines in zone 1 on the right are parabolas with focus $\mathbf{F}_1$ and axis parallel to $\mathbf{F}_1\mathbf{F}_3$. Flow lines in zone 2 are vertical straight lines.

axis parallel to r_2, as can be observed in Figure 5.20a. In zone 1 on the left side, the lines of **J** are symmetrical. In zone 2, a source $\mathbf{F_2F_3}$ is visible, that is, the edges of the visible source are $\mathbf{F_2}$ and $\mathbf{F_3}$. Vector **J** at a point **Q** of zone 2 must then point in the direction of the bisector to the edge rays of the source r_3 and r_2. Therefore, the lines of **J** in this area are vertical. The lines of **J** generated by a Lambertian source $\mathbf{F_2F_1F_3}$ must then be shaped according to what is represented in Figure 5.20b. The parabolic mirrors **AC** and **BD** form a CPC concentrator with a half-acceptance angle θ and an inverted V receiver $\mathbf{AF_1B}$.

A CPC for a straight receiver can also be obtained. Truncating the Lambertian source $\mathbf{F_2F_1F_3}$ at **AB**, we get the result presented in Figure 5.21.[7]

Also in here, the space must be divided into several different zones to analyze the shape of the lines of flow of **J** in each one of them. In zone 1, only the source $\mathbf{BF_3}$ is visible (remember that $\mathbf{F_3}$ is placed at an infinite distance), and therefore, in here, the lines of **J** are parabolas with focus **B** and axis parallel to $\mathbf{F_3B}$. In zone 2, a source $\mathbf{ABF_3}$ is visible. The Lambertian source $\mathbf{ABF_3}$ in these points, behaves as an equivalent source shaped as a straight line with origin at **A** and parallel to $\mathbf{BF_3}$. Therefore, in here, the lines of flux of **J** are parabolas with focus **A** and axis parallel to $\mathbf{BF_3}$. From zone 3, a Lambertian source $\mathbf{F_2F_3}$ is visible, and it behaves as an infinite V-shaped source, therefore making the lines of flow vertical. Finally, in zone 4, only the segment of a straight line **AB** is visible and, therefore, here the lines of flow are shaped as hyperbolas with foci at **A** and **B**. The lines of vector flux **J** from points **A** to **C** and **B** to **D** form a CPC with half-acceptance angle θ and concentrating onto **AB** the radiation falling on **CD**.

To obtain other types of nonimaging devices using the geometrical vector flux, we may need to introduce Lambertian absorbers, which act as sinks

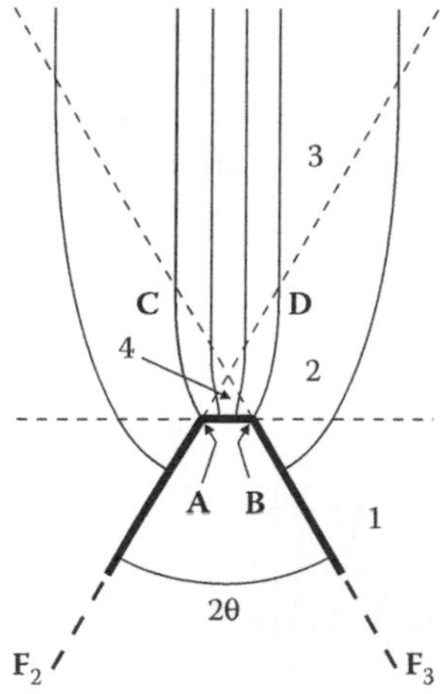

FIGURE 5.21
Cutting the Lambertian source between points **A** and **B** in Figure 5.20, it is possible to obtain a pattern of lines of flow of **J** defining a CPC, placing mirrors along the parts **BD** and **AC** of those lines of flow.

for **J**, in addition to Lambertian radiators that act as sources of **J**. That is the case, for example, of the CEC).[9]

5.7 Examples

The examples presented use expressions for the curves and functions that are derived in Chapter 21.

EXAMPLE 5.1

Calculate the vector flux at point **P** = (0.5,0.35) created by a linear Lambertian source between points **A** = (–0.5,0) and **B** = (0.5,0) emitting in air.

Angle 2θ between the lines **A P** and **B P** is given by

$$\theta = \text{ang}(\mathbf{A}-\mathbf{P},\mathbf{B}-\mathbf{P})/2 = 35.355\,\text{deg}$$

as shown in Figure 5.22.

The vector flux at point **P** = (0.5,0.35) is given by

$$\mathbf{J} = 2\sin\theta\,\text{nrm}\,(\text{nrm}(\mathbf{P}-\mathbf{A}) + \text{nrm}(\mathbf{P}-\mathbf{B})) = (0.66965, 0.943858)$$

since the refractive index for air is $n = 1$.

EXAMPLE 5.2

Calculate the vector flux at point **P** = (1,1.5) inside a CPC for an acceptance angle of $\alpha = 30°$, a small aperture of length 2 centered at the origin and made of a material of refractive index $n = 1.5$.

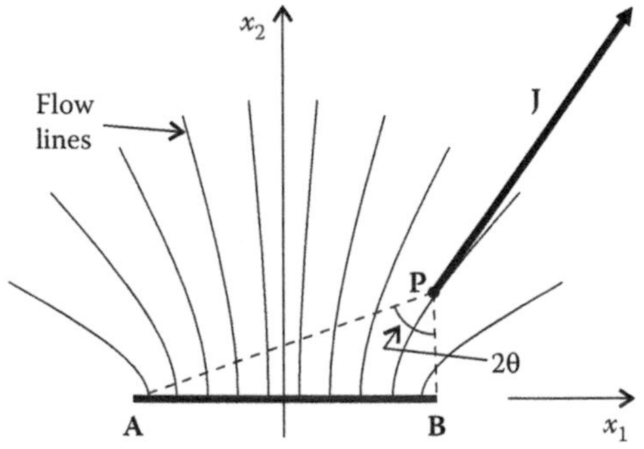

FIGURE 5.22
Flow-lines of a flat Lambertian source and vector flux at point **P**.

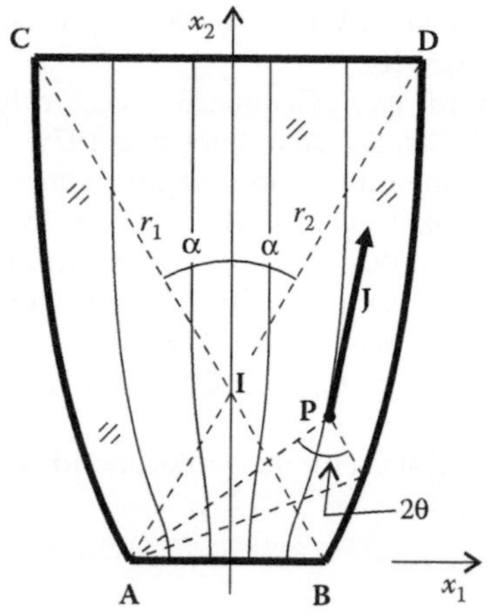

FIGURE 5.23
Flow-lines inside a CPC and vector flux **J** at a point **P**.

The flow lines inside a CPC, as shown in Figure 5.23, are hyperbolas with foci **A** and **B** in the triangle **AIB**, parabolas with focus **A** and axis parallel to r_1 in the area to the right of **DIB** and vertical straight lines in the area above line **CID**. Flow-lines are symmetrical relative to axis x_2, which is the perpendicular bisector of **AB** and **CD**.

The small aperture of the CPC is bounded by points $\mathbf{A} = (-1,0)$ and $\mathbf{B} = (1,0)$. At point **P**, one of the edge rays as direction **AP** and the other is parallel to r_1, that is, to direction $(\cos(\pi/2 + \alpha), \sin(\pi/2 + \alpha)) = (-\sin\alpha, \cos\alpha)$. These two edge rays make an angle 2θ to each other. Angle θ is given by

$$\theta = \text{ang}\,(\mathbf{P} - \mathbf{A}, (-\sin\alpha, \cos\alpha))/2 = 41.56°$$

And vector **J** is given by

$$\mathbf{J} = 2n \sin\theta\, \text{nrm}(\text{nrm}(\mathbf{P} - \mathbf{A}) + (-\sin\alpha, \cos\alpha)) = (0.398825, 1.94984)$$

since the CPC is made of material with refractive index $n = 1.5$.

References

1. Klein, M. V., Furtak, T. E., *Optics*, John Wiley & Sons, New York, 1986.
2. Meyer-Arendt, J. R., *Introduction to Classical and Modern Optics*, Prentice-Hall, New Jersey, 1989.
3. Chandrasekhar, S., *Radiative Transfer*, Dover Publications, Inc., New York, 1960.

4. Gershun, A., The light field, *J. Math. Phys.* XVII, published by the Massachusetts Institute of Technology, 1938.
5. Winston, R. and Welford, W. T., Geometrical vector flux and some new nonimaging concentrators, *J. Opt. Soc. Am. A*, 69, 532, 1979.
6. Miñano, J. C., Two-dimensional nonimaging concentrators with inhomogeneous media: A new look, *J. Opt. Soc. Am. A*, 2, 11, 1826, 1985.
7. Winston, R. and Welford, W. T., Ideal flux concentrators as shapes that do not disturb the geometrical vector flux field: A new derivation of the compound parabolic concentrator, *J. Opt. Soc. Am. A*, 69, 536, 1979.
8. Welford, W. T. and Winston, R., *High Collection Nonimaging Optics*, Academic Press, San Diego, 1989.
9. Greenman, P., Geometrical vector flux sinks and ideal flux concentrators, *J. Opt. Soc. Am.*, 71, 777, 1981.

6

Combination of Primaries with Flow-Line Secondaries

6.1 Introduction

Heat engines are one of the means that generate electricity from a heat source. The efficiency of these engines increases with the temperature of the source, which means that, for high efficiency, we need high temperatures. Solar energy is a clean, renewable source of energy that can generate high heat if highly concentrated. Another possible way to generate clean electricity is by using solar cells. They are, however, expensive, and it may be interesting to replace a large cell by a large optic combined with a small, highly efficient, and less expensive solar cell. This again means a high concentration of solar radiation. These are just two examples of application for a high concentration of radiation with a small angular aperture (such as sunlight).

Nonimaging concentrators, such as the compound parabolic concentrator (CPC), are ideal (in 2-D) for the concentration of radiation. For small acceptance angles, however, they become very tall, and that makes them impractical. Imaging optics, such as lenses or parabolic mirrors, are much more compact, but they cannot achieve high concentrations that nonimaging optics can deliver. It is then interesting to combine them into primary–secondary systems in which, at the focus of the imaging primary, we have a nonimaging secondary that boosts concentration. Further refinements can also be made in these optics: changing the shape of the primary to improve the efficiency or compactness.

Figure 6.1 presents the geometry of a parabolic mirror.[1,2]

This device concentrates the radiation arriving with an angular aperture θ and falling on $\mathbf{P}_1\mathbf{P}_2$ onto a focal zone of (approximate) width $R = 2D \sin \theta / \cos \varphi$, where D is the distance between the focus of the mirror and edge $\mathbf{P}_1$. Considering that $[\mathbf{P}_1, \mathbf{P}_2] = 2D \sin \varphi$, where φ is the rim angle of the parabolic mirror, the maximum concentration is

$$C = \frac{2D \sin\varphi}{2D \sin\theta/\cos\varphi} = \frac{1}{2}\frac{2\sin\varphi\cos\varphi}{\sin\theta} = \frac{1}{2}\frac{\sin(2\varphi)}{\sin\theta} \tag{6.1}$$

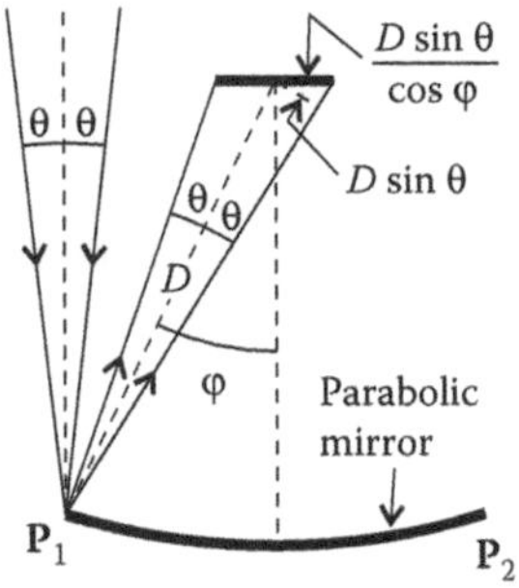

FIGURE 6.1
A parabolic mirror with a flat receiver falls short of the maximum concentration.

The maximum concentration can be obtained for $\phi = \pi/4$:

$$C_{\max} = \frac{1}{2}\frac{1}{\sin\theta} \tag{6.2}$$

which is half of the maximum ideal concentration. If a 3-D parabolic mirror is considered, the maximum concentration is $C_{\max 3D} = 1/(4\sin^2\theta)$, which corresponds to one-fourth of the ideal concentration. Actually, the concentration a parabolic mirror can attain, is slightly higher, if we displace the receiver from the focal plane slightly toward the parabolic mirror. The improvement, however, is negligible, especially for small angles θ. The minimum size spot will be calculated in Section 6.3 when we combine a parabolic mirror with a CEC secondary. The concentration that can be obtained, both in cases, is much lower than the ideal maximum. Similar conclusions can be reached for converging lenses.

Parabolic mirrors may be combined with kaleidoscope secondaries to produce a uniform flux distribution on the receiver.[3,4]

In the case of circular receivers, the concentration produced by a parabolic primary is also lower than the ideal maximum. Figure 6.2 shows the geometry for this case.[5]

The radius of the circular receiver is $R = D\sin\theta$ and the maximum concentration is

$$C = \frac{2D\sin\varphi}{2\pi D\sin\theta} = \frac{\sin\varphi}{\pi\sin\theta} \tag{6.3}$$

The maximum is obtained for $\varphi = \pi/2$,

$$C_{\max} = \frac{1}{\pi\sin\theta} \tag{6.4}$$

which is $1/\pi$ of the maximum ideal concentration.

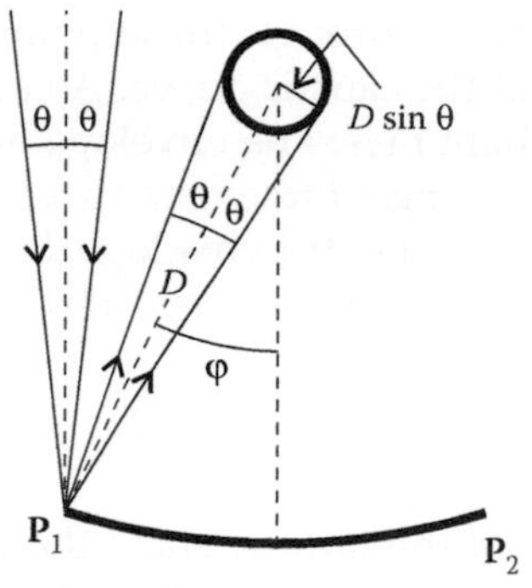

FIGURE 6.2
A parabolic mirror with a circular receiver falls short of the maximum limit of concentration.

6.2 Reshaping the Receiver

One way to increase the concentration of a parabolic primary is to reshape the receiver to better fit all the edge rays reflected by the mirror, as shown in Figure 6.3,[6] which shows a bundle of edge rays tilted by an angle θ to the left, then reflected off the parabolic mirror.

The edges $\mathbf{E}_1$ and $\mathbf{E}_2$ of the receiver are at the intersection of edge rays r_1 and r_2 with the axis of symmetry of the parabola. From there, we design straight sections tangent to the envelope (caustic) of the edge rays. The receiver we obtain captures all the rays reflected off the primary, and is smaller than what a circular receiver would need to be to do the same.

To calculate the shape of the receiver, we first calculate the positions of $\mathbf{E}_1$ and $\mathbf{E}_2$ and then, from the equation of the caustic curve, we determine

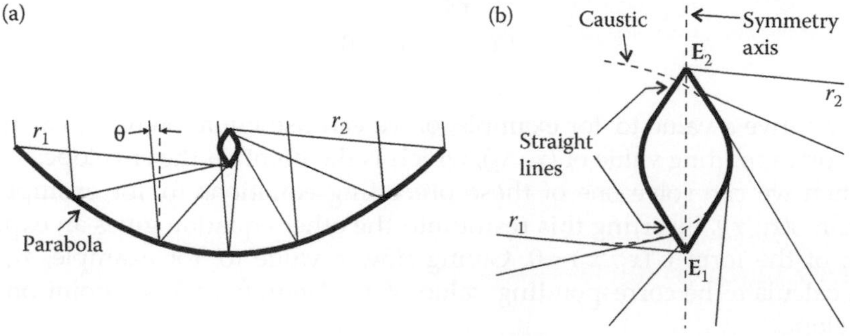

FIGURE 6.3
(a) The receiver of a parabolic mirror can be reshaped to better fit the ray envelope formed by the primary, and to increase the concentration. (b) Detail of the receiver showing some edge rays.

the points whose tangents go through these points. For this, however, we need to know the shape of the caustic curve. A caustic is the envelope of a one-parameter family of light rays. The envelope of a one-parameter family of curves is a curve that is tangent to every curve of the family. Also, each member of the family is tangent to the envelope. In general, a one-parameter family of curves is defined in parametric form by

$$(f(t,\phi),g(t,\phi)) \tag{6.5}$$

where ϕ is the parameter of the family, and t the parameter of each curve. This means that, for a particular value of ϕ, we have a curve parameterized by $(f_\phi(t), g_\phi(t))$ in parameter t. The envelope can be calculated by solving

$$\frac{\partial f}{\partial t}\frac{\partial g}{\partial \phi} - \frac{\partial f}{\partial \phi}\frac{\partial g}{\partial t} = 0 \tag{6.6}$$

By giving a value to ϕ, Equation 6.6 enables us to calculate the corresponding value of t. Or, if we give a value to t, it enables us to obtain ϕ. Introducing then, the pair (t, ϕ) into $(f(t, \phi), g(t, \phi))$, we obtain a point on the envelope.[7]

An alternative way of calculating the caustic is to describe the one-parameter family of curves implicitly by

$$C(x_1,x_2,\phi) = 0 \tag{6.7}$$

where, again, ϕ is the parameter of the family. For a particular value of ϕ we get an expression $C_\phi(x_1, x_2) = 0$, which implicitly defines one curve. The envelope of this family of curves in this case is given by simultaneously solving

$$\begin{gathered} \frac{\partial C}{\partial \phi} = 0 \\ C(x_1,x_2,\phi) = 0 \end{gathered} \tag{6.8}$$

If we give a value to, for example, ϕ, we can use Equation 6.8 to calculate the corresponding value of (x_1, x_2), which is the point on the envelope.[7–9]

Then we can solve one of these preceding equations to, for example, ϕ, obtain $\phi(x_1, x_2)$. Inserting this result into the other equation gives an expression of the form $C^*(x_1, x_2) = 0$. Giving now a value to, for example, x_1, we can calculate the corresponding value of x_2. Again (x_1, x_2) is a point on the envelope.

We may now use these expressions to calculate the caustic of a family of parallel light rays after being reflected by a parabolic mirror. In this case, the family of curves is a family of straight lines, because light rays travel straight (in a homogeneous medium). The caustic is the envelope of this bundle of

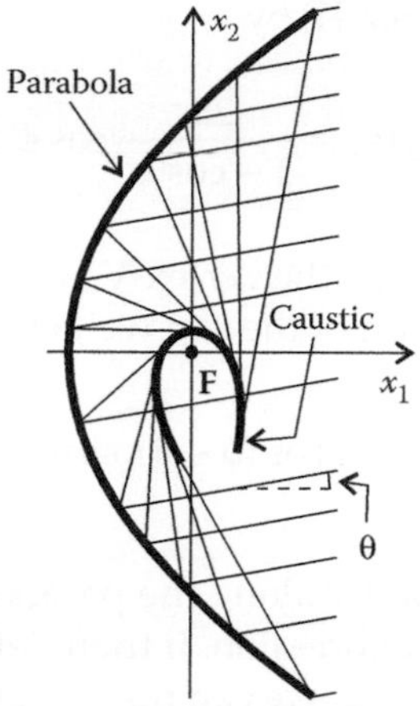

FIGURE 6.4
A caustic of a bundle of parallel rays tilted by an angle θ after being reflected by a parabolic mirror with horizontal axis.

straight lines (rays). Figure 6.4 shows a parabola with a horizontal axis and a bundle of parallel rays tilted by an angle θ to the horizontal. After reflection, these rays form a caustic around the focus **F** of the parabola.

We may now find a parameterization for the family of parallel rays after reflection on the parabola. Figure 6.5 shows a light ray (dashed line) traveling parallel to the axis of the parabola. It is reflected at a point **P** toward a focus **F** = (0,0). Another ray, traveling at an angle θ to the axis, is also reflected at **P** in a direction tangent to the caustic. The latter ray can be parameterized after reflection as

$$\mathbf{P} - t\,R(-\theta)\cdot\mathbf{P} \tag{6.9}$$

where $R(-\theta)$ is a rotation matrix of the angle $-\theta$.

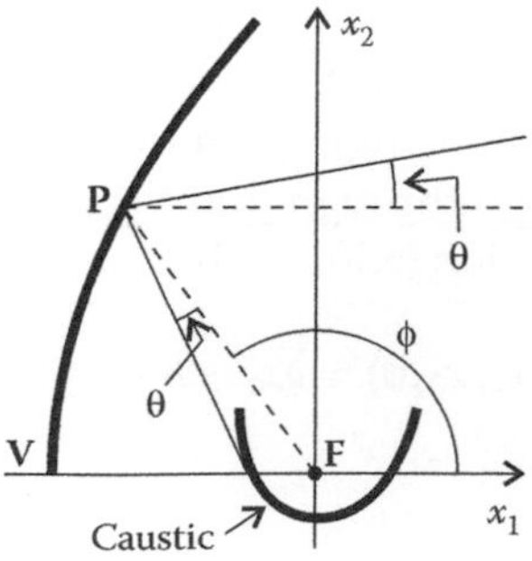

FIGURE 6.5
A light ray tangent to the caustic.

Point **P** is given (see Chapter 21) by

$$\mathbf{P} = (P_1, P_2) = \frac{2d}{1 - \cos\phi}(\cos\phi, \sin\phi) \tag{6.10}$$

where d is the distance between the vertex **V** and the focus **F** of the parabola. Then, for the parameterization of the ray, we get

$$\frac{2d}{1 - \cos\phi}(\cos\phi - t\cos(\phi - \theta), \sin\phi - t\sin(\phi - \theta)) \tag{6.11}$$

For each value of ϕ (each point **P** along the parabola), we have a ray with the parameter t defined by this expression. It then defines a one-parameter family of rays with the parameter ϕ. We can now calculate

$$\frac{\partial f}{\partial t}\frac{\partial g}{\partial \phi} - \frac{\partial f}{\partial \phi}\frac{\partial g}{\partial t} = d^2 \csc^5\left(\frac{\phi}{2}\right)\left(\sin\left(\theta - \frac{\phi}{2}\right) - t\sin\left(\frac{\phi}{2}\right)\right) = 0 \tag{6.12}$$

Solving for t we have,

$$t = \csc\left(\frac{\phi}{2}\right)\sin\left(\frac{\phi}{2} - \theta\right) \tag{6.13}$$

Inserting (6.13) into expression (6.11) we can calculate the points of the involute as a function of ϕ. For a given θ, the involute is then

$$\mathbf{C}(\phi,\theta) = \frac{d}{2}\csc^3\left(\frac{\phi}{2}\right)\left(\sin\left(2\theta - \frac{3\phi}{2}\right) + \sin\left(\frac{3\phi}{2}\right), \cos\left(2\theta - \frac{3\phi}{2}\right) - \cos\left(\frac{3\phi}{2}\right)\right) \tag{6.14}$$

Another possible way to calculate the involute is giving the bundle of rays in an implicit form, such as expression (6.7). A straight line is given by an expression of the form $x_2 = ax_1 + b$. As shown in Figure 6.5, after reflection at point **P**, the light ray makes an angle $\phi - \theta$ to the x_1 axis and we have $a = \tan(\phi - \theta)$. Also, the ray must go through point **P** and we can write $P_2 = aP_1 + b$, where **P** is given by expression (6.10), giving us b. Then, in this case, we get

$$C(x_1, x_2, \phi) = ax_1 + b - x_2 = 0 \tag{6.15}$$

or

$$\frac{x_2 - 2d\sin\phi + x_1\tan(\theta - \phi) - (x_2 + (2d + x_1)\tan(\theta - \phi))\cos\phi}{\cos\phi - 1} = 0 \tag{6.16}$$

This can also be written as

$$x_2 = \frac{-2d\sin\phi + (x_1 - (2d + x_1)\cos\phi)\tan(\theta - \phi)}{\cos\phi - 1} \tag{6.17}$$

Since θ and d are given, expression defines implicitly a straight line defining a ray after reflection by the parabola for each value of ϕ. For the points of the caustic we have

$$\frac{\partial C}{\partial \phi} = \frac{-2\sec^2(\theta - \phi)\sin(\phi/2)}{(\cos\phi - 1)^2} \times (d\sin(2\theta - 3\phi/2) + \sin(\phi/2)(d - x_1 + (2d + x_1)\cos\phi)) = 0 \tag{6.18}$$

which can also be written as

$$x_1 = \frac{d}{2}\csc^3\left(\frac{\phi}{2}\right)\left(\sin\left(2\theta - \frac{3\phi}{2}\right) + \sin\left(\frac{3\phi}{2}\right)\right) \tag{6.19}$$

Inserting this into expression (6.17) gives the corresponding x_2 coordinate for the point of the caustic, which can now be written as in expression (6.14).

The points of a parabola with a horizontal axis and a focus at the origin are given by expression (6.10) (Figure 6.6). Horizontal parallel rays such as r_1 and r_2 are focused to **F**. The distance d between the vertex **V** and the focus **F** is just a scale factor in the equation of the parabola. Points $\mathbf{P}_1$ and $\mathbf{P}_2$ at the edges are obtained for $\phi = \phi_1 = \pi - \varphi$ and $\phi = \phi_2 = \pi + \varphi$, respectively.

We can replace the parameter ϕ by another parameter α, such that $\phi = \pi + \alpha$ with $-\varphi \le \alpha \le \varphi$ as the new parameter range for the parabola. The caustic as a function of this new parameter is

$$\mathbf{C}(\alpha,\theta) = \frac{d}{2}\sec^3\left(\frac{\alpha}{2}\right)\left(\cos\left(\frac{3\alpha}{2} - 2\theta\right) - \cos\left(\frac{3\alpha}{2}\right), \sin\left(\frac{3\alpha}{2} - 2\theta\right) - \sin\left(\frac{3\alpha}{2}\right)\right) \tag{6.20}$$

and $\mathbf{C}(\alpha, \theta)$ is symmetrical to $\mathbf{C}(-\alpha, -\theta)$ in relative to the x_1 axis.

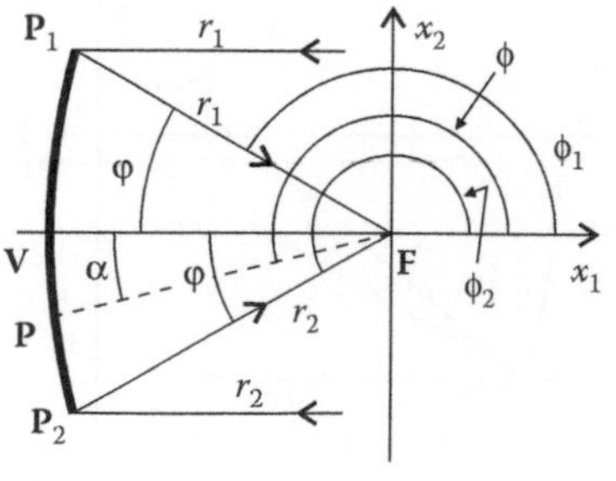

FIGURE 6.6
Parabola with horizontal axis and focus at the origin.

6.3 Compound Elliptical Concentrator Secondary

Two rays r_1 and r_2 parallel to the axis of a parabola are concentrated onto its focus **F**. We now consider another two rays, r_3 and r_4, making an angle θ to the axis of the parabola, as shown in Figure 6.7. After reflection, they intersect at a point **X** of the intersection of the straight lines passing through points $\mathbf{P}_1$ and $\mathbf{P}_2$, with the directions of rays r_3 and r_4 after reflection. It is given by

$$\mathbf{X} = \frac{d}{\cos\varphi + \cos^2\varphi}(-2\sin^2\theta, \sin 2\theta) \tag{6.21}$$

The parabolic mirror $\mathbf{P}_1\mathbf{P}_2$ can be considered as a linear Lambertian source emitting toward the receiver **XY**, where **Y** is symmetrical to **X** relative to the axis x_1. It is represented as a dashed line $\mathbf{P}_1\mathbf{P}_2$ in Figure 6.8. This approximation is not exact. If we had a Lambertian source $\mathbf{P}_1\mathbf{P}_2$ emitting toward **XY**, from midpoint **M**, we would have two edge rays headed toward **X** and **Y**. What we have, instead, are two rays, r_1 and r_2, reflected at the vertex **V** of the parabolic mirror, that do not cross **XY** at the edges.

That same mismatch can be seen when we calculate the étendues. The étendue received by the parabolic mirror is

$$U_P = 2[\mathbf{P}_1,\mathbf{P}_2]\sin\theta = 8d\sin\theta\tan(\varphi/2) \tag{6.22}$$

whereas the étendue emitted by a Lambertian source $\mathbf{P}_1\mathbf{P}_2$ toward a receiver **XY** is given by

$$U_{LS} = 2([\mathbf{P}_2,\mathbf{X}] - [\mathbf{X},\mathbf{P}_1]) = U_P/\cos\varphi \tag{6.23}$$

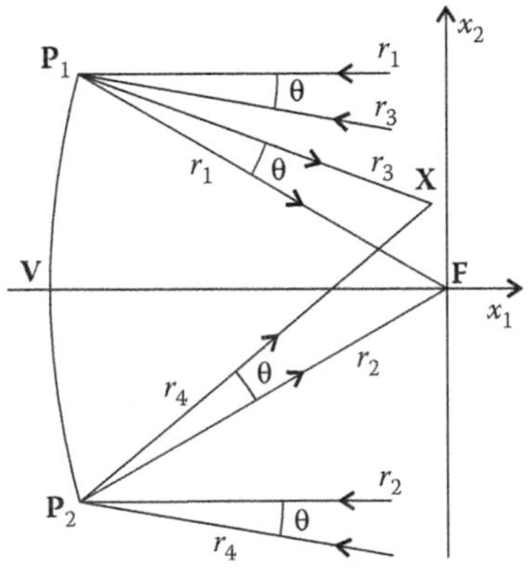

FIGURE 6.7
The rays r_3 and r_4 making an angle θ to the axis of the parabola intersect at a point **X**.

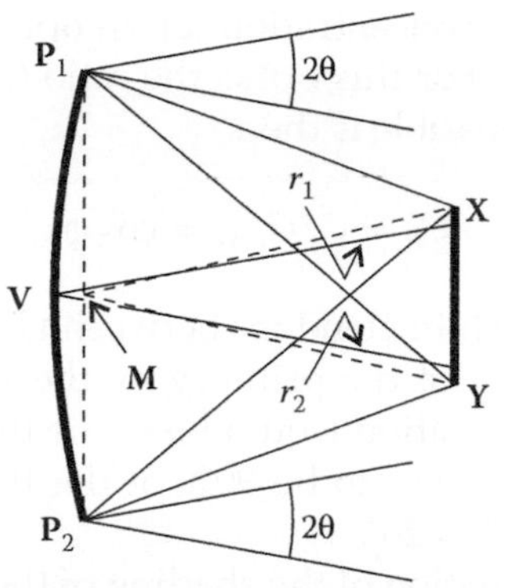

FIGURE 6.8
Lambertian source approximation of a parabolic mirror.

Since U_{LS} is larger than U_P, the concentration we can achieve by approximating the parabola by a Lambertian source is less than the ideal.

To the parabolic primary, we may now add a compound elliptical concentrator (CEC) secondary with an entrance aperture between **X** and its symmetry **Y**, and with a receiver *R*, as shown in Figure 6.9. The CEC is an ideal concentrator for a Lambertian source $\mathbf{P}_1\mathbf{P}_2$.[10–13]

If we want maximum concentration on the receiver, it must be illuminated by radiation with an angular aperture $\pm\pi/2$. The receiver size *R* then fulfills $U_{LS} = 2R \sin(\pi/2)$ or $R = U_{LS}/2$, and the concentration *C* the device achieves is

$$C = [\mathbf{P}_1, \mathbf{P}_2]/R = \cos\varphi/\sin\theta \tag{6.24}$$

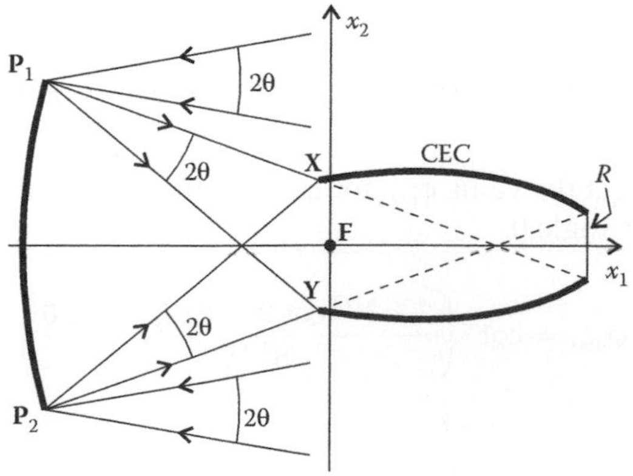

FIGURE 6.9
Parabolic primary and a CEC secondary.

The maximum possible concentration for an optic with a total acceptance angle 2θ is $C_{max} = 1/\sin\theta$. For this optic, the ratio C_R between its concentration and the maximum possible is then

$$C_R = C/C_{max} = \cos\varphi \tag{6.25}$$

which is also the mismatch in étendues between U_P and U_{LS}.

If there was no shading of the primary by the secondary, the smaller φ was, the closer the concentration would be to the theoretical maximum. For example, for the concentration to be 90% of the theoretical maximum, we should have $\varphi = \arccos(0.9) = 26°$.

An optimistic approximation of the shading of the primary by the secondary can be obtained by (note that the CEC is wider than $[\mathbf{X}, \mathbf{Y}]$):

$$S_H = [\mathbf{X},\mathbf{Y}]/[\mathbf{P}_1,\mathbf{P}_2] = \sin 2\theta/\sin 2\varphi \tag{6.26}$$

which increases as φ decreases. Therefore, for smaller values of φ we not only have increased concentration, but also increased shading. The illuminated portion of the primary (not shaded) is given by

$$I_L = 1 - S_H \tag{6.27}$$

The geometrical concentration that the CEC can provide is given by $[\mathbf{P}_1, \mathbf{P}_2]/R$, but the light concentration is affected by the loss mechanism (shading) mentioned previously. It is then approximately given by

$$C = \frac{[\mathbf{P}_1,\mathbf{P}_2]}{R} I_L = \cos\varphi\csc\theta - \cos\theta\csc\varphi \tag{6.28}$$

It will be maximum when its derivative is zero, and that means, for a given value of θ,

$$\frac{dC}{d\varphi} = 0 \Leftrightarrow \cos\theta\cot\varphi\csc\varphi - \csc\theta\sin\varphi = 0 \tag{6.29}$$

Solving for φ we get the value φ_{max} for maximum concentration as a function of the acceptance angle θ:

$$\varphi_{max} = \cot^{-1}\left(\frac{2\times 3^{1/3}\cot\theta - 2^{1/3}\xi^{2/3}\tan\theta}{6^{2/3}\xi^{1/3}}\right) \tag{6.30}$$

where

$$\xi = -9\cot^2\theta\csc^2\theta + \sqrt{12\cot^6\theta + 81\cot^4\theta\csc^4\theta} \tag{6.31}$$

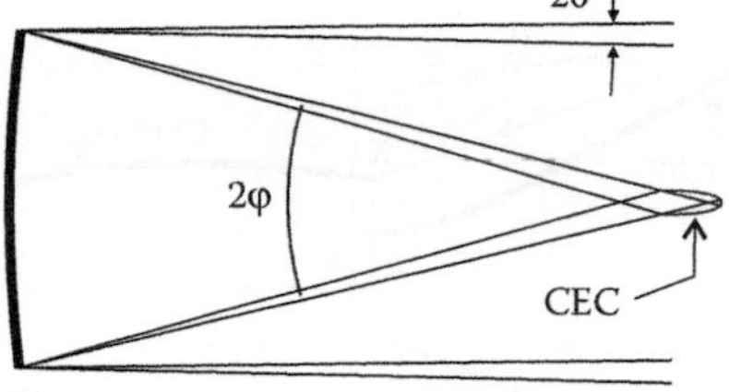

FIGURE 6.10
Parabolic primary and a CEC secondary designed for maximum concentration.

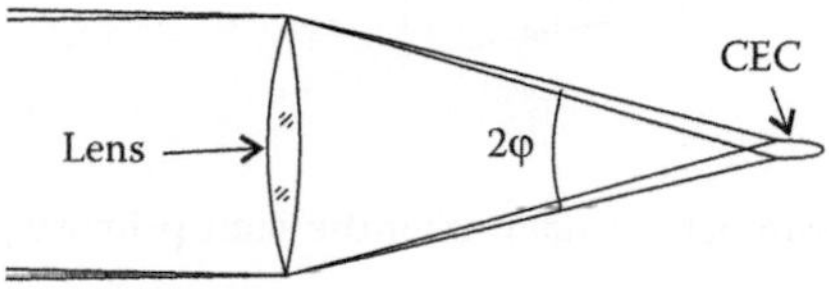

FIGURE 6.11
CEC secondary with a lens as a primary.

For each value of θ, these expressions tell us the value for the rim angle of the primary for maximum concentration.

For example, for an acceptance angle of ±1°, we get a rim angle for the primary of $\varphi_{max} = 14.86°$, a concentration of $C = 51.5$, and a ratio to the maximum concentration of $C/(1/\sin\theta) = 90\%$. Figure 6.10 shows a parabolic primary and a CEC secondary for these parameters.

The parabolic primary may be replaced by a lens, resulting in another primary–secondary optic, as shown in Figure 6.11.[14,15]

6.4 Truncated Trumpet Secondary

Another way to increase the concentration of a parabolic primary is to add a trumpet secondary. Also in this case, we approximate the parabolic primary by a Lambertian source, $\mathbf{P}_1\mathbf{P}_2$, emitting toward **XY**, just as we did for the CEC secondary. The trumpet is composed of two hyperbolic branches with foci **X** and **Y**, as shown in Figure 6.12. The complete trumpet extends all the way from the receiver to the primary, completely shading it. The figure also shows the corresponding CEC as a comparison. Both the complete trumpet and the CEC are ideal for the Lambertian source $\mathbf{P}_1\mathbf{P}_2$, and, therefore, the receiver size R is the same in both the cases.[16–18]

The working principle of the trumpet is as shown in Figure 6.13. Light emitted by the Lambertian source $\mathbf{P}_1\mathbf{P}_2$ toward **XY** bounces back and forth between the reflectors of the trumpet, until it reaches the receiver R. An edge

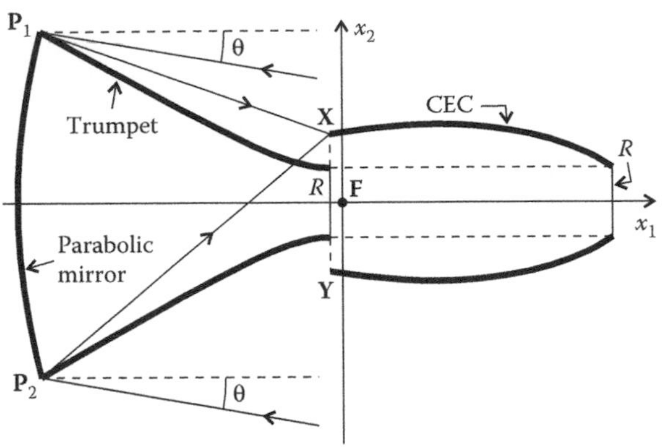

FIGURE 6.12
Comparison of CEC and trumpet secondaries for the same primary parabolic mirror.

ray r headed toward focus **X** of the hyperbola is reflected toward the other focus **Y**. In contrast, an edge ray headed toward the focus **Y** of the hyperbola is reflected toward the other focus **X**.

As the complete trumpet totally shades the primary mirror, it must be truncated to be usable.[19] The truncated trumpet will not capture some of the light reflected by the primary, and will still shade it. For example, if we truncate to the right of point $\mathbf{K}_r$ in Figure 6.13, the ray r will not be captured. Figure 6.14 shows a truncated trumpet. The light losses by the top hyperbola branch for a Lambertian source emitting from $\mathbf{P}_1\mathbf{P}_2$ to **XY** are given by the étendue from $\mathbf{P}_1\mathbf{T}$ to **XK** as

$$U_{LT} = [\mathbf{T},\mathbf{X}] - [\mathbf{T},\mathbf{K}] + [\mathbf{P}_1,\mathbf{K}] - [\mathbf{P}_1,\mathbf{X}] \tag{6.32}$$

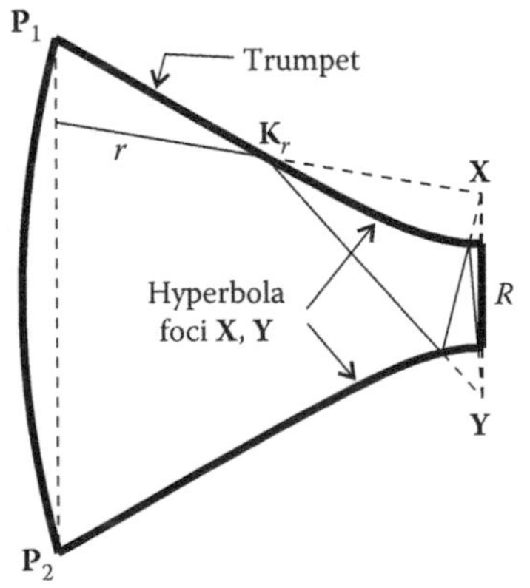

FIGURE 6.13
Working principle of the trumpet secondary.

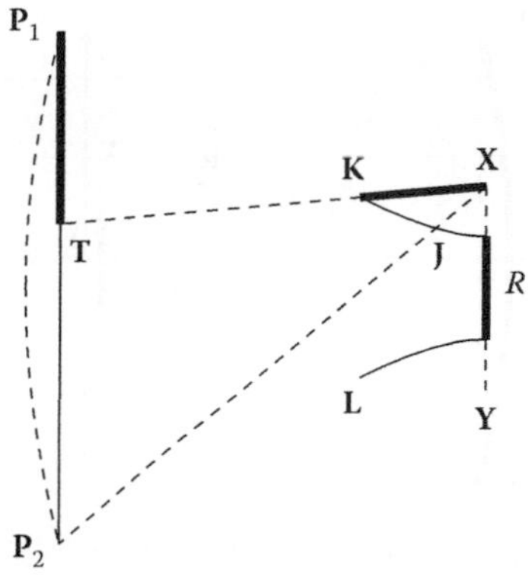

FIGURE 6.14
The trumpet secondary must be truncated to be usable. This, however, leads to some light losses.

where **T** is at the intersection of the straight lines through $\mathbf{P}_1\mathbf{P}_2$ and **XK**, and where the subscript L stands for Lost and T for Top.

If the trumpet was truncated to the right of point **J**, so that point **K** was also to the right of **J**, point **T** would be outside the source $\mathbf{P}_1\mathbf{P}_2$. In that case, the lost étendue for the top hyperbola branch would be the one from $\mathbf{P}_1\mathbf{P}_2$ to **XK** given by the same expression as U_{LT}, only replacing **T** by $\mathbf{P}_2$. The fraction of the étendue lost by the secondary relative to what a Lambertian source $\mathbf{P}_1\mathbf{P}_2$ would emit, is then U_L/U_{LS}, and the fraction captured is

$$c_U = 1 - U_L/U_{LS} \tag{6.33}$$

The other loss of light comes from shading, which is given by

$$S_H = [\mathbf{K},\mathbf{L}]/[\mathbf{P}_1,\mathbf{P}_2] \tag{6.34}$$

Point **L** is symmetrical relative to point **K**. The illuminated portion of the primary is then given by

$$I_L = 1 - S_H \tag{6.35}$$

The geometrical concentration the trumpet can provide is given by $[\mathbf{P}_1, \mathbf{P}_2]/R$, but the light concentration is affected by the two loss mechanisms mentioned previously. It is then approximately given by

$$C = \frac{[\mathbf{P}_1,\mathbf{P}_2]}{R} I_L c_U = \frac{\cos\varphi}{\sin\theta} I_L c_U \tag{6.36}$$

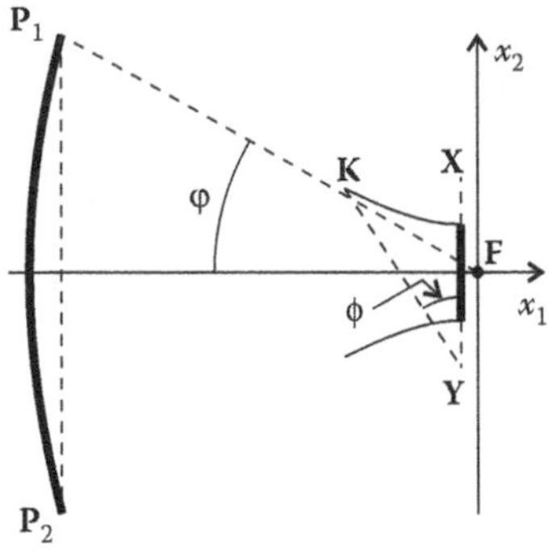

FIGURE 6.15
Parametric definition of the trumpet hyperbola.

If the hyperbola is parameterized by the parameter ϕ, as in Figure 6.15, for a given acceptance angle θ, concentration C can now be obtained as a function of the rim angle φ of the primary and the parameter ϕ of the hyperbola for point **K**.

The hyperbola (and therefore point **K**) can be parameterized as

$$\mathbf{K}(\phi) = \frac{R^2 - f^2}{2R - 2f\cos\phi}(\cos(\phi + \pi/2), \sin(\phi + \pi/2)) + \mathbf{Y} \qquad (6.37)$$

where R is the receiver size given by $R = U_{LS}/2$ and $f = [\mathbf{X}, \mathbf{Y}]$.

The geometry of the system is defined by angles θ, ϕ, and φ. Parameter d appearing in the equation of the parabolic primary (expression (6.10)) is only a scale factor that does not affect the relative sizes.

If we define the acceptance angle θ, and the rim angle φ for the primary, we can plot C as a function of ϕ, which defines the truncation of the primary. This curve has a maximum for a given value of ϕ. By trying different values of φ we can optimize the design. Alternatively, we can give a value to θ, and numerically search for the maximum of $C(\varphi,\phi)$. For $\theta = 1°$, for example, this maximization yields $C = 43.2$ for $\phi = 61.9°$ and $\varphi = 19.4°$. This concentration corresponds to $C \sin\theta = 75.5\%$ of the maximum. For the amount of light striking the receiver, we get $I_L c_U = 80\%$. Ray-tracing with the parabolic primary shows 79% of light striking the receiver. Figure 6.16 shows a parabolic primary and a trumpet concentrator for these parameters, that is, for maximum concentration.

6.5 Trumpet Secondary for a Large Receiver

Another way to designing the trumpet secondary is to do it for a larger receiver R_L, as shown in Figure 6.17. The hyperbola still has foci **X** and **Y** but

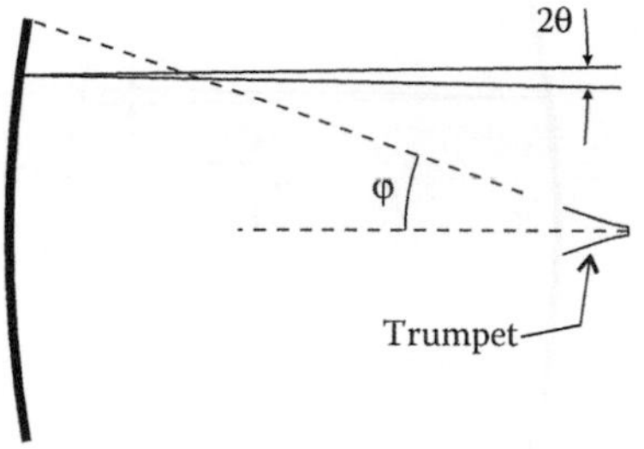

FIGURE 6.16
A parabolic primary and a truncated trumpet secondary designed for maximum concentration.

now intersects line **XY** further away from its center, as needed for a larger receiver.

Now the upper branch of the hyperbola intersects line $\mathbf{XP}_1$ at point **H** instead of at point $\mathbf{P}_1$, as in Figure 6.13. The trumpet, therefore, intersects all the light emitted by the Lambertian source $\mathbf{P}_1\mathbf{P}_2$ and no light is lost. Concentration, however, will be smaller because receiver R_L is now larger.

The working principle is still the same, as discussed earlier. An edge ray r headed toward focus **X** of the hyperbola is reflected toward the other focus **Y**. However, an edge ray headed toward focus **Y** of the hyperbola is reflected toward the other focus **X**. Edge rays keep bouncing back and forth between the hyperbola branches until they reach the receiver.

We may start the design by defining point **H** along the line $\mathbf{P}_1\mathbf{X}$ as $\mathbf{H} = \mathbf{P}_1 + y(\mathbf{X} - \mathbf{P}_1)$ with $0 \le y \le 1$. The shading produced on the primary is

$$S_H = \frac{[\mathbf{H},\mathbf{I}]}{[\mathbf{P}_1,\mathbf{P}_2]} = 1 - y + \frac{y \csc\varphi \sec\varphi \sin(2\theta)}{2d} \tag{6.38}$$

where **I** is symmetrical to **H**. The receiver size is given by

$$R_L = [\mathbf{Y},\mathbf{H}] - [\mathbf{X},\mathbf{H}] \tag{6.39}$$

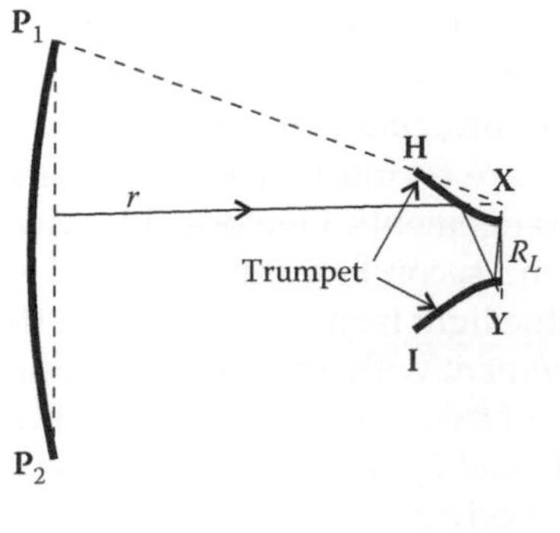

FIGURE 6.17
Trumpet for a larger receiver R_L.

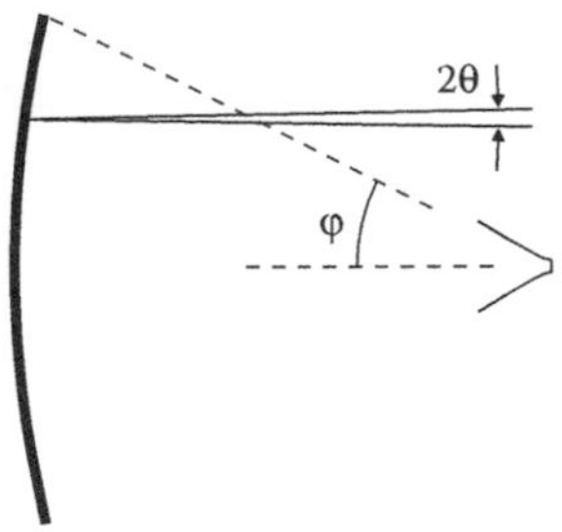

FIGURE 6.18
A parabolic primary and a trumpet secondary for a large absorber, designed for maximum concentration.

And the concentration is approximately given by

$$C = \frac{[\mathbf{P}_1, \mathbf{P}_2]}{R_L} I_L \tag{6.40}$$

where $I_L = 1 - S_H$ is the portion of the primary that is illuminated. Concentration C is a function of parameter y defining the position of **H** along the line $\mathbf{P}_1\mathbf{X}$ and the rim angle φ of the primary. For a half-acceptance angle $\theta = 1°$ and $d = 1$, we have $C = 35.4$ for $y = 0.86$ and $\varphi = 26.7°$. The illuminated portion of the primary is in this case $I_L = 82\%$. The ratio to the maximum possible concentration is $C \sin\theta = 62\%$. Figure 6.18 shows a parabolic primary and a trumpet for a large absorber, designed with these parameters.

6.6 Secondaries with Multiple Entry Apertures

CEC secondaries attain higher concentration when the rim angle φ of the primary is small. This, however, leads to parabolic primaries with a long focal distance, and, therefore, concentrators that are quite long. This is, however, precisely what we are trying to avoid, by going from simple CPCs to primary–secondary arrangements. One possible way around this problem is to divide the primary and secondary into sections, and have each section of the secondary collects the light from one section of the primary.[20] Figure 6.19 shows one such arrangement with an acceptance angle 2θ.

The primary is made of two parabolic sections, $\mathbf{P}_1\mathbf{P}_2$ and $\mathbf{P}_3\mathbf{P}_4$, with a horizontal axis and foci at $\mathbf{F}_1$ and $\mathbf{F}_2$, respectively. The secondary is a combination of two CECs with the receiver $\mathbf{G}_2\mathbf{G}_3$. The top CEC is made of a top ellipse with foci $\mathbf{P}_2$ and $\mathbf{G}_3$, a hyperbola with foci $\mathbf{G}_1$ and $\mathbf{G}_3$ and a bottom ellipse with foci $\mathbf{P}_1$ and $\mathbf{G}_1$. The bottom CEC is symmetrical relative to the top one.

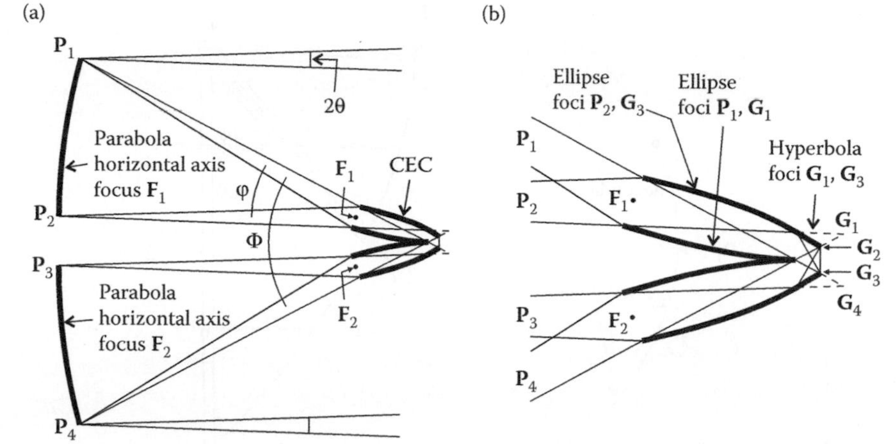

FIGURE 6.19
(a) Primary secondary arrangement with divided primary and secondary. The light from each section of the primary is collected by a corresponding section of the secondary. (b) Detail of the secondary showing the ellipses of the CEC and smaller hyperbolas guiding the light to receiver G_2G_3.

In this concentrator, each one of the CECs collects light from a primary subtending a small angle φ, whereas the whole primary subtends a total angle Φ.

This kind of device has been proposed with combinations of larger number of divisions for the primary and secondary.[21] In this case, the concentrator was designed for a high concentration of solar energy, and, therefore, the acceptance angle 2θ was small (θ = 0.73°). As seen from Figure 6.20, for a small acceptance angle 2θ light rays r_1 and r_2 are almost parallel, and therefore, a CEC secondary was approximated by a CPC.

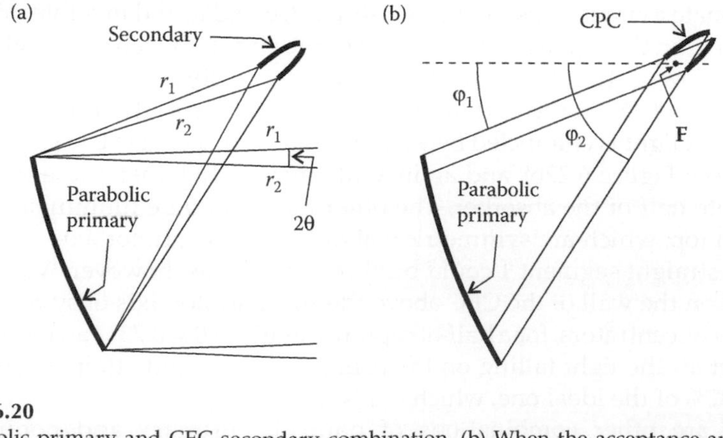

FIGURE 6.20
(a) Parabolic primary and CEC secondary combination. (b) When the acceptance angle 2θ is small, rays r_1 and r_2 are almost parallel, and the secondary can be approximated by a CPC.

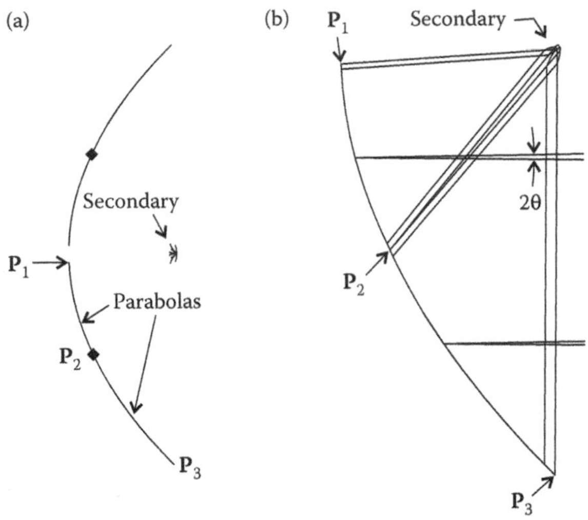

FIGURE 6.21
(a) Combination of two optics, each one with a parabolic primary and a CPC secondary. (b) Bottom half of same optic as in (a), but now showing the edge rays of the CPCs.

The primary is bound by angles φ_1 and φ_2 to the axis of the parabola (which contains its focus **F**), as shown in Figure 6.20b. The concentrator is a combination of four of these shapes, as shown in Figure 6.21a, with the bottom half shown in greater detail in Figure 6.21b.

Angle φ_1 for the parabola $\mathbf{P}_1\mathbf{P}_2$ is 3° to eliminate the central portion of the primary that is shaded by the secondary.[21] The angles ϕ_2 and ϕ_1 are not exactly the same for parabolas $\mathbf{P}_1\mathbf{P}_2$ and $\mathbf{P}_2\mathbf{P}_3$, respectively, to avoid a gap at point $\mathbf{P}_2$ between these two sections of the primary. Figure 6.22 shows the details of the secondary area and also that the exit apertures of the CPCs can be combined to illuminate a circular absorber using straight, circular, and involute mirrors.[21]

The CPC on the top illuminates a part of the circular receiver, its light being channeled by a straight segment 1 (see Figure 6.22b) and a pair of involute mirrors. The CPC at the bottom illuminates another portion of the circular receiver. Its light is channeled by a circular arc with a center **C**, a straight segment 2 (see Figure 6.22b), and an involute mirror. Together, these two CPCs illuminate half of the absorber. The other half would be illuminated by the CPCs on top, which are symmetrical about the concentrator axis of symmetry. The straight segment 1 could be eliminated. This, however, would push point **M** on the wall of the CPC above the concentrator axis of symmetry.[21]

These concentrators, for a half-acceptance angle of $\theta = 0.73°$, can be designed to collect all the light falling on the primary mirror and attain a concentration of 82% of the ideal one, which is $1/\sin\theta$.

There are other combinations of parabolic primary and nonimaging secondaries that attain high concentrations and have a high collection

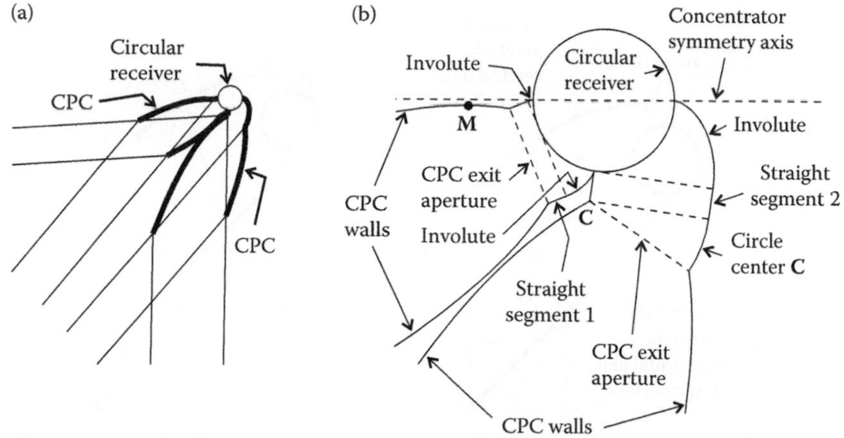

FIGURE 6.22
(a) The two CPC secondaries of Figure 6.21 can be combined to illuminate half of a circular receiver. The other half would be illuminated by two more CPCs symmetrical about the concentrator axis of symmetry. (b) Detail of the secondary close to the receiver.

efficiency.[22] Figure 6.23 shows one such optical system for an acceptance angle 2θ with $\theta = 0.73°$ and a rim angle $\varphi = 90°$.

The right half of the secondary is shown in greater detail in Figure 6.24. The left half (not shown) is symmetrical relative to the vertical line containing the center of the circular receiver *c*.

This secondary optic has a top mirror composed of an involute curve **AB** and a macrofocal parabola **BD** with a vertical axis and having as macrofocus the circular receiver *c*. The bottom mirror is made of a macrofocal parabola **EF** with a horizontal axis and having as macrofocus the circular receiver *c*, and a flat mirror **FI** tangent to **EF** at point **F**. A horizontal light ray r_1 entering the concentrator hits the flat mirror, becoming vertical, and is then reflected by the top macrofocal parabola in a direction tangential to *c*. Another horizontal light ray r_2 entering the concentrator is reflected by the bottom macrofocal parabola, also in a direction tangential to *c*. This optic may be truncated

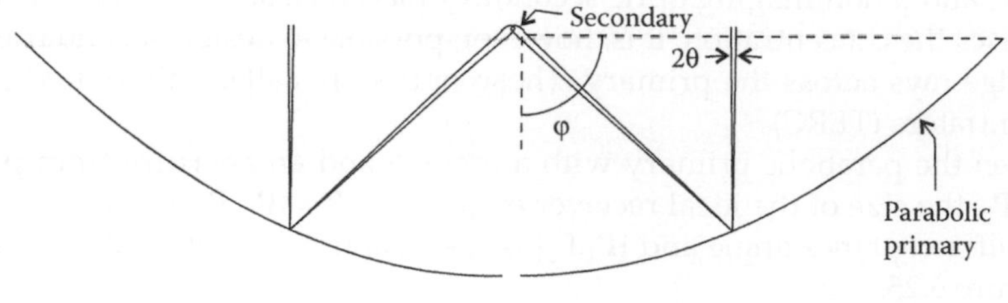

FIGURE 6.23
Parabolic primary and a nonimaging secondary.

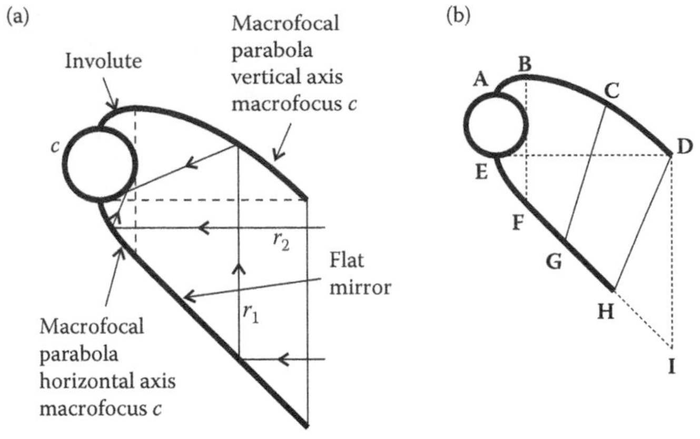

FIGURE 6.24
(a) Secondary for a parabolic primary. (b) Secondary truncated along line **DH**.

at **CG**, with **C** along line **AD**, and **G** along line **EI**, and it still collects all the radiation entering its entrance aperture coming from the primary.

The perimeter of *c* is chosen to match the maximum concentration, and is therefore, given by $d \sin \theta$, where *d* is the size of the entrance aperture defined by the edges of the primary. For this case and the configuration with aperture **DH**, the secondary captures 90% of the radiation reflected by the primary, and therefore, attains 90% of the maximum theoretical concentration.[22]

We may now increase the size of the secondary (optical and circular receiver *c*) so that the perimeter of *c* is chosen to match 82% of the maximum concentration. The secondary now captures 98% of the light reflected by the primary and attains 80% of the maximum theoretical concentration.[22]

6.7 Tailored Edge Ray Concentrators Designed for Maximum Concentration

In the designs presented earlier, the primary was considered as a Lambertian source, and a nonimaging optic secondary placed close to the focus further increases the concentration. It is, however, possible to design secondaries for the edge rays across the primary. These optics are called tailored edge ray concentrators (TERC).[23]

Given the parabolic primary with a focus **F** and an aperture from points $\mathbf{P}_1$ to $\mathbf{P}_2$, the size of the ideal receiver is given by $R = [\mathbf{P}_1, \mathbf{P}_2] \sin \theta$, where θ is the half-acceptance angle and $[\mathbf{P}_1,\mathbf{P}_2]$ is the distance from $\mathbf{P}_1$ to $\mathbf{P}_2$, as shown in Figure 6.25.

We then reflect at point $\mathbf{P}_2$, a ray making an angle θ to the optical axis of the parabola, as shown in Figure 6.25. After reflection, the ray r_1 is headed

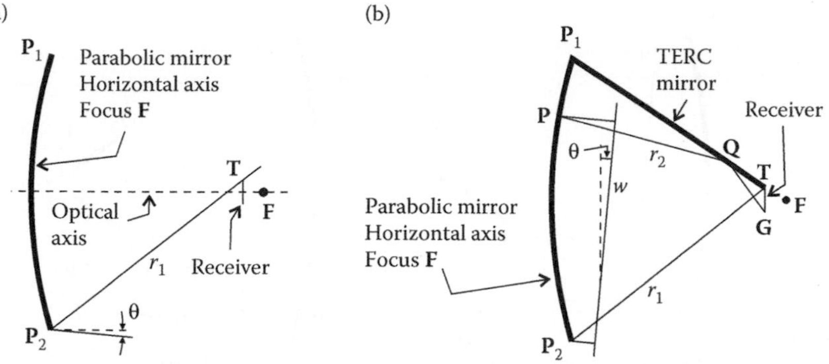

FIGURE 6.25
(a) The path of ray r_1 determines the position of the receiver. (b) Calculation of a TERC mirror as a secondary for a parabolic primary.

toward the optical axis and we search for the point **T** along this ray, but on the other side of the optical axis, at a distance of $R/2$ to the optical axis. Point **T** defines the position of one edge of the receiver. The other edge is point **G**, which is symmetrical to **T**.

All rays perpendicular to the wavefront w, making an angle θ to the vertical, must be focused to point **G**, and the TERC mirror can be defined by the constant optical path length. For a ray r_2, d_{wP} + [**P**, **Q**] + [**Q**, **G**] = S, where d_{wP} is the distance from w to **P** and S is defined by the ray r_1 as $S = d_{wP2}$ + [$\mathbf{P}_2$, **T**] + [**T**, **G**], where d_{wP2} is the distance from w to $\mathbf{P}_2$. This condition defines the position of each point **Q** on the TERC mirror.

The TERC mirror completely shades the primary and must, therefore, be truncated, as shown in Figure 6.26.

Between point $\mathbf{P}_3$ and its symmetrical $\mathbf{P}_4$, all light reflected by the primary is collected by the truncated TERC mirror, but between $\mathbf{P}_3$ and $\mathbf{P}_1$, and

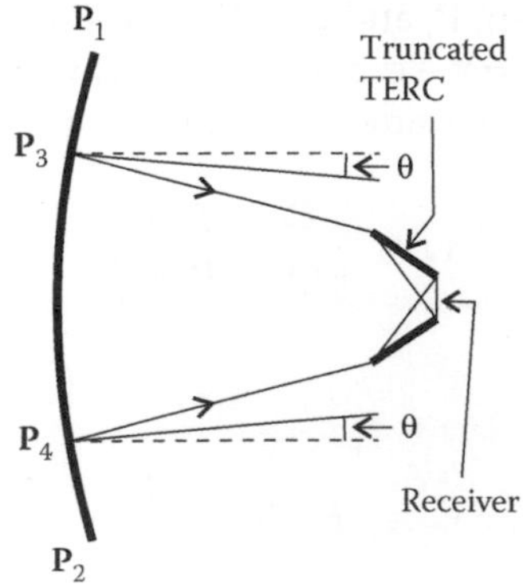

FIGURE 6.26
Truncated TERC.

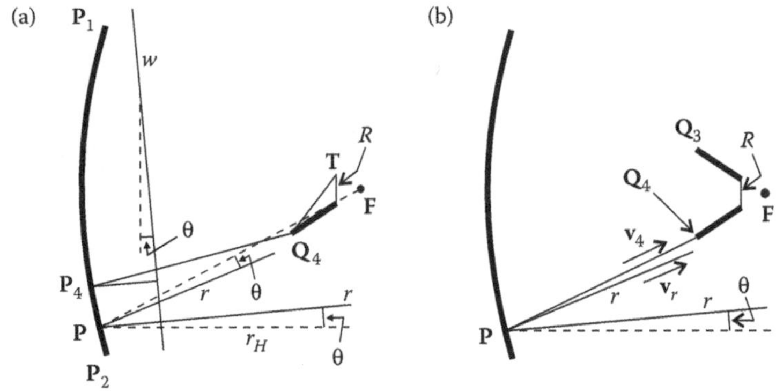

FIGURE 6.27
Étendue losses when the TERC is truncated at a point $\mathbf{Q}_4$. (a) A horizontal incoming ray r_H is reflected at point **P** (between $\mathbf{P}_4$ and $\mathbf{P}_2$) to the focus **F** of the parabola. Another incoming ray r tilted by an angle θ to the horizontal is reflected at **P** missing the TERC secondary and the receiver R. (b) Ray r leaves point **P** in direction $\mathbf{v}_r$. All light leaving point **P** in directions between $\mathbf{v}_4$ and $\mathbf{v}_r$ is lost. Vector $\mathbf{v}_4$ points from **P** to $\mathbf{Q}_4$ where the TERC is truncated.

between $\mathbf{P}_4$ and $\mathbf{P}_2$, there will be some light losses because some rays miss the secondary. Figure 6.27 shows these light losses.

If we place ourselves at point $\mathbf{P}_4$ on the primary mirror, we can calculate the corresponding point $\mathbf{Q}_4$ on the TERC by the constant optical path length from the wavefront w to the edge **T** of the receiver. The edge ray r, tilted θ to the optical axis and reflected at another point **P** between $\mathbf{P}_4$ and $\mathbf{P}_2$, will miss the TERC. All the rays between directions $\mathbf{v}_r$ and $\mathbf{v}_4$ will also miss the TERC, where v_r is the direction of ray r after reflection and $\mathbf{v}_4$ is the direction from **P** to $\mathbf{Q}_4$. The étendue of the light that misses that mirror of the TERC can then be calculated by integration along the parabola from point $\mathbf{P}_4$ to the edge $\mathbf{P}_2$. If focus $\mathbf{F} = (0,0)$, then point **P** on the parabola is given by the expression (6.10), where d can be seen as a scale factor. The parameter range is $\phi_1 < \phi < \phi_2$, where ϕ_1 is the parameter for point $\mathbf{P}_1$ and ϕ_2 is the parameter for point $\mathbf{P}_2$, as shown in Figure 6.6. We have $\phi_1 = \pi - \varphi$ and $\phi_2 = \pi + \varphi$, where φ is the rim angle of the primary. Point $\mathbf{P}_4$ has a parameter ϕ_4. Directions $\mathbf{v}_4$ and $\mathbf{v}_r$ are obtained as

$$\mathbf{v}_4 = \frac{\mathbf{Q}_4 - \mathbf{P}(\phi)}{\| \mathbf{Q}_4 - \mathbf{P}(\phi) \|} \tag{6.41}$$

and

$$\mathbf{v}_r = -R(-\theta) \cdot \frac{\mathbf{P}}{\| \mathbf{P} \|} \tag{6.42}$$

where $R(-\theta)$ is a rotation matrix of an angle $-\theta$. The étendue that is lost by that mirror can then be obtained as

$$U_L(\phi_4) = \int_{\phi 4}^{\phi 2} (\mathbf{v}_r - \mathbf{v}_4) \cdot \frac{d\mathbf{P}}{d\phi}\, d\phi \tag{6.43}$$

and the total étendue lost by both mirrors is $2U_L$.

The étendue balance then starts with that of the light intersected by the concentrator, which is $U_P = 2[\mathbf{P}_1, \mathbf{P}_2] \sin\theta$. The étendue of the light lost to shading is $U_{SH} = 2[\mathbf{Q}_3, \mathbf{Q}_4] \sin\theta$. The remaining light continues toward the primary, is reflected by it, and an additional étendue given by $2U_L$ is lost, because of light that misses the secondary mirrors. The étendue that reaches the receiver is then

$$U_R = 2[\mathbf{P}_1, \mathbf{P}_2]\sin\theta - 2[\mathbf{Q}_3, \mathbf{Q}_4]\sin\theta - 2U_L \tag{6.44}$$

If we define shading as

$$S_H(\phi_4) = \frac{U_{SH}}{U_P} = \frac{[\mathbf{Q}_3, \mathbf{Q}_4]}{[\mathbf{P}_1, \mathbf{P}_2]} \tag{6.45}$$

the portion of the light that gets to receiver R is then given by

$$L_R(\phi_4) = \frac{U_R}{U_P} = 1 - S_H - \frac{U_L}{[\mathbf{P}_1, \mathbf{P}_2]\sin\theta} \tag{6.46}$$

If all the light got to the receiver, concentration would be the maximum possible. Since it is reduced to L_R, concentration is also reduced by the same amount. Giving values to ϕ_4 we can plot $L_R(\phi_4)$ against ϕ_4 and maximize the light getting to the receiver and, therefore, the light concentration. The value of ϕ_{4M} that maximizes $L_R(\phi_4)$ will also give us the point $\mathbf{Q}_4(\phi_{4M})$, where we should truncate the TERC.

For a half-acceptance angle $\theta = 0.01$ radians, the caustic of the edge rays will be above the TERC mirror for rim angles smaller than $\varphi = 36.6°$. For this case, we obtain $\phi_{4M} = 208.5°$, or 28.5° to the optical axis. The optic captures 87.5% of the light, or $L_R = 0.875$. Figure 6.28 shows a TERC designed for these parameters.

The optimum performance for a 3-D optic with a rotational symmetry is also a tradeoff between light captured by the secondary and the shading it produces. Shading, however, is S_H^2 for circular symmetry, which is smaller than that in the 2-D case. In 3-D, it is then possible to increase the size of the TERC slightly, improving the collection efficiency. Increasing ϕ_{4M} by 5° to $\phi_{4M} = 213.5°$, we obtain a concentration with a circular symmetry of 92% of the theoretical maximum, obtained by ray-tracing.[24]

TERC secondaries may also be designed for lens primaries, as shown in Figure 6.29. As described earlier, for the case of a parabolic mirror primary,

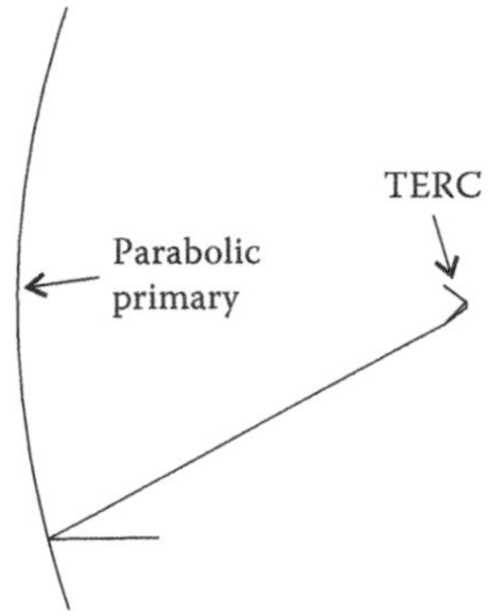

FIGURE 6.28
TERC for maximum light collection.

all rays perpendicular to the wavefront w, making an angle θ to the vertical, must be focused to point **G**, and the TERC mirror can be defined by the constancy of the optical path length. For a ray r_2, $d_{wPL} + n[\mathbf{P}_L, \mathbf{P}_R] + [\mathbf{P}_R, \mathbf{Q}] + [\mathbf{Q}, \mathbf{G}] = S$, where d_{wPL} is the distance from w to $\mathbf{P}_L$ on the left-hand side of the lens, $\mathbf{P}_R$ on the right-hand side of the lens, and S is defined by ray r_1 as $S = d_{wP1} + [\mathbf{P}_1, \mathbf{T}] + [\mathbf{T}, \mathbf{G}]$ where d_{wP1} is the distance from w to $\mathbf{P}_1$. This condition defines the position of each point **Q** on the TERC mirror.

TERC mirrors are flow-line mirrors. If, for example, in Figure 6.25, the optic is made of a material with refractive index n and we replace the parabolic mirror $\mathbf{P}_1\mathbf{P}_2$ by a refractive surface, we obtain a dielectric total internal reflection concentrator (DITRC).

Changing the shape of the primary from a simple parabola in the case of a reflective primary to a more elaborate design, we can improve the optical behavior of the TERC secondary.[24] Figure 6.30 shows one such possibility. The primary is now composed of a central flat portion $\mathbf{V}_1\mathbf{V}_2$ and a parabolic section $\mathbf{V}_2\mathbf{P}_2$ with a focus **F** at the edge of the receiver R and an axis parallel to edge rays r, which make an angle θ to the horizontal.

The portion $\mathbf{Q}_1\mathbf{Q}_2$ of the secondary closer to the receiver receives parallel edge rays in a direction r (parallel to r_1), and it concentrates them to an edge **F** of the receiver R, as shown in Figure 6.31. This curve is, therefore,

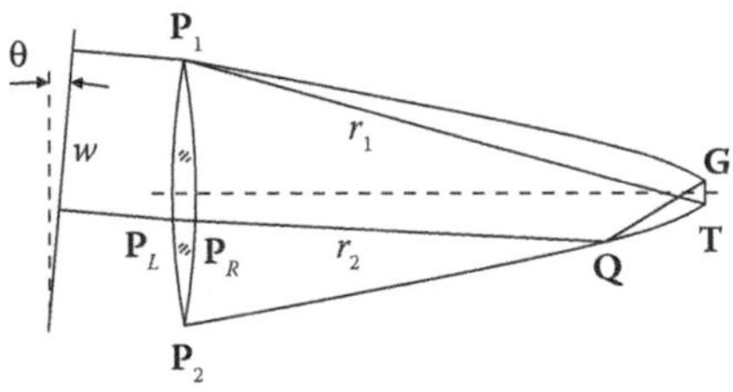

FIGURE 6.29
TERC with a lens as the primary.

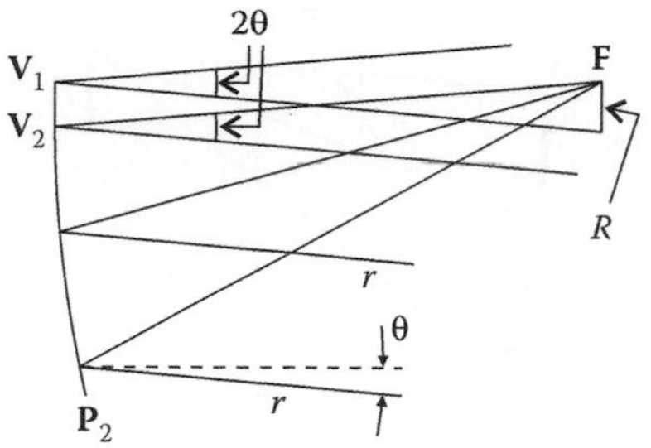

FIGURE 6.30
A compound parabolic primary with a central flat zone.

a parabola with a focus **F** and an axis parallel to *r*. Together with its symmetrical on the other side of the receiver, this portion of the secondary is a highly truncated CPC.

The remainder of the secondary is a TERC mirror calculated by the constant optical path length. For a ray r_2, d_{wP} + [**P**, **Q**] + [**Q**, **F**] = *S*, where d_{wP} is the distance from a flat wavefront *w* to **P**, and *S* is the optical path length defined by ray r_1 as $S = d_{wV2}$ + [$\mathbf{V}_2$, $\mathbf{Q}_2$] + [$\mathbf{Q}_2$, **F**], where d_{wV2} is the distance from *w* to $\mathbf{V}_2$. This condition defines the position of each point **Q** on the TERC mirror.

The design of the whole optic can be simplified if we eliminate the central flat portion of the primary. Figure 6.32 shows the altered configuration, a compound parabolic primary.

This primary is now composed of two symmetrical parabolic arcs. The side of the primary shown in Figure 6.32 is a parabola with a focus **F** and an axis parallel to the edge rays *r*. It then concentrates to the edge **F** of the receiver one bundle of edge rays. The other side of the primary is symmetrical relative to the perpendicular bisector of the receiver *R*. The design of the primary may start by defining the position and the size of the receiver *R*. The acceptance angle 2θ enables us to calculate the point **V** that will be the vertex of the

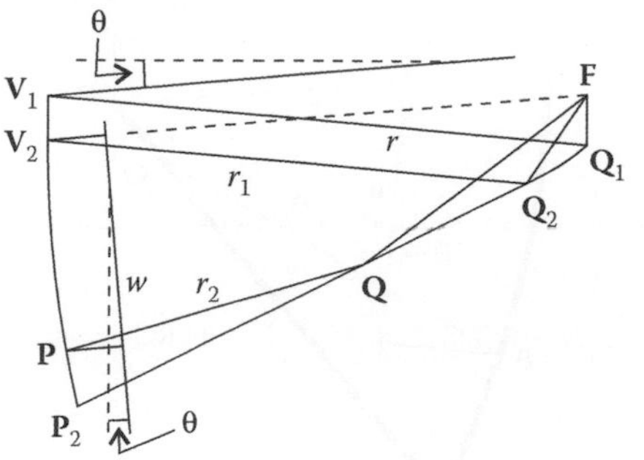

FIGURE 6.31
Central CPC portion of the secondary, and construction method by the constant optical path length of the TERC portion of the secondary.

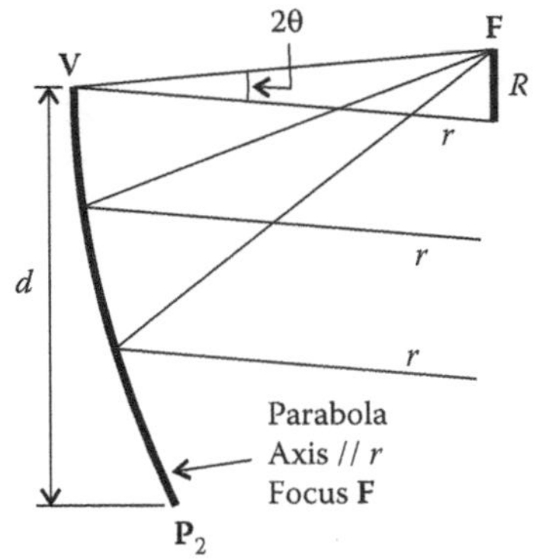

FIGURE 6.32
Definition of a compound parabolic primary reflector for a TERC secondary.

primary. It also defines the direction of rays r that make an angle θ to the horizontal. We now define the bottom parabola as having a focus **F**, an axis parallel to r and passing through **V**. It extends from **V** to a point $\mathbf{P}_2$ at a distance d from the perpendicular bisector of the receiver R, such that $d \sin \theta = R/2$.

The secondary TERC mirror, as described earlier, is now defined so as to reflect the other bundle of edge rays to **F** (Figure 6.33).

All rays perpendicular to the wavefront w, making an angle θ to the vertical, must be focused to point **F**, and the TERC mirror can be defined by the constant optical path length. For a ray r_2, $d_{wP} + [\mathbf{P}, \mathbf{Q}] + [\mathbf{Q}, \mathbf{F}] = S$, where d_{wP} is the distance from w to **P** and S the optical path length defined by ray r_1 as $S = d_{wV} + [\mathbf{V}, \mathbf{T}] + [\mathbf{T}, \mathbf{F}]$, where d_{wV} is the distance from w to **V**. This condition defines the position of each point **Q** on the TERC mirror. This TERC also completely shades the primary and must be truncated to make it usable.

The design without the central flat portion of the primary is not only simpler, but is also more compact, with a larger rim angle for the primary.

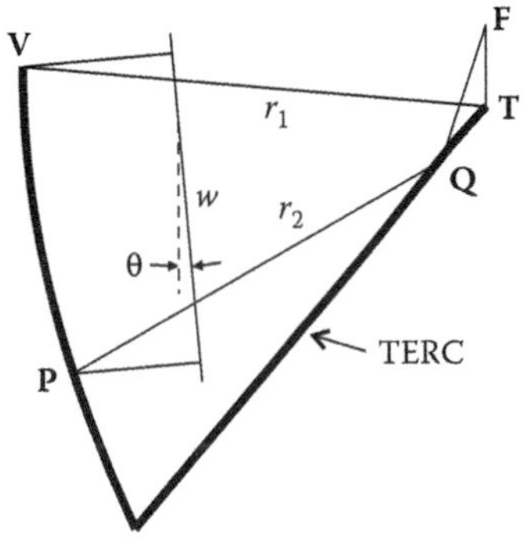

FIGURE 6.33
Calculation of the TERC secondary by the constant optical path length.

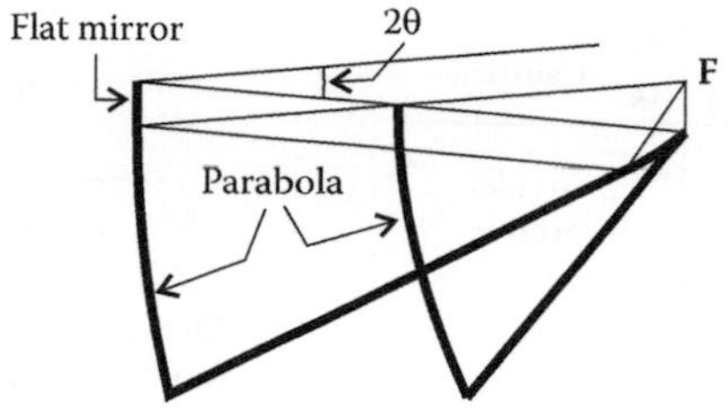

FIGURE 6.34
Comparison of the resulting optic with the central flat portion of the primary and the optic without the central flat portion.

Figure 6.34 shows two of these optics superimposed, for comparison, both with the same acceptance angle and receiver size.

The optic without the central flat portion is, therefore, preferable. The maximum concentration this optic can achieve is similar to that obtained with a parabolic primary. The advantage is that now the primary rim angle is larger and, therefore, the optic is more compact.

For this more compact case in which the primary has no central flat section, if the half-acceptance angle is θ and the vertex **V** of the primary is at **V** = (0,0), we have, for the edge **F** of the receiver,

$$\mathbf{F} = \left(\frac{R}{2\tan\theta}, \frac{R}{2} \right) \tag{6.47}$$

and the points **P** on the parabola from **V** to $\mathbf{P}_2$, as shown in Figure 6.32, are parametrized as

$$\mathbf{P}(\phi) = \left(\frac{\cos^2((\phi - 2\theta)/2)\cot\theta}{1-\cos\phi}, \frac{\sin(\phi - 3\theta) + 3\sin(\phi-\theta) + 2\sin\theta}{8\sin^2(\phi/2)\sin\theta} \right) \tag{6.48}$$

The parameter range is $\pi + 2\theta \le \phi \le \phi_2$, where ϕ_2 is the parameter value for point $\mathbf{P}_2 = (P_{21}, P_{22})$, calculated such that $|P_{22}\sin\theta| = R/2$.

We can now maximize light collection using for this case the same procedure as earlier for the case of a parabolic primary. The étendue that is lost by each mirror as a result of the truncation of the TERC is again given by the expression (6.43), but now $\mathbf{P}_4$ is a point between **V** and $\mathbf{P}_2$. Shading can also be calculated by the expression (6.45) and light collection by the expression (6.46). For a half-acceptance angle $\theta = 0.01$ rad, the rim angle of the primary (angle that the line from the midpoint of receiver R to $\mathbf{P}_2$ makes with the optical axis) is 53.1°. For this case, we obtain $\phi_{4M} = 227.6°$, or 47.6° to the optical axis. The optic captures 85% of the light, or $L_R = 0.85$.

Just as earlier, if we consider the 3-D case in which the optic has rotational symmetry, shading is not so high and the TERC can be extended, increasing

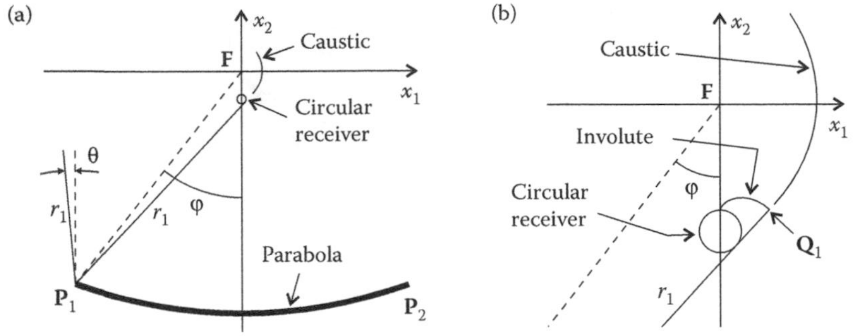

FIGURE 6.35
(a) The circular receiver is centered on the optical axis and is tangential to the edge ray r_1 reflected off the edge point $\mathbf{P}_1$ of the primary. (b) The first part of the secondary is an involute section that must not intersect the caustic of the edge rays.

the light collection. The maximum concentration this optic can provide in 3-D with rotation symmetry is 93% of the theoretical maximum.[24]

TERC secondaries described earlier for linear receivers can also be designed for circular receivers, as shown in Figure 6.35. Just as in the case of a linear receiver, we start with the case of a parabolic primary.

For an acceptance angle of $\pm\theta$, if the TERC is to achieve maximum concentration, the perimeter of the circular receiver must be given by $2\pi R = [\mathbf{P}_1, \mathbf{P}_2] \sin\theta$, where R is now the radius of the receiver. Given the symmetry of the optic, its center must be on the vertical axis x_2. Its position along this axis is such that it is tangent to the ray edge r_1 reflected off the edge of the mirror, as shown in Figure 6.35.

The first section of the secondary is an involute. Its end point $\mathbf{Q}_1$ is defined by the rim angle φ of the parabola. The angle φ must be adjusted such that the caustic formed by the edge rays lies above the mirror, as shown in Figure 6.35. It is now possible to calculate the TERC mirror by the constant optical path length, as shown in Figure 6.36.

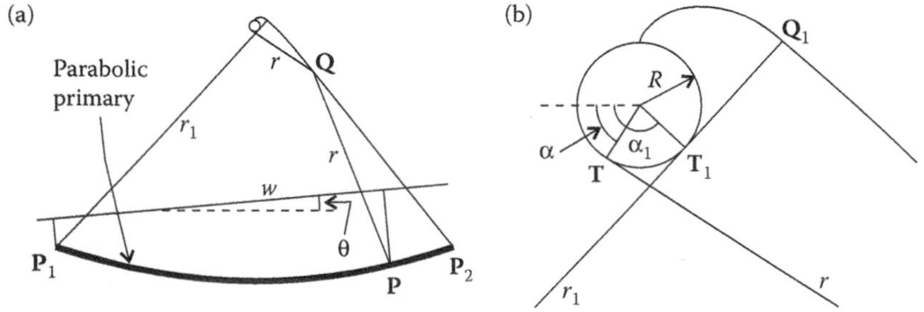

FIGURE 6.36
(a) Calculation of the TERC portion of the secondary. (b) Detail of the secondary close to the circular receiver.

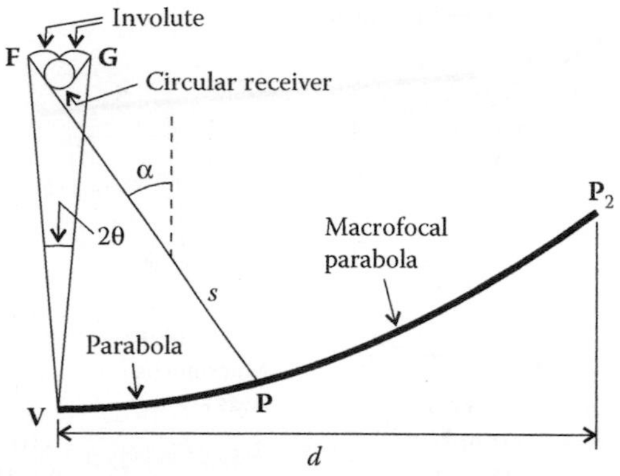

FIGURE 6.37
Compound primary for a circular receiver.

We can then write d_{wP} + [**P**, **Q**] + [**Q**, **T**] + $R\alpha = S$ for ray r, where d_{wP} is the distance from the wavefront w to point **P** and the optical path length S is defined by ray r_1 reflected at point $\mathbf{P}_1$ as $S = d_{wP1}$ + [$\mathbf{P}_1$, $\mathbf{Q}_1$] + [$\mathbf{Q}_1$, $\mathbf{T}_1$] + $R\alpha_1$, where d_{wP1} is the distance from the wavefront w to point $\mathbf{P}_1$. As described earlier, the TERC completely shades the primary and must be truncated to be usable.

Just as in the case of a TERC for a linear receiver, the shape of the primary can be modified to improve the compactness of the optic in the case of a circular receiver also. Figure 6.37 shows a primary made of two parts, a central parabolic arc and an outer macrofocal parabolic arc. The total acceptance angle for the concentrator is 2θ.[25]

The first portion of the secondary is an involute with the end point defined by the line s tilted by an angle α to the vertical and tangent to the circular receiver. This line also defines the point **P** where the two curves of the primary meet. The vertex **V** of the primary mirror is defined by the edge points **F** and **G** of the involute and the acceptance angle θ. Both the primary mirror curves have their axes in the direction of the parallel rays, r, tilted by an angle θ relative to the vertical, as shown in Figure 6.38. The rays, r, are perpendicular to the wavefront w_R. The focus of the parabolic arc is the point **F**, where the left involute ends. The macrofocus of the macrofocal parabola is the circular receiver c.

Once the geometry of the primary and the involutes of the secondary are defined, the first portion of the TERC mirror can be calculated. Figure 6.39 shows an edge ray r_3 perpendicular to the wavefront w_L and incident on a point $\mathbf{P}_3$ of the parabolic portion of the primary. The corresponding point $\mathbf{M}_3$ on the secondary, as shown in Figure 6.40, can be calculated by the constant optical path length.

We have d_{wLP3} + [$\mathbf{P}_3$, $\mathbf{M}_3$] + [$\mathbf{M}_3$, $\mathbf{T}_3$] + $R\alpha_3 = S$, where R is the radius of the circular receiver c, d_{wLP3} is the distance between the wavefront w_L and point $\mathbf{P}_3$, and S the optical path length given by $S = d_{wLV}$ + [**V**, $\mathbf{M}_V$] + [$\mathbf{M}_V$, $\mathbf{T}_V$] + $R\alpha_V$

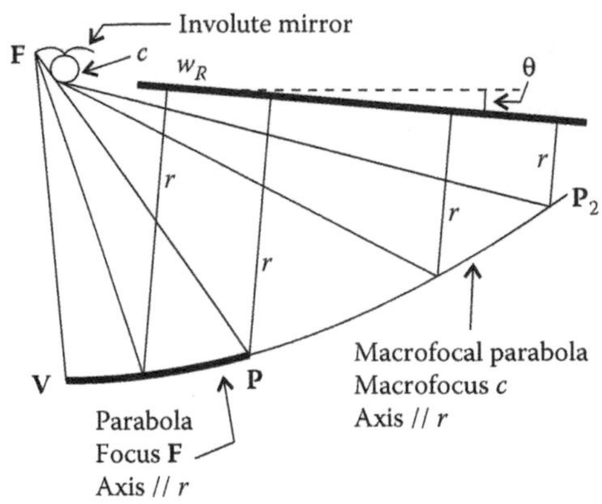

FIGURE 6.38
The central parabolic arc has a focus **F** and the macrofocal parabolic arc has a macrofocus *c*.

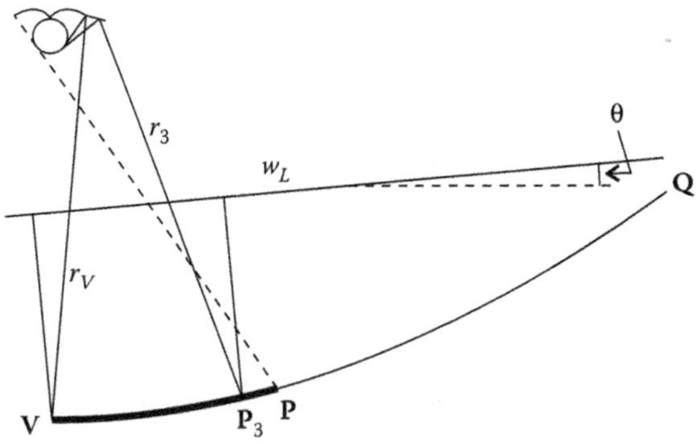

FIGURE 6.39
TERC section for the parabolic portion of the primary calculated by the constant optical path length.

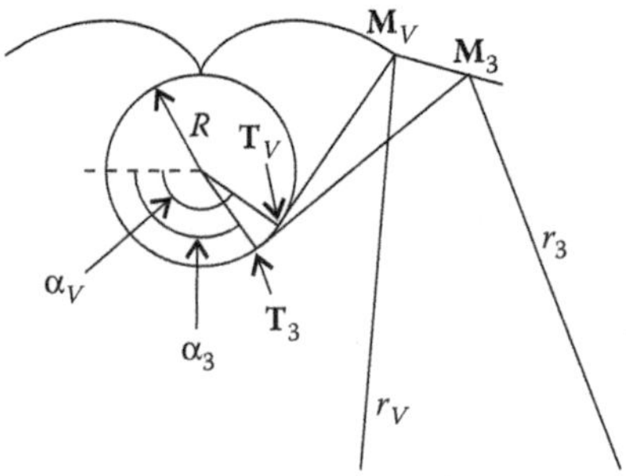

FIGURE 6.40
Details of the receiver zone for the calculation of the first part of the TERC mirror.

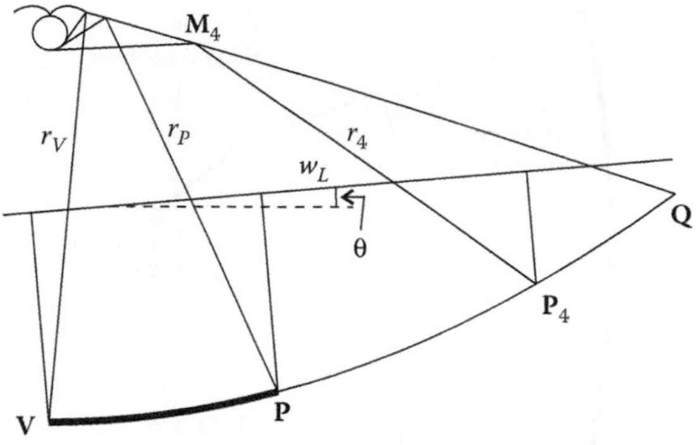

FIGURE 6.41
TERC section for the macrofocal parabolic portion of the primary calculated by the constant optical path length.

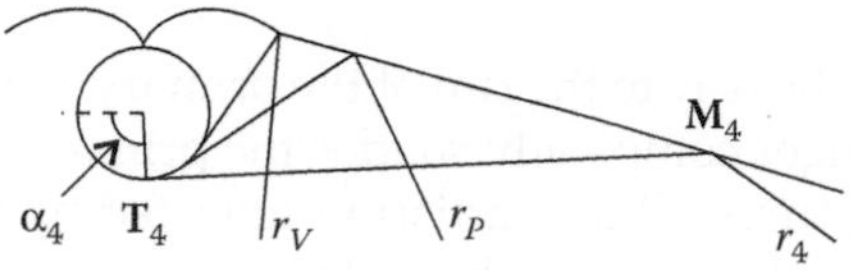

FIGURE 6.42
Details of the receiver zone for the calculation of the second part of the TERC mirror.

defined for ray r_V, where d_{wLV} is the distance between the wavefront w_L and point **V**.

The outermost section of the TERC secondary is calculated in the same way, but using the macrofocal parabolic portion of the primary, as shown in Figure 6.41. We have an edge ray r_4 perpendicular to the wavefront w_L and incident on a point $\mathbf{P}_4$ of the macrofocal parabolic portion of the primary. The corresponding point $\mathbf{M}_4$ on the secondary, shown in more detail in Figure 6.42, can be calculated by the constant optical path length.

We have $d_{wLP4} + [\mathbf{P}_4, \mathbf{M}_4] + [\mathbf{M}_4, \mathbf{T}_4] + R\alpha_4 = S$ for ray r_4, where d_{wLP4} is the distance between the wavefront w_L and point $\mathbf{P}_4$.

As discussed earlier, the TERC completely shades the primary and must be truncated to be usable.

6.8 Tailored Edge Ray Concentrators Designed for Lower Concentration

In all the examples presented earlier, it was always necessary to truncate the TERC secondary to prevent complete shading of the primary. An alternative approach is to increase the size of the receiver. In this case, the TERC mirror

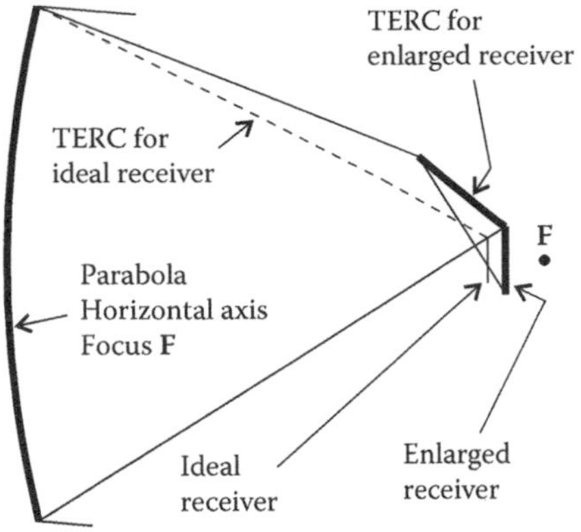

FIGURE 6.43
TERC secondary for a receiver larger than the ideal. The TERC mirror does not extend all the way to the edge of the primary.

no longer extends all the way to the rim of the primary, as shown in Figure 6.43, and, therefore, no longer completely shades the primary.

The secondary captures all the radiation reflected by the primary, but the concentration of the optic is lower than ideal.

Enlarging the size of the receiver to prevent complete shading of the primary can also be done for parabolic primaries with circular receivers.[5] Figure 6.44 shows the left half of a parabolic primary with a tubular secondary. The center of the circular receiver is on the axis of symmetry s and it is tangent to ray r_1 reflected at the edge $\mathbf{P}_1$ of the parabolic primary. Before reflection, ray r_1 makes an angle θ to the left of the vertical.

The size of the circular receiver is larger than the minimum. This means its radius is given by

$$R = \delta[\mathbf{P}_1, \mathbf{P}_6] \sin\theta / 2\pi \tag{6.49}$$

where $\delta > 1$ and points $\mathbf{P}_1$ and $\mathbf{P}_6$ are the edges of the primary, as shown in Figure 6.45. The central portion of the secondary mirror is an involute to the receiver. Although the receiver is tangent to edge ray r_1, this will not be the

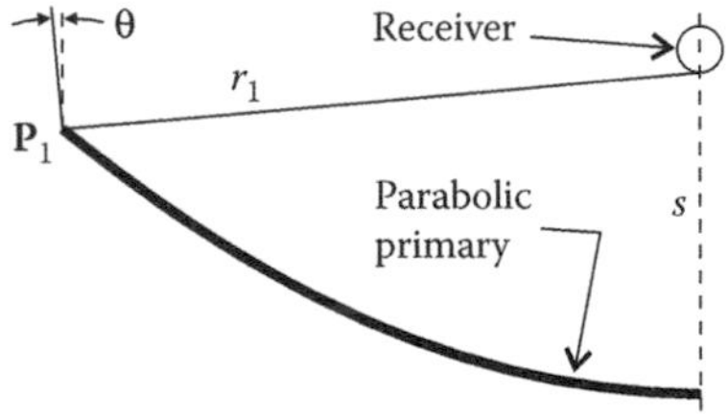

FIGURE 6.44
Position of the circular receiver relative to the parabolic primary.

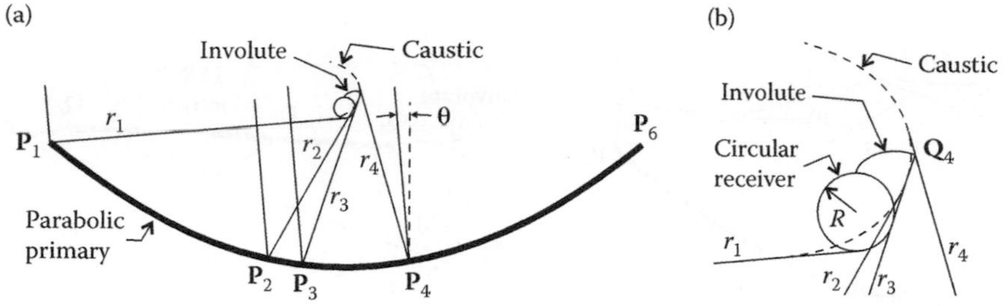

FIGURE 6.45
(a) The central portion of the secondary is an involute to the receiver. (b) Detail close to the receiver.

case for ray r_2 reflected at points (such as $\mathbf{P}_2$) to the right of $\mathbf{P}_1$. Because we are not trying to get maximum concentration, this is not a problem. Only ray r_3 reflected at point $\mathbf{P}_3$ is again tangent to the receiver. The involute starts at the uppermost point of the receiver and ends at point $\mathbf{Q}_4$ where it intersects ray r_3.

In this example, the caustic intersects the involute. There are, therefore, edge rays further to the right of r_3, such as ray r_{34} in Figure 6.46, coming from a point on the primary between $\mathbf{P}_3$ and $\mathbf{P}_4$, that hit the involute and are reflected by it toward the receiver in an uncontrolled way.

Only ray r_4, reflected from point $\mathbf{P}_4$ by the primary, reaches the edge of the involute. Therefore, from $\mathbf{P}_4$ onward, it is possible to design a TERC secondary that redirects the edge rays tangentially to the receiver, as shown in Figure 6.47.

The points of the secondary can be calculated by the constant optical path length and we have $d_{wP5} + [\mathbf{P}_5, \mathbf{Q}_5] + [\mathbf{Q}_5, \mathbf{T}_5] + R\alpha_5 = S$ for ray r_5, where R is the radius of the circular receiver, d_{wP5} is the distance between the wavefront w and point $\mathbf{P}_5$, and S the optical path length, given by $S = d_{wP4} + [\mathbf{P}_4, \mathbf{Q}_4] + [\mathbf{Q}_4, \mathbf{T}_4] + R\alpha_4$ defined for ray r_4, where d_{wP4} is the distance between the wavefront w and point $\mathbf{P}_4$.

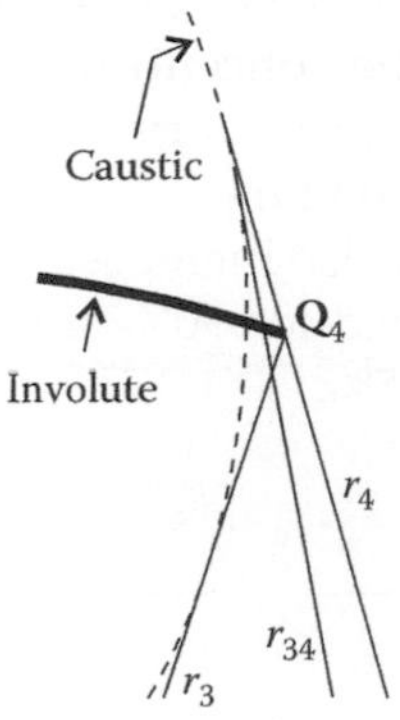

FIGURE 6.46
Details of the rays at the end of the involute.

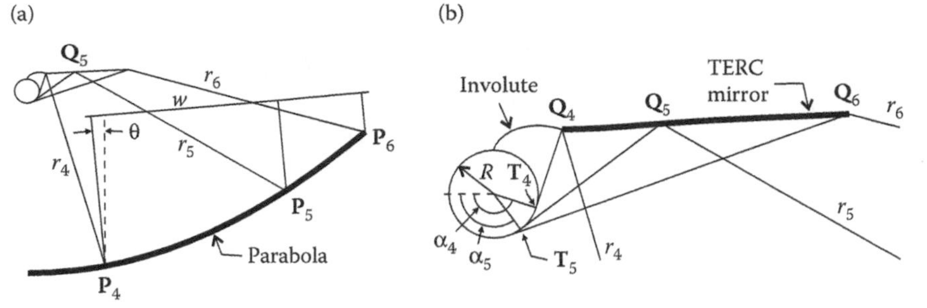

FIGURE 6.47
(a) Design of the TERC mirror to the right of the central involute section. (b) Detail close to the receiver.

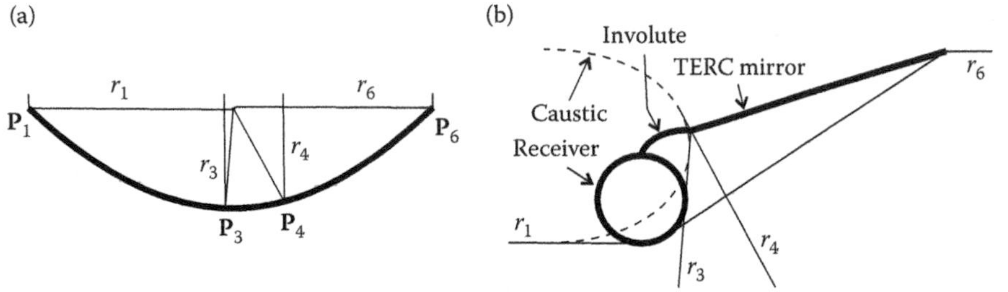

FIGURE 6.48
(a) Secondary for a small acceptance angle and a primary with a large rim angle. (b) Detail close to the receiver.

Different design parameters may lead to different ray assignments between primary and secondary.[5]

This method enables us to design simple secondary optics that attain high concentrations at large primary rim angles. For example, for a rim angle of 90°, an acceptance angle (θ) of 0.007 rad (0.4°) and a concentration of 70% of the ideal maximum, we get the concentrator as shown in Figure 6.48. The shading of the primary by the secondary is about 2%. All the light reflected by the primary reaches the secondary.

Increasing the concentration also increases the shading. If, for example, the same concentrator was designed for 90% of the ideal maximum concentration, the shading would be 15.5%.

6.9 Fresnel Primaries

Parabolic mirrors can get quite big and hard to handle, especially for large rim angles, those that yield the most compact concentrators. One way around

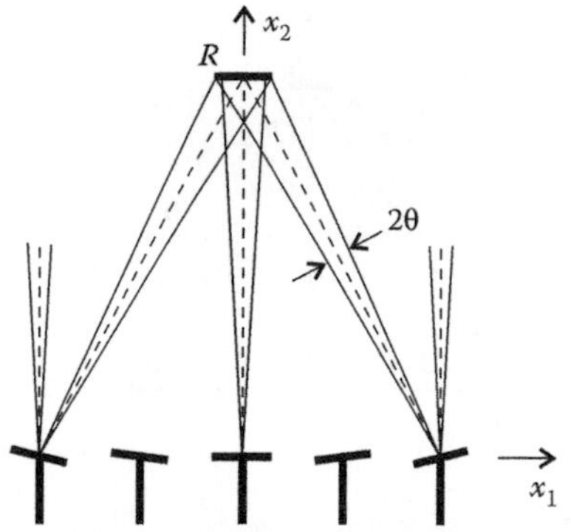

FIGURE 6.49
Field of heliostats reflecting light to a receiver R.

this problem is to Fresnelize the mirror, that is, replace it by a set of small mirrors (heliostats) on a straight line, flats that mimic the optical behavior of the parabolic mirror they replace. Figure 6.49 shows a field of heliostats replacing a parabolic mirror. They reflect the incoming light to the receiver R.

To simplify the analysis of this optic, we consider the limit case in which we have an infinite number of infinitely small heliostats. The heliostat field then becomes a continuous Fresnel primary, as represented in Figure 6.50.

If θ is the half-angular aperture of the radiation, φ the rim angle of the primary, and h the distance (height) from the Fresnel primary to the receiver R, the minimum receiver size to capture all the light is given by

$$R = 2h\sin\theta/\cos^2\varphi \tag{6.50}$$

where $D = h/\cos\varphi$ is the distance from the center of the absorber to the rim of the primary. The maximum étendue this receiver can accept is given by $2R\sin(\pi/2) = 2R$, when it is illuminated by full Lambertian light ($\pm\pi/2$

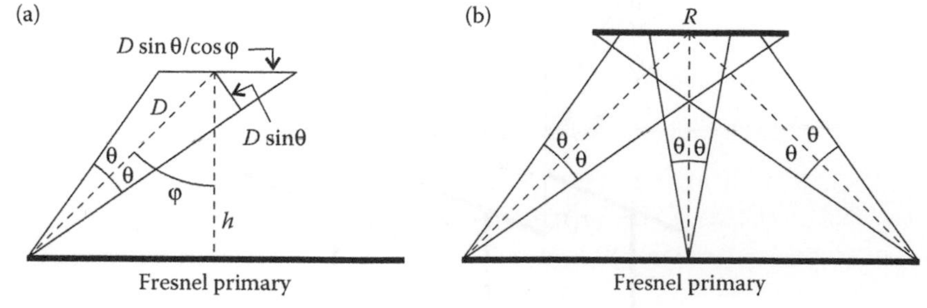

FIGURE 6.50
(a) Minimum absorber size for a Fresnel primary. (b) The receiver captures all light coming from the primary.

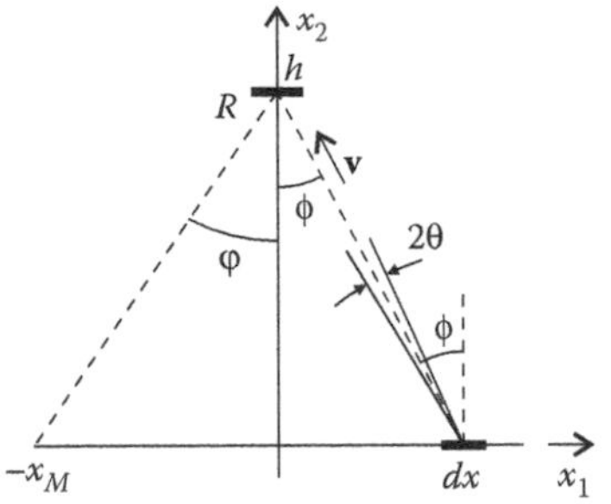

FIGURE 6.51
Étendue calculation of the Fresnel primary.

angle). Concentration relative to the maximum is the ratio between the étendue (amount of light) it receives from the primary and the maximum it can accept,

$$C = U_R/2R \tag{6.51}$$

where U_R is the étendue the receiver R collects from the primary. To calculate the concentration, we need to determine the étendue of the light emitted by the primary. The étendue of a small area dx emitting light within a cone 2θ whose bisector **v** is tilted by an angle ϕ relative to the perpendicular to dx and points to the center of R, as shown in Figure 6.51, is given by

$$dU = 2\,dx \sin\theta \cos\phi \tag{6.52}$$

where the cos ϕ factor is the projection of dx in the direction of **v**.

The factor cos ϕ also corresponds to the shading of the heliostats, as shown in Figure 6.52. The illuminated area A in the direction of the reflected rays

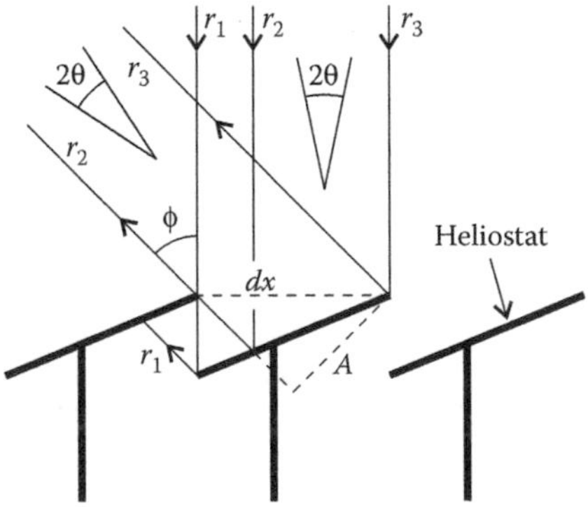

FIGURE 6.52
Heliostat shading produces loss of étendue.

is $A = dx \cos \phi$, and it corresponds to the radiation between the rays r_2 and r_3. The heliostat to the left shades the remaining radiation between the rays r_1 and r_2.

The primary extends from $-x_M$ to x_M with $x_M = h \tan \varphi$, where φ is the rim angle of the primary. The horizontal coordinate of dx is $x = h \tan \phi$, where h is the height of the receiver R relative to the heliostat field. Since

$$\int_0^{\varphi} \cos\phi \, dx = \int_0^{\varphi} \cos\phi \frac{dx}{\phi} d\phi = h \int_0^{\varphi} \frac{1}{\cos\phi} d\phi = h \ln\left(\tan\left(\frac{\pi}{4} + \frac{\varphi}{2}\right)\right) \tag{6.53}$$

the total étendue of the primary is then given by

$$U_P = 2 \times 2h \sin\theta \ln\left(\tan\left(\frac{\pi}{4} + \frac{\varphi}{2}\right)\right) \tag{6.54}$$

where φ is the rim angle of the primary and the factor of 2 is due to the two sides of the primary Fresnel reflector. Since all the light is collected by the receiver R, we have $U_R = U_P$. Concentration can now be written as

$$C = \frac{U_R}{2R} = \cos^2 \varphi \ln\left(\tan\left(\frac{\pi}{4} + \frac{\varphi}{2}\right)\right) \tag{6.55}$$

The maximum is $C_M = 44.8\%$ of the ideal maximum obtained for a primary rim angle of $\varphi_M = 40.4°$.

By lowering collection efficiency, we can increase concentration. Consider, for example, the smaller receiver **AB**, as shown in Figure 6.53, such that there

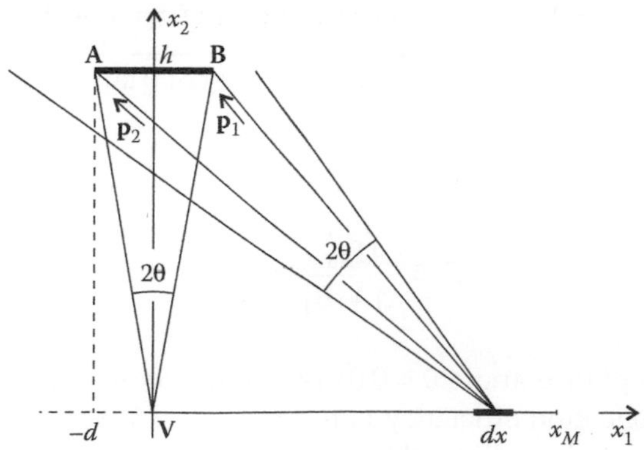

FIGURE 6.53
Concentration of a Fresnel primary can be increased by lowering light collection. Only the right-hand side of the Fresnel primary is shown.

will be light that misses it. All the light outside the angular space between the directions of $\mathbf{p}_1$ and $\mathbf{p}_2$ misses **AB**.

Receiver **AB** is the largest possible width that receives light from all points on the primary. For example, a point to the left of **A** does not receive light from point **V** at the center of the primary. The same thing happens to a point to the right of **B**. Height h acts as a scale factor for the entire optic.

The size of the receiver **AB** is now given by $[\mathbf{A}, \mathbf{B}] = 2d = 2h \tan \theta$, where $\mathbf{A} = (-d, h)$ and $\mathbf{B} = (d, h)$. Since the refractive index is $n = 1$, the optical momentum of the rays from dx to the edges **B** and **A** of the receiver are unit vectors

$$\mathbf{p}_1 = (p_{11}, p_{12}) = \frac{\mathbf{B} - (x,0)}{\| \mathbf{B} - (x,0) \|} \tag{6.56}$$

and

$$\mathbf{p}_2 = (p_{21}, p_{22}) = \frac{\mathbf{A} - (x,0)}{\| \mathbf{A} - (x,0) \|} \tag{6.57}$$

The étendue from the Fresnel primary to the receiver **AB** is then

$$U_{AB} = 2\int_0^{x_M} p_{11} - p_{21}\, dx \tag{6.58}$$

where $x_M = h \tan \varphi$ is the distance from the center to the edge of the heliostat field. We have

$$\begin{aligned} U_{AB} &= 2d\sqrt{\csc^2\theta + \cot\theta \tan\varphi(\cot\theta\tan\varphi + 2)} \\ &\quad - 2d\sqrt{\csc^2\theta + \cot\theta\tan\varphi(\cot\theta\tan\varphi - 2)} \end{aligned} \tag{6.59}$$

Concentration is now given by

$$C = \frac{U_{AB}}{2[\mathbf{A},\mathbf{B}]} = \frac{U_{AB}}{4d} \tag{6.60}$$

For a half-acceptance angle $\theta = 0.01$ rad[26] and $\varphi = \varphi_M$, we get $C = 64.8\%$ of the ideal. The collection efficiency is $\eta = U_{AB}/U_P = 84\%$, which is the energy flux the receiver receives divided by the flux the Fresnel reflector emits.

The rim angle φ_M was calculated as the optimum rim angle for maximum concentration when the receiver collected all the light. Now the situation is different because the receiver is smaller and collects only part of the light emitted by the primary. Increasing the rim angle beyond φ_M will increase the concentration on **AB**, since more light will be directed toward it, but efficiency will decline because more light will also miss it. For example, for an acceptance angle $\theta = 0.01$ rad and a rim angle of $\varphi = 49.6°$, we have $U_P = 2[\mathbf{A}, \mathbf{B}]$ and, therefore, the étendue the receiver **AB** can accept matches the étendue emitted by the primary. For that rim angle, the amount of light captured by the receiver is also the efficiency of the optic, or $C = \eta = U_{AB}/U_P$. We have, in that case, $C = 76.2\%$ of the ideal concentration and also $\eta = 76.2\%$ light collection efficiency.

6.10 Tailored Edge Ray Concentrators for Fresnel Primaries

The TERC secondary mirrors can also be designed for Fresnel primaries.[26] We Fresnelize the primary's continuous curved mirror by replacing it with a set of small mirrors (heliostats) on a straight line, reflecting the light in a way that mimics that of the mirror they replace. We take, for example, the case of the compound parabolic primary, as in Figure 6.32, which results in compact primary–secondary optics. For the heliostats to mimic the optical behavior of this primary, they are oriented so that the edge ray, r_A, on the right-hand side is reflected to the left-hand-side edge of the receiver **A**, as shown in Figure 6.54. The size of the receiver **AB** and TERC, relative to that of the heliostats, is grossly exaggerated, so as to show them on the same scale.

To calculate the shape of the TERC, we may use a method in which we progress along the mirror by very small steps, as shown in Figure 6.55.

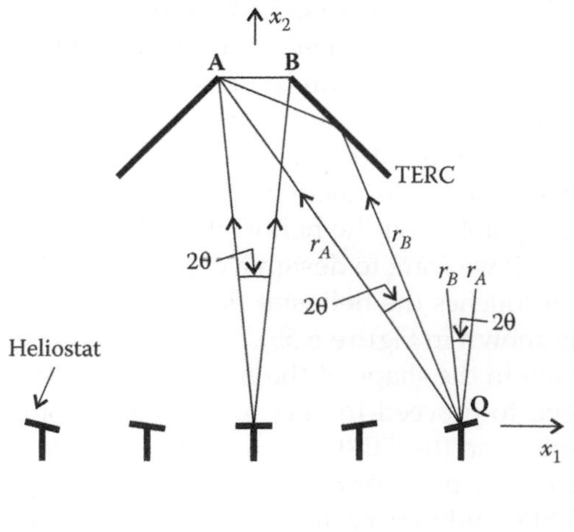

FIGURE 6.54
Aiming strategy of the heliostats for a TERC secondary.

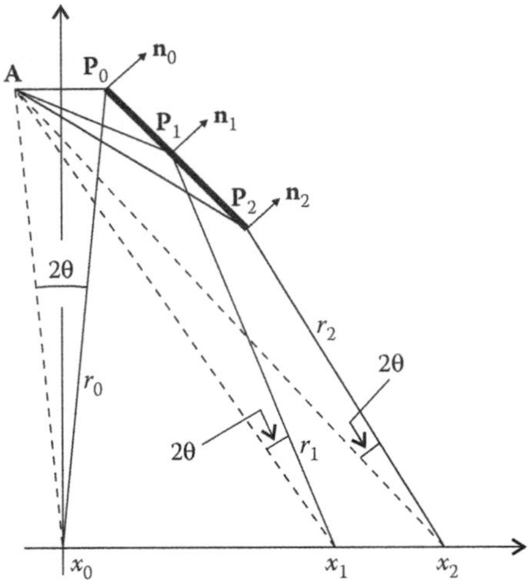

FIGURE 6.55
Constructing the shape of the TERC mirror progressing in small steps.

Again, we consider an infinite number of infinitely small heliostats so that they become a continuum.

We first choose the initial point for the mirror as $\mathbf{P}_0 = \mathbf{B}$, where $\mathbf{B}$ is the left edge of the receiver (see Figure 6.54). We know that ray r_0, coming from the heliostat field at position x_0, must be reflected to edge $\mathbf{A}$ of the receiver. This condition enables us to calculate the normal $\mathbf{n}_0$ to the TERC mirror at point $\mathbf{P}_0$. We now move on to another point at position x_1 on the heliostat field. We know the direction of the ray r_1 and we can intersect it with the straight line passing through point $\mathbf{P}_0$ and tangent to the mirror (perpendicular to $\mathbf{n}_0$) at $\mathbf{P}_0$. This yields the position of point $\mathbf{P}_1$. As before, ray r_1 must be reflected at $\mathbf{P}_1$ toward $\mathbf{A}$, and this enables us to calculate the normal $\mathbf{n}_1$ at point $\mathbf{P}_1$. We now move on to another point at position x_2 on the heliostat field. We know the direction of the ray r_2 and we can intersect it with the straight line passing through point $\mathbf{P}_1$ and tangent to the mirror (perpendicular to $\mathbf{n}_1$) at $\mathbf{P}_1$. This yields the position of point $\mathbf{P}_2$. As earlier, ray r_2 must be reflected at $\mathbf{P}_2$ toward $\mathbf{A}$ and this enables us to calculate the normal $\mathbf{n}_2$ at point $\mathbf{P}_2$. We then move on to another point x_3 on the heliostat field and calculate a new point on the TERC mirror. If we want to design a complete TERC, this process goes on until the mirror touches the heliostat field (crosses the horizontal axis if the geometry is as shown in Figure 6.55).

For good precision in the shape of the mirror using the method described previously, we need to proceed in very small steps along the primary and calculate many points on the TERC mirror. This does not mean we need to save all those points. We may, for example, calculate 10^6 points, but save the position on the TERC only every 10^4 points. This method will provide us with just 100 points on the mirror, but calculated with high precision.

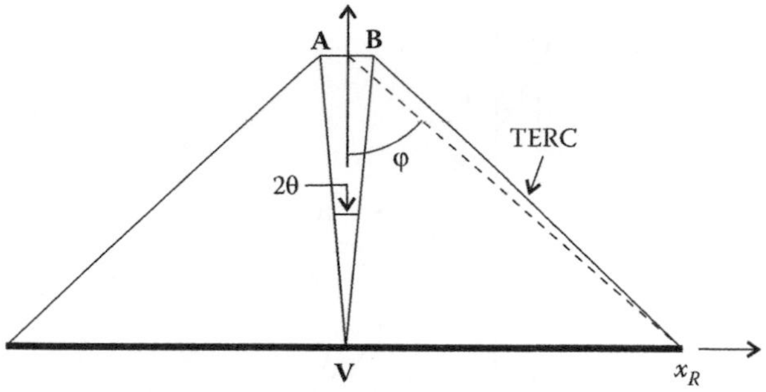

FIGURE 6.56
TERC for an acceptance angle 2θ. It ends at a radius x_R (corresponding to a rim angle φ) for which the étendue of the primary matches what the receiver **AB** can accept.

If the TERC is complete, in the sense that it extends all the way to the primary, it will touch the heliostat field at a point a distance x_R from its center, as shown in Figure 6.56. To calculate the value of x_R, we need to calculate the point of the primary, where its étendue matches that of the receiver **AB**.

For a complete TERC, the illumination of the receiver **AB** will be completely Lambertian and uniform (ignoring the shading of the Fresnel reflector heliostats by the TERC and receiver **AB**). This means that the étendue of the light hitting **AB** is

$$U_{AB} = 2[\mathbf{A},\mathbf{B}] \tag{6.61}$$

where [**A**, **B**] is the distance from **A** to **B** (size of the receiver). This must also be the étendue of the radiation captured by the TERC.

For calculating the étendue of the reflected radiation by the heliostat field, we first consider that we have an infinite number of infinitely small heliostats so that they become a continuum. The étendue of a small area dx emitting light within a cone 2θ with bisector **v** tilted by an angle β relative to the perpendicular to dx, as shown in Figure 6.57, is given as follows

$$dU = 2\,dx\sin\theta\cos\beta \tag{6.62}$$

where the cos β factor is the projection of dx in the direction of **v**.

If **A** is the edge of the receiver and is given by **A** = $(-d, h)$, then

$$\beta = \arctan\left(\frac{x+d}{h}\right) - \theta \tag{6.63}$$

and for the étendue of the heliostat field between x_C and x_D we get

$$U(x_C,x_D) = 2\sin\theta\int_{x_C}^{x_D}\cos\left(\arctan\left(\frac{x+d}{h}\right)-\theta\right)dx = F(x_D) - F(x_C) \tag{6.64}$$

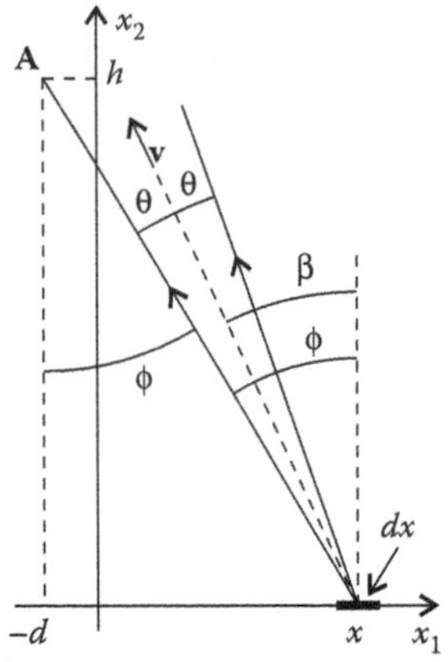

FIGURE 6.57
Étendue of a small area dx emitting a cone 2θ tilted by an angle β to the perpendicular of dx.

with

$$F(x) = 2\sin\theta\left[h\cos\theta\ln\left(d + x + \sqrt{h^2 + (d + x)^2}\right) + \sqrt{h^2 + (d + x)^2}\sin\theta\right] \tag{6.65}$$

Now, making $x_C = 0$ in expression (6.64), we start the integration at the midpoint **V** of the primary and we have

$$U_{xD} = 2U(0, x_D) \tag{6.66}$$

where the factor of 2 in this expression is due to the two sides (left and right) of the heliostat field. By solving numerically,

$$U_{AB} = U_{xD} \tag{6.67}$$

for x_D we get the solution $x_D = x_R$ for the "radius" (length on each side) of the heliostat field, and a value

$$U_P = 2U(0, x_R) = 2[\mathbf{A}, \mathbf{B}] \tag{6.68}$$

for the étendue emitted by the primary that matches that of the receiver. Point $(x_R, 0)$ is where the TERC touches the primary. Figure 6.56 shows a complete TERC for an acceptance angle 2θ.

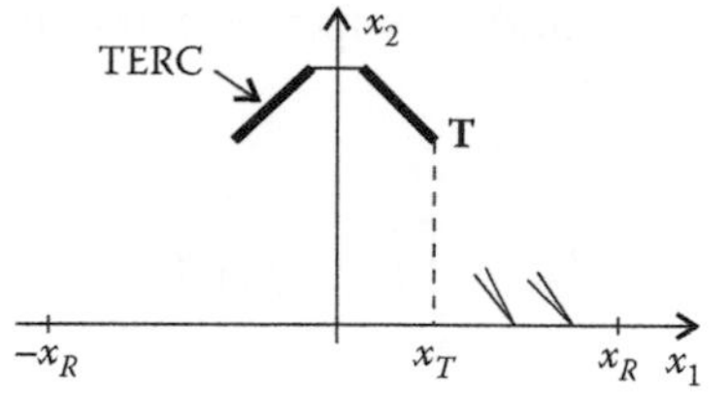

FIGURE 6.58
Shading of a truncated TERC for a Fresnel primary.

As always for TERC secondaries, we need to truncate it to make it usable. By doing this we will have two loss mechanisms: shading and light that is not captured by the secondary.

If we truncate the TERC at a point **T**, as in Figure 6.58, the étendue of the nonshaded portion of the primary can be obtained by $2U(x_T, x_R)$, where the factor of 2 again accounts for the two sides of the optic, and x_T is the horizontal coordinate of point **T**.

We now calculate the étendue of the light that is lost because it misses the secondary. If we use the method described earlier to calculate the TERC, we have a list of points and normals for the mirror. We truncate the TERC at one of its points **T** that has normal $\mathbf{n}_T$. A ray r coming from edge **A** of the source, and reflected at point **T**, intersects the primary at point x_L, as shown in Figure 6.59. From x_L to x_R where the primary ends, there will be losses due to light that misses the secondary. To calculate these losses, we need to consider the edge rays, emitted from a point **Q** on the primary, that have optical momenta $\mathbf{p}_A$ and $\mathbf{p}_B$. These are unit vectors since the refractive index is $n = 1$. Momentum $\mathbf{p}_A$ points in the direction from **Q** to edge **A** of the receiver, and $\mathbf{p}_B$ makes an angle 2θ relative to $\mathbf{p}_A$, where 2θ is the total acceptance of the optic.

The light emitted from **Q** and contained between the directions of $\mathbf{p}_B$ and $\mathbf{p}_T$ is lost, where $\mathbf{p}_T$ is the optical momentum of the light ray emitted from

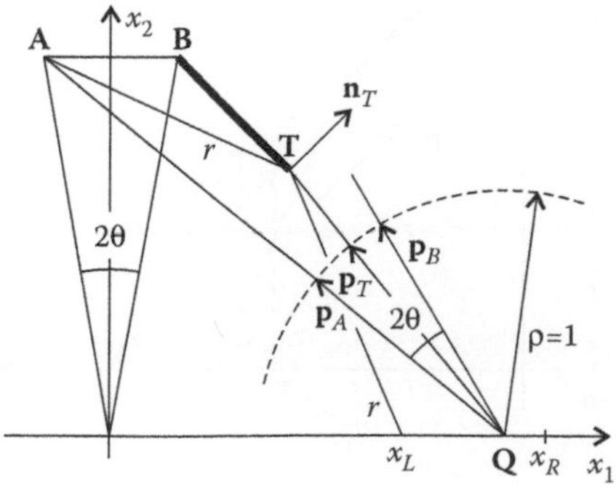

FIGURE 6.59
Étendue lost by truncating the TERC secondary.

Q toward **T**. If $\mathbf{p}_B = (p_{B1}, p_{B2})$ and $\mathbf{p}_T = (p_{T1}, p_{T2})$, the lost étendue can then be written as

$$U_L = \int_{x_L}^{x_R} p_{B1} - p_{T1}\, dx \tag{6.69}$$

as seen from Figure 6.60.

We need to calculate p_{B1} and p_{T1} now. If $\mathbf{A} = (-d, h)$ and $\mathbf{Q} = (x, 0)$, as shown in Figure 6.60, we have

$$\mathbf{p}_A = \frac{\mathbf{A} - \mathbf{Q}}{\|\mathbf{A} - \mathbf{Q}\|} = \frac{1}{\sqrt{h^2 + (d + x)^2}}(-d - x, h) \tag{6.70}$$

Momentum $\mathbf{p}_B$ is also a unit vector, but rotated by an angle 2θ relative to $\mathbf{p}_A$, and therefore,

$$\mathbf{p}_B = R(-2\theta) \cdot \mathbf{p}_A \Rightarrow p_{B1} = \frac{-(d + x)\cos(2\theta) + h\sin(2\theta)}{\sqrt{h^2 + (d + x)^2}} \tag{6.71}$$

where $R(-2\theta)$ is a rotation matrix of the angle -2θ. However, if $\mathbf{T} = (T_1, T_2)$, we have for $\mathbf{p}_T$

$$\mathbf{p}_T = \frac{\mathbf{T} - \mathbf{Q}}{\|\mathbf{T} - \mathbf{Q}\|} = \frac{1}{\sqrt{(T_1 - x)^2 + T_2^2}}(T_1 - x, T_2) \Rightarrow p_{T1} = \frac{T_1 - x}{\sqrt{(T_1 - x)^2 + T_2^2}} \tag{6.72}$$

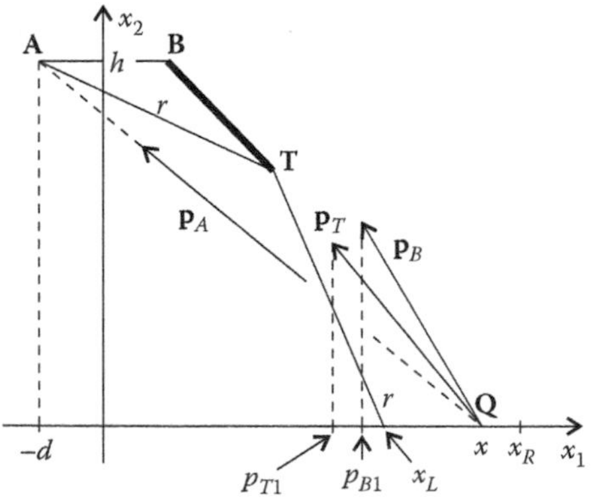

FIGURE 6.60
The étendue lost by radiation missing the secondary is calculated by the integration of $p_{B1} - p_{T1}$ along the length of the heliostat field from x_L to x_R.

And the étendue of the radiation that is lost, because it misses the secondary from position x_L to x_R, is given by

$$U_L(x_L, x_R) = \int_{x_L}^{x_R} p_{B1} - p_{T1}\, dx = G(x_R) - G(x_L) \tag{6.73}$$

with

$$\begin{aligned} G(x) = {} & \sqrt{(T_1 - x)^2 + T_2^2} - \sqrt{h^2 + (d + x)^2}\cos(2\theta) \\ & + h\sin(2\theta)\ln\left(d + x + \sqrt{h^2 + (d + x)^2}\right) \end{aligned} \tag{6.74}$$

The total étendue of the light that reaches the receiver is given by

$$U_R = 2(U(x_T, x_R) - U_L(x_L, x_R)) \tag{6.75}$$

where the factor of 2 accounts for the two sides of the optic. Optimizing the truncation means maximizing U_R relative to position of point **T**.

If we use the method described earlier to calculate the TERC, we have a list of points for the mirror. We may simply calculate U_R for each one of them and take the one that maximizes it. Concentration relative to the maximum possible is given by

$$C = \frac{U_R}{U_{AB}} = \frac{U(x_T, x_R) - U_L(x_L, x_R)}{[\mathbf{A}, \mathbf{B}]} \tag{6.76}$$

since the maximum étendue the receiver **AB** can receive is U_{AB}. The efficiency of the optic is $\eta = U_R/U_P = C$ because $U_P = U_{AB}$, that is, the étendue emitted by the primary is the same as the maximum the receiver can accept.

Figure 6.61 shows the concentration C and the collection efficiency η as a function of the horizontal coordinate x_T of the truncation point of the TERC.

We consider, for example, a receiver of size of $[\mathbf{A}, \mathbf{B}] = 1$ and an acceptance angle of $\theta = 0.01$ rad.[26] The rim angle is now $\varphi = 49.5°$. With no secondary (just the receiver), the ratio to the maximum concentration and the collection efficiency are $C = \eta = 75\%$, whereas, with the truncated TERC, the quantities have increased $\Delta\eta = \Delta C = 13\%$ to $C = \eta = 88\%$. The TERC is truncated at a distance of 2.8 from the optical axis. Figure 6.62 shows a TERC secondary designed for these parameters.

Just as with the case of the continuous compound parabolic primary, in this case of a Fresnel primary also, the TERC mirror may be designed with a CPC portion in the center, as shown in Figure 6.63.

In this case, at the central portion of the heliostat field, from point **C** to **D**, the heliostats are horizontal (as is the central heliostat in Figure 6.54), so

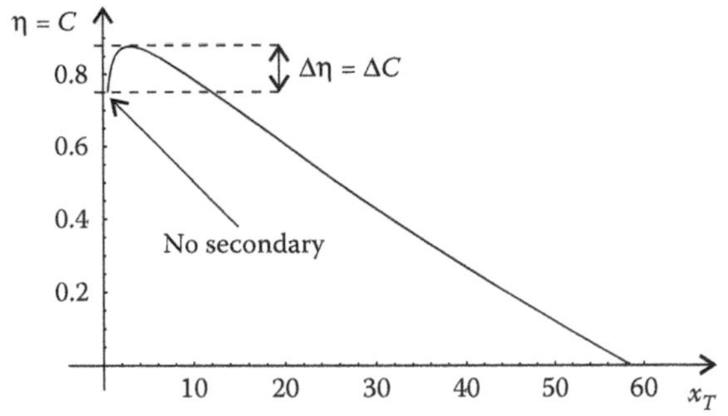

FIGURE 6.61
Concentration and collection efficiency as a function of the TERC truncation.

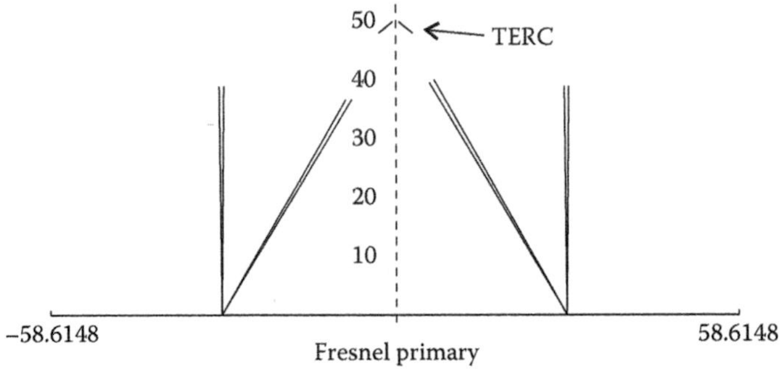

FIGURE 6.62
Truncated TERC for maximum concentration and collection efficiency.

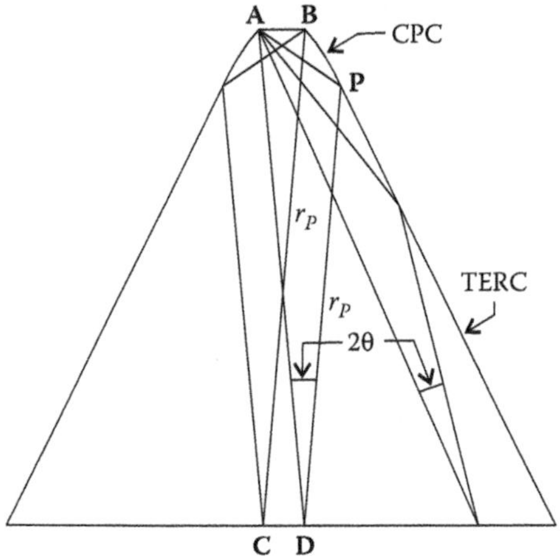

FIGURE 6.63
TERC secondary with a central CPC portion.

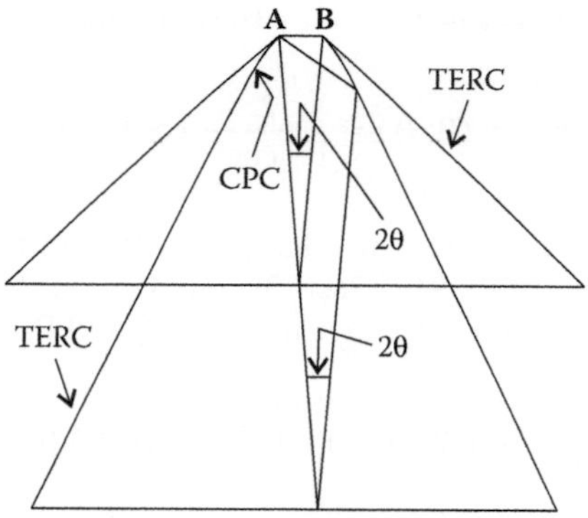

FIGURE 6.64
A comparison of two TERCs for the same receiver size and acceptance angle, but with and without a central CPC portion.

that vertical rays are reflected vertically again. The edge rays r_P hitting the portion **BP** of the mirror are parallel and tilted by an angle θ to the vertical. These rays are reflected to the edge **A** of the receiver, and that makes the portion **BP** of the mirror a parabola with a focus **A** and tilted by an angle θ to the vertical. The other side of the mirror is symmetrical, and, therefore, the top portion of the secondary concentrator is a portion of a CPC.

To the right of point **D**, heliostats are oriented as before. They reflect one of the edge rays to the edge **A** of the receiver, and the other edge ray is reflected by the TERC, also toward **A**. This portion of the mirror can be calculated by the same method as shown in Figure 6.55, starting at point **P**, where the parabolic section ends. Figure 6.64 shows a comparison of two TERCs designed for the same receiver size and acceptance angle, but with and without a central CPC portion. Just as in the case of the continuous primary, in this case also the option in which the secondary has a central CPC portion is less compact than the one that does not.

The TERC secondary extends all the way to the primary and must be truncated to be usable.

6.11 Examples

The following examples use expressions for the curves and functions that are derived in Chapter 21.

EXAMPLE 6.1

Design a CEC secondary for a parabolic primary with a rim angle $\varphi = 15°$ ($\varphi = \pi/12$ rad) and a concentrator acceptance angle of $\theta = \pm 1°$.

The parabolic primary is defined by the equation

$$\mathbf{P}(\phi) = (P_1, P_2) = \frac{2}{1-\cos\phi}(\cos\phi, \sin\phi) \tag{6.77}$$

with $\pi - \pi/12 \le \phi \le \pi + \pi/12$. The edge points $\mathbf{P}_1$ and $\mathbf{P}_2$ of the parabola are given as $\mathbf{P}_1 = (-0.982994,\ 0.260813)$ and $\mathbf{P}_2 = (-0.982994,\ -0.260813)$. Point **X**, where two of the edge rays through $\mathbf{P}_1$ and $\mathbf{P}_2$ meet, is given by

$$\mathbf{X} = \frac{1}{\cos\varphi + \cos^2\varphi}(-2\sin^2\theta, \sin 2\theta) = (-0.000320484, 0.0183605) \tag{6.78}$$

We now have the situation as in Figure 6.65.

The entrance aperture of the CEC will be defined by point **X** and its symmetrical relative to the x_1 axis. The parabolic primary is considered a Lambertian source $\mathbf{P}_1\mathbf{P}_2$ emitting toward the entrance aperture of the CEC. The étendue from this Lambertian source is

$$U_{LS} = 8\sin\theta\tan(\varphi/2)/\cos\varphi = 0.0188372 \tag{6.79}$$

If we want maximum concentration, the receiver will be illuminated completely by Lambertian radiation, with angles ranging between $\pm\pi/2$. The size R of the receiver will then be $2R\sin(\pi/2) = U_{LS}$ or

$$R = U_{LS}/2 = 0.00941861 \tag{6.80}$$

We can now determine the positions of the edge points of the receiver. The lower edge $\mathbf{R}_1$ of the receiver is at the intersection of ray r_1 through **X** coming from $\mathbf{P}_1$ and a horizontal line through the point $(0, -R/2)$,

$$\mathbf{R}_1 = \text{isl}(\mathbf{X}, \mathbf{X} - \mathbf{P}_1, (0, -R/2), (1,0)) = (0.0931829, -0.00470931)$$

as shown in Figure 6.66. Point $\mathbf{R}_2$ is symmetrical to $\mathbf{R}_1$ relative to the axis x_1.

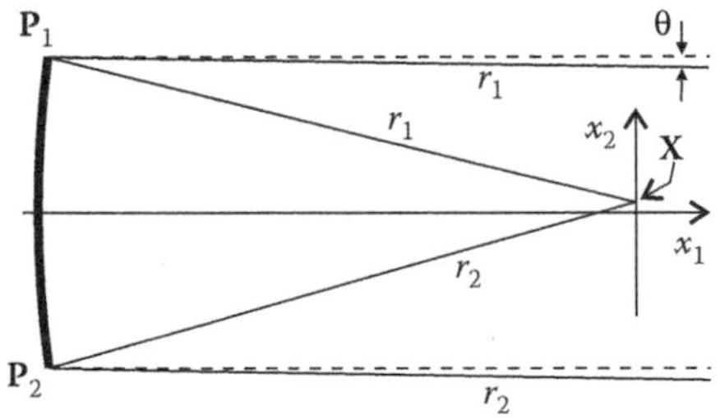

FIGURE 6.65
Parabolic primary.

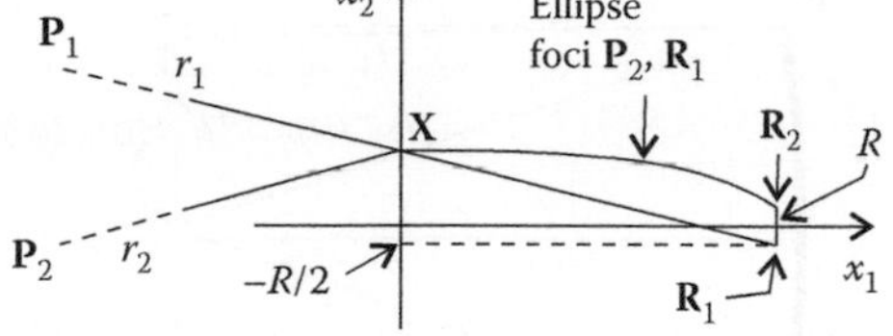

FIGURE 6.66
Determining the position of receiver R and calculating the ellipses that compose the CEC secondary.

We now have the foci $\mathbf{P}_2$ and $\mathbf{R}_1$ for the upper ellipse of the CEC and a point **X** that it must go through. This completely defines the curve, which is given by

$$\text{eli}(\mathbf{P}_2,\mathbf{R}_1,\mathbf{X}) = \left(-0.982994 + \frac{0.0258818\cos(0.23363+\phi)}{2.23574-2.21246\cos\phi},\right.$$
$$\left.-0.260813 + \frac{0.0258818\sin(0.23363+\phi)}{2.23574-2.21246\cos\phi}\right) \tag{6.81}$$

with a parameter range $\alpha_1 \le \phi \le \alpha_2$, where

$$\begin{aligned}\alpha_1 &= \text{ang}(\mathbf{R}_2-\mathbf{P}_2,\mathbf{R}_1-\mathbf{P}_2) = 0.473627^\circ \\ \alpha_2 &= \text{ang}(\mathbf{X}-\mathbf{P}_2,\mathbf{R}_1-\mathbf{R}_2) = 2.47363^\circ\end{aligned} \tag{6.82}$$

The ellipse at the bottom of the CEC is symmetrical to the one at the top relative to the x_1 axis.

EXAMPLE 6.2

Design a trumpet secondary for a parabolic primary with a rim angle $\varphi = 19.5^\circ$ $(19.5\pi/180 \text{ rad})$, a concentrator acceptance angle $\theta = \pm 1^\circ$, and a truncation of the hyperbola at a parameter value of $\phi = 62^\circ$.

The parabolic primary is defined by the equation

$$\mathbf{P}(\phi) = (P_1, P_2) = \frac{2}{1-\cos\phi}(\cos\phi, \sin\phi) \tag{6.83}$$

with $\pi - 19.5\pi/180 \le \phi \le \pi + 19.5\pi/180$. The edge points of the parabola, $\mathbf{P}_1$ and $\mathbf{P}_2$, are given by: $\mathbf{P}_1 = (-0.970474, 0.343663)$ and $\mathbf{P}_2 = (-0.970474, -0.343663)$. The point **X** where two of the edge rays passing through the points $\mathbf{P}_1$ and $\mathbf{P}_2$ meet is given by

$$\mathbf{X} = \frac{1}{\cos\varphi + \cos^2\varphi}(-2\sin^2\theta, \sin 2\theta) = (-0.000332661, 0.0190581) \tag{6.84}$$

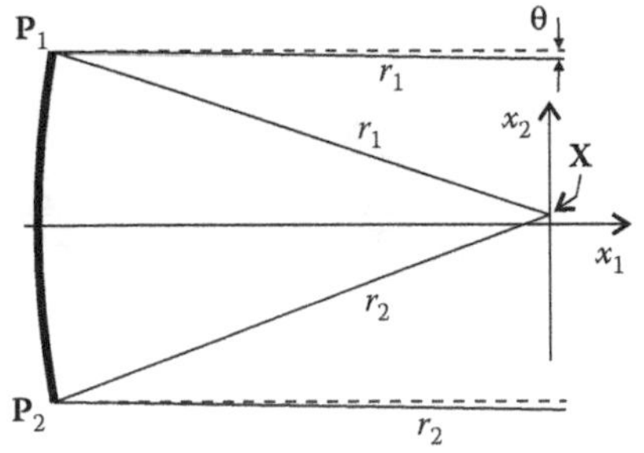

FIGURE 6.67
Parabolic primary.

Its symmetrical **Y** relative to the x_1 axis is

$$\mathbf{Y} = (-0.000332661, -0.0190581) \tag{6.85}$$

We now have the situation as in Figure 6.67.
The receiver size R as shown in Figure 6.68 is given by

$$R = 4\sin\theta\tan(\varphi/2)/\cos\varphi = 0.0127254 \tag{6.86}$$

The hyperbola of the trumpet concentrator is given by

$$\mathbf{K}(\phi) = \frac{R^2 - f^2}{2R - 2f\cos\phi}(\cos(\phi + \pi/2), \sin(\phi + \pi/2)) + \mathbf{Y} \tag{6.87}$$

where $f = [\mathbf{X}, \mathbf{Y}] = 0.0381162$.
The edge point of the truncated hyperbola is obtained as $\mathbf{K} = \mathbf{K}(62°) = (-0.110585, 0.039564)$
The parabolic primary is considered a Lambertian source $\mathbf{P}_1\mathbf{P}_2$ emitting toward **XY**. The étendue from this Lambertian source is

$$U_{LS} = 8\sin\theta\tan(\varphi/2)/\mathrm{coc}\,\varphi = 0.0254508$$

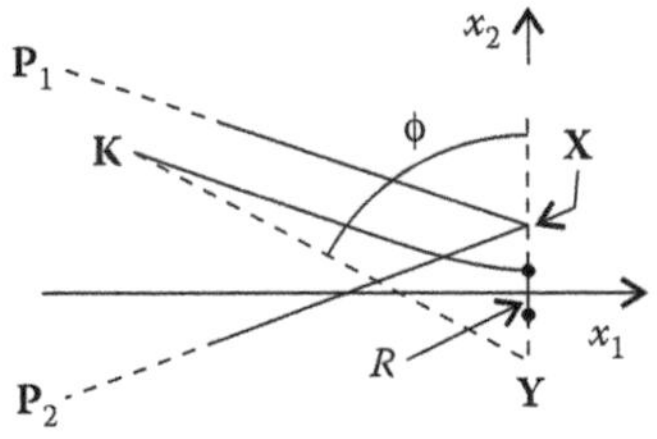

FIGURE 6.68
Truncated hyperbola of the secondary concentrator.

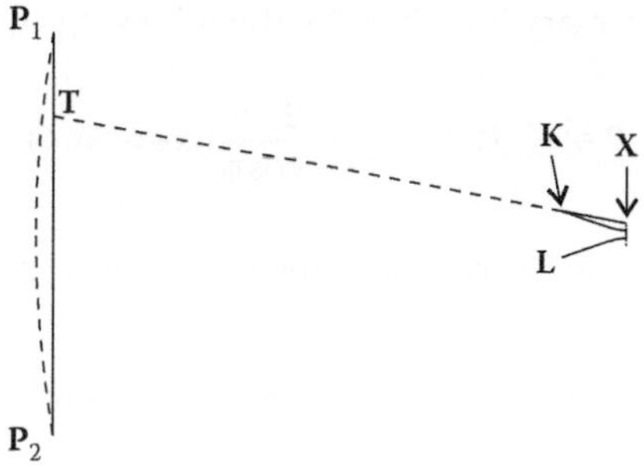

FIGURE 6.69
The parabolic mirror is considered a Lambertian source $\mathbf{P}_1\mathbf{P}_2$. The light emitted by $\mathbf{P}_1\mathbf{T}$ toward **KX** is lost.

The amount of light that is lost due to the truncation of the hyperbola is twice the étendue from $\mathbf{P}_1\mathbf{T}$ to **KX** (Figure 6.69). Point **T** can be obtained by

$$\mathbf{T} = \text{isl}\,(\mathbf{K},\mathbf{K}-\mathbf{X},\mathbf{P}_2,(0,1)) = (-0.970474, 0.199496) \tag{6.88}$$

The étendue lost due to the truncation of the trumpet is then given by

$$U_L = 2([\mathbf{T},\mathbf{X}] - [\mathbf{T},\mathbf{K}] + [\mathbf{P}_1,\mathbf{K}] - [\mathbf{P}_1,\mathbf{X}]) = 0.00242781 \tag{6.89}$$

The fraction of étendue captured by the trumpet is

$$c_U = 1 - U_L/U_{LS} = 0.904608 \tag{6.90}$$

The illuminated portion of the primary is given by

$$I_L = 1 - [\mathbf{K},\mathbf{L}]/[\mathbf{P}_1,\mathbf{P}_2] = 0.884875 \tag{6.91}$$

where **L** is symmetrical to **K** relative to the axis of symmetry of the optic x_1. The concentration the optic provides is then

$$C = \frac{[\mathbf{P}_1,\mathbf{P}_2]}{R} I_L c_U = \frac{\cos\varphi}{\sin\theta} I_L c_U = 43.2348 \tag{6.92}$$

The ratio to the maximum possible concentration is

$$C\sin\theta = 0.754552 \tag{6.93}$$

EXAMPLE 6.3

Design a TERC secondary for a parabolic primary with a rim angle of $\varphi = 36.6°$, a concentrator acceptance angle of $\theta = \pm 0.01$ rad and a flat receiver.

The parabolic primary is defined by the following equation,

$$\mathbf{P}(\phi) = (P_1, P_2) = \frac{2}{1-\cos\phi}(\cos\phi, \sin\phi) \tag{6.94}$$

with $\pi - \varphi \le \phi \le \pi + \varphi$. The caustic of the edge rays is given by

$$\mathbf{C}(\phi,\theta) = \frac{d}{2}\csc^3\left(\frac{\phi}{2}\right)\left(\sin\left(2\theta - \frac{3\phi}{2}\right) + \sin\left(\frac{3\phi}{2}\right),\right.$$
$$\left.\cos\left(2\theta - \frac{3\phi}{2}\right) - \cos\left(\frac{3\phi}{2}\right)\right) \tag{6.95}$$

with $\pi - 36.6\pi/180 \le \phi \le \pi + 36.6\pi/180$ and $\theta = 0.01$ rad.

The edge points of the parabola $\mathbf{P}_1$ and $\mathbf{P}_2$ are

$$\begin{aligned} \mathbf{P}_1 &= \mathbf{P}(\pi - 36.6\pi/180) = (-0.890625, 0.661437) \\ \mathbf{P}_2 &= \mathbf{P}(\pi + 36.6\pi/180) = (-0.890625, -0.661437) \end{aligned} \tag{6.96}$$

The edge points of the caustic are

$$\begin{aligned} \mathbf{C}_1 &= \mathbf{C}(\pi - 36.6\pi/180, 0.01) = (-0.00962638, -0.00662272) \\ \mathbf{C}_2 &= \mathbf{C}(\pi + 36.6\pi/180, 0.01) = (0.00949201, -0.00681391) \end{aligned} \tag{6.97}$$

as shown in Figure 6.70. The size of the ideal receiver is

$$R = [\mathbf{P}_1, \mathbf{P}_2]\sin\theta = 0.0132285 \tag{6.98}$$

The position of the lower edge of the receiver can then be calculated by the intersection of the upper edge ray reflected at $\mathbf{P}_1$ with a horizontal line through the point $(-R/2, 0)$. Note that, since the focus of the parabola

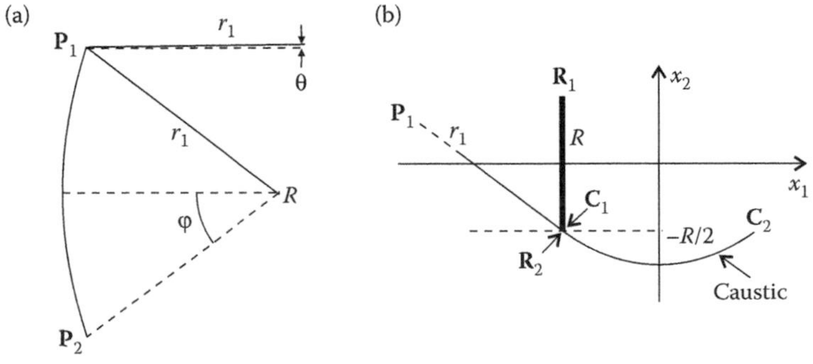

FIGURE 6.70
(a) A parabolic primary, a receiver, and a caustic of the edge rays. (b) Details of the receiver.

is at $\mathbf{F} = (0, 0)$, the vector from $\mathbf{P}_1$ to $\mathbf{F}$ is given by $\mathbf{F} - \mathbf{P}_1 = -\mathbf{P}_1$. A ray parallel to the axis of the parabola is reflected at $\mathbf{P}_1$ toward $\mathbf{F}$, and, therefore, has a direction $-\mathbf{P}_1$ after reflection. For this reason, the reflected top edge ray r_1, making an angle θ to the horizontal before reflection, has a direction $-R(-\theta) \cdot \mathbf{P}_1$ after reflection at point $\mathbf{P}_1$. The edge point $\mathbf{R}_2$ of the receiver is then given by

$$\begin{aligned}\mathbf{R}_2 &= \text{isl}\,(\mathbf{P}_1, -R(-\theta)\cdot\mathbf{P}_1, (0, -R/2), (1, 0)) \\ &= (-0.00963753, -0.00661426)\end{aligned} \tag{6.99}$$

where R is the size of the receiver but $R(-\theta)$ is a rotation matrix of an angle $-\theta$. The other edge $\mathbf{R}_1 = (-0.00963753, 0.00661426)$ of the receiver is symmetrical to $\mathbf{R}_2$ relative to the horizontal axis of symmetry x_1, as shown in Figure 6.70.

The edge points $\mathbf{C}_1$ and $\mathbf{C}_2$ of the caustic are to the right (larger x_1 components) of the edge points $\mathbf{R}_1$ and $\mathbf{R}_2$ of the receiver. Therefore, the caustic does not intersect the receiver or the TERC mirror that will be to the left of R.

We can now define the position of a wavefront w perpendicular to the incoming edge rays, as shown in Figure 6.71. Vectors $\mathbf{v}_1$ (perpendicular to w) and $\mathbf{v}_2$ (parallel to w) are given by

$$\begin{aligned}\mathbf{v}_1 &= (\cos\theta, \sin\theta) \\ \mathbf{v}_2 &= \left(\cos\left(\frac{\pi}{2}+\theta\right), \sin\left(\frac{\pi}{2}+\theta\right)\right) = (-\sin\theta, \cos\theta)\end{aligned} \tag{6.100}$$

We define point $\mathbf{W}_2$ of the wavefront w, for example, as

$$\mathbf{W}_2 = \mathbf{P}_2 + 0.2\mathbf{v}_1 = \mathbf{P}_2 + 0.2(\cos\theta, \sin\theta) = (-0.690635, -0.659437) \tag{6.101}$$

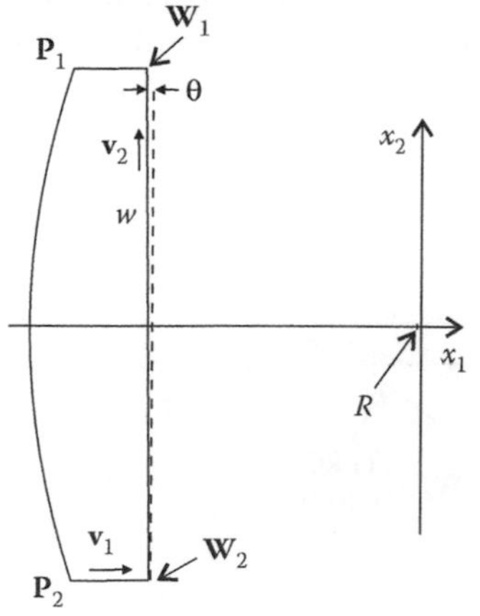

FIGURE 6.71
Definition of a wavefront, w, perpendicular to a set of edge rays.

And point $\mathbf{W}_1$ can then be obtained by the intersection of the straight line through $\mathbf{W}_2$ with a direction $\mathbf{v}_2$ with the straight line through $\mathbf{P}_1$ with direction $\mathbf{v}_1$ as

$$\begin{aligned}\mathbf{W}_1 &= \text{isl}\,(\mathbf{W}_2,\mathbf{v}_2,\mathbf{P}_1,\mathbf{v}_1) = \text{isl}\,(\mathbf{W}_2,(-\sin\theta,\cos\theta),\mathbf{P}_1,(\cos\theta,\sin\theta))\\ &= (-0.703863, 0.663304)\end{aligned} \tag{6.102}$$

We now take points along the parabolic primary and calculate the points on the TERC by the constant optical path length. The edge rays perpendicular to w reflect at the parabola, then on the TERC, and are redirected from there to the top edge $\mathbf{R}_1$ of the receiver. The optical path length S for these light rays is then

$$S = [\mathbf{W}_1,\mathbf{P}_1] + [\mathbf{P}_1,\mathbf{R}_2] + [\mathbf{R}_2,\mathbf{R}_1] = 1.30564 \tag{6.103}$$

We now take a parameter value for the parabolic primary and calculate the corresponding point. For example, for $\phi_4 = \pi + 20\pi/180$, we have

$$\mathbf{P}_4 = \mathbf{P}(\pi + 20\pi/180) = (-0.968909, -0.352654) \tag{6.104}$$

as shown in Figure 6.72.

We can now calculate the corresponding point on the wavefront w by

$$\mathbf{W}_4 = \text{isl}\,(\mathbf{W}_2,\mathbf{v}_2,\mathbf{P}_4,\mathbf{v}_1) = (-0.693731, -0.349902) \tag{6.105}$$

Now, we can calculate the optical path length between points $\mathbf{P}_4$ and $\mathbf{R}_1$ as

$$S_4 = S - [\mathbf{W}_4,\mathbf{P}_4] = 1.03044 \tag{6.106}$$

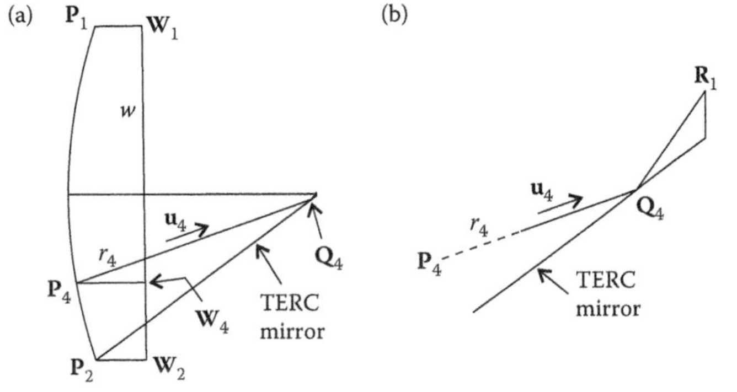

FIGURE 6.72
(a) Calculation of a point $\mathbf{Q}_4$ on the TERC mirror. (b) Detail of the path of ray r_4 at point $\mathbf{Q}_4$ on the TERC mirror.

The edge ray r_4 perpendicular to the wavefront w and reflected at $\mathbf{P}_4$ has a direction $-R(-\theta) \cdot \mathbf{P}_4$ after reflection. Therefore, the direction of the ray after reflection is given by

$$\mathbf{u}_4 = -\text{nrm}\,(R(-\theta) \cdot \mathbf{P}_4) = (0.943066, 0.332606) \tag{6.107}$$

where $\mathbf{u}_4$ is a unit vector. Finally, point $\mathbf{Q}_4$ on the TERC mirror is given by

$$\begin{aligned} \mathbf{Q}_4 &= \text{coptpt}\,(\mathbf{P}_4, \mathbf{u}_4, \mathbf{R}_1, n, S_4) \\ &= (-0.0293346, -0.0212791) \end{aligned} \tag{6.108}$$

where the refractive index is $n = 1$ since the mirrors are in air and the optical path length was also calculated with $n = 1$.

Carrying out the same calculation for other values of parameter ϕ for the parabolic mirror, we get other points on the TERC mirror. The complete mirror extends all the way to the edge of the primary, as shown in Figure 6.72a.

If we want to truncate the TERC at point $\mathbf{Q}_4$, the étendue lost from not being collected by the secondary would be

$$U_L = \int_{\phi_4}^{\pi+\varphi} [\mathbf{v}_r(\phi) - \mathbf{v}_4(\phi)] \cdot \frac{d\mathbf{P}(\phi)}{d\phi} d\phi \tag{6.109}$$

where

$$\begin{aligned} \mathbf{v}_4(\phi) &= \text{nrm}\,(\mathbf{Q}_4 - \mathbf{P}(\phi)) \\ \mathbf{v}_r(\phi) &= \text{nrm}(-R(-\theta) \cdot \mathbf{P}(\phi)) \end{aligned} \tag{6.110}$$

and we get $U_L = 0.00164788$. We now define point $\mathbf{Q}_3$ as symmetrical to $\mathbf{Q}_4$ about the axis of symmetry of the system (axis x_1) as $\mathbf{Q}_3 = (-0.0293346, 0.0212791)$. Truncation of the TERC mirror at point $\mathbf{Q}_3$ results in an additional loss of the étendue also given by U_L (due to the symmetry of the optic). Total étendue loss is then $2U_L$. The shading produced by a TERC truncated at point $\mathbf{Q}_4$ (and the symmetrical mirror at point $\mathbf{Q}_3$) is given by

$$S_H = \frac{[\mathbf{Q}_3, \mathbf{Q}_4]}{[\mathbf{P}_1, \mathbf{P}_2]} = 0.0321711 \tag{6.111}$$

The fraction of light captured by the receiver is then

$$L_R = 1 - S_H - \frac{2U_L}{2[\mathbf{P}_1, \mathbf{P}_2]\sin\theta} = 0.843259 \tag{6.112}$$

If we define the angle α, such that $\phi = \pi + \alpha$ and the parabolic primary is defined by $\mathbf{P}(\alpha)$ with $-\phi \le \alpha \le \phi$, we can plot L_R as a function of truncation angle α_T (in degrees), as shown in Figure 6.73.

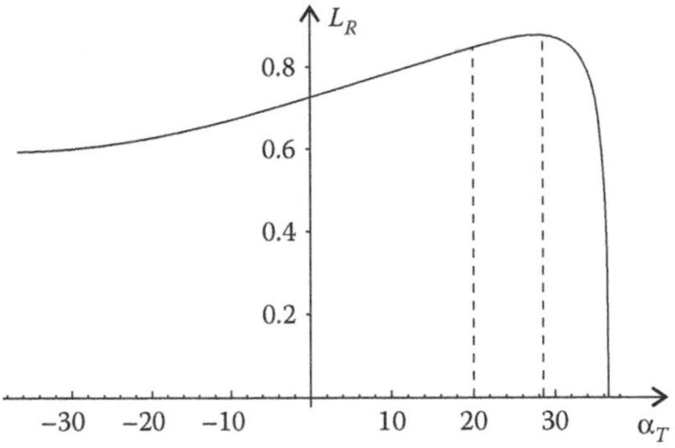

FIGURE 6.73
Étendue of the captured light by the truncated TERC as a function of the truncation parameter α_T.

Truncation at point $\mathbf{P}_4$ corresponds to truncation at $\alpha_T = \phi_4 - \pi = 20°$. The maximum is 0.75 for $\alpha_T = 28.5°$ giving the point where the TERC can be truncated for maximum light collection.

EXAMPLE 6.4

Design a TERC secondary for a parabolic primary with a rim angle of $\varphi = 38°$, a concentrator acceptance angle of $\theta = \pm 5°$ ($\pm\pi/36$ rad) and a circular receiver.

The parabolic primary is defined by the equation

$$\mathbf{P}(\phi) = (P_1, P_2) = \frac{-2}{1+\cos\alpha}(\cos\alpha, \sin\alpha) \tag{6.113}$$

with $-\varphi \le \alpha \le \varphi$. The caustics of the edge rays are given by

$$\begin{aligned}\mathbf{C}(\alpha,\theta) = \frac{d}{2}\sec^3\left(\frac{\alpha}{2}\right)\Bigg(&\cos\left(\frac{3\alpha}{2} - 2\theta\right) - \cos\left(\frac{3\alpha}{2}\right),\\ &\sin\left(\frac{3\alpha}{2} - 2\theta\right) - \sin\left(\frac{3\alpha}{2}\right)\Bigg)\end{aligned} \tag{6.114}$$

with $-\varphi \le \alpha \le \varphi$ and $\theta = \pm\pi/36$ rad. Figure 6.74 shows the primary mirror with a focus at $\mathbf{F} = (0, 0)$ and a rim angle φ. Figure 6.75 shows caustic of the edge rays.

The edge points of the parabola are

$$\begin{aligned}\mathbf{P}_1 &= \mathbf{P}(-\varphi) = (-0.881438, 0.688655)\\ \mathbf{P}_2 &= \mathbf{P}(\varphi) = (-0.881438, -0.688655)\end{aligned} \tag{6.115}$$

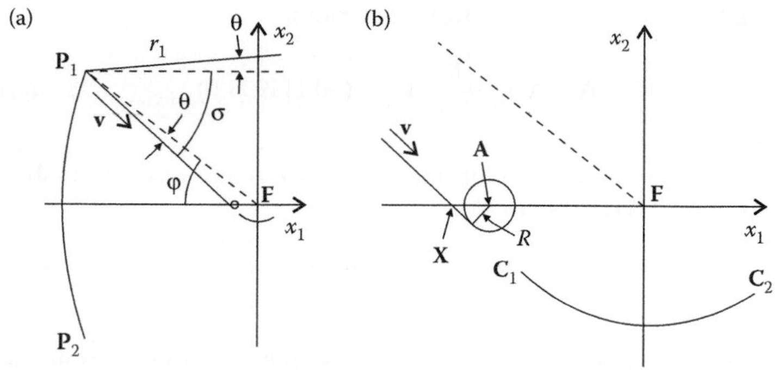

FIGURE 6.74
(a) Parabolic primary with a circular receiver. (b) Detail close to the receiver.

The edge points $\mathbf{C}_1$ and $\mathbf{C}_2$ of the caustic formed by the edge rays are

$$\begin{aligned} \mathbf{C}_1 &= \mathbf{C}(-\varphi,\theta) = (-0.0910376,-0.0484055) \\ \mathbf{C}_2 &= \mathbf{C}(\varphi,\theta) = (0.081249,-0.0634787) \end{aligned} \tag{6.116}$$

A ray r_1, making an angle θ to the horizontal axis, reflects at point $\mathbf{P}_1$ and has a direction $\mathbf{v}$ after reflection. The vector $\mathbf{v}$ makes an angle $\sigma = \varphi + \theta$ to the horizontal and is given by $\mathbf{v} = (\cos(-\sigma), \sin(-\sigma))$. Point $\mathbf{X}$, where ray r_1, after reflection, intersects the axis x_1, is

$$\mathbf{X} = \text{isl}\,(\mathbf{P}_1,\mathbf{v},(0,0),(1,0)) = (-0.142946,0) \tag{6.117}$$

The circular receiver has a radius given by

$$R = \frac{2[\mathbf{P}_1,\mathbf{P}_2]\sin\theta}{2\pi} = 0.019105 \tag{6.118}$$

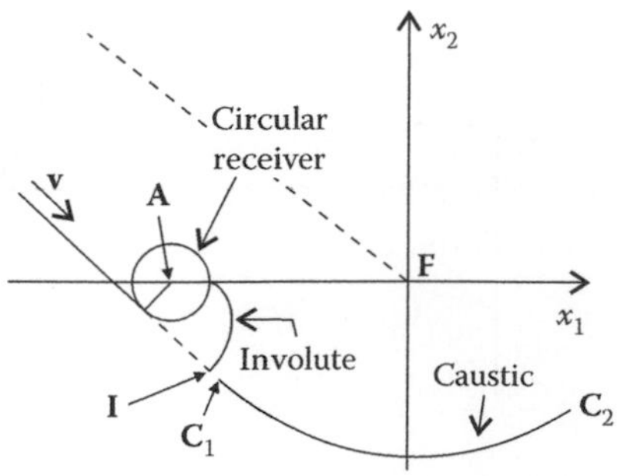

FIGURE 6.75
Circular receiver, an involute mirror, and caustic of one set of edge rays.

The center of the circular receiver is at a point

$$\mathbf{A} = \mathbf{X} + \left(\frac{R}{\sin\sigma}, 0 \right) = (-0.114933, 0) \tag{6.119}$$

We can now design the involute mirror to the circular receiver, as shown in Figure 6.75. It is given by

$$\mathbf{I}_V(\gamma) = R(\cos\gamma, \sin\gamma) - R\gamma(-\sin\gamma, \cos\gamma) + \mathbf{A} \tag{6.120}$$

with $-\pi/2 - \alpha \le \gamma \le 0$.

Its end point $\mathbf{I}$ is on the straight line through $\mathbf{P}_1$ with direction $\mathbf{v}$ and is given by

$$\mathbf{I} = (-0.0955282, -0.044218) \tag{6.121}$$

The caustic (bounded by points $\mathbf{C}_1$ and $\mathbf{C}_2$) is, therefore, to the right of the mirror and does not intersect it.

We may now calculate the TERC mirror. We start by defining the wavefront w perpendicular to a set of edge rays, as shown in Figure 6.76.

We define point $\mathbf{W}_1$ on w as

$$\mathbf{W}_1 = \mathbf{P}_1 + 0.1(\cos\theta, \sin\theta) = (-0.781819, 0.697371) \tag{6.122}$$

and point $\mathbf{W}_2$ as

$$\mathbf{W}_2 = \text{isl}\,(\mathbf{W}_1, (-\sin\theta, \cos\theta), \mathbf{P}_2, (\cos\theta, \sin\theta)) = (-0.662235, -0.669477) \tag{6.123}$$

The ray r_1 reflected at $\mathbf{P}_1$ is perpendicular to the wavefront w and tangential to the circular receiver at point $\mathbf{T}$. It has a direction $\mathbf{v}$ after reflection at $\mathbf{P}_1$. If $\mathbf{t} = R(-\pi/2) \cdot \mathbf{v} = (-\sin\sigma, -\cos\sigma)$, we have

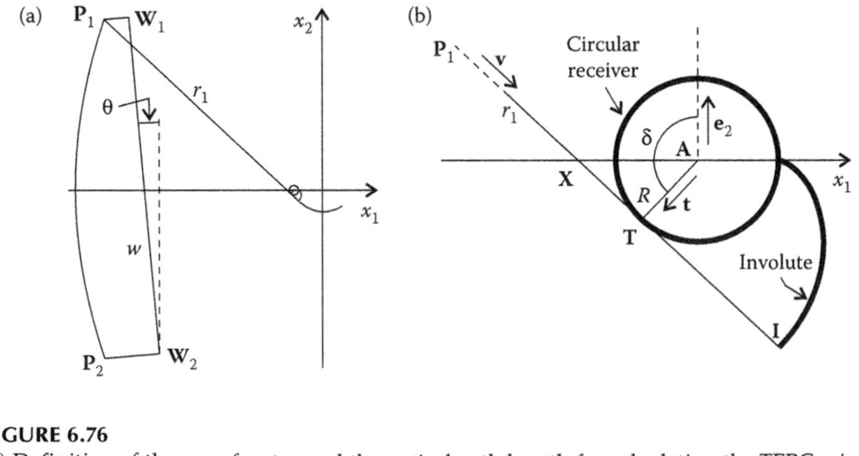

FIGURE 6.76
(a) Definition of the wavefront w and the optical path length for calculating the TERC mirror.
(b) Detail close to the receiver.

$$\mathbf{T} = \mathbf{A} + R\mathbf{t} = (-0.127962, -0.0139725) \tag{6.124}$$

The optical path length is defined as

$$S = [\mathbf{W}_1, \mathbf{P}_1] + [\mathbf{P}_1, \mathbf{I}] + [\mathbf{I}, \mathbf{T}] + R\delta = 1.26463 \tag{6.125}$$

where $\delta = \text{ang}\ (\mathbf{t}, \mathbf{e}_2)$, where $\mathbf{e}_2 = (0, 1)$. We can now choose a point $\mathbf{P}_4$ on the parabolic primary and calculate the corresponding point on the TERC mirror, as shown in Figure 6.77.

For example, taking $\alpha = 20°$, we get

$$\mathbf{P}_4 = \mathbf{P}(20\pi/180) = (-0.950851, -0.443389) \tag{6.126}$$

We can now calculate the corresponding point on the wavefront w as

$$\begin{aligned} \mathbf{W}_4 &= \text{isl}\,(\mathbf{W}_1, (-\sin\theta, \cos\theta), \mathbf{P}_4, (\cos\theta, \sin\theta)) \\ &= (-0.684058, -0.420048) \end{aligned} \tag{6.127}$$

The ray r_4 reflects at $\mathbf{P}_4$ in the direction $\mathbf{u}_4 = -\text{nrm}\,(-R(-\theta) \cdot \mathbf{P}_4) = (0.939693, 0.34202)$. We define the position of point $\mathbf{Q}_4$ as

$$\mathbf{Q}_4 = \mathbf{P}_4 + x\mathbf{u}_4 \tag{6.128}$$

where x is unknown. The Tangent point $\mathbf{T}_4$ on the receiver can be calculated as a function of $\mathbf{Q}_4$, as shown in Figure 6.78.

The angle β is given by

$$\beta = \arccos\left(\frac{R}{[\mathbf{A}, \mathbf{Q}_4]}\right) \tag{6.129}$$

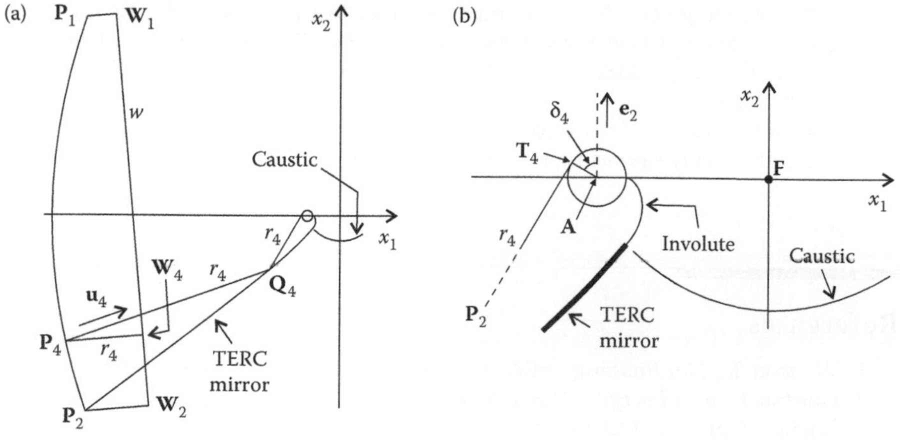

FIGURE 6.77
(a) Construction of the TERC mirror. (b) Detail close to the receiver.

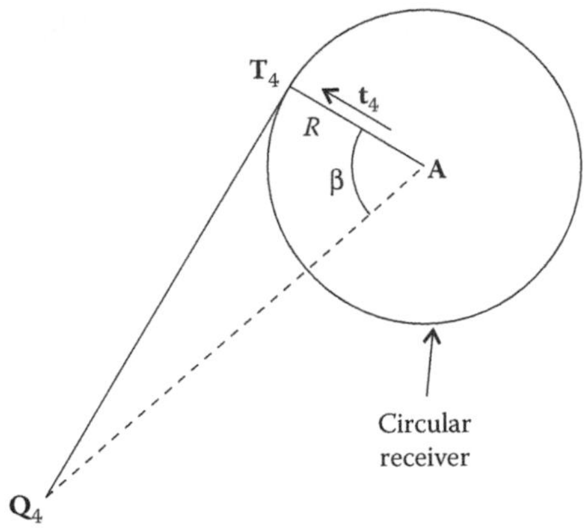

FIGURE 6.78
Calculation of point T_4 on the tangent line to the receiver through point $\mathbf{Q}_4$ on the TERC mirror.

The vector $\mathbf{t}_4$ is then given by

$$\mathbf{t}_4 = R(-\beta) \cdot \text{nrm}(\mathbf{Q}_4 - \mathbf{A}) \tag{6.130}$$

and point $\mathbf{T}_4$ can be obtained as

$$\mathbf{T}_4 = \mathbf{A} + R\mathbf{t}_4 \tag{6.131}$$

Since $\mathbf{Q}_4 = \mathbf{Q}_4(x)$, we have $\mathbf{t}_4 = \mathbf{t}_4(x)$ and $\mathbf{T}_4 = \mathbf{T}_4(x)$. The point $\mathbf{Q}_4$ on the TERC mirror must fulfill the condition of constant optical path length,

$$[\mathbf{W}_4, \mathbf{P}_4] + x + [\mathbf{Q}_4, \mathbf{T}_4] + R\delta_4 = S \tag{6.132}$$

where $\delta_4 = \text{ang}\,(\mathbf{T}_4 - \mathbf{A}, (0, 1))$ and x is the distance between $\mathbf{P}_4$ and $\mathbf{Q}_4$. Solving this equation for x, we get $x = 0.747344$. We can now calculate $\mathbf{Q}_4 = (-0.248578, -0.187783)$ on the TERC mirror and $\mathbf{T}_4 = (-0.131363, 0.00974957)$.

For other values of α we get other points on the parabolic primary mirror and the corresponding points on the TERC mirror.

References

1. Winston, R., Nonimaging optics, *Scientific American*, p. 76, March 1991.
2. Gordon, J., *Solar Energy—The State of the Art, ISES Position Papers*, James & James Science Publishers Ltd, London, 2001.
3. Ries, H., Gordon, J., and Lasken, M., High-flux photovoltaic solar concentrators with kaleidoscope-based optical designs, *Sol. Energy*, 60, 11, 1997.

4. Feuermann, D. and Gordon, J., High-concentration photovoltaic designs based on miniature parabolic dishes, *Sol. Energy*, 70, 423, 2001.
5. Ries, H. and Spirkl, W., Nonimaging secondary concentrators for large rim angle parabolic troughs with tubular absorbers, *App. Opt.*, 35, 2242, 1996.
6. Ries, H. and Spirkl, W., Caustic and its use in designing optimal absorver shapes for 2D concentrators, *Nonimaging Opt.: Maximum Efficiency Light Transfer III*, Proceedings of SPIE, Vol. 2538, 2, 1995.
7. Lawrence, J. D., *A Catalog of Special Plane Curves*, Dover Publications, New York, 1972.
8. Riley, K. F., Hobson, M. P., and Bence, S. J., *Mathematical Methods for Physics and Engineering*, 3rd ed., Cambridge University Press, Cambridge, 2006.
9. Edwards, H. M., *Advanced Calculus, A Differential Forms Approach*, Birkhäuser, Boston, 1969.
10. Rabl, A. and Winston, R., Ideal concentrators for finite sources and restricted exit angles, *Appl. Opt.*, 15, 2880, 1976.
11. Winston, R. and Welford, W. T., Design of nonimaging concentrators as second stages in tandem with image-forming first-stage concentrators, *Appl. Opt.*, 19, 347, 1980.
12. Winston, R., Cone collectors for finite sources, *Appl. Opt.*, 17, 688, 1978.
13. Kritchman, E. M., Second-stage CEC concentrator, *Appl. Opt.*, 21, 751, 1982.
14. Collares-Pereira, M., Rabl, A., and Winston, R., Lens-mirror combinations with maximal concentration, *Appl. Opt.*, 16, 2677, 1977.
15. Collares-Pereira, M., High temperature solar collector with optimal concentration: Non-focusing Fresnel lens with secondary concentrator, *Sol. Energy*, 23, 409, 1979.
16. Winston, R. and Welford, W. T., Geometrical vector flux and some new nonimaging concentrators, *J. Opt. Soc. Am. A*, 69, 532, 1979.
17. O'Gallagher, J., Winston, R., and Welford, W. T., Axially symmetrical nonimaging flux concentrators with the maximum theoretical concentration ratio, *J. Opt. Soc. Am.*, 4, 66, 1987.
18. Kritchman, E. M., Nonimaging second-stage elements: A brief comparison, *Appl. Opt.*, 20, 3824, 1981.
19. Kritchman, E., Optimized second stage concentrator, *Appl. Opt.*, 20, 2929, 1981.
20. Rabl, A., Comparison of solar concentrators, *Sol. Energy*, 18, 93, 1976.
21. Collares-Pereira, M. et al., High concentration two-stage optics for parabolic trough solar collectors with tubular absorber and a large rim angle, *Sol. Energy*, 47, 6, 457, 1991.
22. Mills, D. R., Two-stage solar collectors approaching maximal concentration, *Sol. Energy*, 54, 41, 1995.
23. Friedman, R. P., Gordon, J. M., and Ries, H., New high flux two-stage optical designs for parabolic solar concentrators, *Sol. Energy*, 51, 317, 1993.
24. Friedman, R. P., Gordon, J. M., and Ries, H., Compact high-flux two-stage solar collectors based on tailored edge-ray concentrators, *Sol. Energy*, 56, 607, 1996.
25. Benitez, P. et al., Design of CPC-like reflectors within the simultaneous multiple-surface design method, *Nonimaging Opt.: Maximum Efficiency Light Transfer IV*, SPIE Vol. 3139, 19, 1997.
26. Gordon, J. M. and Ries, H., Tailored edge-ray concentrators as ideal second stages for Fresnel reflectors, *Appl. Opt.*, 32, 2243, 1993.

7

Stepped Flow-Line Nonimaging Optics

7.1 Introduction

Typically, optical devices designed using the flow-line method, are quite large in the case of small acceptance angles, and they also need to touch the edges of the receiver. A possible way around this limitation is to consider a step curve along the flow-lines, with some portions of the curve along the flow-lines and others perpendicular to those lines. The portions along the flow-lines are converted to mirrors, and to those perpendicular to the flow-lines, we add optics. This results in a microstructured optic, with many small optical elements combined into one. Different versions of these devices have numerous applications, such as very compact concentrators, concentrators that do not touch the receiver, backlights and frontlights, and light guides that distribute the light of a source to several receivers, or those that combine the light from several small sources onto a single exit aperture (synthetic large source).

7.2 Compact Concentrators

The compound parabolic concentrator (CPC) can be derived by the flow-line method. For large acceptance angles, the size of the CPC is reasonable, but for small acceptance angles, it becomes very tall. Combining several small CPCs into a single device produces an equivalent, much shorter concentrator.[1]

Figure 7.1 shows a concentrator for an inverted V-shaped receiver **AFB**. Up to the dashed line *s*, the left-hand-side vector flux lines f_i inside this device are parabolas having a focus **F** and an axis parallel to *r*, which is in line with **AF**. Upward from line *s*, they are all straight lines. All of them have the same shape, only scaled upward or downward, as each one of them defines the same concentrator for different receiver sizes. The right-hand-side vector flux lines are symmetrical to ones on the left-hand-side.

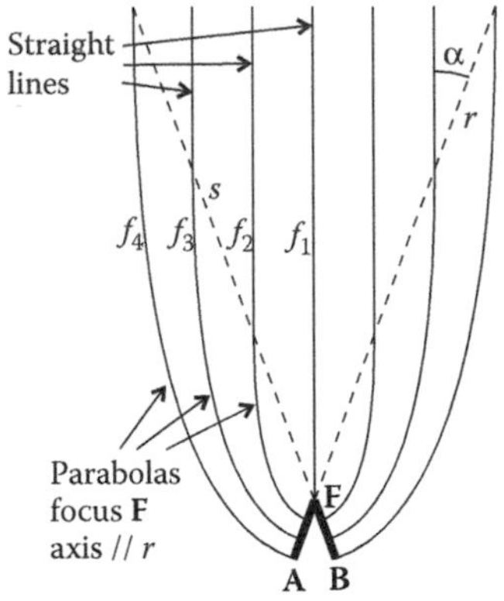

FIGURE 7.1
The vector flux lines inside a CPC for an inverted V-shaped receiver are shaped as parabolas below the dashed lines and straight lines above them.

Taking two of these vector flux lines, we can obtain different concentrators. Figure 7.2 shows three examples. In the case of Figure 7.2a, the radiation is "compressed" by multiple reflections bouncing back and forth between the two parabolas into the receiver R. In the case of Figure 7.2b, the receiver is asymmetric. Figure 7.2c shows a CPC for a tilted receiver. This concentrator has an acceptance angle 2α and results from taking flow-line f_1 in Figure 7.1 as a mirror (which divides in half the CPC for the inverted V-shaped receiver).

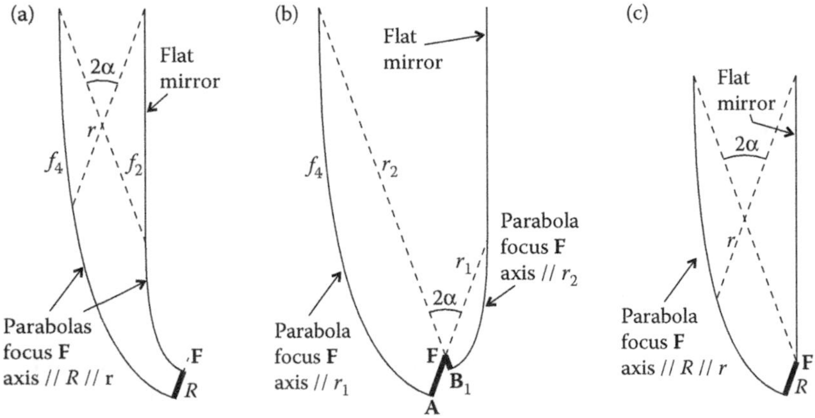

FIGURE 7.2
Different optics can be obtained by taking different flow-lines inside a CPC for an inverted V-shaped receiver. (a) A concentrator with receiver R and composed of two parabolic arcs having the same focus and axis direction. (b) A concentrator for an asymmetrical receiver. (c) CPC for a tilted receiver. It has an acceptance angle 2α, and results from halving a CPC for an inverted V-shaped receiver with a vertical flow-line.

It is possible, however, to take other shapes along the vector flux lines. Figure 7.3a shows one such possibility. In this case, we take a stepped line where the vertical lines, v_i, are along the vector flux lines, but the horizontal ones, h_i, are perpendicular to them.

If **AF** is now a source, its radiation will exit through h_1, h_2, h_3, … and these horizontal lines can now be considered as small "sources," emitting radiation with the same angular aperture 2α as the acceptance angle of the concentrator. A "source" h_i will now have a maximum space, e_i, from $\mathbf{P}_i$ to $\mathbf{Q}_i$, through which its radiation can exit, as shown in Figure 7.3b. This means that we can decrease the angular spread of the radiation to increase its aperture area, and make it exit perpendicular to e_i filling it completely. If h_i and e_i refer now to the sizes, étendue conservation requires that $2h_i \sin\alpha = 2e_i \sin\theta$. Since we have $h_i = e_i \sin\alpha$ we obtain

$$\theta = \arcsin(\sin^2\alpha) \tag{7.1}$$

For the device in Figure 7.3a, we can now put on top of each source h_i a concentrator such as the one shown in Figure 7.2c but turned upside down, as shown in Figure 7.4.

This transforms the source **AF** into a set of smaller collinear sources S_i. Their added widths equal that of **AF**. On top of each one of these exits, S_i, we can now put a CPC with an acceptance angle θ calculated according to expression (7.1). The resulting device is a compact optic shown in Figure 7.5a. This optic can now be seen as a concentrator having an acceptance angle 2θ and a receiver **AF**. Horizontal dashed line **GF** divides the compact optic in

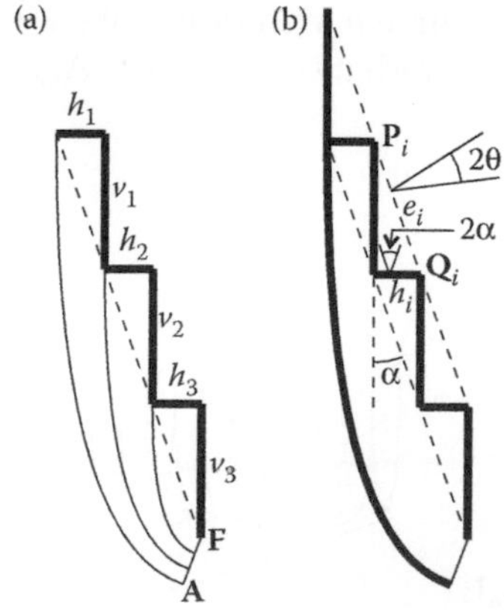

FIGURE 7.3
(a) If **AF** is a source of radiation, its light will be divided into h_1, h_2, and h_3, which can now also be considered as small sources. (b) Each one of these sources h_i now has a distance e_i through which its radiation can exit. This means that the angular aperture of the radiation can be decreased to increase its area and fill e_i.

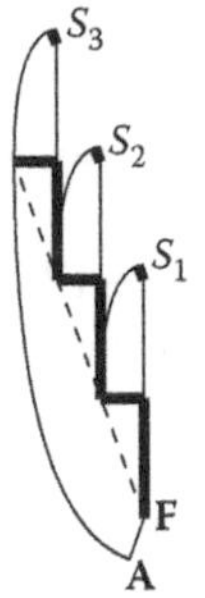

FIGURE 7.4
Placing a CPC for a tilted receiver (as in Figure 7.2c) on top of each horizontal line, we divide the source **AF** into smaller sources S_i.

upper optics (above **GF**) and lower optic (below **GF**). The lower optic **GA** has a parabolic shape and concentrates to **F** the edge rays parallel to r, making an angle 2α to the horizontal. Figure 7.5b shows a similar device, except that it combines a larger number of CPCs.

Now consider a light pipe composed of two vertical parallel flat mirrors, as shown in Figure 7.6a. The length of the light pipe is such that the lines that connect its opposing edges make an angle 2β. If light having an angular aperture 2α is injected at one end, it will bounce around until it exits at the other end with the same angular aperture. The vector flux lines inside this optic are straight, vertical parallel lines. We can now take a stepped line along these lines, just as we did previously. The result is shown in Figure 7.6b.

As described earlier, a source h_i will now have a maximum width e_i from $\mathbf{P}_i$ to $\mathbf{Q}_i$, through which its radiation can exit. Again, we can decrease the

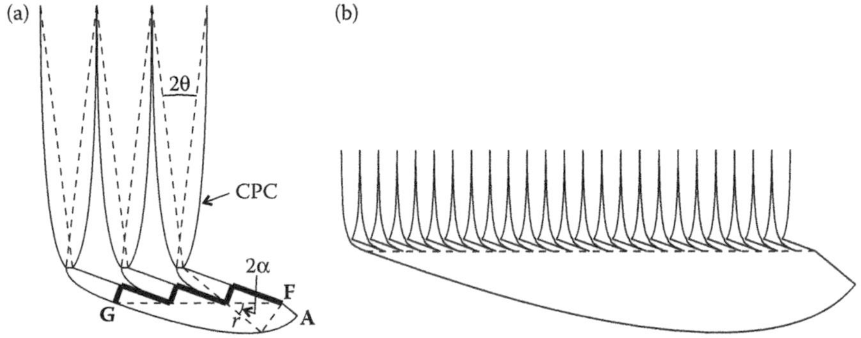

FIGURE 7.5
(a) Compact optics: concentrators that result from a combination of several CPCs into a single receiver. (b) Same optic as in (a), but now with many more upper optics, resulting in a much more compact design.

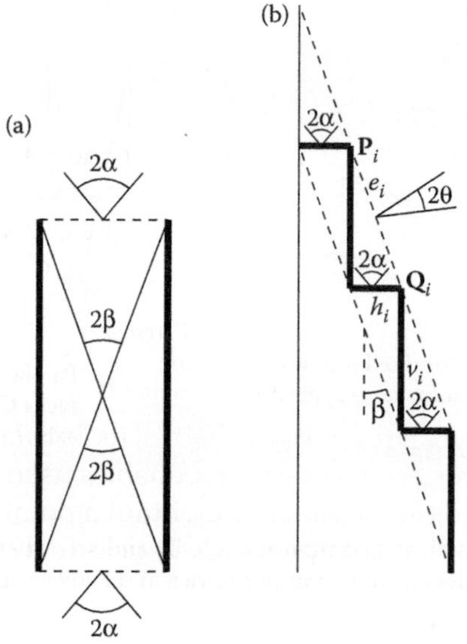

FIGURE 7.6
(a) Light pipe composed of two vertical flat mirrors. If light with angular aperture 2α is injected at one end, it exits at the other end with the same angular extent. (b) The vector flux lines inside this device are vertical straight lines. A stepped vector flux line can also be taken in this case.

angular spread of the radiation to increase its area, and make it exit perpendicular to e_i, filling it completely. If h_i and e_i refer now to the sizes, étendue conservation requires that $2h_i \sin\alpha = 2e_i \sin\theta$. Since $h_i = e_i \sin\beta$ we have

$$\theta = \arcsin(\sin\alpha \sin\beta) \tag{7.2}$$

As described previously, on top of each source h_i we can now put a CPC for a tilted receiver (similar to that of Figure 7.2c, but turned upside down. Because of the difference between angles α and β, we also need circular arcs with an angle α – β to make the exit apertures of these optics collinear. We can now add the CPCs with an acceptance angle 2θ, where θ is determined by expression (7.2). We end up with the device as shown in Figure 7.7a.

Figure 7.7b shows the same device, but rotated so that the CPCs are in the vertical position. This optic can now be seen as an angle transformer having an acceptance angle 2θ and an exit angle 2α. The horizontal dashed line *b* divides the optic into two. The optics above this line are called the upper optics, and the optic below this line is called the lower optic. Figure 7.8 shows an upper optic from Figure 7.7b.

Since these concentrators result from the combination of several small CPCs, they are much more compact than the CPCs themselves.

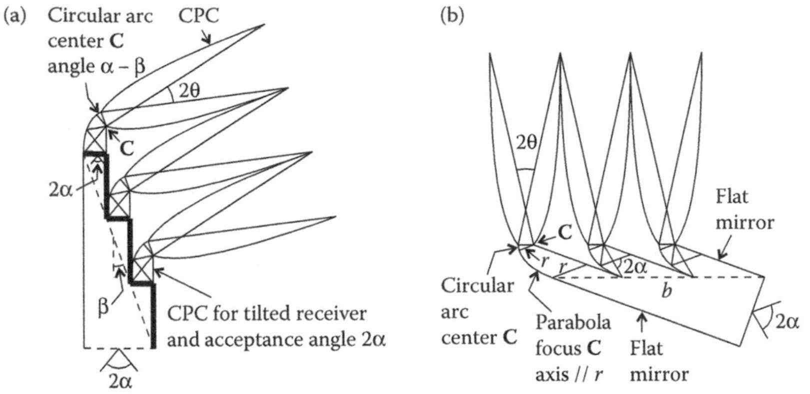

FIGURE 7.7
(a) An angle transformer with an acceptance angle 2θ and an exit angle 2α. (b) The horizontal dashed line *b* divides the device into an upper optics and a lower optic.

Other such designs can combine several light sources into a single exit aperture.[2,3] One such possibility uses a combination of several optics, like the one seen in Figure 7.9.

Figure 7.9 shows a CPC with a small aperture *R* and a large aperture **AM** coupled to an angle rotator. Several conic curves form the sidewalls of this angle rotator. Between points **K** and **J**, light is not confined by a mirror. The optic may then be extended to the right of line **K-J** with a nonoptical surface. This extra material between line **K-J** and the nonoptical surface has no optical function and may be used, for example, for holding the optic.

The wall **ML** of the angle rotator is flat, and reflects edge rays parallel to direction r_2 into a direction parallel to r_3. The portion **LK** is a parabola with a focus **A** and an axis parallel to r_3. On the other side, the curve **AB** is a parabola with a focus **J**, and an axis parallel to r_1. The curve **BC** is an

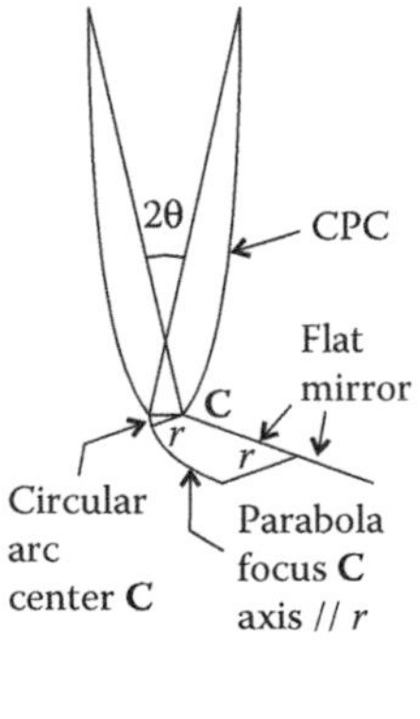

FIGURE 7.8
An upper optic for the device shown in Figure 7.7.

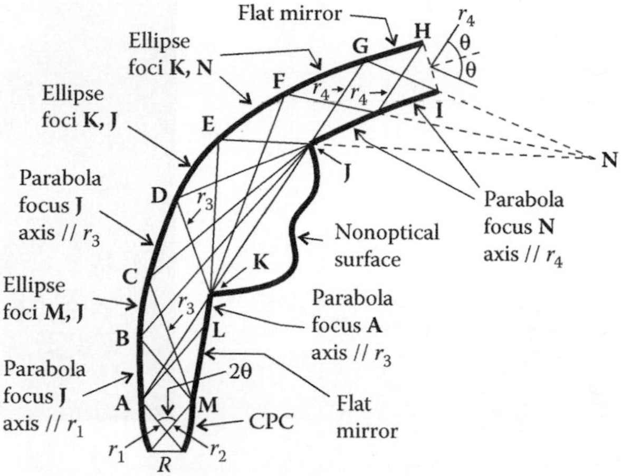

FIGURE 7.9
CPC and an angle rotator. Between points **K** and **J** there is no need for a mirror to confine the light.

ellipse with foci **M** and **J**. The curve **CD** is a parabola with a focus **J**, and an axis parallel to r_3. The curve **DE** is an ellipse with foci **K** and **J**. The curve **EG** is an ellipse with foci **K** and **N**. This side of the optic ends in a flat portion **GH**. On the other side, we have a parabola **IJ** with a focus **N** and an axis parallel to r_4.

Several of these optics can be combined using a stepped flow-line approach, as shown in Figure 7.10.

Note that if the direction of the light is reversed, we can place a large source at the exit aperture, and this optic will distribute its light to several places.

Figure 7.11a shows an optic that combines four light sources into a single exit aperture. Figure 7.11b shows an optic that combines two light sources into a single optical aperture. The two "legs" of the optic have nonoptical surfaces (with holes) for mechanical assembly.

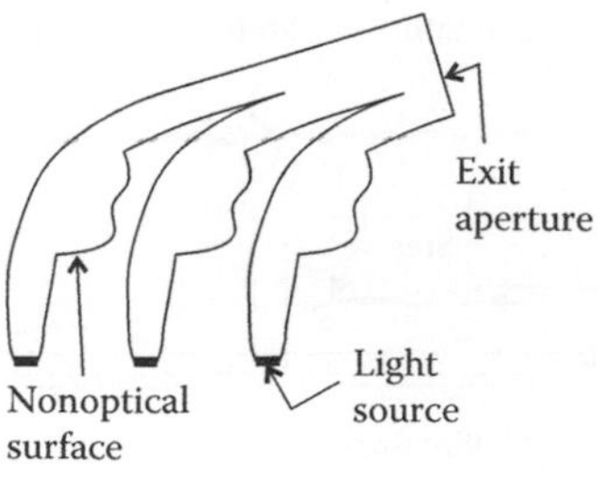

FIGURE 7.10
Combination of several light sources into one single exit aperture by a stepped flow-line optic.

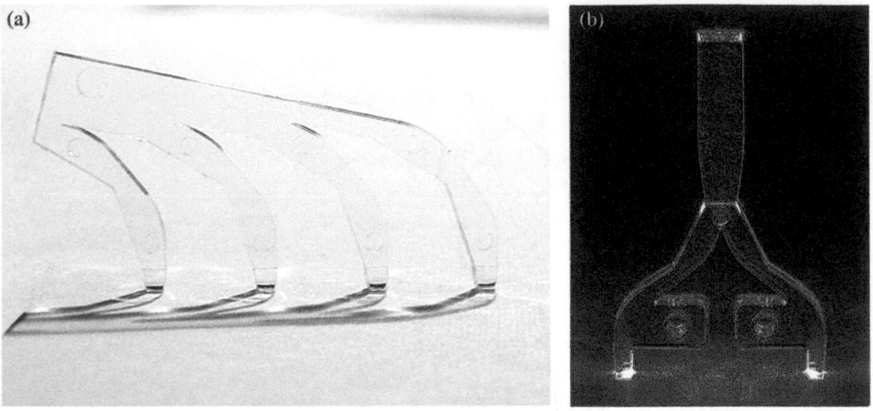

FIGURE 7.11
(a) Combiner optic with four legs. (b) Combiner optic with two legs. Both legs have nonoptical surfaces for mechanical assembly. (Courtesy of Light Prescriptions Innovators.)

Another possible application of stepped flow-line optics is in high efficiency backlights and frontlights.[4] A simple example is to replace the horizontal steps shown in Figure 7.6 by tilted steps, as shown in Figure 7.12, which reflect the light perpendicular to the direction of the light guide.

The resulting optic consists of a light source, a collimator, and a stepped flow-line optic. In the case of a backlight (Figure 7.12a), light is reflected through a transmissive screen, and will be observed on the other side. In the case of a frontlight (Figure 7.12b), light is reflected toward a reflective screen, which is reflected back through the optic and then observed. The size of

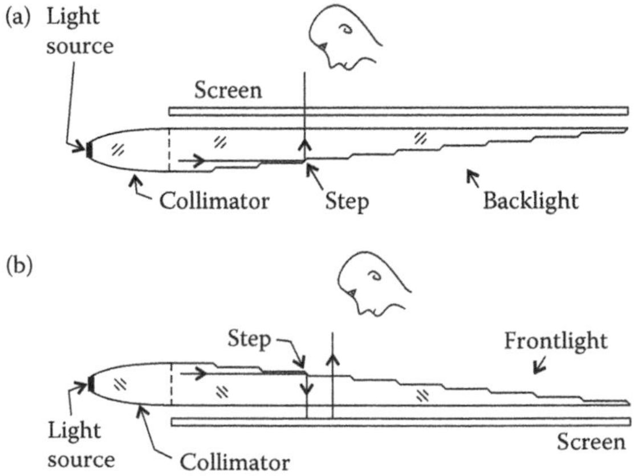

FIGURE 7.12
Stepped flow-line optics can be used in the design of backlights (a) and frontlights (b).

the steps may be made very small compared to the spacing between steps, making them imperceptible to the observer. The frontlight then behaves as a transparent plate that illuminates a reflective screen. More elaborate designs may illuminate a large target with a single small source, or have a thickness that is nearly constant for the whole optic.[4]

7.3 Concentrators with Gaps

Flow-line design methods typically produce designs in which the mirrors touch the source of radiation or the receiver. It is, however, possible to modify these designs to obtain ideal concentrators that do not touch either the source or the receiver.[5]

The CPC in Figure 7.8 can be seen as a concentrator for an infinite source placed at an infinite distance. If, however, the source is now, for example, a circle placed at a finite distance, a tailored concentrator for this new source can also be designed.

Let us then consider a circular Lambertian source of radius r, as shown in Figure 7.13a. Some of its light is captured by an optic having an acceptance angle 2θ, and whose entrance aperture is a circular arc **AB** of radius R. The entrance aperture spans an angle δ at the center of the source, and its arc length is $R\delta$. The étendue entering the optic will then be $U = 2R\,\delta \sin\theta$. Now, R and r are related by $r = R \sin\theta$ and therefore, we can write $U = 2r\delta$. If this radiation is transferred to the exit aperture $\mathbf{F}_1\mathbf{F}_2$ (Figure 7.13b) and concentrated to the maximum (exit angle $\pi/2$), we have $U_2 = 2[\mathbf{F}_1, \mathbf{F}_2]$ for the étendue exiting the device, where $[\mathbf{F}_1, \mathbf{F}_2]$ is the distance between $\mathbf{F}_1$ and $\mathbf{F}_2$. We should

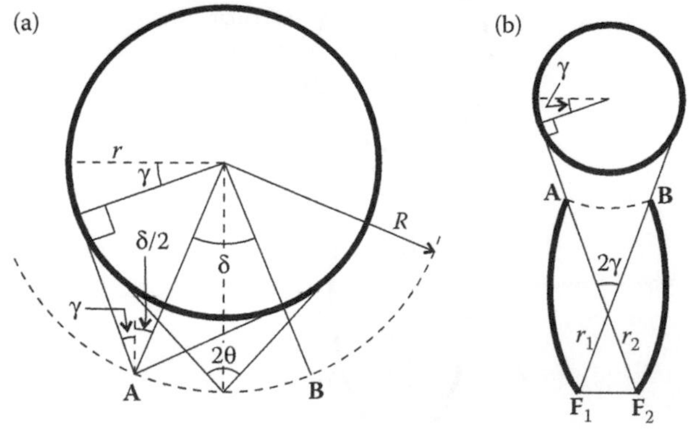

FIGURE 7.13
(a) Circular light source with a radius r and geometry for the entrance aperture **AB** of a concentrator that captures light from this circular source. (b) Concentrator designed for this circular light source.

then have $2r\delta = 2[\mathbf{F}_1, \mathbf{F}_2]$ and, therefore, $[\mathbf{F}_1, \mathbf{F}_2] = r\delta$. The acceptance angle 2θ of the optic is related to the angle 2γ between rays r_1 and r_2 by $\theta = \gamma + \delta/2$. We may then calculate angle 2γ from the values of θ and δ, and, therefore, the directions of rays r_1 and r_2 and, thus, determine the positions of points $\mathbf{F}_1$ and $\mathbf{F}_2$.

Figure 7.14 shows one of these devices for the acceptance angle 2θ. Each one of its points **P** is calculated so that $l_1 + l_2 + r\varphi = Cte$, where *Cte* is a constant.

The whole upper optic as seen in Figure 7.8 can also be adjusted so that several component optics can be placed around the circular source. Figure 7.15a shows one of these modified upper optics. Now the parabola and flat mirror have become elliptical arcs. Figure 7.15b shows how these optics can be put around a circular source. It shows the upper optic in Figure 7.15a, together with a second upper optic obtained by a rotation of an angle δ around the center **C** of the source.

The reason for the angles δ with vertices $\mathbf{P}_1$ and $\mathbf{P}_2$ in Figure 7.15a is that, when this optic is rotated by δ around center **C** to become the next optic in the chain, the line $\mathbf{P}_1\mathbf{P}_3$ of the next optic will be parallel to $\mathbf{P}_1\mathbf{P}_4$ and the line $\mathbf{P}_2\mathbf{P}_5$ of the next optic will be parallel to $\mathbf{P}_2\mathbf{P}_4$.

Let us now consider that $\mathbf{v}_{jk}$ is a unit vector pointing from point $\mathbf{P}_j$ to point $\mathbf{P}_k$, as shown in Figure 7.16. This vector can also be written as $\mathbf{v}_{jk} = (\cos\theta_{jk}, \sin\theta_{jk})$, where θ_{jk} is the angle that line $\mathbf{P}_j\mathbf{P}_k$ makes to the horizontal. Point $\mathbf{P}_k$ can then be obtained from point $\mathbf{P}_j$ as $\mathbf{P}_k = \mathbf{P}_j + x_{jk}\mathbf{v}_{jk}$, where $x_{jk} = [\mathbf{P}_j, \mathbf{P}_k]$ is the distance between $\mathbf{P}_j$ and $\mathbf{P}_k$.

Now for calculating the geometry of the upper optic shown in Figure 7.15, we first note that point $\mathbf{P}_3 = \mathbf{F}_2$, as in Figure 7.14. We can now make $\mathbf{v}_{35} = (\cos\theta_{35}, \sin\theta_{35})$ and $\mathbf{v}_{36} = (\cos\theta_{36}, \sin\theta_{36})$, where angles θ_{35} and θ_{36} are unknown. Since

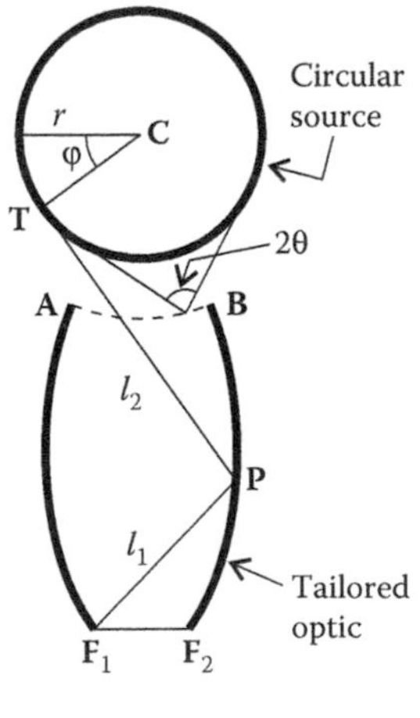

FIGURE 7.14
Concentrator for a circular source of radius r and a receiver $\mathbf{F}_1\mathbf{F}_2$. Its points can be drawn with a string of constant length, that is, $l_1 + l_2 + r\varphi = Cte$, where *Cte* is a constant.

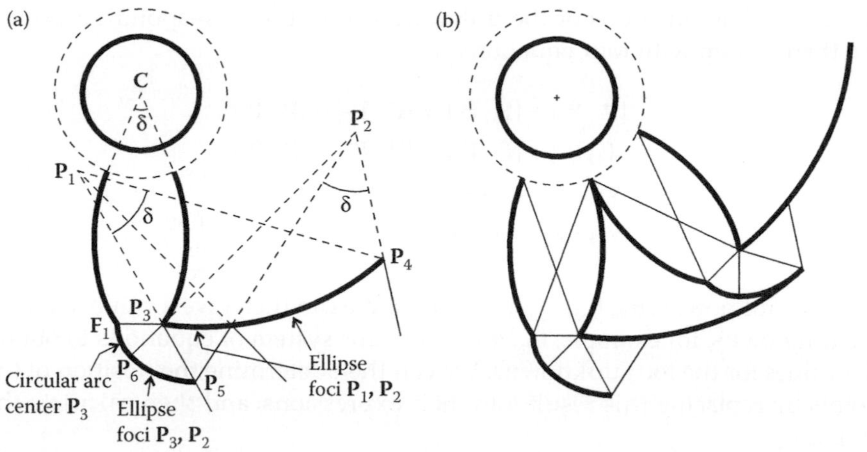

FIGURE 7.15
(a) An upper optic for a circular source. (b) Two of these optics can be combined around the source.

$\mathbf{P}_6$ is connected to point $\mathbf{F}_1$ by a circular arc, it can be obtained as $\mathbf{P}_6 = [\mathbf{F}_1, \mathbf{F}_2]\,\mathbf{v}_{36}$. Point $\mathbf{P}_5$ can be obtained as $\mathbf{P}_5 = \mathbf{P}_3 + x_{35}\mathbf{v}_{35}$, where x_{35} is unknown. Points $\mathbf{P}_1$ and $\mathbf{P}_2$ can be obtained by $\mathbf{P}_1 = \mathbf{P}_3 - x_{31}\mathbf{v}_{35}$ and $\mathbf{P}_2 = \mathbf{P}_3 - x_{32}\mathbf{v}_{36}$, where distances x_{31} and x_{32} are unknown. Point $\mathbf{P}_4$ can be obtained by rotating point $\mathbf{P}_5$ by an angle δ around center $\mathbf{C}$, and, therefore, we have $\mathbf{P}_4 = R(\delta) \cdot \mathbf{P}_5$, where $R(\delta)$ is a rotation matrix of an angle δ.

All the points are now defined as functions of the five unknowns: θ_{35}, θ_{36}, x_{35}, x_{31}, and x_{32}. We can now impose on the system the condition resulting from the curves and angles δ in Figure 7.15a. For the ellipse with foci $\mathbf{P}_2$ and $\mathbf{P}_3$, we have $[\mathbf{P}_3, \mathbf{P}_6] + [\mathbf{P}_6, \mathbf{P}_2] = [\mathbf{P}_3, \mathbf{P}_5] + [\mathbf{P}_5, \mathbf{P}_2]$. For the ellipse with foci $\mathbf{P}_1$ and $\mathbf{P}_2$, we have $[\mathbf{P}_1, \mathbf{P}_3] + [\mathbf{P}_3, \mathbf{P}_2] = [\mathbf{P}_1, \mathbf{P}_4] + [\mathbf{P}_4, \mathbf{P}_2]$. We must also impose the condition that $\mathbf{v}_{35} \cdot \mathbf{v}_{14} = \cos\delta$ and $\mathbf{v}_{25} \cdot \mathbf{v}_{24} = \cos\delta$. Vectors $\mathbf{v}_{14}$, $\mathbf{v}_{25}$, and $\mathbf{v}_{24}$ can

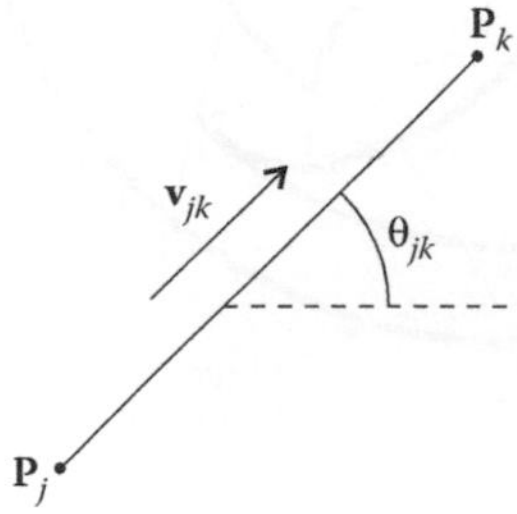

FIGURE 7.16
Definition of vector $\mathbf{v}_{jk}$ and angle θ_{jk} from two points, $\mathbf{P}_j$ and $\mathbf{P}_k$.

be obtained as functions of the unknowns from the corresponding points. We then end up with four equations

$$\begin{aligned} [\mathbf{P}_3,\mathbf{P}_6]+[\mathbf{P}_6,\mathbf{P}_2] &= [\mathbf{P}_3,\mathbf{P}_5]+[\mathbf{P}_5,\mathbf{P}_2] \\ [\mathbf{P}_1,\mathbf{P}_3]+[\mathbf{P}_3,\mathbf{P}_2] &= [\mathbf{P}_1,\mathbf{P}_4]+[\mathbf{P}_4,\mathbf{P}_2] \\ \mathbf{v}_{35}\cdot\mathbf{v}_{14} &= \cos\delta \\ \mathbf{v}_{25}\cdot\mathbf{v}_{24} &= \cos\delta \end{aligned} \tag{7.3}$$

and five unknowns: θ_{35}, θ_{36}, x_{35}, x_{31}, and x_{32}. We can then give a value to one of the unknowns, for example, θ_{36}, and solve this system of equations to obtain the values for the four unknowns. We can then determine the position of the points by replacing this result into their expressions, and then calculate the ellipses.

Once the upper optic has been defined, the corresponding lower optic can be calculated. Figure 7.17 shows an optic having two upper optics and the corresponding lower optic. The lower optic is composed of elliptical arcs that concentrate the light they receive from $\mathbf{P}_1$ and $\mathbf{Q}_1$ to $\mathbf{P}_2$ and $\mathbf{Q}_2$. Just like the whole second upper optic, also points $\mathbf{Q}_1$ and $\mathbf{Q}_2$ are obtained from $\mathbf{P}_1$ and $\mathbf{P}_2$ by a rotation of an angle δ about the center $\mathbf{C}$ of the source.

For the points of the lower optic between $\mathbf{L}_1$ (point $\mathbf{P}_5$ in Figure 6.14a) and $\mathbf{L}_2$, points $\mathbf{P}_1$ and $\mathbf{P}_2$ are "visible." The edge rays at those points on the lower optic are those appearing to come from $\mathbf{P}_1$ and those headed to $\mathbf{P}_2$ and, therefore, require an ellipse having foci $\mathbf{P}_1$ and $\mathbf{P}_2$. Beyond point $\mathbf{L}_2$, point $\mathbf{Q}_2$ becomes "visible."

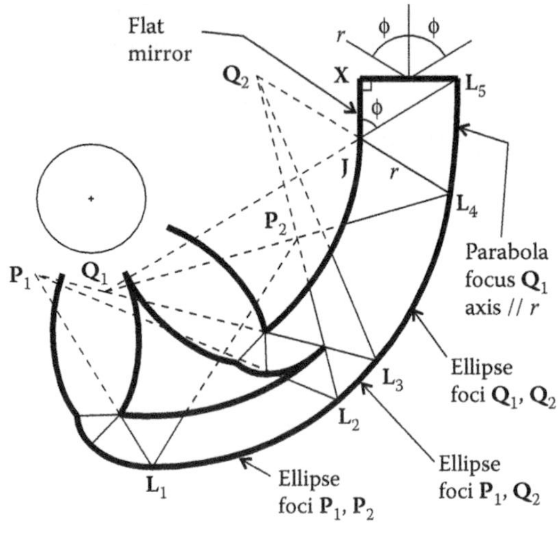

FIGURE 7.17
The lower portion of an optic that captures light coming from a circular source is composed of a set of elliptical arcs that concentrate to $\mathbf{P}_2$ and $\mathbf{Q}_2$, the edge rays it receives from $\mathbf{P}_1$ and $\mathbf{Q}_1$.

Thus, there must be an ellipse with foci $\mathbf{P}_1$ and $\mathbf{Q}_2$. At point $\mathbf{L}_3$, point $\mathbf{P}_1$ stops being visible and point $\mathbf{Q}_1$ becomes visible; hence there is an ellipse with foci $\mathbf{Q}_1$ and $\mathbf{Q}_2$ that extends until point $\mathbf{L}_4$, where $\mathbf{Q}_2$ stops being visible.

The design of the final portion of the optic starts by defining the flat mirror **JX**, which is tangent to the upper optic at point **J**. The angle ϕ between line $\mathbf{Q}_1\mathbf{J}$ and **JX** was chosen as the exit angle of the device. The étendue of the radiation entering the optic coming from the source is known, and therefore the distance [**X**, $\mathbf{L}_5$] can be calculated. Since the exit aperture $\mathbf{XL}_5$ is perpendicular to the flat mirror **JX**, point $\mathbf{L}_5$ can be determined. The design is completed by a parabolic arc from $\mathbf{L}_4$ to $\mathbf{L}_5$.

A large number of upper optics can be placed around the source, completely surrounding it.[6] Figure 7.18 shows this possibility. The upper optics are the same as shown in Figure 7.17, and the design method of the lower optic is also the same.

This device is a gap optic and can also be viewed as a concentrator having an acceptance angle 2ϕ and a circular receiver. In this case, it can be seen that the optic does not touch the central circular receiver.

It is possible to combine a compact optic (Figure 7.5) with a gap optic (Figure 7.18). To the gap optic of Figure 7.18, we remove a portion of the lower optic (above point $\mathbf{L}_{16}$), and adjust the length of the vertical flat mirror, resulting in the optic of Figure 7.19 (shown rotated by 90° relative to the position in Figure 7.18).

To optically connect these two optics, we start with the flat mirror $\mathbf{P}_5\mathbf{C}_1$, which is tangent to the upper optic at point $\mathbf{P}_5$ (defined in Figure 7.15a. Its length defines angle 2α, and is chosen so as to match angle 2α of the upper optics of the compact optic.

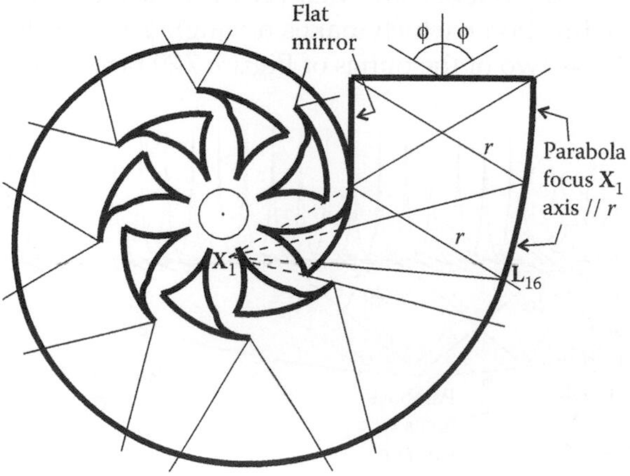

FIGURE 7.18
Gap optic: a concentrator having an acceptance angle 2α that does not touch its circular receiver.

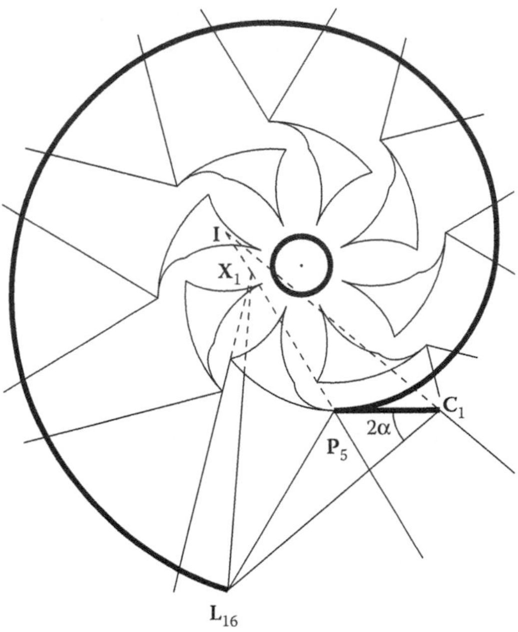

FIGURE 7.19
To optically connect a gap optic and a compact optic, we start from mirror $\mathbf{P}_5\mathbf{C}_1$, tangent to the upper optic at $\mathbf{P}_5$, and whose chosen length defines the angle 2α that matches that for the upper optics of the compact optic.

Figure 7.20 shows a combination of a compact optic and a gap optic. Point **I** is a mirror image of point $\mathbf{L}_{16}$, with $\mathbf{P}_5\mathbf{C}_1$ as the mirror. The upper optics of the compact optic are similar to the ones in Figure 7.5. The lower optic is composed of a set of parabolic arcs that redirect the edge rays coming from the gap optic in direction r, which makes an angle 2α to the horizontal.

Figure 7.21 shows two of the optics of Figure 7.20 placed side by side.

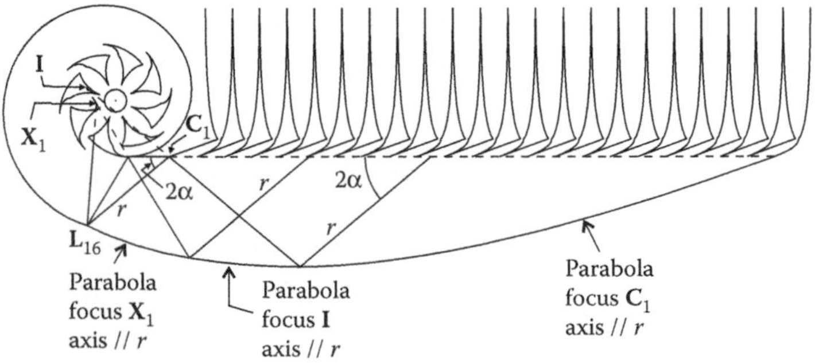

FIGURE 7.20
Combination of a compact optic and a concentrator with a gap between the optic and the receiver.

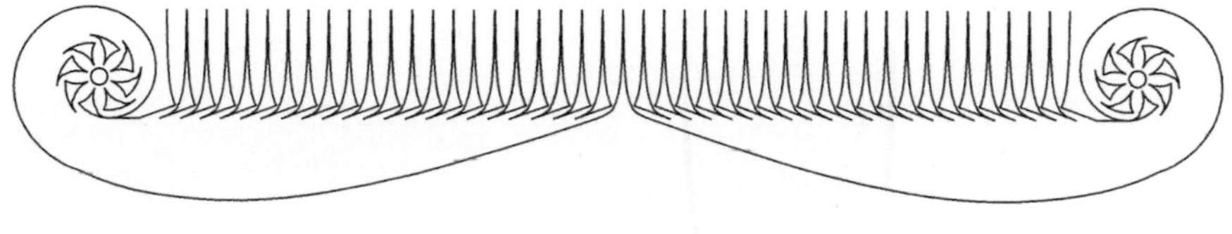

FIGURE 7.21
Two of the optics seen in Figure 7.20 placed side by side.

7.4 Examples

The following examples use expressions for the curves and functions that are derived in Chapter 21.

EXAMPLE 7.1

Design a stepped flow-line concentrator for a half-acceptance angle of $\theta = 20°$.

We start by the parameterization of a CPC. The right-hand-side parabola of a CPC with a half-acceptance angle θ and a small aperture of dimension $2a$ centered at the origin is given by

$$\mathbf{c}_R(\phi) = (c_1(\phi), c_2(\phi))$$
$$= \left(a\frac{1 - \cos(\phi + 2\theta) + 2\sin(\phi + \theta)}{\cos\phi - 1}, a\frac{\cos(\phi + \theta)}{\sin^2(\phi/2)}(1 + \sin\theta) \right) \quad (7.4)$$

with $3\pi/2 - \theta \leq \varphi \leq 2\pi - 2\theta$, as shown in Figure 7.22.

The left-hand-side parabola is symmetrical relative to the vertical axis x_2, and is given by $\mathbf{c}_L(\phi) = (-c_1(\phi), c_2(\phi))$. If the center of the small aperture

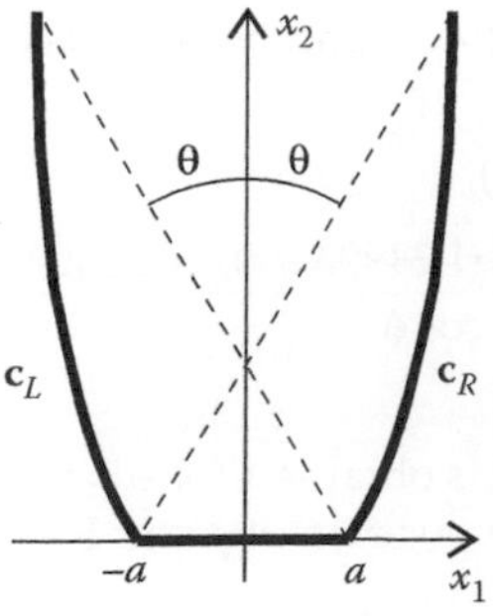

FIGURE 7.22
CPC with a small aperture of dimension $2a$ centered at the origin.

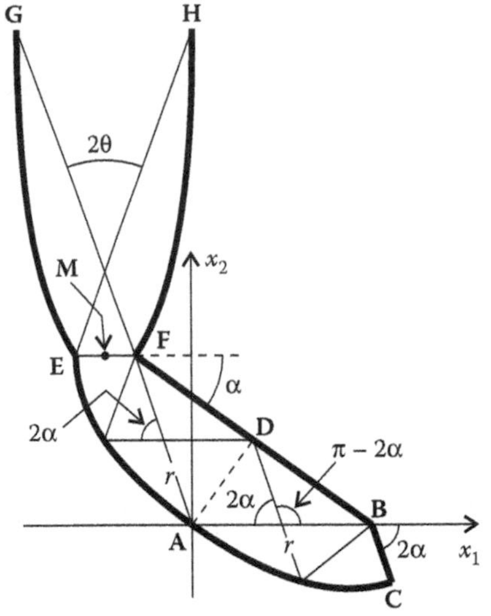

FIGURE 7.23
Stepped flow-line concentrator with an upper optic (above **AB**) and a lower optic (below **AB**).

is now at a point **M** instead of at (0,0), then the parabolas of the CPC are given by $\mathbf{c}_R(\phi) + \mathbf{M}$ and $\mathbf{c}_L(\phi) + \mathbf{M}$ with the same parameter range.

To design the stepped flow-line concentrator, consider that the horizontal line separating the lower optic from the upper optic is bounded by points $\mathbf{A} = (0,0)$ and $\mathbf{B} = (1,0)$, as shown in Figure 7.23.

Begin with the lower optic below line **AB**. Design it as a parabola **AC** tilted by an angle $\pi - 2\alpha$ to the horizontal, that is, with its axis parallel to r. The half-acceptance angle for the optic is $\theta = 20°$ and the angle α that defines the lower optic is given by

$$\alpha = \arcsin(\sqrt{\sin\theta}) = 35.7906\text{ deg} \tag{7.5}$$

The parameterization of lower parabola **AC** is then

$$\begin{aligned}\mathbf{a}(\phi) &= \text{par}(\pi - 2\alpha, \mathbf{B}, \mathbf{A}) \\ &= \left(1 - \frac{0.68404\cos(1.24933 - \phi)}{1 - \cos\phi}, \frac{0.68404\sin(1.24933 - \phi)}{1 - \cos\phi}\right)\end{aligned} \tag{7.6}$$

where $2\alpha \le \varphi \le \pi$. Point **C** is obtained for $\phi = \pi$ as $\mathbf{C} = (1.10806, -0.324499)$. We can now calculate the position of point **F** as

$$\begin{aligned}\mathbf{F} &= \text{isl}(\mathbf{B}, (\cos(-\alpha), \sin(-\alpha)), \mathbf{A}, (\cos(-2\alpha), \sin(-2\alpha))) \\ &= (-0.31596, 0.948773)\end{aligned} \tag{7.7}$$

Parabola **AE** has a horizontal axis and focus **F**, and is given by

$$\mathbf{b}(\phi) = \text{par}(0,\mathbf{F},\mathbf{A}) = \left(-0.31596 + \frac{0.68404\cos\phi}{1-\cos\phi}, 0.948773 + \frac{0.68404\sin\phi}{1-\cos\phi}\right) \quad (7.8)$$

where $\pi \leq \varphi \leq 2\pi - 2\alpha$. Point **E** is obtained for $\phi = \pi$ as $\mathbf{E} = (-0.65798, 0.948773)$. Finally, add the top CPC whose sidewalls are parabolas **EG** and **FH**. The midpoint of its small aperture is $\mathbf{M} = \mathbf{E} + 0.5(\mathbf{F} - \mathbf{E}) = (-0.48697, 0.948773)$, and the parameter a, that defines the small aperture **EF**, is given by $a = [\mathbf{E}, \mathbf{F}]/2 = 0.17101$. The parabolas of the CPC are then

$$\mathbf{c}_R(\phi) + \mathbf{M} = \left(-0.48697 + 0.17101\frac{1-\cos(0.698132+\phi)+2\sin(0.349066+\phi)}{\cos\phi - 1}, 0.948773 + 0.229499\frac{\cos(0.349066+\phi)}{\sin^2(0.5\phi)}\right) \quad (7.9)$$

for the right-hand-side and

$$\mathbf{c}_L(\phi) + \mathbf{M} = \left(-0.48697 - 0.17101\frac{1-\cos(0.698132+\phi)+2\sin(0.349066+\phi)}{\cos\phi - 1}, 0.948773 + 0.229499\frac{\cos(0.349066+\phi)}{\sin^2(0.5\phi)}\right) \quad (7.10)$$

for the left-hand-side, where $3\pi/2 - \theta \leq \varphi \leq 2\pi - 2\theta$.

To combine several upper optics with one single lower optic, scale the lower optic and make an array of upper optics to cover it. For example, to combine 10 upper optics, scale the lower optic by a factor of 10 obtaining a parameterization $10\mathbf{a}(\phi)$ with the same parameter range. The parabolas **AE** of the upper optics are now an array

$$\mathbf{b}(\phi) + i(\mathbf{B} - \mathbf{A}), \quad i = 0,1,\ldots,9 \quad (7.11)$$

with the same parameter range as before for each one of them. The same is true for the parabolas of the CPCs that are now given by

$$\mathbf{c}_R(\phi) + \mathbf{M} + i(\mathbf{B} - \mathbf{A}), \quad i = 0,1,\ldots,9 \quad (7.12)$$

for the right-hand-side parabolas and

$$\mathbf{c}_L(\phi) + \mathbf{M} + i(\mathbf{B} - \mathbf{A}), \quad i = 0,1,\ldots,9 \quad (7.13)$$

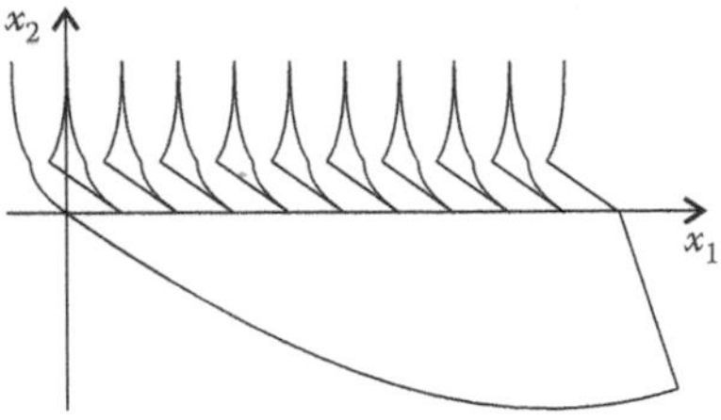

FIGURE 7.24
Combination of several optics into a single concentrator.

for the left-hand-side parabolas. Figure 7.24 shows the resulting optic.

This optic is composed of several upper optics and a lower optic.

References

1. Chaves, J. and Collares-Pereira, M., Ultra flat ideal concentrators of high concentration, *Sol. Energy*, 69, 269, 2000.
2. Chaves, J. et al., Combination of light sources and light distribution using manifold optics, *Nonimaging Opt. Efficient Illumination Systems III*, SPIE Vol. 6338, 63380M, 2006.
3. Dross, O. et al., LED headlight architecture that creates a high quality beam pattern independent of LED shortcomings, *Nonimaging Opt. Efficient Illumination Systems II*, SPIE Vol. 5942, 126, 2005.
4. Miñano, J. C. et al., High-efficiency LED backlight optics designed with the flow-line method, *Nonimaging Opt. Efficient Illumination Systems II*, SPIE Vol. 5942, 6, 2005.
5. Chaves, J. and Collares-Pereira, M., Ideal concentrators with gaps, *Appl. Opt.*, 41, 1267, 2002.
6. Feuermann, D., Gordon, J. M., and Ries, H., Nonimaging optical designs for maximum-power-density remote irradiation, *Appl. Opt.*, 37, 1835, 1998.

8

Luminaires

8.1 Introduction

A luminaire is a mirror (or an optical device) deflecting the light from a source to a receiver, so as to obtain on the receiver a prescribed light distribution. If the dimension of the light source is much smaller than the dimension of the luminaire mirror, such that the light source can be considered as a point source, and the distance from the luminaire to the receiver is also very large, it is possible to obtain the shape of the luminaire mirror by the traditional design methods.[1,2] If the light source is large, it is then necessary to use anidolic optics to obtain the shape of the luminaire's mirrors.[1]

In the previous analysis of nonimaging optics, we analyzed the problem of concentrating the light from a given source. The inverse problem is the illumination, where the source takes the place of the receiver or absorber, and the objective is to produce, at some distance, a given distribution of radiation. The calculation of a luminaire that produces a uniform illumination on a plane is simplified if the plane is considered to be at a large distance from the luminaire. Let us consider initially an infinitesimal source da in a 2-D space illuminating a line, as presented in Figure 8.1.

The illuminance E produced by this source on the line is given (see Chapter 20) by

$$\frac{E}{E_0} = \frac{I}{I_0}\cos^2\theta \tag{8.1}$$

where E_0 is the illuminance produced at the center of the line ($\theta = 0$) and I_0 is the intensity of the radiation in the same direction. To make the illuminance same for all its points, that is, to make $E = E_0$ for all points in the line, we must have

$$I = I_0/\cos^2\theta \tag{8.2}$$

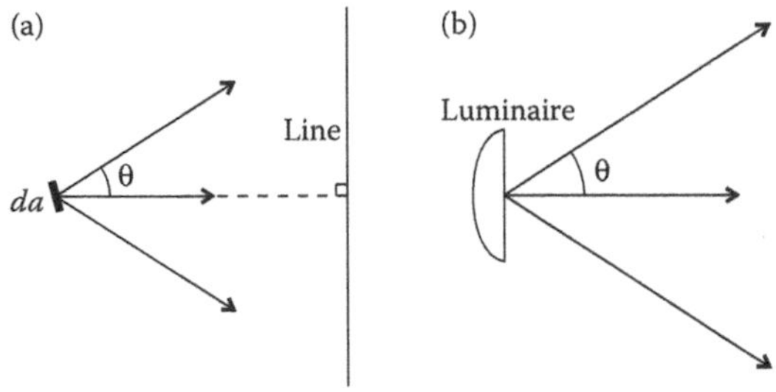

FIGURE 8.1
Figure (a) presents an infinitesimal source *da* illuminating a straight line at a finite distance. (b) If the line is at an infinite distance, the infinitesimal source can be replaced by a luminaire producing the same angular intensity distribution. It will produce on the plane placed at infinity a uniform distribution of radiation.

Since this relation does not depend on the distance from the infinitesimal source *da* to the line, the angular distribution of radiation must also enable us to obtain a constant illuminance on a line placed at an infinite distance. But if the line is now at an infinite distance, the size of the source is no longer an issue, and, therefore, it need not be infinitesimal. We can then replace the point source by a finite dimension one, as presented in Figure 8.1b. A luminaire for uniform light distribution on a line placed at an infinite distance must then produce an angular intensity distribution for the light, which is given by Equation 8.2.

We could consider a compound parabolic concentrator (CPC) with an acceptance angle θ as a candidate for this luminaire if the direction of light is reversed inside of it. This is not, however, a solution to this problem, because if we place a Lambertian source at the smaller aperture of the CPC, through the larger aperture, the radiation will also exit having the characteristics of a Lambertian source but radiating inside an angle θ.

We must then look for another solution. In the presentation of the CPC, we started by placing flat mirrors at each side of the receiver, which were then transformed into the mirrors constituting the CPC. Here, we could start with something similar.

8.2 Luminaires for Large Source and Flat Mirrors

Let us suppose that, at each side of a Lambertian source of light, we place two flat mirrors at an angle $\pi/4$ to the horizontal, as presented in Figure 8.2. The angles are considered positive if measured clockwise, relative to the vertical. Therefore, the angle θ shown in Figure 8.2a is negative, and represented as $-|\theta|$.

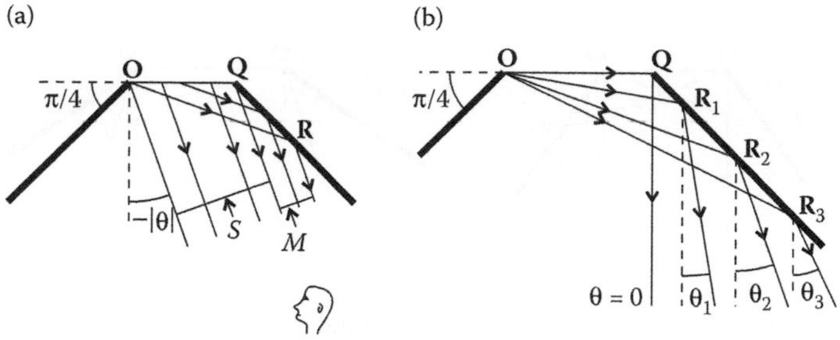

FIGURE 8.2
The power emitted by a Lambertian source **OQ** in a direction making an angle θ to the vertical is proportional to the area S of **OQ** projected in this direction. If mirrors are placed at each side of the source, the portion **QR** of the right-hand-side mirror reflects in direction θ a power proportional to the width M of the source reflection. (a) An observer, from direction θ, sees a total power proportional to $S + M$. (b) As angle θ takes values θ_1, θ_2, and θ_3, the reflection of the source on the mirror extends to points $\mathbf{R}_1$, $\mathbf{R}_2$, and $\mathbf{R}_3$, respectively.

An observer looking into the luminaire from a direction $-|\theta|$ sees the source **OQ**. The intensity produced by the source in direction θ is given by $I_S = L_V[\mathbf{O},\mathbf{Q}]\cos\theta$ or

$$I_S = L_V[\mathbf{O},\mathbf{Q}]\cos\theta = L_V S \tag{8.3}$$

where $S = [\mathbf{O},\mathbf{Q}]\cos\theta$ is the dimension of the source when viewed from direction θ and L_V its luminance.

But the observer also sees an "image" of the source reflected on the mirror. From direction θ, this image has a dimension M. The luminance L_V is conserved on reflection and, therefore, the intensity corresponding to the image on the mirror is given by

$$I_M = L_V M \tag{8.4}$$

An observer looking at the luminaire in the direction θ will then see a source of radiation of width I_S and an image on the mirror of width I_M. This set is equivalent to a source having width $I_M + I_S$.[1,3] The luminaire radiates in the direction θ a power per unit angle proportional to $I_M(\theta) + I_S(\theta)$.

The intensity produced by the luminaire as a function of angle θ is then given by

$$I(\theta) = I_M(\theta) + I_S(\theta) = L(M + S) \tag{8.5}$$

It is important to note that as the absolute value of θ increases, that is, takes values $\theta_1, \theta_2, \theta_3, \ldots$, point **R** moves along the mirror through points $\mathbf{R}_1, \mathbf{R}_2, \mathbf{R}_3, \ldots$ as presented in Figure 8.2b.

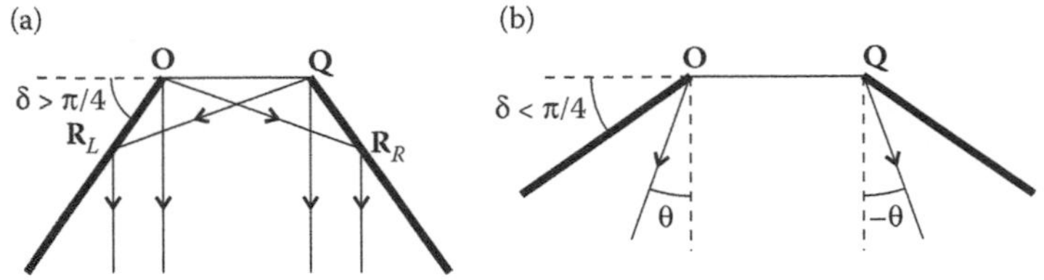

FIGURE 8.3
(a) If the angle δ made by the mirrors with the horizontal is larger than $\pi/4$, for $\theta = 0$ in addition to the source, two reflections $\mathbf{OR}_L$ and $\mathbf{QR}_R$ are visible on the mirrors. (b) If $\delta < \pi/4$, for angles with the vertical between $\pm\theta$ there will be no visible reflections on the mirrors, and the distribution of light between these angles is the one given by the source and cannot be changed.

Figure 8.3a shows the result of the slope δ of the mirrors when it exceeds $\pi/4$. In this case, for $\theta = 0$, in addition to the radiation coming directly from the source, there are two images, one on each mirror, extending through $\mathbf{OR}_L$ on the left-hand-side mirror and through $\mathbf{QR}_R$ on the right-hand-side mirror. For emission angles close to $\theta = 0$ we would then have two images (one on each mirror), complicating the analysis of the system.

However, if we had chosen an angle δ smaller than $\pi/4$, as in Figure 8.3b, a ray coming from **O** would be reflected by the mirror at **Q**, making an angle $-\theta$ with the vertical. By symmetry, a ray coming from **Q** toward **O** would be reflected, making an angle θ with the vertical. Therefore, for angles between $\pm\theta$, there would be no images of the source reflected on the mirrors, and in this interval only the radiation produced by the source would be available; so it would not be possible to change the distribution of light produced by the luminaire to the desired distribution.

The slope of the mirrors (making an angle $\pi/4$ to the vertical) in Figure 8.2 is chosen in such a way that a ray coming from the edge **O** of the source would be reflected at edge **Q** of the mirror and leave the luminaire in the vertical, that is, with $\theta = 0$. Therefore, for $\theta = 0$, the whole source is visible but there are no images on the mirrors. For other values of θ, the light leaving the luminaire comes directly from the source and also from one image of one of the mirrors.

The condition that the light coming from the edge of the source must be reflected by the edge of the mirror, and leave the luminaire in the vertical, will be used later as a boundary condition for the design of luminaires precisely for the same reasons pointed here.

The distribution of radiation that this luminaire produces on a distant target is derived next. From Figure 8.4a we can see that $\psi = \pi/4 - \theta$. Since the internal angles of the triangle **ROQ** add up to π, and considering that $\angle\mathbf{ROQ} = 3\pi/4$, we can calculate $\angle\mathbf{OQR} = \theta$.

Now $d = [\mathbf{O}, \mathbf{Q}]\sin\theta = [\mathbf{R}, \mathbf{O}]\sin\psi$ or

$$[\mathbf{R},\mathbf{O}] = [\mathbf{O},\mathbf{Q}]\frac{\sin\theta}{\sin\psi} \tag{8.6}$$

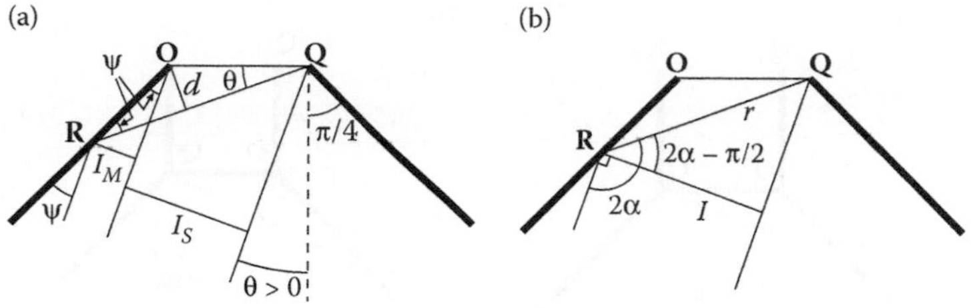

FIGURE 8.4
Calculation of the angular intensity distribution produced by a luminaire made of flat mirrors making an angle π/4 with the horizontal. (a) Separate intensity components I_s coming from the source and I_M coming from the mirror. (b) Total intensity $I = I_S + I_M$ emitted by the luminaire.

But $I_M = L_V$[**R**, **O**]sin ψ and therefore $I_M = L_V$[**O**, **Q**]sin θ. Since $I_S = L_V$[**O**, **Q**] cos θ we get

$$I(\theta) = L_V[\mathbf{O},\mathbf{Q}](\cos\theta + \sin\theta) = I_0(\cos\theta + \sin\theta) \tag{8.7}$$

for the case with image on the left-hand-side mirror with $0 < \theta < \pi/4$ and where $I_0 = L_V$[**O**, **Q**]. For the reflections on the right-hand-side mirror where $-\pi/4 < \theta < 0$ and we get

$$I(\theta) = I_0(\cos\theta - \sin\theta) \tag{8.8}$$

From Figure 8.4b we can also see that the total intensity produced by the luminaire is given by

$$I = r\cos(2\alpha - \pi/2) = r\sin(2\alpha) \tag{8.9}$$

Expression (8.7) can also be obtained from Figure 8.5. In this figure we again have a luminaire similar to the one presented in Figure 8.4, that is, mirrors **AQ** and **BO** make an angle π/4 to the horizontal.

Flat mirrors **AQ** and **OB** then create images $\mathbf{QO}_M$ and $\mathbf{OQ}_M$ of the source, respectively. In the example presented in Figure 8.5, we have an observer placed far away from the luminaire looking at it from an angle θ. This observer is looking at the optical system through the aperture **AB**. What he or she sees is the source **OQ** and the image $\mathbf{OQ}_M$. The intensity produced by the source is given by L_V[**O**, **Q**]cos θ and the intensity produced by the mirror image $\mathbf{OQ}_M$ is L_V[**O**, $\mathbf{Q}_M$]cos ζ = L_V[**O**, **Q**]sin θ since $\theta + \zeta = \pi/2$. We then obtain expression (8.7) for the intensity produced by the luminaire. The image that the observer sees on the mirror extends from **O** to **R**. Note that if the observer is looking from a direction that makes an angle to the vertical larger than β, he will only be able to see part of the image $\mathbf{OQ}_M$, as shown in Figure 8.5b. However, if the observer is looking from the vertical direction

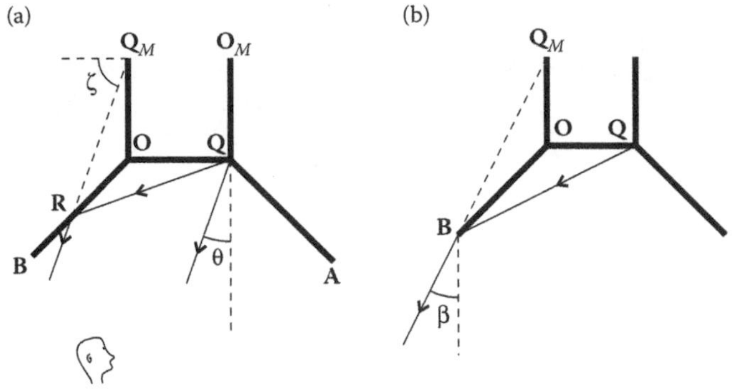

FIGURE 8.5
(a) Mirrors **QA** and **OB** create, respectively, images $\mathbf{QO}_M$ and $\mathbf{OQ}_M$ of the source. When looking through the aperture **AB**, one sees the equivalent source $\mathbf{Q}_M\mathbf{OQO}_M$. (b) For viewing angles larger than β only part of the image $\mathbf{OQ}_M$ will be visible. Therefore, an observer looking into the system from direction θ sees source **OQ** and the image $\mathbf{OQ}_M$, that is, he see sees the equivalent source $\mathbf{QOQ}_M$.

(perpendicular to the aperture **AB**), he will only be able to see the source **OQ** and no images on the mirrors, because the images are perpendicular to this viewing direction and, therefore, cannot be seen.

Since $\cos(\pi/4) = \sin(\pi/4)$ we can write expression (8.8) for the reflections on the right-hand-side mirror with $\theta < 0$, in the following form:

$$I(\theta) = I_0 \frac{\cos\theta\cos(\pi/4) - \sin\theta\sin(\pi/4)}{\cos(\pi/4)} = I_0 \frac{\cos(\pi/4 + \theta)}{\cos(\pi/4)} \tag{8.10}$$

We then see that the intensity produced in direction θ with $\theta < 0$ is proportional to $\cos(\pi/4 + \theta)$ since $I_0/\cos(\pi/4)$ is a constant. Replacing I/I_0 in Equation 8.1 we get

$$E = \frac{E_0}{\cos(\pi/4)} \cos\left(\frac{\pi}{4} + \theta\right)\cos^2\theta \Rightarrow E \propto \cos\left(\frac{\pi}{4} + \theta\right)\cos^2\theta \tag{8.11}$$

and, therefore, we can conclude that the illuminance on a distant target is proportional to $\cos(\pi/4 + \theta)\cos^2\theta$.[1] As seen, the illuminance on a distant plane is not constant and, therefore, a luminaire for this purpose cannot be built in V-shape.

Let us now consider that the flat mirror of the luminaire does not touch the source of light. This situation is presented in Figure 8.6.

If, as mentioned earlier, for $\theta = 0$ we want no images on the mirrors, and for $\theta \neq 0$ we want just one image on one mirror, the edge point $\mathbf{R}_0$ of the mirror must have an angle such that the ray coming from **O** is reflected vertically, as shown in Figure 8.6. The reasons for choosing this condition are the same as given earlier, for the case in which the mirrors touched the edges of the

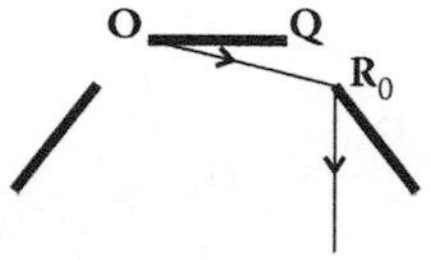

FIGURE 8.6
Luminaire consisting of two flat mirrors that do not touch the source. The initial point $\mathbf{R}_0$ of the mirror is such that a ray of light coming from the edge **O** of the source is reflected vertically.

source. That is, this condition guarantees that for $\theta \neq 0$ there is an image of the source in one mirror so that we can tailor the light distribution. But we never have a situation in which we have images in both mirrors.

A point on a mirror can be described in a coordinate system, as presented in Figure 8.7a, that is, using $r(\phi)$. As seen in Figure 8.7b, for different angles ϕ, we have different rays reflected on the mirror. That is, for different values of ϕ, for example, ϕ_1 and ϕ_2, the corresponding ray of light exiting from **O** is reflected on the mirror at points $\mathbf{R}_1$, $\mathbf{R}_2$, … and leave the luminaire making angles θ_1, θ_2, respectively, to the vertical. And also, the incident and reflected rays at $\mathbf{R}_1$ and $\mathbf{R}_2$ make angles $2\alpha_1$ and $2\alpha_2$, respectively. It is, therefore, possible for each shape of the mirror to establish functions $\phi(\theta)$, $\alpha(\theta)$, and $r(\phi(\theta)) = r(\theta)$.

The size of the source **OQ** is merely a scale factor for the design of the luminaire. For a source twice the size, the luminaire would have to be twice as large. We can, therefore, make $[\mathbf{O}, \mathbf{Q}] = 1$ without loss of generality. Now suppose we want, for example, a constant illuminance on a distant target. We can say that the source has a luminance (brightness) L_V and that we need the luminaire to produce an intensity $I(\theta) = L_V/\cos^2\theta$. Or we can make $L_V = 1$ for the source and say that we need the luminaire to produce an intensity

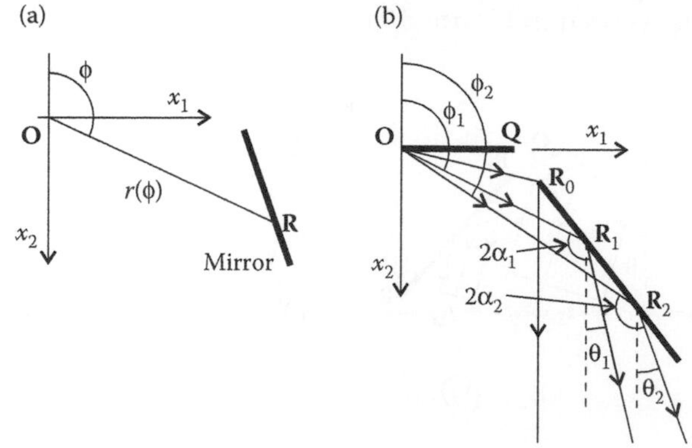

FIGURE 8.7
(a) The points of a mirror can be described by coordinate $r(\phi)$. (b) For different angles ϕ, the light rays are reflected at different points $\mathbf{R}_i$ on the mirror in different directions θ_i. The incident and reflected rays also make different angles α_i and, in this case, it is possible to define functions $\phi(\theta)$, $\alpha(\theta)$, and $r(\phi(\theta)) = r(\theta)$.

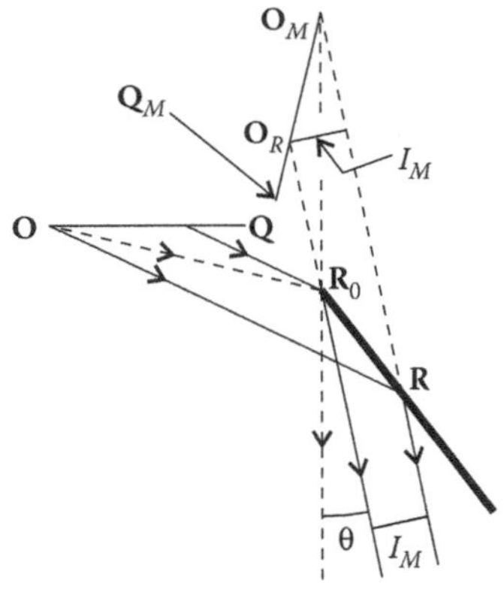

FIGURE 8.8
The mirror of the luminaire creates an image $\mathbf{Q}_M\mathbf{O}_M$ of the source and, in direction θ, it produces an intensity given by I_M.

$I(\theta) = 1/\cos^2\theta$. This does not affect the shape of the luminaire. Without loss of generality, therefore, we can make $L_V = 1$ for the design of the luminaire. For a Lambertian source, we then have $I_S(\theta) = I_0 \cos\theta$ with $I_0 = 1$ or $I_S(\theta) = \cos\theta$. From now on we will then use $[\mathbf{O}, \mathbf{Q}] = 1$ and $L_V = 1$ for the source.

The image of the source produced by the mirror in Figure 8.7b is presented in Figure 8.8.

The mirror of the luminaire creates an image $\mathbf{Q}_M\mathbf{O}_M$ of the source. From direction θ, only the portion $\mathbf{O}_M\mathbf{O}_R$ of the image of the source can be seen, and it produces the intensity I_M in the direction θ. For different values of θ, point **R** is at different positions on the mirror and the visible portion $\mathbf{O}_M\mathbf{O}_R$ of the source image varies. Again the intensity produced by the luminaire is given by $I_M + I_S$, as seen in Figure 8.9a.

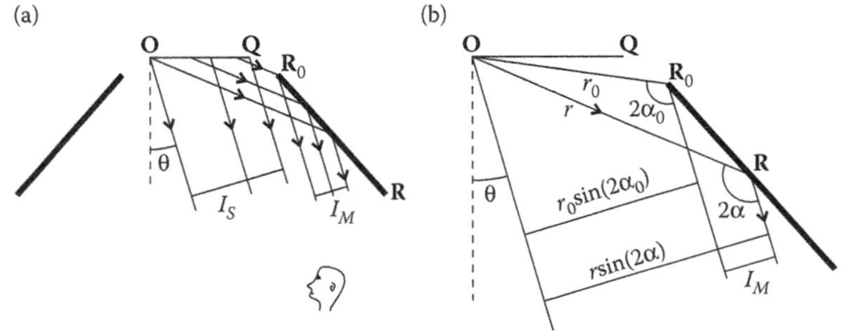

FIGURE 8.9
The mirrors of a luminaire do not necessarily have to touch the edges of the light source. (a) A luminaire where the mirrors do not touch the source. Intensity in direction θ is still given by $I_S + I_M$. (b) The intensity of the reflection on the mirror is given by $I_M = r\sin(2\alpha) - r_0 \sin(2\alpha_0) = p - p_0$, with p given by $p = r\sin(2\alpha)$.

For obtaining the intensity produced by the mirror, we first define a function $p(\theta)$ given by

$$p = r\sin(2\alpha) \tag{8.12}$$

The intensity produced by the mirror can now be obtained from the construction presented in Figure 8.9b. For an angle θ, the image of the source in the mirror extends from point $\mathbf{R}_0$ to point $\mathbf{R}$. The intensity produced by the mirror in direction θ can then be obtained from[4]

$$I_M = r\sin(2\alpha) - r_0\sin(2\alpha_0) = p - p_0 \tag{8.13}$$

Note that line r in Figure 8.9b represents in fact a ray of light exiting point **O**, being reflected at **R**, and exiting the luminaire making an angle θ to the vertical. Line r_0 does not, however, represent a ray of light because, as we have seen in Figure 8.6, a ray of light exiting **O** and reflected at $\mathbf{R}_0$ leaves the luminaire vertically, that is, with an angle to the vertical $\theta = 0$. For point $\mathbf{R}_0$, that is, we then have $p_0(\theta) = r_0\sin(2\alpha_0(\theta))$ where r_0 is the distance between points **O** and $\mathbf{R}_0$, that is, $r_0 = [\mathbf{O},\mathbf{R}_0]$. For point **R** we have $p(\theta) = r(\theta)\sin(2\alpha(\theta))$, so that

$$I_M(\theta) = p(\theta) - p_0(\theta) \tag{8.14}$$

The total intensity produced by the luminaire is then given by

$$I(\theta) = I_M(\theta) + I_S(\theta) = r(\theta)\sin(2\alpha(\theta)) - r_0\sin(2\alpha_0(\theta)) + I_S(\theta) \tag{8.15}$$

This expression is still valid even in the case where the mirrors are not flat. Note that $I_M(\theta)$ is the contribution of only one mirror because we never have images of the source on both mirrors at the same time.

Another example of a luminaire with flat mirrors is presented in Figure 8.10, where the mirror of the luminaire is defined in such a way that a ray reflected vertically at the lower edge point $\mathbf{R}_0$ of the mirror is coming from the near edge **O** of the source.

Then, for a different angle θ to the vertical, the observer will see an image on the mirror that extends from $\mathbf{R}_0$ to **R**, as seen in Figure 8.10b. The mirror of the luminaire creates an image $\mathbf{Q}_M\mathbf{O}_M$ of the source, as seen in Figure 8.11.

From direction θ, only the portion $\mathbf{O}_M\mathbf{O}_R$ of the image of the source can be seen, and it produces the intensity I_M in the direction θ For different values of θ, point **R** is at different positions on the mirror and the visible portion $\mathbf{O}_M\mathbf{O}_R$ of the source image varies.

In this case, as seen in Figure 8.10b, the intensity produced by the mirror will be proportional to

$$I_M(\theta) = r_0\sin(2\alpha_0) - r\sin(2\alpha) = p_0(\theta) - p(\theta) \tag{8.16}$$

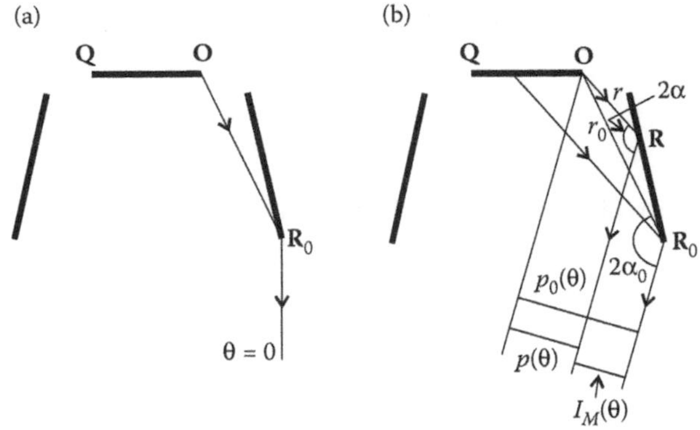

FIGURE 8.10
(a) A luminaire with flat mirrors. A ray coming from the near edge (to the mirror) of the source is reflected vertically at $\mathbf{R}_0$. (b) The width of the image of the source on the mirror is given by $I_M = r_0\sin(2\alpha_0) - r\sin(2\alpha) = p_0 - p$, with p given by $p = r\sin(2\alpha)$.

In contrast to the previous example, where the intensity produced by the mirror is given by expression (8.14), the total intensity produced by the luminaire in this case is given by

$$I(\theta) = I_M(\theta) + I_S(\theta) = r_0\sin(2\alpha_0) - r\sin(2\alpha) + I_S(\theta) \tag{8.17}$$

Note that in this case also, $I_M(\theta)$ is the contribution of only one mirror because we never have images of the source on both mirrors at the same time.

As stated earlier, the position of a point $\mathbf{R}$ on the mirror can be defined using the coordinates r and ϕ defined in Figure 8.7. The position of the edge

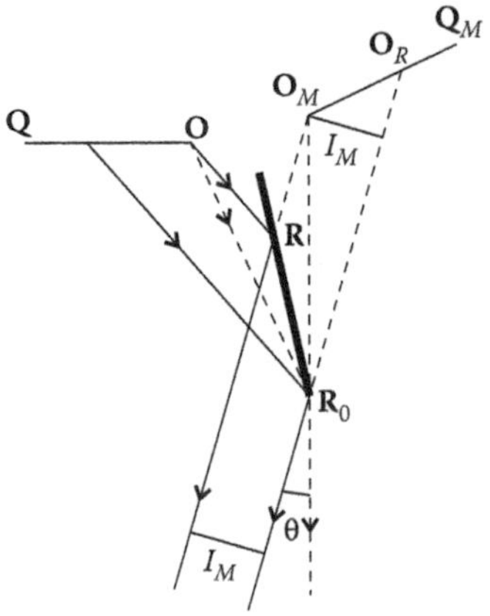

FIGURE 8.11
The mirror of the luminaire creates an image $\mathbf{Q}_M\mathbf{O}_M$ of the source and, in the direction θ, it produces the intensity I_M.

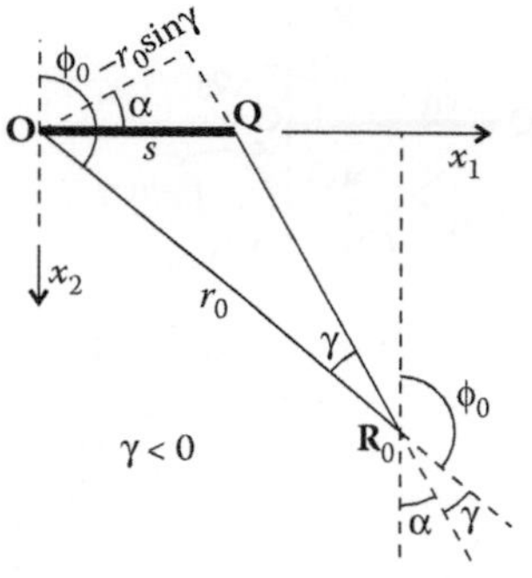

FIGURE 8.12
Calculation of the angle subtended by the source at the edge point $\mathbf{R}_0$ of the mirror when the origin is at the far edge of the source.

point $\mathbf{R}_0$ of the mirror can, however, be also defined using the angle γ subtended by the source at $\mathbf{R}_0$ and the distance from $\mathbf{R}_0$ to the origin as parameters. If the origin $\mathbf{O}$ is the source edge further away from $\mathbf{R}_0$, we have the far-edge case, as presented in Figure 8.12.

From Figure 8.12, we can see that $\alpha = \pi - (\phi_0 - \gamma)$. In this figure, $\gamma < 0$ and therefore $\alpha = \pi - (\phi_0 + |\gamma|)$. We have $\cos\alpha = -\cos(\phi_0 - \gamma)$. Although we are considering sources for which $[\mathbf{O}, \mathbf{Q}] = 1$, in this particular calculation we will make $[\mathbf{O}, \mathbf{Q}] = s$. We then have

$$-s\cos(\phi_0 - \gamma) = -r_0 \sin\gamma \Leftrightarrow \cos(\phi_0 - \gamma) = \frac{r_0}{s}\sin\gamma \tag{8.18}$$

From Equation 8.18, ϕ_0 can be obtained as

$$\phi_0 = \gamma + \arccos\left(\frac{r_0}{s}\sin\gamma\right) \tag{8.19}$$

If γ and r_0 (or r_0/s) are given, the angle ϕ_0 can be determined and therefore $\mathbf{R}_0$ can be defined as a function of these parameters.

Let us now consider the case in which the origin $\mathbf{O}$ is the source edge closer to $\mathbf{R}_0$. We then have the near-edge case as presented in Figure 8.13.

From Figure 8.13 we can see that

$$[\mathbf{Q},\mathbf{C}] = \sqrt{s^2 - r_0^2\sin^2\gamma} \tag{8.20}$$

and, therefore,

$$\begin{aligned} s\sin\alpha &= [\mathbf{A},\mathbf{B}] - [\mathbf{A},\mathbf{O}] = [\mathbf{Q},\mathbf{C}]\cos\gamma - [\mathbf{A},\mathbf{O}] \\ &= \cos\gamma\sqrt{s^2 - r_0^2\sin^2\gamma} - r_0\sin^2\gamma \end{aligned} \tag{8.21}$$

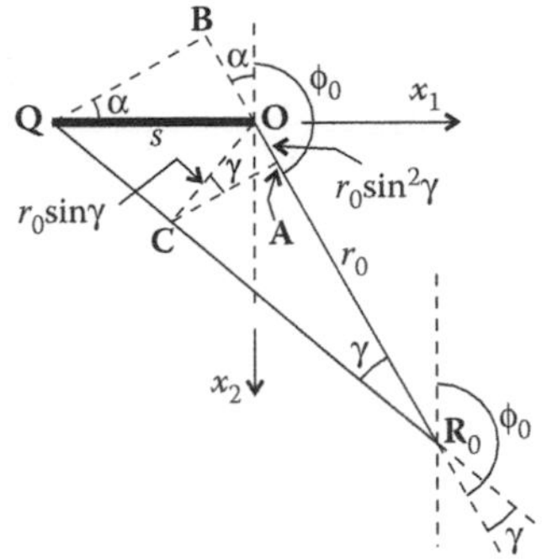

FIGURE 8.13
Calculation of the angle subtended by the source at the edge point $\mathbf{R}_0$ of the mirror when the origin is at the near edge of the source.

Since $\alpha = \pi - \varphi_0$, we can write for ϕ_0

$$\phi_0 = \pi - \arcsin\left(\cos\gamma\sqrt{1-\left(\frac{r_0}{s}\sin\gamma\right)^2} - \frac{r_0}{s}\sin^2\gamma\right) \tag{8.22}$$

As mentioned earlier, if γ and r_0 (or r_0/s) are given, angle ϕ_0 can be determined and, therefore, the position of $\mathbf{R}_0$ can be defined. By using this result, instead of defining the edge point $\mathbf{R}_0$ in terms of its coordinates $\mathbf{R}_0 = (R_{01}, R_{02})$, we can define it by the values of r_0 and γ.

Angle γ is important because it defines the maximum angle θ for which the image of the source on the mirror extends to point $\mathbf{R}_0$, as seen in Figure 8.14. In Figure 8.14a we have the near-edge case. In this case, for $\theta > \gamma$ the image on

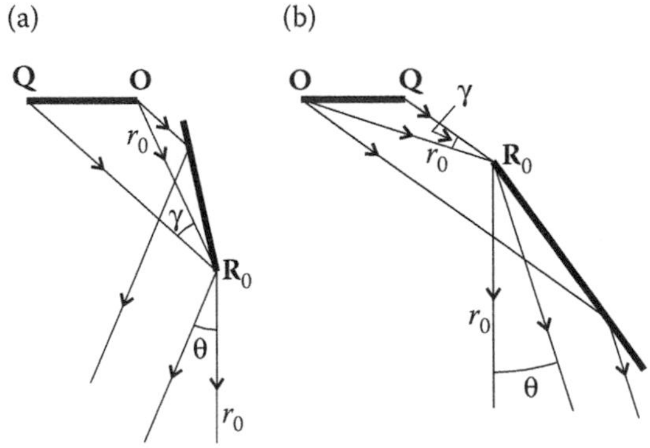

FIGURE 8.14
Angle γ defines the maximum angle θ for which the image of the source on the mirror extends to point $\mathbf{R}_0$. In both cases, ray r_0 coming from the origin $\mathbf{O}$ is reflected vertically at $\mathbf{R}_0$. If the image on the mirror was to extend to $\mathbf{R}_0$ for angles θ larger than γ, the light would have to come from points at the left side of $\mathbf{Q}$ in (a) and at the right side of $\mathbf{Q}$ in (b), which is not possible.

the mirror will no longer extend to $\mathbf{R}_0$ because if a light ray was to leave $\mathbf{R}_0$ with an angle $\theta > \gamma$, it would have to come from a point on the source further to the left-hand-side than **Q**. But the source ends at **Q** and, therefore, this is not possible.

Something similar happens in the far-edge case presented in Figure 8.14b. Also in this case, for $|\theta| > \gamma$ the image in the mirror will no longer extend to $\mathbf{R}_0$ because if a light ray was to leave $\mathbf{R}_0$ with an angle $|\theta| > \gamma$, it would have to come from a point on the source further to the right side than **Q**. But the source ends at **Q** and, therefore, this is not possible.

Expressions (8.14) and (8.16) are obtained for the intensity produced by the mirror considering that the image on the mirror extends from a point **R** to the edge point $\mathbf{R}_0$. These expressions are, therefore, valid only for values of $|\theta|$ smaller than γ. This will be important in the following analysis of luminaires with curved mirrors.

8.3 The General Approach for Flat Sources

We can now move forward to the analysis of curved luminaire mirrors. First, we need a coordinate system for the mirror parameterizations. Figure 8.15a presents the customary coordinate system, although this coordinate system is usually presented with the x_2-axis pointing downward, as seen in Figure 8.15b.[1,4]

As can be seen

$$2\alpha = \phi - \theta \tag{8.23}$$

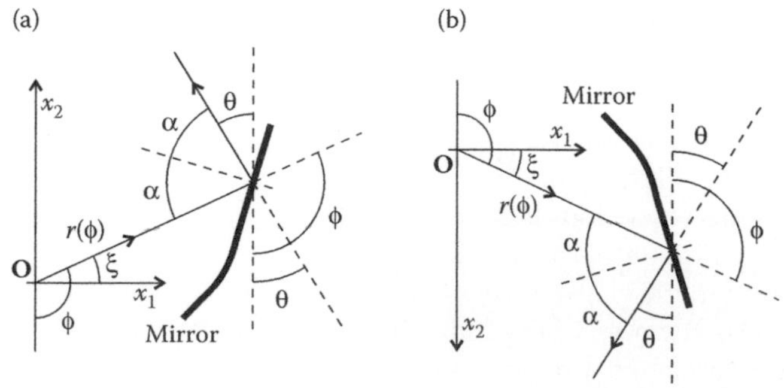

FIGURE 8.15
(a) The coordinate system used in the parameterization of the luminaire mirrors. (b) The same coordinate system, but with the x_2-axis pointing downward. This is the usual orientation of the axes for the presentation of the luminaires.

If, as discussed earlier, for different values of ϕ, we have different values of θ, we can define the function $\phi(\theta)$ and therefore $r(\phi) = r(\phi(\theta)) = r(\theta)$. Also α is a function of θ designated by $\alpha(\theta)$. Now $p(\theta)$ can be written as

$$p(\theta) = r(\phi)\sin(\phi - \theta) \Leftrightarrow p(\theta) = r(\theta)\sin(2\alpha(\theta)) \quad (8.24)$$

Then the equation of the mirror, that is, $r(\phi)$ can now be obtained from

$$r(\phi) = p(\theta)/\sin(\phi - \theta) \Leftrightarrow r(\theta) = p(\theta)/\sin(2\alpha(\theta)) \quad (8.25)$$

There are two possible expressions for $p(\theta)$, which can be obtained from expression (8.15)

$$p(\theta) = I(\theta) + r_0 \sin(\phi_0 - \theta) - I_S(\theta) \quad (8.26)$$

or from expression (8.17)

$$p(\theta) = -I(\theta) + r_0 \sin(\phi_0 - \theta) + I_S(\theta) \quad (8.27)$$

where $r_0 = r(\phi_0)$ is a constant. Note that α_0 is obtained for $\phi = \phi_0$ and is a function of θ as seen in Figures 8.9b and 8.10b. Therefore, $\alpha_0(\theta) = (\phi_0 - \theta)/2$ or $2\alpha_0(\theta) = \phi_0 - \theta$. In these expressions $I(\theta)$ is the desired intensity distribution for the luminaire.

The expression for $p_0(\theta)$ can be obtained from the initial conditions for the design of the mirror. As shown in Figure 8.16, if the position of the edge point $\mathbf{R}_0$ of the mirror is given, r_0 and ϕ_0 can be calculated.

Since the desired intensity distribution $I(\theta)$ is given, the expression for $p(\theta)$ can now be calculated. If the starting point $\mathbf{R}_0$ for the design of the mirror is close to the source, we have a situation similar to the one in Figure 8.9 and $p(\theta)$ is given by expression (8.26). If the starting point for the design of the

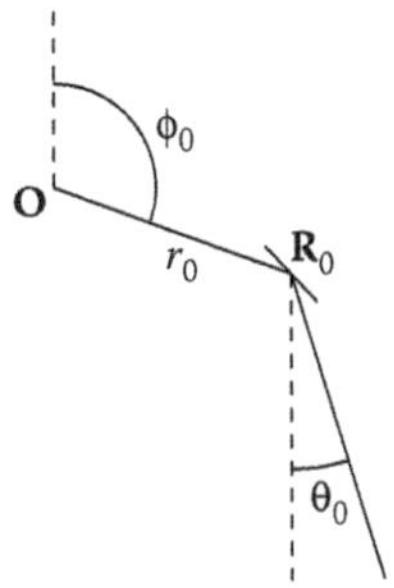

FIGURE 8.16
The initial conditions for the design of the mirror are the position of the initial point $\mathbf{R}_0$ (which gives us the values of r_0 and ϕ_0) and the direction of the reflected ray coming from **O** at $\mathbf{R}_0$ (given by angle θ_0). In previous examples, $\theta_0 = 0$.

mirror is away from the source, we have a situation similar to the one in Figure 8.10, and $p(\theta)$ is given by expression (8.27).

Since ϕ can be expressed as a function of θ, that is, $\phi(\theta)$ we can write expression (8.25) as $r(\phi(\theta)) = p(\theta)/\sin(\phi(\theta) - \theta)$ or $r(\theta) = p(\theta)/\sin(2\alpha(\theta))$. However, this Equation 8.25 in this form cannot be solved because we still do not have the relation $\alpha(\theta)$. We then need another equation enabling us to obtain $\alpha(\theta)$.

Let us then consider an infinitesimal portion of the mirror. This situation is presented in Figure 8.17. When $d\phi \to 0$, **OA** and **OC** tend to be parallel and Figure 8.17a tends to the situation depicted in Figure 8.17b.

From Figures 8.17a and 8.17b we can verify that **CB** = dr and **AC** = $rd\phi$, and we can therefore write[4]

$$\frac{dr}{rd\phi} = \tan\alpha \Leftrightarrow \frac{1}{r}\frac{dr}{d\phi} = \tan\alpha \Leftrightarrow \frac{d\ln r}{d\phi} = \tan\alpha \tag{8.28}$$

Equation 8.28 can be solved for $\alpha(\theta)$ using expression (8.23) to relate α and ϕ and Equation 8.24 to replace r by p (see Appendix A) resulting in

$$\alpha(\theta) = \arctan\left(\frac{p(\theta)}{P(\theta) - C_m}\right) \tag{8.29}$$

where $P(\theta)$ the primitive of $p(\theta)$ given by

$$P(\theta) = \int p(\theta)d\theta = \int I(\theta)d\theta + r_0\cos(\theta - \phi_0) - \int I_S(\theta)d\theta \tag{8.30}$$

if $p(\theta)$ is given by expression (8.26) and by

$$P(\theta) = \int p(\theta)d\theta = -\int I(\theta)d\theta + r_0\cos(\theta - \phi_0) + \int I_S(\theta)d\theta \tag{8.31}$$

if $p(\theta)$ is given by expression (8.27).

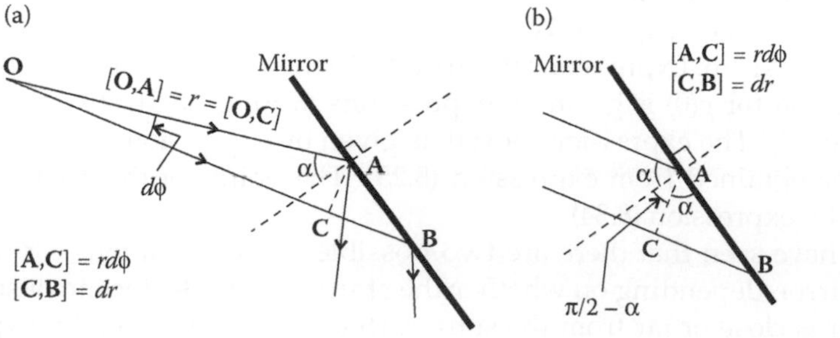

FIGURE 8.17
(a) To obtain the differential equation describing the mirror, consider a small part of it. The origin of the coordinate system is point **O**. As $d\phi \to 0$, **OA** and **OC** tend to become parallel and (a) becomes (b), for which $dr/(r\, d\phi) = \tan\alpha$.

In expression (8.29), C_m is a constant of integration that must be determined from the initial conditions. Figure 8.16 shows the initial conditions for the design of the mirror. If the position of initial point $\mathbf{R}_0$ is given, the values of ϕ_0 and r_0 can be determined (r_0 is the distance between **O** and $\mathbf{R}_0$). The expression for $p(\theta)$ can then be obtained from expressions (8.26) or (8.27), and for $P(\theta)$ from expressions (8.30) or (8.31). Let us further suppose that at point $\mathbf{R}_0$, the ray coming from **O** is reflected making an angle θ_0 to the vertical. We then have the initial conditions necessary to start designing the mirror. Constant C_m can be determined for the given value θ_0. Note that α_0 is related to ϕ_0 and θ_0 by expression (8.23), that is, $\alpha_0 = (\phi_0 - \theta_0/2)$. Expression (8.29) can be solved for constant C_m, which can then be obtained from the following expression:

$$C_m = P(\theta_0) - \frac{p(\theta_0)}{\tan((\phi_0 - \theta_0)/2)} = P(\theta_0) - \frac{p(\theta_0)}{\tan(\alpha(\theta_0))} \tag{8.32}$$

The points on the mirror can now be calculated. If **O** is located at $\mathbf{O} = (O_1, O_2)$, the points in the mirror then have coordinates of the form (see Figure 8.15):

$$\begin{aligned}\mathbf{R}(\theta) &= \mathbf{O} + r(\theta)(\cos\xi, \sin\xi) = \mathbf{O} + r(\theta)(\cos(\phi)(\theta) - \pi/2), \sin(\phi(\theta) - \pi/2)\\ &= (O_1, O_2) + r(\theta)(\sin(2\alpha(\theta) + \theta), -\cos(2\alpha(\theta) + \theta))\end{aligned} \tag{8.33}$$

If we consider that **O** is located at the origin, that is, $\mathbf{O} = (0, 0)$, we have

$$\mathbf{R}(\theta) = r(\theta)(\sin(2\alpha)(\theta) + \theta), -\cos(2\alpha)(\theta) + \theta) \tag{8.34}$$

and expression (8.25) gives us $r(\theta)$.

The process can now be summarized. As initial conditions, we must give the source size **OQ**, the desired angular distribution of intensity $I(\theta)$, the initial point $\mathbf{R}_0$ for the mirror, and the direction of the reflected ray on $\mathbf{R}_0$, which comes from the origin **O**. With these initial conditions, the values of r_0, ϕ_0, and θ_0 are given. Expressions (8.26) or (8.27) then enable us to obtain $p(\theta)$. The expression for $p(\theta)$ is given by expressions (8.30) or (8.31). Expression (8.32) gives us C_m. The expression for $\alpha(\theta)$ is given by expression (8.29) and $r(\theta)$ can now be obtained from expression (8.25). The points of the mirror are then given by expression (8.34).

We have seen that there are two possible sets of equations for designing the mirror, depending on whether the starting point $\mathbf{R}_0$ for the design of the mirror is close or far from the source, that is, if $p(\theta)$ is given by expressions (8.26) or (8.27). However, the position for the origin of the coordinate system can also be chosen to be on the edge of the source closer to the mirror or away from the mirror. This creates a total of four possible configurations in the design of a luminaire for a flat source.

8.4 Far-Edge Diverging Luminaires for Flat Sources

We now have the tools needed to calculate the shape of a luminaire, which produces a constant illuminance on a distant plane.

We start with a luminaire with initial point $\mathbf{R}_0$ for the design of the mirror close to the source, so we have a situation similar to that in Figure 8.9. This case is called far-edge diverging. "Far-edge," because the edge of the source chosen for origin is the one further away from the mirror, and "diverging," because the rays coming from the origin **O** diverge after reflecting off the mirror. In this case, the caustic of the edge rays coming from the origin falls behind the reflector. Then, expression (8.26) is to be used for $p(\theta)$. The luminaire to be designed is presented in Figure 8.18.

The process summarized earlier can now be used. Consider a source **OQ** of luminance L_V. If we want a constant illuminance on a distant plane, the intensity produced by the source and the luminaire must be given by expression (8.2), where I_0 is the intensity for $\theta = 0$. But for $\theta = 0$, there are no images of the source on the mirrors, so I_0 is only from the source contribution in this direction and can be obtained from expression (8.3) with $\theta = 0$. We, therefore, have $I_0 = L_V[\mathbf{O}, \mathbf{Q}]$. The source contribution is given by expression (8.3). Replacing these results in expression (8.26) we then have

$$p(\theta) = \frac{L_V[\mathbf{O},\mathbf{Q}]}{\cos^2\theta} + r_0 \sin(\phi_0 - \theta) - L_V[\mathbf{O},\mathbf{Q}]\cos(\theta) \tag{8.35}$$

When we have $[\mathbf{O}, \mathbf{Q}] = 1$ and $L_V = 1$, we can write

$$p(\theta) = \frac{1}{\cos^2\theta} + r_0 \sin(\phi_0 - \theta) - \cos(\theta) \tag{8.36}$$

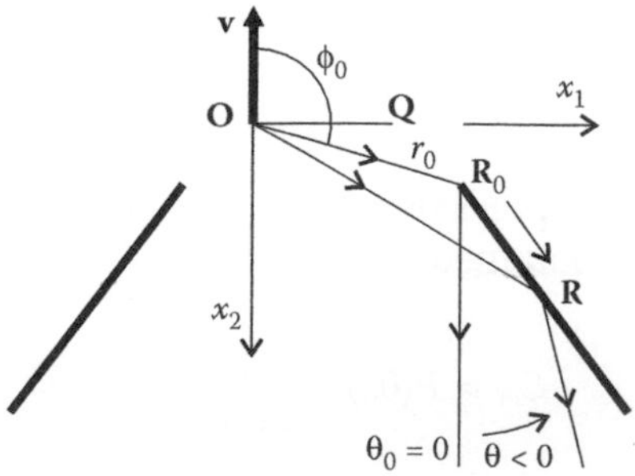

FIGURE 8.18
Far-edge diverging luminaire. The design starts at point $\mathbf{R}_0$ and the light ray coming from **O** is reflected vertically at this point. The design then starts at $\theta = 0$ and, as θ evolves to negative values, the design evolves to points on the mirror further away from the source.

and $P(\theta)$ can be obtained from expression (8.30) as

$$P(\theta) = \int p(\theta)d\theta = \tan\theta + r_0\cos(\theta - \phi_0) - \sin\theta \tag{8.37}$$

It is assumed that points **O** and **Q** and the initial point for the mirror $\mathbf{R}_0$ are given. If the coordinates of **O** and $\mathbf{R}_0$ are given by $\mathbf{O} = (O_1, O_2)$ and $\mathbf{R}_0 = (R_{01}, R_{02})$, ϕ_0 can be calculated (as in Figure 8.18) by

$$\phi_0 = \arccos\left(\frac{\mathbf{v}\cdot\mathbf{r}_0}{\sqrt{(\mathbf{v}\cdot\mathbf{v})(\mathbf{r}_0\cdot\mathbf{r}_0)}}\right) \tag{8.38}$$

with $\mathbf{v} = (0, -1)$ and $\mathbf{r}_0 = (R_{01}, R_{02}) - (O_1, O_2)$. If $\mathbf{O} = (0, 0)$ we get $\mathbf{r}_0 = (R_{01}, R_{02})$. We can also obtain r_0 by

$$r_0 = [\mathbf{O}, \mathbf{R}_0] = \sqrt{\mathbf{r}_0\cdot\mathbf{r}_0} \tag{8.39}$$

We can now impose the boundary condition $\theta = \theta_0 = 0$ for $\phi = \phi_0$, that is, the ray coming from **O** reflects at $\mathbf{R}_0$ leaving the luminaire in the vertical direction. This is again the boundary condition used earlier. This condition guarantees that for $\theta \neq 0$, there is an image of the source on one mirror, enabling a tailoring of the intensity pattern by adjusting the shape of the mirror. We never, however, had a situation in which the images of the source appear on both mirrors at the same time. In this case, expressions (8.36) and (8.37) can be written as

$$p(\theta_0) = r_0\sin\phi_0 \tag{8.40}$$

and

$$P(\theta_0) = r_0\cos\phi_0 \tag{8.41}$$

with ϕ_0 and r_0 given by expressions (8.38) and (8.39). From expression (8.32) constant C_m can now be obtained,

$$C_m = P(\theta_0) - \frac{p(\theta_0)}{\tan(\phi_0/2)} \tag{8.42}$$

$\alpha(\theta)$ is given by expression (8.29). From expression (8.25) it is now possible to obtain $r(\theta)$. The mirror points are given by expression (8.34). In this case, the design starts with $\theta = 0$ and evolves to negative values of θ, as seen in Figure 8.18.

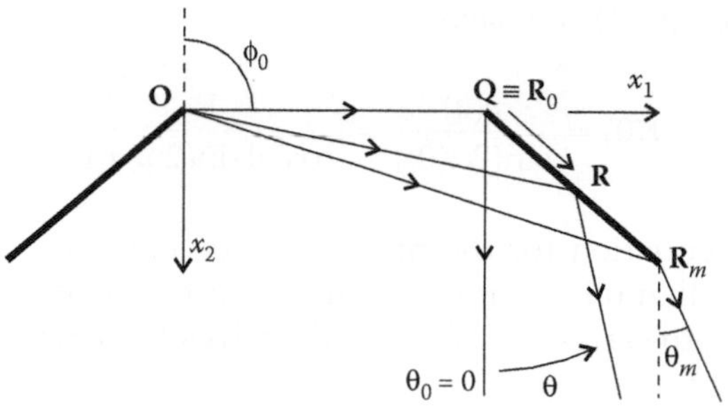

FIGURE 8.19
Luminaire that produces a constant illuminance at a distant plane. Since the whole device should extend to infinity, it has to be truncated. The constant illuminance of a distant plane is then only achieved for $|\theta| < |\theta_m|$.

A particular case occurs when point $\mathbf{R}_0$ is chosen to be on the edge $\mathbf{Q}$ of the source. This situation is presented in Figure 8.19. In this case, the equations describing the mirror are simpler.

Since the mirror starts at point $\mathbf{Q}$, we then have $\mathbf{R}_0 = \mathbf{Q}$, $r_0 = 1$, and $\phi_0 = \pi/2$. Expressions (8.36) and (8.37) can then be simplified to

$$p(\theta) = \frac{1}{\cos^2 \theta} \tag{8.43}$$

and

$$P(\theta) = \tan \theta \tag{8.44}$$

From the boundary condition $\phi_0 = \pi/2$ for $\theta_0 = 0$, we can also conclude that $\alpha_0 = (\phi_0 - \theta_0)/2 = \pi/4$. For $\theta = \theta_0 = 0$ we then have

$$p(\theta_0) = 1 \quad \text{and} \quad P(\theta_0) = 0 \tag{8.45}$$

These results can also be obtained from expressions (8.40) and (8.41), with $r_0 = 1$ and $\phi_0 = \pi/2$. From expression (8.42) we then obtain

$$C_m = 0 - \frac{1}{\tan(\pi/4)} = -1 \tag{8.46}$$

From expression (8.29), we get

$$\alpha(\theta) = \arctan\left(\frac{1/\cos^2 \theta}{\tan \theta + 1}\right) = \arctan\left(\frac{1}{\cos \theta (\cos \theta + \sin \theta)}\right) \tag{8.47}$$

From expression (8.25), we have

$$r(\theta) = \frac{p(\theta)}{\sin(2\alpha(\theta))} = \frac{1}{\cos^2\theta \sin(2\alpha(\theta))} \tag{8.48}$$

and, finally, the equation for the mirror can be obtained from expression (8.34). Remember that θ is positive when measured clockwise. Therefore, in this case, the design starts with $\theta = 0$ and evolves to negative values of θ, as seen in Figure 8.19.

From expression (8.47), when $\theta \to -\pi/4$, we see that $\cos\theta + \sin\theta \to 0$, and, therefore, $\alpha(\theta) \to \pi/2$, so that $r(\theta) \to +\infty$. This means that the mirrors extend from the source of light to infinity. They must then be truncated for a realistic design to be obtained. Therefore, the points in the mirrors are given by expression (8.34) with $\theta_m < \theta < 0$, where $-\pi/4 < \theta_m < 0$.

The shape of the mirrors resulting from these calculations is presented in Figure 8.19. As can be verified, the obtained mirrors are almost flat.[1] As seen earlier, these do not produce a constant illuminance on a distant plane.

It can also be noted from the obtained result that to design a complete luminaire, it should extend to infinity. This means that, ideally, the mirrors should touch the plane at infinity. This is similar to what happens in the design of tailored edge ray concentrators (TERCs) as secondary concentrators for Fresnel primary reflectors, since the TERC should ideally extend to the primary completely covering it. The same method can, in fact, be used in both cases.[1]

Truncating does not affect the uniformity inside the truncating angle, that is, does not affect the uniformity for $\theta_m < \theta < 0$. Outside the truncating angle, the intensity pattern is uncontrolled.

It can be shown that these luminaires, such as the one in Figure 8.19, can be designed in the same way as TERC mirrors for Fresnel primaries.[1]

Three-dimensional rotationally symmetric luminaires with a cross section similar to the one in Figure 8.19, can also be designed to enable uniform illuminance of a distant target.[5]

8.5 Far-Edge Converging Luminaires for Flat Sources

We now present a luminaire with initial point $\mathbf{R}_0$ being away from the source, with the origin **O** of the coordinate system at the edge of the source and further away from the mirror. This case is called far-edge converging. "Far-edge," because the edge of the source chosen for the origin is the one further away from the mirror, and "converging," because the rays coming from the origin **O** converge after reflecting on the mirror. In this case, the caustic of the edge rays coming from the origin falls in front of the reflector.

Since the starting point $\mathbf{R}_0$ for the design of the mirror is far from the source, the expression (8.27) is to be used for $p(\theta)$ as seen in Figure 8.20b. The luminaire to be designed is presented in Figure 8.20a.

Let us again suppose that we are interested in designing a luminaire producing a constant illuminance on a distant plane. In this case, $I(\theta) = 1/\cos^2\theta$. The intensity produced by the source is given by $I_S(\theta) = \cos\theta$. In this case, expression (8.27) can be written as

$$p(\theta) = -\frac{1}{\cos^2\theta} + r_0 \sin(\phi_0 - \theta) + \cos\theta \tag{8.49}$$

The primitive of this function is given by expression (8.31),

$$P(\theta) = -\tan\theta + r_0 \cos(\theta - \phi_0) + \sin\theta \tag{8.50}$$

It is assumed that points **O** and **Q** and the initial point for the mirror $\mathbf{R}_0$ are given. If the coordinates of **O** and $\mathbf{R}_0$ are given by $\mathbf{O} = (0, 0)$ and $\mathbf{R}_0 = (R_{01}, R_{02})$, ϕ_0 can be calculated by expression (8.38) with $\mathbf{v} = (0, -1)$ and $\mathbf{r}_0 = (R_{01}, R_{02})$, as seen in Figure 8.20a. In the same way, r_0 is given by expression (8.39).

We can now impose the boundary condition $\theta = \theta_0 = 0$ for $\phi = \phi_0$, that is, the ray coming from **O** reflects on $\mathbf{R}_0$, leaving the luminaire in the vertical. This is again the boundary condition used earlier. In this case, expressions (8.49) and (8.50) can also be written as expressions (8.40) and (8.41). From expression (8.42) constant C_m can now be obtained.

The expression for $\alpha(\theta)$ is given by expression (8.29) with $p(\theta)$, where $P(\theta)$ is given by expressions (8.49) and (8.50). As mentioned earlier, from expression

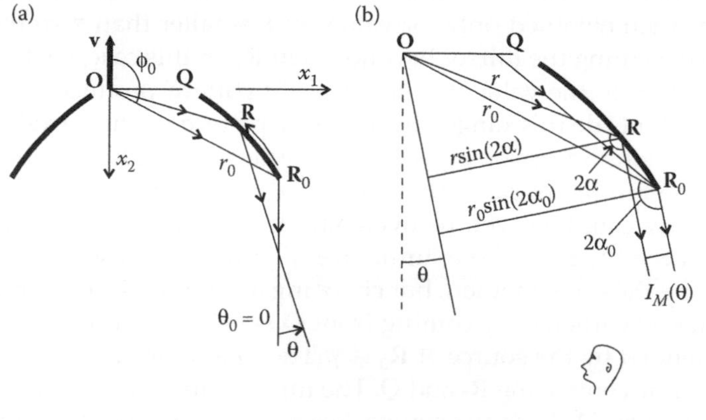

FIGURE 8.20
Far-edge converging luminaire. (a) The design starts at point $\mathbf{R}_0$ and the light ray coming from **O** is reflected vertically at this point. The design then starts at $\theta = 0$ and, as θ evolves to negative values, evolves toward the source. (b) The power emitted by the luminaire in direction θ corresponds to the reflection of the source on the mirror between points $\mathbf{R}_0$ and **R**.

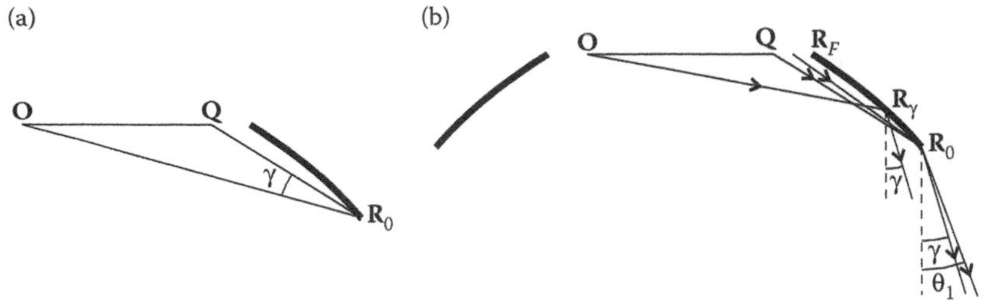

FIGURE 8.21
Far-edge converging luminaire. (a) The design of the mirror starts at $\mathbf{R}_0$ and ends at the horizontal line through the source. (b) The intensity produced can only be tailored for values of $|\theta| < \gamma$ because, for larger angles, the image on the mirror no longer extends to $\mathbf{R}_0$.

(8.25) it is now possible to obtain $r(\theta)$. The mirror points are again given by expression (8.34). In this case, the design starts with $\theta = 0$ and evolves to negative values of θ, as seen in Figure 8.20a. Figure 8.21 presents a luminaire calculated using the method just described.

The initial point for the mirror is $\mathbf{R}_0$, and a ray coming from **O** is reflected vertically at this point. These are the initial conditions for the design of the luminaire. The angle subtended by the source at this point is γ, as shown in Figure 8.21a.

The design of the mirror starts at point $\mathbf{R}_0$ and should end, at most, at the horizontal line through **O** and **Q**, that is, at point $\mathbf{R}_F$ because beyond this point the source is not visible from the mirror as seen in Figure 8.21. From point $\mathbf{R}_0$, however, it is not possible for light to exit at an angle θ_1 larger than γ, since this ray of light would have to come from a point between **Q** and $\mathbf{R}_F$, and this is not possible, as seen in Figure 8.21b. The desired intensity distribution is then obtained only for values of θ smaller than γ, so there is no interest in designing the mirror beyond point $\mathbf{R}_\gamma$. In this case, we have $\theta_m = \gamma$. For angles θ up to the value θ_m, the intensity pattern produced will be the desired one. Outside this range, the intensity pattern is uncontrolled.

This is a serious limitation of the design because the light distribution produced by the luminaire can only be tailored for a narrow range of angles. This difficulty can, however, be overcome if we allow multiple reflections to occur on the mirrors of the luminaire. Figure 8.22 presents a luminaire calculated by the same method, but choosing a different starting point $\mathbf{R}_0$.

As discussed earlier, a ray coming from **O** is reflected vertically at $\mathbf{R}_0$. The angle subtended by the source at $\mathbf{R}_0$ is γ. The final point $\mathbf{R}_m$ of the mirror is now on the line connecting $\mathbf{R}_0$ and **Q**. The mirror cannot be extended beyond $\mathbf{R}_m$ because it would shade the source. Point $\mathbf{R}_m$ can be calculated by solving the equation

$$\mathbf{R}(\theta) = \mathbf{Q} + x(\mathbf{R}_0 - \mathbf{Q}) \tag{8.51}$$

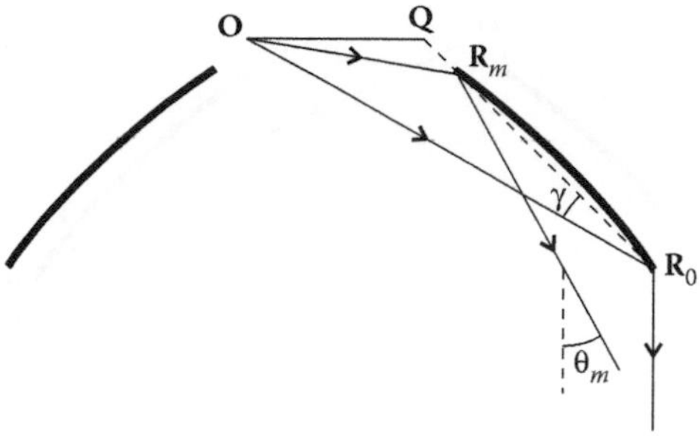

FIGURE 8.22
Far-edge converging luminaire designed by choosing the initial point in such a way that the mirror now ends at point $\mathbf{R}_m$ on the line connecting **Q** and $\mathbf{R}_0$. The mirror cannot be extended beyond $\mathbf{R}_m$ because it would shade the source.

This is a set of two equations with two unknowns, θ and x. The value obtained for x tells us the position of $\mathbf{R}_m$ along line $\mathbf{QR}_0$ and the obtained value θ_m for θ tells us the maximum value for which the luminaire tailors the radiation pattern. For angles θ up to the value θ_m, the intensity pattern produced will be the desired one. Outside this range, the intensity pattern is uncontrolled.

From what was said earlier, the image on the mirror should extend to $\mathbf{R}_0$ only for θ smaller than γ. This would mean that the design method would only make sense for values of θ smaller than γ, as in the example presented earlier. This is not, however, the case for this luminaire because of multiple reflections on the mirror.

When θ reaches the value of γ, we have the situation presented in Figure 8.23. Now, the ray of light reflected at $\mathbf{R}_0$ is coming from the edge **Q** of the source.

For values of θ larger than the value of γ, there are multiple reflections on the mirror, and the reflections extend the image of the source to $\mathbf{R}_0$, as presented in Figure 8.24. The mirror can then be designed for θ beyond γ using the same equations.

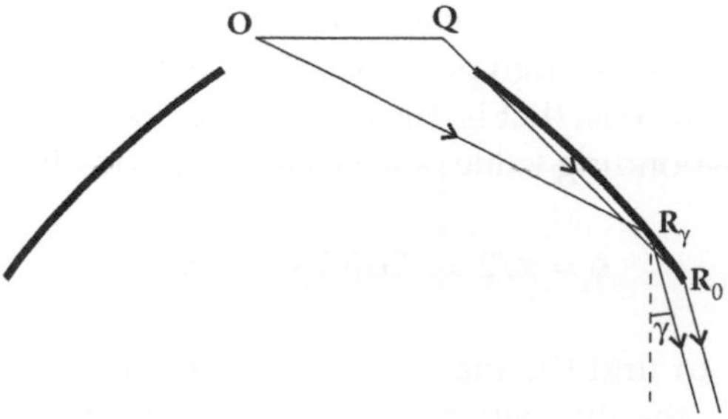

FIGURE 8.23
If only one reflection on the mirror is allowed for the light rays before exiting the luminaire, the image on the mirror only extends to $\mathbf{R}_0$ for values of θ up to γ.

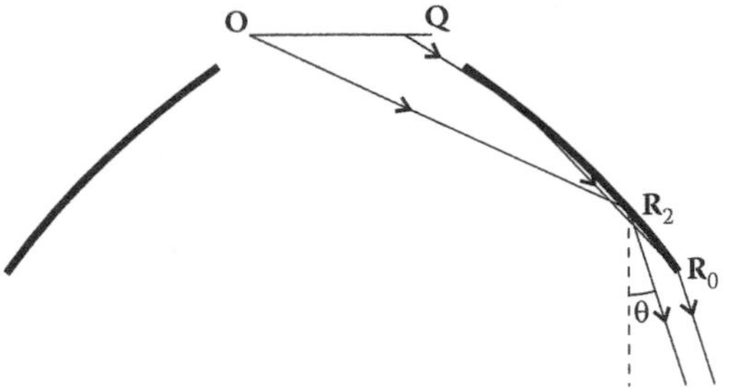

FIGURE 8.24
If multiple reflections are allowed for the light rays on the mirrors, then it is possible to extend the image of the source to $\mathbf{R}_0$, even for angles θ larger than γ.

The maximum angle for which we can see light coming out of point $\mathbf{R}_0$ is the one that is tangent to the mirror at this point.

8.6 Near-Edge Diverging Luminaires for Flat Sources

We now present a luminaire with its initial point $\mathbf{R}_0$ away from the source and the origin **O** of the coordinate system at the edge of the source closer to the mirror. This case is called near-edge diverging. "Near-edge," because the edge of the source chosen for origin is the one closer to the mirror, and "diverging," because the rays coming from the origin **O** diverge after reflecting on the mirror. In this case, the caustic of the edge rays coming from the origin falls behind the reflector. Since the starting point $\mathbf{R}_0$ for the design of the mirror is far from the source, the expression (8.27) is to be used for $p(\theta)$, as seen in Figure 8.25b. The luminaire to be designed is presented in Figure 8.25a. Note that now the positions of **O** and **Q** are inverted, since the origin **O** must now be on the edge closer to the mirror to be designed.

The equations describing this luminaire are then the same as the ones used for the far-edge converging case presented earlier, but the design starts at $\theta = 0$ and progresses through positive values of θ. The mirror must extend until it touches the x_1-axis, that is, the horizontal line through the source. To calculate the corresponding value of θ we can numerically solve the equation

$$\phi = \pi/2 \Leftrightarrow 2\alpha(\theta) + \theta = \pi/2 \tag{8.52}$$

This enables us to find the maximum value θ_m for angle θ. For angles θ up to the value θ_m, the intensity pattern produced will be the desired one. Outside this range, the intensity pattern is uncontrolled.

For these luminaires, and depending on the position chosen for the initial point $\mathbf{R}_0$, the mirrors can partially shade the light coming directly from the

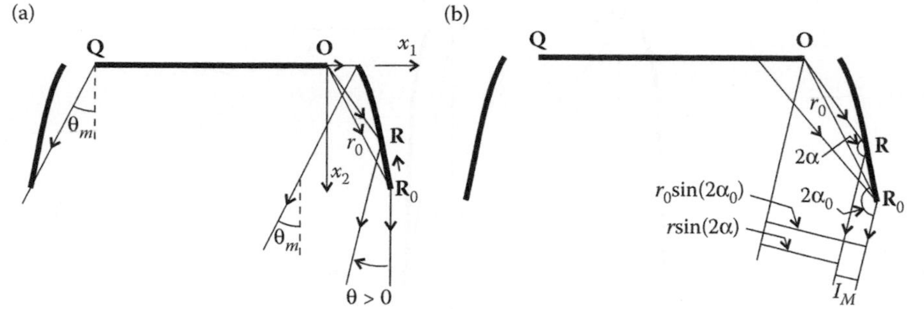

FIGURE 8.25
Near-edge diverging luminaire. (a) A light ray coming from the origin **O** is reflected vertically at initial point $\mathbf{R}_0$, at a distance r_0 from **O**. Angle θ starts at 0 and, as it evolves to positive values, the design of the mirror evolves toward the source. (b) The power emitted by the luminaire in direction θ corresponds to the reflection of the source on the mirror between points $\mathbf{R}_0$ and **R**.

source for a given angular interval. Figure 8.26 shows a luminaire where this shading occurs. This shading effect is presented in Figure 8.27.

The method of design for these luminaires is similar to the previous one, with the same equations. The difference is in the expression used for the contribution of the source for the illumination. Depending on the angle θ, the expressions differ for this contribution.

For values of θ between $\pm\theta_1$, the source of light is completely visible. Its intensity must then be given by $I_S(\theta) = [\mathbf{O}, \mathbf{Q}]\cos\theta$.

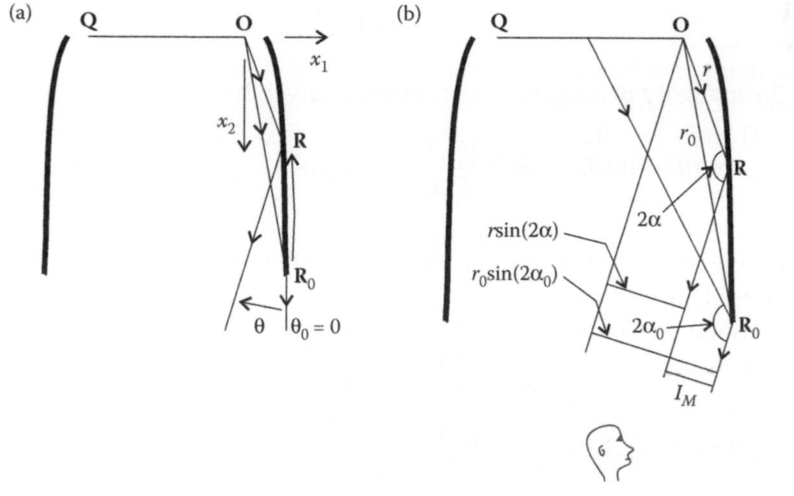

FIGURE 8.26
The initial point $\mathbf{R}_0$ for the design of a near-edge diverging luminaire can be chosen so that the mirror shades the source for some values of θ. (a) The design of the mirror evolves toward the source as θ evolves from 0 to positive values. (b) The contribution of the mirror to the luminaire's intensity.

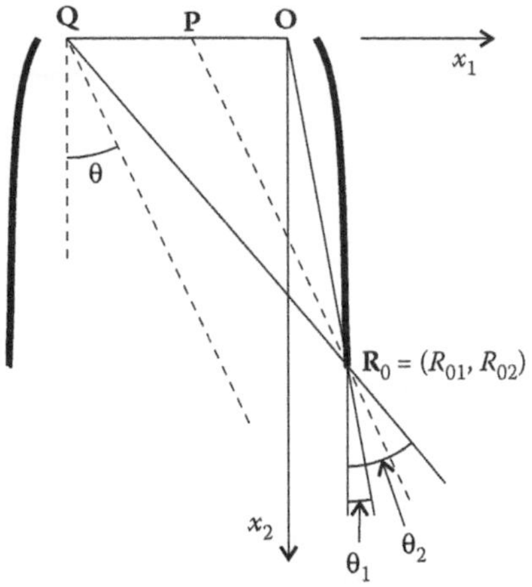

FIGURE 8.27
For angles θ with the vertical larger than θ_1, the mirrors shadow the source and, therefore, only the portion **QP** is visible. For values of θ larger than θ_2, the source is no longer visible.

For a value of θ between θ_1 and θ_2, only the part **PQ** of the source is visible. Therefore, only this part of the source contributes to the luminaire's intensity in this direction. The expression $I_M(\theta) = [\mathbf{O}, \mathbf{Q}]\cos\theta$ must then be multiplied by $[\mathbf{P}, \mathbf{Q}]/[\mathbf{O}, \mathbf{Q}]$ for these values of θ. The portion of the source visible for angle θ is then given by

$$f(\theta) = \frac{[\mathbf{Q},\mathbf{P}]}{[\mathbf{O},\mathbf{Q}]} \tag{8.53}$$

and the intensity produced by the source is given by

$$I_S(\theta) = [\mathbf{Q},\mathbf{P}]\cos\theta = \frac{[\mathbf{Q},\mathbf{P}]}{[\mathbf{O},\mathbf{Q}]}[\mathbf{O},\mathbf{Q}]\cos\theta = f(\theta)[\mathbf{O},\mathbf{Q}]\cos\theta \tag{8.54}$$

for $\theta_1 < |\theta| < \theta_2$. Making $\mathbf{Q} = (Q_1, 0)$, $\mathbf{P} = (P_1, 0)$ and $\mathbf{R}_0 = (R_{01}, R_{02})$, we can now calculate

$$\begin{aligned} \tan\theta_2 &= \frac{R_{01} - Q_1}{R_{02}} \\ \tan\theta &= \frac{R_{01} - P_1}{R_{02}} \end{aligned} \tag{8.55}$$

and conclude that

$$\tan(\theta_2) - \tan\theta = \frac{R_{01} - Q_1}{R_{02}} - \frac{R_{01} - P_1}{R_{02}} = \frac{P_1 - Q_1}{R_{02}} = \frac{[\mathbf{Q},\mathbf{P}]}{R_{02}} \tag{8.56}$$

Therefore, making $R_{02} = y_0$, we get

$$f(\theta) = (\tan\theta_2 - \tan\theta)y_0 / [\mathbf{O},\mathbf{Q}] \tag{8.57}$$

for $\theta_1 < |\theta| < \theta_2$.

Since the source of light is completely visible for $|\theta_1 < \theta_1$ and completely invisible for $|\theta| > \theta_2$, and considering $[\mathbf{O}, \mathbf{Q}] = 1$, we can write

$$f(\theta) = \begin{cases} 1 & |\theta| < \theta_1 \\ (\tan\theta_2 - \tan\theta)y_0 & \theta_1 < |\theta| < \theta_2 \\ 0 & |\theta| > \theta_2 \end{cases} \tag{8.58}$$

The design of the luminaire must then be divided into parts, according to the branches of $f(\theta)$.

For constant illuminance, we must have $I(\theta) = 1/\cos^2\theta$, as mentioned earlier. Since the starting point $\mathbf{R}_0$ for the design of the mirror is far from the source, the expression (8.27) is to be used for $p(\theta)$, as seen in Figure 8.26b.

$$p(\theta) = -\frac{1}{\cos^2\theta} + r_0\sin(\phi_0 - \theta) + f(\theta)\cos\theta \tag{8.59}$$

For $\theta < \theta_1$ we have $f(\theta) = 1$ and, therefore, the expressions to be used in this case are the same as used in the previous example for the far-edge converging luminaire, since $p(\theta)$ is the same as obtained in expression (8.49); but in this case, the design starts at $\theta = 0$ and evolves to positive values of θ, as seen in Figure 8.26a. The point of the mirror $\mathbf{R}_1$ obtained for $\theta = \theta_1$ is now used as boundary condition for the next section of the mirror, since it must be continuous. The boundary condition to be used in the next section of mirror is that, for $\theta = \theta_1$, we must have $\alpha = \alpha_1 = \alpha(\theta_1)$. The part of the mirror from $\mathbf{R}_0$ to $\mathbf{R}_1$ is presented in Figure 8.28.

In the new section of the mirror, we have $\theta_1 < |\theta| < \theta_2$. In this case, from expression (8.59) we obtain

$$p(\theta) = -\frac{1}{\cos^2\theta} + r_0\sin(\phi_0 - \theta) + (\tan\theta_2 - \tan\theta)y_0\cos\theta \tag{8.60}$$

where r_0 and ϕ_0 still have the same value as the previous section of the mirror, because the image of the source on the mirror still extends to $\mathbf{R}_0$. Integrating expression (8.60) we get

$$P(\theta) = r_0\cos(\phi_0 - \theta) + y_0\frac{\cos(\theta_2 - \theta)}{\cos\theta_2} - \tan\theta \tag{8.61}$$

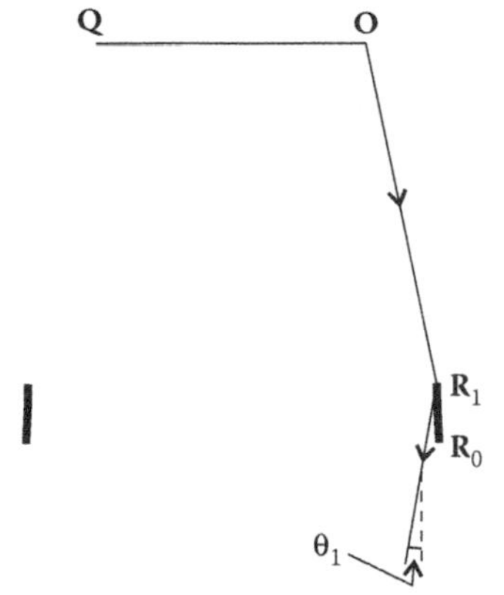

FIGURE 8.28
For angles θ with the vertical smaller than θ_1, the mirrors do not shadow the source. For this range of angles, the mirrors extend from point $\mathbf{R}_0$ to $\mathbf{R}_1$ and are symmetrical.

Constant C_m can be obtained from expression (8.32) by making $\theta_0 = \theta_1$ and $\alpha = \alpha_1$:

$$C_m = P(\theta_1) - \frac{p(\theta_1)}{\tan\alpha_1} \tag{8.62}$$

Note that, in expression (8.62), $p(\theta)$ and $P(\theta)$ are calculated for the new section of mirror and given by expressions (8.60) and (8.61); but $\alpha_1 = \alpha(\theta_1)$ for $\theta = \theta_1$, is obtained from the previous section of mirror, that is, from the function $\alpha(\theta)$ for the previous section of the mirror.

We now have $p(\theta)$, $P(\theta)$, and C_m for the new section of the mirror, and a new expression for the function $\alpha(\theta)$ can then be calculated as done earlier by expression (8.29) and a new expression for $r(\theta)$ by expression (8.25). The mirror points are again given by expression (8.34).

The mirror must extend until it touches the x_1-axis. To calculate the corresponding value of θ we can numerically solve Equation 8.52. This enables us to find the maximum value θ_m of the angle θ. In the case presented in Figure 8.27, the second section of the mirror extends through the interval $\theta_1 < \theta < \theta_m$.

A comparison between the luminaires obtained with and without shading is presented in Figure 8.29. As seen, the luminaire without shading is smaller for the same exit aperture and maximum angle θ_m. Since the luminaire with shading is also more complex to calculate, there is no point in using it.

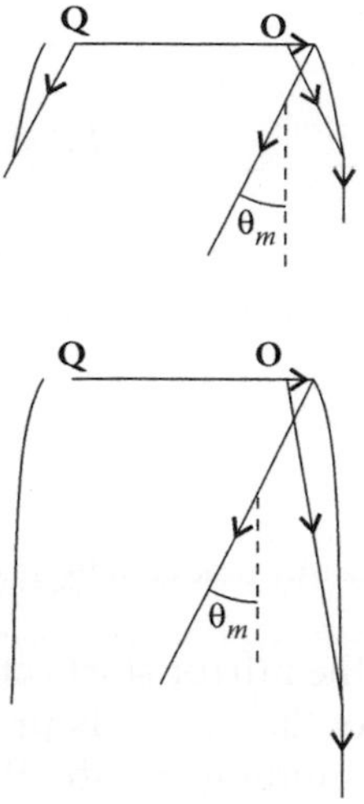

FIGURE 8.29
Comparison of two near-edge diverging luminaires. The top one is designed without shading, and the bottom one with shading. Both have the same exit aperture and the same maximum angle θ_m. As seen, the luminaire designed with shading is larger (and also more complex to design).

8.7 Near-Edge Converging Luminaires for Flat Sources

We finally present a luminaire with initial point $\mathbf{R}_0$ close to the source and with the origin **O** of the coordinate system at the edge of the source, closer to the mirror. This case is called near-edge converging. "Near-edge," because the edge of the source chosen for origin is the one closer to the mirror, and "converging," because the rays coming from the origin **O** converge after reflecting off the mirror. In this case, the caustic of the rays coming from the origin falls in front of the reflector. Since the starting point $\mathbf{R}_0$ for the design of the mirror is close to the source, the expression (8.26) is to be used for $p(\theta)$ as seen in Figure 8.30. The equations used to design these luminaires are then the same as those used to design the far-edge diverging case, but the design starts at $\theta = 0$ and evolves to positive values of θ, as seen in Figure 8.31a.

These luminaires may or may not have shading, as in the case of near-edge diverging designs presented earlier, but the designs with shading do not perform better than the ones without shading, and they are larger.[6] This result is similar to that presented earlier for the near-edge diverging luminaires. Therefore, there are no apparent advantages in using near-edge converging luminaires with shading. Besides, this design method is much more

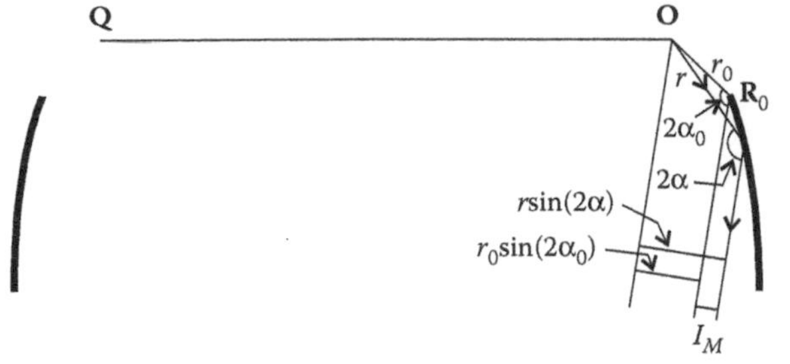

FIGURE 8.30
Contribution of the mirror to the intensity produced by a near-edge converging luminaire.

complex, since the design of the mirror starts at $\mathbf{R}_0$, close to the source, and, therefore, the final point $\mathbf{R}_m$ of the mirror is unknown. For this reason, it is impossible to know at the beginning the shading that the mirror will produce. An iterative method is then necessary in this design. We must try to guess the end position $\mathbf{R}_m$ of the mirror and perform the calculations using the shading that the mirror would produce if it started at $\mathbf{R}_0$ and ended at $\mathbf{R}_m$. If, after designing the luminaire, we verify that the mirror does not end at $\mathbf{R}_m$, its position must be changed and a new mirror calculated. This iterative process must continue until a coherent solution is found, that is, until the mirror profile terminates at the chosen point $\mathbf{R}_m$.

The designs with no shading are much simpler to calculate. Figure 8.31 presents one such luminaire. The mirror starts at $\mathbf{R}_0$. As mentioned earlier, the initial condition for the design is that a ray of light coming from **O** is reflected at $\mathbf{R}_0$ and exits the luminaire vertically. The end-point of the mirror is where the mirror becomes vertical. For this point, the light coming from the source, and making an angle θ_m to the vertical, still leaves the luminaire

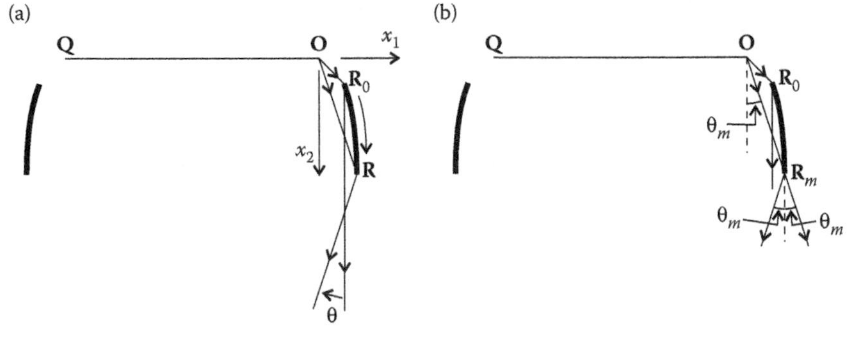

FIGURE 8.31
Near-edge converging luminaire. (a) The initial point $\mathbf{R}_0$ for this design is such that the light ray coming from **O** is reflected vertically at this point. As θ evolves to positive values, the design evolves to points further away from the source. (b) The maximum angle θ_m is where the mirror becomes vertical. Extending the mirror beyond this point will produce shading.

without blocking, and the light reflected at $\mathbf{R}_m$ leaves the luminaire also making an angle θ_m to the vertical. This point can be obtained by noting that, for point $\mathbf{R}_m$, we have $\phi_m + \theta_m = \pi$, and solving the equation:

$$\pi - \phi = \theta \Leftrightarrow \pi - (2\alpha(\theta) + \theta) = \theta \tag{8.63}$$

This enables us to find the maximum value θ_m for angle θ. For angles θ up to the value θ_m, the intensity pattern produced will be the desired one. Outside this range, the intensity pattern is uncontrolled.

We then have four possibilities for the design of the luminaires.[4] As for the choice of the edge of the source used as a basis for the design, we have two possibilities—near-edge and far-edge. As for the shape of mirrors, we also have two possibilities—converging or diverging.

Note that in the far-edge configurations, the design of the mirror starts at $\theta = 0$ and advances through negative values of θ. In the near-edge configurations, the design of the mirror also starts at $\theta = 0$ but now advances through positive values of θ.

The equations used for far-edge diverging designs and near-edge converging configurations are the same, and the equations used for far-edge converging designs and near-edge diverging configurations are also the same.

8.8 Luminaires for Circular Sources

The designs presented earlier are based on a linear Lambertian light source. It is, however, interesting to have luminaires for other forms of light sources. A case also studied is for tubular sources.[7–9] A widely used type of these sources is the fluorescent tube.

The designs presented earlier can be immediately applied to tubes (or other kinds of source shapes) if this new source is transformed into a Lambertian linear source by means of an involute mirror.[8] Figure 8.32 shows a tubular light source and the corresponding involute mirrors that transform it into a virtual linear Lambertian source **OQ**.

FIGURE 8.32
A tubular source of radiation can be transformed into an apparent linear source **OQ** by means of two involutes. The design techniques developed for linear sources can now be applied to the source **OQ**.

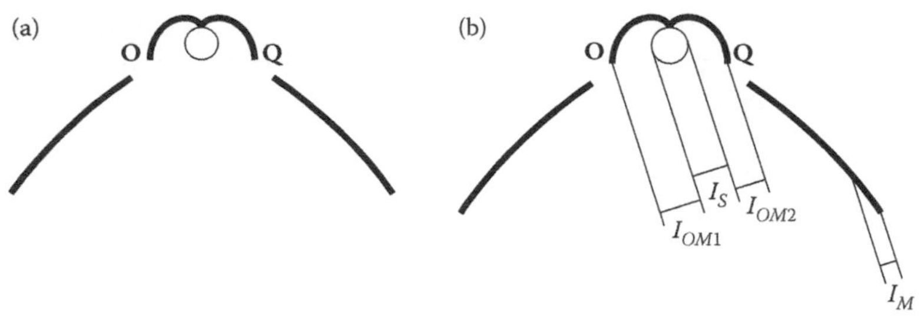

FIGURE 8.33
A tubular source is transformed into an equivalent flat source **OQ** by two involute arcs. (a) A far-edge converging luminaire for flat source can tailor the intensity pattern. (b) The contribution to the luminaire's intensity by the source is I_S, the mirror is I_M, and the two involute arcs are I_{OM1} and I_{OM2}.

These involute mirrors can be built for any source of light, not just circular ones.

This arrangement can be directly applied to the luminaires for flat sources, as presented earlier. For example, the flat source of the luminaire presented in Figure 8.22 can be replaced by the source presented in Figure 8.32. Figure 8.33 presents one such arrangement.

It is, nevertheless, possible to develop the theory presented earlier directly for tubes as generalizations of Equations 8.28 and 8.24 for tubular sources.[9]

In the case of luminaires for linear sources presented earlier, edge rays are emitted from a point to obtain the curves (see Figure 8.34a). There are two possibilities—the edge rays could come from the near edge or the far edge of

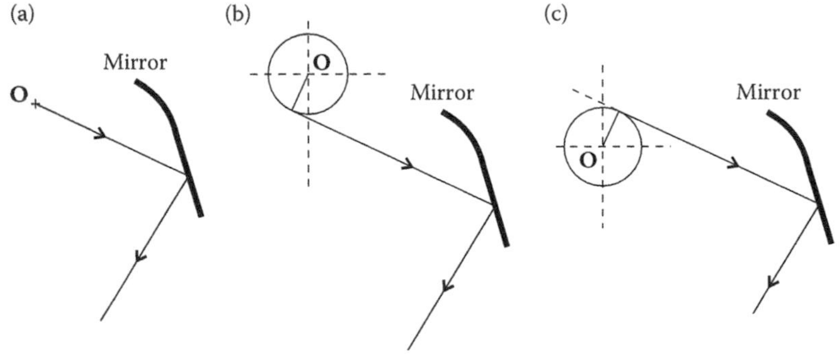

FIGURE 8.34
(a) In the case of linear sources of radiation, the design of the luminaires is based on one of the edges of the source, which is a point. For tubular sources, the situation is different, and two cases can be considered. (b) The edge rays on which the design of the luminaire is based are those coming from the lower part of the tube. This case is called "far-edge." (c) The edge rays on which the design is based are those coming from the upper part of the tube. This case is called "near-edge."

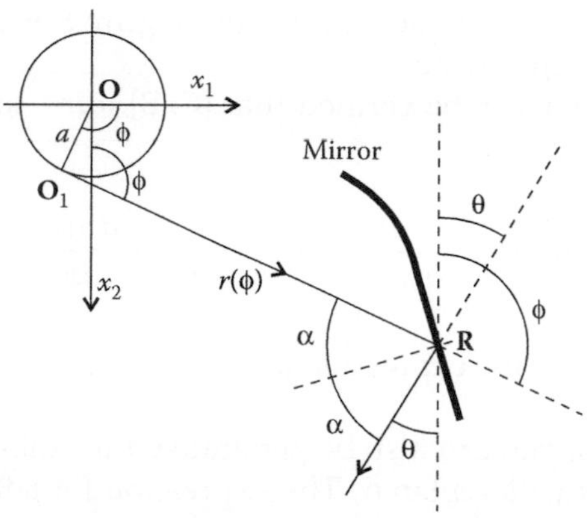

FIGURE 8.35
Coordinate system used to define the mirror of the luminaire in the far-edge case.

the source. In the case of tubular sources, this analysis is similar, although more complex.

The edge rays to be considered are tangents to the tube and there are two possibilities for the design—they can come from the lower part of the tube (see Figure 8.34b) or from the upper part of the tube (see Figure 8.34c). The case in which the edge rays come from the lower part of the tube is called the far edge; the case in which the edge rays come from the upper part of the tube is called the near edge.

To analyze the far-edge case, the coordinates in Figure 8.35 will be used. As seen, expression (8.23) is still valid in this case.

The equation for the shape of the mirror can now be obtained with the help of Figure 8.36.

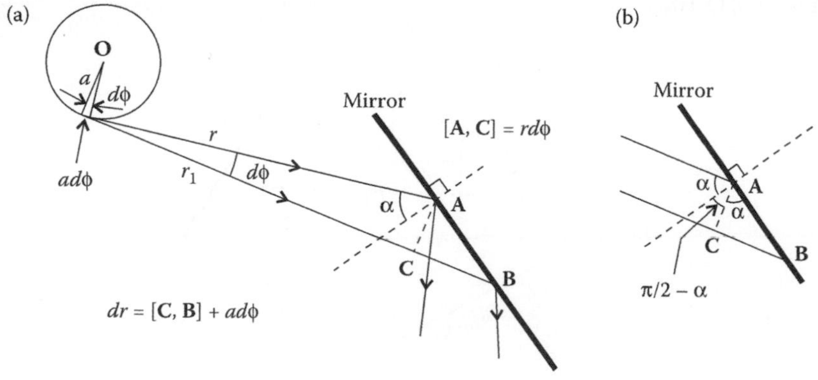

FIGURE 8.36
To find the differential equation describing the mirror, consider a small portion **AB**. The origin of the coordinate system is point **O**. When $d\phi \to 0$, r and r_1 become parallel, (a) tends to the situation (b), for which $(dr - a\,d\phi)/(r\,d\phi) = \tan\alpha$.

When $d\phi \rightarrow 0$, r and r_1 become parallel and Figure 8.36a becomes the situation presented in Figure 8.36b.

From these figures it can be verified that $[\mathbf{C}, \mathbf{B}] = dr - ad\phi$ and $[\mathbf{A}, \mathbf{C}] = rd\phi$ and we can write

$$\frac{dr - ad\phi}{rd\phi} = \tan\alpha \Leftrightarrow \frac{1}{r}\frac{dr}{d\phi} = \tan\alpha + \frac{a}{r} \Leftrightarrow \frac{d\ln r}{d\phi} = \tan\alpha + \frac{a}{r} \tag{8.64}$$

As seen, when $a \rightarrow 0$, the expression (8.64) tends to Equation 8.28 obtained earlier.

The expression for $p(\theta)$ can also be generalized for tubular sources. From Figure 8.37 we see that $b = a/\tan\alpha$. The expression for $p(\theta)$ can now be generalized to

$$p(\theta) = \left(r + \frac{a}{\tan\alpha}\right)\sin(2\alpha) \tag{8.65}$$

When the radius of the tube approaches zero, $a \rightarrow 0$, and the expression (8.65) becomes Equation 8.24.

As mentioned earlier, the expression for $p(\theta)$ can be used to calculate the contribution of the mirror to the intensity produced by the luminaire. Two different possibilities for the design can also be considered in this case, similar to the luminaires for flat sources. Figure 8.38 presents the case in which the edge rays do not intersect after reflection off the luminaire mirror. This is the diverging case, and the contribution of the mirror is given by

$$I_M = (r + b)\sin(2\alpha) - (r_0 + b_0)\sin(2\alpha_0) = p(\theta) - p_0(\theta) \tag{8.66}$$

with $b_0 = a/\tan\alpha_0$.

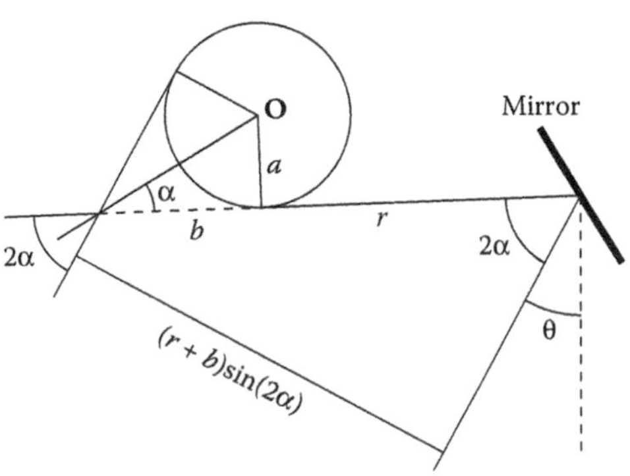

FIGURE 8.37
As in luminaires for linear sources, for tubular sources also, the intensity produced by the mirrors is defined at the cost of function $p(\theta)$.

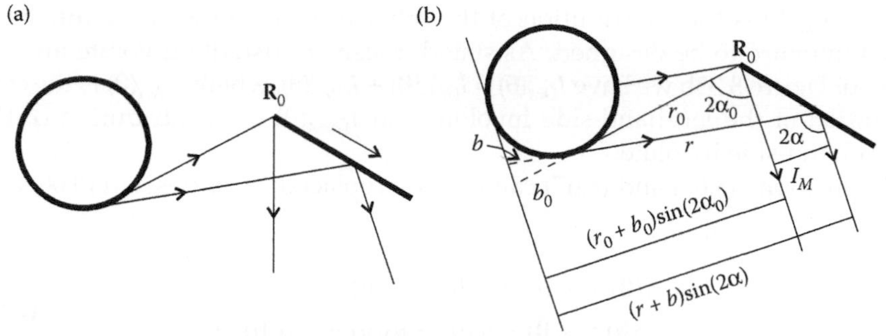

FIGURE 8.38
Near-edge diverging configuration. (a) The edge (tangent) rays to the cylinder diverge after reflection on the mirror as presented. (b) The contribution of the mirror to the intensity is given by $I_M = p(\theta) - p_0(\theta)$ with $p(\theta)$ given by the expression (8.65).

Figure 8.39 presents the case where the edge rays intersect after reflection off the luminaire mirror. This is the converging case and the contribution of the mirror is

$$I_M = (r_0 + b_0)\sin(2\alpha_0) - (r + b)\sin(2\alpha) = p_0(\theta) - p(\theta) \qquad (8.67)$$

with $b_0 = a/\tan\alpha_0$.

In the case of linear sources, the intensity produced by the luminaire is given by expression (8.5). For tubular sources, besides the luminaire mirror we have other mirrors, usually involute mirrors, as in Figure 8.33. The expression for the intensity of the luminaire is then given by

$$I(\theta) = I_S(\theta) + I_M(\theta) + I_{OM}(\theta) \qquad (8.68)$$

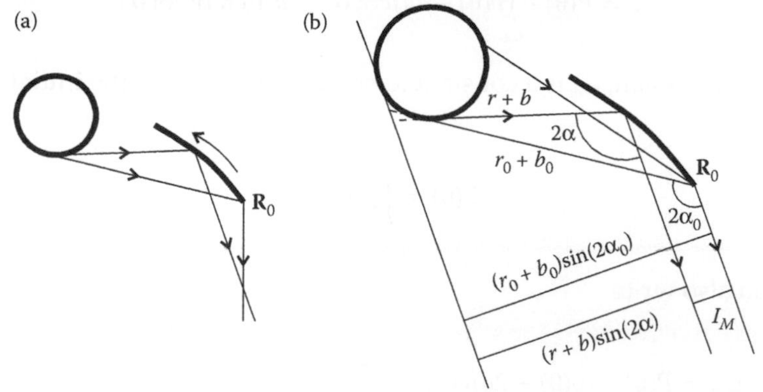

FIGURE 8.39
Near-edge converging configuration. (a) The edge (tangent) rays to the cylinder converge after reflection on the mirror. (b) In this case, the contribution of the mirror to the intensity is given by $I_M = p_0(\theta) - p(\theta)$ with $p(\theta)$ given by expression (8.65).

where $I_{OM}(\theta)$ is the contribution of the other mirrors, besides the mirror of the luminaire to be designed. As stated, these are usually involute arcs. In case of Figure 8.33b we have $I_{OM}(\theta) = I_{OM1}(\theta) + I_{OM2}(\theta)$, where $I_{OM1}(\theta)$ is the contribution of the left-hand-side involute and $I_{OM2}(\theta)$ the contribution of the right-hand-side involute.

Expressions (8.66) and (8.67) can now be replaced in expression (8.68) and we get

$$\begin{aligned} I(\theta) &= I_S(\theta) + p(\theta) - p_0(\theta) + I_{OM}(\theta) \\ p(\theta) &= I(\theta) - I_S(\theta) + p_0(\theta) - I_{OM}(\theta) \end{aligned} \tag{8.69}$$

or

$$\begin{aligned} I(\theta) &= I_S(\theta) + p_0(\theta) - p(\theta) + I_{OM}(\theta) \\ p(\theta) &= I_S(\theta) + p_0(\theta) + I_{OM}(\theta) - I(\theta) \end{aligned} \tag{8.70}$$

In any case, we can obtain $p(\theta)$ if the desired intensity for the luminaire $I(\theta)$, intensity produced by the source, and intensity produced by the other mirrors of the system $I_{OM}(\theta)$ are given. These other mirrors are defined as a starting point for the design. They are not calculated, so their contribution for the intensity must be determined.

As with flat sources, also in this case we must give a starting point $\mathbf{R}_0$ for the design of the mirror.

The expression for $r(\theta)$ can be obtained from expression (8.65) if $\alpha(\theta)$ is given. Expressions (8.64) and (8.65), together with $2\alpha = \phi - \theta$, result in expression (8.71) (see Appendix B), which defines $\alpha(\theta)$ implicitly.

$$C_m = P(\theta) - (p(\theta) - 2a)\cot\alpha - 2a\arctan(\cot\alpha) \tag{8.71}$$

In this expression, C_m is a constant to be determined from the initial conditions and

$$P(\theta) = \int p(\theta)d\theta \tag{8.72}$$

We can also write

$$C_m = P(\theta) - (p(\theta) - 2a)\cot\left(\frac{\phi - \theta}{2}\right) - 2a\arctan\left(\cot\left(\frac{\phi - \theta}{2}\right)\right) \tag{8.73}$$

Given an initial point for the mirror, we can calculate ϕ_0. Given also a value of $\theta = \theta_0$ for this initial point, we can obtain C_m.

Now, giving values to θ, the corresponding value for α can be obtained by solving Equation 8.71. Repeating the process for different values of θ, we can obtain $\alpha(\theta)$. The expression for $r(\theta)$ can now be obtained from expression (8.65) as

$$r(\theta) = \frac{p(\theta)}{\sin(2\alpha(\theta))} - \frac{a}{\tan(\alpha(\theta))} \tag{8.74}$$

The parameterization for the points of the mirror can now be obtained. From Figure 8.35 it is seen that $\mathbf{O}_1 = a(\cos\phi, \sin\phi)$ and from expression (8.33) (in which point **O** is now considered to be at position $\mathbf{O}_1$) we have

$$\mathbf{R} = \mathbf{O}_1 + r(\theta)(\sin(2\alpha(\theta)+\theta), -\cos(2\alpha(\theta)+\theta)) \tag{8.75}$$

or

$$\mathbf{R} = a(\cos(2\alpha(\theta)+\theta), \sin(2\alpha(\theta)+\theta)) + r(\theta)(\sin(2\alpha(\theta)+\theta), -\cos(2\alpha(\theta)+(\theta)) \tag{8.76}$$

A luminaire designed for a tubular source starts with an involute. This involute may or may not touch the source. If it does, we have two possibilities—total or partial involute. Figure 8.32 shows a complete involute. In this case, the tubular source is completely transformed into a linear source and the solutions found for this kind of source can be immediately applied. The solutions with partial involute are designed for truncated involutes.[8] Figure 8.40 presents one of these involutes truncated for an angle μ with the vertical. The case of a complete involute can be obtained for $|\mu| = \pi/2$.

The method of design for the luminaires with partial involutes is similar to the one described for linear sources. It is, however, necessary to remember that the light source has a different geometry, so the equations must adapt to this new situation.

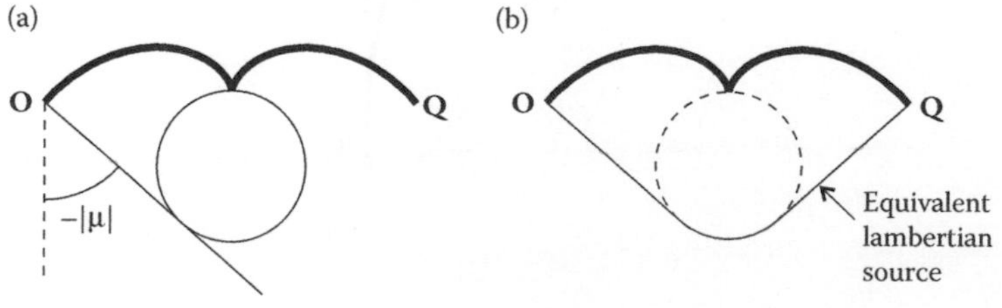

FIGURE 8.40
(a) In tubular sources, the design of a luminaire can start by two mirrors shaped as involutes touching the source. In this case, for angles with the vertical smaller than μ, this set behaves as a Lambertian source of width **OQ**. (b) The equivalent Lambertian source is shaped as a rounded wedge with edges **O** and **Q**.

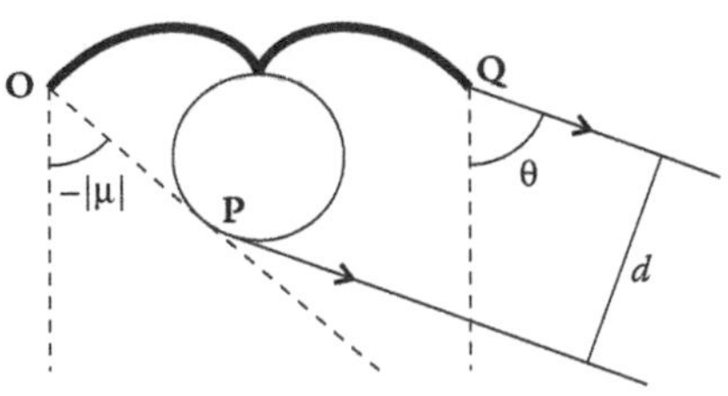

FIGURE 8.41
For the case presented in Figure 8.40, if the exit angle of light is larger than μ, the tubular source plus mirrors behave as a luminous source of width d defined by the extremity **Q** of the mirror and tangent to the source at point **P**.

For θ between ±μ (both edges **O** and **Q** are visible), the source behaves as a Lambertian emitter of width **OQ**. But for $|\theta| > |\mu|$ (only **Q** is visible), the width d visible for the source is defined by an edge point of the involute **Q** and by a tangent point **P** to the tube,[8] as presented in Figure 8.41.

Similarly, the equations to be used for defining the mirrors will depend on this behavior and the equations for flat source or tubular source must be used accordingly.

If a far-edge luminaire is to be designed, the edge of the source serving as a basis for the design can be the involute edge **O** or the tangent to the tube. Figure 8.42 presents this situation. For the part of the mirror between points $\mathbf{R}_B$ and $\mathbf{R}_T$, the edge **O** is visible. The differential equation defining the mirror is then Equation 8.28. For the points beyond $\mathbf{R}_T$, the edge of the source corresponds to the tangent to the tube and the differential equation to which the mirror must obey is expression (8.64).[8]

This problem does not exist in the design of a near-edge luminaire since, in this case, the edge **Q** is always visible from the mirror, as seen in Figure 8.42.

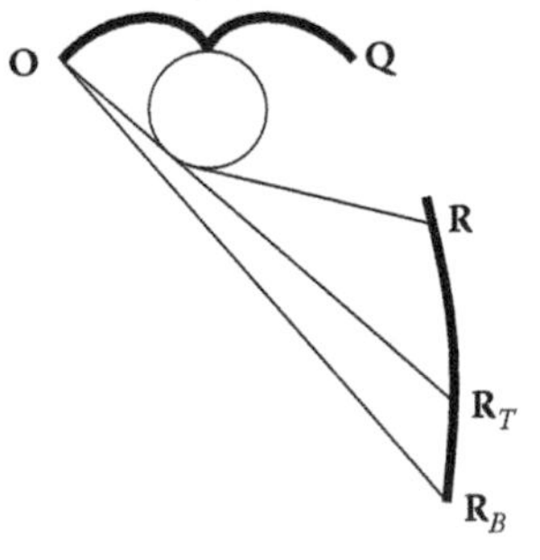

FIGURE 8.42
If the far edge **O** is used as the basis for the design of the luminaire, the equations used to define the mirror must be different for $|\theta| < |\mu|$ and $|\theta| > |\mu|$. For the portion $\mathbf{R}_B\mathbf{R}_T$ of the mirror, which is $\theta < \mu$, the edge **O** of the involute is visible and the mirror is described by differential equation (8.28) for linear sources. For the portion $\mathbf{R}_T\mathbf{R}$ of the mirror, the point **O** is no longer visible, so the differential equation defining it is now expression (8.64).

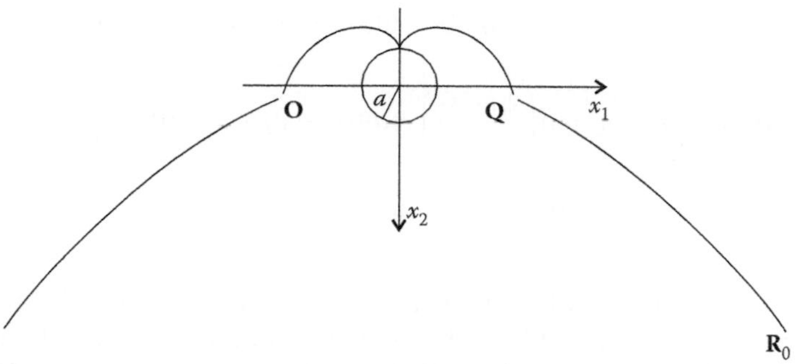

FIGURE 8.43
Far-edge converging luminaire. It consists of two involute arcs starting at the highest point on the tubular source and ending at **O** and **Q**, and a mirror starting at $\mathbf{R}_0$ and extending toward the source.

Figure 8.43 shows a far-edge converging luminaire. Its design can start by the involute. Once this is done, we have the coordinates for points **O** and **Q** and, therefore, the size of the apparent source is $s = [\mathbf{O}, \mathbf{Q}]$.

We now choose a point $\mathbf{R}_0$ to start the design of the mirror. As seen in Figure 8.44, the edge **O** of the involute can be seen from $\mathbf{R}_0$. Therefore, the first part of the design is done in the same way as the one for a far-edge converging luminaire for a linear source **OQ**. This enables us to design the portion of the mirror from $\mathbf{R}_0$ to $\mathbf{R}_1$. At point $\mathbf{R}_1$, the edge ray from the source becomes tangent to the tube, as seen in Figure 8.44a. Now the equations for a tubular source must be used. Also for this new portion of the mirror, the image of the source extends to $\mathbf{R}_0$.

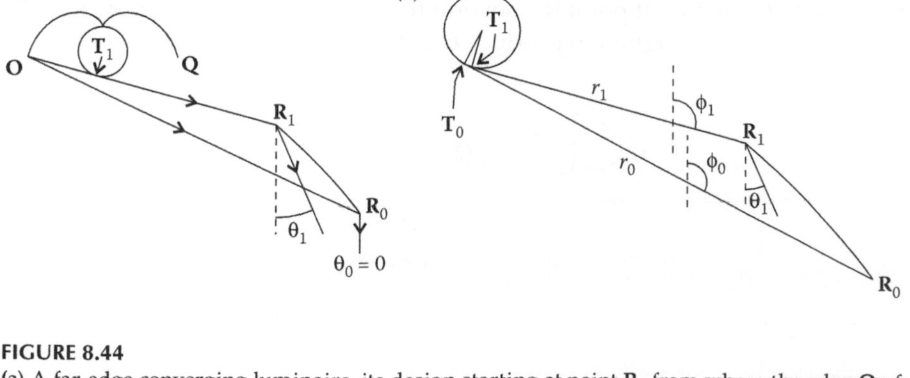

FIGURE 8.44
(a) A far-edge converging luminaire, its design starting at point $\mathbf{R}_0$ from where the edge **O** of the involute can be seen. The portion $\mathbf{R}_0\mathbf{R}_1$ of the mirror is then designed as if we have a linear source **OQ**. Beyond point $\mathbf{R}_1$, the equations for a tubular source must be used. (b) The initial conditions for the design beyond point $\mathbf{R}_1$.

The expression for $p(\theta)$ can then be obtained from expression (8.70) as

$$p(\theta) = \left(r_0 + a \Big/ \tan\left(\frac{\phi_0 - \theta}{2}\right)\right) \sin(\phi_0 - \theta) + s\cos\theta - \frac{s}{\cos^2\theta}$$
$$= a + a\cos(\theta - \phi_0) - r_0 \sin(\theta - \phi_0) + s\cos\theta - \frac{s}{\cos^2\theta} \quad (8.77)$$

where $s = [\mathbf{O}, \mathbf{Q}]$ and r_0 and ϕ_0 are as indicated in Figure 8.44b. Distance r_0 is now the distance from point $\mathbf{R}_0$ to $\mathbf{T}_0$ on the tangent to the source through $\mathbf{R}_0$. Angle ϕ_0 is defined by line $\mathbf{T}_0\mathbf{R}_0$ and the vertical. The expression (8.77) can now be integrated to obtain

$$P(\theta) = a\theta + \cos\theta(r_0\cos\phi_0 - a\sin\phi_0) + \sin\theta(s + a\cos\phi_0 + r_0\sin\phi_0) - s\tan\theta \quad (8.78)$$

Constant C_m given by expression (8.73) can now be obtained from point $\mathbf{R}_1$ where the first part of the mirror ends. This is the initial point for the new section of the mirror. From angle θ_1, which the light makes to the vertical when coming from $\mathbf{O}$ and reflecting at $\mathbf{R}_1$, we can obtain $p(\theta_1)$ and $P(\theta_1)$. Replacing also the values for ϕ_1 and θ_1 in expression (8.73), we can calculate C_m. Now, for different values of θ, we can obtain the corresponding values of α by solving Equation 8.71 numerically. These pairs (θ, α) can now be introduced in expression (8.74) and $r(\theta)$ is obtained. Finally, the mirror points can be calculated by expression (8.76). Figure 8.43 shows the complete luminaire.

Like the design for a far-edge diverging luminaire for a linear source **OQ**, also in this case the design can continue beyond angle γ subtended by the source at initial point $\mathbf{R}_0$. This angle is indicated in Figure 8.45. Also in this case, this is due to multiple reflections on the mirror. The maximum angle that the light can come out of point $\mathbf{R}_0$ corresponds to the direction of the tangent to the mirror at point $\mathbf{R}_0$. Beyond this direction, the luminaire mirror can no longer be designed to produce the desired intensity pattern.

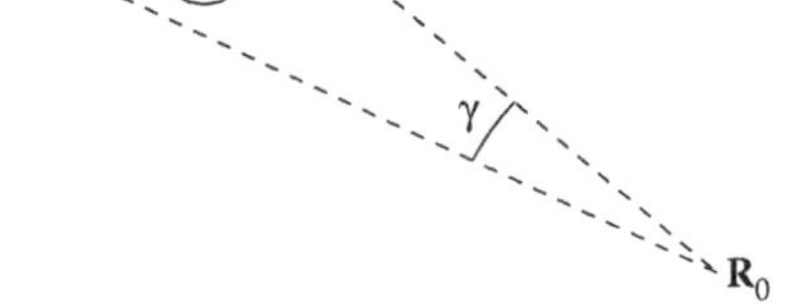

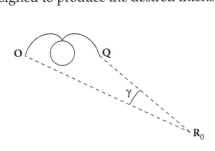

FIGURE 8.45
From initial point $\mathbf{R}_0$, the source subtends an angle γ. The luminaire can, however, be designed for angles larger than γ due to multiple reflections on the mirror, as in far-edge converging luminaire for a flat source.

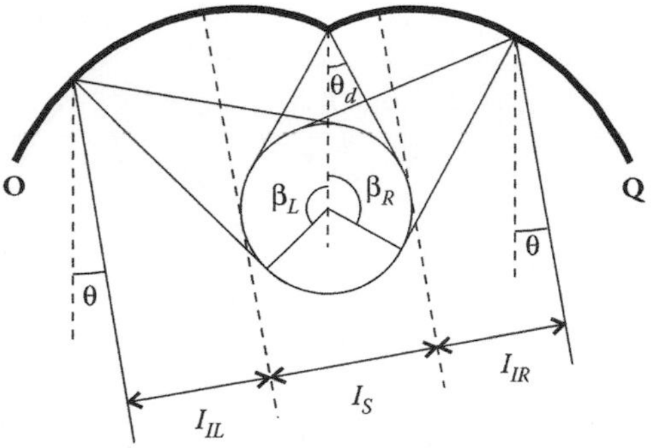

FIGURE 8.46
If the central part of the luminaire is made of two involute arcs, there will be reflections of the source in direction θ in both portions of the involute. Therefore, the intensity produced by the optic in direction θ is given by the intensity of the source plus the two parts resulting from reflections on both the mirrors.

Let us now consider a different situation in which the central part of the luminaire consists of two involute arcs that do not touch the source, as presented in Figure 8.46. Their optical behavior is completely different from the partial involutes presented earlier in Figure 8.40. It no longer behaves as a Lambertian source with edges **O** and **Q**. To see this, a brief presentation of the optical characteristics of these involute arcs is made in Figure 8.46.

Figure 8.46 presents the contributions of the source I_S and images I_{IL} and I_{IR} on the left-hand- and right-hand-side involutes, respectively.

The intensity of each of the involutes for the intensity of the luminaire can be determined from its geometry, as presented in Figure 8.47.

From Figure 8.47, we have $\delta = \beta_R - \theta - \pi/2$. We then have $d_1 = a - a \cos \delta = a - a \sin(\beta_R - \theta)$ and $d_2 = a + a \cos \delta = a + a \sin(\beta_R - \theta)$ and also $d_3 = d_2 - d_1 = 2a \sin(\beta_R - \theta)$. This is the contribution of the right-hand-side involute. In the same way, the contribution of the left-hand-side involute is given by $2a \sin(\beta_L + \theta)$. Therefore, the sum of the contributions of source I_S and the involutes $I_I = I_{IL} + I_{IR}$ is given by

$$I_S + I_I = 2a + 2a\sin(\beta_R - \theta) + 2a\sin(\beta_L + \theta) \tag{8.79}$$

Nevertheless, it should be noted that this expression is valid only for θ in the interval $\pm\theta_d$, θ_d being defined in Figure 8.46. For values of θ outside this range, it is necessary to recalculate this expression, since the images are disjointed.[9] The total intensity of the luminaire can then be calculated by

$$I(\theta) = I_S(\theta) + I_I(\theta) + I_M(\theta) \tag{8.80}$$

where $I(\theta)$ is the desired intensity, $I_I(\theta)$, the intensity produced by the involutes, and $I_M(\theta)$ the intensity of the mirror of the luminaire to be designed.

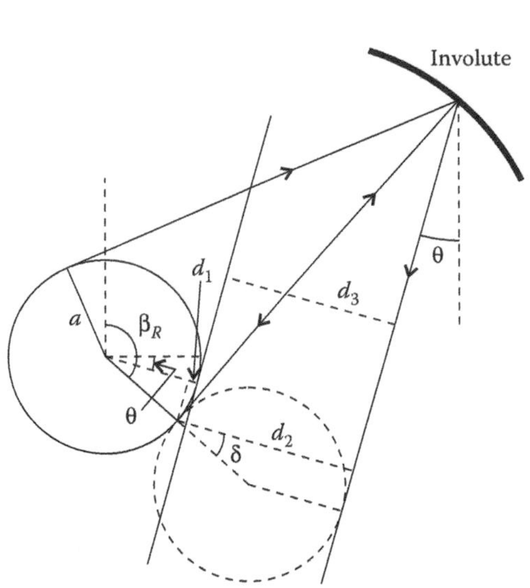

FIGURE 8.47
Contribution of reflection on each of the involute mirrors for the illuminance of the luminaire.

The analytical method described earlier can be used to design the luminaire mirror for this kind of involute. Let us then suppose that the mirror starts at a point $\mathbf{R}_0$ and that, at this point, a light ray tangent to the lower part of the source is reflected vertically, as presented in Figure 8.48a. This boundary condition is similar to the one used in the examples of flat-source case. The points $\mathbf{R}_i$ of the mirror can now be obtained for $0 < \theta_i < \gamma$ and, therefore, the mirror can be obtained for points between $\mathbf{R}_0$ and $\mathbf{R}\gamma$. For these angles θ_i, the image of the source in the mirror extends from $\mathbf{R}_0$ to the point $\mathbf{R}_i$. The maximum value γ is the one for which the ray reflected at $\mathbf{R}_0$ is tangent to the upper part of the source, as presented in Figure 8.48c. For angles to the vertical larger than γ, the shape of the mirror can no longer be obtained, since no light ray coming from the source can be reflected at $\mathbf{R}_0$ and leave the luminaire in these directions. For this reason, the image of the source on

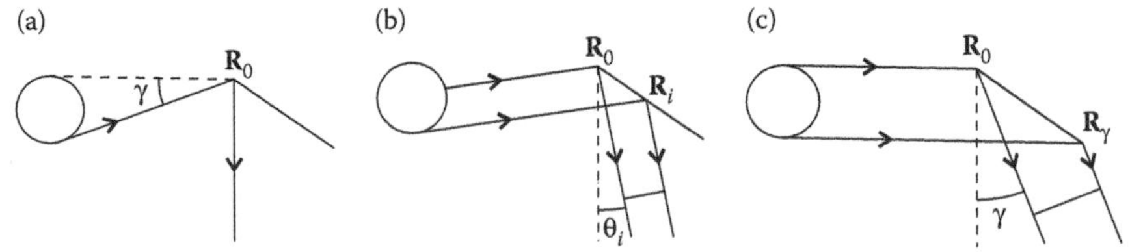

FIGURE 8.48
(a) At point $\mathbf{R}_0$, the ray of light tangent to the lower part of the tube is reflected vertically. (b) For $\theta \neq 0$, the image extends from $\mathbf{R}_0$ to another point $\mathbf{R}_i$ on the mirror. However, this is valid only for values of $|\theta|$ smaller than γ. (c) Limit case in which $|\theta| = \gamma$. For larger values of $|\theta|$, the image of the tube on the mirror no longer extends to $\mathbf{R}_0$.

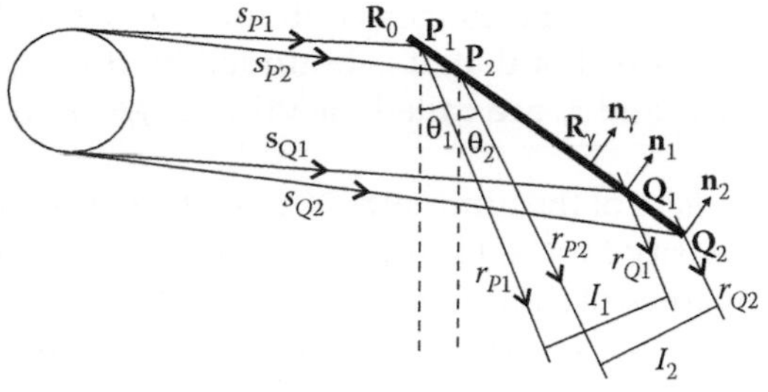

FIGURE 8.49
For angles θ_1, θ_2, ... larger than γ, the image of the source on the mirror extends only to $\mathbf{P}_1$, $\mathbf{P}_2$, ... and not to $\mathbf{R}_0$. Knowing the desired intensity in these directions, it is possible to calculate new points $\mathbf{Q}_1$, $\mathbf{Q}_2$, ... of the mirror based on the part of the mirror already calculated.

the mirror no longer extends to $\mathbf{R}_0$ and therefore the intensity that this image produces can no longer be obtained by $I(\theta) = p(\theta) - p_0(\theta)$.

The mirror can, however, be extended beyond point $\mathbf{R}\gamma$ using a different design method. The part of the mirror between $\mathbf{R}_0$ and $\mathbf{R}\gamma$ calculated earlier is used in this extension. Let us then suppose that a light ray s_{p1} tangent to the upper part of the source hits a point $\mathbf{P}_1$ on the mirror $\mathbf{R}_0\mathbf{R}\gamma$, as presented in Figure 8.49. This ray will be reflected with an angle to the vertical given by θ_1, that is, it is reflected as ray r_{P1}. If the desired intensity I_1 produced by the mirror in this direction is known, line r_{Q1} can be obtained. This line is parallel to r_{P1} and the distance between them is I_1. Since point $\mathbf{R}\gamma$ of the mirror is known, the normal $\mathbf{n}_\gamma$ to the mirror at $\mathbf{R}_\gamma$ is also known. Line r_{Q1} can then be intersected with the tangent to the mirror at point $\mathbf{R}_\gamma$, and point $\mathbf{Q}_1$ obtained. The normal $\mathbf{n}_1$ of the mirror at point $\mathbf{Q}_1$ can also be obtained, since at this point the mirror must reflect ray s_{Q1} tangent to the source into reflected ray r_{Q1} leaving the luminaire.

We now consider a light ray s_{P2} tangent to the upper part of the source and hitting a point $\mathbf{P}_2$ on the known part of the mirror $\mathbf{R}_0\mathbf{Q}_1$. This ray will be reflected with an angle to the vertical given by θ_2, that is, it is reflected as ray r_{P2}. If the desired intensity I_2 produced by the mirror in this direction is known, line r_{Q2} can be obtained. This line is parallel to r_{P2} and the distance between them is I_2. Point $\mathbf{Q}_1$ of the mirror is known, and the normal $\mathbf{n}_1$ to the mirror at $\mathbf{Q}_1$ is also known. Line r_{Q2} can then be intersected with the tangent to the mirror at point $\mathbf{Q}_1$ and point $\mathbf{Q}_2$ is obtained. The normal $\mathbf{n}_2$ of the mirror at point $\mathbf{Q}_2$ can also be obtained, since, at this point, the mirror must reflect ray s_{Q2} tangent to the source into reflected ray r_{Q2} leaving the luminaire.

This process can now continue for more points on the mirror. To obtain a good approximation to the shape of the mirror, it is important to proceed in very small steps.

The image of the source in the mirror now extends from $\mathbf{P}_1$ to $\mathbf{Q}_1$ for angle θ_1 and from $\mathbf{P}_2$ to $\mathbf{Q}_2$ for angle θ_2. For these θ angles ($|\theta| > \gamma$), the image of the source in the mirror is no longer contained between $\mathbf{R}_0$ and another point of the mirror. Instead of that, it is bounded by the edge rays s_P and s_Q of the source. Rays s_Q and s_P are called "leading edge" and "trailing edge," respectively.[7]

Note that the intensity of the luminaire is given by expression (8.68), where I is the desired intensity for the luminaire, I_S the intensity produced by the source itself, I_M the intensity produced by the mirror whose construction is described earlier, and I_{OM} the contribution of other mirrors that the luminaire may contain (involute arcs).[9]

Figure 8.50 shows an example of the application of design methods presented earlier. The side mirrors are calculated from point $\mathbf{R}_0$ to point $\mathbf{R}_\gamma$, using the analytical method, and beyond point $\mathbf{R}_\gamma$ using the numerical method presented in Figure 8.49. The slope of the mirror at the initial point $\mathbf{R}_0$ must be such that ray r_1 coming from the source is reflected and exits the luminaire in the vertical direction. This is the initial condition also used in the examples presented earlier for linear sources.

The analytical method described earlier is valid for $|\theta| < \gamma$, where γ is the angle that the tubular source subtends, when seen from $\mathbf{R}_0$, as presented in Figure 8.50a. In this case, the image of the source on the mirror extends from $\mathbf{R}_0$ to $\mathbf{R}_1(\theta)$, and the intensity I_M for $|\theta| = |\theta_1| < \gamma$ is represented by d_1. This analytical method can then be applied in the design of the mirror between $\mathbf{R}_0$ and $\mathbf{R}\gamma$ and the points of the mirror obtained from expression (8.76).

From this point onward (i.e., for $|\theta| > \gamma$), the situation is different, as shown in Figure 8.50b, wherein it is impossible for the light to exit $\mathbf{R}_0$ in these directions. Therefore, the image of the tubular source, when seen from $|\theta| = |\theta_2| > \gamma$, extends from $\mathbf{R}_T(\theta)$ to $\mathbf{R}_2(\theta)$. For each θ_2, we can then determine $\mathbf{R}_T$ on the part of the mirror already calculated. Based on this point, we can calculate a new point $\mathbf{R}_2$ ahead. We then see that the new portion of the mirror to be calculated is based on the part of the mirror already calculated. As

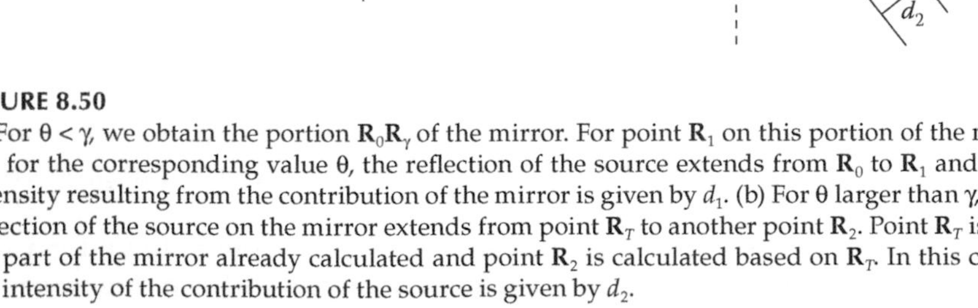

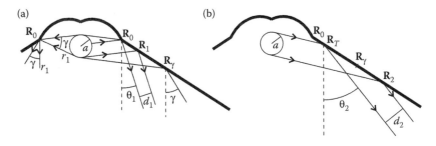

FIGURE 8.50
(a) For $\theta < \gamma$, we obtain the portion $\mathbf{R}_0\mathbf{R}_\gamma$ of the mirror. For point $\mathbf{R}_1$ on this portion of the mirror, for the corresponding value θ, the reflection of the source extends from $\mathbf{R}_0$ to $\mathbf{R}_1$ and the intensity resulting from the contribution of the mirror is given by d_1. (b) For θ larger than γ, the reflection of the source on the mirror extends from point $\mathbf{R}_T$ to another point $\mathbf{R}_2$. Point $\mathbf{R}_T$ is on the part of the mirror already calculated and point $\mathbf{R}_2$ is calculated based on $\mathbf{R}_T$. In this case, the intensity of the contribution of the source is given by d_2.

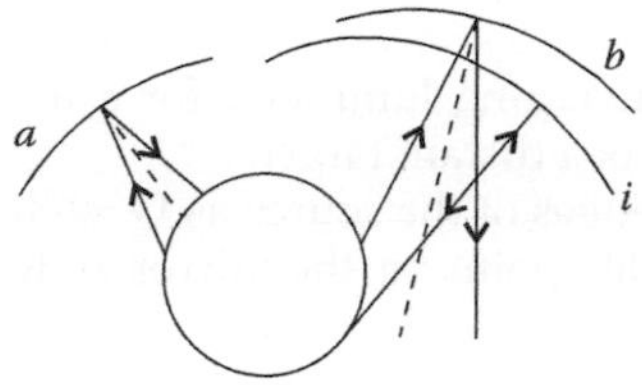

FIGURE 8.51
The central part of the luminaire has the shape of an involute (curve *i*) since this is the curve that allows more compact designs possible with no light reflected back to the source. In the case of curve *a*, the normal to the curve intersects the tubular source so that there will be light reflected back to the tube instead of to the target. In the case of curve *b*, the normal passes far from the tube and, therefore, a mirror designed according to one of these curves would be bigger than the one shaped as an involute.

one ray goes through the part of the mirror already calculated, the other one enables us to calculate a new portion of the mirror.[7,9] In some cases, in the design of luminaires, there is an added complexity of the "shade" produced by the mirrors as in the luminaire of Figure 8.27.

In the design of a luminaire, an important aspect is to avoid radiation coming from the source being redirected back to the source. The goal is to make all the radiation to exit from the luminaire so that the exit power is maximized. However, one intends that the luminaire should be as compact as possible. The central part of the luminaire is then designed by placing two involute arcs side by side.

Then, the normal to the mirror cannot intersect the tube, because some radiation will be reflected back to the source (the case of curve *a* in Figure 8.51). If the normal to the curve passes far off from the source (case of curve *b*), this problem no longer exists and the light is reflected far from the source. Nevertheless, the mirror obtained in this case is bigger than necessary. Involute-shaped curve *i* is the one that, avoiding the reflection of radiation back to the source, enables us to design the smallest possible mirror.

In the design method presented earlier, the central part of the luminaire does not necessarily have to consist of two arcs of involute.[9] However, this is the solution that enables the design of the most compact devices.

Also, note that the mirror must have a wedge point. This results from the fact that the normal to the mirror cannot intersect the tube.

8.9 Examples

The examples presented as follows use expressions for the curves and functions that are derived in Chapter 21.

EXAMPLE 8.1

Design a far-edge converging luminaire for a unit length source and uniform illumination of a distant target.

We first define the edges of the source as $\mathbf{O} = (-0.5, 0)$ and $\mathbf{Q} = (0.5, 0)$. We now define the edge point for the mirror as $\mathbf{R}_0 = (1.82, 1.3)$. We can now calculate

$$\begin{aligned} \phi_0 &= \text{ang}(\mathbf{R}_0 - \mathbf{O},(0,-1)) = 2.08155 \\ r_0 &= [\mathbf{R}_0,\mathbf{O}] = 2.6594 \end{aligned} \tag{8.81}$$

And we get

$$\begin{aligned} p(\theta) &= -\frac{1}{\cos^2\theta} + r_0\sin(\phi_0 - \theta) + \cos\theta \\ &= \cos\theta - \sec^2\theta + 2.6594\sin(2.08155 - \theta) \\ P(\theta) &= -\tan\theta + r_0\cos(\theta - \phi_0) + \sin\theta \\ &= 2.6594\cos(2.08155 - \theta) + \sin\theta - \tan\theta \end{aligned} \tag{8.82}$$

We set that the ray coming from $\mathbf{O}$ and reflected at $\mathbf{R}_0$ exits the luminaire in the vertical direction, and we have $\theta_0 = 0$. We then get

$$\begin{aligned} p(\theta_0) &= r_0\sin\phi_0 = 2.32 \\ P(\theta_0) &= r_0\cos\phi_0 = -1.3 \end{aligned} \tag{8.83}$$

Constant C_m is now given by

$$C_m = P(\theta_0) - \frac{p(\theta_0)}{\tan(\phi_0/2)} = -2.6594 \tag{8.84}$$

and

$$\alpha(\theta) = \arctan\left(\frac{p(\theta)}{P(\theta) - C_m}\right) \tag{8.85}$$

Finally, the points of the mirror are given by

$$\mathbf{R}(\theta) = \mathbf{O} + \frac{p(\theta)}{\sin(2\alpha(\theta))}(\sin(2\alpha(\theta)+\theta), -\cos(2\alpha(\theta)+\theta)) \tag{8.86}$$

To find the maximum value for θ, we numerically solve the equation

$$\mathbf{R}(\theta) = \mathbf{Q} + x(\mathbf{R}_0 - \mathbf{Q}) \tag{8.87}$$

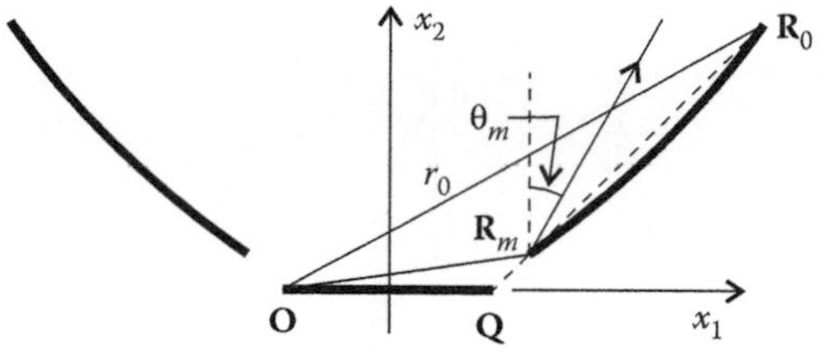

FIGURE 8.52
Far-edge converging luminaire.

and get $\theta = \theta_m = -0.512409$ rad $= -29.3588°$ and $x = x_m = 0.13299$. This also enables us to obtain $\mathbf{R}_m = \mathbf{R}(\theta_m)$ which is on the line connecting $\mathbf{Q}$ and $\mathbf{R}_0$. The mirror is finally given by $\mathbf{R}(\theta)$ with $\theta_m \le \theta \le 0$. Figure 8.52 shows the resulting luminaire.

The illuminance pattern on a distant target can now be determined by ray tracing. Figure 8.53 shows the geometry of luminaire and target.

The illuminance pattern on the target as a function of the angle θ is shown in Figure 8.54.

A note could now be added about how to ray trace these optics. The first thing is to generate a ray set. One way of doing it is by using the Monte Carlo integration.[10]

The integral of a function $I(\theta)$ over an interval $\Delta\theta$ from θ_1 to θ_2, where $\Delta\theta = \theta_2 - \theta$, can be approximately calculated using Monte Carlo integration:

$$\int_{\theta_1}^{\theta_2} I(\theta) d\theta \approx \Delta\theta \frac{1}{N} \sum_{i=1}^{N} I(\theta_i) \tag{8.88}$$

It is given by the product of the interval $\Delta\theta$ by the mean value of the function in that interval. Its mean value is approximated by generating random values θ_i uniformly distributed in the interval $\Delta\theta$, calculating the value of the function at the points $I(\theta_i)$ and then dividing by the total number of points.

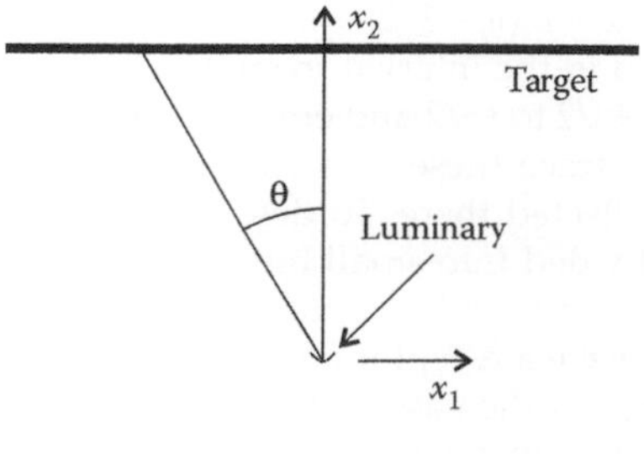

FIGURE 8.53
Geometry of far-edge converging luminaire and distant target.

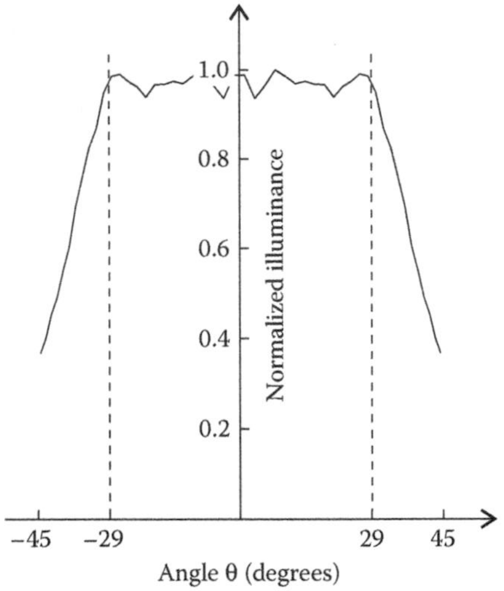

FIGURE 8.54
Normalized illuminance pattern on a distant target as a function of angle θ in degrees.

Using expression (8.88) we can generate random rays that can be used to simulate a source in a computer ray trace. Let us consider that we have a uniform 2-D Lambertian source of length L_S, for example, line source **OQ** in Figure 8.52, and, therefore, $L_S = [\mathbf{O}, \mathbf{Q}]$. The intensity in each direction produced by this uniform source is given by $I(\theta) = I_0 \cos\theta$. The total flux it emits is given by

$$\Phi = \int_{-\pi/2}^{\pi/2} I(\theta)d\theta = \int_{-\pi/2}^{\pi/2} I_0 \cos\theta d\theta \approx \sum_{i=1}^{N} \frac{\pi I_0}{N} \cos\theta_i \tag{8.89}$$

We can then generate a set of rays, each of them defined by a point $\mathbf{P}_i = (x_i, 0)$, a direction $\mathbf{v}_i = (\cos\theta_i, \sin\theta_i)$, and a power given by $p_i = \pi I_0 \cos\theta_i/N$, where N is the number of rays in the ray set. The rays in the rayset are then defined by $(\mathbf{P}_i, \mathbf{v}_i, p_i)$ with $i = 1,\ldots,N$. The values of x_i and θ_i can be obtained by $x_i = L_s(y_i - 1/2)$ and $\theta_i = \pi(z_i - 1/2)$ where y_i and z_i are randomly generated in the interval from 0 to 1. This simulates a source that extends from $-L_s/2$ to $L_S/2$ and emits in angles from $-\pi/2$ to $\pi/2$.

We may now ray trace these rays through the system. They hit the receiver and are collected there. To determine the flux distribution on the receiver, it is divided into small bins by a process called, naturally enough, binning.

Suppose that we have a receptor defined by a parameter α that varies between α_m and α_M. In the case of Figure 8.53, the receptor would be the target and the parameter α could be the horizontal coordinate x_1. We divide this parameter space into bins of, for example, equal length, as shown in Figure 8.55. In this particular case, we have seven bins,

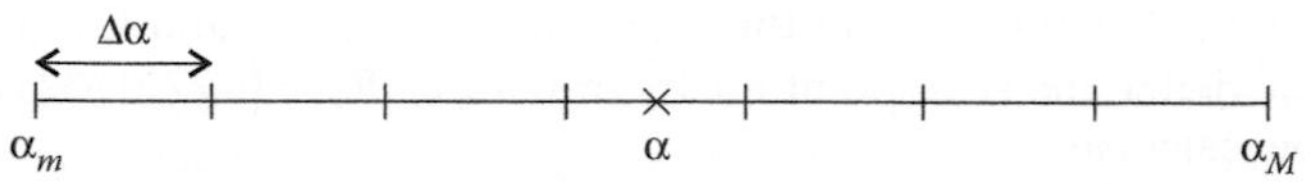

FIGURE 8.55
Receiver divided into bins.

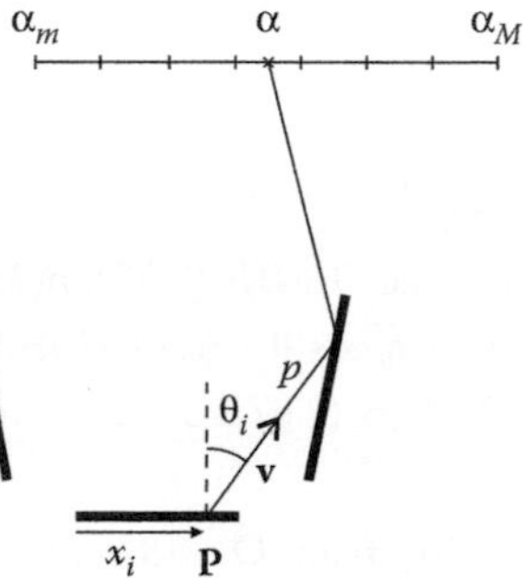

FIGURE 8.56
A ray with power p leaves the source from point **P** in a direction **v**, goes through the optic, and hits the receiver at a position defined by parameter value α.

each of a length Δα; and, in general, we have N bins, each of a length $\Delta\alpha = (\alpha_M - \alpha_m)/N$.

If a ray hits the receiver now at a point with a parameter value α, and we want to determine to which bin it corresponds, we will calculate $b = (\alpha - \alpha_m)/\Delta\alpha$. In the particular case of Figure 8.55, b is 3.5, and the point is in bin number 4. We define sig(b) as a function that returns the smallest integer greater than or equal to b. For example, sig(3.1) = 4, sig(3.8) = 4, and sig(4) = 4. This function sig(b) will give us the bin number for any parameter value $\alpha > \alpha_m$. For $\alpha = \alpha_m$ the bin number is 1.

Figure 8.56 shows a ray with power p exiting the source at a point **P** in a direction **v**. It goes through the optic and hits the target at a point with parameter α.

Every time a ray hits one of these bins, we add the power of the ray to that bin. In the end, we will know how much power falls in each bin and, therefore, the power distribution on the receiver.

If we are interested in the intensity pattern instead, we would divide the angular space into bins and count how much flux falls into each bin. These bins could be, for example, at angular intervals to the optical axis from 0° to 5°, from 5° to 10° and so on. For each ray leaving the optic, we could then check the angle it makes to the optical axis, and add its power to the corresponding bin. In the end, we would have a distribution of power as a function of direction to the optical axis.

EXAMPLE 8.2

Design a near-edge diverging luminaire for a unit length source and uniform illumination of a distant target.

We first define the edges of the source as $\mathbf{O} = (-0.5, 0)$ and $\mathbf{Q} = (0.5, 0)$. We now define the edge point for the mirror as $\mathbf{R}_0 = (0.77, 0.5324)$. We can now calculate

$$\begin{aligned}\phi_0 &= \text{ang}(\mathbf{R}_0 - \mathbf{O},(0,-1)) = 153.109\,\text{deg} \\ r_0 &= [\mathbf{R}_0,\mathbf{O}] = 0.579957\end{aligned} \tag{8.90}$$

And we get

$$\begin{aligned}p(\theta) &= -\frac{1}{\cos^2\theta} + r_0\sin(\phi_0 - \theta) + \cos\theta \\ &= \cos\theta - \sec^2\theta + 0.579957\sin(2.67225 - \theta) \\ P(\theta) &= -\tan\theta + r_0\cos(\theta - \phi_0) + \sin\theta \\ &= 0.579957\cos(2.67225 - \theta) + \sin\theta - \tan\theta\end{aligned} \tag{8.91}$$

We set that the ray coming from $\mathbf{O}$ and reflected at $\mathbf{R}_0$ exits the luminaire in the vertical direction, and we have $\theta_0 = 0$. We then get

$$\begin{aligned}p(\theta_0) &= r_0\sin\phi_0 = 0.262314 \\ P(\theta_0) &= r_0\cos\phi_0 = -0.517244\end{aligned} \tag{8.92}$$

Constant C_m is now given by

$$C_m = P(\theta_0) - \frac{p(\theta_0)}{\tan(\phi_0/2)} = -0.579957 \tag{8.93}$$

and

$$\alpha(\theta) = \arctan\left(\frac{p(\theta)}{P(\theta) - C_m}\right) \tag{8.94}$$

Finally, the points of the mirror are given by

$$\mathbf{R}(\theta) = \mathbf{O} + \frac{p(\theta)}{\sin(2\alpha(\theta))}(\sin(2\alpha(\theta) + \theta), -\cos(2\alpha(\theta) + \theta)) \tag{8.95}$$

In order to find the maximum value for θ, we numerically solve the equation,

$$2\alpha + \theta = \pi/2 \tag{8.96}$$

and get $\theta = \theta_m = 0.462083$ rad $= 26.4754°$ which terminates the mirror at the horizontal (x_1) axis. The mirror is finally given by $\mathbf{R}(\theta)$ with $0 \le \theta \le \theta_m$. Figure 8.57 shows the resulting luminaire.

The illuminance pattern on a distant target can now be determined by ray tracing. Figure 8.58 shows the geometry of luminaire and target.

The illuminance pattern on the target as a function of the angle θ is shown in Figure 8.59.

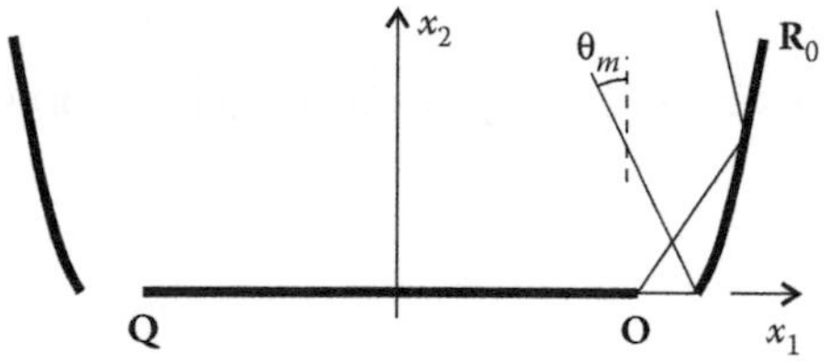

FIGURE 8.57
Near-edge diverging luminaire.

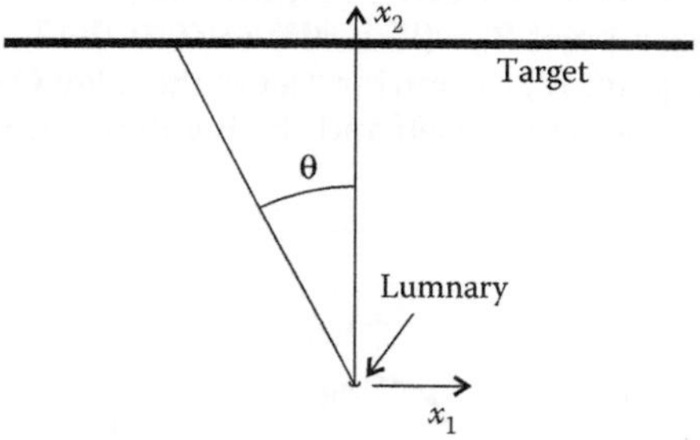

FIGURE 8.58
Geometry of near-edge diverging luminaire and distant target.

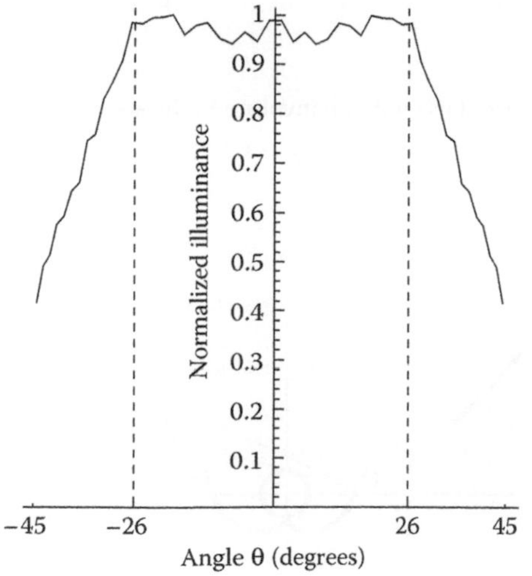

FIGURE 8.59
Normalized illuminance pattern on a distant target.

EXAMPLE 8.3

Design a far-edge converging luminaire for a tubular source of unit radius $a = 1$ and uniform illumination of a distant target.

The design of the luminaire starts with an involute to the source given by

$$\mathrm{inv}(\xi) = a(\cos(\xi - \pi/2), \sin(\xi - \pi/2)) + \xi a(\cos(\xi - \pi), \sin(\xi - \pi)) \tag{8.97}$$

with $a = 1$ and

$$-(\pi/2 + \mu) \leq \xi \leq \pi/2 + \mu \tag{8.98}$$

We choose $\mu = 75°$ and get $\mathbf{Q} = \mathrm{inv}(\pi/2 + \mu) = (3.04049, 0.22058)$ and **O** as its symmetrical point, as shown in Figure 8.60.

We now choose a point $\mathbf{R}_0 = (10.4, 605)$ to start the luminaire mirror. From point $\mathbf{R}_0$ to point $\mathbf{R}_1$, the mirror "sees" the edge **O** of the apparent source formed by the source itself and the involute mirrors as shown in Figure 8.61.

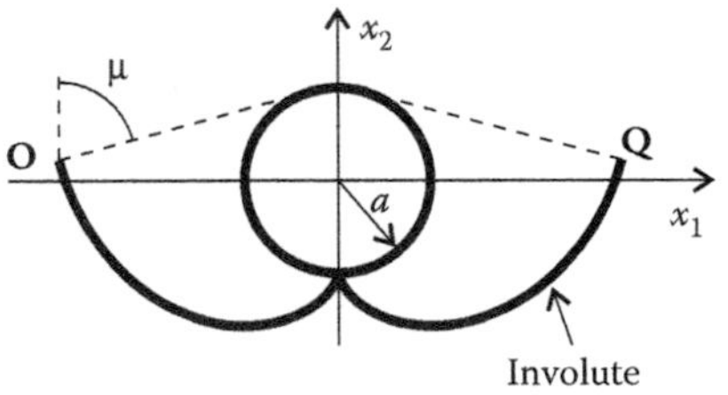

FIGURE 8.60
The design of the luminaire starts with an involute to the source.

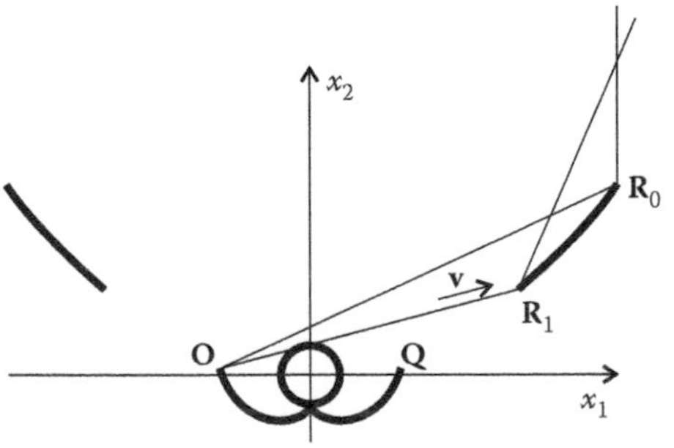

FIGURE 8.61
We choose the position of initial point $\mathbf{R}_0$. The first part $\mathbf{R}_0\mathbf{R}_1$ of the mirror is designed as a far-edge converging luminaire for a linear source **OQ**.

This portion of the mirror is then calculated as a far-edge converging luminaire for a linear source **OQ**. We calculate

$$\begin{aligned} s &= [\mathbf{O},\mathbf{Q}] = 6.08097 \\ \phi_0 &= \text{ang}(\mathbf{R}_0 - \mathbf{O},(0,-1)) = 115.042° \\ r_0 &= [\mathbf{R}_0,\mathbf{O}] = 14.835 \end{aligned} \tag{8.99}$$

and we get

$$\begin{aligned} p(\theta) &= -\frac{s}{\cos^2\theta} + r_0\sin(\phi_0 - \theta) + s\cos\theta \\ &= 6.08097(\cos\theta - \sec^2\theta) + 14.835\sin(2.00786 - \theta) \\ P(\theta) &= -s\tan\theta + r_0\cos(\theta - \phi_0) + s\sin\theta \\ &= 14.835\cos(2.00786 - \theta) + 6.08097(\sin\theta - \tan\theta) \end{aligned} \tag{8.100}$$

We set that the ray coming from **O** and reflected at $\mathbf{R}_0$ exits the luminaire in the vertical direction, and we have $\theta_0 = 0$. We obtain

$$\begin{aligned} p(\theta_0) &= r_0\sin\phi_0 = 13.4405 \\ P(\theta_0) &= r_0\cos\phi_0 = -6.27942 \end{aligned} \tag{8.101}$$

Constant C_m is now given by

$$C_m = P(\theta_0) - \frac{p(\theta_0)}{\tan(\phi_0/2)} = -14.835 \tag{8.102}$$

and

$$\alpha(\theta) = \arctan\left(\frac{p(\theta)}{P(\theta) - C_m}\right) \tag{8.103}$$

Finally, the points of the mirror are given by

$$\mathbf{R}(\theta) = \mathbf{O} + \frac{p(\theta)}{\sin(2\alpha(\theta))}(\sin(2\alpha(\theta) + \theta), -\cos(2\alpha(\theta) + \theta)) \tag{8.104}$$

Vector **v** in Figure 8.61 is given by $\mathbf{v} = (\cos(\pi/2 - \mu), \sin(\pi/2 - \mu))$ and point $\mathbf{R}_1$ can be obtained by numerically solving the pair of equations (note that **O**, **v**, and **R** are two-dimensional),

$$\mathbf{O} + d\mathbf{v} = \mathbf{R}(\theta) \tag{8.105}$$

and we obtain $\theta = \theta_1 = -23.5297°$ and $d = d_1 = 10.4278$. The portion of the mirror $\mathbf{R}_0\mathbf{R}_1$ is then given by $\mathbf{R}(\theta)$ for $\theta_1 \le \theta \le 0$.

From point $\mathbf{R}_1$ downward, point **O** is no longer visible and the mirror is calculated as a far-edge converging luminaire for a tubular source. The image on the mirror for this second part of the luminaire also extends all the way to $\mathbf{R}_0$, so $p(\theta)$ and $P(\theta)$ are calculated relative to this position. For this new section of the mirror, we first calculate point $\mathbf{P}_0$ on the source that is also on the tangent line to the source through point $\mathbf{R}_0$, as shown in Figure 8.62.

We have $\mathbf{P}_0 = (-0.45909, 0.88839)$ and then

$$\begin{aligned} r_0 &= [\mathbf{P}_0, \mathbf{R}_0] = 12.2233 \\ \phi_0 &= \text{ang}(\mathbf{R}_0 - \mathbf{P}_0, (0,-1)) = 117.328^\circ \end{aligned} \tag{8.106}$$

and also

$$\begin{aligned} p(\theta) &= a + a\cos(\theta - \phi_0) - r_0 \sin(\theta - \phi_0) + s\cos\theta - s/\cos^2\theta \\ &= 1 + 16.481\cos\theta - 6.08097\sec^2\theta + 6.5\sin\theta \\ P(\theta) &= a\theta + \cos\theta(r_0\cos\phi_0 - a\sin\phi_0) + \sin\theta(s + a\cos\phi_0 + r_0\sin\phi_0) \\ &= \theta - 6.5\cos\theta + 16.481\sin\theta - 6.08097\tan\theta \end{aligned} \tag{8.107}$$

To calculate the value of the integration constant C_m, we use the values relative to point $\mathbf{R}_1$ where this new portion of the mirror starts. We have already obtained θ_1 and ϕ_1 is given by (Figure 8.62)

$$\phi_1 = \text{ang}(\mathbf{R}_1 - \mathbf{O}, (0,-1)) = 105^\circ \tag{8.108}$$

We also have

$$\begin{aligned} p(\theta_1) &= 6.28175 \\ P(\theta_1) &= -10.302 \end{aligned} \tag{8.109}$$

Constant C_m is now given by

$$\begin{aligned} C_m &= P(\theta_1) - \cot\left(\frac{\phi_1 - \theta_1}{2}\right)(p(\theta_1) - 2a) - 2a\arctan\left(\cot\left(\frac{\phi_1 - \theta_1}{2}\right)\right) \\ &= -13.2642 \end{aligned} \tag{8.110}$$

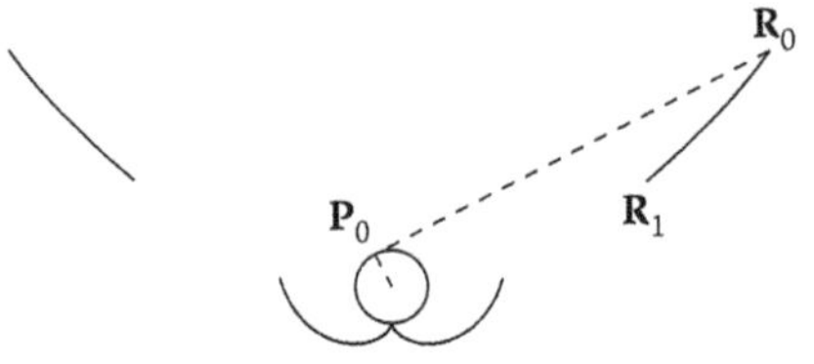

FIGURE 8.62
Point $\mathbf{P}_0$ of the source is also on the tangent line to the source that goes through point $\mathbf{R}_0$.

We now give a maximum value of $\theta_m = -31°$ to θ. For different values of $\theta_m \le \theta \le \theta_1$ we can find the corresponding values of α by numerically solving the equation:

$$C_m = P(\theta) - \cot\alpha(p(\theta) - 2a) - 2a\arctan(\cot\alpha) \tag{8.111}$$

We obtain, for example, for the pairs (θ_i, α_i): ((−31,0.94981), (−30,0.997741), (−29,1.0334), (−28,1.06044), (−27,1.08113), (−26,1.09701), (−25,1.10912), (−24,1.11822)).

For each of these pairs, we can calculate

$$r_i = \frac{p(\theta_i)}{\sin(2\alpha_i)} - \frac{a}{\tan\alpha_i} \tag{8.112}$$

The points of the mirror are finally given by

$$\mathbf{R}_i = a\left(\cos(2\alpha_i + \theta_i), \sin(2\alpha_i + \theta_i)\right) + r_i\left(\sin(2\alpha_i + \theta_i), -\cos(2\alpha_i + \theta_i)\right) \tag{8.113}$$

as

$\mathbf{R}_i$ = ((7.67964,8.43946), (71.0586, −260.689), (20.1508, −5.73802), (19.5833,19.7013), (−27.6671,94.4873), (−9.60706,3.42271), (9.64254,10.9493), (4.85006, −25.3726)).

Figure 8.63 shows a complete luminaire.

The illuminance pattern on a distant target can now be determined by ray tracing. Figure 8.64 shows the geometry of luminaire and target.

The illuminance pattern on a distant target is shown in Figure 8.65 as a function of angle θ as defined in Figure 8.64.

The pattern is uniform within the design angle.

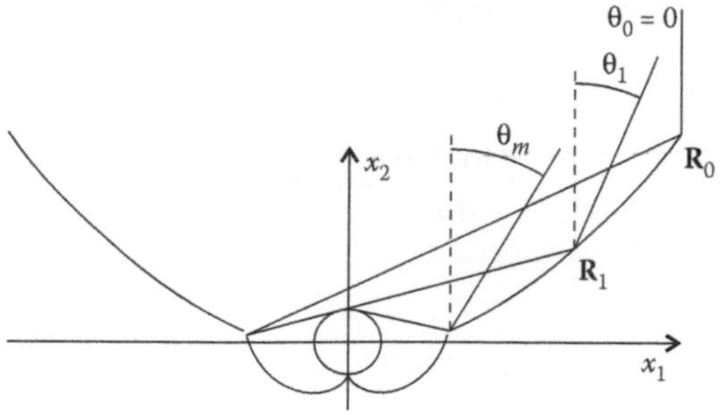

FIGURE 8.63
Complete luminaire.

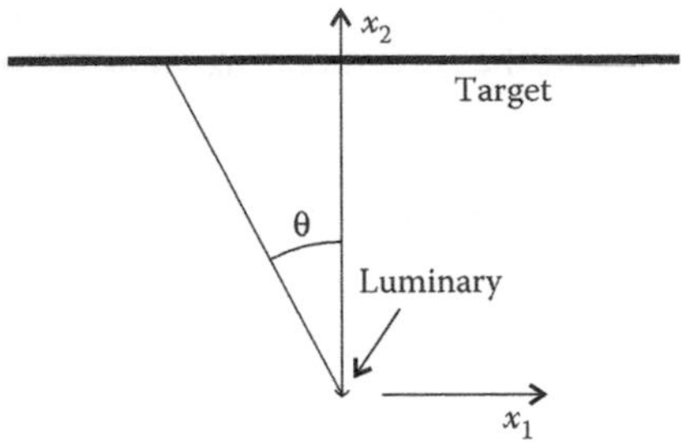

FIGURE 8.64
Geometry of distant target and far-edge converging luminaire for a tubular source.

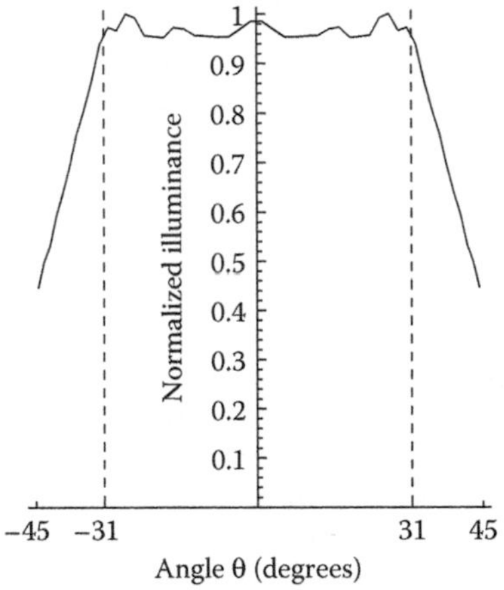

FIGURE 8.65
Illuminance pattern on the target as a function of angle θ.

Appendix A: Mirror Differential Equation for Linear Sources

Solve the equation:

$$\frac{d\ln r(\phi)}{d\phi} = \tan\alpha(\theta) \tag{A.1}$$

for $\alpha(\theta)$ where

$$\begin{aligned} \phi &= 2\alpha(\theta) + \theta \\ r(\theta) &= p(\theta)/\sin(2\alpha) \end{aligned} \tag{A.2}$$

in which $p(\theta)$ is known.

to solve Equation A.1 for $\alpha(\theta)$ we start by calculating the derivative of ln r and we get

$$\frac{d\ln r}{d\theta} = \frac{d\ln r}{d\phi}\frac{d\phi}{d\theta} \Leftrightarrow \frac{d\ln r}{d\theta} = \frac{1}{r}\frac{dr}{d\phi}\frac{d\phi}{d\theta} \tag{A.3}$$

From the first expression of (A.2) we get

$$\frac{d\phi}{d\theta} = 2\frac{d\alpha}{d\theta} + 1 \tag{A.4}$$

and therefore we have

$$\frac{d\ln r}{d\theta} = \tan\left(\alpha(\theta)\right)\left(2\frac{d\alpha}{d\theta} + 1\right) \tag{A.5}$$

which is a differential equation for $\alpha(\theta)$. From the second expression of (A.2) we have

$$\ln(p(\theta)) = \ln(r(\phi(\theta))) + \ln(\sin(2\alpha(\theta))) \tag{A.6}$$

Calculating the θ derivative, we get

$$\frac{d\ln p}{d\theta} = \frac{1}{r}\frac{dr}{d\phi}\frac{d\phi}{d\theta} + 2\frac{\cos(2\alpha)}{\sin(2\alpha)}\frac{d\alpha}{d\theta} \tag{A.7}$$

Considering Equations A.3 and A.5, we get

$$\begin{aligned}\frac{d\ln p}{d\theta} &= \tan\alpha\left(2\frac{d\alpha}{d\theta} + 1\right) + \frac{\cos^2\alpha - \sin^2\alpha}{\sin\alpha\cos\alpha}\frac{d\alpha}{d\theta} \\ &= \left(\frac{2\sin^2\alpha}{\sin\alpha\cos\alpha} + \frac{\cos^2\alpha - \sin^2\alpha}{\sin\alpha\cos\alpha}\right)\frac{d\alpha}{d\theta} + \frac{\sin\alpha}{\cos\alpha}\end{aligned} \tag{A.8}$$

which is a differential equation for $\alpha(\theta)$ since $p(\theta)$ is known. We can now write[4]

$$\frac{d\alpha}{d\theta} = \sin\alpha\cos\alpha\frac{d\ln(p(\theta))}{d\theta} - \sin^2\theta \tag{A.9}$$

Dividing Equation A.9 by $\sin^2\alpha$ we get

$$-\frac{1}{\sin^2\alpha}\frac{d\alpha}{d\theta} + \frac{1}{\tan\alpha}\frac{d\ln(p(\theta))}{d\theta} = 1 \tag{A.10}$$

Equation A.10 is an equation for $\alpha(\theta)$, which can be solved by making the change of variables[4]

$$u = 1/\tan\alpha \tag{A.11}$$

resulting in

$$\frac{du}{d\theta} + u\frac{d\ln(p(\theta))}{d\theta} = 1 \Leftrightarrow \frac{d(u\,p(\theta))}{d\theta} = p(\theta) \tag{A.12}$$

Integrating both sides of this equation, we get

$$u\,p = \int p(\theta)d\theta - C_m \Leftrightarrow u(\theta) = \frac{P(\theta) - C_m}{p(\theta)} \tag{A.13}$$

where C_m is the integration constant and $P(\theta)$ the primitive of $p(\theta)$ given by

$$P(\theta) = \int p(\theta)d\theta \tag{A.14}$$

We can now obtain α from Equation A.11

$$\alpha(\theta) = \arctan\left(\frac{1}{u}\right) = \arctan\left(\frac{p(\theta)}{P(\theta) - C_m}\right) \tag{A.15}$$

where C_m is a constant to be determined from the initial conditions of the problem.

Appendix B: Mirror Differential Equation for Circular Sources

Obtain an expression for $\alpha(\theta)$ from the expression

$$\frac{1}{r}\frac{dr}{d\phi} = \tan\alpha + \frac{a}{r} \Leftrightarrow \frac{d\ln r(\phi)}{d\phi} = \tan\alpha(\theta) + \frac{a}{r} \tag{B.1}$$

where a is a constant and this expression is subject to

$$\begin{aligned} \phi &= 2\alpha(\theta) + \theta \\ p(\theta) &= \left(r + \frac{a}{\tan\alpha}\right)\sin(2\alpha) \end{aligned} \tag{B.2}$$

where the first expression relates ϕ, α, and θ, and the second expression relates r to a known quantity p.

We start by calculating the logarithm of p, given by

$$\ln(p(\theta)) = \ln\left(r + \frac{a}{\tan\alpha}\right) + \ln(\sin(2\alpha)) \tag{B.3}$$

calculating the θ derivative, we get

$$\begin{aligned}\frac{d\ln p}{d\theta} &= \frac{1}{r + a/\tan\alpha}\left(\frac{dr}{d\phi}\frac{d\phi}{d\theta} - a\frac{1/\cos^2\alpha}{\tan^2\alpha}\frac{d\alpha}{d\theta}\right) + 2\frac{\cos(2\alpha)}{\sin(2\alpha)}\frac{d\alpha}{d\theta}\\ &= \frac{r\sin(2\alpha)}{p}\frac{1}{r}\frac{dr}{d\phi}\frac{d\phi}{d\theta} - \frac{\sin(2\alpha)}{p}\frac{a}{\sin^2\alpha}\frac{d\alpha}{d\theta} + 2\frac{\cos(2\alpha)}{\sin(2\alpha)}\frac{d\alpha}{d\theta}\end{aligned} \tag{B.4}$$

considering expression (B.1) and the first expression of (B.2) we can write

$$\begin{aligned}\frac{d\ln p}{d\theta} &= \frac{r\sin(2\alpha)}{p}\left(\tan\alpha + \frac{a}{r}\right)\left(2\frac{d\alpha}{d\theta} + 1\right)\\ &\quad - \frac{\sin(2\alpha)}{p}\frac{a}{\sin^2\alpha}\frac{d\alpha}{d\theta} + 2\frac{\cos(2\alpha)}{\sin(2\alpha)}\frac{d\alpha}{d\theta}\\ &= \tan\alpha\left(\frac{r\sin(2\alpha)}{p} + \frac{a\sin(2\alpha)}{p\tan\alpha}\right)\left(2\frac{d\alpha}{d\theta} + 1\right)\\ &\quad - \frac{\sin(2\alpha)}{p}\frac{a}{\sin^2\alpha}\frac{d\alpha}{d\theta} + 2\frac{\cos(2\alpha)}{\sin(2\alpha)}\frac{d\alpha}{d\theta}\end{aligned} \tag{B.5}$$

The second expression of (B.2) can now be written as

$$r\sin(2\alpha) + \frac{a\sin(2\alpha)}{\tan\alpha} = p \Leftrightarrow \frac{r\sin(2\alpha)}{p} + \frac{a\sin(2\alpha)}{p\tan\alpha} = 1 \tag{B.6}$$

and, therefore,

$$\begin{aligned}\frac{d\ln p}{d\theta} &= \tan\alpha\left(2\frac{d\alpha}{d\theta} + 1\right) - \frac{\sin(2\alpha)}{p}\frac{a}{\sin^2\alpha}\frac{d\alpha}{d\theta} + 2\frac{\cos(2\alpha)}{\sin(2\alpha)}\frac{d\alpha}{d\theta}\\ &= \frac{2\sin^2 a}{\sin\alpha\cos\alpha}\frac{d\alpha}{d\theta} - \frac{(2a/p)\cos^2\alpha}{\sin\alpha\cos\alpha}\frac{d\alpha}{d\theta} + \frac{\cos^2\alpha - \sin^2\alpha}{\sin\alpha\cos\alpha}\frac{d\alpha}{d\theta} + \frac{\sin^2\alpha}{\sin\alpha\cos\alpha}\end{aligned} \tag{B.7}$$

so it can be concluded that

$$\left(1 - \frac{2a\cos^2\alpha}{p(\theta)}\right)\frac{d\alpha}{d\theta} = \sin\alpha\cos\alpha\frac{d\ln(p(\theta))}{d\theta} - \sin^2\theta \tag{B.8}$$

As seen, when $a \to 0$, Equation B.8 tends to Equation A.9, presented earlier for the case of a linear source. Dividing Equation B.8 by $\sin^2 \alpha$ and making $d \ln p(\theta)/d\theta = (1/p)dp/d\theta$, we get

$$\frac{1}{\sin^2 \theta} \frac{d\alpha}{d\theta} - \frac{2a}{p} \cot^2 \alpha \frac{d\alpha}{d\theta} = \cot \alpha \frac{1}{p} \frac{dp}{d\theta} - 1 \tag{B.9}$$

To solve Equation B.9, we can now make

$$\cot \alpha = \tan u \tag{B.10}$$

Note that, from expression (B.10), one can obtain $\cos u \cos \alpha - \sin u \sin \alpha = 0$ or $\cos(u + \alpha) = 0$ and therefore $u + \alpha = \pi/2 + n\pi$ or $u = (2n + 1)\pi/2 - \alpha$ where n is an integer. Squaring both terms of expression (B.10) and considering that, for any angle β, we have $\sin^2 \beta + \cos^2 \beta = 1$, we get

$$\frac{1 - \sin^2 \alpha}{\sin^2 \alpha} = \frac{1 - \cos^2 u}{\cos^2 u} \Leftrightarrow \frac{1}{\sin^2 \alpha} = \frac{1}{\cos^2 u} \tag{B.11}$$

Calculating the θ derivative of expression (B.10) and considering expression (B.11), we get

$$\frac{d \cot \alpha}{d\theta} = \frac{d \tan u}{d\theta} \Leftrightarrow -\frac{1}{\sin^2 \alpha} \frac{d\alpha}{d\theta} = \frac{1}{\cos^2 u} \frac{du}{d\theta} \Leftrightarrow -\frac{d\alpha}{d\theta} = \frac{du}{d\theta} \tag{B.12}$$

Replacing expressions (B.11) and (B.12) in expression (B.9), we get

$$\tan u \frac{dp}{d\theta} + \frac{p}{\cos^2 u} \frac{du}{d\theta} = p + 2a \tan^2 u \frac{du}{d\theta} \tag{B.13}$$

which is equivalent to

$$\frac{d}{d\theta}(p \tan u) = p + 2a \tan^2 u \frac{du}{d\theta} \tag{B.14}$$

Using $\tan^2 u = -(1 - 1/\cos^2 u)$, Equation B.14 can now be integrated with θ, and we get

$$p \tan u = \int p(\theta) d\theta - 2a(u - \tan u) - C_m \tag{B.15}$$

where C_m is the integration constant. This expression can be rewritten as

$$\tan u(p - 2a) + 2au = P(\theta) - C_m \tag{B.16}$$

With $P(\theta)$ given by expression $P(\theta) = \int p(\theta)d\theta$.

From expressions (B.16) and (B.10), we have

$$C_m = P(\theta) - (p(\theta) - 2a)\cot\alpha - 2a\arctan(\cot\alpha) \tag{B.17}$$

Since from expression (B.2) we have $2\alpha = \phi - \theta$ we can therefore write

$$C_m = P(\theta) - \left(p(\theta) - 2a\right)\cot\left(\frac{\phi - \theta}{2}\right) - 2a\arctan\left(\cot\left(\frac{\phi - \theta}{2}\right)\right) \tag{B.18}$$

Given the initial values, ϕ_0 and θ_0, we can obtain C_m from expression (B.18). Now, giving values to θ, the corresponding value for α can be obtained by solving Equation B.17. This enables us to calculate $\alpha(\theta)$.

References

1. Winston, R. and Ries, H., Nonimaging reflectors as functionals of the desired irradiance, *J. Opt. Soc. Am. A*, 10, 1902, 1993.
2. Elmer, W.B., *The Optical Design of Reflectors*, John Wiley & Sons, New York, 1980.
3. Rabl, A., Edge-ray method for analysis of radiation transfer among specular reflectors, *Appl. Opt.*, 33, 1248, 1994.
4. Rabl, A. and Gordon, J.M., Reflector design for illumination with extended sources: The basic solutions, *Appl. Opt.*, 33, 6012, 1994.
5. Gordon, J.M. and Rabl, A., Reflectors for uniform far-field irradiance: Fundamental limits and example of an axisymmetric solution, *Appl. Opt.*, 37, 44, 1998.
6. Ong, P.T. et al., Tailored edge-ray designs for uniform illumination of distant targets, *Opt. Eng.*, 34, 1726, 1995.
7. Ries, H. and Winston, R., Tailored edge-ray reflectors for illumination. *J. Opt. Soc. Am. A*, 11, 1260, 1994.
8. Ong, P.T., Gordon, J.M., and Rabl, A., Tailored lighting reflectors to prescribed illuminance distributions: Compact partial-involute designs, *Appl. Opt.*, 34, 7877, 1995.
9. Ong, P.T., Gordon, J.M., and Rabl, A., Tailored edge-ray designs for illumination with tubular sources, *Appl. Opt.*, 35, 4361, 1996.
10. Fournier, F., *Freeform Reflector Design with Extended Sources*, PhD thesis, College of Optics and Photonics at the University of Central Florida Orlando, 2010.

9

Miñano–Benitez Design Method (Simultaneous Multiple Surface)

9.1 Introduction

This chapter describes a nonimaging optics design method known in the field as the simultaneous multiple surface (SMS) or the Miñano–Benitez design method. The abbreviation SMS comes from the fact that it enables the simultaneous design of multiple optical surfaces.[1] The original idea came from Miñano. The design method itself was initially developed in 2-D by Miñano and later also by Benítez. The first generalization to 3-D geometry came from Benítez. It was then much further developed by contributions of Miñano and Benítez. Other people have worked initially with Miñano and later with Miñano and Benítez on programming the method.

We have seen in previous chapters that in the Winston–Welford (or flow-line) design method, the nonimaging optic is obtained by using the edge ray principle, in which the light rays coming from the edge of the source are redirected to the edge of the receiver. The edge rays are reflected by mirrors that channel the light, where each mirror reflects only one set of edge rays. In the Miñano–Benitez method, the situation is different, and the surfaces sequentially reflect or refract both sets of edge rays.

SMS surfaces (in 2-D geometry) are piecewise curves made of several portions of Cartesian ovals, so that some of their characteristics are first detailed.

First consider, for example, that we have a point source (emitter) **E** in air ($n = 1$) and we want to perfectly focus its light onto another point **R** (receiver) immersed in a medium of refractive index n, as shown in Figure 9.1a. If we choose an optical path length S between **E** and **R**, we can design a surface that concentrates on **R** the light emitted by **E**. This surface is called a Cartesian oval, after Descartes, who solved the problem for spherical wavefronts (it was Levi-Civita who solved the general problem in 1900). Although it is possible to obtain an analytical expression for this curve[2] (or see Chapter 21), a numerical method is presented here because it will be useful for the

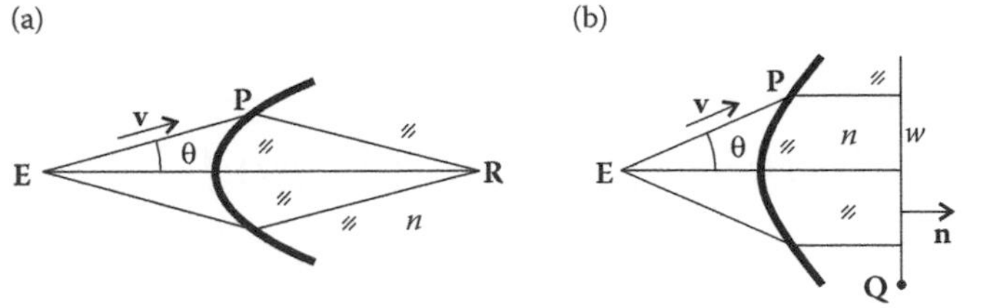

FIGURE 9.1
(a) The light emitted by a point source **E** is concentrated on to a point **R** inside a medium of refractive index n. (b) The light emitted by a point source **E** is made parallel after entering a medium of refractive index n.

Miñano–Benitez design method presented in this chapter. If **v** is a given unit vector, point **P** on the Cartesian oval can be obtained by

$$\mathbf{P} = \mathbf{E} + t\,\mathbf{v} \tag{9.1}$$

where t is the distance between **E** and **P** and $\mathbf{v} = (\cos\theta, \sin\theta)$. The distance between **P** and **R** can now be obtained by

$$d_{PR} = \sqrt{(\mathbf{R}-\mathbf{P})\cdot(\mathbf{R}-\mathbf{P})} \tag{9.2}$$

The distance t can be obtained by solving the equation

$$t + n\sqrt{(\mathbf{E}+t\mathbf{v}-\mathbf{R})\cdot(\mathbf{E}+t\mathbf{v}-\mathbf{R})} = S \tag{9.3}$$

and, therefore, point **P** can be obtained. Point **P** is given by **P** = ccoptpt(**E**, 1, **v**, **R**, n, S), where ccoptpt is defined in Chapter 21. Note that we are considering that point **E** is in air ($n = 1$) and point **R** is in a medium of refractive index n. By doing this for different direction vectors **v**, we can completely define the Cartesian oval curve. For each value of θ we get a point **P** on the curve.

Figure 9.1b presents another situation. Now the light rays emitted by **E** are made parallel after entering the medium of refractive index n. These rays will be perpendicular to wavefront w defined by point **Q** and normal **n**. Point **P** can now be calculated as **P** = coptsl(**E**, 1, **v**, **Q**, n, **n**, S), where S is the optical path length between **E** and w (see Chapter 21). Note that we are considering that point **E** is in air ($n = 1$) and wavefront w is in a medium of refractive index n. As seen earlier, varying angle θ gives different points **P** on the curve (optical surface).

Generally, we have a situation as shown in Figure 9.2. Here, we have two given wavefronts w_1 and w_2, and we have to calculate the refractive curve c, separating two media of refraction indices n_1 and n_2, that refracts w_1 to w_2. Curve c is a generalized Cartesian oval.

This curve can be obtained by the constant optical path length S. For each point $\mathbf{W}_1$ on w_1, we know the direction $\mathbf{t}_1$ of the light ray perpendicular to w_1.

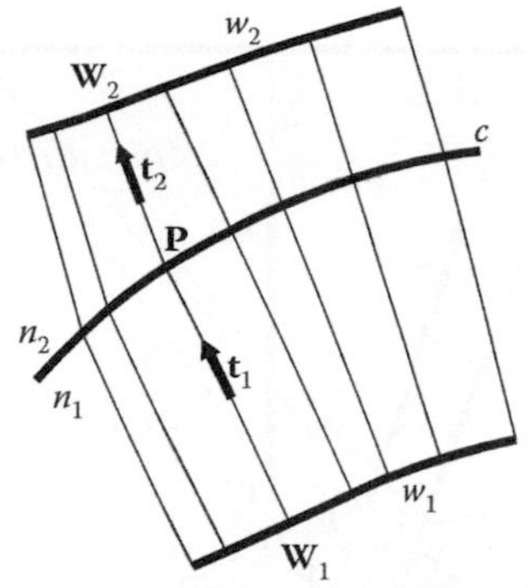

FIGURE 9.2
Generalized Cartesian oval.

We can then calculate the position of point **P** along this light ray such that $n_1[\mathbf{W}_1, \mathbf{P}] + n_2[\mathbf{P}, \mathbf{W}_2] = S$. Repeating the process for different points on w_1, we obtain different points on c. The light ray refracted at **P** intersects the wavefront w_2 at point $\mathbf{W}_2$ and points the direction $\mathbf{t}_2$ perpendicular to w_2.

The same method could be used to calculate a mirror that reflects one wavefront to another.

9.2 The RR Optic

The ideas presented previously about the Cartesian ovals can now be used to design an SMS optic. We start with a lens, that is, an optic with two refractive surfaces. The procedure described here calculates the refractive surfaces point by point, and is related to the algorithm used by Schulz in the design of aspheric imaging lenses.[3,4]

Figure 9.3 shows an SMS chain, the basic construction of an SMS optic. We start, for example, by defining the two point sources $\mathbf{E}_1$ and $\mathbf{E}_2$. Here, they are shown as the edges of a flat source. We also define two point receivers, $\mathbf{R}_1$ and $\mathbf{R}_2$. Here, they are shown as the edges of a flat receiver. We assume that the system is symmetrical so that the perpendicular bisector b of $\mathbf{E}_1\mathbf{E}_2$ is also the perpendicular bisector of $\mathbf{R}_1\mathbf{R}_2$. We want to concentrate the light emitted by point $\mathbf{E}_2$ on point $\mathbf{R}_1$ and the light emitted by $\mathbf{E}_1$ on point $\mathbf{R}_2$.

Now choose the refractive index n of the lens we want to design. Consider, for example, that $\mathbf{E}_1\mathbf{E}_2$ and $\mathbf{R}_1\mathbf{R}_2$ are both in air ($n = 1$). Now choose a point $\mathbf{P}_0$ and its normal $\mathbf{n}_0$. Given the symmetry of the system, choose point $\mathbf{P}_0$ on the bisector line b of $\mathbf{E}_1\mathbf{E}_2$ and $\mathbf{R}_1\mathbf{R}_2$ and its normal $\mathbf{n}_0$ as the vertical (perpendicular to both $\mathbf{E}_1\mathbf{E}_2$ and $\mathbf{R}_1\mathbf{R}_2$). Point $\mathbf{P}_0$ is on the top surface of the lens to be designed. Refract a ray r_1 coming from $\mathbf{E}_2$ at $\mathbf{P}_0$. Choose a point $\mathbf{P}_1$ along the

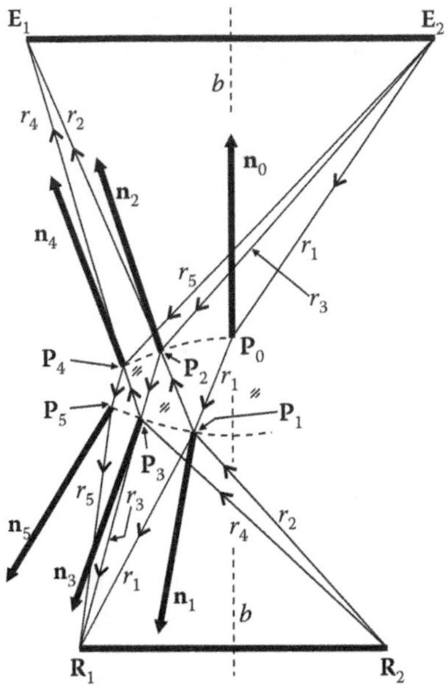

FIGURE 9.3
SMS chain.

refracted ray. This point is on the bottom surface of the lens. Force the ray r_1 to be refracted to $\mathbf{R}_1$, this condition gives us the direction $\mathbf{n}_1$ of the normal at point $\mathbf{P}_1$. Also, from the path of ray r_1 we can obtain the optical path length between $\mathbf{E}_2$ and $\mathbf{R}_1$ as

$$S = [\mathbf{E}_2, \mathbf{P}_0] + n[\mathbf{P}_0, \mathbf{P}_1] + [\mathbf{P}_1, \mathbf{R}_1] \tag{9.4}$$

Given the symmetry of the system, this is also the optical path length between $\mathbf{E}_1$ and $\mathbf{R}_2$. Now, refract at $\mathbf{P}_1$ a ray r_2 coming from point $\mathbf{R}_2$. Since we know the optical path length S between $\mathbf{R}_2$ and $\mathbf{E}_1$, we can determine the optical path length between $\mathbf{P}_1$ and $\mathbf{E}_1$ as

$$S_1 = S - [\mathbf{R}_2, \mathbf{P}_1] \tag{9.5}$$

Since we know the direction of r_2 after refraction at $\mathbf{P}_1$ and the optical path length between $\mathbf{P}_1$ and $\mathbf{E}_1$, we can calculate the position of another point $\mathbf{P}_2$ on the top surface of the lens. Calculating the position of $\mathbf{P}_2$ is similar to calculating a point $\mathbf{P}$ in the case of a Cartesian oval, as shown in Figure 9.1a, but now the ray is moving from a high refractive index material into a low one. From the direction of the incident and refracted rays at point $\mathbf{P}_2$, we can calculate its normal $\mathbf{n}_2$.

Now refract a ray r_3 coming from $\mathbf{E}_2$ at point $\mathbf{P}_2$. Using the same procedure as described earlier, calculate a new point $\mathbf{P}_3$ and its normal $\mathbf{n}_3$ on the bottom surface of the lens. Again, refracting at $\mathbf{P}_3$ a ray coming from $\mathbf{R}_2$ gives point $\mathbf{P}_4$ and its normal $\mathbf{n}_4$ on the top surface of the lens. Refracting at $\mathbf{P}_4$ a ray coming from $\mathbf{E}_2$ gives point $\mathbf{P}_5$ and its normal $\mathbf{n}_5$ on the bottom surface of the lens. This process goes on and on with the calculation of alternate points on the top and bottom surfaces of the lens, building an SMS chain of points and normals.

This process does not completely define the surfaces of the lens, but gives us only two sets of isolated points. A way around this limitation is by using the construction shown in Figure 9.4.

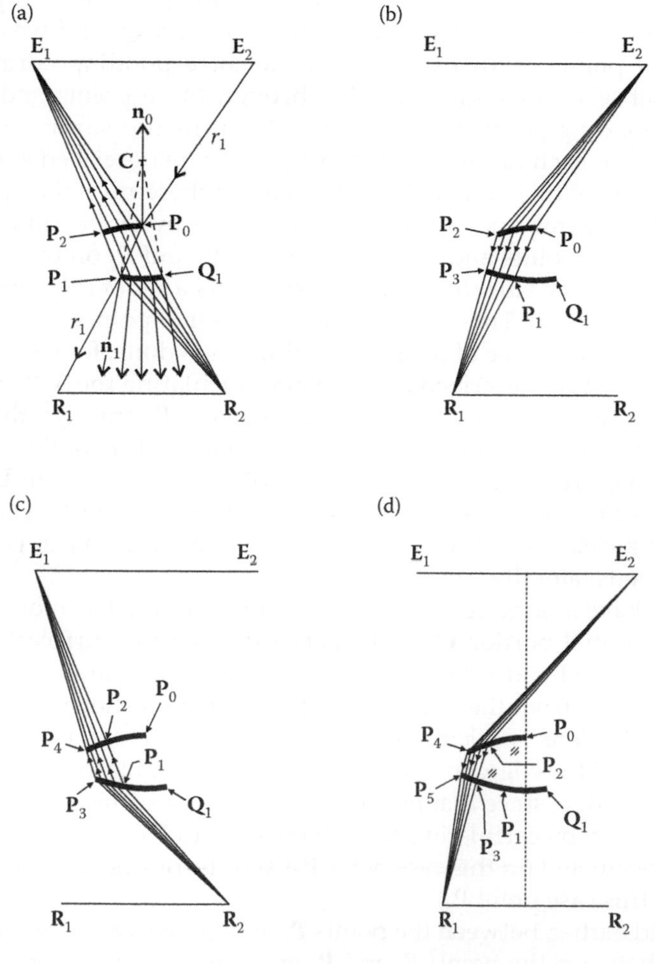

FIGURE 9.4
Definition of an SMS lens by the alternate addition of portions of the top and bottom surfaces of the lens. Calculation is done from the center to the edge. (a) Rays propagated from $\mathbf{R}_2$ through the bottom circular section $\mathbf{P}_1\mathbf{Q}_1$ define top section $\mathbf{P}_0\mathbf{P}_2$. (b) Rays propagated from $\mathbf{E}_2$ through $\mathbf{P}_0\mathbf{P}_2$ define a new section $\mathbf{P}_1\mathbf{P}_3$. (c) Rays propagated from $\mathbf{R}_2$ through $\mathbf{P}_1\mathbf{P}_3$ define a new section $\mathbf{P}_2\mathbf{P}_4$. (d) Rays propagated from $\mathbf{E}_2$ through $\mathbf{P}_2\mathbf{P}_4$ define a new section $\mathbf{P}_3\mathbf{P}_5$.

As earlier, we choose a point $\mathbf{P}_0$ on the axis of symmetry of the system and its normal $\mathbf{n}_0$ as a vertical vector. Refract a ray r_1 coming from $\mathbf{E}_2$ at $\mathbf{P}_0$. Choose the position of a point $\mathbf{P}_1$ along the direction of the refracted ray. Force this ray r_1 to be refracted at $\mathbf{P}_1$ toward $\mathbf{R}_1$ and this gives us the normal $\mathbf{n}_1$ at $\mathbf{P}_1$. Now, given the symmetry of the system, define a point $\mathbf{Q}_1$ symmetrical to $\mathbf{P}_1$. The normal at $\mathbf{Q}_1$ is also symmetrical to $\mathbf{n}_1$ at point $\mathbf{P}_1$. Now choose a curve that goes through $\mathbf{P}_1$ and $\mathbf{Q}_1$ and is perpendicular to the normals at these two points. In the case of Figure 9.4a make this curve as a circle with center $\mathbf{C}$. The position of $\mathbf{C}$ is obtained by intersecting a straight line through $\mathbf{P}_1$ with direction $\mathbf{n}_1$ with another straight line through $\mathbf{P}_0$ with direction $\mathbf{n}_0$. Now we have a curve between $\mathbf{P}_1$ and $\mathbf{Q}_1$ and we can calculate a set of points on that curve with the corresponding normals. Now launch a set of rays coming from $\mathbf{R}_2$ through these points and calculate the corresponding points on a portion $\mathbf{P}_0\mathbf{P}_2$ of the top surface of the lens. This is done for each ray, using the same procedure described earlier when calculating the SMS chain of Figure 9.3. For all these rays, the optical path length S is the same and is calculated as earlier by expression (9.4). Now we have a set of points and normals between $\mathbf{P}_0$ and $\mathbf{P}_2$ on the top surface of the lens. We now launch through these points a set of rays coming from point $\mathbf{E}_2$ (Figure 9.4b). These rays define a new set of points between $\mathbf{P}_1$ and $\mathbf{P}_3$ on the bottom surface of the lens. Again, this is done, for each ray, using the same procedure described earlier when calculating the SMS chain as in Figure 9.3. Now launch a set of rays coming from $\mathbf{R}_2$ through these points and calculate the corresponding points on a portion $\mathbf{P}_2\mathbf{P}_4$ of the top surface of the lens (Figure 9.4c). Next launch a set of rays coming from $\mathbf{E}_2$ through these points and calculate the corresponding points on a portion $\mathbf{P}_3\mathbf{P}_5$ of the bottom surface of the lens (Figure 9.4d). This process can go on as we extend the lens laterally.

Now, we have a large set of points on both surfaces. The more points we pick on the initial portion $\mathbf{Q}_1\mathbf{P}_1$, the better defined the surfaces of the lens will be. The right-hand-side of the lens is obtained by symmetry.

As can be seen from the sequence of the calculation shown in Figure 9.4a, b, c, and d, this lens is calculated starting from the center and growing the surfaces toward the edge. Sometimes it may be preferable to start the calculation from the edge, since that may lead to smoother surfaces. The process is similar. We start by calculating an SMS chain as in Figure 9.3. Now take the last point calculated (in this case point $\mathbf{P}_5$) and the one next to it on the same surface (in this case point $\mathbf{P}_3$).

As we did earlier, between the points $\mathbf{P}_1$ and $\mathbf{Q}_1$, we can now interpolate a curve $\mathbf{c}(x)$ between the points $\mathbf{P}_5$ and $\mathbf{P}_3$ ensuring that it is perpendicular to $\mathbf{n}_5$ at point $\mathbf{P}_5$ and perpendicular to $\mathbf{n}_3$ at point $\mathbf{P}_3$, as shown in Figure 9.5a. We may choose, for example, a third-degree polynomial of the form

$$p(x) = a + bx + cx^2 + dx^3 \tag{9.6}$$

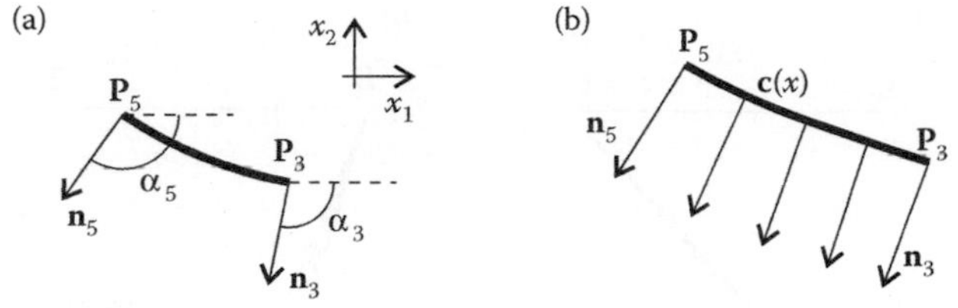

FIGURE 9.5
(a) Interpolation of a curve between points $\mathbf{P}_5$ and $\mathbf{P}_3$. The curve is normal to $\mathbf{n}_5$ at point $\mathbf{P}_5$ and is perpendicular to $\mathbf{n}_3$ at point $\mathbf{P}_3$. (b) A set of points and normals may be extracted from the curve interpolation.

Now, if points $\mathbf{P}_5$ and $\mathbf{P}_3$ are given by $\mathbf{P}_5 = (P_{51}, P_{52})$ and $\mathbf{P}_3 = (P_{31}, P_{32})$ we have

$$\begin{aligned} P_{52} &= a + bP_{51} + cP_{51}^2 + dP_{51}^3 \\ P_{32} &= a + bP_{31} + cP_{31}^2 + dP_{31}^3 \end{aligned} \tag{9.7}$$

Also, for $p(x)$ to have the right derivatives at points $\mathbf{P}_5$ and $\mathbf{P}_3$, we must have

$$\begin{aligned} \frac{dp(P_{51})}{dx} &= \tan\left(\alpha_5 + \frac{\pi}{2}\right) \Leftrightarrow b + 2cP_{51} + 3dP_{51}^2 = \tan\left(\alpha_5 + \frac{\pi}{2}\right) \\ \frac{dp(P_{31})}{dx} &= \tan\left(\alpha_3 + \frac{\pi}{2}\right) \Leftrightarrow b + 2cP_{31} + 3dP_{31}^2 = \tan\left(\alpha_3 + \frac{\pi}{2}\right) \end{aligned} \tag{9.8}$$

where α_5 and α_3 are the angles that $\mathbf{n}_5$ and $\mathbf{n}_3$ make to the horizontal. In this particular case, we have $\alpha_5 < 0$ and $\alpha_3 < 0$. Equations 9.7 and 9.8 give us the values of a, b, c, and d and define $p(x)$. The curve is then parameterized by

$$\mathbf{c}(x) = (x, p(x)) \tag{9.9}$$

with $P_{51} < x < P_{31}$. The normal to the curve is given by

$$\mathbf{n}_C(x) = (dp(x)/dx, -1) \tag{9.10}$$

for the same parameter range. The expression for $\mathbf{n}_C(x)$ can also be normalized to make its length unity for all values of parameter x.

Now we use the same procedure as earlier to calculate the surfaces of the lens. We first define a set of points and normals on the portion of curve $\mathbf{P}_5\mathbf{P}_3$ we just defined, as shown in Figure 9.5b.

Now, using the same optical path length S, as was used to calculate the SMS chain of Figure 9.3, launch a set of rays from $\mathbf{R}_1$ through the points of $\mathbf{P}_5\mathbf{P}_3$ and calculate a set of points $\mathbf{P}_4\mathbf{P}_2$ on the top surface, as shown in Figure 9.6a.

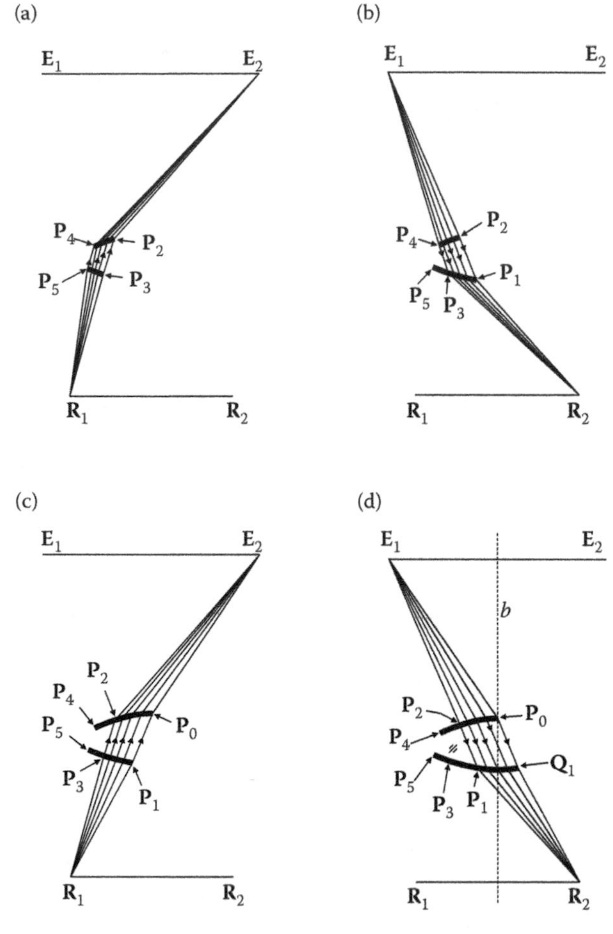

FIGURE 9.6
Definition of an SMS lens by the alternate addition of portions of surface on the top and bottom surfaces of the lens. Calculation is done from the edge to the center. (a) Rays propagated from $\mathbf{R}_1$ through P_3P_5 define a new section $\mathbf{P}_2\mathbf{P}_4$. (b) Rays propagated from $\mathbf{E}_1$ through $\mathbf{P}_2\mathbf{P}_4$ define a new section $\mathbf{P}_1\mathbf{P}_3$. (c) Rays propagated from $\mathbf{R}_1$ through $\mathbf{P}_1\mathbf{P}_3$ define a new section $\mathbf{P}_0\mathbf{P}_2$. (d) Rays propagated from $\mathbf{E}_1$ through $\mathbf{P}_0\mathbf{P}_2$ define a new section $\mathbf{P}_1\mathbf{Q}_1$.

Now launch a set of rays coming from $\mathbf{E}_1$ through the new points between $\mathbf{P}_4$ and $\mathbf{P}_2$ and calculate a new set of points on the bottom surface between the points $\mathbf{P}_3$ and $\mathbf{P}_1$ (Figure 9.6b). Next, launch a set of rays from $\mathbf{R}_1$ through the points of $\mathbf{P}_3\mathbf{P}_1$ and calculate a set of points $\mathbf{P}_2\mathbf{P}_0$ on the top surface (Figure 9.6c). Then launch a set of rays coming from $\mathbf{E}_1$ through the new points between $\mathbf{P}_2$ and $\mathbf{P}_0$ and calculate a new set of points on the bottom surface between points $\mathbf{P}_1$ and $\mathbf{Q}_1$ (Figure 9.6d) The calculation ends when the surfaces cross the axis of symmetry b of the system. The right-hand side of the lens is obtained by symmetry.

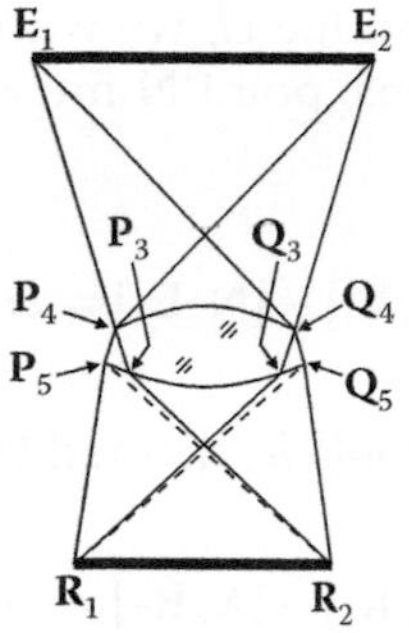

FIGURE 9.7
An SMS lens with an entrance aperture $\mathbf{P_4Q_4}$ that redirects to $\mathbf{R_1R_2}$ all the light it receives from $\mathbf{E_1E_2}$. The lens is not symmetrical in the sense that if $\mathbf{R_1R_2}$ is now the emitter, not all the light that hits the entrance aperture $\mathbf{P_5Q_5}$ will end up on $\mathbf{E_1E_2}$.

Figure 9.7 shows the complete lens. All the light emitted by $\mathbf{E_1}$ toward the entrance aperture of the lens $\mathbf{P_4Q_4}$ will be redirected to $\mathbf{R_2}$ and the light emitted by $\mathbf{E_2}$ toward the entrance aperture of the lens $\mathbf{P_4Q_4}$ will be redirected to $\mathbf{R_1}$.

Another possibility for calculating an SMS lens is by imposing the condition that the étendues from emitter to the lens and from the lens to the receiver are equal and also that the lens has a thick edge.[5] Now we choose the points at the edges of the surfaces of the lens by using the condition of matching the étendues from $\mathbf{E_1E_2}$ toward the lens and from the lens toward $\mathbf{R_1R_2}$. Suppose that we already have one such lens as shown in Figure 9.8.

The étendue from $\mathbf{E_1E_2}$ to the entrance aperture **NM** of the lens is given by

$$U = 2([\mathbf{N},\mathbf{E}_1] - [\mathbf{N},\mathbf{E}_2]) \tag{9.11}$$

Also, the étendue from the exit aperture of the lens **XY** to $\mathbf{R_1R_2}$ is given by

$$U = 2([\mathbf{X},\mathbf{R}_1] - [\mathbf{X},\mathbf{R}_2]) \tag{9.12}$$

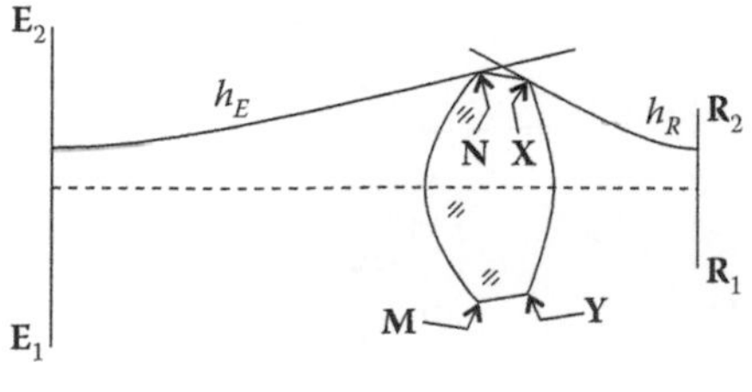

FIGURE 9.8
An SMS lens in which the étendue from $\mathbf{E_1E_2}$ to the entrance aperture of the lens **NM** matches the étendue of the exit aperture **XY** of the lens to $\mathbf{R_1R_2}$.

If we give a value to the étendue U, we want to couple the light between $\mathbf{E}_1\mathbf{E}_2$ and $\mathbf{R}_1\mathbf{R}_2$ through the lens; point **N** must then be on the hyperbola h_E defined by

$$[\mathbf{N},\mathbf{E}_1]-[\mathbf{N},\mathbf{E}_2]=U/2 \tag{9.13}$$

and point **X** on another hyperbola h_R defined by

$$[\mathbf{X},\mathbf{R}_1]-[\mathbf{X},\mathbf{R}_2]=U/2 \tag{9.14}$$

These hyperbolas can be obtained by using the function hyp(**F**, **G**, U, n) as defined in Chapter 21, in which **F** and **G** are either $\mathbf{E}_1$ and $\mathbf{E}_2$ or $\mathbf{R}_1$ and $\mathbf{R}_2$ and $n = 1$ since these hyperbolas are considered to be in air. The design of the lens then begins by choosing a value for U and choosing point **N** on the hyperbola h_E and point **X** on the hyperbola h_R as shown in Figure 9.9.

Consider a ray r_1 emitted from $\mathbf{E}_1$ toward point **N**. There it refracts toward point **X** and from there it is redirected to point $\mathbf{R}_1$. Since now we know the directions of incident and refracted rays at point **N**, we can calculate its normal $\mathbf{n}_N$. Also, since we know the directions of incident and refracted rays at point **X**, we can calculate its normal $\mathbf{n}_X$.

Point **N** will receive the light emitted from $\mathbf{E}_1$ to $\mathbf{E}_2$, as shown in Figure 9.10.

Knowing the direction of the normal at point **N**, refract at that point a ray r_2 coming from $\mathbf{E}_2$. This ray is redirected at some point $\mathbf{X}_1$ on the right-hand-side surface of the lens toward $\mathbf{R}_1$. We now have a situation in which the light coming out of point **N**, confined between rays r_1 and r_2 is headed toward portion $\mathbf{XX}_1$ of the lens and must be concentrated on to point $\mathbf{R}_1$. The portion $\mathbf{XX}_1$ must then be a Cartesian oval that concentrates on $\mathbf{R}_1$ these light rays coming from **N**. Since we know the path of ray r_1, we can calculate the optical path length between **N** and $\mathbf{R}_1$ as

$$S_N = n[\mathbf{N},\mathbf{X}]+[\mathbf{X},\mathbf{R}_1] \tag{9.15}$$

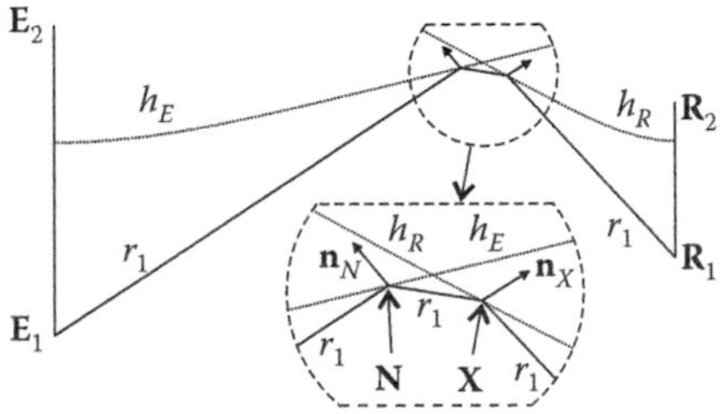

FIGURE 9.9
Path of a ray r_1 from $\mathbf{E}_1$ to **N**, then to **X** and finally to $\mathbf{R}_1$. Its path gives the normals at points **N** and **X**.

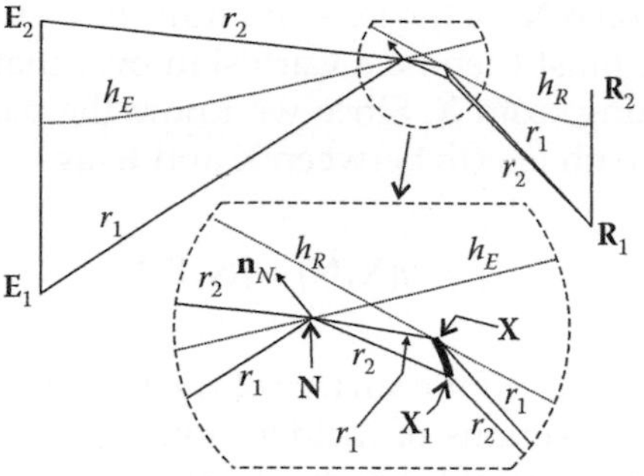

FIGURE 9.10
Portion $\mathbf{XX_1}$ of the lens concentrates to point $\mathbf{R_1}$ the light confined between the rays r_1 and r_2 coming from point **N**.

This defines portion $\mathbf{XX_1}$ of the surface of the lens. Note that for each point of $\mathbf{XX_1}$ we have the directions of incident and refracted rays and we can, therefore, calculate the normal for each one of its points.

Now we have the whole path of ray r_2 defined as $\mathbf{E_2}$-$\mathbf{N}$-$\mathbf{X_1}$-$\mathbf{R_1}$. We can, therefore, calculate the optical path length between $\mathbf{E_2}$ and $\mathbf{R_1}$ as

$$S = [\mathbf{E}_2,\mathbf{N}] + n[\mathbf{N},\mathbf{X}_1] + [\mathbf{X}_1,\mathbf{R}_1] \tag{9.16}$$

The optic is symmetrical in the sense that if $\mathbf{R_1R_2}$ was now the source of light and $\mathbf{E_1E_2}$ the receiver, all the light emitted from $\mathbf{R_1R_2}$ toward the lens would have to be redirected to $\mathbf{E_1E_2}$. We can then use the same reasoning to build a portion of the surface on the left-hand side of the lens (Figure 9.11).

Since we know the direction of the normal at point **X**, we now refract at that point a ray r_3 coming from $\mathbf{R_2}$. This ray is redirected at some point $\mathbf{N_1}$ on the left-hand-side surface of the lens toward $\mathbf{E_1}$. We now have a situation in which the light coming out of point **X** confined between rays r_1 and r_3 is

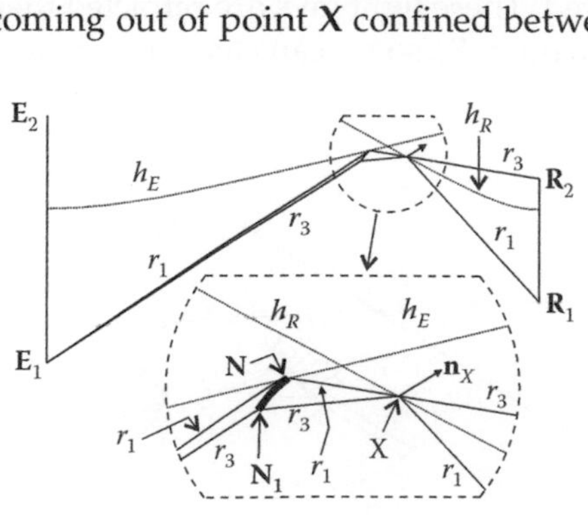

FIGURE 9.11
Portion $\mathbf{NN_1}$ of the lens concentrates on point $\mathbf{E_1}$ the light confined between rays r_1 and r_3 coming from point **X**.

headed toward portion $\mathbf{NN}_1$ of the lens and must be concentrated on to point $\mathbf{E}_1$. The portion $\mathbf{NN}_1$ must then be a Cartesian oval that concentrates on $\mathbf{E}_1$ these light rays coming from **X**. Since we know the path of ray r_1, we can calculate the optical path length between **X** and $\mathbf{E}_1$ as

$$S_X = n[\mathbf{X},\mathbf{N}] + [\mathbf{N},\mathbf{E}_1] \tag{9.17}$$

This defines portion $\mathbf{NN}_1$ of the surface of the lens. Note that for each point of $\mathbf{NN}_1$ we have the directions of incident and refracted rays and we can, therefore, calculate the normal for each one of its points.

Now we have the whole path of ray r_3 defined as $\mathbf{R}_2$-$\mathbf{X}$-$\mathbf{N}_1$—$\mathbf{E}_1$. We can, therefore, calculate the optical path length between $\mathbf{R}_2$ and $\mathbf{E}_1$ as

$$S = [\mathbf{R}_2,\mathbf{X}] + n[\mathbf{X},\mathbf{N}_1] + [\mathbf{N}_1,\mathbf{E}_1] \tag{9.18}$$

Given the symmetry of the system, the optical path lengths calculated by Equation 9.18 (between $\mathbf{E}_1$ and $\mathbf{R}_2$) and Equation 9.16 (between $\mathbf{E}_2$ and $\mathbf{R}_1$) are equal.

We now have a situation, as in Figure 9.12, that shows the curve $\mathbf{NN}_1$ with a set of points and their normals and the curve $\mathbf{XX}_1$ also with a set of points and their normals.

Now, we can calculate the remaining portions of the surfaces of the SMS lens using the same method as earlier for calculating the lens of Figure 9.6. As shown in Figure 9.13a, launch a set of rays from $\mathbf{E}_2$ through the portion $\mathbf{NN}_1$ of the lens already calculated. These light rays are refracted there and we know the optical path length to point $\mathbf{R}_1$, so we can calculate a new portion of lens $\mathbf{X}_1\mathbf{X}_2$ on the other side.

We can also launch a set of rays from $\mathbf{R}_2$ through the portion $\mathbf{XX}_1$ of the lens already calculated. These light rays are refracted there and we know the optical path length to point $\mathbf{E}_1$, so we can calculate a new portion of lens $\mathbf{N}_1\mathbf{N}_2$ on the other side.

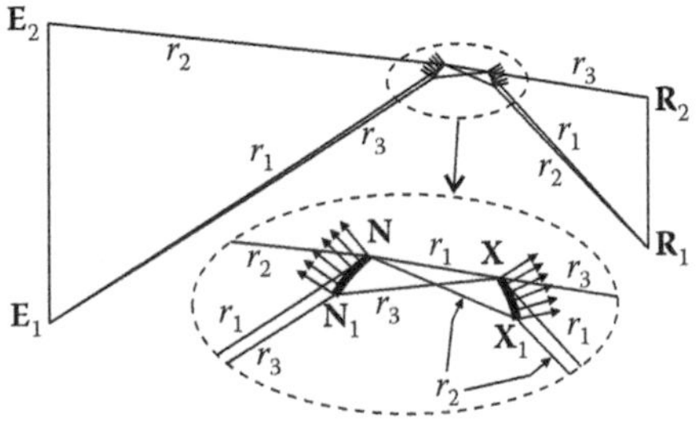

FIGURE 9.12
The design of the surfaces of the SMS lens starts at the edge with two Cartesian ovals $\mathbf{NN}_1$ and $\mathbf{XX}_1$.

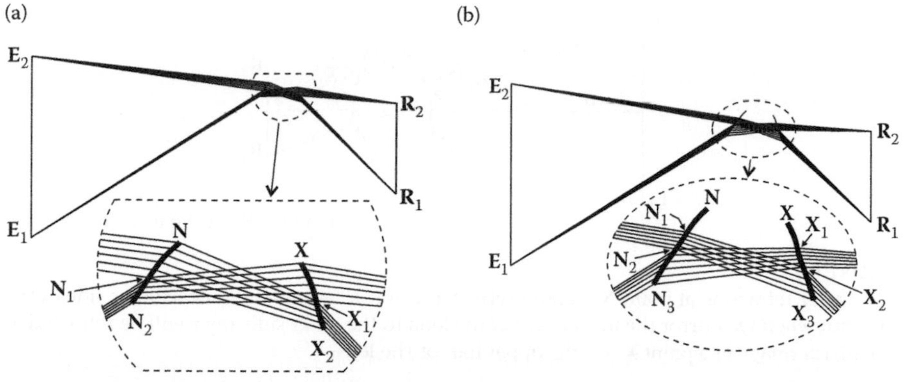

FIGURE 9.13
By launching rays through the portions of the lens already calculated, we can calculate a new portion of lens on the other side. (a) Rays through **N**–$\mathbf{N}_1$ define a new section $\mathbf{X}_1\mathbf{X}_2$ and rays through **X**–$\mathbf{X}_1$ define a new section $\mathbf{N}_1\mathbf{N}_2$. (b) Rays through $\mathbf{N}_1\mathbf{N}_2$ define a new section $\mathbf{X}_2\mathbf{X}_3$ and rays through $\mathbf{X}_1\mathbf{X}_2$ define a new section $\mathbf{N}_2\mathbf{N}_3$.

We now repeat the process for the portions of lens we just calculated, as shown in Figure 9.13b. We can launch a set of rays from $\mathbf{E}_2$ through the portion $\mathbf{N}_1\mathbf{N}_2$ of the lens already calculated. These light rays are refracted there and we know the optical path length to point $\mathbf{R}_1$, so we can calculate a new portion of lens $\mathbf{X}_2\mathbf{X}_3$ on the other side. We can also launch a set of rays from $\mathbf{R}_2$ through the portion $\mathbf{X}_1\mathbf{X}_2$ of the lens already calculated. These light rays are refracted there and we know the optical path length to point $\mathbf{E}_1$, so we can calculate a new portion of lens $\mathbf{N}_2\mathbf{N}_3$ on the other side. This process continues to the optical axis. Figure 9.14 shows the last step of this calculation, when the surfaces of the lens cross the axis of symmetry *b*.

Since the system is symmetrical relative to line *b*, we take only the portion of the lens above *b* and mirror it on the other side, completing the lens.

When we do this, we are replacing the portion of the lens we calculated below the line *b* with the mirror image of the lens above the line *b*. This makes the lens nonideal at the center because we are replacing what we should have (what we calculated below line *b*) with something else (the mirror image of

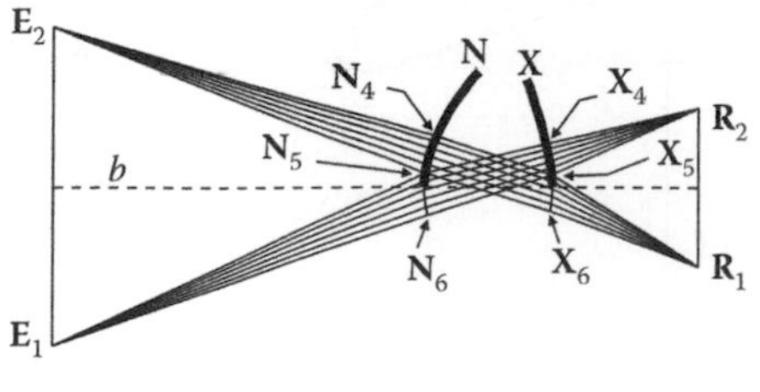

FIGURE 9.14
The design of the lens ends when the surfaces cross the axis of symmetry *b*.

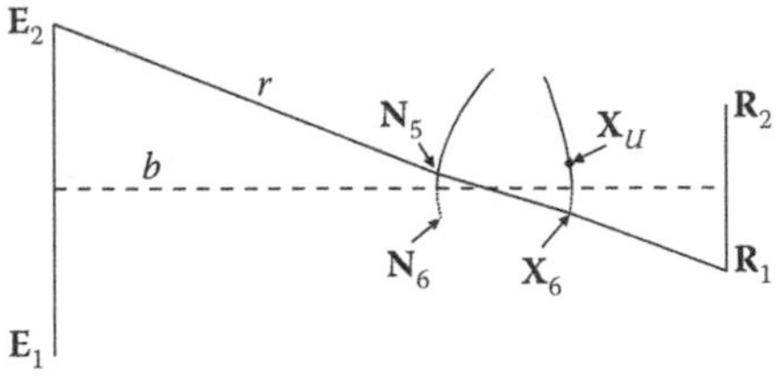

FIGURE 9.15
Ray r after refraction at point $\mathbf{N}_5$ should refract at point $\mathbf{X}_6$ calculated for the other side of the lens, but when we mirror the top surface of the lens to the other side, ray r will be refracted at the mirror image of a point $\mathbf{X}_U$ on the upper half of the lens.

the surfaces of the lens above line b). Figure 9.15 shows this effect for a ray r. The path of the ray is calculated such that it refracts at point $\mathbf{N}_5$ on the left-hand-side surface of the lens and then at a point $\mathbf{X}_6$ on the other side. When we mirror the surfaces of the lens above axis of symmetry b, ray r no longer refracts at point $\mathbf{X}_6$, but at the mirror image of another point $\mathbf{X}_U$ on the upper portion of the lens.

The regions of the surfaces of the lens for which perfect focusing of the light onto $\mathbf{R}_1$ and $\mathbf{R}_2$ cannot be guaranteed are shown in Figure 9.16. It is defined by the edge rays crossing the center of the lens. such minor blurring is quite acceptable in numerous situations.

An edge ray emitted from $\mathbf{E}_1$ toward $\mathbf{C}_L$ on the center point of the left-hand side surface of the lens defines point $\mathbf{X}_C$ on the other surface. Also, an edge ray emitted from $\mathbf{R}_1$ toward $\mathbf{C}_R$ on the center point of the right-hand side surface of the lens defines point $\mathbf{N}_C$ on the other surface. Ideality cannot be guaranteed for the rays crossing the lens between $\mathbf{N}_C$ and $\mathbf{C}_L$ and between $\mathbf{C}_R$ and $\mathbf{X}_C$.

This nonideality is, however, quite small and for practical applications the lens behaves quite well. Figure 9.17 shows a complete lens, obtained by mirroring the top portion of the lens above the axis of symmetry b to the other side.

Although we have presented the lens as being perpendicular to the optical axis, this does not have to be the case. If we just take points $\mathbf{X}$ and $\mathbf{N}$ according to the conservation of étendue and design the lens, we may get a lens

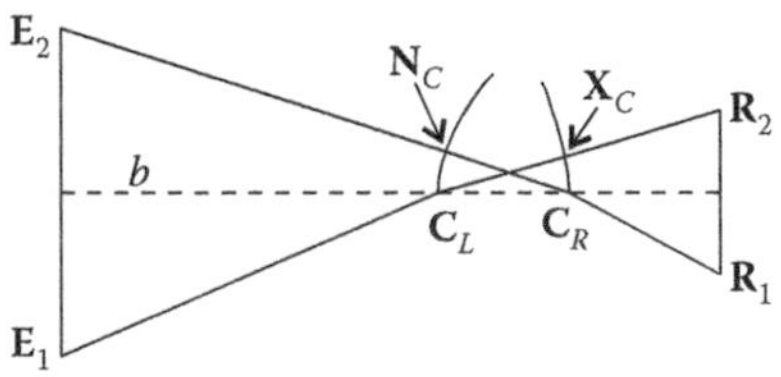

FIGURE 9.16
The region of the center of the lens for which no perfect focusing can be guaranteed is confined between points $\mathbf{N}_C$ and $\mathbf{C}_L$ for the left-hand-side surface and between $\mathbf{X}_C$ and $\mathbf{C}_R$ for the right-hand-side.

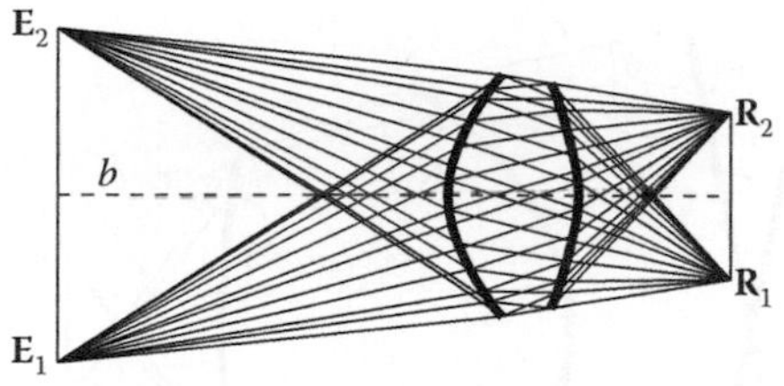

FIGURE 9.17
An SMS lens that focuses onto $\mathbf{R}_1$ and $\mathbf{R}_2$ the light rays emitted by $\mathbf{E}_2$ and $\mathbf{E}_1$.

with a peak at the center. To avoid this, the positions of points **N** and **X** must be moved until the desired result is obtained. First point **N** is kept fixed and point **X** moved along the hyperbola h_R until the left-hand-side lens surface is perpendicular to the optical axis. Then point **X** is kept fixed and point **N** is moved along the hyperbola h_E until the right-hand-side surface of the lens is perpendicular to the optical axis. This process is repeated until both the surfaces are perpendicular to the axis of symmetry.

We now go back to the lens shown in Figure 9.7. These lenses can also be designed for infinite sources at an infinite distance. Figure 9.18 shows one such lens. This is the limit case in which the edges $\mathbf{E}_1$ and $\mathbf{E}_2$ of the emitter are moved to an infinite distance away from the lens. The rays coming from the edges of the emitter are now two sets of parallel rays perpendicular to wavefronts w_1 and w_2. The lens focuses at point $\mathbf{R}_1$ the rays perpendicular to w_2 and at $\mathbf{R}_2$ those perpendicular to w_1. The SMS chains are calculated in the same way as earlier.

We now consider the lens shown in Figure 9.17. These lenses can also be designed for infinite sources at an infinite distance. Figure 9.19 shows such a lens.

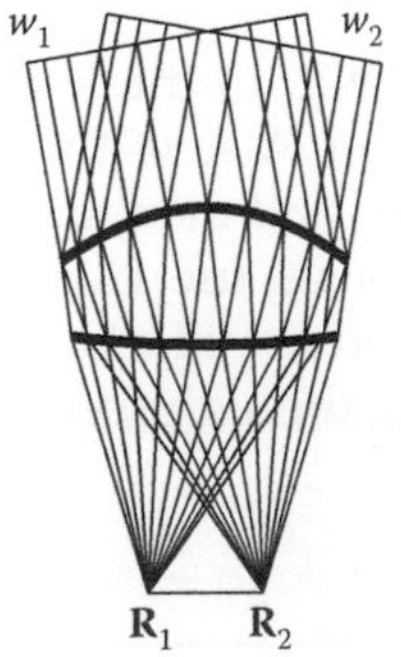

FIGURE 9.18
An SMS lens designed for the case in which the emitter is infinitely large at an infinite distance. In this case, the lens focuses at points $\mathbf{R}_1$ and $\mathbf{R}_2$ the rays perpendicular to the flat wavefronts w_2 and w_1.

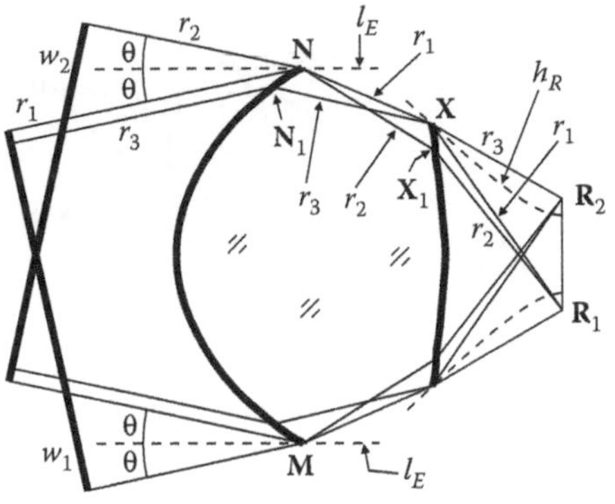

FIGURE 9.19
An SMS lens with a thick edge designed for the case in which the emitter is infinitely large at an infinite distance. In this case, the lens focuses at points $\mathbf{R}_1$ and $\mathbf{R}_2$ the rays perpendicular to the flat wavefronts w_2 and w_1.

The edge point **X** of the lens is on the hyperbola h_R, as seen earlier, but the edge point **N** is now on a horizontal straight line l_E. The distance between the top and bottom lines l_E is such that, if we choose points **N** and **M** on those lines, the étendue the lens intercepts (given by $U = 2[\mathbf{N}, \mathbf{M}]\sin\theta$) matches that defined by the hyperbola h_R.

The design of the lens is as described earlier. An edge ray r_1 perpendicular to a flat wavefront w_1 makes an angle θ to the horizontal and refracts at point **N** toward **X**. It then refracts again at **X** toward $\mathbf{R}_1$. The deflections required along the path of this ray give us the normals at points **N** and **X**. Now refract at point **N** the rays with directions between those of r_1 and r_2 and calculate the Cartesian oval $\mathbf{XX}_1$ that concentrates them on point $\mathbf{R}_1$. Refract at point **X** the rays emitted from the receiver toward **X**. These rays are contained between r_1 and r_3. Then calculate the Cartesian oval $\mathbf{NN}_1$ that makes them perpendicular to wavefront w_1.

Next use portions $\mathbf{NN}_1$ and $\mathbf{XX}_1$ of the lens to calculate the SMS chains, using the same method as earlier. Also in this case, the lens surfaces are calculated as far as the optical axis and then mirrored to the other side.

The lenses described previously use two refractions. Refractions in the Miñano–Benitez design method are identified as R. These lenses are then RR devices because light going through them undergoes two refractions.

9.3 SMS with a Thin Edge

The lens in Figure 9.7 is not symmetrical, in the sense that if we reverse the direction of the light, the behavior of the lens is different. Suppose then that

$\mathbf{R}_1$ is an emitter. All the light received by the lens between the points $\mathbf{P}_5$ and $\mathbf{Q}_3$ is redirected to point $\mathbf{E}_2$. The light received by portion $\mathbf{Q}_3\mathbf{Q}_5$, however, is not redirected. If we want the light coming from $\mathbf{R}_1$ and refracted at $\mathbf{Q}_3\mathbf{Q}_5$ to be redirected to $\mathbf{E}_2$, we would need a further portion of the top surface of the lens to the right of $\mathbf{Q}_4$. We could design this new portion of the top surface, but then we would have the same issue for the light coming from $\mathbf{E}_1$ and refracted at this new portion of the top surface. This light would not be refracted now toward $\mathbf{R}_2$.

The same thing happens for the light emitted by $\mathbf{R}_2$ toward the portion $\mathbf{P}_3\mathbf{P}_5$ of the bottom surface of the lens. This light is also not concentrated on to the point $\mathbf{E}_1$.

There are two ways around this situation. One of them is to continue building the lens toward the edges. As we do that, the points of the SMS chains get closer and the asymmetry decreases, as shown in Figure 9.20.

As we calculate more points toward the edges of the lens and these points become closer, the étendue of the light emitted from $\mathbf{E}_1\mathbf{E}_2$ toward the entrance aperture of the lens also gets closer to the étendue from the exit aperture of the lens toward $\mathbf{R}_1\mathbf{R}_2$.

Another possibility for calculating one of these lenses is to start both surfaces from a point, as shown in Figure 9.21.

Once we have defined the positions of $\mathbf{E}_1$ and $\mathbf{E}_2$ for the emitter and $\mathbf{R}_1$ and $\mathbf{R}_2$ for the receiver, we can define the position of point $\mathbf{X}$ where the surfaces of the lens start. Since the rays crossing the lens at point $\mathbf{X}$ do not enter it, the optical path length between $\mathbf{E}_1$ and $\mathbf{R}_2$, which is the same as from $\mathbf{E}_2$ to $\mathbf{R}_1$, is given by

$$S_1 + S_2 = S_3 + S_4 \tag{9.19}$$

as shown in Figure 9.22a.

This condition can also be written as

$$S_1 - S_3 = S_4 - S_2 \tag{9.20}$$

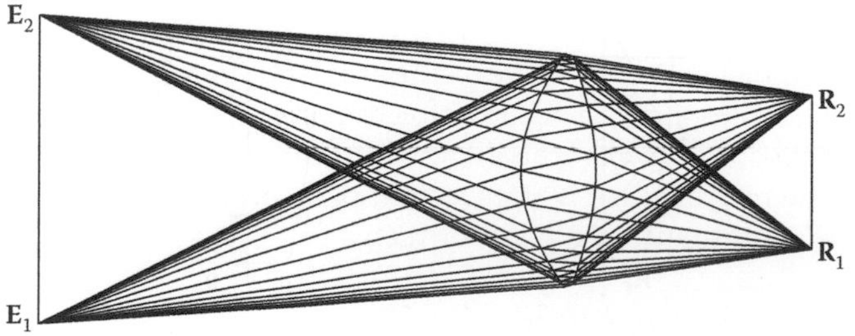

FIGURE 9.20
As we calculate more and more points on the SMS chains, these points get closer toward the edges of the lens.

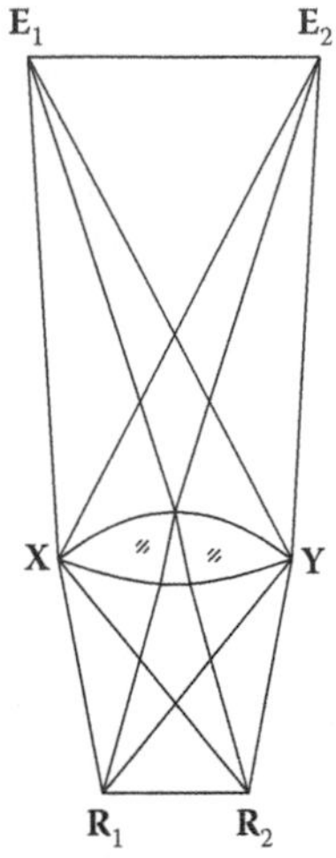

FIGURE 9.21
RR SMS lens whose surfaces touch at the end points **X** and **Y**.

which is the condition that states that the étendue from the emitter $\mathbf{E}_1\mathbf{E}_2$ to the entrance aperture **XY** of the lens is the same as the étendue from the lens to the receiver $\mathbf{R}_1\mathbf{R}_2$.

Now, we can define, for example, point **X** as $\mathbf{X} = (X_1, y)$, where X_1 is chosen by us and y is calculated using expression (9.19). Once we have the position of point **X**, we also have the directions of vectors $\mathbf{v}_1$, $\mathbf{v}_3$, $\mathbf{v}_4$ and $\mathbf{v}_6$ defined by the edges of the emitter and receiver and by point **X**, as shown in Figure 9.22b.

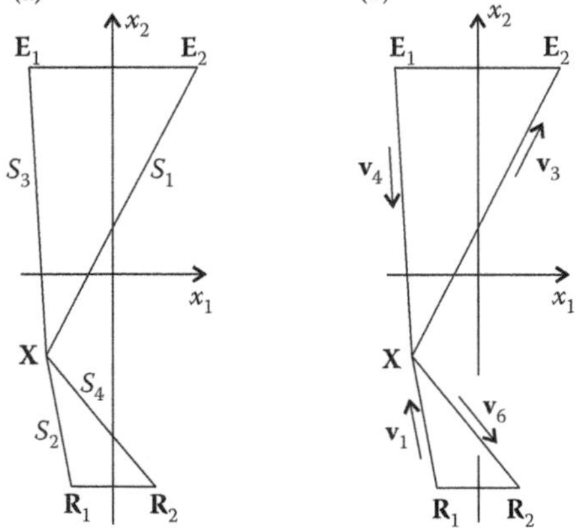

FIGURE 9.22
(a) Starting point **X** of the surfaces of the RR SMS lens and optical path lengths to the edges of emitter $\mathbf{E}_1\mathbf{E}_2$ and receiver $\mathbf{R}_1\mathbf{R}_2$. (b) Same point **X** with directions of edge rays $\mathbf{v}_1$, $\mathbf{v}_3$, $\mathbf{v}_4$, and $\mathbf{v}_6$ crossing it.

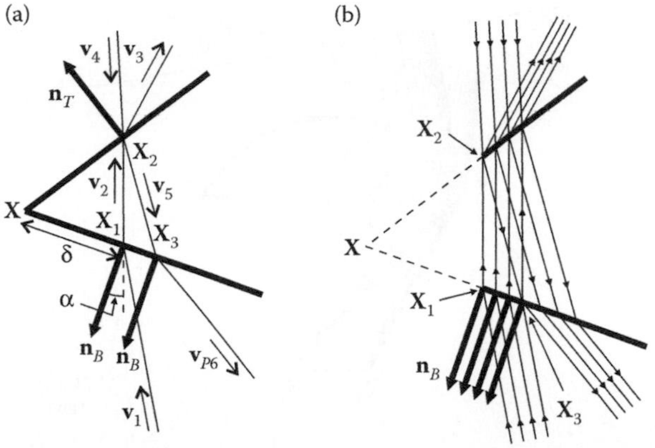

FIGURE 9.23
(a) Calculation of the directions of the normals $\mathbf{n}_T$ and $\mathbf{n}_B$ to the top and bottom surfaces at point **X**. (b) SMS chains for generating an RR lens from a thin edge toward the center.

Consider that the bottom surface, of the lens starting at **X** has normal $\mathbf{n}_B$ making an unknown angle α to the vertical, as shown in Figure 9.23a. consider also a point $\mathbf{X}_1$ on the bottom surface a very small distance δ away from **X**.

As $\delta \to 0$, vector $\mathbf{v}_1$ for a light ray emitted from point $\mathbf{R}_1$ is the same as in Figure 9.22b. We then refract the light ray with direction $\mathbf{v}_1$ at point $\mathbf{X}_1$ and calculate the direction of vector $\mathbf{v}_2$ inside the lens. This ray, after refraction of the top surface, is headed toward $\mathbf{E}_2$, and again as $\delta \to 0$, this ray has direction $\mathbf{v}_3$ as defined in Figure 9.22b. Since we have the directions of an incident $\mathbf{v}_2$ and refracted $\mathbf{v}_3$ rays on the top surface, we can determine its normal $\mathbf{n}_T$. Next, refract a ray with direction $\mathbf{v}_4$ on the top surface with normal $\mathbf{n}_T$ calculating the direction $\mathbf{v}_5$ of the light inside the lens. Then refract a light ray with direction $\mathbf{v}_5$ on the bottom surface with normal $\mathbf{n}_B$ to obtain the direction of ray $\mathbf{v}_{P6}$. We iterate on the value of α until $\mathbf{v}_{P6}$ is parallel to $\mathbf{v}_6$ (the angle between them becomes zero) as in Figure 9.22b. This gives us the directions of the normals $\mathbf{n}_B$ and $\mathbf{n}_T$ to the bottom and top surfaces at point **X**.

Consider that the bottom and top surfaces in the neighborhood of point **X** are flat with normals $\mathbf{n}_B$ and $\mathbf{n}_T$. Consider again a point $\mathbf{X}_1$ very close to point **X**, but at a distance $\delta > 0$ away. We refract at point $\mathbf{X}_1$ a ray with the direction of $\mathbf{v}_1$ on the bottom surface and calculate point $\mathbf{X}_2$ on the top surface. Refract at point $\mathbf{X}_2$ a ray with direction $\mathbf{v}_4$ and calculate point $\mathbf{X}_3$ on the bottom surface, as shown in Figure 9.23a. Between points $\mathbf{X}_1$ and $\mathbf{X}_3$, consider the bottom surface to be flat with normal $\mathbf{n}_B$, as shown in Figure 9.23b. We now calculate the SMS chains as earlier, starting from the portion of the bottom surface $\mathbf{X}_1\mathbf{X}_3$. Figure 9.23b shows the first step of this calculation. As earlier, the surfaces are calculated to the axis of symmetry and then mirrored to the other side. A complete lens is shown in Figure 9.21. This process of filling the spaces between the SMS points with an interpolation $\mathbf{X}_1\mathbf{X}_3$ and

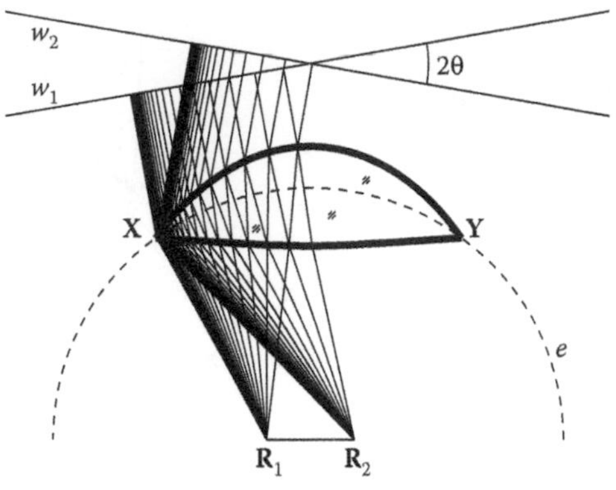

FIGURE 9.24
SMS lens with thin edges at **X** and **Y** designed for flat input wavefronts w_1 and w_2 and point receiver wavefronts $\mathbf{R}_1$ and $\mathbf{R}_2$. Edges **X** and **Y** are on a curve *e*.

corresponding SMS chains is called skinning (contrary to the usual meaning of the word, which means removing the skin).

It is also possible to use the same design method when the input (emitter) wavefonts are flat, as shown in Figure 9.24. Here, a lens with thin edges at points **X** and **Y** focuses to $\mathbf{R}_2$ parallel rays perpendicular to w_1 and focuses to $\mathbf{R}_1$ parallel rays perpendicular to w_2. The total acceptance angle of this optic is 2θ.

As before, the calculation of the optical surfaces starts at edge **X** whose normal vectors $\mathbf{n}_T$ to the top surface and $\mathbf{n}_B$ to the bottom surface are obtained from the directions of vectors $\mathbf{v}_1$, $\mathbf{v}_3$, $\mathbf{v}_4$ and $\mathbf{v}_6$, as shown in Figure 9.25. Point X verifies the condition (9.19).

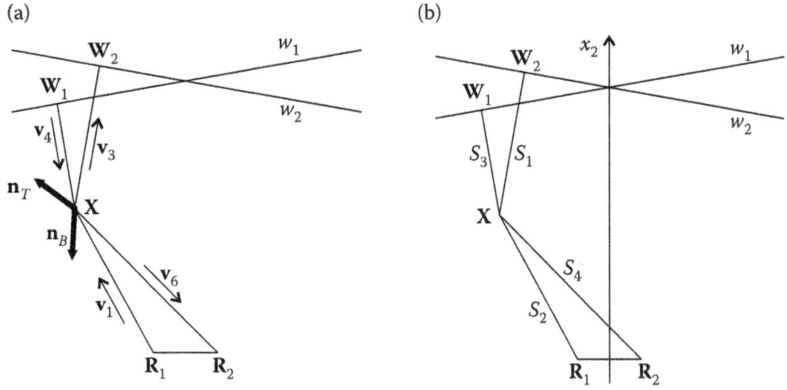

FIGURE 9.25
Starting conditions for the design of an SMS lens with thin edge for flat input wavefronts w_1 and w_2 and point receiver wavefronts $\mathbf{R}_1$ and $\mathbf{R}_2$. (a) Determining the orientations of normals $\mathbf{n}_T$ to the top surface and $\mathbf{n}_B$ to the bottom surface at edge **X** of the lens. (b) The optical path length between w_2 and $\mathbf{R}_1$ is $S_1 + S_2$ and that between w_1 and $\mathbf{R}_2$ is $S_3 + S_4$. The optic verifies $S_1 + S_2 = S_3 + S_4$.

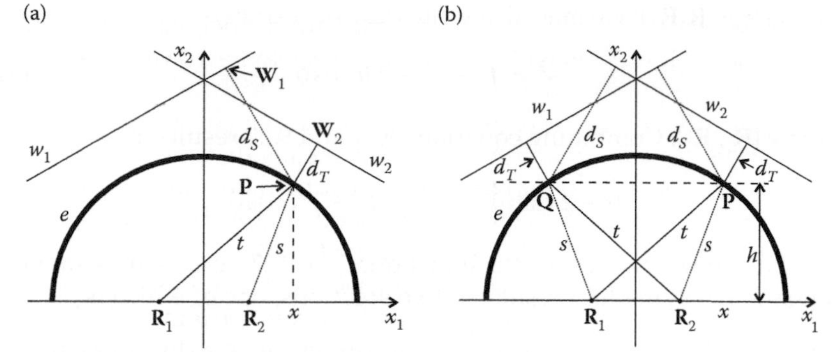

FIGURE 9.26
(a) Equal optical path length between w_1 and $\mathbf{R}_2$ and between w_2 and $\mathbf{R}_1$ defines the shape of a curve e. (b) Cutting curve e with a horizontal line at height h defines points **P** and **Q**. Étendue is conserved from w_1, w_2 to **PQ** and to $\mathbf{R}_1\mathbf{R}_2$.

Point **X** and its symmetrical point **Y** are on a curve e shown in Figure 9.24. We now determine the shape of this curve. Consider curve e in Figure 9.26a whose points **P** verify the condition of equal of optical path length from wavefront w_1 to $\mathbf{R}_2$ and from w_2 to $\mathbf{R}_1$. This condition may be written as $S = [\mathbf{R}_1, \mathbf{P}] + [\mathbf{P}, \mathbf{W}_2] = [\mathbf{R}_2, \mathbf{P}] + [\mathbf{P}, \mathbf{W}_1]$, or $S = t + d_T = s + d_S$.

Referring now to Figure 9.26b, if we cut this curve e with a horizontal line at height h, this defines the position of two points **P** and **Q**. The étendue from w_1, w_2 to **PQ** is $U_1 = 2d_S - 2d_T$ and the entendue from **PQ** to $\mathbf{R}_1\mathbf{R}_2$ is $U_2 = 2t - 2s$. If étendue is conserved, we get $U_1 = U_2$, or $t + d_T = s + d_S$ which is the condition of equal optical path length from w_1 to $\mathbf{R}_2$ and from w_2 to $\mathbf{R}_1$.

In order to obtain the shape of curve e, consider now the geometry in Figure 9.27, from which we get $t + d\sin\theta = s + (2x + d)\sin\theta$, or

$$s = t - 2x\sin\theta \tag{9.21}$$

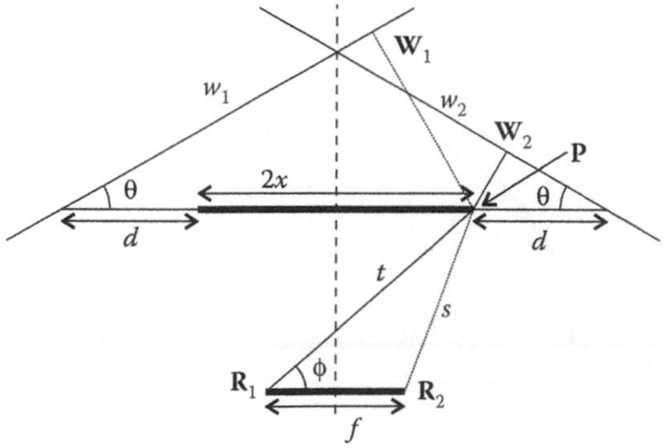

FIGURE 9.27
Calculation of point **P** for which there is equal optical path length between w_1 and $\mathbf{R}_2$ and between w_2 and $\mathbf{R}_1$.

For triangle $\mathbf{R}_1\mathbf{R}_2\mathbf{P}$ we may also write (law of cosines):

$$s^2 = f^2 + t^2 - 2ft\cos\phi \tag{9.22}$$

where $f = [\mathbf{R}_1, \mathbf{R}_2]$. Combining equations 9.21 and 9.22 results in

$$(t - 2x\sin\theta)^2 = f^2 + t^2 - 2ft\cos\phi \tag{9.23}$$

Replacing now $x = t\cos\phi - f/2$ into Equation 9.23 results in a quadratic equation in t. Solving for t results in two solutions, one of which is

$$t(\phi) = \frac{f - f\sin^2\theta}{2\sin\theta - 2\sin^2\theta\cos\phi} \tag{9.24}$$

Multiplying numerator and denominator by $f/\sin^2\theta$ we get

$$t(\phi) = \frac{(f/\sin\theta)^2 - f^2}{2(f/\sin\theta) - 2f\cos\phi} = \frac{K^2 - f^2}{2K - 2f\cos\phi} \tag{9.25}$$

where $K = f/\sin\theta$. Point $\mathbf{P}$ is given by $\mathbf{P} = \mathbf{R}_1 + t(\phi)(\cos\phi, \sin\phi)$ and is therefore on an ellipse (see Chapter 21) with semi-major axis

$$a = K/2 = f/(2\sin\theta) \tag{9.26}$$

and semi-minor axis b which may be obtained from the geometry in Figure 9.28:

$$(f/2)^2 + b^2 = (K/2)^2 \tag{9.27}$$

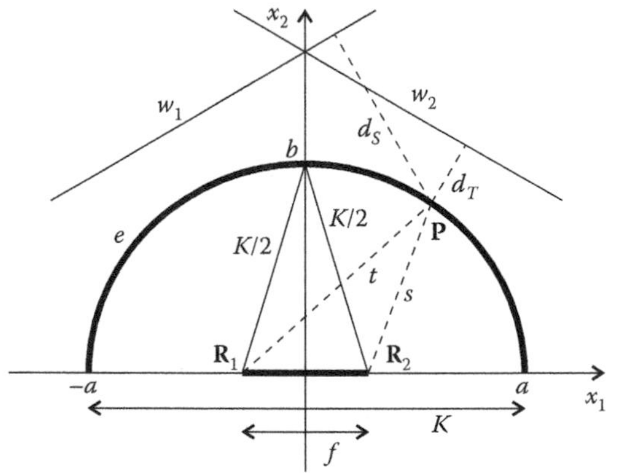

FIGURE 9.28
The points of ellipse e verify the consition of equal optical path length between w_1 and $\mathbf{R}_2$ and between w_2 and $\mathbf{R}_1$, that is, $t + d_T = s + d_S$.

Replacing $K = f/\sin\theta$ and solving for b results in

$$b = f/(2\tan\theta) \tag{9.28}$$

If $\mathbf{R}_1\mathbf{R}_2$ centered at (0, 0), the equation of the ellipse may be rewritten as

$$\left(\frac{x_1}{a}\right)^2 + \left(\frac{x_2}{b}\right)^2 = 1 \Leftrightarrow \left(\frac{x_1}{f/(2\sin\theta)}\right)^2 + \left(\frac{x_2}{f/(2\tan\theta)}\right)^2 = 1$$

$$\Leftrightarrow x_1^2\sin^2\theta + x_2^2\tan^2\theta = \left(\frac{f}{2}\right)^2 \tag{9.29}$$

This ellipse e may be reparameterized as

$$\mathbf{e}(\varphi) = (a\cos\varphi, b\sin\varphi) = \frac{f}{2}\left(\frac{\cos\varphi}{\sin\theta}, \frac{\sin\varphi}{\tan\theta}\right) \tag{9.30}$$

with $0 < \varphi < 2\pi$. If $f = [\mathbf{R}_1, \mathbf{R}_2] = 2$, we get[6,7]

$$x_1^2\sin^2\theta + x_2^2\tan^2\theta = 1 \tag{9.31}$$

The points **P** of the ellipse in Figure 9.28 verify $t + d_T = s + d_S$.

Points **X** and **Y** at the edges of the optic in Figure 9.24 are on curve e at a given height h, as was the case of points **Q** and **P** in Figure 9.26b.

In the limit case in which f goes to zero, the acceptance angle θ also goes to zero and ellipse e becomes a circle c centered at a point **R**, as shown in Figure 9.29.

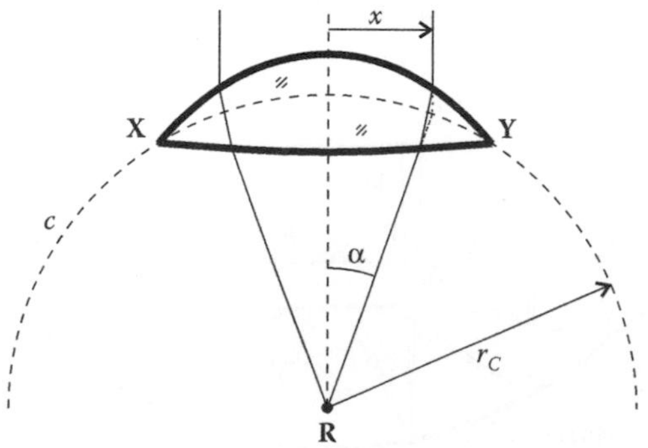

FIGURE 9.29
The limit case when the size of the receiver goes to zero and becomes a point **R**. This is called the aplanatic limit.

For this optic we may write

$$x = r_C \sin\alpha \tag{9.32}$$

where r_C is the radius of circle c and is, therefore, constant for all values of α. The étendue now becomes infinitesimal and this is called the aplanatic limit (see Chapter 11).

9.4 The XR, RX, and XX Optics

The RR lens presented earlier has two refractive surfaces. Other types of SMS optics can also be calculated using the Miñano–Benitez design method. For example, the case in which one of the surfaces is reflective (a mirror) and the other refractive. The resulting optic has the geometry as shown in Figure 9.30. In the Miñano–Benitez design method, refractions are denoted by R, reflections by X (from the Spanish reflexión) and total internal reflections (TIRs) by I. This new optic is, therefore, an XR because light is first reflected and then refracted before reaching the receiver. The rays perpendicular to wavefront w_1 are concentrated on to the edge $\mathbf{R}_2$ of the receiver and the edge rays perpendicular to w_2 are concentrated on to the receiver edge $\mathbf{R}_1$. The receiver is immersed in a medium of refractive index n.

The algorithm for calculating this optic is the same as for the RR lens shown in Figure 9.4. During the design process we ignore the shadow the secondary (refractive element) produces on the primary (mirror). Start with the positions of the end points $\mathbf{R}_1$ and $\mathbf{R}_2$ for the receiver. Now choose a point $\mathbf{P}_0$ and its normal $\mathbf{n}_0$, as shown in Figure 9.31. Next, refract at $\mathbf{P}_0$ a ray r_1 coming from the edge $\mathbf{R}_2$ of the receiver and choose a point $\mathbf{P}_1$ along the refracted ray. This edge ray must be reflected in a direction perpendicular to

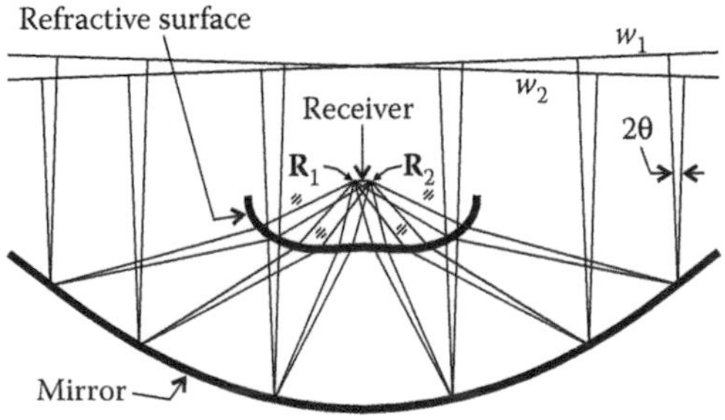

FIGURE 9.30
An XR SMS optic. One on the surfaces is a mirror (X) and the other surface is refractive (R). The receiver is immersed in a dielectric of refractive index n.

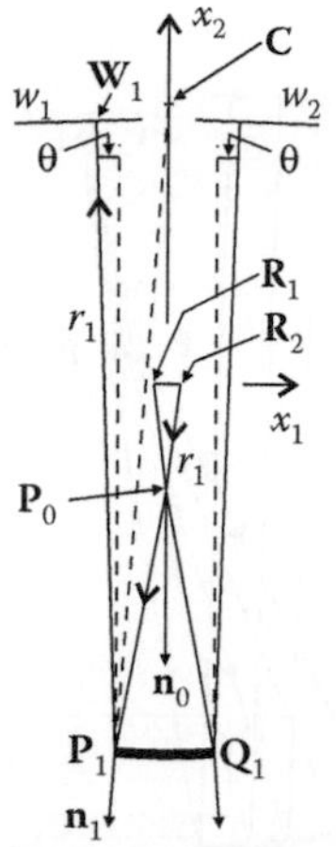

FIGURE 9.31
Calculating the first portion of the mirror of an XR optic from which the rest of the surfaces will be derived.

the wavefront w_1 and this gives us the direction of the normal $\mathbf{n}_1$ at point $\mathbf{P}_1$. The ray r_1 intersects the wavefront w_1 at point $\mathbf{W}_1$. Point $\mathbf{Q}_1$ is symmetrical to point $\mathbf{P}_1$ and also has a symmetrical normal. The intersection of the axis of symmetry x_2 by the straight line defined by point $\mathbf{P}_1$ and its normal $\mathbf{n}_1$ gives us point **C**, which we take as the center of a circular arc from $\mathbf{P}_1$ to $\mathbf{Q}_1$ with normals at $\mathbf{P}_1$ and $\mathbf{Q}_1$ matching those calculated for those points.

We now calculate the SMS chains as we did for the RR lens. Figure 9.32 shows some of the steps in those calculations.

The path of the ray r_1 gives us the optical path length between $\mathbf{R}_2$ and w_1, which is also the optical path length between $\mathbf{R}_1$ and w_2. It is given by

$$S = n[\mathbf{R}_2, \mathbf{P}_0] + [\mathbf{P}_0, \mathbf{P}_1] + [\mathbf{P}_1, \mathbf{W}_1] \tag{9.33}$$

Reflect off the mirror a set of rays perpendicular to wavefront w_2 between points $\mathbf{P}_1$ and $\mathbf{Q}_1$ as shown in Figure 9.32a. By constant optical path length, calculate a portion $\mathbf{P}_0\mathbf{P}_2$ of the refractive surface. Next refract a set of rays coming from $\mathbf{R}_2$ on that surface and calculate, again by constant optical path length, the points on the mirror between $\mathbf{P}_1$ and $\mathbf{P}_3$ that reflect these rays in a direction perpendicular to wavefront w_1 (Figure 9.32b). Again take a set of rays perpendicular to wavefront w_2, reflect them on the portion $\mathbf{P}_1\mathbf{P}_3$ of the mirror and, by the constant optical path length, calculate a portion $\mathbf{P}_2\mathbf{P}_4$ of the refractive surface (Figure 9.32c). The process continues to calculate alternate portions of the reflective (mirror) and refractive surfaces.

Just as with the case of the RR optic in Figure 9.8, also in this case also we may design an optic that matches the étendues from the emitter to the optic

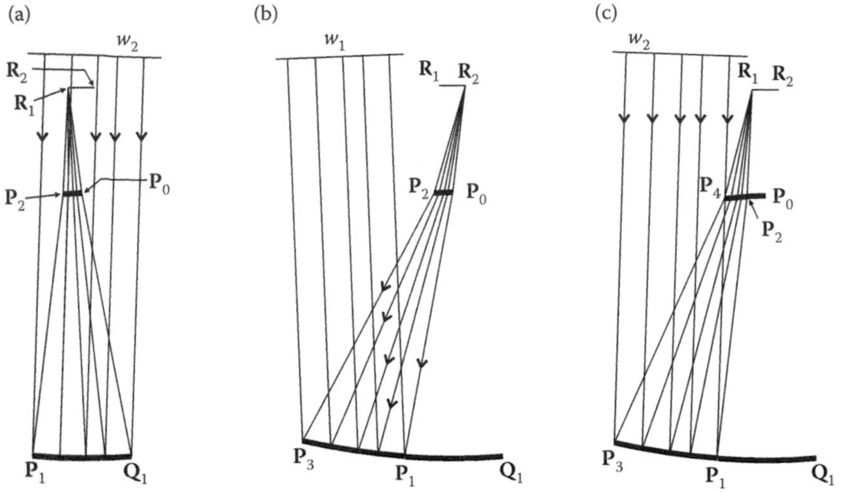

FIGURE 9.32
SMS chains for calculating the shape of the optical surfaces of an XR optic. (a) Rays propagated from w_2 through $\mathbf{Q}_1\mathbf{P}_1$ on the mirror define a new section $\mathbf{P}_0\mathbf{P}_2$ on the refractive surface. (b) Rays propagated from $\mathbf{R}_2$ through $\mathbf{P}_0\mathbf{P}_2$ define a new section $\mathbf{P}_1\mathbf{P}_3$ on their way to w_1. (c) Rays propagated from w_2 through $\mathbf{P}_1\mathbf{P}_3$ define a new section $\mathbf{P}_2\mathbf{P}_4$.

and from the optic to the receiver, as shown in Figure 9.33 for the case in which the emitter is a finite source $\mathbf{E}_1\mathbf{E}_2$.[5]

The design method used to obtain this device is similar to the one described for the RR lens. It starts by defining the positions for the starting points **N** and **X**. The étendue from the source $\mathbf{E}_1\mathbf{E}_2$ to the mirror is given by $U = 2([\mathbf{N}, \mathbf{E}_1] - [\mathbf{N}, \mathbf{E}_2])$. In the same way, the étendue from $\mathbf{R}_1\mathbf{R}_2$ to the refractive surface

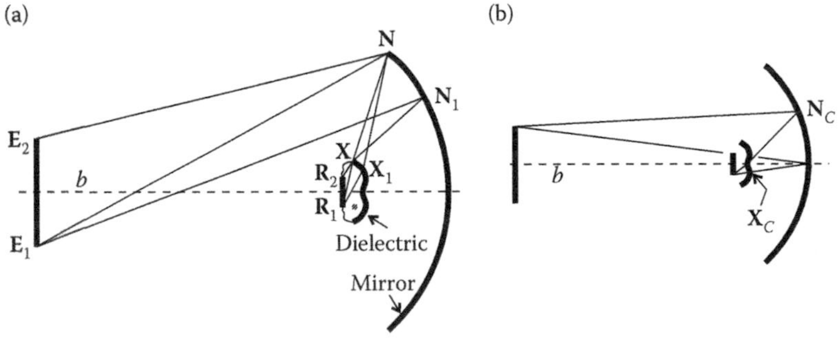

FIGURE 9.33
(a) XR optic that matches the étendues from the emitter to the optic and from the optic to the receiver. It comprises a mirror collecting the light from the emitter and a dielectric piece in contact with the receiver. (b) Also in this case, the Miñano–Benitez design method does not guarantee the convergence of all the rays for the portion of the mirror comprised between $\mathbf{N}_C$ and its symmetrical point relative to axis b.

is given by $U = 2([[\mathbf{X}, \mathbf{R}_1]] - [[\mathbf{X}, \mathbf{R}_2]])$. Note that in the latter case, the optical path lengths $[[\mathbf{X}, \mathbf{R}_1]]$ and $[[\mathbf{X}, \mathbf{R}_2]]$ are calculated in a medium of refractive index n, and therefore, they are the product of this refractive index and the distance between the points. The étendue arriving at the mirror from $\mathbf{E}_1\mathbf{E}_2$ must be the same as that exiting the refractive surface toward $\mathbf{R}_1\mathbf{R}_2$. We can then choose a value for U. Point **N** must be on the line defined by $[\mathbf{N}, \mathbf{E}_1] - [\mathbf{N}, \mathbf{E}_2] = U/2$, that is, on an hyperbola having foci $\mathbf{E}_1$ and $\mathbf{E}_2$. Point **X** must be on a line defined by $[[\mathbf{X}, \mathbf{R}_1]] - [[\mathbf{X}, \mathbf{R}_2]] = U/2$, that is, on a hyperbola having foci $\mathbf{R}_1$ and $\mathbf{R}_2$.

Similar to the RR lens shown in Figure 8.16, in this case also the design of the curves begins with portions $\mathbf{NN}_1$ of the first surface (now a mirror) and $\mathbf{XX}_1$ of the second (refractive) surface. Portion $\mathbf{NN}_1$ of the first surface (mirror) focuses to **X** the rays coming from $\mathbf{E}_1$. Now it is an ellipse with foci $\mathbf{E}_1$ and **X**, and passing through point **N**. Portion $\mathbf{XX}_1$ of the second surface focuses to $\mathbf{R}_1$ the rays coming from **N**. It is a Cartesian oval with foci **N** and $\mathbf{R}_1$, and passing through point **X**. From the path of ray $\mathbf{E}_2$-**N**-$\mathbf{X}_1$-$\mathbf{R}_1$, we can calculate the optical path length between $\mathbf{E}_2$ and $\mathbf{R}_1$. Also, from the path of ray $\mathbf{R}_2$-**X**-$\mathbf{N}_1$-$\mathbf{E}_1$ we can calculate the optical path length between $\mathbf{R}_2$ and $\mathbf{E}_1$. Given the symmetry of the design, these optical path lengths should be equal to one another. The design of the XR optic now continues in a way similar to that shown in Figure 8.17 for the RR lens. Launch a set of rays from $\mathbf{E}_2$ through the portion $\mathbf{NN}_1$ of the first surface (mirror) already calculated. We know the optical path length to point $\mathbf{R}_1$, therefore, we can calculate a new portion $\mathbf{X}_1\mathbf{X}_2$ of the second (refractive) surface. We can also launch a set of rays from $\mathbf{R}_2$ through the portion $\mathbf{XX}_1$ of the second surface (refractive) already calculated. We know the optical path length to point $\mathbf{E}_1$, therefore, we can calculate a new portion $\mathbf{N}_1\mathbf{N}_2$ on the first surface (mirror). as in the case of the RR lens, in this case also this iterative process is continued until the surfaces reach the optical (symmetry) axis.

Also in this case, there will be a central zone for both surfaces for which no perfect focusing of the rays is guaranteed. In the case of the mirror, this zone is between point $\mathbf{N}_C$ and its symmetrical point relative to axis b. Point $\mathbf{N}_C$ is defined by the edge ray emitted from the edge $\mathbf{R}_1$ of the receiver through the center of the refractive surface. In the case of the refractive surface, this zone is between $\mathbf{X}_C$ and its symmetrical point. Point $\mathbf{X}_C$ is defined by the edge ray emitted from the edge $\mathbf{E}_2$ of the emitter through the center of the mirror.

Calculating an XR optic can often result in loops on the surface profiles caused by caustics formed between the optical surfaces. Adjusting the parameters to avoid those loops may prove to be a difficult task. One option is to calculate the optic with the offending loops and then, when defining the optical surfaces, take only those points that define smooth surfaces, removing all the loops. The resulting optical surfaces, although not ideal, may still work for some applications.

It is also possible to design an XR optic for an infinite source at an infinite distance subtending an angle 2θ, as shown in Figure 9.34. This optic can be used as a solar concentrator with acceptance angle 2θ and may be given rotational symmetry to make it into a 3-D optic.

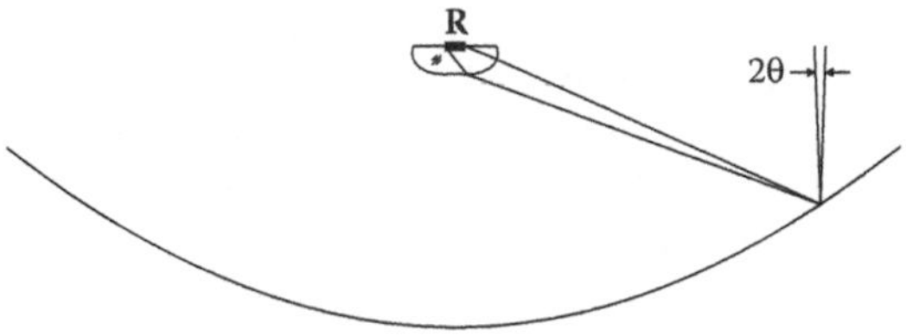

FIGURE 9.34
XR optic designed for an acceptance angle 2θ.

For practical reasons, solar concentrators are usually assembled side by side in arrays. Trimming the primary mirror in a square shape is therefore desireable, as shown in Figure 9.35.

In Figure 9.36, the dashed and dotted lines show the rotational optic resulting from the 2-D design in Figure 9.34. These lines show a circular primary (dashed lines) and a circular receiver (dotted line). The square trimming (solid lines) of the primary is inscribed in the circular primary. The resulting primary mirror captures less light, reducing concentration. If the square receiver R circumscribes the circular receiver, it will capture all the light that would reach the circular receiver. This, however, further reduces concentration.[8]

Another possibility for an SMS optic is to have the first surface as refractive (R) and the second reflective (X), as shown in Figure 9.37. This optic is called RX.[9,10] The receiver $\mathbf{R}_1\mathbf{R}_1$ is immersed in a medium of refractive index n and faces down (it is illuminated from below).

The calculation method is again the same shown in Figure 9.4 for an RR lens. By symmetry, start by choosing a point $\mathbf{P}_0$ on the perpendicular bisector of the receiver $\mathbf{R}_1\mathbf{R}_2$ and its vertical normal $\mathbf{n}_0$. We refract at that point a ray r_1 perpendicular to wavefront w_2 and choose a point $\mathbf{P}_1$ along the refracted ray, as shown in Figure 9.38. Calculate the normal $\mathbf{n}_1$ at $\mathbf{P}_1$ such that this ray

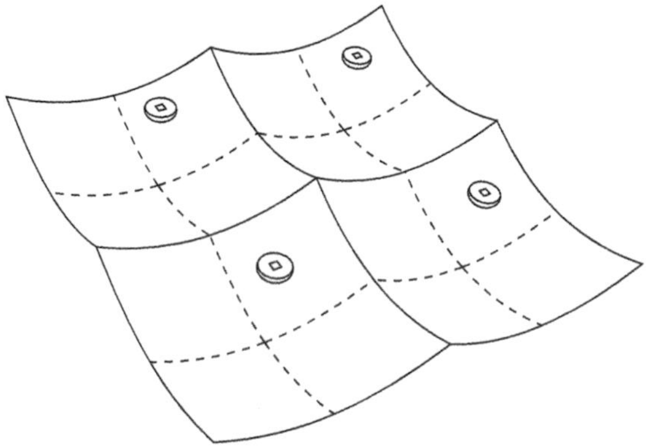

FIGURE 9.35
XR optic trimmed in a square shape so that several of those optics can be placed side by side.

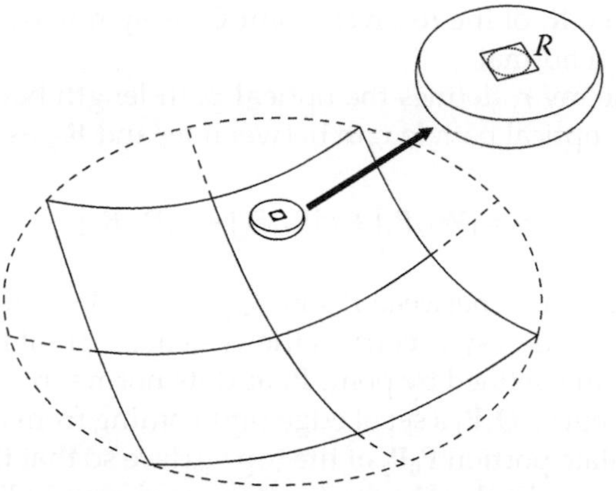

FIGURE 9.36
XR square concentrator with a square primary inscribed in the circular primary (rotationally symmetric) and a square receiver that circumscribes the circular receiver (rotationally symmetric).

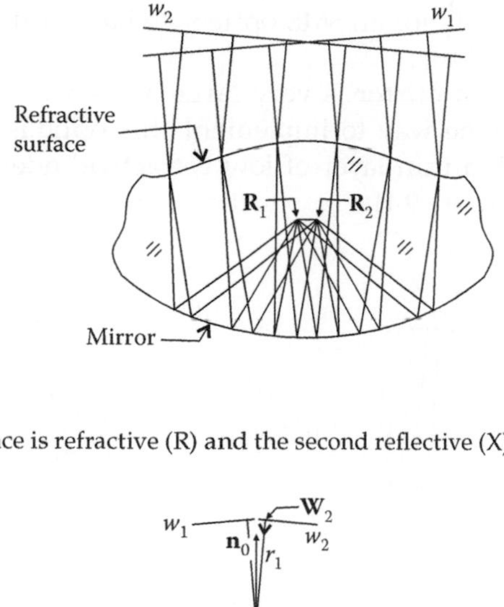

FIGURE 9.37
RX optic. The first surface is refractive (R) and the second reflective (X).

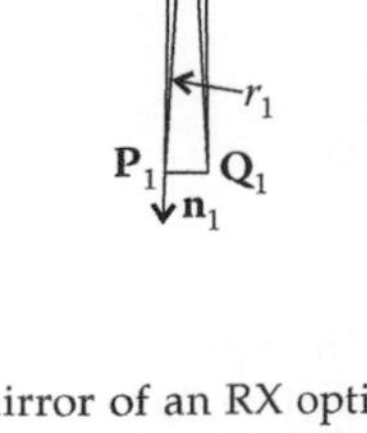

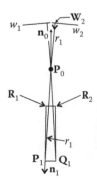

FIGURE 9.38
Calculating the first portion of the mirror of an RX optic from which the rest of the surfaces will be derived.

is reflected to edge $\mathbf{R}_1$ of the receiver. Point $\mathbf{Q}_1$ is symmetrical to $\mathbf{P}_1$ and also has a symmetrical normal.

The path of the ray r_1 defines the optical path length between w_2 and $\mathbf{R}_1$, which is also the optical path length between w_1 and $\mathbf{R}_2$, as

$$S = [\mathbf{W}_2,\mathbf{P}_0] + n[\mathbf{P}_0,\mathbf{P}_1] + n[\mathbf{P}_1,\mathbf{R}_1] \tag{9.34}$$

Next, choose the curve between $\mathbf{P}_1$ and $\mathbf{Q}_1$ as a circle whose center is at the intersection of the axis of symmetry of the system (or point $\mathbf{P}_0$ and normal $\mathbf{n}_0$) and the straight line defined by point $\mathbf{P}_1$ and its normal $\mathbf{n}_1$.

Reflect off that curve $\mathbf{Q}_1\mathbf{P}_1$ a set of edge rays coming from the edge $\mathbf{R}_2$ of the source and calculate portion $\mathbf{P}_0\mathbf{P}_2$ of the top surface so that they are refracted in a direction perpendicular to wavefront w_1, as shown in Figure 9.39.

Refract a set of rays perpendicular to wavefront w_2 on $\mathbf{P}_0\mathbf{P}_2$ and calculate portion $\mathbf{P}_1\mathbf{P}_3$ of the mirror so that these rays are concentrated on to the edge $\mathbf{R}_1$ of the receiver. The process goes on to calculate alternate portions of the reflective and refractive surfaces.

Yet another possibility for an SMS optic is to have both surfaces reflective, as shown in Figure 9.40.

In this optic, the top mirror is very large and covers the bottom mirror almost completely. One way to implement this optic is to make it as two dielectric parts with a thin layer of low refractive index material between them, as shown in Figure 9.41.

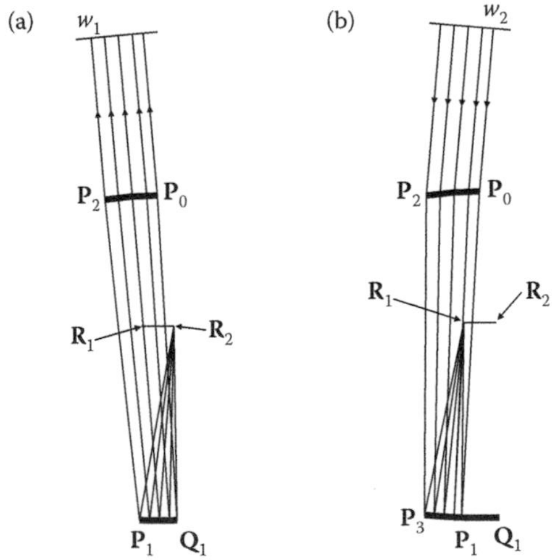

FIGURE 9.39
SMS chains for calculating the shape of the optical surfaces of an RX optic. (a) A set of rays from $\mathbf{R}_2$ reflected on $\mathbf{P}_1\mathbf{Q}_1$ determine a new section $\mathbf{P}_0\mathbf{P}_2$ of the top surface. (b) A set of rays from w_2 refracted on $\mathbf{P}_0\mathbf{P}_2$ determine a new section $\mathbf{P}_1\mathbf{P}_3$ of the mirror.

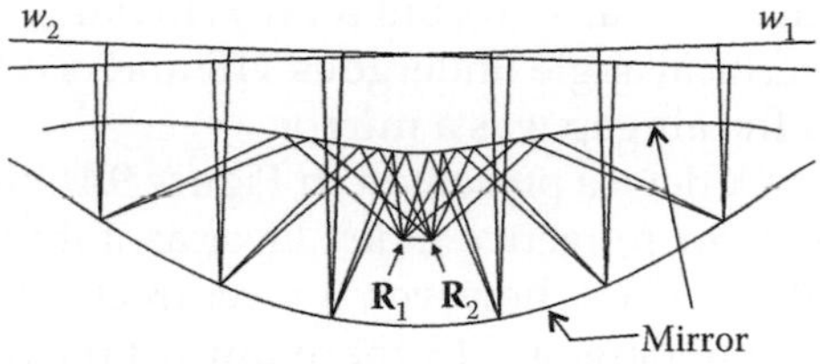

FIGURE 9.40
An XX optic. The first surface is reflective (X) and the second is also reflective (X). When designing the optic we ignore the shading the top mirror produces on the bottom mirror.

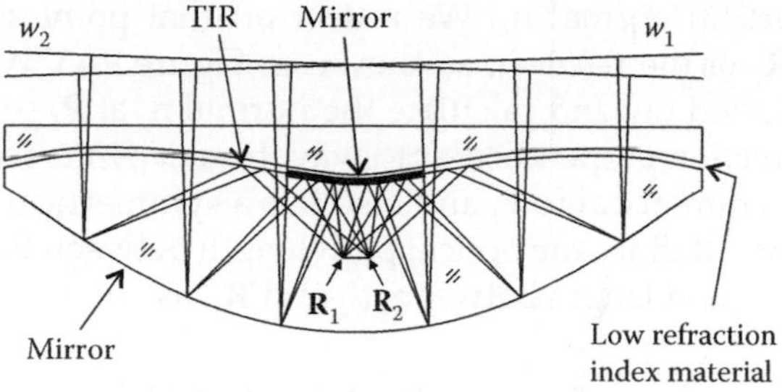

FIGURE 9.41
The XX optic can be implemented as two parts or refractive index n separated by a thin layer of low refractive index material (such as air). Light goes through this layer, reflects off the bottom mirror and undergoes TIR off the top mirror toward the receiver. The TIR condition is not attained along the whole top surface, where a small central mirror is required.

The way this layer works is shown in Figure 9.42. A ray r_1 traveling in the medium of refractive index n and making an angle to the normal smaller than the critical angle α_C is refracted into the low refractive index layer and then again from that layer to the medium of refractive index n and continues in the same direction. Therefore, for this ray, it is as if the air gap was not

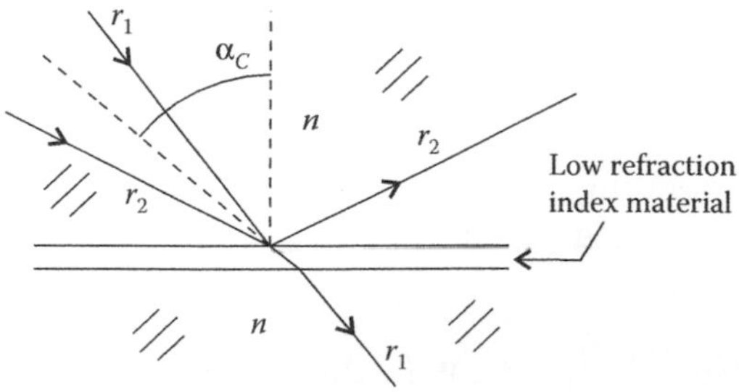

FIGURE 9.42
A thin layer of low refractive index material (such as air) separating two parts of a high refractive index material. This layer behaves as a mirror for light, making an angle larger than the critical angle to the vertical, but lets the light through if the angle to the normal is smaller than the (large) critical angle.

there (except for a small lateral shift), but a ray r_2 making an angle to the vertical larger than the critical angle undergoes TIR and is reflected back. Thus, for this ray, it is as if the air gap was a mirror.

For the case of the XX device presented in Figure 9.41, the incoming radiation goes through the low refractive index layer as if it was not there. After reflecting on the bottom mirror, however, it is redirected toward the top mirror at large angles and undergoes TIR there toward the receiver. The condition of TIR is not achieved along all the top surface of the XX, and there must be a small central mirror.

The calculation method for the XX is the same as earlier. By symmetry, we start by choosing a point $\mathbf{P}_0$ on the perpendicular bisector of the receiver $\mathbf{R}_1\mathbf{R}_2$ and its vertical normal $\mathbf{n}_0$. We reflect off that point a ray r_1 coming from the edge $\mathbf{R}_2$ of the receiver, as shown in Figure 9.43. We choose point $\mathbf{P}_1$ along the reflected ray and calculate the normal $\mathbf{n}_1$ at $\mathbf{P}_1$ so that this ray is reflected in a direction perpendicular to wavefront w_1, intersecting it at point $\mathbf{W}_1$. Point $\mathbf{Q}_1$ is symmetrical to $\mathbf{P}_1$ and also has a symmetrical normal.

The path of ray r_1 defines the optical path length between $\mathbf{R}_2$ and w_1, which is also the optical path length between w_2 and $\mathbf{R}_1$, as

$$S = [\mathbf{R}_2, \mathbf{P}_0] + [\mathbf{P}_0, \mathbf{P}_1] + [\mathbf{P}_1, \mathbf{W}_1] \tag{9.35}$$

Choose the curve between $\mathbf{P}_1$ and $\mathbf{Q}_1$ as a circle whose center is defined by the intersection of the axis of symmetry of the system (or point $\mathbf{P}_0$ and normal $\mathbf{n}_0$) and the straight line defined by point $\mathbf{P}_1$ and its normal $\mathbf{n}_1$.

Reflect off that curve $\mathbf{Q}_1\mathbf{P}_1$ a set of edge rays perpendicular to wavefront w_2 and calculate portion $\mathbf{P}_0\mathbf{P}_2$ of the top surface so that they are reflected toward edge $\mathbf{R}_1$ of the receiver, as shown in Figure 9.44.

Reflect off $\mathbf{P}_0\mathbf{P}_2$ a set of rays coming from edge $\mathbf{R}_2$ of the receiver and calculate portion $\mathbf{P}_1\mathbf{P}_3$ of the bottom mirror, so that these rays are redirected in a

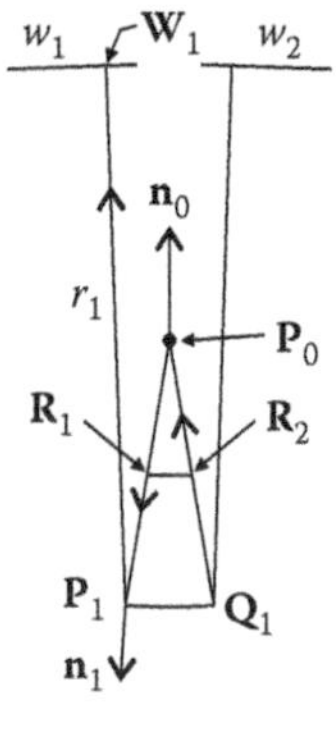

FIGURE 9.43
Calculating the first portion of the mirror of an XX optic from which the rest of the surfaces will be derived.

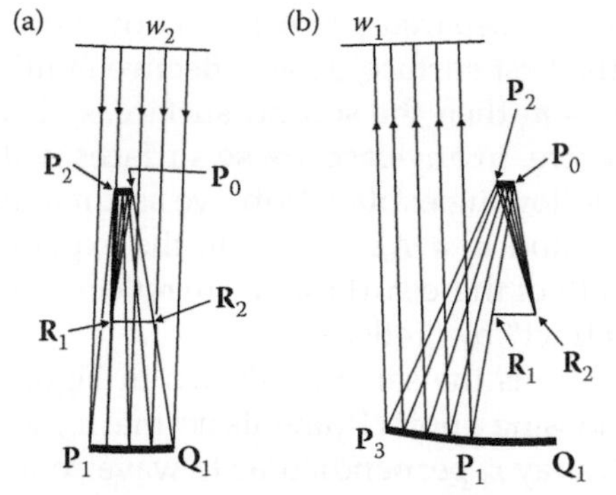

FIGURE 9.44
SMS chains for calculating the shape of the optical surfaces of an XX optic. (a) Rays from w_2 reflected on Q_1P_1 define a new portion P_0P_2 of the top mirror. (b) Rays from R_2 reflected on P_0P_2 define a new portion P_1P_3 of the bottom mirror.

direction perpendicular to wavefront w_1. The process goes on as we calculate alternating portions of the reflective and refractive surfaces.

9.5 The Miñano–Benitez Design Method with Generalized Wavefronts

The Miñano–Benitez method can be used with generalized input and output wavefronts. Figure 9.45 shows a more general situation with two input wavefronts w_1 and w_2 and two output wavefronts w_3 and w_4. We want to design an optic that couples w_1 with w_4 and w_2 with w_3. This optic has two optically active surfaces s_1 and s_2. Surface s_1 separates two media of refractive indices

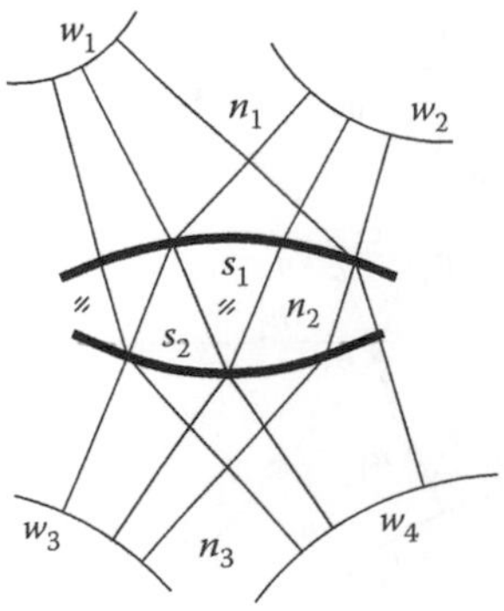

FIGURE 9.45
The Miñano–Benitez in the more general case in which an SMS optic couples two generalized input wavefronts w_1 and w_2 and two generalized output wavefronts w_4 and w_3.

n_1 and n_2 and surface s_2 separates two media of refraction indices n_2 and n_3.[11] If $n_1 = n_2$ then the first surface, s_1, is reflective (a mirror), otherwise it is refractive. Also, if $n_2 = n_3$ then the second surface, s_2, is reflective (a mirror), otherwise it is refractive. In any case, these surfaces deflect (either reflect or refract) light. In the following explanation, we assume that both the surfaces are refractive ($n_1 \neq n_2$ and $n_2 \neq n_3$). However, the explanation would still be the same if one or both of these surfaces were reflective ($n_1 = n_2$ or $n_2 = n_3$), by simply replacing "refract" by "reflect."

The design procedure is the same as shown in Figure 9.3, Figure 9.4, and Figure 9.6. Start by choosing a point $\mathbf{P}_0$ and its normal $\mathbf{n}_0$, as shown in Figure 9.46.

Refract at point $\mathbf{P}_0$ a ray r_1 perpendicular to wavefront w_2. Choose the optical path length S_{23} between w_2 and w_3. With that value, calculate point $\mathbf{P}_1$ and its normal $\mathbf{n}_1$ so that ray r_1 is deflected at $\mathbf{P}_1$ in a direction perpendicular to wavefront w_3. Then refract at point $\mathbf{P}_1$ a ray r_2 perpendicular to wavefront w_4. We choose the optical path length S_{14} between w_1 and w_4. With this value, we can calculate point $\mathbf{P}_2$ and its normal $\mathbf{n}_2$ so that ray r_2 is deflected at $\mathbf{P}_2$ in a direction perpendicular to w_1.

Interpolate a curve $\mathbf{c}$ between points $\mathbf{P}_2$ and $\mathbf{P}_0$ such that it is perpendicular to $\mathbf{n}_0$ at $\mathbf{P}_0$ and to $\mathbf{n}_2$ at $\mathbf{P}_2$. For this curve, calculate a set of points and normals between $\mathbf{P}_0$ and $\mathbf{P}_2$. Refract at those points a set of rays perpendicular to wavefront w_2, as shown in Figure 9.47a. Since we have the optical path length S_{23} between w_2 and w_3, we can calculate a new portion $\mathbf{P}_1\mathbf{P}_3$ of surface s_2 of the optic. Refract at those new points a set of rays perpendicular to w_4, as shown in Figure 9.47b. Since we have the optical path length S_{14} between w_4 and w_1, we can calculate a new portion $\mathbf{P}_2\mathbf{P}_4$ of surface s_1 of the optic.

This process builds the surfaces leftward from the initial point $\mathbf{P}_0$. The same process can also build the surfaces s_1 and s_2 to the right of $\mathbf{P}_0$. As earlier, calculate a set of points and normals on curve $\mathbf{c}$ between $\mathbf{P}_0$ and $\mathbf{P}_2$ (e.g., the same set of points as earlier). Refract at these points a set of rays perpendicular to wavefront w_1, as shown in Figure 9.48a. Since we have the optical

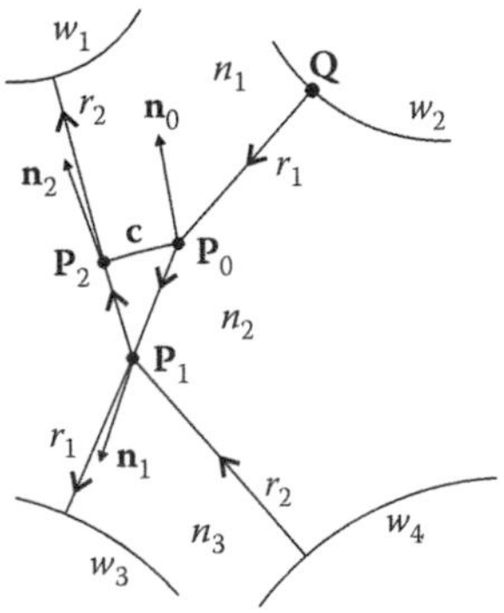

FIGURE 9.46
The design process starts by choosing a point $\mathbf{P}_0$ and its normal $\mathbf{n}_0$. We then build the first steps of an SMS chain and interpolate a curve $\mathbf{c}$ between points $\mathbf{P}_0$ and $\mathbf{P}_2$.

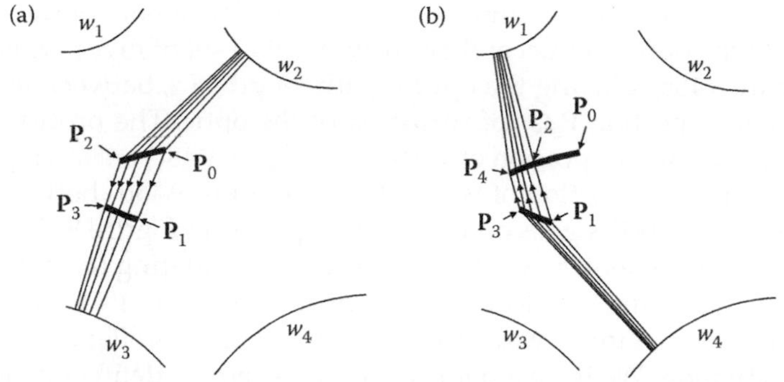

FIGURE 9.47
Starting at curve **c** between $\mathbf{P}_0$ and $\mathbf{P}_2$, the design of the optical surfaces proceeds to the left as we calculate alternate portions of the top (s_1) and bottom (s_2) surfaces of the optic. (a) Rays propagated from w_2 through $\mathbf{P}_0\mathbf{P}_2$ on s_1 define a new section $\mathbf{P}_1\mathbf{P}_3$ on s_2. (b) Rays propagated from w_4 through $\mathbf{P}_1\mathbf{P}_3$ on s_2 define a new section $\mathbf{P}_2\mathbf{P}_4$ on s_1.

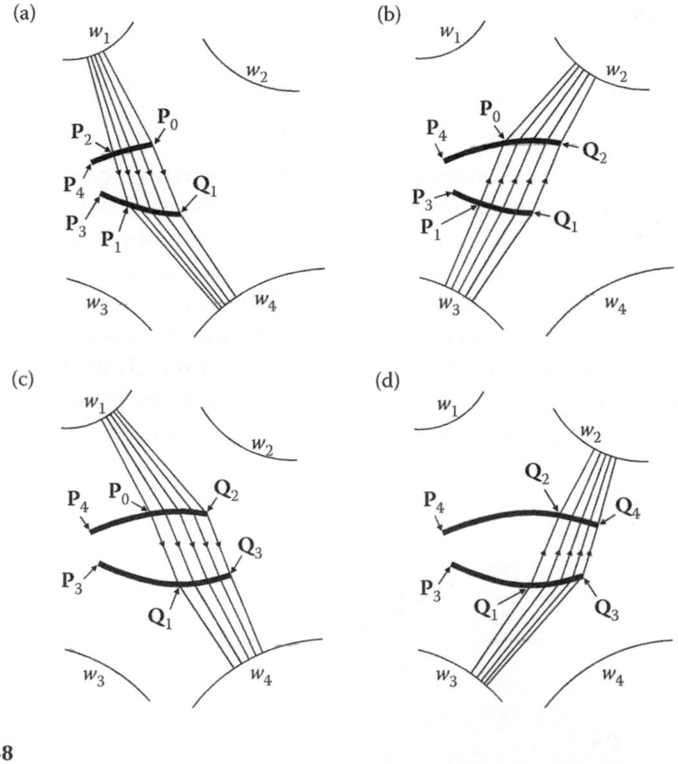

FIGURE 9.48
Starting at curve **c** between $\mathbf{P}_0$ and $\mathbf{P}_2$, the design of the optical surfaces proceeds to the right as we calculate alternate portions of the top (s_1) and bottom (s_2) surfaces of the optic. (a) Rays propagated from w_1 through $\mathbf{P}_0\mathbf{P}_2$ on s_1 define a new section $\mathbf{Q}_1\mathbf{P}_1$ on s_2. (b) Rays propagated from w_3 through $\mathbf{Q}_1\mathbf{P}_1$ define a new section $\mathbf{Q}_2\mathbf{P}_0$. (c) Rays propagated from w_1 through $\mathbf{Q}_2\mathbf{P}_0$ define a new section $\mathbf{Q}_3\mathbf{Q}_1$. (d) Rays propagated from w_3 through $\mathbf{Q}_1\mathbf{Q}_3$ define a new section $\mathbf{Q}_2\mathbf{Q}_4$.

path length, S_{14}, between w_1 and w_4, we can calculate a new portion $\mathbf{P}_1\mathbf{Q}_1$ of surface s_2 of the optic. Refract at those new points a set of rays perpendicular to w_3 (Figure 9.48b). Having the optical path length, S_{23}, between w_3 and w_2, calculate a new portion $\mathbf{P}_0\mathbf{Q}_2$ of surface s_1 of the optic. The process goes on as we calculate another portion of surface s_2 (Figure 9.48c) between points $\mathbf{Q}_1$ and $\mathbf{Q}_3$ and another portion of top surface s_1 (Figure 9.48d) between points $\mathbf{Q}_2$ and $\mathbf{Q}_4$. This process goes on to give the optic as in Figure 9.45.

Now take a closer look at what happens when calculating the path of a ray, such as ray r_1 in Figure 9.46. Knowing the position of point $\mathbf{P}_0$, we must determine from which point of wavefront w_2 ray r_1 is coming. Figure 9.49 shows a similar situation. We have a point $\mathbf{P}$ and a wavefront defined by a parameterization $\mathbf{w}(\sigma)$. We want to determine for what point $\mathbf{Q}$ the perpendicular to $\mathbf{w}(\sigma)$ goes through the point $\mathbf{P}$.

We now consider a possible way to determine the position of point $\mathbf{Q}$. Take, for example, a point $\mathbf{Q}(\sigma)$ on the wavefront and determine its tangent a at that point. It has unit normal vector $\mathbf{n}(\sigma)$ and unit tangent vector $\mathbf{t}(\sigma)$. For a straight line a, we can determine point $\mathbf{R}(\sigma)$ on the line perpendicular to a through $\mathbf{P}$ as (see Chapter 21),

$$\mathbf{R}(\sigma) = \text{isl}(\mathbf{P}, \mathbf{n}(\sigma), \mathbf{Q}(\sigma), \mathbf{t}(\sigma)) = \mathbf{P} + \frac{(\mathbf{Q} - \mathbf{P}) \cdot \mathbf{n}}{\mathbf{n} \cdot \mathbf{n}} \mathbf{n} \tag{9.36}$$

Now varying the parameter σ on wavefront $\mathbf{w}(\sigma)$ we can determine its value so that $(\mathbf{R}(\sigma) - \mathbf{Q}(\sigma)) \cdot \mathbf{t}(\sigma) = 0$, or that the distance between $\mathbf{R}(\sigma)$ and $\mathbf{Q}(\sigma)$ is zero $[\mathbf{R}(\sigma), \mathbf{Q}(\sigma)] = 0$. This gives us the position of point $\mathbf{Q}$.

In the case of ray r_1 in Figure 9.46, we could now refract it at point $\mathbf{P}_0$ since we know its normal $\mathbf{n}_0$. The optical path length between w_2 and w_3 is S_{23} and we can calculate the optical path length between $\mathbf{P}_0$ and w_3 as $S_{03} = S_{23} - n_1[\mathbf{P}_0, \mathbf{Q}]$.

We now consider the situation in Figure 9.50, in which we have a point $\mathbf{F}$ (in the case of ray r_1 in Figure 9.46, this point would be $\mathbf{P}_0$) emitting a light ray in the direction of the unit vector $\mathbf{v}$ (in the case of ray r_1, this would be

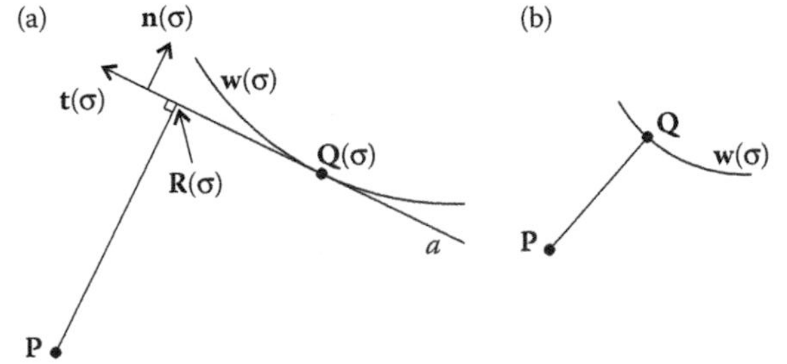

FIGURE 9.49
(a) A possible way to find the ray through point **P** that is perpendicular to a wavefront described by parameterization **w**(σ). (b) Point **Q** on **w**(σ) for which this condition is met: the normal to **w**(σ) at point **Q** crosses **P**.

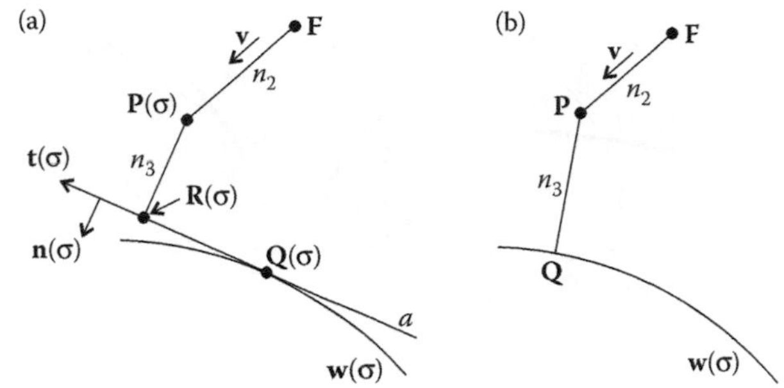

FIGURE 9.50
(a) A possible way to find the ray path from a point **F** to a wavefront parameterized by **w**(σ), given the direction **v** of the ray at **F** and the optical path length *S* between **F** and **w**(σ). (b) Position of point **Q** after solving [**R**(σ),**Q**(σ)] = 0 for parameter σ.

the direction of the ray after refraction at $\mathbf{P}_0$), and we know the optical path length *S* between **F** and wavefront **w**(σ), which is a curve with parameter σ (in the case of ray r_1, this wavefront would be w_3). The refractive index before deflection is n_2 and is n_3 after deflection (in the case of ray r_1, this deflection occurs at point $\mathbf{P}_1$). We want to determine the path of this light ray. Note that this situation is also similar to that in Figure 9.2 in which we want to calculate point **P** on curve *c*.

Let us now consider a possible way to determine the path of this light ray. Take a point **Q**(σ) on the wavefront and determine its tangent *a* at that point. It has unit normal vector **n**(σ) and unit tangent vector **t**(σ). For a straight line *a*, we can determine a point **P**(σ) that matches the optical path length between **F** and *a* as (see Chapter 21),

$$\mathbf{P}(\sigma) = \text{coptsl}(\mathbf{F}, n_2, \mathbf{v}, \mathbf{Q}(\sigma), n_3, \mathbf{n}(\sigma), S) = \mathbf{F} + \frac{S - n_3(\mathbf{Q} - \mathbf{F}) \cdot \mathbf{n}}{n_2 - n_3 \mathbf{v} \cdot \mathbf{n}} \mathbf{v} \tag{9.37}$$

Point **R**(σ) is given by an intersection of the straight lines defined by **Q**(σ) and **t**(σ) and by **P**(σ), and **n**(σ) as (see Chapter 21),

$$\mathbf{R}(\sigma) = \text{isl}(\mathbf{P}(\sigma), \mathbf{n}(\sigma), \mathbf{Q}(\sigma), \mathbf{t}(\sigma)) = \mathbf{P} + \frac{(\mathbf{Q} - \mathbf{P}) \cdot \mathbf{n}}{\mathbf{n} \cdot \mathbf{n}} \mathbf{n} \tag{9.38}$$

Now varying the parameter σ on wavefront **w**(σ) we can determine its value so that $(\mathbf{R}(\sigma) - \mathbf{Q}(\sigma)) \cdot \mathbf{t}(\sigma) = 0$, or that the distance between **R**(σ) and **Q**(σ) is zero [**R**(σ), **Q**(σ)] = 0. This gives us the position of point **Q** as shown in Figure 9.50b.

The tangent vector **t**(σ) to curve **w**(σ) is given by

$$\mathbf{t}(\sigma) = \frac{d\mathbf{w}(\sigma)/d\sigma}{\| d\mathbf{w}(\sigma)/d\sigma \|} \tag{9.39}$$

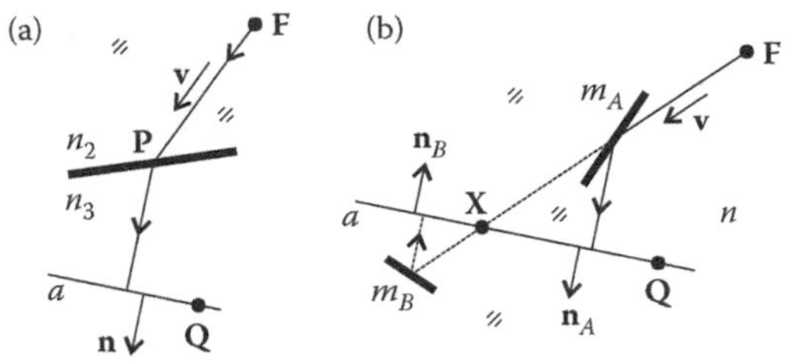

FIGURE 9.51
Determining the right normal direction to straight line a for the case of refraction (a) and reflection (b).

and the normal $\mathbf{n}(\sigma)$ to a is obtained by rotating this vector by either $\pi/2$ or $-\pi/2$.

For $\mathbf{P}(\sigma)$ to be calculated properly with expression (9.37), the normal to straight line a must point in the direction of the light ray as it crosses it. In the case of refraction, this means that the normal $\mathbf{n}$ to a must fulfill $\mathbf{v}\cdot\mathbf{n} > 0$, as in Figure 9.51a. If it does not, its direction must be inverted.

Note that this construction is also valid in the case in which $n_2 = n_3 = n$. In this case, as in Figure 9.50b, the light ray would reflect at point $\mathbf{P}$ in a direction perpendicular to the wavefront parameterized by $\mathbf{w}(\sigma)$. To determine the right normal direction to a in this case, we may start by calculating point $\mathbf{X}$ at the intersection of the straight line a and the light ray defined by point $\mathbf{F}$ and direction $\mathbf{v}$, as shown in Figure 9.51b. We calculate the optical path length between $\mathbf{F}$ and $\mathbf{X}$ as $S_{fX} = n[\mathbf{F}, \mathbf{X}]$. If $S_{fX} < S$, then the reflection is on a mirror m_B further away than point $\mathbf{X}$ and we have normal $\mathbf{n}_B$ to a, which fulfills $\mathbf{v}\cdot\mathbf{n}_B < 0$. In this case, point $\mathbf{P}$ would be on m_B. However, if $S_{fX} > S$, then the reflection is on a mirror m_A closer than point $\mathbf{X}$, and we have normal $\mathbf{n}_A$ to a, which fulfills $\mathbf{v}\cdot\mathbf{n}_A > 0$. In this case, point $\mathbf{P}$ would be on m_A.

Unless we already know that we are choosing the right normal direction to curve $\mathbf{w}(\sigma)$, we should verify it before iterating in parameter σ.

9.6 The RXI Optic: Iterative Calculation

The optical surfaces of an RXI look similar to those of an XX, but it can be implemented as a single piece, instead of two pieces with an air gap between them. The RXI is a compact concentrator (or collimator) made of a material with a refractive index n. If used as a concentrator, light refracts at the top surface s_1, then is reflected at the (mirrored) bottom surface s_2, and again undergoes TIR at the top surface s_1 and from there redirected to the receiver **AB**, which is immersed in the medium of refractive index n.[12,10] Its name comes from this path of the light, along which there are deflections by a

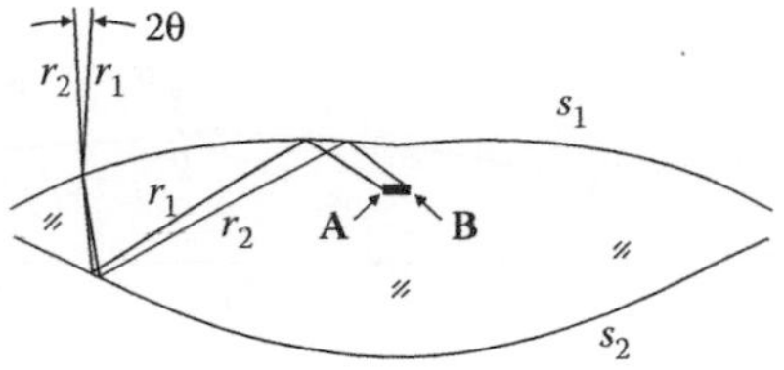

FIGURE 9.52
Paths of two edge rays inside an RXI.

refraction (R), reflection (X), and TIR (I). The center of the top surface s_1 is mirror-coated so that TIR fails in that surface portion. Figure 9.52 shows two light rays r_1 and r_2 and their paths inside an RXI.

The ray r_1 makes an angle $-\theta$ to the vertical before entering the CPC. It is redirected to left-hand-side edge **A** of the receiver. The ray r_2 makes an angle $+\theta$ to the vertical before entering the CPC. It is redirected to right-hand-side edge **B** of the receiver. Figure 9.53 shows the two input flat wavefronts w_1 and w_2. The rays perpendicular to these wavefronts are concentrated on to edges **A** and **B** of the receiver, respectively.

The design process starts with the definition of the receiver size. For example, it is centered at the origin and has a length of 2 units so that $\mathbf{A} = (-1, 0)$ and $\mathbf{B} = (1, 0)$. We choose also the refractive index n of the optic and its half-acceptance angle θ. With these values, we can calculate the width of the RXI, which is given by conservation of étendue as

$$w_{RXI} = n[\mathbf{A},\mathbf{B}]/\sin\theta \tag{9.40}$$

Now choose a top curve s_1. With it we will calculate the bottom curve s_2 and then recalculate the top curve s_1. With the new top curve s_1, calculate the bottom curve s_2 and recalculate again the top curve s_1. This process goes on until the variation in the curves from one iteration to the other is small enough. Figure 9.54 shows the initial top curve used to generate the RXI in Figure 9.53.

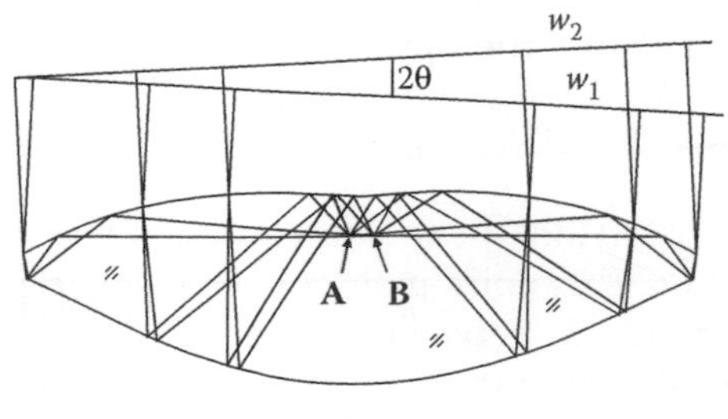

FIGURE 9.53
The rays perpendicular to wavefront w_1 are concentrated to point **A** and those perpendicular to wavefront w_2 are concentrated on to point **B**.

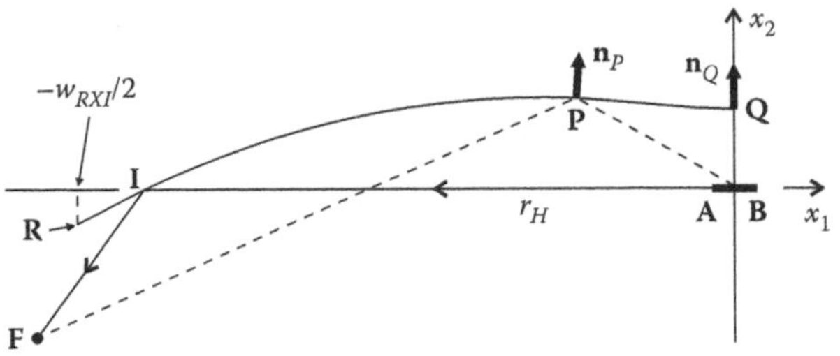

FIGURE 9.54
A possible initial top curve for calculating an RXI.

In this particular case, we may first choose the position of point **R**. It must be of the form **R** = $(-w_{RXI}/2, -y)$ where y is chosen. Then choose point **I** on the horizontal axis x_1. There is no special rule for choosing portion **RI** of the upper curve, so in this example it is a straight line. We can now reflect off point **I** a horizontal ray, r_H, coming from the receiver **AB** and choose a point **F** on the reflected ray. We now choose portion **IP** of the top curve as an ellipse with foci **F** and (0,0), which is the midpoint of the receiver **AB**. Portion **PQ** is defined as a third-degree polynomial through points **P** and **Q** (on the vertical axis x_2) that has normal $\mathbf{n}_P$ at point **P** (the normal to the ellipse at that point) and normal $\mathbf{n}_Q$ (vertical) at point **Q**.

Having a possible top curve, calculate the bottom curve. The fist step is to calculate its left-hand-side end point **X**. Refract at point **R** a ray r_1 making an angle $-\theta$ to the vertical, as shown in Figure 9.55. Intersect the refracted ray r_1 with the reflection of ray r_H at point **I** and obtain point **X**, the first point of the bottom surface.

Now we can calculate the first iteration of the bottom surface s_2. First define the position of wavefront w_1. We may do that, for example, by choosing a point $\mathbf{W}_1$ along ray r_1. Since we know the path of ray r_1, which is $\mathbf{W}_1$-**R**-**X**-**I**-**A**, we can calculate the optical path length S_1 between w_1 and edge **A** of the receiver, as shown in Figure 9.56.

Take a point $\mathbf{P}_3$ on the top surface s_1 and determine the position $\mathbf{W}_3$ on wavefront w_1 of the ray through $\mathbf{P}_3$, in this case perpendicular to w_1 through $\mathbf{P}_3$. Now determine the optical path length between $\mathbf{P}_3$ and **A** as

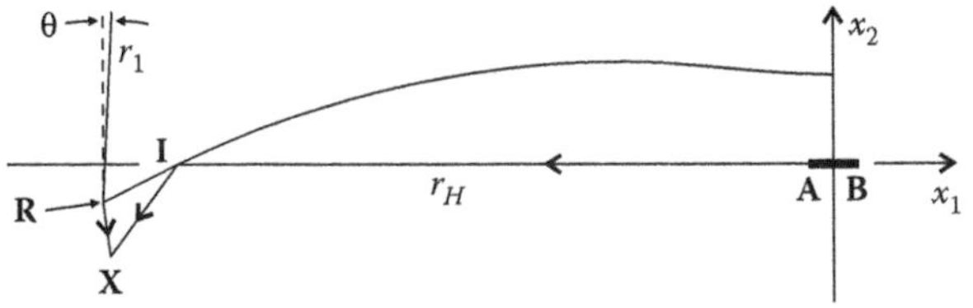

FIGURE 9.55
Calculation of the first point **X** of the bottom surface.

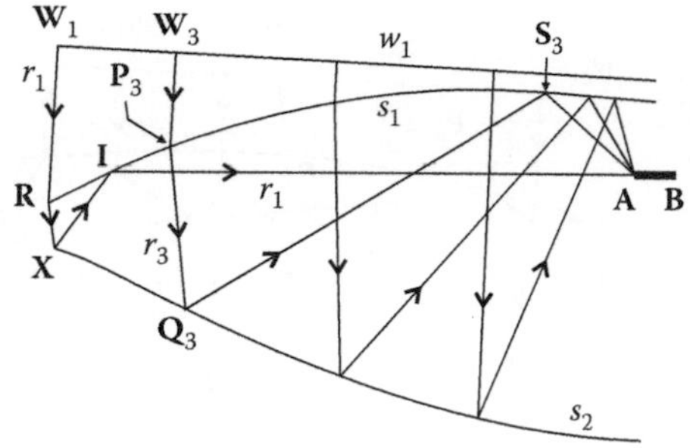

FIGURE 9.56
Calculation of the bottom curve of the RXI.

$S_{P3} = S_1 - [\mathbf{W}_3, \mathbf{P}_3]$. then determine point $\mathbf{Q}_3$ on the bottom surface s_2, such that ray r_3 from $\mathbf{W}_3$ refracts at $\mathbf{P}_3$, reflects at $\mathbf{Q}_3$, then reflects at some point $\mathbf{S}_3$ on the top surface s_1. From there it is redirected to the edge **A** of the receiver. Repeating this process for a set of points on s_1, we can determine a set of points and normals for s_2.

With this new bottom surface s_2, we can recalculate the top surface s_1. First refract at point **R** a set of rays contained between the edge rays r_1 and r_2. We concentrate these rays to edge **B** of the receiver. We now consider that ray r_1, instead of ending at point **A**, continues to point **B**. Since we know the path of ray r_1, which is now **R-X-I-B**, we can determine the optical path length between **R** and **B**, as shown in Figure 9.57.

The rays refracted at point **R** are reflected off the bottom surface s_2 and, using the constant optical path length, we can determine a new portion **IJ** of the top surface that concentrates these rays to the edge **B** of the receiver.

Define the position of the wavefront w_2, as shown in Figure 9.58. Since we know the path of ray r_2 (from the step in Figure 9.57), which is now

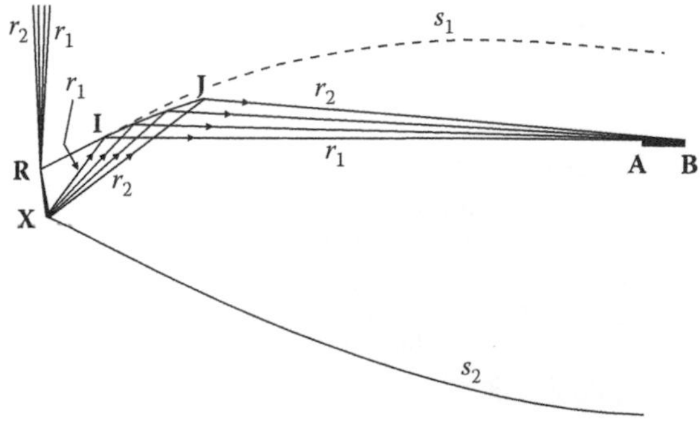

FIGURE 9.57
Refracting at point **R** of the top surface a set of rays between the directions of r_1 and r_2 enables the recalculation of the part of the top surface between points **I** and **J**.

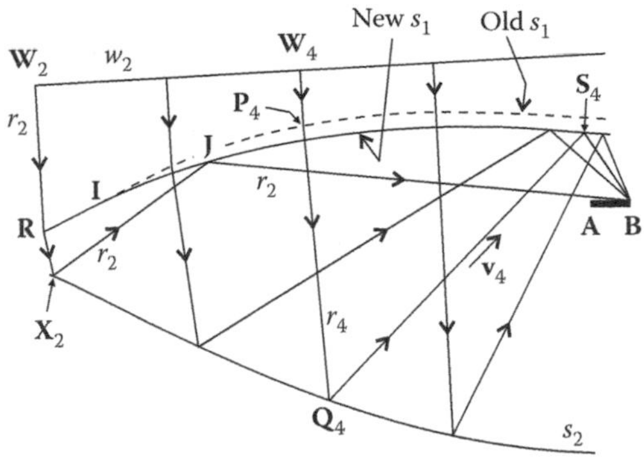

FIGURE 9.58
Refracting the rays perpendicular to wavefront w_2 on the old top surface s_1 and reflecting them on the bottom surface s_2, we can recalculate the top surface s_1 to the right of point **J**.

$\mathbf{W}_2$-$\mathbf{R}$-$\mathbf{X}_2$-$\mathbf{J}$-$\mathbf{B}$, we can determine the optical path length S_2 between the wavefront w_2 and the edge **B** of the receiver.

Determine the path $\mathbf{W}_4$-$\mathbf{P}_4$-$\mathbf{Q}_4$ of a ray r_4 from wavefront w_2 to a point $\mathbf{Q}_4$ on the bottom surface, as shown in Figure 9.58.

The optical path length between $\mathbf{Q}_4$ and **B** is now $S_{Q4} = S_2 - [\mathbf{W}_4, \mathbf{P}_4] - n[\mathbf{P}_4, \mathbf{Q}_4$. Since we know the direction $\mathbf{v}_4$ or ray r_4 after reflection at point $\mathbf{Q}_4$, we can determine the position of point $\mathbf{S}_4$ on the new top surface s_1. Repeating this process for a set of points on s_2, we can determine a new set of points and normals for s_1.

With this new top surface s_1, we repeat the process: calculate a new bottom surface s_2 and recalculate the top surface s_1. The process ends when the latest s_1 is close enough to the previous s_1.

Just as in the case of the XX, in the case of the RXI also the central portion of the top surface cannot reflect light by TIR to the receiver. Figure 9.59 shows this central portion, which has to be mirrored. In Figure 9.59, α_C is the critical angle inside the material of the RXI. This central mirror on the top surface completes the design of the RXI.

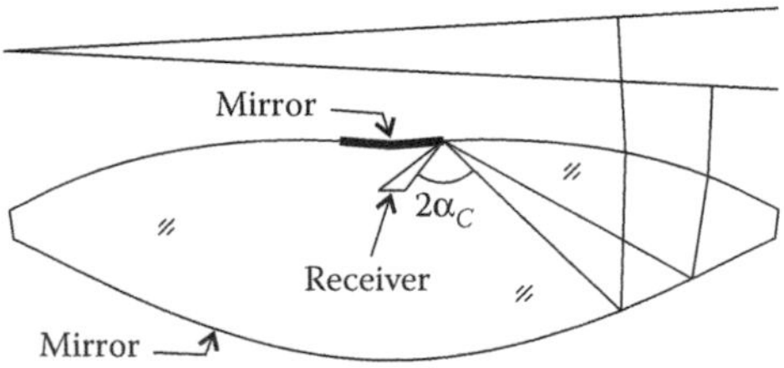

FIGURE 9.59
The central portion of the top surface of the RXI needs to be mirrored because light reflected by the bottom mirror should not undergo TIR there.

There are several ways to calculate the points of the bottom and top surfaces. We now consider some possible ways to do it. In the first iteration we define the top curve, but in the following iterations it is defined by a set of points and normals. These points and normals may, for example, be interpolated by a piecewise curve. Between each pair of points we may interpolate a third-degree polynomial (similar to what we did in Figure 9.5), which completely defines the top curve s_1.

Now, for a given point $\mathbf{P}_3$ on the top surface s_1, we know the direction of the refracted ray r_3 and the optical path length between $\mathbf{P}_3$ and $\mathbf{A}$. We have, therefore, the situation as in Figure 9.60, which shows a ray r_3 with a geometry similar to that of ray r_3 in Figure 9.56 after refraction at point $\mathbf{P}_3$.

A light ray emitted from a point $\mathbf{P}_3$ in a direction $\mathbf{d}$ (the direction of the refracted ray r_3 at point $\mathbf{P}_3$) reflects off an unknown point $\mathbf{Q}_3$ toward a point $\mathbf{S}_3$ on a given curve $\mathbf{s}(\sigma)$ (in our case s_1), and from there it is reflected to a point $\mathbf{A}$. We know the optical path length S_{P3} between $\mathbf{P}_3$ and $\mathbf{A}$. Curve $\mathbf{s}(\sigma)$ is described as a function of the parameter σ. The ray paths are inside in a medium of refractive index n.

To calculate the position of point $\mathbf{Q}_3$, first define a straight line w through point $\mathbf{P}_3$ and perpendicular to $\mathbf{d}$. The tangent to w is given by unit vector $\mathbf{t}$ and its normal by $\mathbf{n} = -\mathbf{d}$.

We now choose a point $\mathbf{S}(\sigma)$ on the curve $\mathbf{s}(\sigma)$. Reflect off $\mathbf{S}(\sigma)$ the light ray coming from $\mathbf{A}$. After reflection, this ray has the direction of unit vector $\mathbf{v}(\sigma)$. The optical path length between $\mathbf{S}(\sigma)$ and w is

$$S_{Sw}(\sigma) = S_{P3} - n[\mathbf{A}, \mathbf{S}(\sigma)] \tag{9.41}$$

The situation now simplifies to a ray r emitted from a point $\mathbf{S}(\sigma)$ with direction $\mathbf{v}(\sigma)$ that we want to reflect off a point $\mathbf{Q}(\sigma)$ in a direction perpendicular

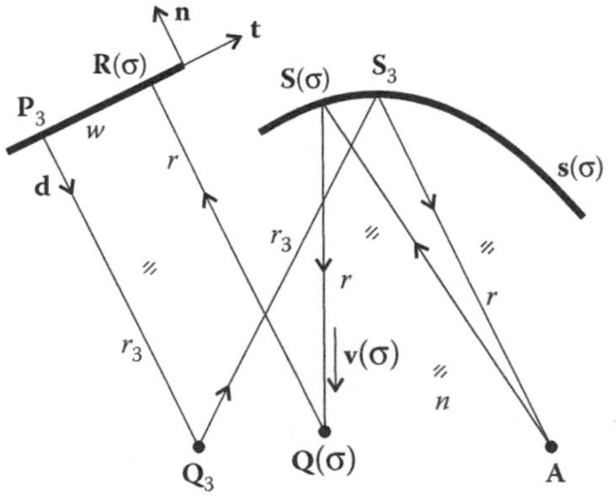

FIGURE 9.60
A possible method to calculate the points of the bottom surface of the RXI.

to w, knowing the optical path length S_{sw} between $\mathbf{S}(\sigma)$ and w. The analytical solution is (see Chapter 21)

$$\mathbf{Q}(\sigma) = \text{coptsl}(\mathbf{S}(\sigma), n, \mathbf{v}(\sigma), \mathbf{P}_3, n, \mathbf{n}, S_{Sw}(\sigma)) = \mathbf{S} + \frac{S_{Sw} - n(\mathbf{P}_3 - \mathbf{S}) \cdot \mathbf{n}}{n - n\mathbf{v} \cdot \mathbf{n}} \mathbf{v} \quad (9.42)$$

Point $\mathbf{R}(\sigma)$ on w is on the perpendicular to w through $\mathbf{Q}(\sigma)$. Varying the parameter σ on the curve $\mathbf{s}(\sigma)$ determines its value, so that $(\mathbf{R}(\sigma) - \mathbf{P}_3) \cdot \mathbf{t} = 0$, namely, that the distance between $\mathbf{R}(\sigma)$ and $\mathbf{P}_3$ is zero $[\mathbf{R}(\sigma), \mathbf{P}_3] = 0$. When this happens, $\mathbf{Q}(\sigma)$ converges to the point $\mathbf{Q}_3$ to be determined.

For the calculation of the new top surface, and referring back to Figure 9.58, we may choose a point $\mathbf{P}_4$ on the existing top surface, calculate the corresponding point $\mathbf{W}_4$ on wavefront w_2, refract the ray at $\mathbf{P}_4$, calculate the intersection point $\mathbf{Q}_4$ with the interpolated bottom surface s_2, and reflect it there. Then the calculation of point $\mathbf{S}_4$ uses the function $S_4 = \text{coptpt}(\mathbf{Q}_4, \mathbf{v}_4, \mathbf{B}, n, S_{Q4})$ as defined in Chapter 21. This same function can also be used to calculate the points of the top surface between **I** and **J**.

Another possible way to calculate points $\mathbf{S}_4$ on the top surface is to use the points and normal we have for the bottom surface s_2. In this case we do not need to interpolate the whole bottom surface, but only a small portion between points **X** and $\mathbf{X}_2$, which is used to calculate portion **IJ** of the top surface s_2. In this alternative way, we take a point $\mathbf{Q}_4$ that we previously calculated for the bottom surface s_2. Now we choose a point $\mathbf{P}(\sigma)$ on the top curve, parameterized as $\mathbf{s}(\sigma)$. The normal to $\mathbf{s}(\sigma)$ at point $\mathbf{P}(\sigma)$ is $\mathbf{n}(\sigma)$, as shown in Figure 9.61.

We refract at $\mathbf{P}(\sigma)$ the ray coming from $\mathbf{Q}_4$. We vary the parameter σ until the refracted ray at $\mathbf{P}(\sigma)$ makes an angle $\alpha(\sigma) = \pi/2 + \theta$ to the horizontal, fixing

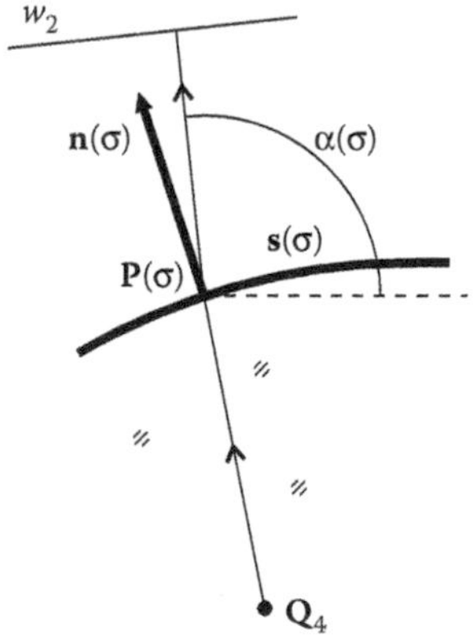

FIGURE 9.61
Possible way to calculate the path of a light ray from wavefront w_2 to a given point $\mathbf{Q}_4$ on the bottom surface.

the position of point $\mathbf{P}_4$. Point $\mathbf{W}_4$ is on the perpendicular to w_2 through $\mathbf{P}_4$, as shown in Figure 9.58. The calculation of $\mathbf{S}_4$ in this case is as described earlier.

9.7 The RXI Optic: Direct Calculation

It is also possible to calculate an RXI optic with a direct method, without iterations in the calculation of the optical surfaces. The design may also be modified so that the emitter does not have to be immersed in a medium of refractive index n, but may be in air.[13]

Start with a receiver $\mathbf{R}_1\mathbf{R}_2$, a virtual entrance aperture $\mathbf{QQ}^*$ and the acceptance angle 2θ, as shown in Figure 9.62. The distance $[\mathbf{Q},\mathbf{Q}^*]$ is obtained from the conservation of étendue $2[\mathbf{Q},\mathbf{Q}^*]\sin\theta = 2[\mathbf{R}_1,\mathbf{R}_2]$. The RXI concentrator will capture the light headed to the virtual aperture $\mathbf{QQ}^*$ inside the acceptance angle 2θ, defined by input wavefronts w_1 and w_2, as in Figure 9.63.

Choose a point $\mathbf{A}$ along direction $\mathbf{v}_2$ perpendicular to wavefront w_2 through point $\mathbf{Q}$. Now choose also the position of point $\mathbf{C}$ and choose a shape for the top surface of the optic between $\mathbf{A}$ and $\mathbf{C}$, for example, a straight line or a low order polynomial (preferably with no inflection points).

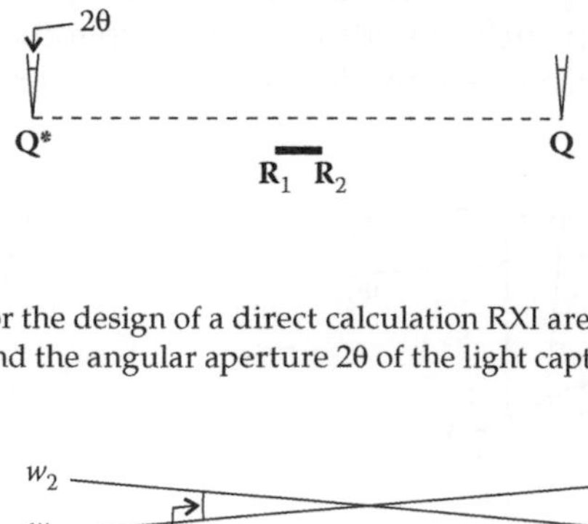

FIGURE 9.62
The starting conditions for the design of a direct calculation RXI are the receiver $\mathbf{R}_1\mathbf{R}_2$, virtual entrance aperture $\mathbf{QQ}^*$, and the angular aperture 2θ of the light captured.

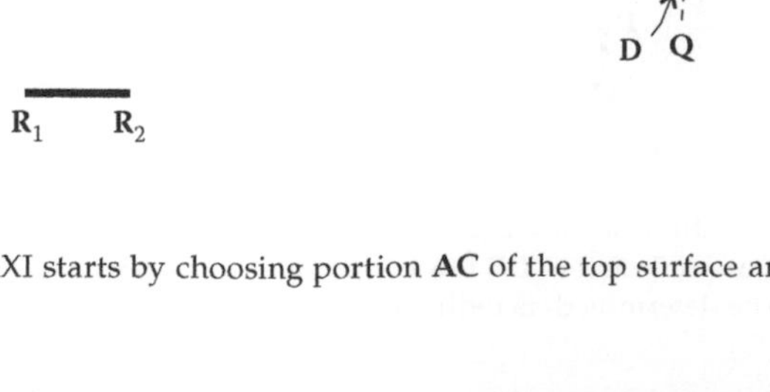

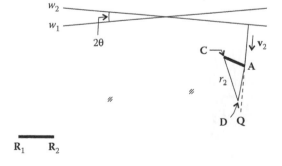

FIGURE 9.63
The design of the RXI starts by choosing portion $\mathbf{AC}$ of the top surface and a point $\mathbf{D}$ on the bottom surface.

We may now calculate the shape of a small portion **DE** of the bottom surface, as shown in Figure 9.64a. The cone of rays with angular aperture 2θ contained between r_2 (perpendicular to w_2) and r_1 (perpendicular to w_1) and headed to point **Q** is refracted at portion **AB** of the top surface **AC** and then reflected by the bottom surface **DE**, being focused to point **C**. In order to determine the shape of **DE**, refract ray r_2 at point **A** and choose a point **D** along it. This will be the starting point of surface **DE**. Now calculate the optical path length between **C** and **Q** as $S_{CQ} = n$[**C**,**D**] + n[**D**,**A**] – [**A**,**Q**]. Note that **Q** is in air ($n = 1$) and is a virtual point from which rays appear to diverge (hence the negative sign in the last term of S_{CQ}, see Chapter 21). Surface **DE** may now be calculated as a Cartesian oval with focus **C** and virtual focus **Q**. Take, for example, the path of a ray headed to **Q** with direction **u**, as in Figure 9.64b. It crosses the top surface at point **I**, is reflected by the bottom surface at point **J** (to be determined) from which it is redirected to point **C**. The optical path length between **C** and **Q** for this ray is n[**C**,**J**] + n[**J**,**I**] – [**I**,**Q**] = S_{CQ} which can also be written as $S_I = S_{CQ}$ + [**I**,**Q**] = n[**C**,**J**] + n[**J**,**I**], where S_I is the optical path length for portion **I-J-C** of the ray. Now, we may refract direction **u** at point **I** of the top surface and obtain direction **v** of the refracted ray. Point **J** on the bottom surface may now be obtained from **J** = ccoptpt(**I**,n,**v**,**C**,n,S_I), where n is the refractive index of the optic and function ccoptpt(…) is defined at the end of Chapter 21. Using the same procedure and moving point **I** along **AB** we obtain the complete portion **DE** of the bottom surface. Point **B** is at the intersection of the top surface with the line perpendicular to wavefront w_1 through point **Q**. The cone of light defined by **DE** and reflected at **C** is bound

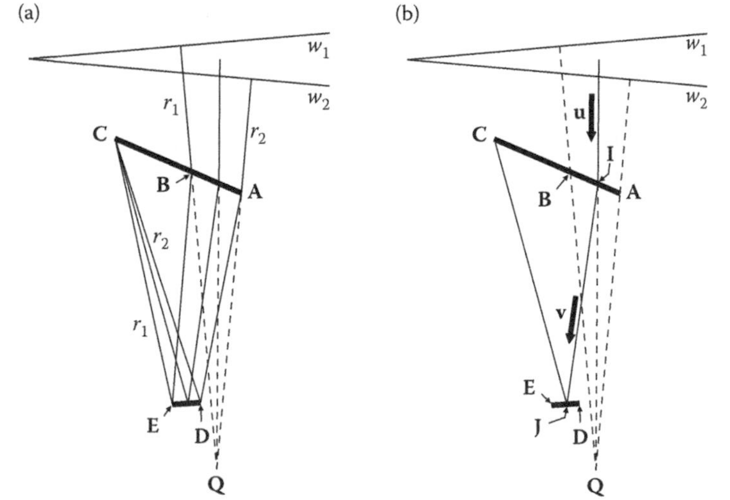

FIGURE 9.64
(a) Portion **DE** on the bottom surface is calculated such that rays headed to (virtual) point **Q** and refracted at **AB**, are reflected at **DE** toward point **C**. (b) A ray headed to **Q**, refracted at **I** and reflected at **J** (to be determined) is redirected to **C**.

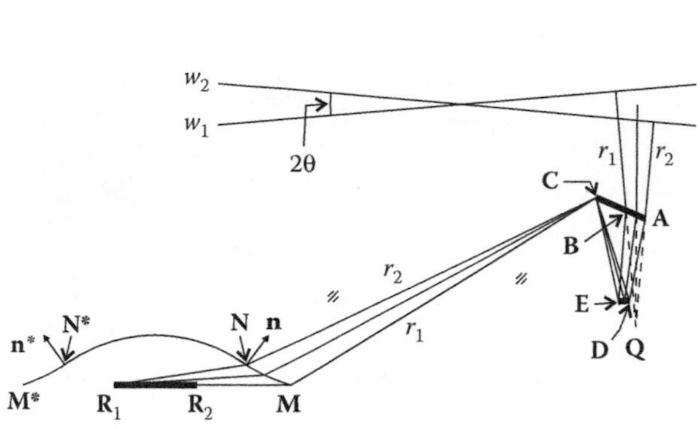

FIGURE 9.65
The cone of rays contained between rays r_1 and r_2 and reflected at point **C** is redirected to $\mathbf{R}_1$ by Cartesian oval **MN**. Curve **MN** and its symmetrical **M*N*** defines the edges of the receiver cavity. Choosing the shape of curve **NN*** completes the design of the cavity.

by rays r_1 and r_2, as shown in Figure 9.65. Now, a portion **MN** of the receiver cavity may be determined in such a way that the cone of rays reflected at **C** are redirected to edge $\mathbf{R}_1$ of the receiver. Surface **MN** may be calculated as a Cartesian oval by constant optical path length between **C** and $\mathbf{R}_1$. In particular, this determines the position of point **N** and its normal **n** and also of its symmetric **N*** and corresponding normal **n***. We may now choose the shape of a curve between **N** and **N*** matching the normals **n** and **n*** at the edges. This completes the shape of the cavity **M-N-N*-M*** over the receiver $\mathbf{R}_1\mathbf{R}_2$.

Now that we have the shape of the cavity, we may propagate rays from $\mathbf{R}_1$ through it and determine the shape of wavefront w_3 by constant optical path length from $\mathbf{R}_1$ to w_3, as shown in Figure 9.66. Wavefront w_4 results from the propagation of rays from $\mathbf{R}_2$ through the cavity and is symmetrical to w_3 (if we use the same optical path length to calculate it).

Now consider the ray r_2 through points $\mathbf{W}_1$-**A-D-C**-$\mathbf{W}_3$ whose path is known, as shown in Figure 9.67. One may determine the optical path length S_{23} for this ray from w_2 to w_3. Reflecting now different rays on **DE** coming

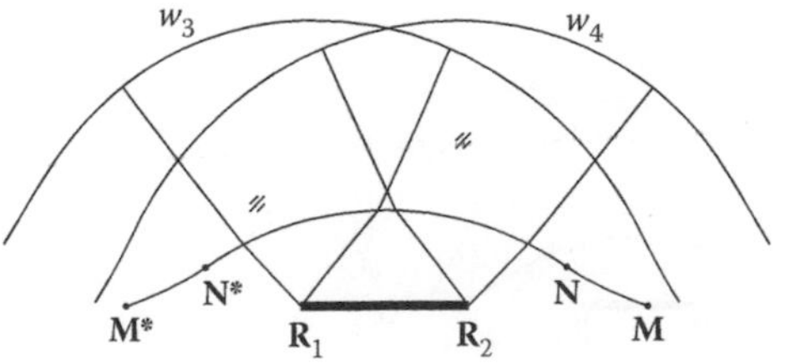

FIGURE 9.66
Wavefronts w_3 and w_4 are the propagation of rays from $\mathbf{R}_1$ and $\mathbf{R}_2$ respectively through the cavity.

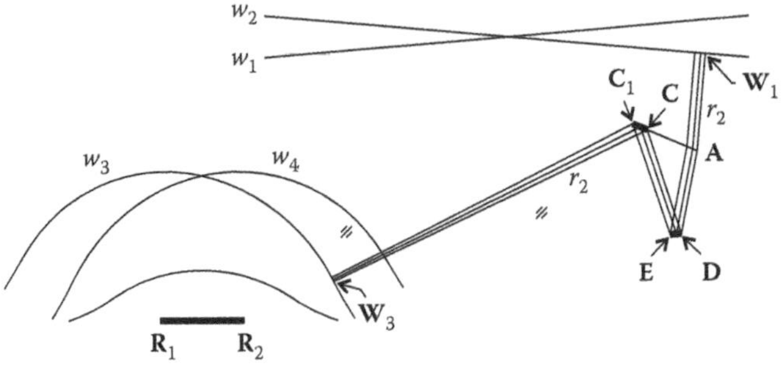

FIGURE 9.67
Reflecting rays on **DE** coming from w_2 and enforcing constant optical path length S_{23} from w_2 to w_3, we may calculate a new portion $\mathbf{CC_1}$ of the top surface.

from w_2 and enforcing constant optical path length S_{23}, one may determine a new portion $\mathbf{CC_1}$ on the top surface.

Now consider the ray r_1 through points $\mathbf{W_2}$-**B**-**E**-**C**-$\mathbf{W_2}$ whose path is known, as shown in Figure 9.68. One may determine the optical path length S_{14} for this ray from w_1 to w_4. Reflecting now different rays on $\mathbf{CC_1}$ coming from w_4 and enforcing constant optical path length S_{14}, one may determine a new portion $\mathbf{EE_1}$ on the bottom surface.

Now, we may reflect a set of rays on $\mathbf{EE_1}$ coming from w_2, and calculate a new portion $\mathbf{C_1C_2}$ of the top surface by enforcing constant optical path length S_{23} from w_2 to w_3, as show in Figure 9.69. The SMS calculation continues as we calculate new portions of top surface s_1 and bottom surface s_2, as shown in Figure 9.69.

Figure 9.70 shows a complete RXI optic. The central portion m_T of the top surface ending at point **M** needs to be mirrored, because TIR fails in that region. The edge ray r_3 from $\mathbf{R_2}$ reaches point **M** at the critical angle α_C to its

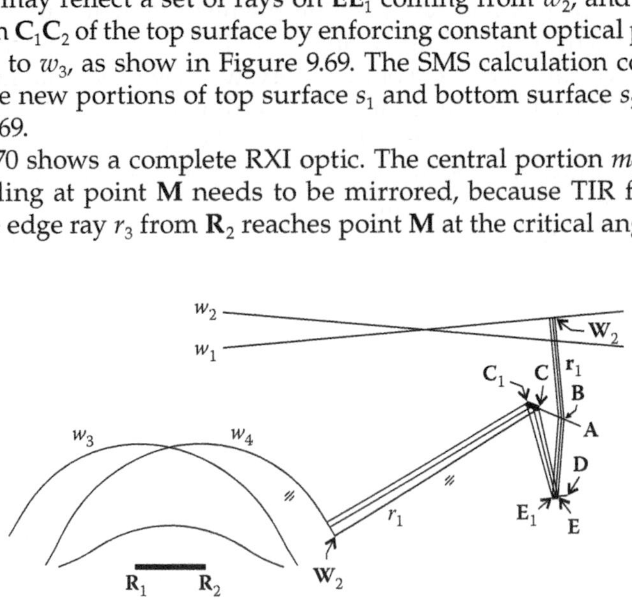

FIGURE 9.68
Reflecting rays on $\mathbf{CC_1}$ coming from w_4 and enforcing a constant optical path length S_{14} from w_4 to w_1, we may calculate a new portion $\mathbf{EE_1}$ of the bottom surface.

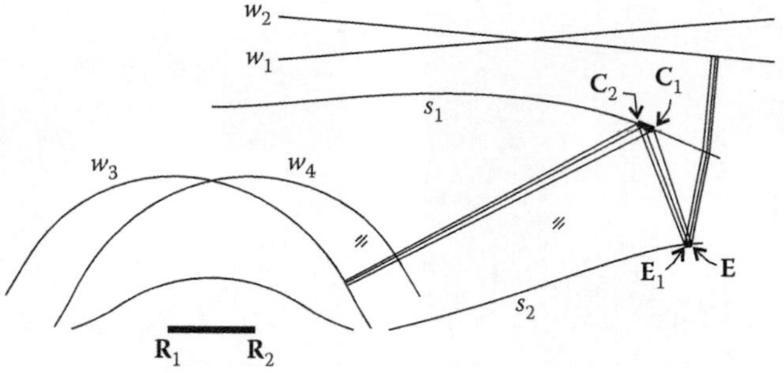

FIGURE 9.69
A set of rays from w_2 reflected at $\mathbf{EE}_1$ determines a new portion $\mathbf{C}_1\mathbf{C}_2$ on the top surface by imposing a constant optical path length S_{23} between w_2 and w_3.

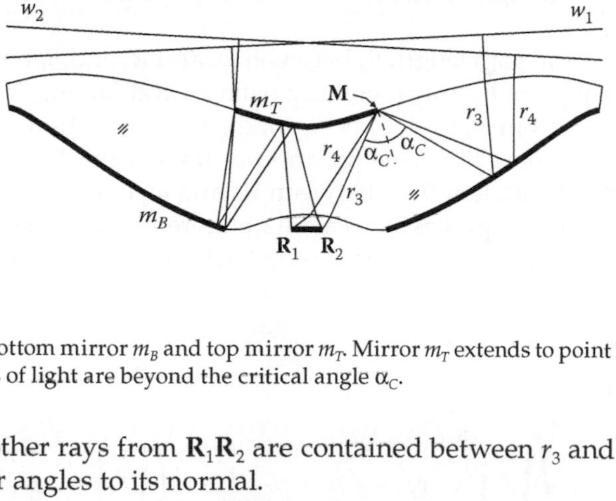

FIGURE 9.70
RXI optic with bottom mirror m_B and top mirror m_T. Mirror m_T extends to point **M** where all the incidence angles of light are beyond the critical angle α_C.

normal. All other rays from $\mathbf{R}_1\mathbf{R}_2$ are contained between r_3 and r_4 and reach **M** with wider angles to its normal.

The back surface m_B of the RXI is also mirrored since light cannot be reflected there by TIR.

The RXI can also be used as a collimator. Where we had the receiver now we have an emitter, and the direction of the light inside the optic is reversed.

Figure 9.71 shows a flashlight based on an RXI optic. Figure 9.71b shows the RXI optic separated from the remaining assembly.

9.8 SMS Optical Path Length Adjustment

In Figure 9.72 we have a 2-D configuration with two given input asymmetric wavefronts w_1 and w_2 immersed in a medium of refractive index n_1 and two asymmetric output wavefronts (points in this example) $\mathbf{R}_1$ and $\mathbf{R}_2$ immersed

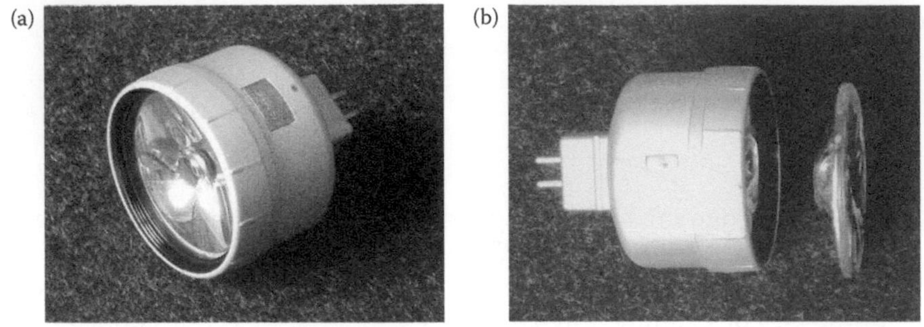

FIGURE 9.71
(a) RXI collimator lamp. (b) The same, but with the RXI optic extracted.

in a medium of refractive index n_3. Also given is point $\mathbf{P}_0$ and its normal $\mathbf{n}_0$. The goal is to design an RR SMS lens made of a material with refractive index n_2.

Choose an optical path length S_A between w_2 and $\mathbf{R}_1$ and S_B between w_1 and $\mathbf{R}_2$. Refracting at point $\mathbf{P}_0$ a ray r_1 coming from w_2 and forcing the optical path length of this ray from w_2 to $\mathbf{R}_1$ to be S_A, one gets the position of point $\mathbf{P}_1$ and its normal. Now, doing the same for another ray r_2 from $\mathbf{R}_2$ through $\mathbf{P}_1$ and forcing an optical path length S_B between $\mathbf{R}_2$ and w_1 one gets point $\mathbf{P}_2$ and its normal. Repeating the process we calculate more points and corresponding normals on the surfaces of the lens. Going back to point $\mathbf{P}_0$ and considering a

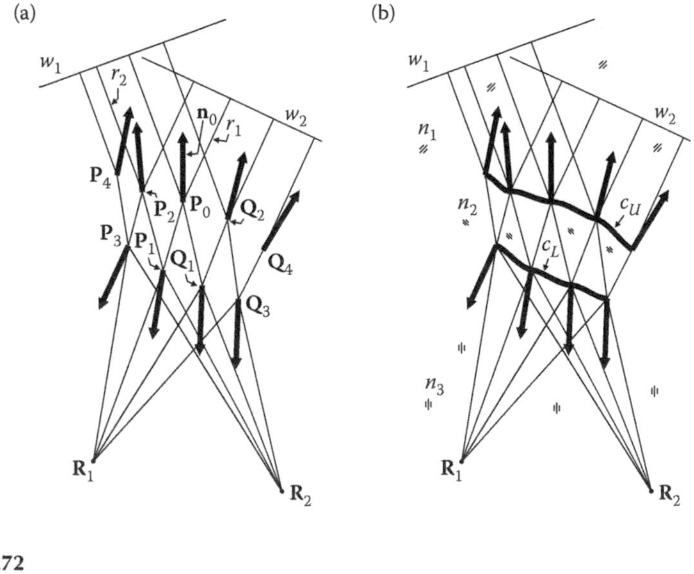

FIGURE 9.72
(a) SMS 2-D chain calculated for asymmetric input wavefronts w_1, w_2 and output (point) wavefronts $\mathbf{R}_1$ and $\mathbf{R}_2$. (b) When interpolating the SMS points and normals, the resulting curves c_U and c_L may have undulations.

ray from w_1 we calculate a new point $\mathbf{Q}_1$. Repeating the process, more points $\mathbf{Q}_2$, $\mathbf{Q}_3$, … are calculated to the right. The result is shown in Figure 9.72a.

If we now interpolate the points and normals just calculated, we may get two curves c_U and c_L as shown in Figure 9.72b. In general, however, these curves have undulations.

Another option is to consider an interpolation c_T of the top points …, $\mathbf{P}_4$, $\mathbf{P}_2$, $\mathbf{P}_0$, $\mathbf{Q}_2$, $\mathbf{Q}_4$, … as in Figure 9.73. This interpolation uses only the points and not the normals calculated by the SMS chain. The result is a top curve c_T clearly smoother than c_U. However, when comparing the normal $\mathbf{n}_0$ from the SMS method at point $\mathbf{P}_0$ with normal $\mathbf{m}_0$ to curve c_T, also at point $\mathbf{P}_0$, one can see that they do not match. The same also happens at the other points, where the SMS normals (solid arrows) point in directions which are clearly different from the normals to curve c_T (dashed arrows).

The same is true for the points of the bottom surface …, $\mathbf{P}_3$, $\mathbf{P}_1$, $\mathbf{Q}_1$, $\mathbf{Q}_3$, … An interpolation curve c_B through these points (SMS normals not considered in the interpolation) is smoother than curve c_L. However, its normals (dashes arrows) point in directions which are clearly different from those calculated by the SMS. An optic built with optical surfaces c_T and c_B will not work.

The mismatch between normal $\mathbf{n}_0$ from the SMS method and normal $\mathbf{m}_0$ to curve c_T may be eliminated by adjusting the value(s) of optical path length(s) S_A or (and) S_B. We may, for example, choose a value for S_A and allow S_B to vary. The angle β between normals $\mathbf{n}_0$ and $\mathbf{m}_0$ will then be a function of S_B, and we may calculate the value of S_B that verifies $\beta(S_B) = 0$. The result is shown in Figure 9.74. Now the normal to curve c_T at point $\mathbf{P}_0$ is the same as in the SMS method.

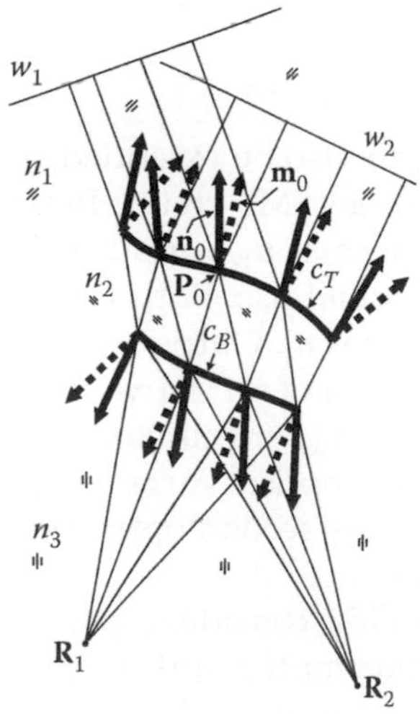

FIGURE 9.73
Interpolating only points and not the normals results in smoother curves c_T and c_B, but their normals to not match those calculated by the SMS method.

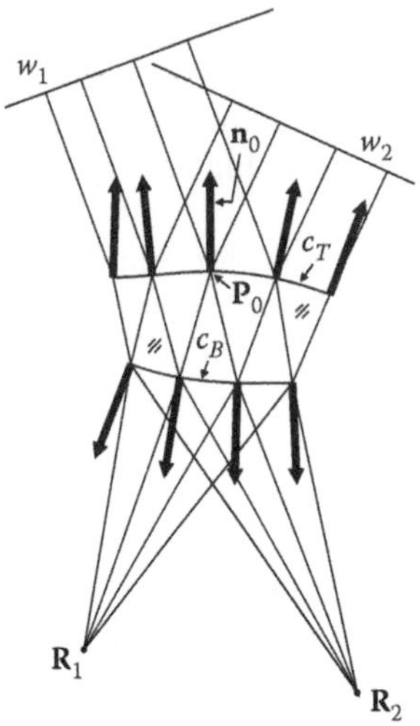

FIGURE 9.74
Adjusting the optical path lengths between w_1 and $\mathbf{R}_2$ and between w_2 and $\mathbf{R}_1$ results in smooth curves c_T and c_B that couple the input and output wavefronts.

This method may be used when calculating asymmetrical optics for adjusting the optical path lengths S_A and S_B. These values of S_A and S_B are then used to calculate the whole SMS optic as shown in Figures 9.46 through 9.48.

9.9 SMS 3-D

The SMS design method can also be extended to 3-D geometry.[14–17] Figure 9.75 shows the first steps of an SMS chain. Input wavefront w_1 is flat and defined by point $\mathbf{W}_1$ and normal $\mathbf{n}_{W1}$. Input wavefront w_2 is also flat and defined by point $\mathbf{W}_2$ and normal $\mathbf{n}_{W2}$. Both input wavefronts are immersed in a medium of refractive index n_1. Output wavefronts $\mathbf{R}_1$ and $\mathbf{R}_2$ are points and are immersed in a medium of refractive index n_3. All points $\mathbf{W}_1$, $\mathbf{W}_2$, $\mathbf{R}_1$ and $\mathbf{R}_2$ and normals $\mathbf{n}_{W1}$ and $\mathbf{n}_{W2}$ are contained in a plane v. The goal is to design an RR optic whose first optical surface separates two media of refractive indices n_1 and n_2 and whose second optical surface separated two media of refractive indices n_2 and n_3.

Referring now to Figure 9.75a, consider a point $\mathbf{P}_0$ with its normal $\mathbf{n}_0$, both in plane v, and choose a value for the optical path length S_A between w_2 and $\mathbf{R}_1$. Take the ray r_{11} perpendicular to w_2 through point $\mathbf{P}_0$. It comes from point $\mathbf{W}_0$ on w_2 given by

$$\mathbf{W}_0 = \text{islp}(\mathbf{P}_0, \mathbf{n}_{W2}, \mathbf{W}_2, \mathbf{n}_{w2}) \tag{9.43}$$

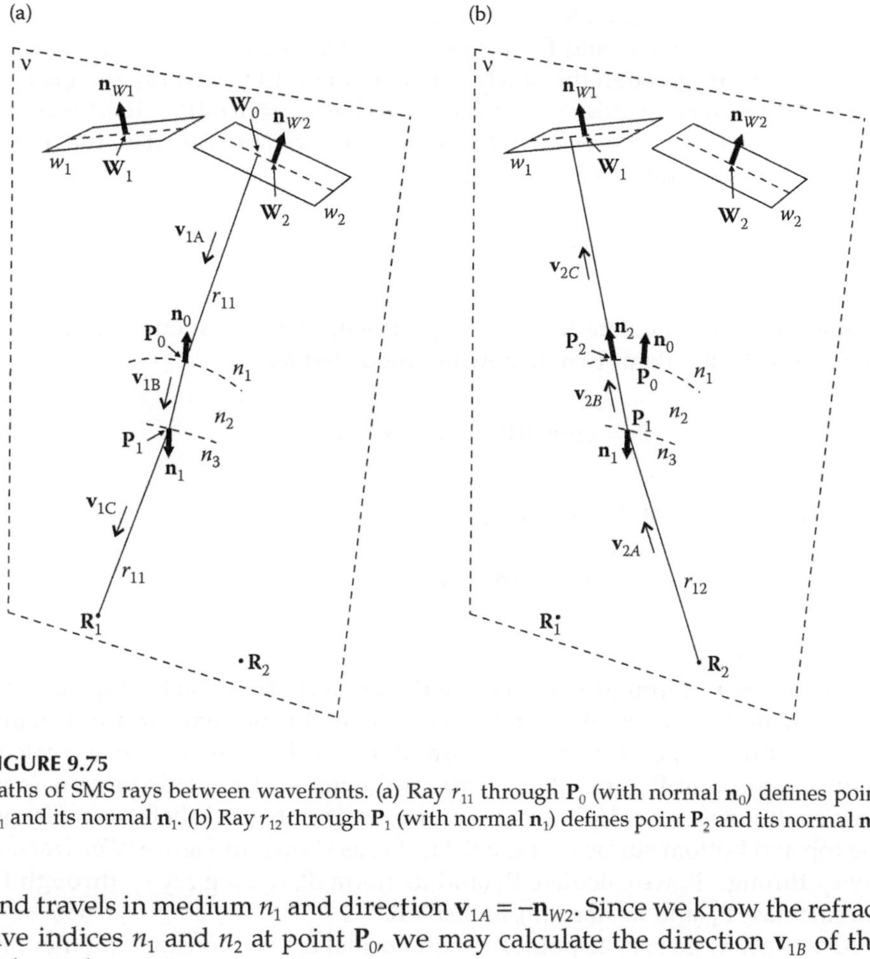

FIGURE 9.75
Paths of SMS rays between wavefronts. (a) Ray r_{11} through $\mathbf{P}_0$ (with normal $\mathbf{n}_0$) defines point $\mathbf{P}_1$ and its normal $\mathbf{n}_1$. (b) Ray r_{12} through $\mathbf{P}_1$ (with normal $\mathbf{n}_1$) defines point $\mathbf{P}_2$ and its normal $\mathbf{n}_2$.

and travels in medium n_1 and direction $\mathbf{v}_{1A} = -\mathbf{n}_{W2}$. Since we know the refractive indices n_1 and n_2 at point $\mathbf{P}_0$, we may calculate the direction $\mathbf{v}_{1B}$ of the refracted ray into n_2 as

$$\mathbf{v}_{1B} = \text{rfr}(\mathbf{v}_{1A}, \mathbf{n}_0, n_1, n_2) \tag{9.44}$$

Also, we may calculate the optical path length S_1 between $\mathbf{P}_0$ and $\mathbf{R}_1$ as $S_1 = S_A - n_1[\mathbf{W}_0, \mathbf{P}_0]$. Point $\mathbf{P}_1$ may now be calculated as

$$\mathbf{P}_1 = \text{ccoptpt}(\mathbf{P}_0, n_2, \mathbf{v}_{1B}, \mathbf{R}_1, n_3, S_1) \tag{9.45}$$

Direction $\mathbf{v}_{1C}$ of the ray after refraction is given by $\mathbf{v}_{1C} = \text{nrm}(\mathbf{R}_1 - \mathbf{P}_1)$. Normal $\mathbf{n}_1$ at point $\mathbf{P}_1$ is given by

$$\mathbf{n}_1 = \text{rfrnrm}(\mathbf{v}_{1B}, \mathbf{v}_{1C}, n_2, n_3) \tag{9.46}$$

Referring now to Figure 9.75b, choose a value S_B for the optical path length between wavefronts w_1 and $\mathbf{R}_2$. We may calculate the path of another ray r_{12} coming from $\mathbf{R}_2$ through the newly calculated point $\mathbf{P}_1$. This ray travels in a medium of refractive index n_3 and has direction $\mathbf{v}_{2A} = \text{nrm}(\mathbf{P}_1 - \mathbf{R}_2)$. Since we know the refractive indices n_3 and n_2 at point $\mathbf{P}_1$, we may calculate the direction $\mathbf{v}_{2B}$ of the ray refracted into n_2 as

$$\mathbf{v}_{2B} = \text{rfr}(\mathbf{v}_{2A}, \mathbf{n}_1, n_3, n_2) \tag{9.47}$$

Also, we may calculate the optical path length S_2 between $\mathbf{P}_1$ and w_1 as $S_2 = S_B - n_3[\mathbf{R}_2,\mathbf{P}_1]$. Point $\mathbf{P}_2$ may now be calculated as

$$\mathbf{P}_2 = \text{coptsl}(\mathbf{P}_1, n_2, \mathbf{v}_{2B}, \mathbf{W}_1, n_1, \mathbf{n}_{W1}, S_2) \tag{9.48}$$

Normal $\mathbf{n}_2$ to point $\mathbf{P}_2$ is given by

$$\mathbf{n}_2 = \text{rfrnrm}(\mathbf{v}_{2B}, \mathbf{v}_{2C}, n_2, n_1) \tag{9.49}$$

where $\mathbf{v}_{2C} = \mathbf{n}_{W1}$.

Tracing ray r_{11} through a given point $\mathbf{P}_0$ (and its normal) on the top surface, we calculated a new point $\mathbf{P}_1$ (and its normal) on the bottom surface. Tracing a ray r_{12} through point $\mathbf{P}_1$ (and its normal) on the bottom surface, we calculated a new point $\mathbf{P}_2$ (and its normal) on the top surface. This process may now continue as we calculate more and more points (and their normals) on the top and bottom surfaces of the RR optic, as shown in Figure 9.76. Tracing ray r_{13} through $\mathbf{P}_2$ we calculate $\mathbf{P}_3$ (and its normal), tracing ray r_{14} through $\mathbf{P}_3$ we calculate $\mathbf{P}_4$ (and its normal), and so on.

Also, we may go back to point $\mathbf{P}_0$ (with normal $\mathbf{n}_0$) and consider another ray r_{21}, this time coming from wavefront w_1. Using the same procedure as before, and the same optical path length S_B between w_1 and $\mathbf{R}_2$, the ray r_{21} defines a new point $\mathbf{Q}_1$ (and its normal) on the bottom surface. Tracing ray r_{22} through $\mathbf{Q}_1$ we calculate $\mathbf{Q}_2$, tracing ray r_{23} through $\mathbf{Q}_2$ we calculate $\mathbf{Q}_3$, tracing ray r_{24} through $\mathbf{Q}_3$ we calculate $\mathbf{Q}_4$. In the end we have a list of points (and normals) …, $\mathbf{P}_4$, $\mathbf{P}_2$, $\mathbf{P}_0$, $\mathbf{Q}_2$, $\mathbf{Q}_4$, … on the top surface of the RR lens and another list of points (and normals) …, $\mathbf{P}_3$, $\mathbf{P}_1$, $\mathbf{Q}_1$, $\mathbf{Q}_3$, … on the bottom surface.

All the points and normals referred above are contained in plane v so this calculation results is a two-dimensional SMS chain contained in that same plane. The same procedure, however, may be used to calculate points on the surfaces of the lens outside plane v. Figure 9.77 shows one such configuration. Now, the SMS chain starts at a point $\mathbf{T}_0$ on a curve c_0 that crosses plane v at point $\mathbf{P}_0$. Point $\mathbf{T}_0$ is, therefore, away from plane v. Normal $\mathbf{u}_0$ to point $\mathbf{T}_0$ is perpendicular to curve c_0. The direction of $\mathbf{u}_0$ is chosen on the plane perpendicular to c_0 at point $\mathbf{T}_0$.

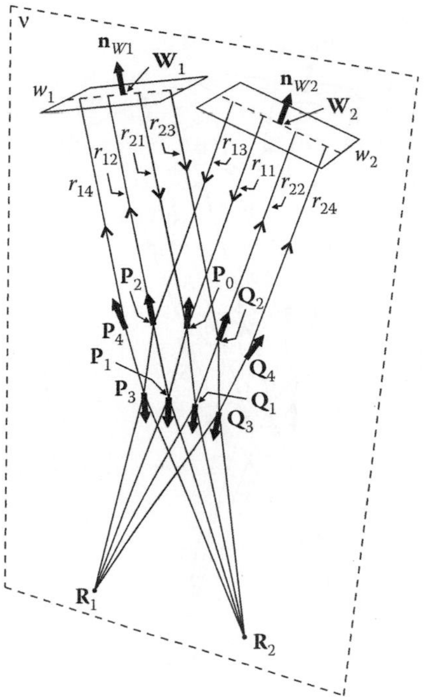

FIGURE 9.76
SMS chain. All rays, points and normals are contained in plane v and, for that reason, this is a two-dimensional SMS chain.

The same procedure described above to calculate the SMS chain starting at point $\mathbf{P}_0$ may now be used to calculate a new SMS chain starting at point $\mathbf{T}_0$. The optical path lengths S_B between wavefronts w_1 and $\mathbf{R}_2$ and S_A between w_2 and $\mathbf{R}_1$ are the same as before. The path of ray r_{31} through $\mathbf{T}_0$ on the top surface of the lens defines a new point $\mathbf{T}_1$ (and its normal) on its bottom surface. The calculations to determine the path of ray r_{31} are similar to those in expressions (9.43) through (9.46). Tracing ray r_{32} through $\mathbf{T}_1$ we calculate $\mathbf{T}_2$ and its normal. The calculations to determine the path of ray r_{32} are similar to those in expressions (9.47) through (9.49). Tracing ray r_{33} through $\mathbf{T}_2$ we calculate $\mathbf{T}_3$ and so on. Going back to point $\mathbf{T}_0$, tracing ray r_{41} through point $\mathbf{T}_0$ we calculate point $\mathbf{S}_1$, tracing ray r_{42} we calculate point $\mathbf{S}_2$ and so on. In the end we have a list of points (and normals) …, $\mathbf{T}_4$, $\mathbf{T}_2$, $\mathbf{T}_0$, $\mathbf{S}_2$, $\mathbf{S}_4$, … on the top surface of the RR lens and another list of points (and normals) …, $\mathbf{T}_3$, $\mathbf{T}_1$, $\mathbf{S}_1$, $\mathbf{S}_3$, … on the bottom surface.

Each one of the SMS chains (and corresponding rays), as in Figure 9.76 or Figure 9.77, may be seen as "layers" in generating the complete lens. As we move $\mathbf{T}_0$ along curve c_0, more and more of these "layers" are generated. The SMS points in these different "layers" define the shape of the lens.

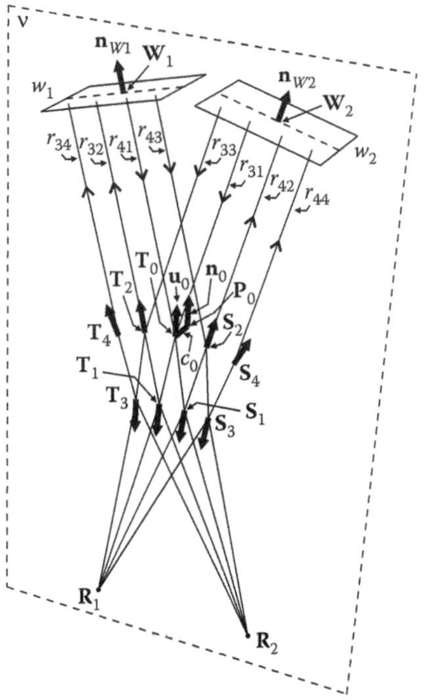

FIGURE 9.77
SMS chain. Point $\mathbf{T}_0$ is on a curve c_0 that crosses plane v at point $\mathbf{P}_0$ and, therefore, is not on plane v. The resulting SMS chain is therefore also not on v and is three-dimensional.

Now, if one moves in small steps from point $\mathbf{P}_0$ to point $\mathbf{T}_0$ along a curve c_0 (with normals moving from $\mathbf{n}_0$ to $\mathbf{u}_0$), point $\mathbf{P}_1$ moves to point $\mathbf{T}_1$ along a curve c_1, point $\mathbf{P}_2$ moves to point $\mathbf{T}_2$ along curve c_2 and so on. Also, point $\mathbf{Q}_1$ moves to point $\mathbf{S}_1$ along a curve e_1, point $\mathbf{Q}_2$ moves to $\mathbf{S}_2$ along e_2, and so on, as shown in Figure 9.78.

This process may then continue by extending the curve c_0 and using more of its points to calculate new SMS chains. The result is shown in Figure 9.79 with several curves …, c_4, c_2, c_0, e_2, e_4, … on the top surface of the lens and …, c_3, c_1, e_1, e_3, … on the bottom surface of the lens. Each one of these two sets of curves looks like a set of ribs, and for that reason they are called ribs.

An alternative way of doing the same calculation is to start with a set of points (and normals) on curve c_0, as shown in Figure 9.80a. By tracing rays coming from w_2 through these points and focusing them to $\mathbf{R}_1$, we calculate a new set of points and normals on curve c_1. As before, for each one of these rays, the calculations are similar to those in expressions (9.43) through (9.46). The optical path length S_A between w_2 and $\mathbf{R}_1$ is also the same as before for all these rays. Now, a new set of rays coming from $\mathbf{R}_2$ may be traced through the points and normals of curve c_1. By making these rays perpendicular to wavefront w_1, a new set of points and normals may be calculated on curve c_2.

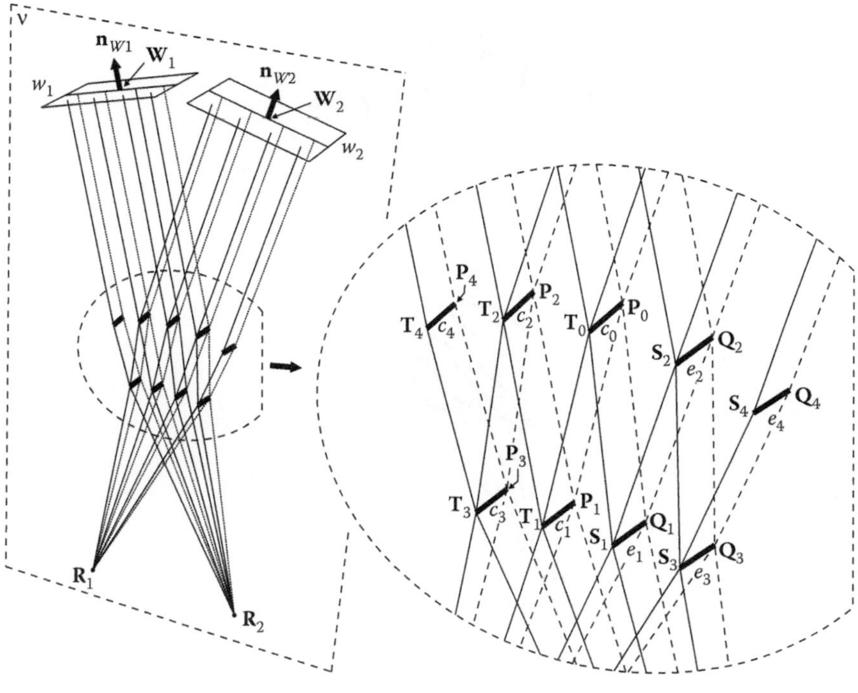

FIGURE 9.78
Moving the starting point $\mathbf{P}_0$ to $\mathbf{T}_0$ along curve c_0 and calculating the corresponding SMS chains generates a set of curves on both the top and bottom surfaces of the lens.

As before, for each one of these rays, the calculations are similar to those in expressions (9.47) through (9.49). The optical path length S_B between wavefronts w_1 and $\mathbf{R}_2$ is also the same as before for all these rays.

By tracing rays coming from w_2 through the points and normals of c_2 and focusing them to $\mathbf{R}_1$, we calculate a new set of points and normals on curve c_3, as shown in Figure 9.81a. Now, a new set of rays coming from $\mathbf{R}_2$ may be traced through the points and normals of curve c_3. By making these rays perpendicular to wavefront w_1, a new set of points and normals may be calculated on curve c_4, as shown in Figure 9.81b. Repeating this process, one obtains the curves shown in Figure 9.79.

At this point we have a set of curves on the top and bottom surfaces of the lens, but the surface of the optic between these curves is not defined. Now create a surface between two neighboring curves on the same surface. The normals to the surface at these curves must coincide with the normals calculated by the SMS method. Figure 9.82 shows one such surface s_0 between curves c_1 and e_1 on the bottom surface of the lens. Now, we may take a set of points and normals on s_0. Tracing rays from $\mathbf{R}_2$ through these points of s_0 and making them perpendicular to w_1 defines the points (and normals) of a surface s_1 connecting the curves c_0 and c_2 on the top surface of the lens. Again,

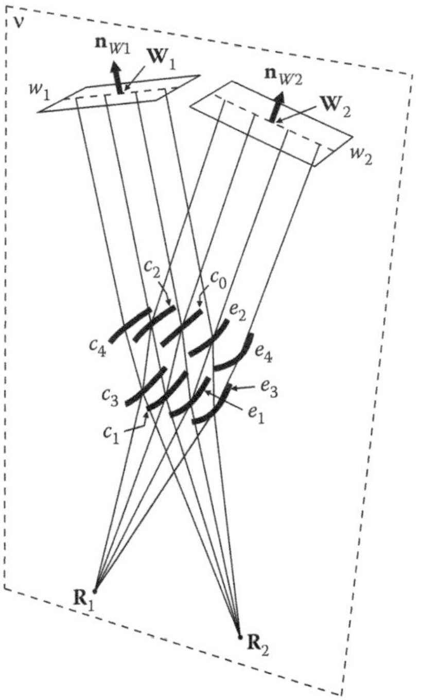

FIGURE 9.79
Set of curves on the top and bottom surfaces of a three-dimensional SMS lens resulting from the SMS chains.

the procedure for calculating these rays and the optical path length between w_1 and $\mathbf{R}_2$ is the same as before.

By tracing the rays from w_2 through the points and normals of s_1 we can calculate the points and normals of a surface s_2 connecting the curves c_1 and c_3 on the bottom surface, as shown in Figure 9.83a. Again, the procedure for calculating these rays and the optical path length between w_2 and $\mathbf{R}_1$ are the same as before. By tracing the rays from $\mathbf{R}_2$ through the points and normals of s_2 we can calculate the points and normals of a new surface s_3 between the curves c_2 and c_4 on the top surface, as shown in Figure 9.83b.

Repeating this process, the complete top and bottom surfaces of the lens may be defined, as shown in Figure 9.84. The process of adding "skin" to the ribs is called skinning (contrary to the usual meaning of the word, which means removing the skin).

Figure 9.85 shows another view of the ribs …, c_4, c_2, c_0, e_2, e_4, … on the top surface of the lens and …, c_3, c_1, e_1, e_3, … on the bottom surface of the lens. Referring back to Figure 9.76, points …, $\mathbf{P}_4$, $\mathbf{P}_2$, $\mathbf{P}_0$, $\mathbf{Q}_2$, $\mathbf{Q}_4$, … define another curve t_3 crossing the ribs of the top surface. The same is true for points …, $\mathbf{P}_3$, $\mathbf{P}_1$, $\mathbf{Q}_1$, $\mathbf{Q}_3$, … that define a curve b_3 crossing the ribs of the bottom surface. Also, referring back to Figure 9.77, points …, $\mathbf{T}_4$, $\mathbf{T}_2$, $\mathbf{T}_0$, $\mathbf{S}_2$, $\mathbf{S}_4$, … define a curve

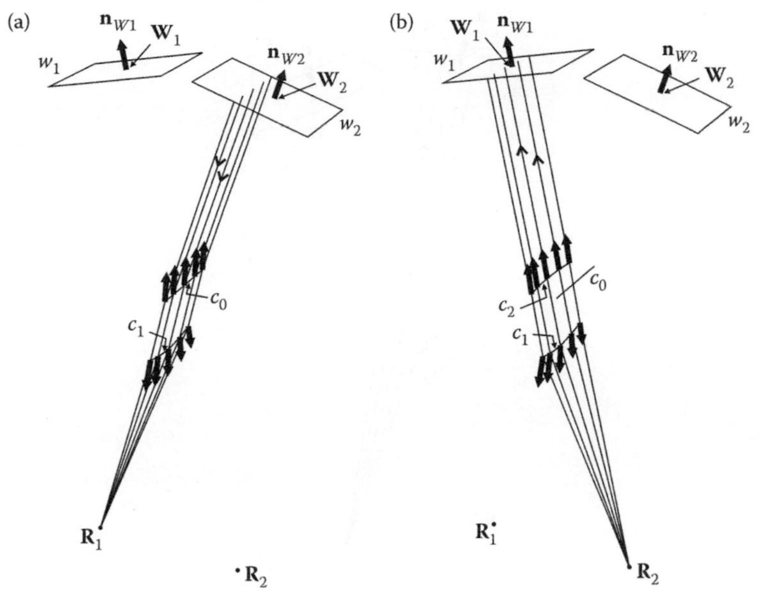

FIGURE 9.80
(a) Tracing rays from w_2 through a curve c_0 (with its normals) and focusing these rays to $\mathbf{R}_1$ generates a curve c_1 and its normals. (b) Rays from $\mathbf{R}_2$ through c_1 generate c_2.

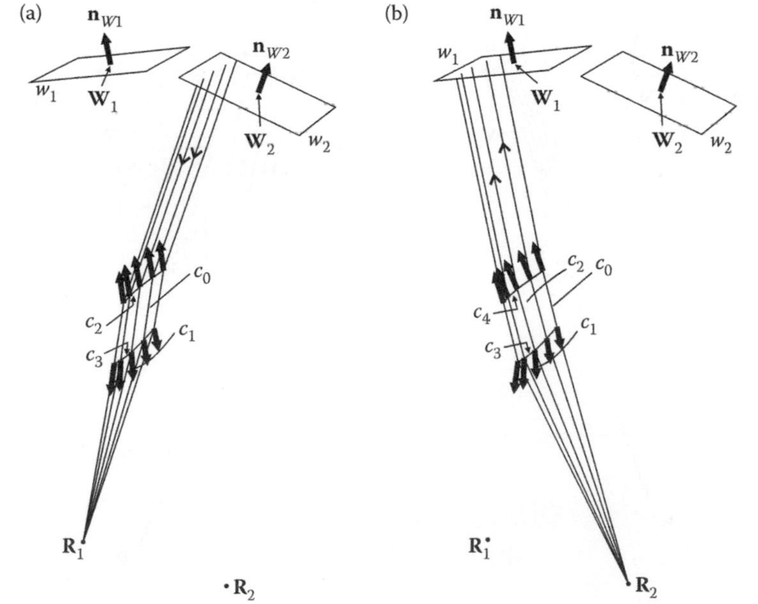

FIGURE 9.81
(a) Rays from w_2 through c_2 generate c_3. (b) Rays from $\mathbf{R}_2$ through c_3 generate c_4.

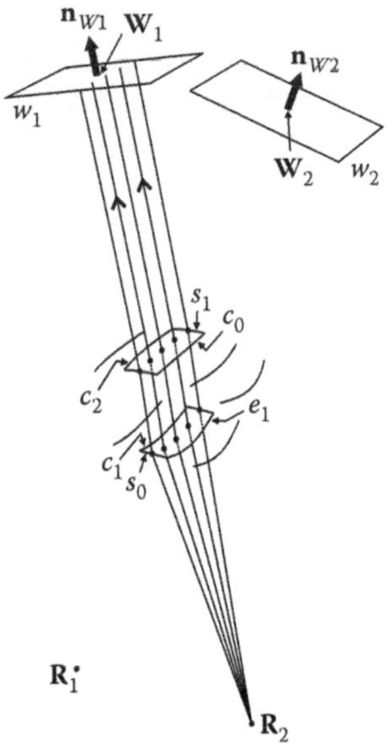

FIGURE 9.82
Creating a surface s_0 between c_1 and e_1 on the bottom surface of the lens and tracing through it rays from $\mathbf{R}_2$ to w_1 generates a surface s_1 between c_0 and c_2 on the top surface of the lens.

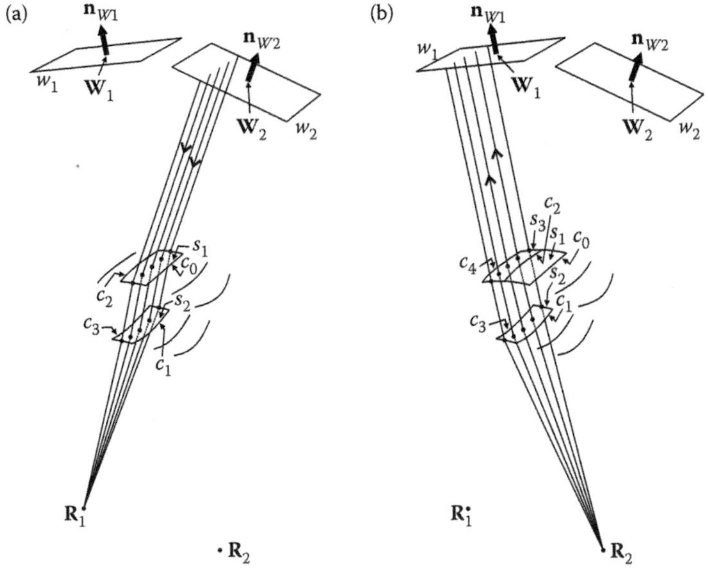

FIGURE 9.83
(a) Ray tracing from w_2 to $\mathbf{R}_1$ through surface s_1 on the top surface of the lens generates surface s_2 on the bottom surface of the lens. (b) Ray tracing from $\mathbf{R}_2$ to w_1 through s_2 generates s_3.

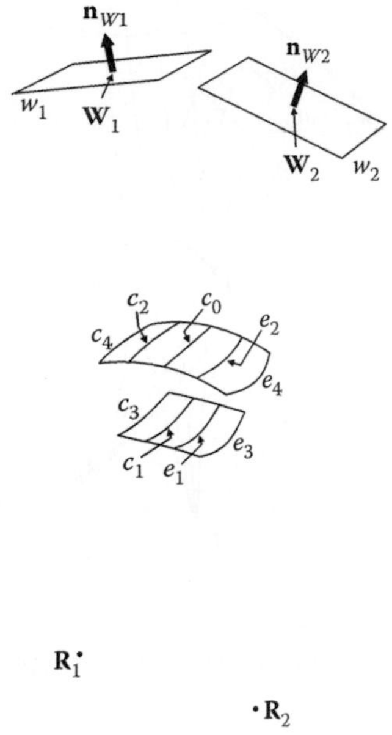

FIGURE 9.84
Complete top and bottom surfaces of the RR SMS 3-D lens.

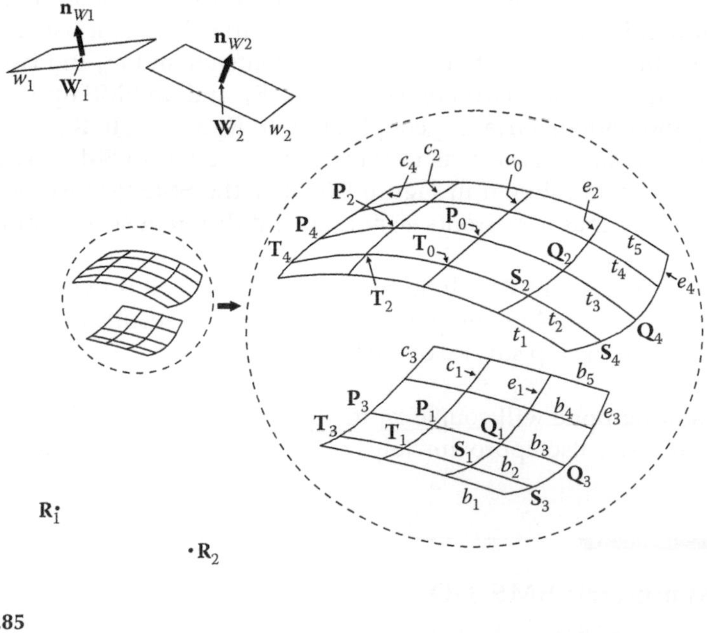

FIGURE 9.85
Spines $t_1, \ldots, t_5$ cross the ribs $c_4, \ldots e_4$ on the top surface of the lens. Spines $b_1, \ldots, b_5$ cross the ribs $c_3, \ldots e_3$ on the bottom surface of the lens.

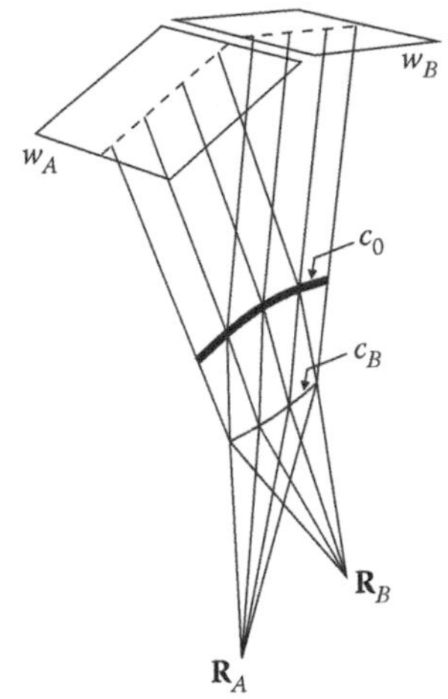

FIGURE 9.86
Initial curve c_0 may be calculated with the SMS method coupling wavefronts w_A and w_B onto $\mathbf{R}_B$ and $\mathbf{R}_A$ respectively.

t_2 crossing the ribs of the top surface. And points ..., $\mathbf{T}_3$, $\mathbf{T}_1$, $\mathbf{S}_1$, $\mathbf{S}_3$, ... define a curve b_2 crossing the ribs of the bottom surface. In general, it is possible to define a set of curves called spines which cross the ribs. These are ..., t_1, t_2, t_3, t_4, t_5, ... on the top surface and ..., b_1, b_2, b_3, b_4, b_5, ... on the bottom surface.

Initial curve c_0 may have different shapes, for example a straight line or a circular arc. However, another possibility is to calculate it using the SMS method. Figure 9.86 shows two input wavefronts w_A and w_B and two output wavefronts (in this case two points) $\mathbf{R}_A$ and $\mathbf{R}_B$ and an SMS optic with top surface c_0 and bottom surface c_B coupling w_A to $\mathbf{R}_B$ and w_B to $\mathbf{R}_A$.

We may now take curve c_0 as the starting curve for an SMS 3-D optic (in this case a lens), as shown in Figure 9.87. For the SMS 3-D we now use a different set of input wavefronts w_1 and w_2 and different set of output wavefronts $\mathbf{R}_1$ and $\mathbf{R}_2$.

Starting at each point $\mathbf{P}$ with normal $\mathbf{n}_P$ on curve c_0, it is possible to calculate a 3-D SMS chain. Moving point $\mathbf{P}$ along curve c_0 results in curves ..., c_4, c_2, c_0, e_2, e_4, ... on the top surface of the lens and ..., c_3, c_2, e_1, e_3, ... on its bottom surface.

The resulting optic will couple w_1 to $\mathbf{R}_2$ and w_2 to $\mathbf{R}_1$, and because of the shape of curve c_0, also approximately couple w_A to $\mathbf{R}_B$ and w_B to $\mathbf{R}_A$.

9.10 Asymmetric SMS 3-D

The design method described above may be applied to optics that are highly asymmetric, as the one in Figure 9.88. It is an XR, composed of a primary

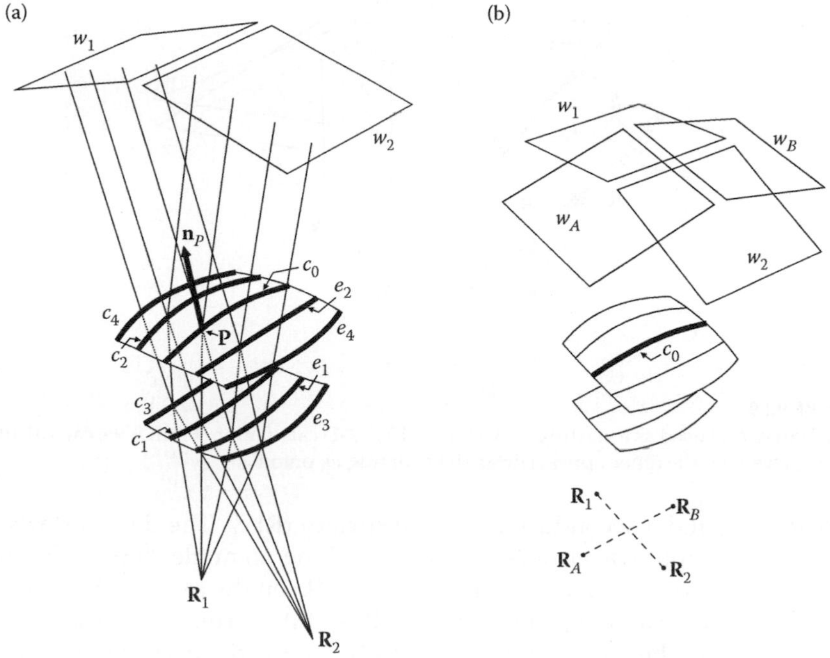

FIGURE 9.87
(a) Curve c_0 calculated by the SMS method is used as a starting curve for the SMS 3-D calculation. (b) Wavefronts w_A, w_B, $\mathbf{R}_A$, $\mathbf{R}_B$ for the initial curve c_0 and wavefronts w_1, w_2, $\mathbf{R}_1$, $\mathbf{R}_2$ for the SMS 3-D calculation.

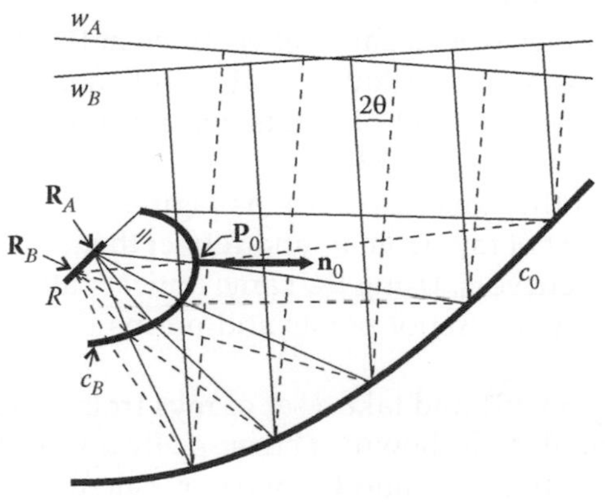

FIGURE 9.88
Initial curve c_0 of an asymmetric design may be calculated with the SMS method coupling wavefronts w_A and w_B onto $\mathbf{R}_B$ and $\mathbf{R}_A$, respectively.

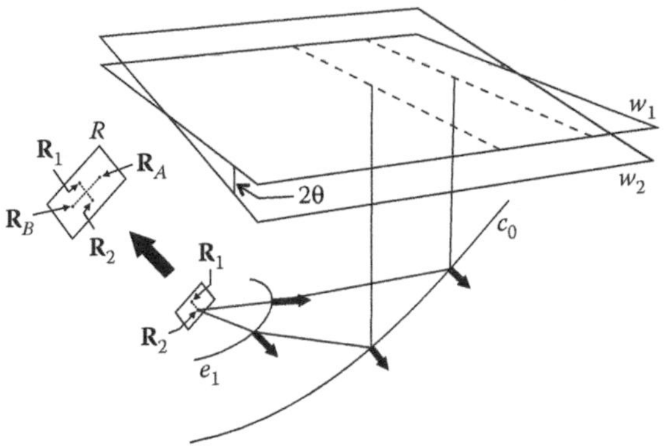

FIGURE 9.89
Initial curve c_0 is used as a starting curve for the SMS 3-D calculation. From it we can calculate a new curve e_1 on the other optical surface by coupling w_1 onto $\mathbf{R}_2$.

mirror (X) c_0 and a secondary refractive surface (R) c_B. The design starts by choosing flat input wavefronts w_A and w_B making an angle 2θ to each other, and choosing point output wavefronts $\mathbf{R}_A$ and $\mathbf{R}_B$ on the receiver R. One may now choose, for example, a starting point $\mathbf{P}_0$ and its normal $\mathbf{n}_0$ on the secondary surface and then adjust the optical path lengths S_{AB} between w_A and $\mathbf{R}_B$ and S_{BA} between w_B and $\mathbf{R}_A$ to get smooth optical surfaces.

Now, we may take, for example, curve c_0 as the starting point for an SMS 3-D design, as shown in Figure 9.89. We then pick two new flat wavefronts w_1 and w_2 making an angle 2θ to each other and two new point wavefronts $\mathbf{R}_1$ and $\mathbf{R}_2$ on receiver R. One possibility is to obtain w_1 and w_2 by a 90° rotation of w_A, w_B around the vertical axis and obtain $\mathbf{R}_1$ and $\mathbf{R}_2$ also be a 90° rotation of $\mathbf{R}_A$ and $\mathbf{R}_B$ around the normal to R through its center point. In that case, the optical path length for the 3-D SMS calculation may be chosen, for example, as $S_{3D} = (S_{AB} + S_{BA})/2$.

Take now a set of rays from w_1 and reflect them at the points and normals of curve c_0. By imposing a constant optical path length S_{3D} between w_1 and $\mathbf{R}_2$, one may calculate a new set of points and normals of a curve e_1 on the secondary, as shown in Figure 9.89.

Figure 9.90 shows the same geometry as Figure 9.89 but from a different perspective. Take a set of rays from $\mathbf{R}_1$ and refract them at the points and normals (not shown) of curve e_1. Using the same optical path length S_{3D} between $\mathbf{R}_1$ and w_2 calculate a new set of points and normals of curve e_2 on the primary mirror.

Now, refer to Figure 9.91 and take a set of rays from w_1 and reflect them at the points and normals (not shown) of curve e_2. By imposing a constant optical path length S_{3D} between w_1 and $\mathbf{R}_2$, one may calculate a new set of points and normals of a curve e_3 on the secondary.

The process now continues as we calculate more curves on the primary mirror and secondary refractive surface.

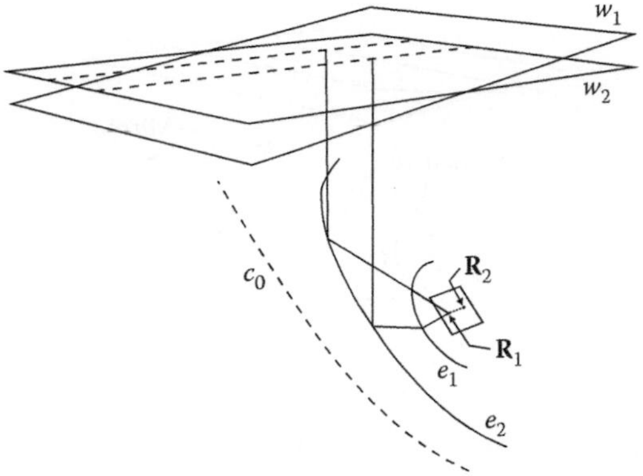

FIGURE 9.90
Coupling point wavefront R_1 onto w_2 we may calculate a new curve e_2 on the mirror based on curve e_1 on the secondary surface.

In this example we have taken the output wavefronts $\mathbf{R}_A$, $\mathbf{R}_B$, $\mathbf{R}_1$, $\mathbf{R}_2$ inside the receiver and not at its edges. This results in curves e_1, e_2, e_3, … closer to each other, easing the skinning process (or even eliminating its need).

A set of curves on the primary and secondary optical elements is shown in Figure 9.92.

Figure 9.93 shows a more realistic design of one of these concentrators designed for a geometrical concentration of 1000× and a 1 cm^2 cell. Using this concept an acceptance angle of 1.3° has been demonstrated.[18] These XR SMS optics may be coupled to an ultrashort prism integrated on the secondary optic to smoothen the irradiance distribution on the cell.[19]

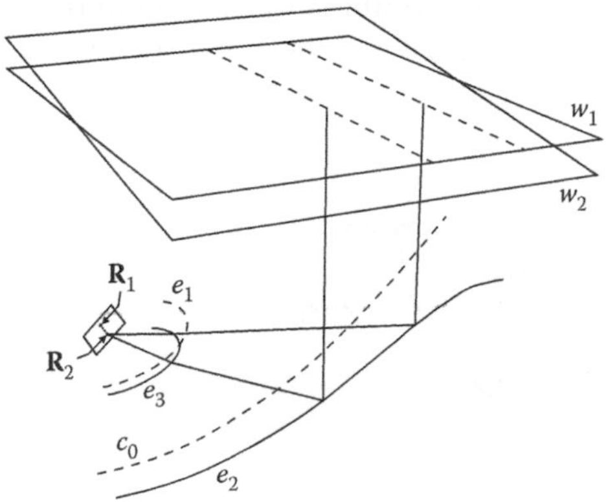

FIGURE 9.91
Coupling wavefront w_1 onto $\mathbf{R}_2$ we may calculate a new curve e_3 on the secondary surface based on curve e_2 on the mirror.

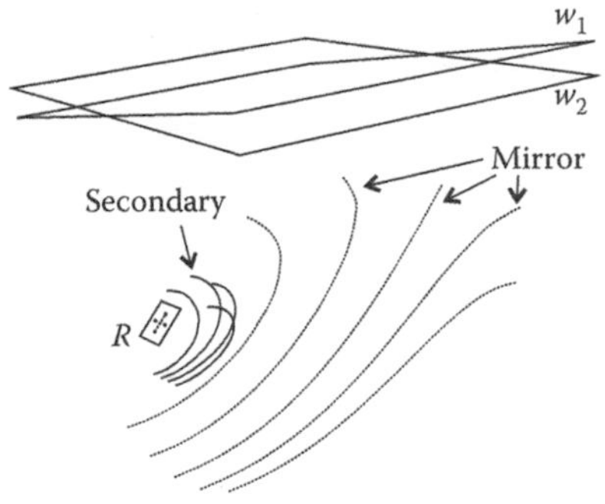

FIGURE 9.92
Set of curves on the mirror and set of curves on the secondary optical surface. These curves define the optical surfaces.

Figure 9.94 shows how these optics can be assembled into an array. The geometry is such that each secondary can be hidden behind the mirror of the next concentrator. This avoids the shading that the secondary would otherwise cast on the primary mirror. When used for photovoltaic concentration, the heat sink of each solar cell (receiver *R*) is also hidden behind the mirror of the next concentrator and, therefore, also does not cast shading on the primary mirror.

Figure 9.95 shows a secondary of a XR free-form concentrator. Aside form the free-form optical surface (Figure 9.95a), the part has holding features and a short prism at the bottom (Figure 9.95b) to homogenize the solar cell illuminance.

An array of six of these concentrators is show in Figure 9.96a. At the top, the photo shows the heat pipes and fins that extract the heat from the solar cells. Figure 9.96b shows a module composed of several of these sets.

The heat sink of each concentrator is hidden behind the mirror of the concentrator below it. Therefore, when looking at the module, its whole surface appears to be mirrored, as seen in Figure 9.96b.

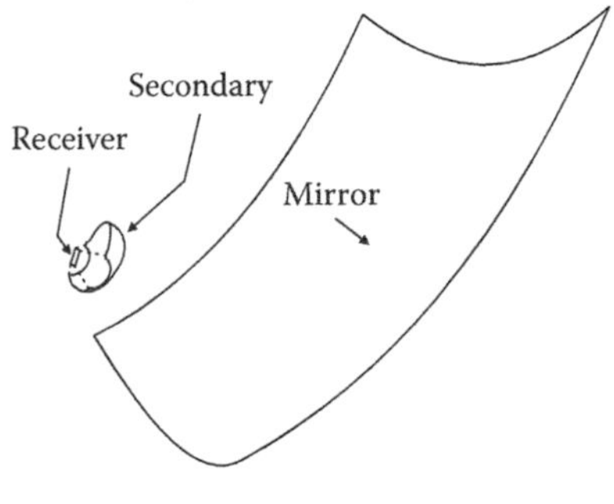

FIGURE 9.93
XR free-form concentrator.

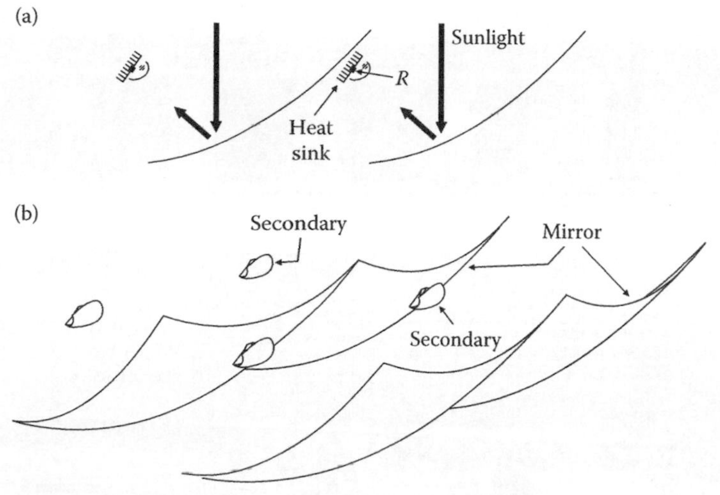

FIGURE 9.94
(a) Array of XR free-form concentrators. Each secondary is hidden behind the mirror of the next concentrator, avoiding shading the mirror. (b) 3-D view of the same.

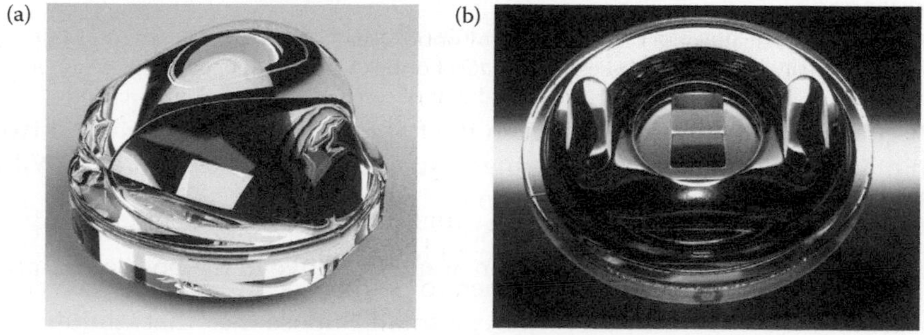

FIGURE 9.95
Secondary free-form concentrator. (a) Top view showing the free-form refractive surface. (b) Bottom view showing a short prism at the center (the two side lobes on the base are holding features). (Courtesy of Light Prescriptions Innovators.)

9.11 SMS 3-D with a Thin Edge

One possibility to design an SMS 2-D optic is to start with a thin edge, as is the case in Figure 9.24. These designs can be extended to 3-D geometry.[6,7] Figure 9.97 shows a lens with a thin edge along curve e_H that focuses to point $\mathbf{R}_2$ the rays perpendicular to flat wavefront w_1 and focuses to $\mathbf{R}_1$ the (symmetrical) rays perpendicular to flat wavefront w_2. The design of the lens starts with the calculation of edge curve e_H.

(a)

(b)

FIGURE 9.96
(a) Array of XR free-form concentrators. (b) Module composed of several of these sets. (Courtesy of Light Prescriptions Innovators.)

Equal optical path length between the flat wavefronts w_1 and w_2 and two points $\mathbf{R}_2$ and $\mathbf{R}_1$, respectively, defines a surface that verifies $[\mathbf{R}_1,\mathbf{P}] + [\mathbf{P},\mathbf{W}_2] = [\mathbf{R}_2,\mathbf{P}] + [\mathbf{P},\mathbf{W}_1]$, or $t + d_T = s + d_S$, as shown in Figure 9.98.

The normal vectors to flat wavefronts w_1 and w_2 are contained in plane x_1x_3. Wavefronts w_1 and w_2 are tilted by an angle θ relative to the horizontal, and

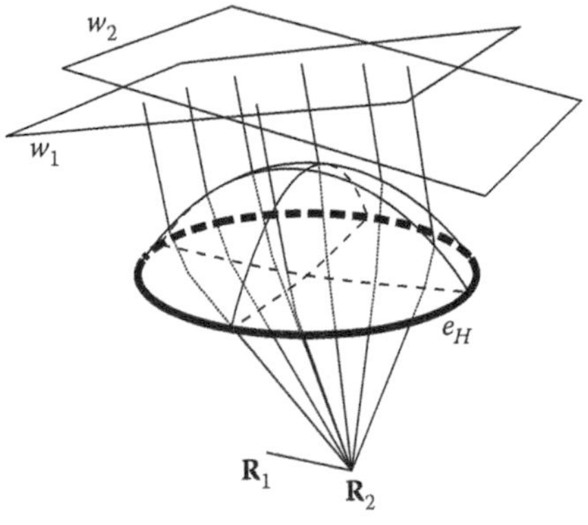

FIGURE 9.97
SMS 3-D lens with thin edge along curve e_H designed for input flat wavefronts w_1 and w_2 and point receiver wavefronts $\mathbf{R}_1$ and $\mathbf{R}_2$.

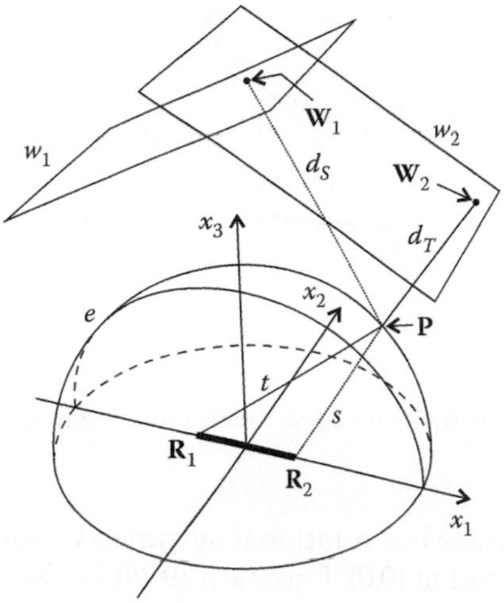

FIGURE 9.98
Equal optical path length between w_1 and $\mathbf{R}_2$ and between w_2 and $\mathbf{R}_1$ defines the shape of a surface e.

are symmetrical relative to plane x_2x_3. Points $\mathbf{R}_1$ and $\mathbf{R}_2$ are also symmetrical relative to plane x_2x_3. Conservation of optical path length in Figure 9.99 may be written as $[\mathbf{R}_2,\mathbf{P}] + [\mathbf{P},\mathbf{W}_1] = [\mathbf{R}_1,\mathbf{P}] + [\mathbf{P},\mathbf{W}_2]$, or $s + (2x + d)\sin\theta = t + d\sin\theta$, which may be rewritten as (9.21), or $s = t - 2x\sin\theta$.

Rotating point $\mathbf{P}$ around axis $\mathbf{R}_1\mathbf{R}_2$ along a circle c does not change t, s, or x, as seen from Figure 9.100, and Equation 9.21 still holds. It may then be concluded that the surface has rotational symmetry around axis $\mathbf{R}_1\mathbf{R}_2$ and the same cross section defined by Equation 9.29 or Equation 9.30. It is then an ellipsoid with foci $\mathbf{R}_1$ and $\mathbf{R}_2$ and rotational symmetry relative to the axis $\mathbf{R}_1\mathbf{R}_2$.

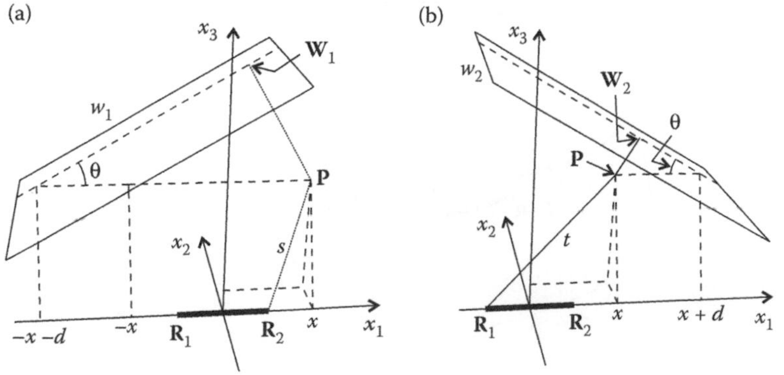

FIGURE 9.99
(a) Ray from w_1 through $\mathbf{P}$ to $\mathbf{R}_2$. (b) Ray from w_2 through $\mathbf{P}$ to $\mathbf{R}_1$. Both flat wavefronts w_1 and w_2 are tilted by an angle θ relative to the horizontal.

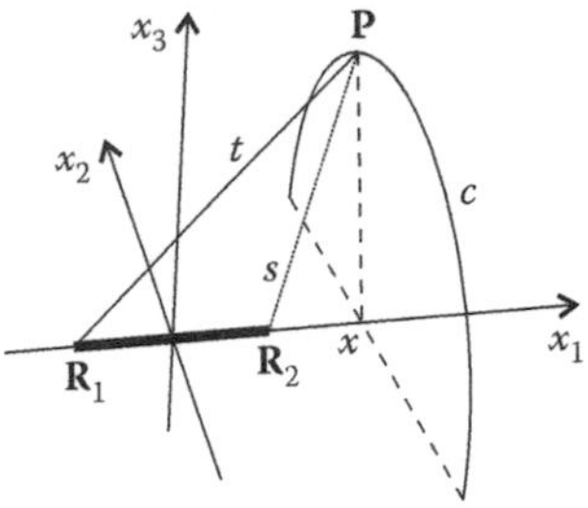

FIGURE 9.100
Rotating point **P** around axis $\mathbf{R}_1\mathbf{R}_2$ on a circle c maintains distances t, s, and x.

Since in 3-D the surface has rotational symmetry around $\mathbf{R}_1\mathbf{R}_2$ and assuming that $\mathbf{R}_1\mathbf{R}_2$ is centered at (0,0), Equation (9.29) is also valid in both the x_1x_2 and x_1x_3 planes. The surface may then be parameterized as

$$\begin{aligned}\mathbf{e}(\varphi,\zeta) &= (a\cos\varphi\sin\zeta, b\sin\varphi\sin\zeta, b\cos\zeta)\\ &= \frac{f}{2}\left(\frac{\cos\varphi\sin\zeta}{\sin\theta}, \frac{\sin\varphi\sin\zeta}{\tan\theta}, \frac{\cos\zeta}{\tan\theta}\right)\end{aligned} \tag{9.50}$$

with $0 < \varphi < 2\pi$ and $0 < \zeta < \pi$.

We now intersect the ellipsoid with a horizontal plane at a height $x_3 = h$, as shown in Figure 9.101. This results in the intersection curve e_H, an ellipse with semi-major axis c and semi-minor axis d.

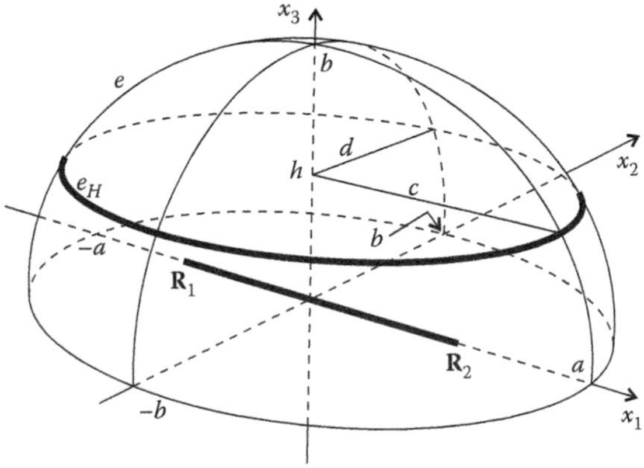

FIGURE 9.101
Ellipsoid surface e intersected with a horizontal plane $x_3 = h$ defines an elliptical curve e_H with semi-major axis c and semi-minor axis d.

Point $\mathbf{P} = (c, 0, h)$ on plane x_1x_3 and verifies Equation 9.29, which for point $\mathbf{P}$ is

$$c^2 \sin^2\theta + h^2 \tan^2\theta = (f/2)^2 \tag{9.51}$$

as seen from Figure 9.102. Semi-minor axis d may be obtained from $d^2 + h^2 = b^2$. Replacing this expression and b from Equation 9.28 into Equation 9.51 and and solving for d we get

$$d = c\cos\theta \tag{9.52}$$

Ellipse e_H is then defined by

$$\left(\frac{x_1}{c}\right)^2 + \left(\frac{x_2}{d}\right)^2 = 1 \Leftrightarrow \left(\frac{x_1}{c}\right)^2 + \left(\frac{x_2}{c\cos\theta}\right)^2 = 1 \tag{9.53}$$

at $x_3 = h$. In parametric form ellipse e_H is given by

$$\mathbf{e}_H(\varphi) = (c\cos\varphi, d\sin\varphi, h) = (c\cos\varphi, c\cos\theta\sin\varphi, h) \tag{9.54}$$

with $0 < \varphi < 2\pi$ and c is obtained from the value of h from Equation 9.51 as

$$c = \frac{\sqrt{f^2 - 4h^2\tan^2\theta}}{2\sin\theta} \tag{9.55}$$

where $0 < h < b = f/(2\tan\theta)$.

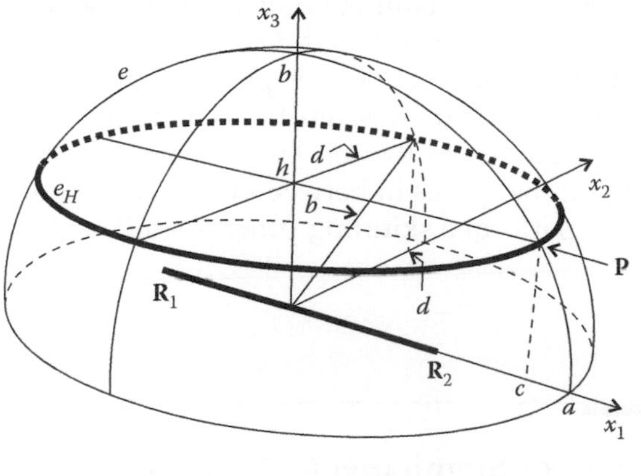

FIGURE 9.102
Calculation of semi-minor axis d from the value of semi-major axis c of elliptical curve e_H resulting from the intersection of ellipsoid e with a horizontal plane $x_3 = h$.

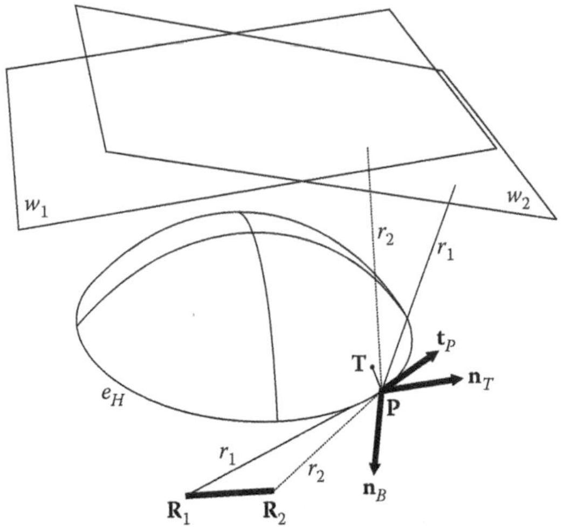

FIGURE 9.103
Given a point **P** on edge curve e_H and its tangent vector $\mathbf{t}_P$, it is possible to calculate the directions of normal $\mathbf{n}_T$ to the top surface and $\mathbf{n}_B$ to the bottom surface. Vectors $\mathbf{n}_T$ and $\mathbf{n}_B$ are perpendicular to $\mathbf{t}_P$.

Curve e_H is the thin edge of the SMS 3-D lens in Figure 9.97. The design of the lens starts by choosing a point **P** on curve e_H and calculating its tangent vector $\mathbf{t}_P$, as shown in Figure 9.103.

We may now restrict the normal $\mathbf{n}_T$ to the top surface and $\mathbf{n}_B$ to the bottom surface to lay in the plane through point **P** with normal $\mathbf{t}_P$. The paths of rays r_1 and r_2 through **P** then define the orientations of $\mathbf{n}_T$ and $\mathbf{n}_B$ in a calculation process similar to that illustrated in Figure 9.23a, only now in 3-D. Once we know $\mathbf{n}_T$ we can now move a very small distance into the top surface of the lens along direction $\mathbf{n}_T \times \mathbf{t}_P$ obtaining point **T**. We may give point **T** the same normal $\mathbf{n}_T$. Since the optical path length for rays r_1 and r_2 is known, we can now calculate an SMS chain that defines points on the top and bottom surfaces of the lens. This calculation is similar to that illustrated in Figure 9.77. These points define curves on the top and bottom surfaces but may be far apart. In order to fill the spaces in between them and completely define the curves, we need to perform a skinning process similar to that illustrated in Figure 9.23b.

9.12 Other Types of Simultaneous Multiple Surface Optics

The Miñano–Benitez design method can also be used to design a large variety of other types of optics.[20] Other examples of application include, afocal lenses,[21] TIR-R lenses,[22] that is, TIR lenses with a secondary covering the receiver, or primary–secondary concentrators for tubular receivers.[23,24] In the latter case, one of the advantages of the SMS optics is that the secondary mirror does not have to touch the receiver, as in the case of the flow-line optics in Chapter 6.

Another possibility is to combine different geometries to form new optics, such as, for example, a combination of an RX and an RXI in one single optic.[25] It is also possible to combine the SMS optics with flow-line mirrors in a single device.[25]

The Miñano–Benitez method can also be used in the design of SMS optics with imaging applications.[2,26]

9.13 Examples

The following examples use expressions for the curves and functions that are derived in Chapter 21.

EXAMPLE 9.1

Calculate an RR SMS lens that concentrates on the edge points of the receiver, $\mathbf{R}_1 = (-0.5, -2)$ and $\mathbf{R}_2 = (0.5, -2)$, the light coming from the edges of the emitter, $\mathbf{E}_1 = (-1, 2)$ and $\mathbf{E}_2 = (1, 2)$. The refractive index of the lens is $n = 1.5$.

First decide how much étendue to couple between the emitter and receiver, say $U = 1$. If the dimensions were, for example, in millimeters, than the étendue would be $U = 1$ mm. The edges of the entrance aperture of the lens must be on a hyperbola that has foci $\mathbf{E}_1$ and $\mathbf{E}_2$ and each point $\mathbf{P}$ on it fulfills $[\mathbf{P}, \mathbf{E}_1] - [\mathbf{P}, \mathbf{E}_2] = U/2$. Also, the exit aperture of the lens must be on a hyperbola that has foci $\mathbf{R}_1$ and $\mathbf{R}_2$ and each point $\mathbf{P}$ on it fulfills $[\mathbf{P}, \mathbf{R}_1] - [\mathbf{P}, \mathbf{R}_2] = U/2$. These hyperbolas are given by

$$\begin{aligned} h_E(\phi) &= \mathrm{hyp}(\mathbf{E}_1, \mathbf{E}_2, U, n) = -\frac{15}{4(1-4\cos(\phi))}(\cos(\phi), \sin(\phi)) + (-1, 2) \\ h_R(\phi) &= \mathrm{hyp}(\mathbf{R}_1, \mathbf{R}_2, U, n) = -\frac{0.75}{1-2\cos(\phi)}(\cos(\phi), \sin(\phi)) + (-0.5, -2) \end{aligned} \tag{9.56}$$

for the emitter and receiver, respectively. Now choose two points, one on each of these hyperbolas, as starting points for the surfaces of the lens. We choose

$$\begin{aligned} \mathbf{N} &= h_E(303.3\,\mathrm{deg}) = (0.721303, -0.620433) \\ \mathbf{X} &= h_R(42.7\,\mathrm{deg}) = (0.673162, -0.917437) \end{aligned} \tag{9.57}$$

Design of the lens starts by defining the path of ray r_1 as $\mathbf{E}_1$-$\mathbf{N}$-$\mathbf{X}$-$\mathbf{R}_1$. This ray path enables us to determine the normals to the lens $\mathbf{n}_N$ and $\mathbf{n}_X$ at points **N** and **X**, respectively, as shown in Figure 9.104.

We start with the normal to the lens at point **N**. The ray r_1 emitted by the edge point $\mathbf{E}_1$ of the source refracts at point **N** toward point **X** as shown in Figure 9.104. This enables us to calculate the normal to the surface $\mathbf{n}_N$ at point **N** as

$$\mathbf{n}_N = \text{rfrnrm}(\mathbf{s}_1, \mathbf{t}_1, 1, n) = (0.774292, 0.632829) \tag{9.58}$$

where $\mathbf{s}_1 = \text{nrm}(\mathbf{N} - \mathbf{E}_1)$ and $\mathbf{t}_1 = \text{nrm}(\mathbf{X} - \mathbf{N}) = (-0.16, -0.987117)$. We may now refract the ray r_2 coming from $\mathbf{E}_2$ toward **N** at this last point obtaining the direction $\mathbf{t}_2$ as

$$\mathbf{t}_2 = \text{rfr}(\mathbf{s}_2, \mathbf{n}_N, 1, n) = (-0.387389, -0.921916) \tag{9.59}$$

where $\mathbf{s}_2 = \text{nrm}(\mathbf{N} - \mathbf{E}_2)$. The path $\mathbf{E}_1$-$\mathbf{N}$-$\mathbf{X}$-$\mathbf{R}_1$ of the ray r_1 is known and the optical path length between points **N** and $\mathbf{R}_1$ may be calculated as

$$S_{NR1} = n[\mathbf{N}, \mathbf{X}] + [\mathbf{X}, \mathbf{R}_1] = 2.04764 \tag{9.60}$$

This enables us to calculate the Cartesian oval between points **X** and $\mathbf{X}_1$ that focuses to point $\mathbf{R}_1$ the rays coming from **N**. We now decide the number of points N_P that we want to calculate for the portion $\mathbf{X}$-$\mathbf{X}_1$. For

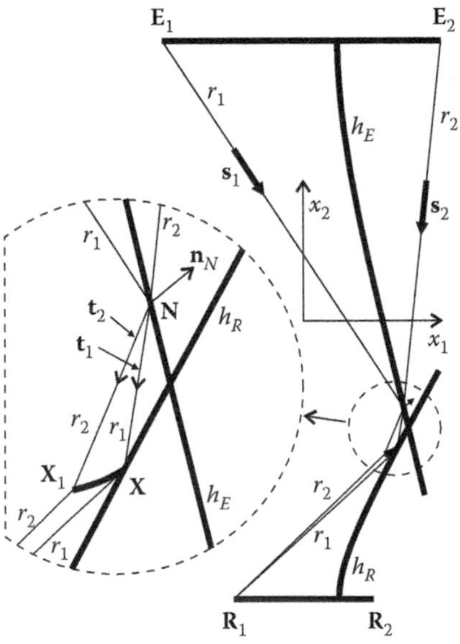

FIGURE 9.104
The design of the RR SMS lens starts with a Cartesian oval $\mathbf{X}$-$\mathbf{X}_1$ on the lower surface of the lens.

example, $N_P = 5$. Intermediate directions between $\mathbf{t}_1$ and $\mathbf{t}_2$ may be calculated, for example, as

$$\mathbf{t} = \text{nrm}(x\mathbf{t}_2 + (1 - x)\mathbf{t}_1) \tag{9.61}$$

with $0 \le x \le 1$ varying in steps of $\Delta x = 1/N_P$. The points on the Cartesian oval **X**-$\mathbf{X}_1$ are obtained as

$$\text{ccoptpt}(\mathbf{N}, n, \mathbf{t}, \mathbf{R}_1, 1, S_{NR1}) \tag{9.62}$$

for each value of **t**. The list of points for this portion of the curve is: CO_1 = ((0.673162,–0.917437,0.524912,–0.851157), (0.656641,–0.926966,0.473593, –0.880744), (0.638962,–0.935759,0.416074,–0.909331), (0.620269,–0.943542,0.351424,–0.936216), (0.60077,–0.95003,0.278588,–0.960411), (0.58074,–0.954948, 0.196453,–0.980513)), where the first two coordinates of each point represent the position and the second two the normal.

The same method can now be used to calculate the first portion of the upper surface of the lens. We first calculate the normal at point **X** as

$$\mathbf{n}_X = \text{rfrnrm}(\text{nrm}(\mathbf{X} - \mathbf{R}_1), \text{nrm}(\mathbf{N} - \mathbf{X}), 1, n) = (0.524912, -0.851157) \tag{9.63}$$

which yields the same resulting normal vector calculated for the first point (point **X**) of the Cartesian oval co_1 for portion **X**-$\mathbf{X}_1$ of the lens. We can also calculate the optical path length between **X** and $\mathbf{E}_1$ as

$$S_{XE1} = n[\mathbf{X}, \mathbf{N}] + [\mathbf{N}, \mathbf{E}_1] = 3.58653 \tag{9.64}$$

We calculate also

$$\begin{aligned} \mathbf{t}_1 &= \text{nrm}(\mathbf{N} - \mathbf{X}) \\ \mathbf{t}_2 &= \text{rfr}(\text{nrm}(\mathbf{X} - \mathbf{R}_2), \mathbf{n}_X, 1, n) = (-0.102201, 0.994764) \end{aligned} \tag{9.65}$$

Again $\mathbf{t} = \text{nrm}(x\mathbf{t}_2 + (1 - x)\mathbf{t}_1)$ with $0 \le x \le 1$ varying in steps of $\Delta x = 1/N_P$. The points on the Cartesian oval **N**-$\mathbf{N}_1$ are obtained as

$$\text{ccoptpt}(\mathbf{X}, n, \mathbf{t}, \mathbf{E}_1, 1, S_{XE1}) \tag{9.66}$$

for each value of **t**. The list of points for this portion of the curve is: CO_2 = ((0.721303,–0.620433,0.774292,0.632829), (0.707238,–0.604229,0.735292,0.677751), (0.69149,–0.5882,0.690403,0.723425), (0.674094,–0.572702, 0.638771,0.769397), (0.655161,–0.558134,0.579425,0.815025), (0.63489,–0.544914,0.511301,0.859401)), where, again, the first two coordinates of each point represent the position and the second two the normal. The two portions of the lens calculated so far are shown in Figure 9.105.

We are now ready to start calculating the SMS chains that extend the surfaces of the lens to the optical axis. Start by calculating the optical

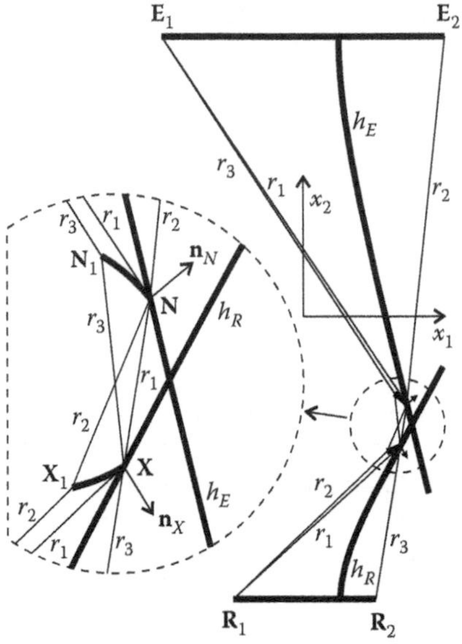

FIGURE 9.105
The first two portions of the lens are Cartesian oval curves **X-X**$_1$ and **N-N**$_1$.

path length between points $\mathbf{R}_2$ and $\mathbf{E}_1$ as $S_{RE1} = [\mathbf{R}_2, \mathbf{X}] + n[\mathbf{X}, \mathbf{N}_1] + [\mathbf{N}_1, \mathbf{E}_1] = 4.68286$ and also the optical path length between points $\mathbf{E}_2$ and $\mathbf{R}_1$, which is $S_{ER2} = S_{RE1}$ by symmetry.

Let us then take, for example, the third point of the first portion of the lower surface of the lens co_1 and calculate the corresponding point on the upper surface of the lens. This point has coordinates $\mathbf{X}_{13} = (0.638962, -0.935759)$ and normal $\mathbf{n}_X 13 = (0.416074, -0.909331)$. The optical path length between $\mathbf{X}_{13}$ and $\mathbf{E}_1$ is given by $S_{13} = S_{RE1} - [\mathbf{R}_2, \mathbf{X}_{13}] = 3.60958$. The direction of the refracted ray at $\mathbf{X}_{13}$ coming from $\mathbf{R}_2$ is given by $\mathbf{t}_{13} = \text{rfr}(\text{nrm}(\mathbf{X}_{13} - \mathbf{R}_2), \mathbf{n}_{X13}, 1, n) = (-0.0677209, 0.997704)$. We can now calculate the corresponding point on the upper surface of the lens as $\mathbf{N}_{13} = \text{ccoptpt}(\mathbf{X}_{13}, n, \mathbf{t}_{13}, \mathbf{E}_1, 1, S_{13}) = (0.611426, -0.530086)$. The normal at $\mathbf{N}_{13}$ is given by $\mathbf{n}_N 13 = \text{rfrnrm}(\mathbf{t}_{13}, \text{nrm}\ (\mathbf{E}_1 - \mathbf{N}_{13}), n, 1) = (0.554893, 0.831922)$. This same process must be repeated for all the points of **X-X**$_1$ resulting in a portion of the lens to the left of $\mathbf{N}_1$. Accordingly, the same process is repeated for all the points of **N-N**$_1$ resulting in a new portion of the lens to the left of $\mathbf{X}_1$. This same process is now repeated for the new points of the lens just calculated, resulting in new portions of the lens while moving toward the axis of symmetry of the emitter and the receiver. After five of these iterations, we reach the axis in this example. The design process grows the surfaces to the left of the vertical axis. We take only the points to the right of it. A complete list of points for both surfaces is then ((0.721303, −0.620433), (0.707238, −0.604229), (0.69149, −0.5882), (0.674094, −0.572702), (0.655161, −0.558134), (0.63489, −0.544914), (0.623548, −0.537957), (0.611426, −0.530086), (0.598783, −0.521426), (0.585999, −0.51221), (0.573593,

−0.5028), (0.551205, −0.485926), (0.526741, −0.468696), (0.500375, −0.451458), (0.472458, −0.434645), (0.44354, −0.418756), (0.415674, −0.404434), (0.385449, −0.389485), (0.353339, −0.374315), (0.320157, −0.359443), (0.287099, −0.345448), (0.242335, −0.328331), (0.193708, −0.312698), (0.142053, −0.299603), (0.0887252, −0.290082), (0.0355541, −0.284913)) for the upper surface and ((0.673162, −0.917437), (0.656641, −0.926966), (0.638962, −0.935759), (0.620269, −0.943542), (0.60077, −0.95003), (0.58074, −0.954948), (0.568433, −0.957662), (0.555116, −0.961122), (0.541026, −0.965332), (0.526515, −0.970238), (0.512055, −0.975706), (0.490197, −0.984131), (0.466905, −0.992482), (0.44246, −1.00056), (0.417302, −1.00813), (0.39205, −1.01494), (0.364609, −1.02189), (0.334999, −1.02938), (0.303642, −1.03725), (0.271216, −1.04528), (0.238675, −1.05321), (0.201091, −1.06175), (0.161133, −1.06961), (0.1196, −1.07627), (0.0776509, −1.08122), (0.036766, −1.08407)) for the lower surface. The left half of the lens is obtained by symmetry relative to the central axis. The complete lens is shown in Figure 9.106.

We may now define a lens by interpolating the points using, for example, a spline. A ray-trace is as shown in Figure 9.107.

A detailed analysis of the focus at point $\mathbf{R}_1$ shows that it is not a point, but has some finite size. There are two waists to be considered. A smaller waist, w_1, is due to the fact that we used a less number of points (only five points per SMS section: $N_P = 5$). The more points we calculate, the smaller w_1 will be. The larger waist, w_2, is produced by the rays that cross the center of the lens for which this SMS calculation method cannot guarantee convergence to a point. Both waists are, however, very small when compared to the size of the lens, as can be seen in Figure 9.107.

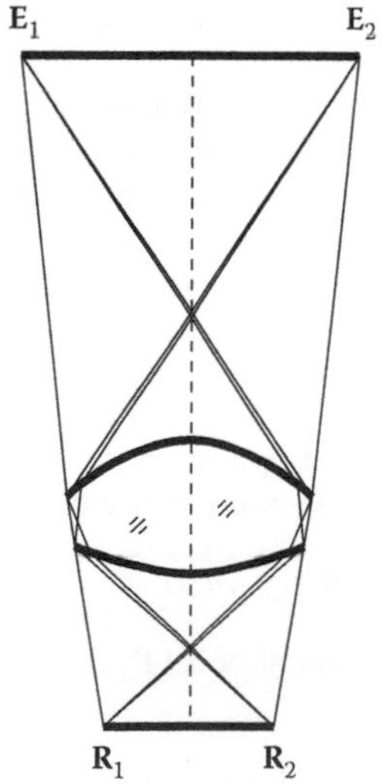

FIGURE 9.106
The complete RR SMS lens.

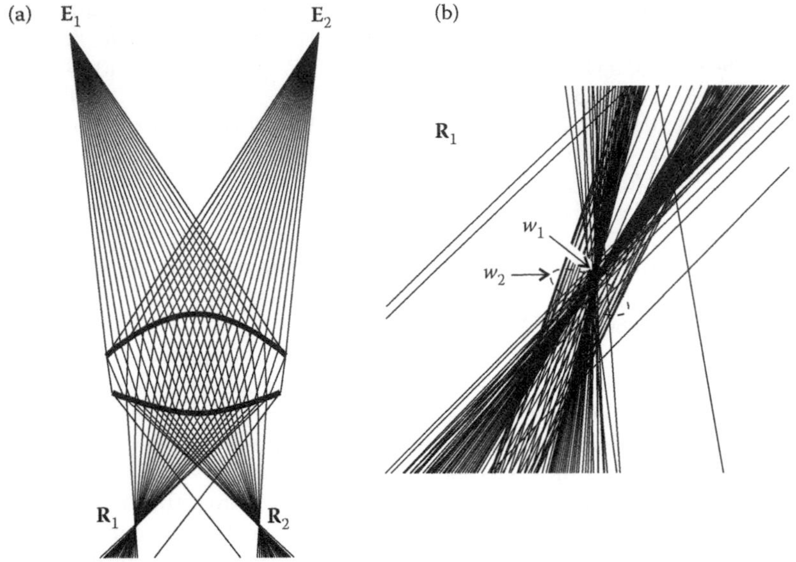

FIGURE 9.107
(a) Ray-tracing of an RR SMS lens. (b) Detail of the rays crossing point $\mathbf{R}_1$ showing that there is a large concentration of rays at a small waist w_1, but some rays spread over a larger waist w_2.

EXAMPLE 9.2

Design an XR SMS optic that concentrates to the edge points of the receiver $\mathbf{R}_1 = (-0.5, 0)$ and $\mathbf{R}_2 = (0.5, 0)$ light reaching the reflective surface with an angular spread of $\theta = 2°$, half-angle. The refractive index of the refractive element is $n = 1.5$.

The XR SMS optic can be designed starting at the edges of the mirror and the refractive surface with Cartesian ovals and then calculating the SMS chains to build the surfaces toward the center, just as with the RR SMS lens in Example 1. The other way to design the optic is to start, for example, at the center, with a prescribed curve, and then calculate the SMS chains toward the edges. This example uses the second method.

Start by specifying a point, for example, $\mathbf{P}_0 = (0, -4)$ and its normal $\mathbf{n}_0 = (0, -1)$ on the refractive surface, as shown in Figure 9.108.

We can now refract at point $\mathbf{P}_0$ a ray r_1 coming from the edge $\mathbf{R}_2$ of the source. After refraction it is headed in the direction

$$\mathbf{t} = \text{rfr}(\text{nrm}(\mathbf{P}_0 - \mathbf{R}_2), \mathbf{n}_0, n, 1) = (-0.186052, -0.98254) \tag{9.67}$$

We now choose the position of point $\mathbf{P}_1$ in the direction of the refracted ray as

$$\mathbf{P}_1 = \mathbf{P}_0 + 10\mathbf{t} = (-1.86052, -13.8254) \tag{9.68}$$

Its normal can also be calculated because we know that, after reflection on the mirror, this ray must be parallel to $\mathbf{s}_1$ with $\mathbf{s}_1 = (\cos(\pi/2 + \theta), \sin(\pi/2 + \theta))$. The normal at $\mathbf{P}_1$ is then given by

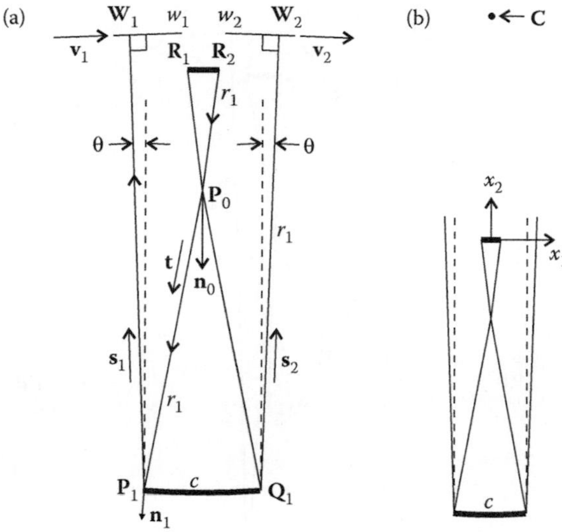

FIGURE 9.108
(a) The design of an XR optic may start by specifying a central portion c and then calculating the SMS points based on this curve. (b) In this example, the central portion c is chosen as a circle with center **C**.

$$\mathbf{n}_1 = \text{rfxnrm}(\mathbf{t},\mathbf{s}_1) = (-0.0760445, -0.997104) \tag{9.69}$$

By symmetry, we can also define another point $\mathbf{Q}_1$ on the other side of the mirror, which will also have a normal symmetric to that of $\mathbf{P}_1$. We now choose the shape of the mirror between the points $\mathbf{P}_1$ and $\mathbf{Q}_1$ as a circle c whose center **C** is at the intersection of the axis of symmetry x_2 and the straight line defined by the point $\mathbf{P}_1$ and its normal $\mathbf{n}_1$

$$\mathbf{C} = \text{isl}((0,0),(0,1),\mathbf{P}_1,\mathbf{n}_1) = (0, 10.57) \tag{9.70}$$

We must also define the flat wavefronts w_1 and w_2 perpendicular to the two bundles of incoming parallel rays. The plane wavefront w_1 is defined by a straight line passing through point $\mathbf{W}_1$ that we choose to be at position $\mathbf{W}_1 = (W_{11}, W_{12}) = \mathbf{P}_1 + 15\mathbf{s}_1$ and tangent vector $\mathbf{v}_1 = (\cos\theta, \sin\theta)$. The wavefront w_2 is defined by point $\mathbf{W}_2 = (-W_{11}, W_{12})$ and tangent vector $\mathbf{v}_2 = (\cos(-\theta), \sin(-\theta))$.

Now we can calculate the optical path length between w_1 and $\mathbf{R}_2$ as

$$S = [\mathbf{W}_1,\mathbf{P}_1] + [\mathbf{P}_1,\mathbf{P}_0] + n[\mathbf{P}_0,\mathbf{R}_2] = 31.0467 \tag{9.71}$$

By symmetry this is also the optical path length between w_2 and $\mathbf{R}_1$.

We now have all the ingredients to start building the SMS chains. We can take a set of, for example, $N_P = 5$ points, on c at equiangular spacing between $\mathbf{Q}_1$ and $\mathbf{P}_1$ and the corresponding normals to c. We may drop

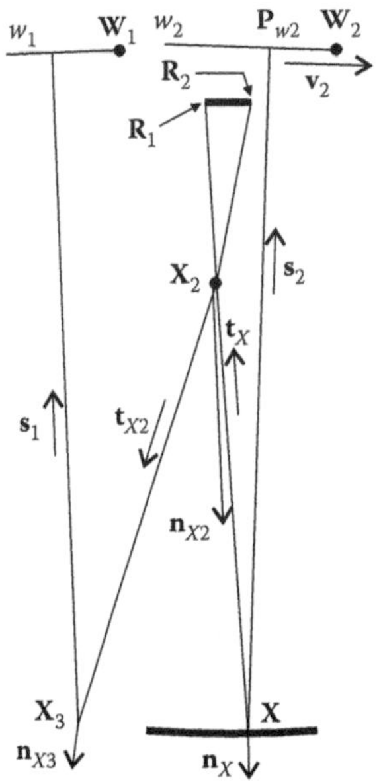

FIGURE 9.109
An SMS chain. A ray coming from $\mathbf{P}_{w2}$ reflects at $\mathbf{X}$ toward $\mathbf{X}_2$ and there it is refracted to $\mathbf{R}_1$. Another ray coming from $\mathbf{R}_2$ refracts at $\mathbf{X}_2$, and then reflects off $\mathbf{X}_3$ in direction $\mathbf{s}_1$.

the last point $\mathbf{P}_1$ of this list to avoid repeated points in the SMS chains. These points are ((1.86052, –13.8254, 0.0760445, –0.997104), (1.117, –13.8707, 0.0456549, –0.998957), (0.372449, –13.8934, 0.015223, –0.999884), (–0.372449, –13.8934, –0.015223, –0.999884), (–1.117, –13.8707, –0.0456549, –0.998957)), where the first two components are the position and the second two the normal.

We now take one of the points on c to exemplify the calculation of the SMS chain. For example, take point $\mathbf{X}$ = (0.372449, –13.8934) and the corresponding normal $\mathbf{n}_X$ = (0.015223, –0.999884). We can now reflect at this point a ray perpendicular to w_2. to do this, we need to first determine for what point $\mathbf{P}_{w2}$ of w_2, the corresponding light ray passes through $\mathbf{X}$ (see Figure 9.109). We have

$$\mathbf{P}_{w2} = \text{isl}(\mathbf{X},\mathbf{s}_2,\mathbf{W}_2,\mathbf{v}_2) = (0.900126, 1.21728) \tag{9.72}$$

The optical path length between point $\mathbf{X}$ and $\mathbf{R}_1$ is then $S_X = S - [\mathbf{X}, \mathbf{P}_{w2}] = 15.9268$. The reflected ray at point $\mathbf{X}$ has direction

$$\mathbf{t}_X = \text{rfx}(-\mathbf{s}_2,\mathbf{n}_X) = (-0.0653073, 0.997865) \tag{9.73}$$

We need to now calculate the point on the Cartesian oval that focuses to point $\mathbf{R}_1$ the ray coming from point $\mathbf{X}$ in direction $\mathbf{t}_X$

$$\mathbf{X}_2 = \text{ccoptpt}(\mathbf{X},1,\mathbf{t}_X,\mathbf{R}_1,n,S_X) = (-0.274686, -4.00547) \tag{9.74}$$

The normal at this point can now be calculated as

$$\mathbf{n}_{X2} = \text{rfrnrm}(\mathbf{t}_X, \text{nrm}(\mathbf{R}_1 - \mathbf{X}_2), 1, n) = (0.0378641, -0.999283) \quad (9.75)$$

We can now repeat the process for a ray coming from $\mathbf{R}_2$, refracted at $\mathbf{X}_2$ and calculate the corresponding point on the mirror to reflect it in the direction of

$$\mathbf{s}_1 = (\cos(\pi/2 + \theta), \sin(\pi/2 + \theta)) \quad (9.76)$$

The optical path length between $\mathbf{X}_2$ and w_1 is $S_X2 = S - n[\mathbf{R}_2, \mathbf{X}_2] = 24.9272$. We now refract at $\mathbf{X}_2$ the ray coming from $\mathbf{R}_2$ as

$$\mathbf{t}_{X2} = \text{rfr}(\text{nrm}(\mathbf{X}_2 - \mathbf{R}_2), \mathbf{n}_{X2}, n, 1) = (-0.304544, -0.952498) \quad (9.77)$$

Next calculate the point on the Cartesian oval that makes the rays coming from point $\mathbf{X}_2$ to become perpendicular to the straight line w_1, for a particular direction $\mathbf{t}_{X2}$

$$\mathbf{X}_3 = \text{coptsl}(\mathbf{X}_2, 1, \mathbf{t}_{X2}, \mathbf{W}_1, 1, \mathbf{s}_1, S_{X2}) = (-3.36293, -13.6643) \quad (9.78)$$

Finally, we calculate the normal to the mirror at $\mathbf{X}_3$

$$\mathbf{n}_{X3} = \text{frxnrm}(\mathbf{t}_{X2}, \mathbf{s}_1) = (-0.136846, -0.990592) \quad (9.79)$$

This process must now be repeated for all other points of c and then for the new points we calculate. A complete list of points is ((−0.372449, −13.8934), (−1.117, −13.8707), (−1.86052, −13.8254), (−2.60783, −13.7569), (−3.36293, −13.6643), (−4.12466, −13.5471), (−4.89286, −13.4046), (−5.66858, −13.2358), (−6.45408, −13.0391), (−7.24813, −12.8138), (−8.04916, −12.5595), (−8.85805, −12.2751), (−9.6786, −11.9579), (−10.5035, −11.6097), (−11.3162, −11.2377), (−12.1135, −10.8449), (−12.8973, −10.4318), (−13.6756, −9.99512), (−14.424, −9.5502), (−15.1094, −9.1216), (−15.7396, −8.70992), (−16.3319, −8.30809), (−16.9082, −7.90342), (−17.4391, −7.51904), (−17.8916, −7.18315), (−18.2898, −6.88175), (−18.6611, −6.59628), (−19.0324, −6.30671), (−19.3709, −6.03937), (−19.6433, −5.82238), (−19.8769, −5.63545), (−20.099, −5.45717), (−20.3351, −5.26718), (−20.5505, −5.09347), (−20.7113, −4.9639), (−20.8434, −4.85785), (−20.972, −4.75509), (−21.1199, −4.63747), (−21.2519, −4.53291), (−21.3344, −4.46803), (−21.3905, −4.42431), (−21.4416, −4.38492)) for the mirror and ((0,−4.), (−0.135644, −4.0014), (−0.274686, −4.00547), (−0.416515, −4.01198), (−0.560526, −4.02071), (−0.706103, −4.03147), (−0.856721, −4.04432), (−1.01593, −4.05894), (−1.18322, −4.07455), (−1.35811, −4.09036), (−1.54021, −4.10571), (−1.733, −4.12013), (−1.93955, −4.13243), (−2.15893, −4.14094), (−2.39036, −4.1441), (−2.6335, −4.14049), (−2.88953, −4.1286), (−3.15683, −4.1064), (−3.4319, −4.07221), (−3.71228, −4.0248), (−3.9971, −3.96305), (−4.28097, −3.88722), (−4.55523, −3.79897), (−4.81814, −3.69882), (−5.07093, −3.58644), (−5.31653, −3.46034), (−5.54775, −3.32438), (−5.75639, −3.18424),

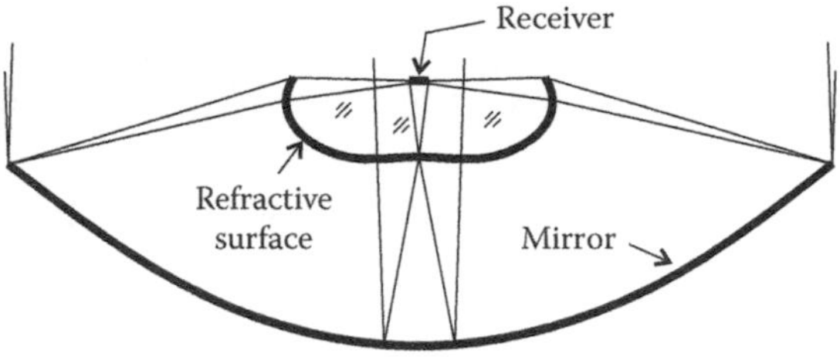

FIGURE 9.110
An XR optic.

(−5.94625, −3.03872), (−6.12191, −2.8849), (−6.28754, −2.71858), (−6.43621, −2.54643), (−6.56186, −2.37674), (−6.66913, −2.20519), (−6.76167, −2.02585), (−6.84127, −1.83188), (−6.90277, −1.63261), (−6.94319, −1.43863), (−6.96476, −1.24271), (−6.96715, −1.03605), (−6.9467, −0.808906), (−6.90058, −0.572656), (−6.82955, −0.339663), (−6.72835, −0.098585), (−6.58436, 0.165038)) for the lens. The profile of the optical surfaces may be obtained by the interpolation of these points using, for example, a spline fit. The optic is shown in Figure 9.110.

If we ray-trace this optic with sets of parallel rays tilted $\pm\theta$ to the vertical we will not get perfect focusing onto points $\mathbf{R}_1$ and $\mathbf{R}_2$. That is due to the small number of points ($N_P = 5$) calculated on the initial curve c. The higher the N_P, the more accurate will be the focusing.

EXAMPLE 9.3

Calculate the SMS 3-D chains of an RR optic with refractive index $n = 1.5$. The calculation starts by specifying flat input wavefront w_A defined by point $\mathbf{A} = (0.5, 0, 3.5)$ and normal $\mathbf{n}_A = (\cos 75°, 0, \sin 75°) = (0.258819, 0, 0.965926)$, flat wavefront w_B defined by point $\mathbf{B} = (-0.5, 0, 3.5)$ and normal $\mathbf{n}_B = (\cos 105°, 0, \sin 105°)$. Both wafefronts are, therefore, tilted by 15° relative to the vertical axis x_3. The output wavefronts are two points at positions $\mathbf{R}_A = (1.3, 0, -5)$ and $\mathbf{R}_B = (-1.3, 0, -5)$, as shown in Figure 9.111a. The optical path length between w_A and $\mathbf{R}_B$ and between w_B and $\mathbf{R}_A$ is specified as $S = 9.7$. We also specify point $\mathbf{P}_0 = (0, 0, 1)$ and its normal $\mathbf{n}_0 = (0, 0, 1)$.

We start by calculating the path of ray $\mathbf{B}_0$-$\mathbf{P}_0$-$\mathbf{H}_0$-$\mathbf{R}_A$. Point $\mathbf{B}_0$ on wavefront w_B is obtained from $\mathbf{B}_0 = \text{islp}(\mathbf{P}_0, \mathbf{n}_B, \mathbf{B}, \mathbf{n}_B) = (-0.658494, 0, 3.45753)$. The optical path length between $\mathbf{P}_0$ and $\mathbf{R}_A$ is given by $S_1 = S\text{-}[\mathbf{P}_0, \mathbf{B}_0] = 7.15578$. Vector $\mathbf{v}_1$ of the ray refracted at $\mathbf{P}_0$, coming from w_B, is given by $\mathbf{v}_1 = \text{rfr}(-\mathbf{n}_B, \mathbf{n}_0, 1, n) = (0.172546, 0, -0.985001)$. Point $\mathbf{H}_0$ on the bottom surface of the lens is now given by $\mathbf{H}_0 = \text{ccoptpt}(\mathbf{P}_0, n, \mathbf{v}_1, \mathbf{R}_A, 1, S_1) = (0.349972, 0, -0.997861)$. The normal to point $\mathbf{H}_0$ is given by $\mathbf{h}_0 = \text{rfrnrm}(\mathbf{v}_1, \text{nrm}(\mathbf{R}_A\text{-}\mathbf{H}_0), n, 1) = (0.055129, 0, -0.998479)$.

We may now calculate the path of ray $\mathbf{R}_B$-$\mathbf{H}_0$-$\mathbf{T}_0$-$\mathbf{A}_0$. The optical path length between $\mathbf{H}_0$ and w_A is given by $S_2 = S\text{-}[\mathbf{R}_B, \mathbf{H}_0] = 5.37108$. Vector $\mathbf{v}_2$ of the ray refracted at $\mathbf{H}_0$, coming from $\mathbf{R}_B$, is given by $\mathbf{v}_2 = \text{rfr}(\text{nrm}(\mathbf{H}_0\text{-}\mathbf{R}_B), \mathbf{h}_0, 1, n) = (0.234457, 0, 0.972127)$. Point $\mathbf{T}_0$ is now given by $\mathbf{T}_0 = \text{coptsl}(\mathbf{H}_0, n,$

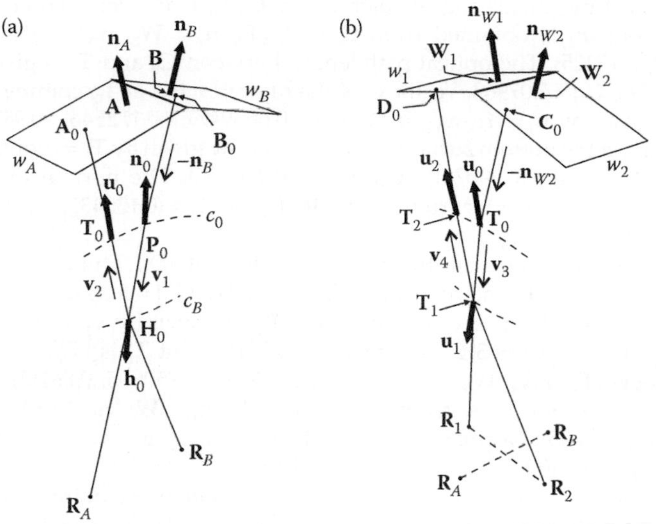

FIGURE 9.111
(a) Initial curve c_0 is calculated with wavefronts w_A, w_B, $\mathbf{R}_A$, and $\mathbf{R}_B$. (b) Wavefronts w_1, w_2, $\mathbf{R}_1$, and $\mathbf{R}_2$ are used to calculate new SMS points based on the points (and normals) of c_0.

$\mathbf{v}_2$, $\mathbf{A}$, 1, $\mathbf{n}_A$, S_2) = (0.812803, 0, 0.921173). Point $\mathbf{A}_0$ on wavefront w_A is given by $\mathbf{A}_0$ = islp($\mathbf{T}_0$, $\mathbf{n}_A$, $\mathbf{A}$, $\mathbf{n}_A$) = (1.43656, 0, 3.24905). Normal $\mathbf{u}_0$ to point $\mathbf{T}_0$ is given by $\mathbf{u}_0$ = rfrnrm(nrm($\mathbf{T}_0$-$\mathbf{H}_0$), $\mathbf{n}_A$, n, 1) = (0.185381, 0, 0.982667).

This process may now continue as we calculate more points and normals on the top curve c_0 and bottom curve c_B of the RR SMS lens.

A list of points and normals on the top curve c_0 is ((1.46339, 0, 0.764853, 0.260348, 0, 0.965515), (0.812803, 0, 0.921173, 0.185381, 0, 0.982667), (0, 0, 1, 0, 0, 1), (−0.812803, 0, 0.921173, −0.185381, 0, 0.982667), (−1.46339, 0, 0.764853, −0.260348, 0, 0.965515)).

A list of points and normals on the bottom curve c_B is ((1.01022, 0, −0.922831, 0.17688, 0, −0.984232), (0.349972, 0, −0.997861, 0.055129, 0, −0.998479), (−0.349972, 0, −0.997861, −0.055129, 0, −0.998479), (−1.01022, 0, −0.922831, −0.17688, 0, −0.984232)).

Each element in these lists has the position (first three coordinates) and the normal calculated from the SMS method at that position (last three cordinates).

We now take point $\mathbf{T}_0$ on the top SMS curve c_0 calculated above and use it to calculate new SMS 3-D points, as shown in Figure 9.111b. Wavefronts w_1, w_2, $\mathbf{R}_1$ and $\mathbf{R}_2$ are obtained by rotating wavefronts w_A, w_B, $\mathbf{R}_A$, and $\mathbf{R}_B$ by −90° around the x_3 axis. Wavefront w_1 = (0, −0.5, 3.5, 0, −0.258819, 0.965926), where the first three components are point $\mathbf{W}_1$ and the last three components are normal $\mathbf{n}_{W1}$. Wavefront w$_2$ = (0, 0.5, 3.5, 0, 0.258819, 0.965926), where the first three components are point $\mathbf{W}_2$ and the last three components are normal $\mathbf{n}_{W2}$. Wavefronts $\mathbf{R}_1$ = (0, −1.3, −5) and $\mathbf{R}_2$ = (0, 1.3, 5).

We start by calculating the path of ray $\mathbf{C}_0$-$\mathbf{T}_0$-$\mathbf{T}_1$-$\mathbf{R}_1$. Point Point $\mathbf{C}_0$ on wavefront w_2 is obtained from $\mathbf{C}_0$ = islp($\mathbf{T}_0$, $\mathbf{n}_{W2}$, $\mathbf{W}_2$, $\mathbf{n}_{W2}$) = (0.812803, 0.6782, 3.45225). The optical path length between $\mathbf{C}_0$ and $\mathbf{T}_0$ is given by $S_3 = S$-[$\mathbf{T}_0$, $\mathbf{C}_0$] = 7.07963. Vector $\mathbf{v}_3$ of the ray refracted at $\mathbf{T}_0$, coming from w_2, is given by $\mathbf{v}_3$ = rfr(-$\mathbf{n}_{W2}$, $\mathbf{u}_0$, 1, n) = (−0.0639477, −0.172546, −0.982923). Point $\mathbf{T}_0$ on the bottom surface of the lens is now given by $\mathbf{T}_1$ = ccoptpt($\mathbf{T}_0$, n, $\mathbf{v}_3$, $\mathbf{R}_1$, 1, S_3) = (0.690781, −0.329246, −0.954406). The normal to point $\mathbf{T}_1$ is given by $\mathbf{u}_1$ = rfrnrm($\mathbf{v}_3$, nrm($\mathbf{R}_1$-$\mathbf{T}_1$), n, 1) = (0.130432, −0.055039, −0.989928).

We may now calculate the path of ray $\mathbf{R}_2$-$\mathbf{T}_1$-$\mathbf{T}_2$-$\mathbf{D}_0$. The optical path length between $\mathbf{T}_1$ and w_1 is given by $S_4 = S$-[$\mathbf{R}_2$, $\mathbf{T}_1$] = 5.28429. Vector $\mathbf{v}_4$ of the ray refracted at $\mathbf{T}_1$, coming from $\mathbf{R}_2$, is given by $\mathbf{v}_4$ = rfr(nrm($\mathbf{T}_1$-$\mathbf{R}_2$), $\mathbf{u}_1$, 1, n) = (0.0566312, −0.225866, 0.972511). Point $\mathbf{T}_2$ is now given by $\mathbf{T}_2$ = coptsl($\mathbf{T}_1$, n, $\mathbf{v}_4$, $\mathbf{W}_1$, 1, $\mathbf{n}_{W1}$, S_4) = (0.796503, −0.750905, 0.86113). Point $\mathbf{D}_0$ on wavefront w_1 is given by $\mathbf{D}_0$ = islp($\mathbf{T}_2$, $\mathbf{n}_{W1}$, $\mathbf{W}_1$, $\mathbf{n}_{W1}$) = (0.796503, −1.39382, 3.2605). Normal $\mathbf{u}_2$ to point $\mathbf{T}_2$ is given by $\mathbf{u}_2$ = rfrnrm(nrm($\mathbf{T}_2$-$\mathbf{T}_1$), $\mathbf{n}_{W1}$, n, 1) = (0.167726, −0.157919, 0.973103).

This process may now continue as we calculate more points and normals on the top and bottom surfaces of the RR SMS lens using the points of the initial curve c_0 calculated above. A list of points and normals on the top surface is (((1.39029, −1.13396, 0.689373, 0.142787, −0.111239, 0.983482), (0.756508, −1.36698, 0.740989, 0.116338, −0.218387, 0.968903), (0, −1.46339, 0.764853, 0, −0.260348, 0.965515), (−0.756508, −1.36698, 0.740989, −0.116338, −0.218387, 0.968903), (−1.39029, −1.13396, 0.689373, −0.142787, −0.111239, 0.983482)), ((1.44351, −0.609109, 0.738966, 0.232015, −0.0931498, 0.968242), (0.796503, −0.750905, 0.86113, 0.167726, −0.157919, 0.973103), (0, −0.812803, 0.921173, 0, −0.185381, 0.982667), (−0.796503, −0.750905, 0.86113, −0.167726, −0.157919, 0.973103), (−1.44351, −0.609109, 0.738966, −0.232015, −0.0931498, 0.968242)), ((1.46339, 0, 0.764853, 0.260348, 0, 0.965515), (0.812803, 0, 0.921173, 0.185381, 0, 0.982667), (0, 0, 1, 0, 0, 1), (−0.812803, 0, 0.921173, −0.185381, 0, 0.982667), (−1.46339, 0, 0.764853, −0.260348, 0, 0.965515)), ((1.44351, 0.609109, 0.738966, 0.232015, 0.0931498, 0.968242), (0.796503, 0.750905, 0.86113, 0.167726, 0.157919, 0.973103), (0, 0.812803, 0.921173, 0, 0.185381, 0.982667), (−0.796503, 0.750905, 0.86113, −0.167726, 0.157919, 0.973103), (−1.44351, 0.609109, 0.738966, −0.232015, 0.0931498, 0.968242)), ((1.39029, 1.13396, 0.689373, 0.142787, 0.111239, 0.983482), (0.756508, 1.36698, 0.740989, 0.116338, 0.218387, 0.968903), (0, 1.46339, 0.764853, 0, 0.260348, 0.965515), (−0.756508, 1.36698, 0.740989, −0.116338, 0.218387, 0.968903), (−1.39029, 1.13396, 0.689373, −0.142787, 0.111239, 0.983482))).

A list of points and normals on the bottom surface is (((1.3185, −0.821472, −0.762582, 0.30345, −0.180023, −0.935687), (0.691184, −0.955168, −0.882298, 0.145044, −0.177442, −0.973384), (0., −1.01022, −0.922831, 0., −0.17688, −0.984232), (−0.691184, −0.955168, −0.882298, −0.145044, −0.177442, −0.973384), (−1.3185, −0.821472, −0.762582, −0.30345, −0.180023, −0.935687)), ((1.31575, −0.280455, −0.829336, 0.277182, −0.0561894, −0.959173), (0.690781, −0.329246, −0.954406, 0.130432, −0.055039, −0.989928), (0., −0.349972, −0.997861, 0., −0.055129, −0.998479), (−0.690781, −0.329246, −0.954406, −0.130432, −0.055039, −0.989928), (−1.31575, −0.280455, −0.829336, −0.277182, −0.0561894, −0.959173)), ((1.31575, 0.280455, −0.829336, 0.277182, 0.0561894, −0.959173), (0.690781, 0.329246, −0.954406, 0.130432, 0.055039,

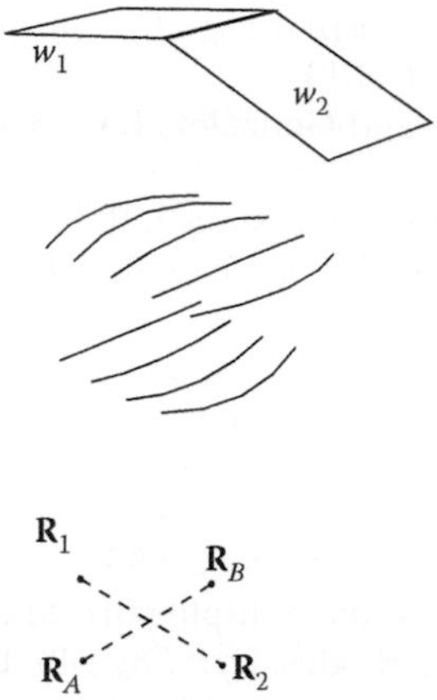

FIGURE 9.112
Sets of curves defining the top and bottom surfaces of the SMS 3-D lens.

−0.989928), (0., 0.349972, −0.997861, 0., 0.055129, −0.998479), (−0.690781, 0.329246, −0.954406, −0.130432, 0.055039, −0.989928), (−1.31575, 0.280455, −0.829336, −0.277182, 0.0561894, −0.959173)), ((1.3185, 0.821472, −0.762582, 0.30345, 0.180023, −0.935687), (0.691184, 0.955168, −0.882298, 0.145044, 0.177442, −0.973384), (0., 1.01022, −0.922831, 0., 0.17688, −0.984232), (−0.691184, 0.955168, −0.882298, −0.145044, 0.177442, −0.973384), (−1.3185, 0.821472, −0.762582, −0.30345, 0.180023, −0.935687))).

These points form curves on the top and bottom surfaces of the SMS lens, as shown in Figure 9.112.

These curves may now be interpolated to generate the SMS surfaces.

References

1. Miñano, J. C. et al., High efficiency non-maging optics, *United States Patent 6.639.733*, 2003.
2. Stavroudis, O. N., *The Optics of Rays, Wave Fronts, and Caustics*, Academic Press, New York, 1972.
3. Schulz, G., Aspheric surfaces, In *Progress in Optics* (Wolf, E., ed.), Vol. XXV, North-Holland, Amsterdam, 351, 1988.
4. Schulz, G., Achromatic and sharp real imaging of a point by a single aspheric lens. *Appl. Opt.*, 22, 3242, 1983.
5. Miñano, J. C. and González, J. C., New method of design of nonimaging concentrators, *Appl. Opt.*, 31, 3051, 1992.
6. Winston, R. et al., *Nonimaging Optics*, Elsevier Academic Press, Amsterdam, 2005.
7. Benitez, P., *Advanced concepts of non-imaging optics: design and manufacture*, PhD thesis, Polytechnic University of Madrid, 1998.

8. Koshel, R. J., *Illumination Engineering: Design with Nonimaging Optics*, Wiley IEEE Press, Piscataway, NJ, 2013.
9. Miñano, J. C., Benítez, P., and González, J. C., RX: A nonimaging concentrator, *Appl. Opt.*, 34, 2226, 1985.
10. Miñano, J. C., Gonzalez, J. C., and Benitez, P., New non-imaging designs: The RX and the RXI concentrators, *Nonimaging optics: Maximum-Efficiency Light Transfer II*, SPIE Vol., 2016, 120, 1993.
11. Gonzalez, J. C. and Miñano, J. C., Design of optical systems which transform one bundle of incoherent light into another, *Nonimaging Optics: Maximum-Efficiency Light Transfer II*, SPIE Vol. 2016, 109, 1993.
12. Miñano, J. C., González, J. C. and Benítez, P., A high-gain, compact, nonimaging concentrator: RXI, *Appl. Opt.*, 34, 7850, 1985.
13. Muñoz, F. et al., Simultaneous multiple surface design of compact air-gap collimators for light-emitting diodes, *Opt. Eng.*, 43, 1522, 2004.
14. Benitez, P. et al., SMS design method in 3D geometry: Examples and applications, *Nonimaging Optics: Maximum Efficiency Light Transfer VII*, SPIE Vol. 5185, 18, 2004.
15. Benitez, P. et al., Simultaneous multiple surface optical design method in three dimensions, *Opt. Eng.*, 43, 1489, 2004.
16. Benitez, P., Mohedano, R., and Miñano, J. C., Design in 3D geometry with the simultaneous multiple surface design method of nonimaging optics, *Nonimaging Optics: Maximum Efficiency Light Transfer V*, SPIE Vol. 3781, 12, 1999.
17. Miñano, J. C. et al., Free-form integrator array optics, *Nonimaging Optics and Efficient Illumination Systems II*, SPIE Vol. 5942, 114, 2005.
18. Plesniak, A. et al., Demonstration of high performance concentrating photovoltaic module designs for utility scale power generation. *Proc. ICSC-5*, Palm Desert, CA, USA, 2008.
19. Cvetkovic, A. et al., The free form XR photovoltaic concentrator: A high performance SMS3D design, *High and Low Concentration for Solar Electric Applications III, Proc. SPIE*, Vol. 7043, August 2008.
20. Dross, O. et al., Review of SMS design methods and real-world applications, *Nonimaging Optics and Efficient Illumination Systems*, SPIE Vol. 5529, 35, 2004.
21. Chaves, J., Miñano, J. C., and Benitez, P., Afocal video-pixel lens for tricolor LEDs, *Nonimaging Optics and Efficient Illumination Systems II*, SPIE Vol. 5942, 18, 2005.
22. Alvarez, J. L. et al., TIR-R concentrator: A new compact high-gain SMS design, *Nonimaging Optics: Maximum Efficiency Light Transfer VI*, SPIE Vol. 4446, 32, 2001.
23. Benitez, P., Garcia R., and Miñano, J. C., Contactless efficient two-stage solar concentrator for tubular absorber, *Appl. Opt.*, 36, 7119, 1997.
24. Benitez, P. et al., Contactless two-stage solar concentrators for tubular absorber, *Nonimaging Optics: Maximum Efficiency Light Transfer IV*, SPIE Vol. 3139, 205, 1997.
25. Benitez, P. et al., New nonimaging static concentrators for bifacial photovoltaic solar cells, *Nonimaging Optics: Maximum Efficiency Light Transfer V*, SPIE Vol. 3781, 22, 1999.
26. Benitez, P. and Miñano, J. C., Ultrahigh-numerical-aperture imaging concentrator, *J. Opt. Soc. Am. A*, 14, 1988, 1997.

10

Wavefronts for Prescribed Output

10.1 Introduction

Nonimaging optics are well suited to transfer light from an emitter to a receiver. However, in many applications, efficient light transfer is not sufficient, and a given irradiance pattern is also desired. In those situations, simply using the edge ray principle to couple the edges of the emitter onto the edges of the receiver, will, typically, not generate the desired result.

Shaping the output wavefronts of a nonimaging optic provides extra degrees of freedom that allow the light output to also be shaped. This process defines how much of the exit aperture of the optic illuminates a given point on the receiver and this, in turn, defines the light distribution on it. Now the edge rays from the emitter are no longer redirected to the edges of the receiver. Instead, they are redirected to points inside it, according to the desired irradiance pattern. This same principle may also be used to shape the output intensity pattern produced by an optic.

By shaping the output wavefronts, nonimaging optics may be designed to generate a prescribed irradiance pattern on a receiver, or a prescribed intensity pattern on the far field.

10.2 Wavefronts for Prescribed Intensity

An optic O_1 captures light from an emitter E and redirects it in varying directions, as shown in Figure 10.1. For simplicity, below we assume that the emitter E is immersed in air ($n = 1$), and that the optic also emits into air ($n = 1$). However, the same methods still apply when the emitter is immersed in different refractive index material, or when the optic emits into a medium of different refractive index.

The exit aperture of the optic extends from x_L on the left to x_R on the right. At a position, x_A, on its aperture, the light emitted is contained between rays r_{A1} and r_{A2}. These make angles θ_{A1} and θ_{A2} to the vertical (normal to the optic

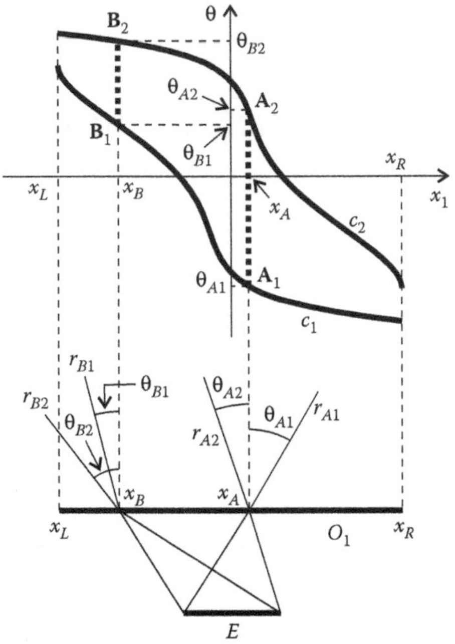

FIGURE 10.1
Optic O_1 captures light from emitter E and redirects it in varying directions. Each edge ray of the emitted radiation may be represented as a point on a (x_1,θ) plot as shown at the top. The two sets of edge rays result in two curves c_1 and c_2 in the (x_1,θ) space.

aperture). Edge ray r_{A1} may be represented as a point $\mathbf{A}_1 = (x_A,\theta_{A1})$ in a (x_1,θ) plot, as shown on top. Also, edge ray r_{A2} may be represented as another point $\mathbf{A}_2 = (x_A,\theta_{A2})$ in the same plot. The same may be done for another position x_B on the optic aperture. The edge rays r_{B1} and r_{B2} at this position make angles θ_{B1} and θ_{B2} to the vertical, and may be represented as points $\mathbf{B}_1 = (x_B,\theta_{B1})$ and $\mathbf{B}_2 = (x_B,\theta_{B2})$ in plot (x_1,θ). Repeating this process for all positions from x_L to x_R, two curves (functions) $c_1(x)$ and $c_2(x)$ are obtained describing the edge rays of the emitted light across the whole optic aperture. The angles that the edge rays make to the vertical, may be obtained from these functions: $\theta_{A1} = c_1(x_A)$, $\theta_{A2} = c_2(x_A)$, $\theta_{B1} = c_1(x_B)$, $\theta_{B2} = c_2(x_B)$.

The directions of the edge rays across optic O_1 also define the étendue of the light emitted. From the geometry in Figure 10.2, we have

$$U_O = \int_{x_L}^{x_R} (\sin\theta_2 - \sin\theta_1)dx = \int_{x_L}^{x_R} (\sin(c_2(x)) - \sin(c_1(x)))dx \qquad (10.1)$$

since the optic emits into air ($n = 1$).

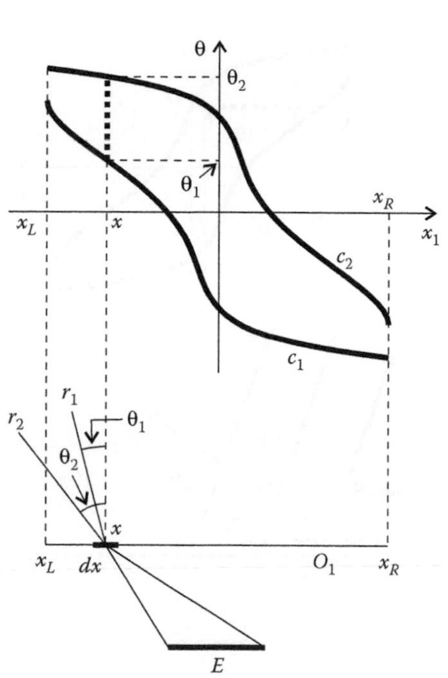

FIGURE 10.2
The étendue of the light emitted by optic O_1 may be obtained integrating the étendue through an infinitesimal portion dx across the whole aperture from x_L to x_R.

It is possible to choose an emitting direction **v** and determine how much intensity the optic emits in that direction, as shown in Figure 10.3. Direction **v** makes an angle θ_I to the vertical. A horizontal line at height $\theta = \theta_I$ in the (x_1,θ) plot intersects curves c_1 and c_2 at points **C** and **D**, with horizontal coordinates x_C and x_D, respectively. The aperture of optic O_1 between horizontal coordinates x_C and x_D emits light in direction **v**, while points outside x_C to x_D do not. Take, for example, a point at horizontal coordinate x_E on optic O_1. It emits light inside an angle α_E, which includes direction **v**. Therefore, this point is lit when seen from direction **v**. At position x_C, the emission angle α_C has edge ray r_{C1} pointing in direction **v**. Therefore, this point is also lit when seen from direction **v**. However, to the left of x_C, another point at x_G has emission angle α_G, which does not include direction **v** and, therefore, this point is seen as dark from direction **v**. Accordingly, at position x_D, the emission angle α_D has edge ray r_{D2} pointing in direction **v**. Therefore, this point is also lit when seen from direction **v**. However, to the right of x_D, another point at x_F has emission angle α_F, which does not include direction **v** and, therefore, this point is seen as dark from direction **v**. Therefore, when seen from direction **v**, optic O_1 has a lit area extending from x_C to x_D, which is the same as the distance from point **C** to point **D** in the (x_1,θ) plot. The intensity emitted

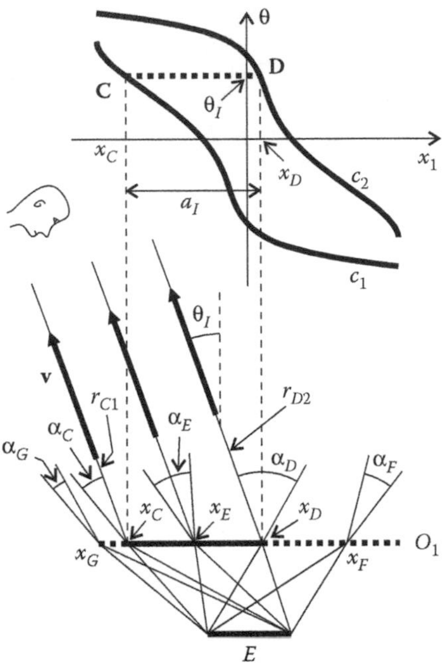

FIGURE 10.3
When seen from direction **v** making an angle θ_I to the vertical, optic O_1 has a lit area extending from x_C to x_D. The intensity I_I produced by optic O_1 in direction **v** may be obtained from the distance from points **C** do **D** in the (x_1,θ) plot and it is proportional to **[C,D]** cos θ_I.

in direction **v** is then proportional to **[C,D]** cos θ_I where **[C,D]** $= x_D - x_C$ is the distance between points **C** and **D**.

The intensity produced by a two-dimensional optic is given by

$$I = \frac{d\Phi}{d\theta} = L\frac{dU}{d\theta} = L\frac{a\cos\theta\, d\theta}{d\theta} = La\cos\theta \tag{10.2}$$

where L is the radiance, $d\Phi = LdU$ is the light flux, and dU is the étendue, both $d\Phi$ and dU for the light emitted inside $d\theta$. Expression (10.2) may be rewritten as

$$I = \frac{d\Phi}{d\theta} = L\frac{dU}{d\theta} = Lv \tag{10.3}$$

where $v = dU/d\theta$ is the étendue per unit angle in direction θ and, therefore, the intensity I in direction θ is proportional to the étendue per unit angle v emitted in that direction. Combining Equation 10.3 with Equation 10.2 results in

$$v = a\cos\theta \tag{10.4}$$

If the optic emits light within angles $\theta_L < \theta < \theta_R$, the total étendue of the light emitted is given by

$$U = \int_{\theta_L}^{\theta_R} v(\theta)d\theta \tag{10.5}$$

A uniform Lambertian emitter E emits light with the same (constant) radiance L from all of its points, and in all directions. The radiance L of the light emitted by optic O_1 in direction θ is then also a constant, independent of where the light is coming from (assuming no losses in the system). Since radiance L is constant, from Equation 10.3 we can see that defining the intensity distribution $I(\theta)$ for the emission angle $\theta_L < \theta < \theta_R$ is equivalent to defining the distribution of the étendue per unit angle $v(\theta)$.

In the case of the optic in Figure 10.3, the intensity in direction **v** is given by $I_I = La_I \cos\theta_I$, where $a_I = x_D - x_C$ is the lit area in direction **v**.

In the (x_1,θ) plot, the horizontal distance a_A between curves c_1 and c_2 for a given emission angle θ_A is the area of the optic lit in that direction, as shown in Figure 10.4. Also, the vertical distance α_B between curves c_1 and c_2, for a given emission position x_B, is the total angle of emission at that position in the optic aperture.

A curve, such as c_2, in the (x_1,θ) plot defines a bundle of edge rays on optic O_1, as shown in Figure 10.5. For each horizontal coordinate x_A, curve c_2 gives the angle $\theta_A = c_2(x_A)$ that the ray r_A makes to the vertical, determining its direction. This bundle of rays is perpendicular to a wavefront w_2.

In order to determine the shape of wavefront w_2, one may start at position x_1 in optic O_1, as shown in Figure 10.6. From curve c_2, it is possible to determine the direction of ray r_1 at x_1, similarly to what is shown in Figure 10.5 for position x_A. Now choose a point $\mathbf{W}_1$ along ray r_1. Move to another position x_2 in optic O_1 and determine the direction of ray r_2. Intersect r_2 with the perpendicular to ray r_1 through point $\mathbf{W}_1$, determining a new point $\mathbf{W}_2$ on

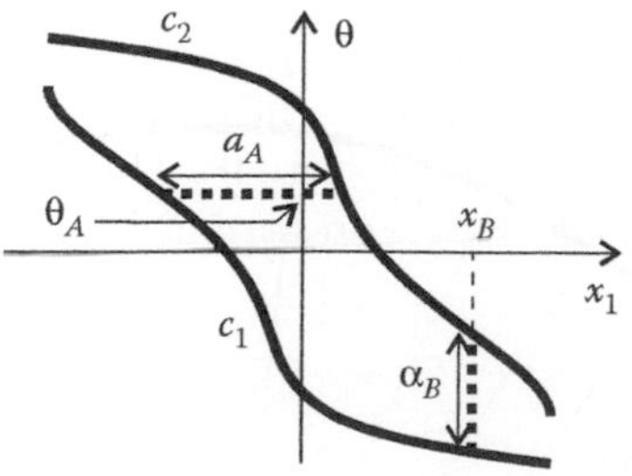

FIGURE 10.4
The horizontal distance a_A between curves c_1 and c_2 at height θ_A is the lit area in direction θ_A. The vertical distance α_B between curves c_1 and c_2 at horizontal coordinate x_B is the total emission angle at position x_B in the optic.

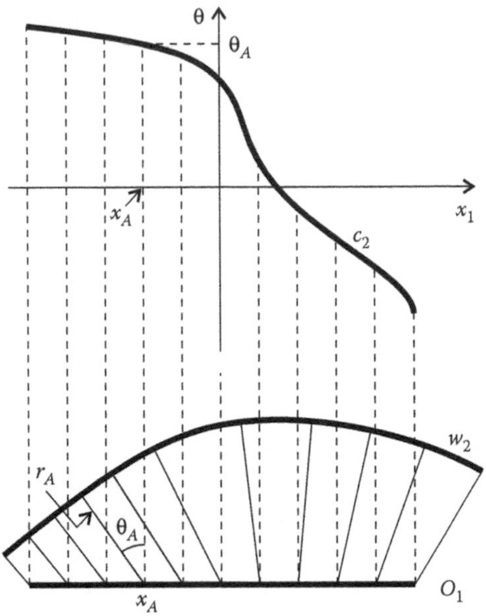

FIGURE 10.5
Edge rays defined by curve c_2 in a (x_1, θ) plot are perpendicular to wavefront w_2.

wavefront w_2. Then move to another position x_3 in the exit aperture of optic O_1 and determine the direction of ray r_3. Intersect r_3 with the perpendicular to ray r_2 through point $\mathbf{W}_2$, determining a new point $\mathbf{W}_3$ on wavefront w_2. Continue this process all the way to the end of the exit aperture of optic O_1. Proceeding in very small steps will give a good approximation of the wavefront shape.

We now describe an example of a prescribed intensity optic using the method outlined above. In this example, the optic is symmetric relative to the vertical axis, and extends from $-x_O$ to x_O, so that $x_L = -x_O$ and $x_R = x_O$.

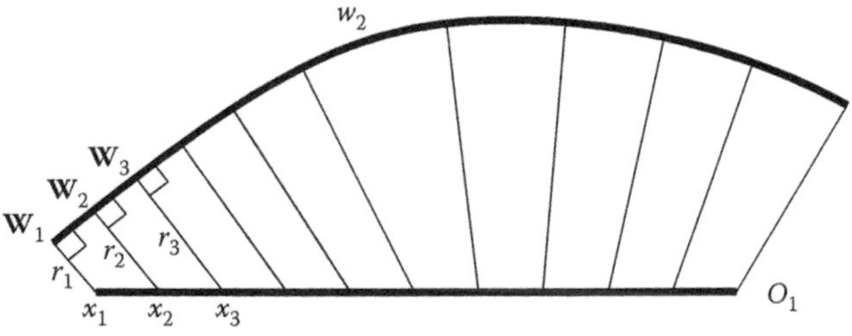

FIGURE 10.6
Wavefront w_2 may be obtained from the paths of the rays through optic O_1.

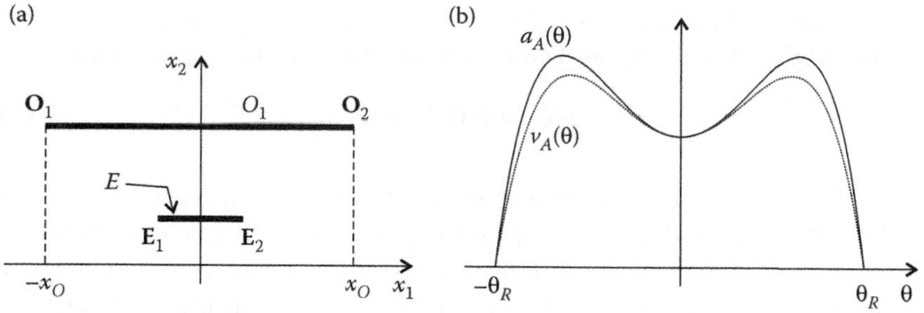

FIGURE 10.7
(a) Emitter E from $\mathbf{E}_1$ to $\mathbf{E}_2$ and optic O_1 from $\mathbf{O}_1$ to $\mathbf{O}_2$, extending from $-x_O$ to x_O. (b) Étendue per unit angle $v_A(\theta)$ in arbitrary units defines the shape of the intensity emission pattern across the emission angle of optic O_1, ranging from $-\theta_R$ to θ_R. The lit area a_A is given by $a_A(\theta) = v_A(\theta)/\cos\theta$.

Let us now assume we want to design an optic $\mathbf{O}_1\mathbf{O}_2$ for an emitter $\mathbf{E}_1\mathbf{E}_2$, and that the optic produces a prescribed intensity pattern. Figure 10.7a shows the overall configuration of one such system with emitter E and optic O_1. The optic extends horizontally from $-x_O$ to x_O.

Since radiance L is constant, from Equation 10.3 we can see that defining the intensity distribution $I(\theta)$ for an emission angle $-\theta_R < \theta < \theta_R$ is equivalent to defining the distribution of the étendue per unit angle $v(\theta)$. Figure 10.7b shows an example of a desired étendue per unit angle v_A within the emission range $-\theta_R < \theta < \theta_R$ in arbitrary units. At this stage in the design process, only the shape of the curve $v_A(\theta)$ is important, not its values. Also shown is lit area $a_A(\theta) = v_A(\theta)/\cos\theta$, obtained from expression (10.4).

The étendue of the light emitted from $\mathbf{E}_1\mathbf{E}_2$ and captured by $\mathbf{O}_1\mathbf{O}_2$ is given by $U = 2([\mathbf{E}_1,\mathbf{O}_2] - [\mathbf{E}_2,\mathbf{O}_2])$. The étendue emitted by the optic and defined by $v_A(\theta)$ is

$$U_A = \int_{-\theta_R}^{\theta_R} v_A(\theta)d\theta \tag{10.6}$$

Ideally, the étendue is conserved and we should have $U_A = U$. We may now define

$$v(\theta) = v_A(\theta)\frac{U}{U_A} \tag{10.7}$$

which verifies

$$U = \int_{-\theta_R}^{\theta_R} v(\theta)d\theta \tag{10.8}$$

We then take $v(\theta)$ as the desired étendue per unit angle distribution inside emission angle $-\theta_R < \theta < \theta_R$, and calculate the corresponding lit area as

$$a(\theta) = v(\theta)/\cos\theta \tag{10.9}$$

Figure 10.8, in its top right corner, shows the plot for $a(\theta)$ given by expression (10.9), but now with the θ axis pointing up. From $a(\theta)$, one may determine curves c_1 and c_2 in the (x_1,θ) plot. These curves now allow us to determine the shape of output wavefronts w_1 and w_2 of optic O_1 extending from $\mathbf{O}_1$ to $\mathbf{O}_2$. This optic will produce in direction θ_I a desired intensity defined by the lit area a_I in that direction.

We now calculate a possible shape for curves c_1 and c_2 from curve $a(\theta)$.

For a given curve $a(\theta)$, curves c_1 and c_2 are not unique, and therefore, the output wavefronts w_1 and w_2 and, thus, optic O_1, are not unique either. Figure 10.9 shows one such possibility. First, we choose the angle α_E for the emission angle at the edge $\mathbf{O}_1$.

Point $\mathbf{C}_1$, where curve c_2 starts, is at position $\mathbf{C}_1 = (-x_O,\theta_R)$, where $-x_O$ is the horizontal coordinate of edge $\mathbf{O}_1$ of the optic, and θ_R is the maximum emission

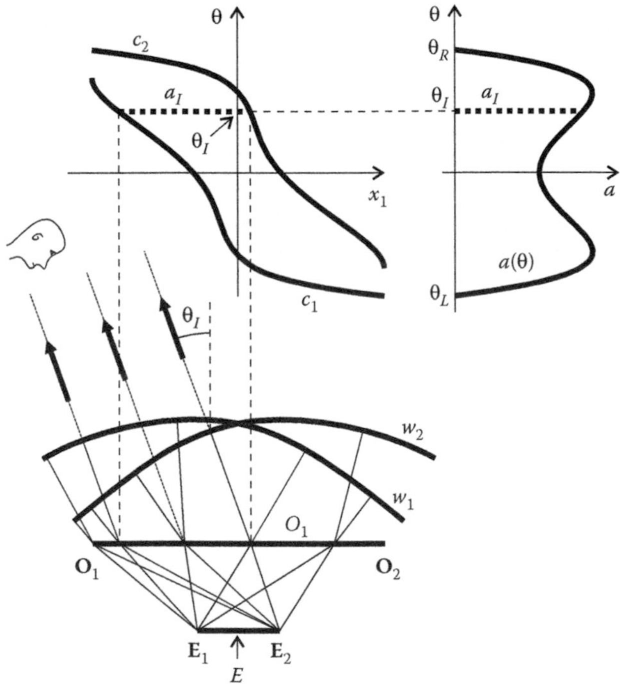

FIGURE 10.8
The desired lit area $a(\theta)$ in each direction θ may be used to determine curves c_1 and c_2 (which are not unique), which are used to determine output wavefronts w_1 and w_2, which are then used to calculate optic O_1.

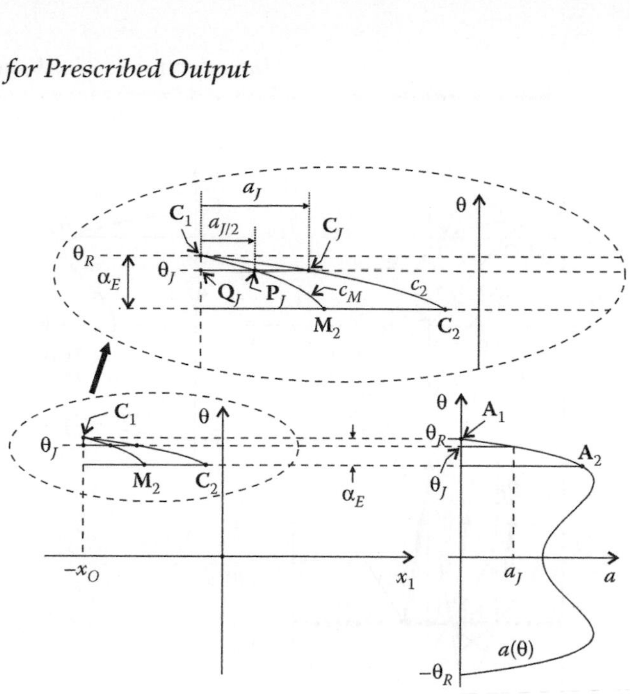

FIGURE 10.9
From point $\mathbf{C}_1$ to point $\mathbf{C}_2$, curve c_2 is the same as $a(\theta)$ from point $\mathbf{A}_1$ to point $\mathbf{A}_2$.

angle. From $\mathbf{C}_1$ to $\mathbf{C}_2$, curve c_2 is the same as curve $a(\theta)$ from point $\mathbf{A}_1$ to point $\mathbf{A}_2$; that is, for an emission angle θ_J, curve c_2 is at point $\mathbf{C}_J = (-x_O + a_J, \theta_J)$ with $a_J = a(\theta_J)$.

We may now define a mid-curve c_M whose points, for an emission angle θ_J, are at $\mathbf{P}_J = (-x_O + a_J/2, \theta_J)$ with $a_J = a(\theta_J)$. Point $\mathbf{P}_J$ is at the mid-point of $\mathbf{Q}_J\mathbf{C}_J$ where $\mathbf{Q}_J = (-x_O, \theta_J)$. This method defines curve c_M from point $\mathbf{C}_1$ to point $\mathbf{M}_2$.

We may now consider point $\mathbf{M}_3$ at the origin $(x_1, \theta) = (0,0)$ and choose an extension of curve c_M from $\mathbf{M}_2$ to $\mathbf{M}_3$, as shown in Figure 10.10. Now take a point $\mathbf{P}_K = (x_K, \theta_K)$ on curve c_M. For this emission angle θ_K, the lit area is $a_K = a(\theta_K)$. A point $\mathbf{L}_K = (x_K - a_K/2, \theta_K)$ may now be defined on curve c_1, and another point $\mathbf{R}_K = (x_K + a_K/2, \theta_K)$ may also be defined on curve c_2. Moving point $\mathbf{P}_K$ along curve c_M between points $\mathbf{M}_2$ and $\mathbf{M}_3$, defines curves c_1 and c_2 above axis x_1. Using the same method (or symmetry, if the emission pattern is symmetrical) allows us to define curves c_1 and c_2 below axis x_1.

Curve $c_M(\theta_K)$ defines the mid-position x_K of the lit area a_K in optic O_1 inside its aperture $-x_O$ to x_O for emission direction θ_K. Therefore, in this case, to define the output wavefronts, two functions of the emission angle θ are given: the lit area $a(\theta)$, and also $c_M(\theta)$, defining the central position of the lit area inside the optic aperture $-x_O$ to x_O.

At this point, we have the shape of curves $c_1(x)$ and $c_2(x)$ for $-x_O < x < x_O$ and we may calculate the shape of wavefronts w_1 and w_2 in a process similar to that shown in Figure 10.6. Also, the values of $-x_O$ and x_O are the horizontal coordinates of edges $\mathbf{O}_1$ and $\mathbf{O}_2$ of the optic to be designed, as defined in Figure 10.7a.

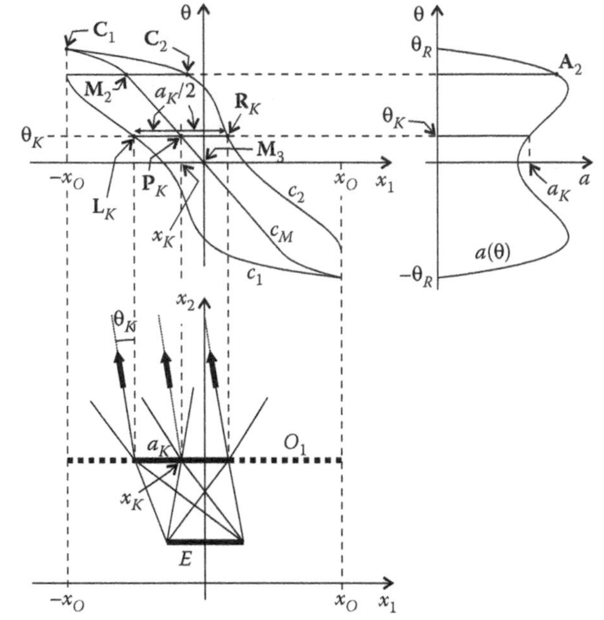

FIGURE 10.10
Curves c_1 and c_2 are obtained from the mid-curve c_M, displacing its points by $a(\theta)/2$ to the left and the right. Curve c_M defines the position of lit area a_K inside optic O_1 aperture from $-x_O$ to x_O.

Given the output wavefronts w_1 and w_2 and an emitter E, an optic O_1 may now be designed,[1] as shown in Figure 10.11. The design starts at the edge $\mathbf{O}_1$, and the calculations are similar to those in Figure 9.23 (or Figure 10.24). The input wavefronts for optic O_1 are the edges of emitter E (limit case of circular wavefronts collapsing to two points at the edges $\mathbf{E}_1$ and $\mathbf{E}_2$ of E).

This optic O_1, extending from $\mathbf{O}_1$ to $\mathbf{O}_2$, may also be coupled with a CEC, as in Figure 10.12, allowing it to be used with a fully Lambertian source $\mathbf{E}_3\mathbf{E}_4$. The CEC is composed of elliptical arc e_2 with foci $\mathbf{E}_3$ and $\mathbf{O}_1$, and symmetrical elliptical arc e_1.

We now consider a second example of a prescribed intensity optic.

Figure 10.13 shows another example of an intensity pattern for an optic extending from x_L to x_R. The lit area $a(\theta)$ in each direction is as shown on the right, and the corresponding curves c_1 and c_2 in a (x_1,θ) plot on the left. In this example, curve c_2 is the same as $a(\theta)$ for $\theta > 0$ and curve c_1 is symmetrical to $a(\theta)$ for $\theta < 0$.

Wavefronts w_1 and w_2 may be calculated from curves c_1 and c_2 and an optic designed for those wavefronts, as shown in Figure 10.14. Figure 10.14a shows a CPC-type device, while Figure 10.14b shows a dielectric total internal reflection concentrator (DTIRC). The étendue of the emitter $\mathbf{E}_1\mathbf{E}_2$ must match the étendue defined by output wavefronts w_1 and w_2 integrated across the exit aperture $\mathbf{O}_1\mathbf{O}_2$ of the optic. The DTIRC is shorter than the CPC-type

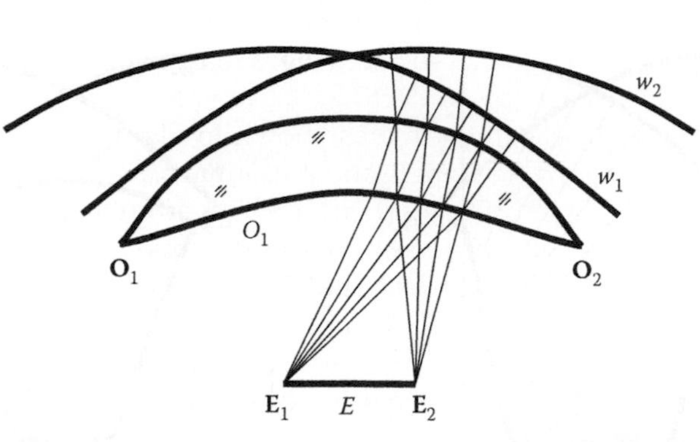

FIGURE 10.11
Optic O_1, extending from $\mathbf{O}_1$ to $\mathbf{O}_2$, designed for an emitter $\mathbf{E}_1\mathbf{E}_2$ and output wavefronts w_1 and w_2. This optic produces a desired intensity pattern defined by the output wavefronts w_1 and w_2.

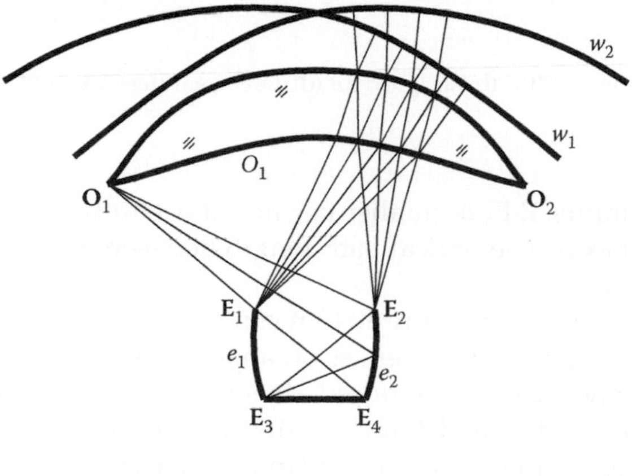

FIGURE 10.12
A CEC made of elliptical arcs e_1 and e_2 coupled to an optic O_1, takes the light from a Lambertian source $\mathbf{E}_3\mathbf{E}_4$, and produces a desired intensity pattern defined by output wavefronts w_1 and w_2.

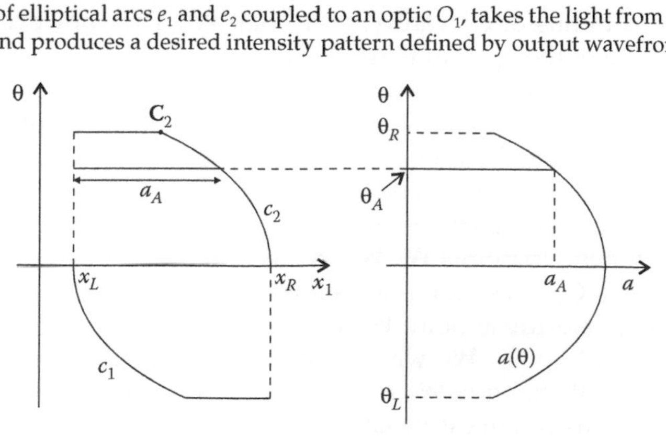

FIGURE 10.13
Right: lit area in each emission direction θ. Left: curves c_1 and c_2 used to calculate the optic output wavefronts.

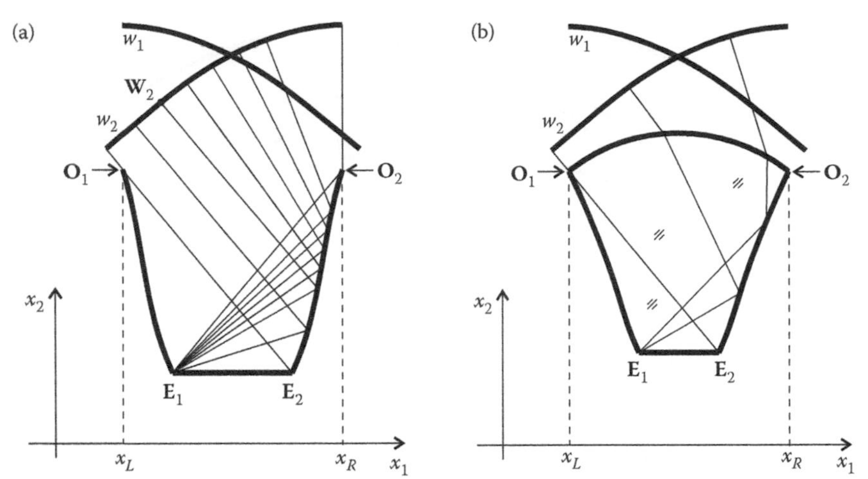

FIGURE 10.14
CPC-type optic (a) and DTIRC (b) designed for (different) emitters $\mathbf{E}_1\mathbf{E}_2$ and output wavefronts w_1 and w_2.

optic, and its emitter $\mathbf{E}_1\mathbf{E}_2$ is smaller because it is emitting inside a material of refractive index n. The optical path length between w_2 and $\mathbf{E}_1$ is constant for both designs.

From curves c_1 and c_2 in Figure 10.13 it can be seen that the aperture of the optic is fully lit from x_L to x_R for emission angle $\theta = 0$. For that reason, the exit aperture size cannot be further reduced. Reducing the aperture size $x_L x_R$ would reduce the lit area for $\theta = 0$ and, therefore, the intensity emitted in that direction.

Curve c_2 in Figure 10.13 has a discontinuous derivative (kink) at point $\mathbf{C}_2$. This translates to a wavefront w_3 with a discontinuous second derivative (discontinuous radius of curvature) at a point $\mathbf{W}_2$. This, however, does not affect the design of the optic in Figure 10.14.

An extreme case of the configuration shown in Figures 10.13 and 10.14 happens in the situation depicted in Figure 10.15, in which the lit area is the same in all directions. Here, the intensity pattern is that of a Lambertian source (falls with cos θ) between θ_L and θ_R. Curve c_2 in the (x_1,θ) plot is constant from x_L to x_R, and then falls sharply to zero at x_R. Points $\mathbf{C}_1$, $\mathbf{C}_2$, $\mathbf{C}_3$, $\mathbf{C}_4$, and $\mathbf{C}_5$, in curve c_2, correspond to points $\mathbf{W}_1$, $\mathbf{W}_2$, $\mathbf{W}_3$, $\mathbf{W}_4$, and $\mathbf{W}_5$, respectively, in wavefront w_2. At point $\mathbf{C}_3$, curve c_2 has a discontinuous derivative. This translates to a change in curvature at point $\mathbf{W}_3$ in wavefront w_2.

From point $\mathbf{W}_1$ to point $\mathbf{W}_3$, wavefront w_2 is flat (infinite radius of curvature). From point $\mathbf{W}_3$ to point $\mathbf{W}_5$, it is a circle with center $\mathbf{O}_2$ and finite radius $[\mathbf{W}_3,\mathbf{O}_2]$. The radius of curvature of w_2 is, therefore, discontinuous at point $\mathbf{W}_3$. However, w_2 is still a smooth curve (continuous derivative).

The optic in Figure 10.15 designed for wavefront w_2 (and its symmetrical w_1, not shown) is a CPC with emitter $\mathbf{E}_1\mathbf{E}_2$, and exit aperture $\mathbf{O}_1\mathbf{O}_2$. Note that

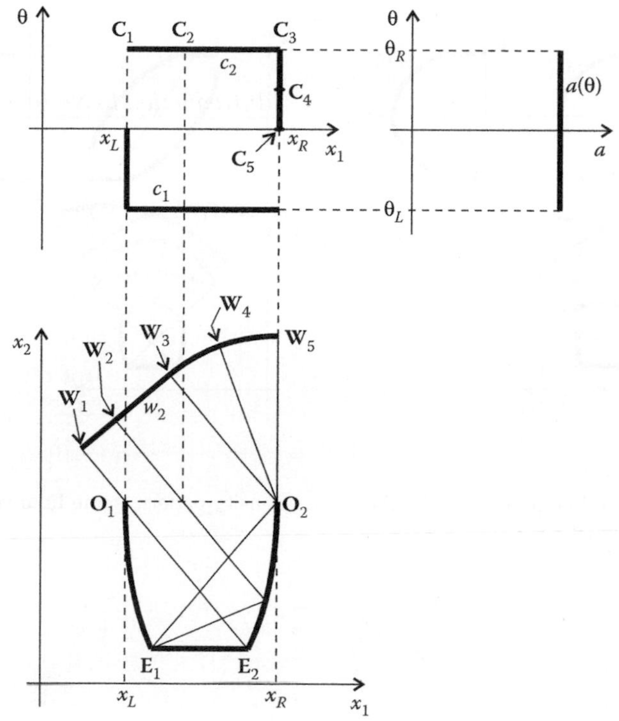

FIGURE 10.15
A constant lit area $a(\theta)$ for all emission angles θ results in a CPC optic.

from $\mathbf{W}_3$ to $\mathbf{W}_5$, all rays perpendicular to w_2 converge at edge $\mathbf{O}_2$ of the aperture and, therefore, this portion of the wavefront does not generate a mirror surface above $\mathbf{O}_2$.

A given intensity pattern may be produced by different optics, with different output wavefronts. Figure 10.16 shows one such example. The optics in Figure 10.16a and Figure 10.16b produce the same intensity pattern (same lit area $a(\theta)$ for each emission direction θ), but have different sizes. Curves c_1 and c_2 in Figure 10.16a are generated using the method in Figure 10.13, while curves c_1 and c_2 in Figure 10.16b are generated using the method in Figure 10.10.

The optic in Figure 10.16a has the smallest possible aperture area $x_L x_R$ since it is fully lit for $\theta = 0$. Reducing the aperture area would reduce the intensity for $\theta = 0$, and the desired pattern would not be produced.

Figure 10.17 compares the optics in Figure 10.16a (mirror m_1 and lens l_1) and in Figure 10.16b (mirror m_2 and lens l_2).

Similar methods may also be used to generate a prescribed irradiance on a target at a finite distance.

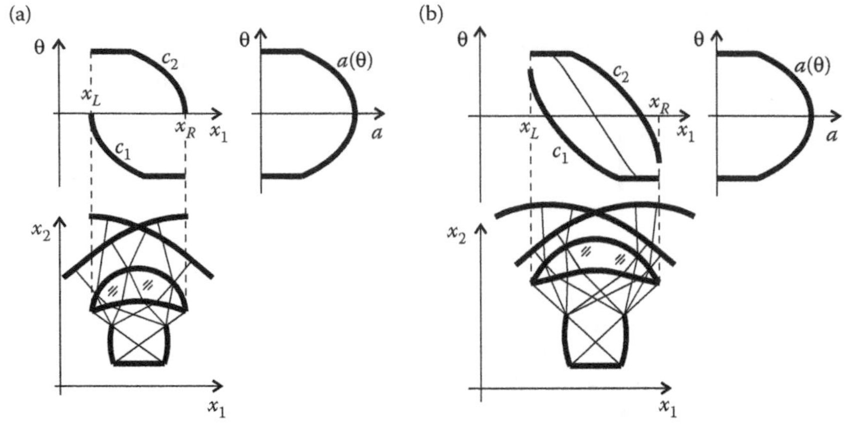

FIGURE 10.16
Small optic (a) and large optic (b) for the same intensity pattern (same lit area $a(\theta)$ for each emission direction θ).

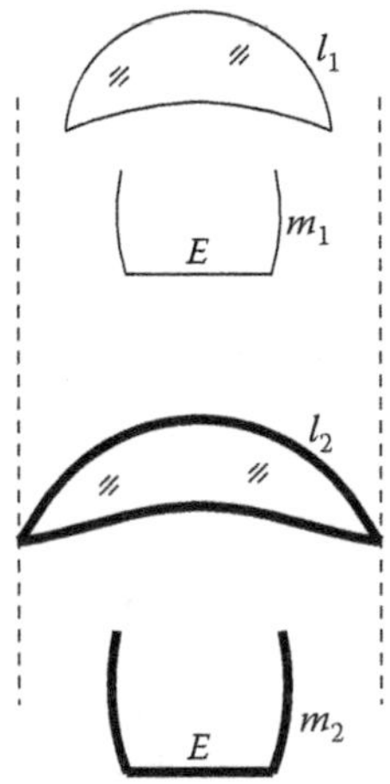

FIGURE 10.17
Small optic (m_1 and l_1 top) and large optic (m_2 and l_2 bottom) for the same emitter E and same intensity pattern produced.

10.3 Wavefronts for Prescribed Irradiance

Figure 10.3 shows how an optic O_1 may produce a desired intensity in a given direction θ_I. Something similar happens in the case of the irradiance on a receiver R placed at a finite distance, as shown in Figure 10.18a. Point **P**

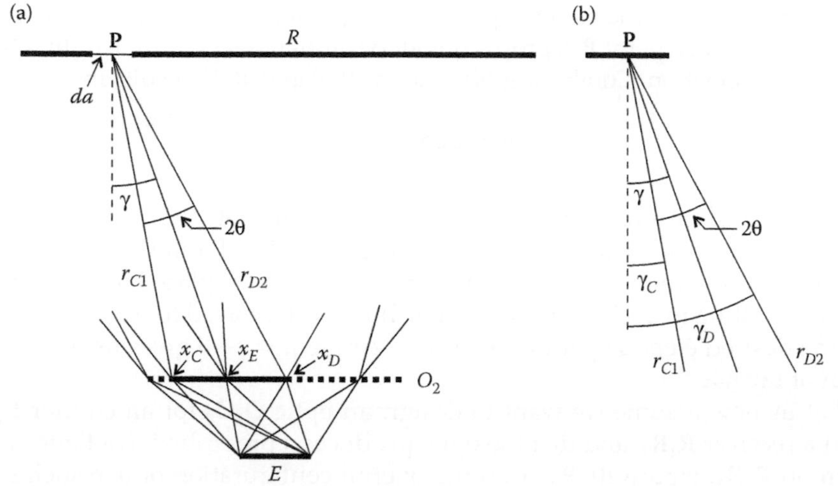

FIGURE 10.18
(a) When seen from point **P** on the receiver R, optic O_2 has a lit area extending from x_C to x_D. The irradiance E_P produced by optic O_2 at position **P** may be obtained from $E_P = 2L \cos\gamma \sin\theta$. (b) Given the direction of ray r_{C1} and the irradiance E_P at point **P**, we may determine the direction of ray r_{D2}.

on receiver R receives light coming from the aperture of optic O_2 between horizontal coordinates x_C and x_D, confined between rays r_{C1} and r_{D2}. This light comes originally from an emitter E. For simplicity, in the example below, we assume that both the emitter E and receiver R are immersed in air ($n = 1$). However, the same methods still apply when emitter and receiver are immersed in different refractive index materials.

A uniform Lambertian emitter E emits light with the same (constant) radiance L from all of its points and in all directions. The radiance L of the light reaching point **P** on the receiver is then also a constant, independent of where the light is coming from (assuming no losses in the system). The light flux $d\Phi$ captured by an area da at point **P** is given by $d\Phi = 2daL \cos\gamma \sin\theta$, where γ is the angle that the bisector of r_{C1} and r_{D2} makes to the normal to R, 2θ is the angular aperture defined by rays r_{C1} and r_{D2}, and L is the radiance. The irradiance E_P at point **P** is then given by

$$E_P = \frac{d\Phi}{da} = L\frac{dU}{da} = L\frac{2da\cos\gamma\sin\theta}{da} = 2L\cos\gamma\sin\theta \qquad (10.10)$$

where $d\Phi = LdU$ is the light flux falling on da, and dU is the étendue of the light crossing da. Expression (10.10) may be rewritten as

$$E_P = \frac{d\Phi}{da} = L\frac{dU}{da} = Lu \qquad (10.11)$$

where $u = dU/da$ is the étendue per unit length at point **P** and, therefore, the irradiance E_P at a point **P** is proportional to the étendue per unit length u falling at that position. Combining Equations 10.10 and 10.11 results in

$$u = 2\cos\gamma\sin\theta \tag{10.12}$$

Note that if we already know ray r_{C1}, we may calculate the direction of ray r_{D2} from expression (10.12), as shown in Figure 10.18b. From the direction or ray r_{C1}, we also know angle γ_C. Angle $\gamma = \gamma_C + \theta$ may be replaced into Equation 10.12, resulting in $u = 2\cos(\gamma_C + \theta)\sin\theta$, that may be solved for θ as a function of the desired étendue per unit length u. Angle $\gamma_D = \gamma + \theta$ gives us the direction of ray r_{D2}.

Let us now assume we want to design an optic $\mathbf{O}_1\mathbf{O}_2$ for an emitter $\mathbf{E}_1\mathbf{E}_2$ and a receiver $\mathbf{R}_1\mathbf{R}_2$, and that the optic produces a prescribed irradiance pattern on $\mathbf{R}_1\mathbf{R}_2$. Figure 10.19 shows the overall configuration of one such system. A point $\mathbf{R}_A$ on the receiver has horizontal coordinate x_{RA} along axis x_1.

Since radiance L is constant, from Equation 10.11 we can see that defining the irradiance distribution $E_P(x_1)$ along receiver $\mathbf{R}_1\mathbf{R}_2$ for $-x_R < x_1 < x_R$ is equivalent to defining the distribution of the étendue per unit length $u(x_1)$. Figure 10.20 shows the desired étendue per unit length along the receiver in arbitrary units, u_A. At this stage in the design process, only the shape of the curve $u_A(x_1)$ is important, not its values.

The étendue of the light emitted from $\mathbf{E}_1\mathbf{E}_2$ and captured by $\mathbf{O}_1\mathbf{O}_2$ is given by $U = 2([\mathbf{E}_1,\mathbf{O}_2] - [\mathbf{E}_2,\mathbf{O}_2])$. The étendue reaching the receiver $\mathbf{R}_1\mathbf{R}_2$ and defined by $u_A(x_1)$ is

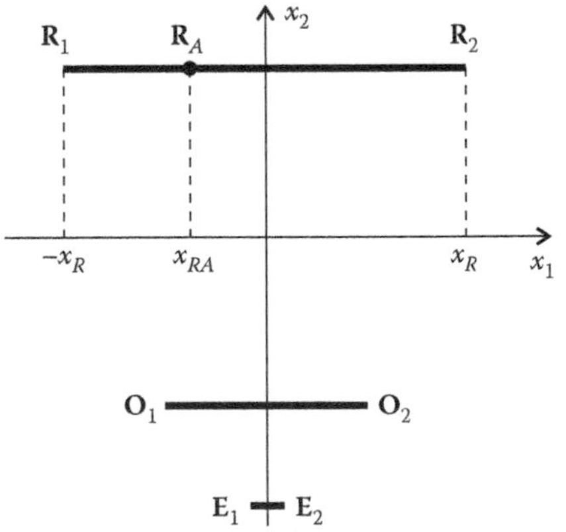

FIGURE 10.19
Overall configuration of an optical system with emitter $\mathbf{E}_1\mathbf{E}_2$, optic $\mathbf{O}_1\mathbf{O}_2$, and receiver $\mathbf{R}_1\mathbf{R}_2$, extending along the x_1 axis from $-x_R$ to x_R. A point $\mathbf{R}_A$ on the receiver has horizontal coordinate x_{RA}.

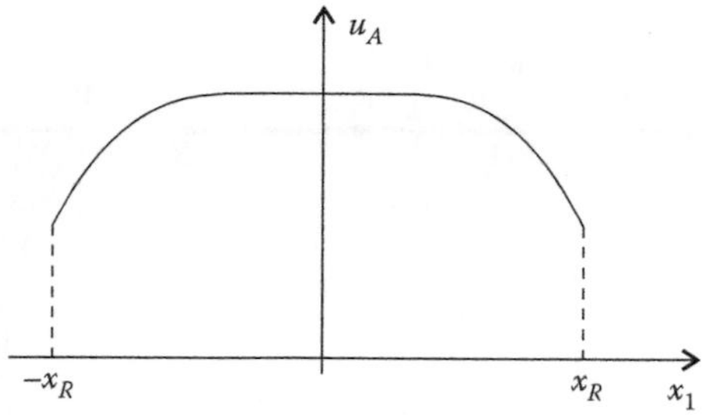

FIGURE 10.20
Étendue per unit length distribution across the receiver in arbitrary units. These values must then be normalized with the total étendue crossing the optic.

$$U_A = \int_{-x_R}^{x_R} u_A(x)dx \tag{10.13}$$

Ideally, if the étendue is conserved we should have $U_A = U$. We may now define

$$u(x_1) = u_A(x_1)\frac{U}{U_A} \tag{10.14}$$

which verifies

$$U = \int_{-x_R}^{x_R} u(x)dx \tag{10.15}$$

We then take $u(x_1)$ as the desired étendue per unit length distribution on receiver $\mathbf{R}_1\mathbf{R}_2$.

There are now different options for calculating different output wavefronts that produce the desired irradiance pattern on $\mathbf{R}_1\mathbf{R}_2$. As a possibility for calculating a pair of such wavefronts, start by defining the acceptance angle θ_E of optic $\mathbf{O}_1\mathbf{O}_2$ at its edge, as shown in Figure 10.21a. Light emitted from edge $\mathbf{O}_1$ of the optic will then illuminate region $\mathbf{R}_1\mathbf{R}_3$ of the receiver, where $\mathbf{R}_3$ is defined by ray r_3 tilted by θ_E relative to the ray from $\mathbf{O}_1$ to $\mathbf{R}_1$.

Since the whole system is symmetric, at the center $\mathbf{R}_0$ of the receiver ($x_1 = 0$), the left and right edge rays r_{0L} and r_{0R} are also symmetric, relative to the vertical, and make an angle $2\theta_0$ to each other. Since the bisector to r_{0L} and r_{0R} is vertical, $\gamma = 0$ (see Figure 10.18), and we get from Equation 10.12 $\theta_0 = \arcsin(u(0)/2)$, which determines the directions of r_{0L} and r_{0R}.

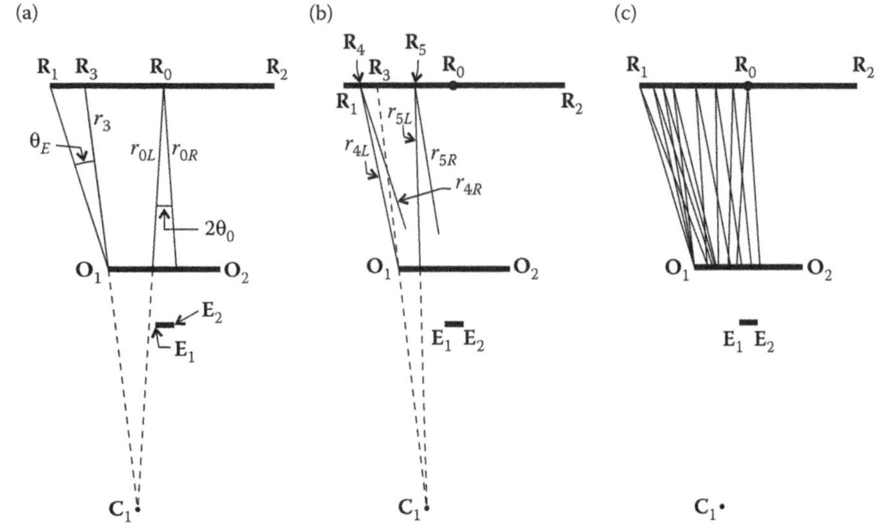

FIGURE 10.21
(a) Choice of the acceptance angle θ_E at the edge of the optic and calculation of the left and right edge rays r_{0L} and r_{0R} at the center of the receiver. (b) Calculation of the edge rays reaching the receiver. (c) A complete set of edge rays of the light illuminating the left half $\mathbf{R}_1\mathbf{R}_0$ of the receiver.

Consider now a point $\mathbf{R}_4$ on the receiver between points $\mathbf{R}_1$ and $\mathbf{R}_3$, as shown in Figure 10.21b. The light it receives from the optic is bound to the left by ray r_{4L} from point $\mathbf{O}_1$, and to the right by r_{4R} to be determined. Point $\mathbf{R}_4$ has horizontal coordinate x_{R4} (similar notation to the horizontal coordinate x_{RA} of point $\mathbf{R}_A$ in Figure 10.19). The étendue per unit length at $\mathbf{R}_4$ is $u(x_{R4})$, and from expression (10.12) we may determine the direction of r_{4R}, using the same method shown in Figure 10.18, and described above. Moving point $\mathbf{R}_4$ along segment $\mathbf{R}_1\mathbf{R}_3$ of the receiver, one may calculate the set of edge rays reaching $\mathbf{R}_1\mathbf{R}_3$.

Going back to Figure 10.21a, the intersection of rays r_3 and r_{0L} determines the position of point $\mathbf{C}_1$. This point may now be used as a pivot point for the left edge rays r_{5L} of points $\mathbf{R}_5$ between $\mathbf{R}_3$ and $\mathbf{R}_0$. This is just one option for defining this bundle of rays, and other options are also possible.

Consider then point $\mathbf{R}_5$ in Figure 10.21b. We have chosen that left edge ray r_{5L} at $\mathbf{R}_5$ points towards $\mathbf{C}_1$. Point $\mathbf{R}_5$ has horizontal coordinate x_{R5} (similar notation to the horizontal coordinate x_{RA} of point $\mathbf{R}_A$ in Figure 10.19). The étendue per unit length at $\mathbf{R}_5$ is $u(x_{R5})$, and from expression (10.12) we may determine the direction of r_{5R}, using the same method shown in Figure 10.18, and described above. Moving point $\mathbf{R}_5$ along segment $\mathbf{R}_3\mathbf{R}_0$ of the receiver, one may calculate the set of edge rays reaching $\mathbf{R}_3\mathbf{R}_0$.

The set of edge rays thus calculated for the left half of the receiver $\mathbf{R}_1\mathbf{R}_0$ is shown in Figure 10.21c. Those same edge rays are shown separately in

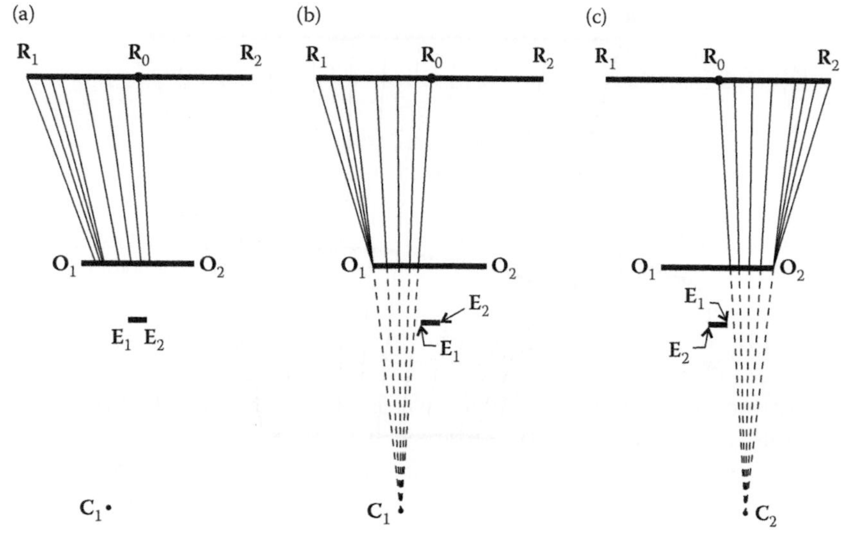

FIGURE 10.22
(a) Right edge rays at portion $\mathbf{R}_1\mathbf{R}_0$ of the receiver. (b) Left edge rays at portion $\mathbf{R}_1\mathbf{R}_0$ of the receiver. (c) Rays symmetric to those in (b).

Figure 10.22a for the right rays, and Figure 10.22b for the left rays. Figure 10.22c shows the symmetric to the left rays in Figure 10.22b.

Figure 10.23 combines the edge rays in Figure 10.22a with those in Figure 10.22c, for a complete set of right edge rays across the whole receiver $\mathbf{R}_1\mathbf{R}_2$. Additional rays crossing $\mathbf{W}_{1A}\mathbf{W}_{1B}$ diverge from edge $\mathbf{R}_1$ of the receiver. These added rays illuminate point $\mathbf{R}_1$, creating there the desired irradiance. The shape of wavefront w_1 may now be determined as a curve perpendicular to all these rays, in a procedure similar to that in Figure 10.5.

Wavefront w_2 is obtained by symmetry, and an optic O_1 may now be calculated between points $\mathbf{O}_1$ and $\mathbf{O}_2$, using the SMS method. The design starts at the edge $\mathbf{O}_1$, as shown in Figure 10.24, and the calculations are similar to those in Figure 9.23.

Figure 10.25 shows how the rays crossing the optic illuminate the receiver $\mathbf{R}_1\mathbf{R}_2$. It shows the illumination cone at two points: $\mathbf{R}_4$ and $\mathbf{R}_6$. In the case of point $\mathbf{R}_4$, the illumination cone is bound by one left r_{4L} ray coming from edge $\mathbf{O}_1$ of the optic, and a right edge ray r_{4R} coming from edge $\mathbf{E}_2$ of the emitter. In the case of point $\mathbf{R}_6$, the illumination cone is bound by one left r_{6L} ray coming from edge $\mathbf{E}_1$ of the emitter and a right edge ray r_{6R} coming from edge $\mathbf{E}_2$ of the emitter.

A CEC optic with mirrors e_1 and e_2 allows optic O_1 to be used with a Lambertian emitter $\mathbf{E}_3\mathbf{E}_4$, where now $\mathbf{E}_1\mathbf{E}_2$ is the exit aperture of the CEC.

Figure 10.25b shows another example of a different optic O_4 designed for the same output aperture $\mathbf{O}_1\mathbf{O}_2$ and the same exit wavefronts w_1 and w_2,

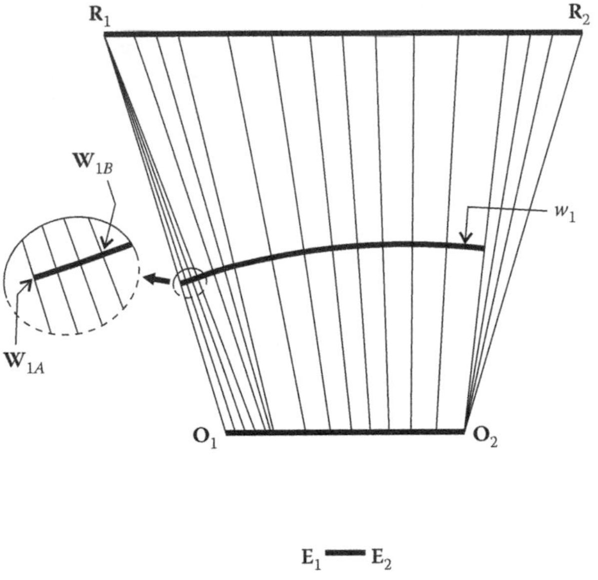

FIGURE 10.23
A complete set of rays covering the whole receiver $\mathbf{R}_1\mathbf{R}_2$ defines wavefront w_1. Rays crossing portion $\mathbf{W}_{1A}\mathbf{W}_{1B}$ of w_1 diverge from edge $\mathbf{R}_1$ of the receiver.

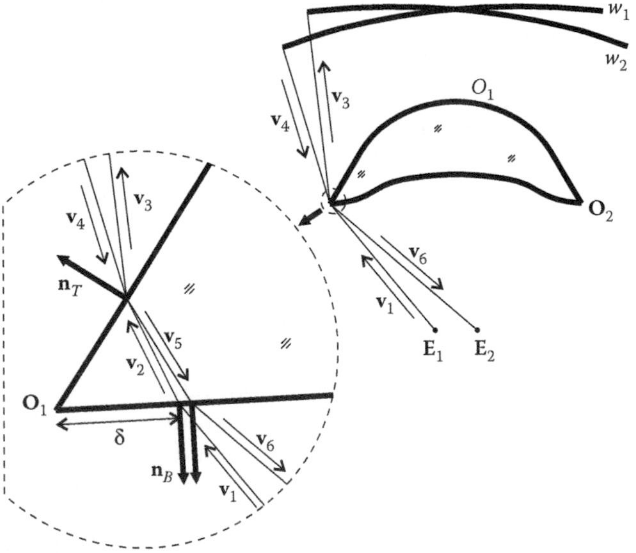

FIGURE 10.24
The design of optic O_1 starts at the edge $\mathbf{O}_1$. The design method is the same as illustrated in Figure 9.23.

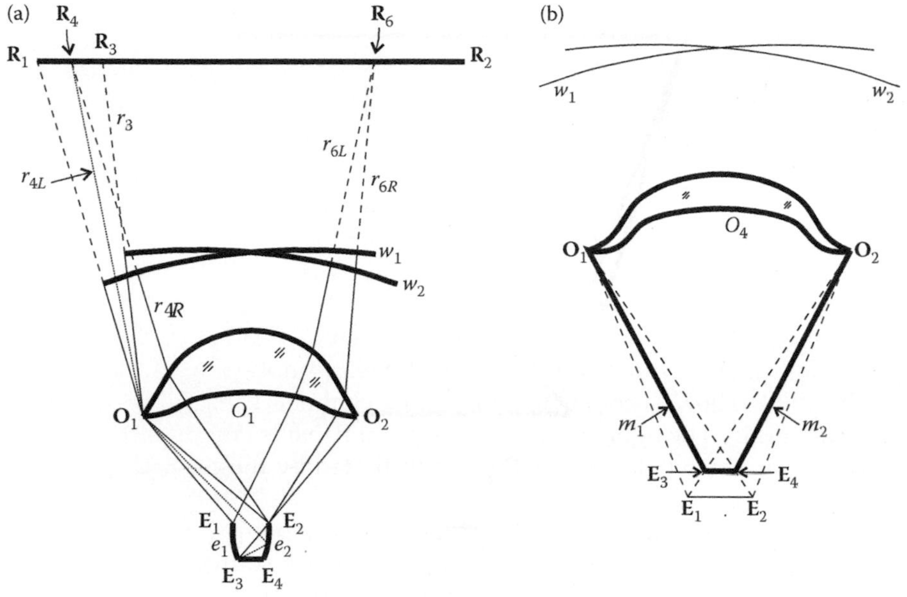

FIGURE 10.25
(a) Optic designed for emitter $\mathbf{E}_1\mathbf{E}_2$ and output wavefronts w_1 and w_2. Also shown are some light rays. A CEC with elliptical mirrors e_1 and e_2 allows the optic O_1 to be used with a Lambertian emitter $\mathbf{E}_3\mathbf{E}_4$. (b) Optic O_4 is similar to optic O_1 in (a), but designed for a different emitter $\mathbf{E}_1\mathbf{E}_2$ and coupled with flat mirrors m_1 and m_2, so that O_4 can be used with Lambertian emitter $\mathbf{E}_3\mathbf{E}_4$.

therefore producing the same irradiance pattern on the same target $\mathbf{R}_1\mathbf{R}_2$ as in Figure 10.25a. This optic has a different emitter $\mathbf{E}_1\mathbf{E}_2$, but was designed using the same method utilized to get Figure 10.25a. Flat mirrors m_1 from $\mathbf{E}_3$ to $\mathbf{O}_1$ and m_2 from $\mathbf{E}_4$ to $\mathbf{O}_2$ allow optic O_4 to be used with a Lambertian emitter $\mathbf{E}_3\mathbf{E}_4$. In this configuration, distance $[\mathbf{E}_3,\mathbf{E}_1] = [\mathbf{E}_3,\mathbf{E}_4]$. Mirror m_1 bisects $\mathbf{E}_1\mathbf{E}_3\mathbf{E}_4$, so that $\mathbf{E}_1$ is the mirror image of $\mathbf{E}_4$. Also, point $\mathbf{O}_1$ is at the intersection of mirror m_1 with the straight line through $\mathbf{E}_2$ and $\mathbf{E}_4$. From the geometry of this configuration, we can see that $U = 2([\mathbf{O}_1,\mathbf{E}_2] - [\mathbf{O}_1,\mathbf{E}_1]) = 2([\mathbf{O}_1,\mathbf{E}_4] + [\mathbf{E}_4,\mathbf{E}_2] - [\mathbf{O}_1,\mathbf{E}_1]) = 2[\mathbf{E}_3,\mathbf{E}_4]$ since $[\mathbf{O}_1,\mathbf{E}_4] = [\mathbf{O}_1,\mathbf{E}_1]$ and $[\mathbf{E}_4,\mathbf{E}_2] = [\mathbf{E}_3,\mathbf{E}_4]$. The etendue U of a Lambertian source $\mathbf{E}_3\mathbf{E}_4$ is then the same as that exchanged between a virtual emitter $\mathbf{E}_1\mathbf{E}_2$ and a receiver $\mathbf{O}_1\mathbf{O}_2$.

Other designs for the optic are also possible. As another example of a possible design, consider the same target irradiance shown in Figure 10.20, same emitter $\mathbf{E}_1\mathbf{E}_2$, same optic end points $\mathbf{O}_1$ and $\mathbf{O}_2$, and same receiver $\mathbf{R}_1\mathbf{R}_2$, as in Figure 10.19. In this example, we prescribe a flat shape for the bottom surface of the optic from $\mathbf{O}_1$ to $\mathbf{O}_2$, as in Figure 10.26. Other shapes are also possible.

Points $\mathbf{R}_1$ and $\mathbf{O}_1$ define the direction of ray r_{1L}, and the étendue per unit length $u(-x_R)$ at point $\mathbf{R}_1$ allows us to calculate the direction of ray r_{1R}. Now calculate portion $\mathbf{O}_1\mathbf{X}_1$ of the top surface (bound by vectors r_{1L} and r_{1R}) as

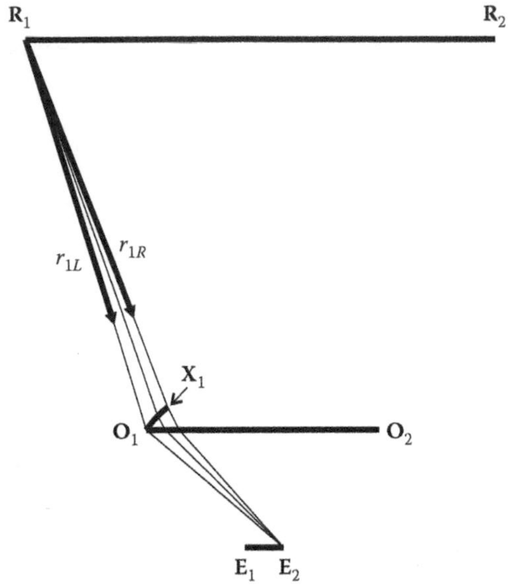

FIGURE 10.26
Design of a lens for a prescribed irradiance on receiver $\mathbf{R}_1\mathbf{R}_2$. Bottom surface from $\mathbf{O}_1$ to $\mathbf{O}_2$ chosen as flat (could be a different shape). Edge $\mathbf{O}_1\mathbf{X}_1$ of the top surface focuses the rays from $\mathbf{E}_2$ to $\mathbf{R}_1$. The cone of rays bound by edge rays r_{1L} and r_{1R} defines the irradiance at $\mathbf{R}_1$.

a Cartesian oval that focuses the rays from $\mathbf{E}_2$ to $\mathbf{R}_1$. These rays are also refracted at the flat bottom surface $\mathbf{O}_1\mathbf{O}_2$, as shown in Figure 10.26.

Now take a point $\mathbf{E}_4$ on $\mathbf{E}_1\mathbf{E}_2$ and calculate the path of a ray form $\mathbf{E}_4$ as it crosses point $\mathbf{O}_1$, as shown in Figure 10.27. This ray has two refractions at point $\mathbf{O}_1$, at the bottom and top surfaces. This ray continues up as ray r_{4L}, and intersects $\mathbf{R}_1\mathbf{R}_2$ at point $\mathbf{R}_4$. From the étendue per unit length $u(x_{R4})$ at point $\mathbf{R}_4$, we may calculate the direction of r_{4R}. We already have point $\mathbf{X}_1$ and its normal $\mathbf{n}_1$ from the previous calculation, so we can intersect ray r_{4R} with the plane defined by point $\mathbf{X}_1$ and normal $\mathbf{n}_1$, defining point $\mathbf{X}_2$. Ray r_{4R} must be redirected to point $\mathbf{E}_2$ and this condition defines its normal $\mathbf{n}_2$.

Moving point $\mathbf{E}_4$ from $\mathbf{E}_2$ to $\mathbf{E}_1$ in small steps, and proceeding as described, we can calculate a new section of the top surface of the lens between points $\mathbf{X}_1$ and $\mathbf{X}_3$, as shown in Figure 10.28.

Now we take a point $\mathbf{X}$ in the portion of top surface already calculated, and calculate its path from $\mathbf{E}_1$ through $\mathbf{X}$. This ray exits the lens as r_{5L} and impacts the receiver at point $\mathbf{R}_5$, as shown in Figure 10.29. From the étendue per unit length $u(x_{R5})$ at point $\mathbf{R}_5$, we may calculate the direction of r_{5R}. We already have point $\mathbf{X}_3$ and its normal $\mathbf{n}_3$ from the previous calculation, so we can intersect ray r_{5R} with the plane defined by point $\mathbf{X}_3$ and normal $\mathbf{n}_3$, defining point $\mathbf{X}_4$. Ray r_{5R} must be redirected to point $\mathbf{E}_2$ and this condition defines its normal $\mathbf{n}_4$.

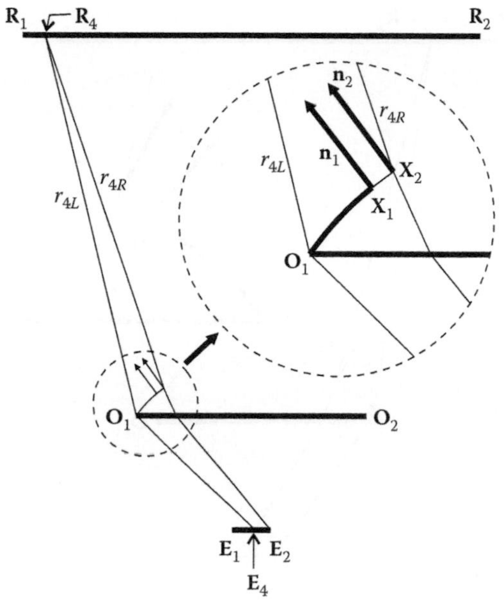

FIGURE 10.27
A ray emitted from point $\mathbf{E}_4$ inside $\mathbf{E}_1\mathbf{E}_2$ and refracted (at the bottom and top surfaces) at point $\mathbf{O}_1$, impacts $\mathbf{R}_1\mathbf{R}_2$ at point $\mathbf{R}_4$ and defines a new point $\mathbf{X}_2$ on the top surface, based on a previously calculated point $\mathbf{X}_1$.

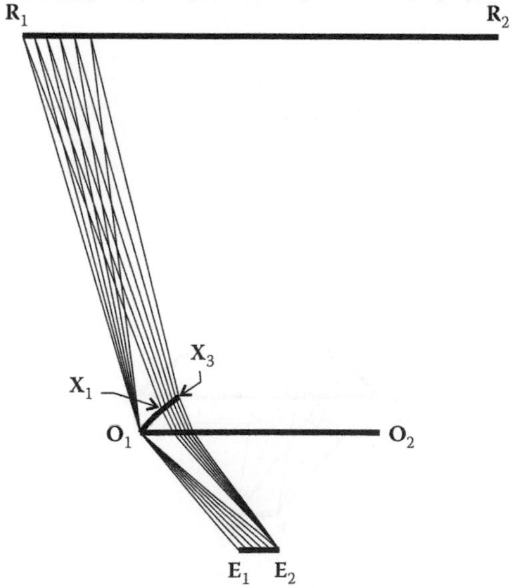

FIGURE 10.28
The rays emitted from points inside $\mathbf{E}_1\mathbf{E}_2$ and crossing the lens at point $\mathbf{O}_1$ (refracted on the bottom and top surfaces), define a new section $\mathbf{X}_1\mathbf{X}_2$ of the top surface of the lens.

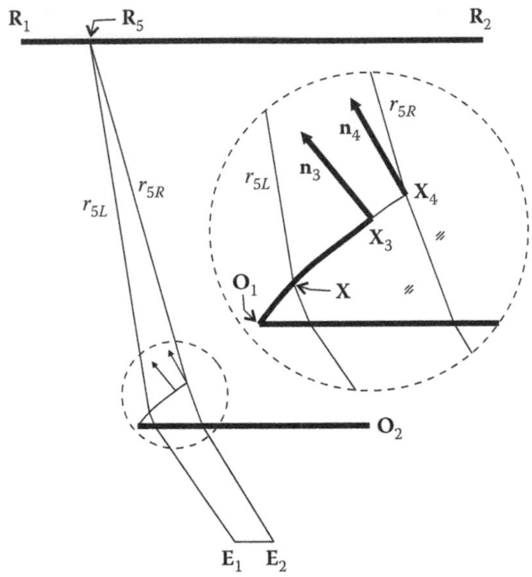

FIGURE 10.29
A ray coming from point $\mathbf{E}_1$ and crossing a known point **X** of the lens impacts $\mathbf{R}_1\mathbf{R}_2$ at point $\mathbf{R}_5$ and allows us to calculate a new point $\mathbf{X}_4$, based on a previously calculated point $\mathbf{X}_3$.

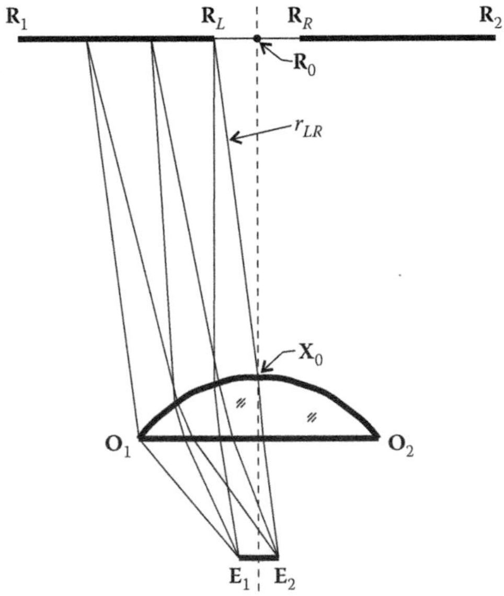

FIGURE 10.30
Complete lens. Ray r_{LR} reaches the center $\mathbf{X}_0$ of the lens before point $\mathbf{R}_L$ reaches $\mathbf{R}_0$. This means that the irradiance is not controlled in region $\mathbf{R}_L\mathbf{R}_R$ of the receiver ($\mathbf{R}_R$ is symmetrical relative to $\mathbf{R}_L$).

Moving point **X** along the points of the lens previously calculated, and proceeding as described above, new points on the top surface of the lens may be calculated. The complete lens is shown in Figure 10.30.

Ray r_{LR} reaches the center $\mathbf{X}_0$ of the lens before point $\mathbf{R}_L$ reaches the center $\mathbf{R}_0$ of the receiver. This means that the irradiance in the central region $\mathbf{R}_L\mathbf{R}_R$ of the receiver is not controlled (point $\mathbf{R}_R$ is symmetrical to $\mathbf{R}_L$).

10.4 Bundle Coupling and Prescribed Irradiance

There are two main groups of design problems in nonimaging optics. The first group is called "bundle-coupling" and its goal is to maximize the light power transferred between emitter (source) and receiver (target). The second group is called "prescribed irradiance" or "prescribed intensity" and its goal is to produce a desired irradiance pattern on a receiver (target).[2–4]

Given an optic O_1 with edges $\mathbf{O}_1$ and $\mathbf{O}_2$, and an emitter $\mathbf{E}_1\mathbf{E}_2$, we may define a bundle of rays b_1 from Lambertian source $\mathbf{E}_1\mathbf{E}_2$ to $\mathbf{O}_1\mathbf{O}_2$ as the set of all rays that cross both $\mathbf{E}_1\mathbf{E}_2$ and $\mathbf{O}_1\mathbf{O}_2$, as shown in Figure 10.31a. Bundle b_1 has étendue U_1. Accordingly, consider that $\mathbf{R}_1\mathbf{R}_2$ is a Lambertian source, emitting down in all directions (emission angle $\pm\pi/2$). The subset of rays that

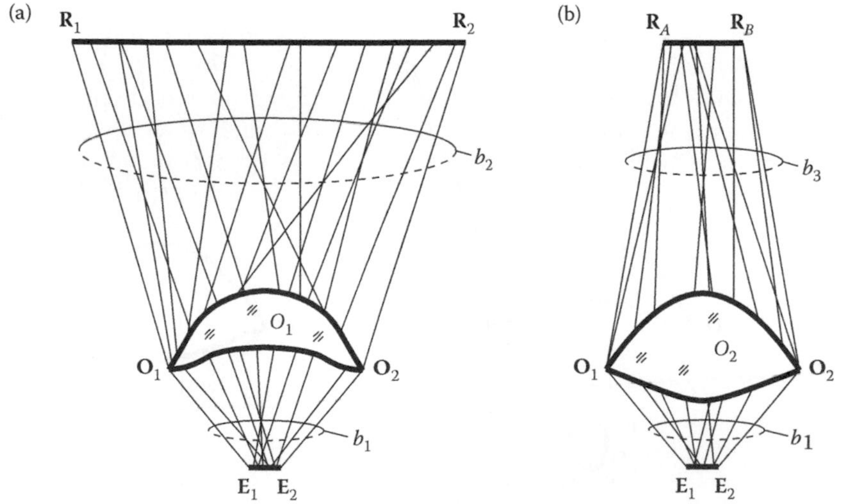

FIGURE 10.31
(a) Bundle of rays b_1 crossing both $\mathbf{E}_1\mathbf{E}_2$ and $\mathbf{O}_1\mathbf{O}_2$, and bundle of rays b_2 crossing both $\mathbf{O}_1\mathbf{O}_2$ and $\mathbf{R}_1\mathbf{R}_2$. (b) Bundle of rays b_1 crossing both $\mathbf{E}_1\mathbf{E}_2$ and $\mathbf{O}_1\mathbf{O}_2$ – same as in (a) – and bundle of rays b_2 crossing both $\mathbf{O}_1\mathbf{O}_2$ and $\mathbf{R}_A\mathbf{R}_B$.

intersects optic $\mathbf{O}_1\mathbf{O}_2$ is another bundle of rays b_2. This is the subset of all rays that cross both $\mathbf{O}_1\mathbf{O}_2$ and $\mathbf{R}_1\mathbf{R}_2$. Bundle b_2 has étendue U_2.

Now consider another optic O_2, with the same edges, $\mathbf{O}_1$ and $\mathbf{O}_2$, and same emitter $\mathbf{E}_1\mathbf{E}_2$, as in Figure 10.31b. The bundle of rays, b_1, from $\mathbf{E}_1\mathbf{E}_2$ to $\mathbf{O}_1\mathbf{O}_2$, is the same as before and, therefore, has the same étendue U_1. This optic O_2 has a different receiver $\mathbf{R}_A\mathbf{R}_B$. We may also consider that $\mathbf{R}_A\mathbf{R}_B$ is a Lambertian source, emitting down in all directions (emission angle $\pm\pi/2$). The subset of rays that intersects optic $\mathbf{O}_1\mathbf{O}_2$ is another bundle of rays b_3. This is the subset of all rays that cross both $\mathbf{O}_1\mathbf{O}_2$ and $\mathbf{R}_A\mathbf{R}_B$. Bundle b_3 has étendue U_3.

Figure 10.32b shows the same optic O_2 of Figure 10.31b, with emitter $\mathbf{E}_1\mathbf{E}_2$ and receiver $\mathbf{R}_A\mathbf{R}_B$. Each point $\mathbf{R}_I$ on the receiver is illuminated by the whole optic $\mathbf{O}_1\mathbf{O}_2$. The cone of rays at $\mathbf{R}_I$ is bound by the left edge ray r_{IL} coming from edge $\mathbf{O}_1$ of the optic, and right edge ray r_{IR} coming from edge $\mathbf{O}_2$ of the optic. Optic O_2 takes all the rays of bundle b_1 and transforms them to the rays of bundle b_3. Is is said that optic O_2 couples bundles b_1 and b_3, and this is an example of bundle coupling. The étendue U_1 of bundle b_1 matches étendue U_3 of bundle b_3, that is $U_1 = U_3$.

Figure 10.32a shows the same optic O_1 as Figure 10.25a and Figure 10.31a. This optic is designed for emitter $\mathbf{E}_1\mathbf{E}_2$ and produces a desired irradiance pattern on receiver $\mathbf{R}_1\mathbf{R}_2$. At each point $\mathbf{R}_I$ on the receiver, the incident light is

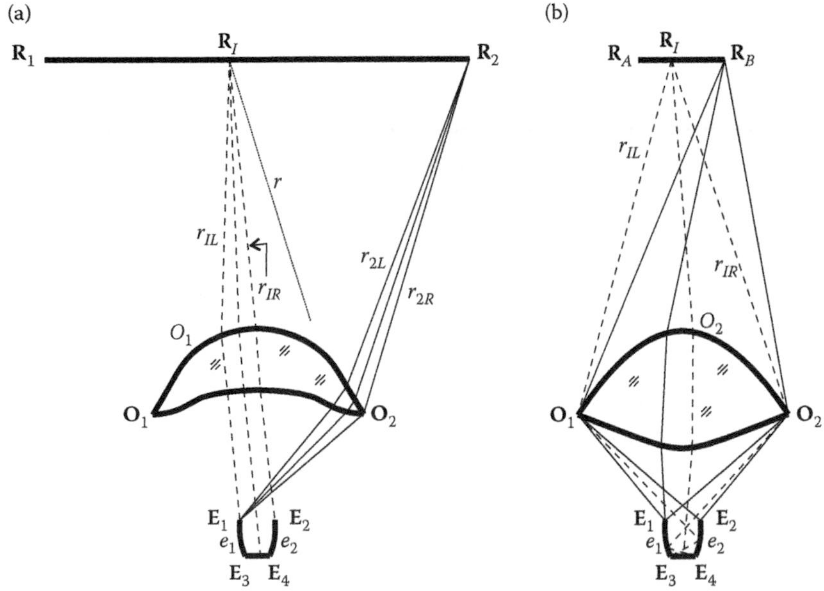

FIGURE 10.32
(a) Optic O_1 illuminates large receiver $\mathbf{R}_1\mathbf{R}_2$ producing there a desired irradiance pattern (prescribed irradiance). (b) Optic O_2 couples the étendue of the source with that of the receiver $\mathbf{R}_1\mathbf{R}_2$ (bundle coupling).

bound by a left edge ray r_{IL} and a right edge ray r_{IR}. In particular, at the edge $\mathbf{R}_2$ the edge incident rays are r_{2L} and r_{2R}.

Optic O_1 captures bundle b_1 emitted by $\mathbf{E}_1\mathbf{E}_2$ toward $\mathbf{O}_1\mathbf{O}_2$, but it does not fill bundle b_2. For example, ray r belongs to bundle b_2, but it is not part of the rays emitted by O_1 toward $\mathbf{R}_1\mathbf{R}_2$. We may, however, consider the bundle of rays b_4 consisting of all rays reaching $\mathbf{R}_1\mathbf{R}_2$ coming from $\mathbf{O}_1\mathbf{O}_2$. Bundle b_4 has étendue U_4. At each point R_I on the receiver, this bundle b_4 consists of all rays inside edge rays r_{IL} and r_{IR}. Bundle b_4 is a subset of b_2, and the étendue U_4 of bundle b_4 is much smaller than the étendue U_2 of bundle b_2, that is $U_1 = U_4 \ll U_2$.

When the étendue U_1 of the emitter is much smaller than the étendue U_2 of the receiver, it is possible to design an optic that will produce a prescribed irradiance pattern on the receiver. That optic is not unique, and different distributions of light on the receiver (coming from different directions) will produce the same irradiance pattern. These are called prescribed irradiance optics.

References

1. Gonzalez, J.C. and Miñano, J.C., Design of optical systems which transform one bundle of incoherent light into another, *Nonimaging Opt.: Maximum-Efficiency Light Transfer II*, SPIE Vol., 2016, 109, 1993.
2. Winston, R. et al., *Nonimaging Optics*, Elsevier Academic Press, Amsterdam, 2005.
3. Hernandez, M. et al., High-performance Köhler concentrators with uniform irradiance on solar cell, *Proc. SPIE 7059, Nonimaging Optics and Efficient Illumination Systems V*, San Diego, California, USA, September 2, 2008.
4. Buljan, M., *Free-Form Optical Systems for Nonimaging Applications*, PhD thesis, Technical University of Madrid, 2014.

11

Infinitesimal Étendue Optics

11.1 Introduction

The Simultaneous Multiple Surface (SMS) design method allows us to design two optical surfaces simultaneously that couple two input wavefronts onto two output wavefronts. Here, we consider the limit case, in which the étendue of the light captured by an optic goes to zero. When that happens, the emitter and receiver sizes become infinitesimal. In the case of a solar concentrator, the acceptance angle goes to zero.

In this limit case, we can still design two optical surfaces simultaneously, but the design process is simpler, more stable, and does not have loops that sometimes appear in the SMS. These characteristics make this an interesting design method when the receiver is very small, or when having a nonoptimal optic is acceptable. These designs may also serve as a starting point for further optimization using the SMS method.

11.2 Infinitesimal Étendue Optics

When we have a light emitter E and a receiver R, as shown in Figure 11.1a, the étendue of the light exchanged between the two is given by

$$U = [[\mathbf{R}_2, \mathbf{E}_1]] + [[\mathbf{R}_1, \mathbf{E}_2]] - [[\mathbf{R}_2, \mathbf{E}_2]] - [[\mathbf{R}_1, \mathbf{E}_1]] \tag{11.1}$$

where $[[\mathbf{X},\mathbf{Y}]]$ is the optical path length between $\mathbf{X}$ and $\mathbf{Y}$.

However, as the sizes of the emitter E and receiver R go to zero, we have the situation shown in Figure 11.1b, in which the infinitesimal étendue U_I received by R goes to

$$U_I = 2nR\cos\alpha\sin\left(\frac{d\alpha}{2}\right) \tag{11.2}$$

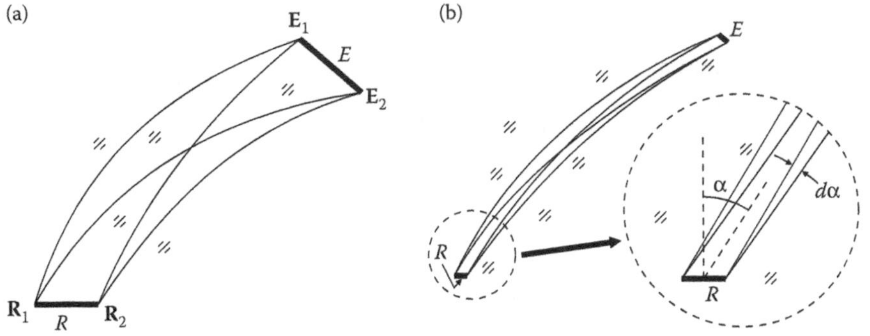

FIGURE 11.1
(a) Large emitter *E* and large receiver *R*. (b) Infinitesimal emitter *E* and infinitesimal receiver *R*.

where n is the refractive index in which the receiver is immersed, and $d\alpha$ is the angular aperture spanned by the emitter E when seen from receiver R. In expression (11.2) R represents the size of the receiver. We now apply this result to an optic with receiver R. Figure 11.2 shows a 2-D vertical cut of an optic O. Some edge rays defining a half-acceptance angle θ (total acceptance 2θ) are also shown diagrammatically. Conservation of étendue may be written as

$$2d\rho \sin\theta = 2nR\cos\alpha \sin\left(\frac{d\alpha}{2}\right) \tag{11.3}$$

Since, $\sin\zeta \to \zeta$ when $\zeta \to 0$ we may write $\sin(d\zeta) = d\zeta$ (angle ζ in radians). Now, making $\sin(d\alpha/2) = d\alpha/2$, we get

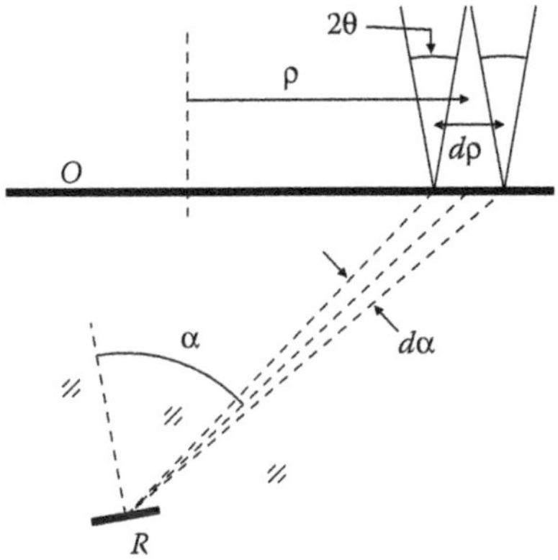

FIGURE 11.2
Étendue conservation through an optic O (in two dimensions) for an infinitesimal receiver R.

$$\frac{d\rho}{d\alpha} = \frac{n}{2\sin\theta} R\cos\alpha \tag{11.4}$$

This expression assumes that ρ increases as α increases. However, the opposite may also be true: ρ decreases as α increases. In that case, we would have

$$\frac{d\rho}{d\alpha} = -\frac{n}{2\sin\theta} R\cos\alpha \tag{11.5}$$

or, in general,

$$\frac{d\rho}{d\alpha} = \pm\frac{n}{2\sin\theta} R\cos\alpha \tag{11.6}$$

Integrating this differential equation we get

$$\rho = \pm\frac{nR}{2\sin\theta}\sin\alpha + C \tag{11.7}$$

or

$$\rho = \pm f\sin\alpha + C \tag{11.8}$$

where $f = nR/(2\sin\theta)$ and C is a constant of integration. In the particular case in which $C = 0$, that is, $\rho = 0$ when $\alpha = 0$, we get from Equation 11.8 $\rho = f\sin\alpha$.

Now, we consider the case of a receiver with an arbitrary shape. It will have a projected area $p(\alpha)$ in direction α, as shown in Figure 11.3. This projected area can also be written as $p(\alpha) = Rb(\alpha)$, where $b(\alpha)$ is a function of angle α. The receiver in the example of Figure 11.3 has a projected area of dimension $p(0) = R$ for direction $\alpha = 0$.

Expression (11.3) can be written in this case as

$$2d\rho\sin\theta = 2nRb(\alpha)\sin\left(\frac{d\alpha}{2}\right) \tag{11.9}$$

and the equivalent expression to Equation 11.6 becomes

$$\frac{d\rho}{d\alpha} = \pm\frac{nR}{2\sin\theta} b(\alpha) \tag{11.10}$$

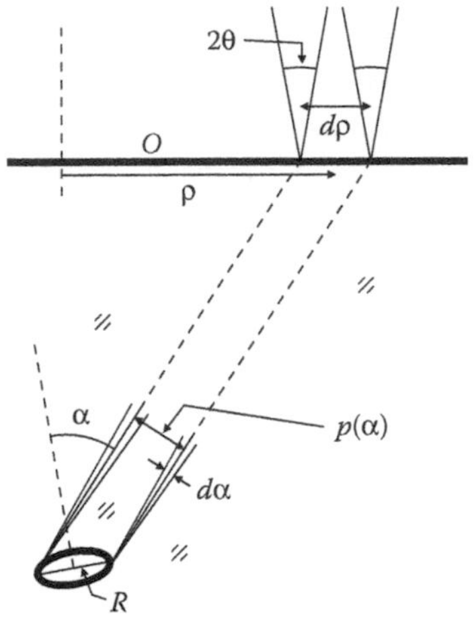

FIGURE 11.3
Infinitesimal étendue optic for an arbitrary shape receiver.

Integrating this differential equation we get

$$\rho(\alpha) = \pm\frac{nR}{2\sin\theta}B(\alpha) + C \tag{11.11}$$

where $B(\alpha) = \int b(\alpha)$ and C is a constant of integration.

If the receiver is flat with size R, as in Figure 11.2, $b(\alpha) = \cos\alpha$ and, therefore, $B(\alpha) = \sin\alpha$. If the receiver is circular with diameter R, $b(\alpha) = 1$ and $B(\alpha) = \alpha$, and we get for a circular receiver

$$\rho(\alpha) = \pm\frac{nR}{2\sin\theta}\alpha + C \tag{11.12}$$

Now we consider infinitesimal étendue optics with circular symmetry.[1] Making $\rho = 0$ for $\alpha = 0$, we get $C = 0$. This is the case when the vertical ray through the receiver ($\alpha = 0$) leaves the optic at $\rho = 0$. That situation is shown in Figure 11.4 for an optic O_C with circular symmetry and meridian acceptance angle θ_M. Expression (11.8) is now

$$\rho = \frac{nR}{2\sin\theta_M}\sin\alpha \tag{11.13}$$

Figure 11.5 shows a three-dimensional view of the same concentrator O_C with circular symmetry, and also the paths of two sagittal rays s_1 and s_2

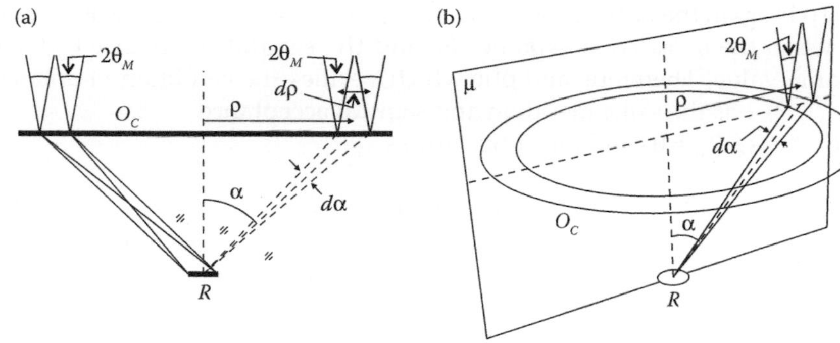

FIGURE 11.4
(a) Meridian rays in a concentrator with circular symmetry. (b) 3-D view of same, showing meridian plane μ.

(shown diagrammatically) contained on sagittal plane σ. If these are edge rays of the incoming radiation, they define a sagittal acceptance angle θ_G. For a given point on the entrance aperture of the concentrator, these rays have the highest value of skewness. They must arrive at the receiver also with the highest value of skewness and, therefore, at its edge. In particular, ray s_1 arrives at the receiver, also as a sagittal ray, making an angle α to the vertical (here the optical axis is vertical). Conservation of skew is then

$$\rho \sin\theta_G = n\frac{R}{2}\sin\alpha \Leftrightarrow \rho = \frac{nR}{2\sin\theta_G}\sin\alpha \tag{11.14}$$

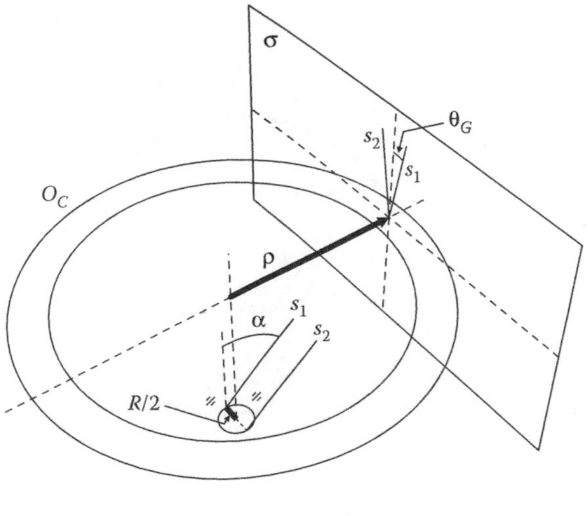

FIGURE 11.5
Saggital rays in a concentrator with circular symmetry.

where ρ is again the radial coordinate. Equation 11.14 is the same as Equation 11.13 when the meridian acceptance θ_M and the sagittal acceptance θ_G have the same value. Therefore, an optic which verifies the condition of aplanatism (11.13), has the same meridian and sagittal acceptance.

In that case, $\theta_M = \theta_G$ and the optic verifies

$$\rho = f \sin\alpha \tag{11.15}$$

with

$$f = \frac{nR}{2\sin\theta} \tag{11.16}$$

These optics that verify Equation 11.15 are called aplanatic,[2] and they are the limit case of SMS optics when the acceptance angle goes to zero.[3] Figure 11.6a shows an SMS optic with acceptance angle 2θ and receiver R. Figure 11.6b shows the same optic crossed by two symmetric flow lines f_1 and f_2. The étendue of the radiation contained by those two flow-lines is given by $U = 2(2\rho)\sin\theta = 2(d_1 - d_2)$. Now, when the size of R goes to zero, we have the situation shown in Figure 11.6c, in which d_1 and d_2 become parallel, and $d_1 - d_2 = R\sin\alpha$. The étendue becomes, in this case, $2\rho\sin\theta = R\sin\alpha$, which is the same as Equations 11.15 and 11.16, for an aplanatic optic. In this example, the receiver R is immersed in a medium of refractive index $n = 1$ (air), but the same argument could be used if $n \neq 1$. It can then be concluded that the SMS optic tends to an aplanatic optic in the limit case in which $R \to 0$.

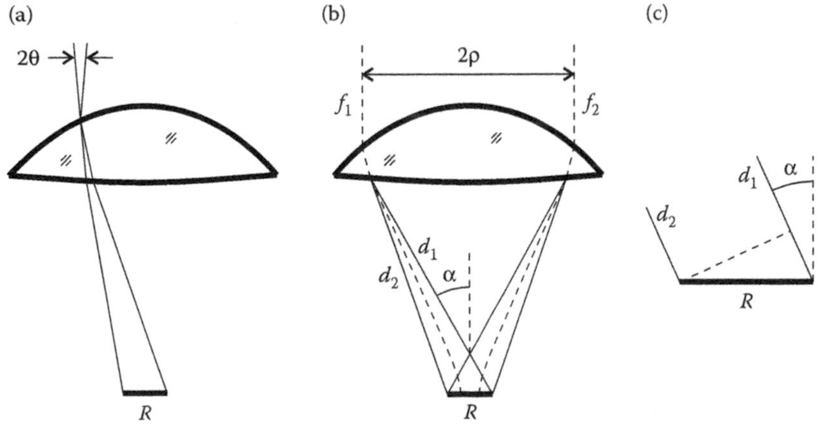

FIGURE 11.6
(a) SMS optic with acceptance angle 2θ and receiver R. (b) Flow lines f_2 and f_1 crossing the SMS lens. (c) Limit case in which the size of R goes to zero.

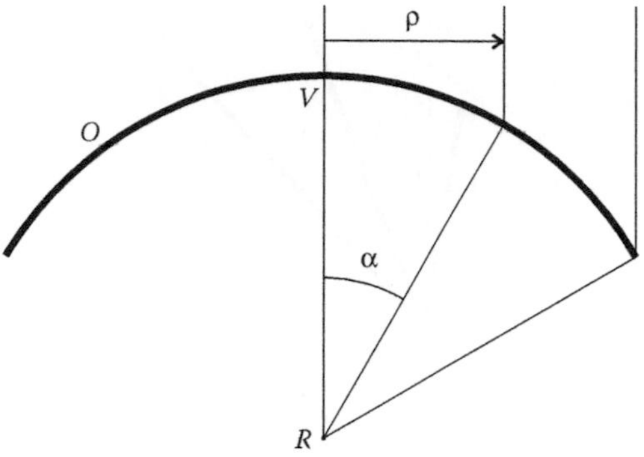

FIGURE 11.7
Infinitesimal étendue (aplanatic) optic.

Consider now a hypothetical circular optic O centered on receiver **R** and that collects a set of parallel vertical rays and concentrates them to **R**, as in Figure 11.7. This optic verifies $\rho = [\mathbf{V},\mathbf{R}] \sin \alpha$, where $[\mathbf{V},\mathbf{R}]$ is the distance from optic vertex **V** to point receiver **R** and is, therefore, aplanatic. Distance $[\mathbf{V},\mathbf{R}]$ is the radius of optic O.

In three dimensions, this optic will be a spherical cap, as shown in Figure 11.8. We may now cut this optic with a vertical plane μ at $x_1 = p_X$. This defines circular curve c on O.

Consider now the particular case in which $[\mathbf{V},\mathbf{R}] = n$, where n is the refractive index in which **R** is immersed. The vectors pointing from **R** to points

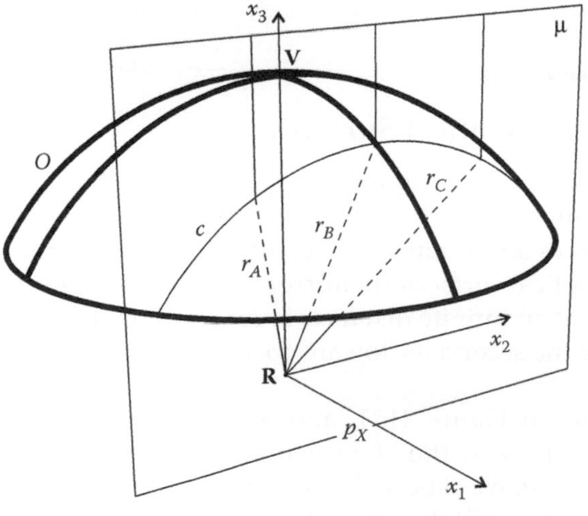

FIGURE 11.8
Cut of a spherical apalanatic optic O by a vertical plane $x_1 = p_X$.

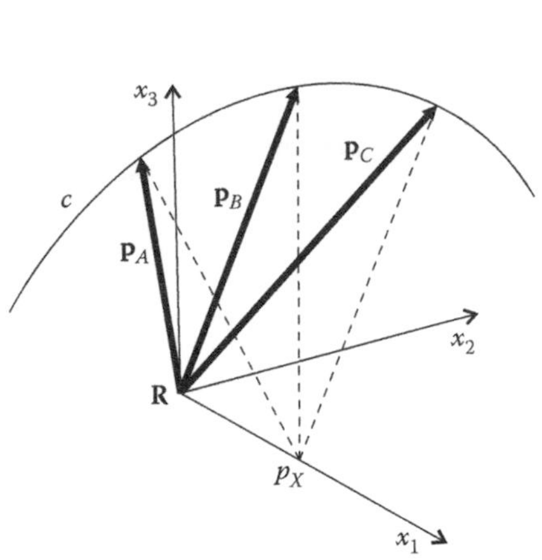

FIGURE 11.9
All rays pointing at curve c have optical momentum with $p_1 = p_X$ and are on a cone of $p_1 =$ constant.

along curve c are the optical momenta of the (reverse) rays emitted from **R** and crossing curve c, as shown in Figure 11.9.

The momenta of all these rays are on the surface of a cone with vertex **R**. For example, rays r_A, r_B, and r_C have momenta $\mathbf{p}_A$, $\mathbf{p}_B$, and $\mathbf{p}_C$. Optical momentum $\mathbf{p}_A = (p_{A1}, p_{A2}, p_{A3})$ with $p_{A1} = p_X$. Also, $p_{B1} = p_X$ and $p_{C1} = p_X$. Therefore, all rays in the cone crossing c have $p_1 =$ constant. These rays leave the optic in the direction of axis x_3 in a plane μ with $x_1 =$ constant, as shown in Figure 11.8. This is a general characteristic of aplanatic systems. A set of rays with $p_1 =$ constant leaves the optic in a plane $x_1 =$ constant.

11.3 Continuous Optical Surfaces

Aplanatic optics may be designed using a simple method in which we proceed in very small steps, building both optical surfaces simultaneously. Figure 11.10 shows an example of one of these optics for the particular case in which both optical surfaces are mirrors. An incoming ray with direction **v** (from an object at an infinite distance) is reflected by the first optical surface m_B, and then by the second m_T, toward focus **R** (where an image of the object is formed).

Referring now to Figure 11.11, and according to the above expression (11.15), we can define a method for calculating the points of the mirrors of an aplanatic optic. First, we choose the size of receiver R (should be small when compared to distance [**P**,**Q**]), and pick a value for the small half-acceptance angle θ. With the values of R and θ, we may calculate magnification f using expression (11.16).

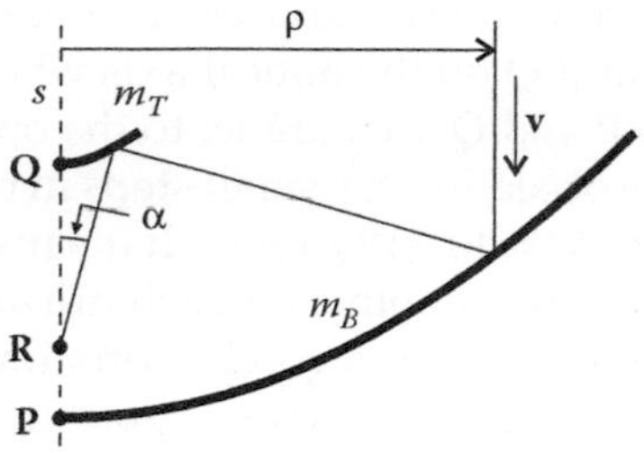

FIGURE 11.10
Aplanatic XX optic with two mirrors m_B and m_T and point receiver **R**.

Suppose now we already know the positions of points $\mathbf{P}_1$ and its normal $\mathbf{n}_1$, and $\mathbf{Q}_1$ and its normal $\mathbf{m}_1$, that is, the path of ray r_1 is already known. We now launch a ray r_2, shifted by a distance $\Delta\rho$ relative to ray r_1 and in the direction of vector **v**. We intersect ray r_2 with the the plane defined by point $\mathbf{P}_1$ and normal $\mathbf{n}_1$, and determine the position of point $\mathbf{P}_2$. From expression (11.15) we can determine $\alpha_2 = \arcsin(\rho_2/f)$.

Launching from **R** a ray tilted by an angle α_2 to the optical axis, and intersecting it with the plane defined by point $\mathbf{Q}_1$ and normal $\mathbf{m}_1$, we can determine the position of point $\mathbf{Q}_2$. We now have the complete path of ray r_2. From the direction of the incident and reflected rays at $\mathbf{P}_2$, we can determine its normal $\mathbf{n}_2$. Also, from the directions of the incident and reflected rays at $\mathbf{Q}_2$, we can determine its normal $\mathbf{m}_2$. Repeating the same process for another ray to the right of r_2, we can determine two new points on m_B and m_T to the right of $\mathbf{P}_2$ and $\mathbf{Q}_2$. This process continues as we determine more and more points,

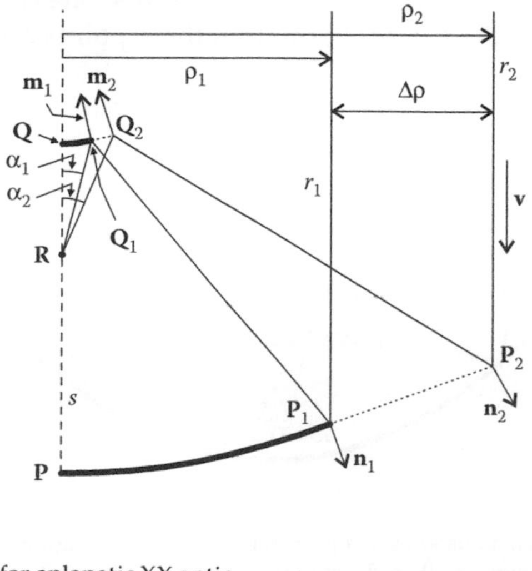

FIGURE 11.11
Construction method for aplanatic XX optic.

and completely define both optical surfaces. The design starts by defining the position of points **P** and **Q** on the optical axis where the surfaces m_B and m_T start. The normals to **P** and **Q** are parallel to the optical axis *s*.

Note that we need to proceed in very small steps in order to get good precision optical surfaces (e.g., $\Delta\rho = 10^{-6}[\mathbf{P},\mathbf{Q}]$), which means calculating hundreds of thousands or even millions of points for each mirror. However, we do not need to save all those points to get a good approximation of the mirror. A few points (say 100 or so for each mirror) are, typically, enough. Therefore, as the calculation proceeds and we calculate more and more points very close to each other, we can save only one point for every 10.000 points calculated, for example.

Aplanatic optics, like the one in Figure 11.10, only work perfectly when used with an infinitesimal acceptance angle, and an infinitesimal size (point) receiver **R**. However, in practice, these optics are used as concentrators with a finite acceptance angle 2θ and a finite size receiver *R* centered at the design position **R**, as shown in Figure 11.12. In that case, the ideal position of the receiver *R* is no longer centered at the design position **R**, but at position R_E, slightly displaced toward the secondary mirror, as shown in Figure 11.12.

This is similar to what happens with the combinations of parabolic primaries and CEC secondaries for which the ideal position for the entrance aperture of the CEC is not centered at the focus of the parabolic mirror, but slightly displaced toward it (see Chapter 6).

In general, this method allows us to design two optical surfaces in which n_1 is the refractive index before the first surface, n_2 is the refractive index between surfaces, and n_3 is the refractive index after the second surface. If $n_1 = n_2$ then the first surface is reflective, otherwise refractive. Also, if $n_2 = n_3$ the second surface is reflective, otherwise refractive. As seen above, as we step through the construction method, we define the path of a new ray using expression (11.15), and determine two new points, one on each optical surface. The corresponding surface normals at these points can be calculated by

$$\text{dflnrm}(\mathbf{i},\mathbf{r},n_A,n_B) = \frac{n_A\mathbf{i} - n_B\mathbf{r}}{\|n_A\mathbf{i} - n_B\mathbf{r}\|} \tag{11.17}$$

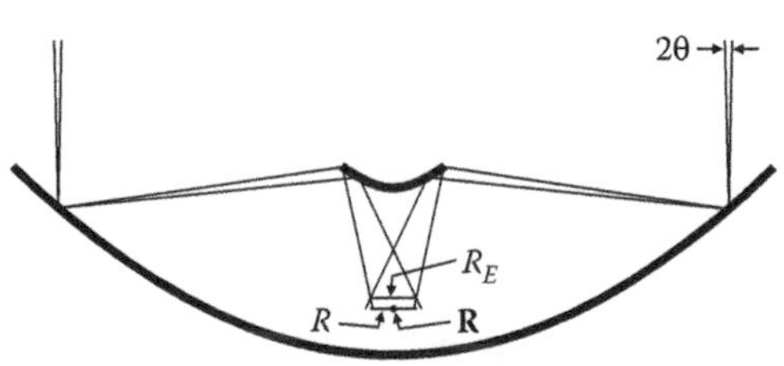

FIGURE 11.12
When an aplanatic optic is used as a concentrator for a finite angular acceptance angle, the ideal position of the receiver will be closer to the secondary mirror.

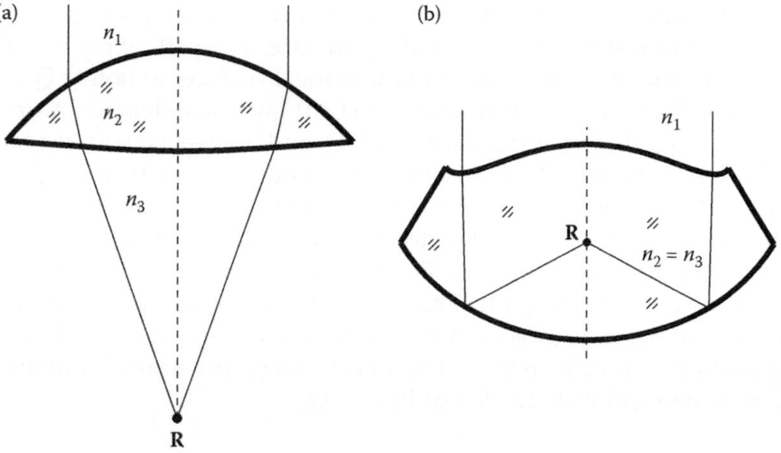

FIGURE 11.13
(a) RR aplanatic optic. (b) RX aplanatic optic.

where $||\mathbf{i}|| = ||\mathbf{r}|| = 1$, and **i** is the direction of the incident ray, and **r** is the direction of the deflected ray, and n_A and n_B are the refractive indices before and after deflection (see Chapter 16). If $n_A \neq n_B$ the ray is refracted, and if $n_A = n_B$ the ray is reflected. This method can then be used to design different types of aplanatic optics, such as RR, XR, RX, XX, or RXI. Figure 11.13 shows examples of an RR and an RX aplanatic optics.

Figure 11.14 shows another example, now an RXI optic. Its design starts by choosing section **AB** which may, for example, be a straight section. Section **AB** verifies $\rho_B = f \sin \alpha_A$, where f is the magnification of the aplanat.

Now take ray r_A coming from receiver **R** and reflect it at **A**. Take also a vertical ray r_B and refract it at **B**. Intersecting these two rays defines point **C** at the

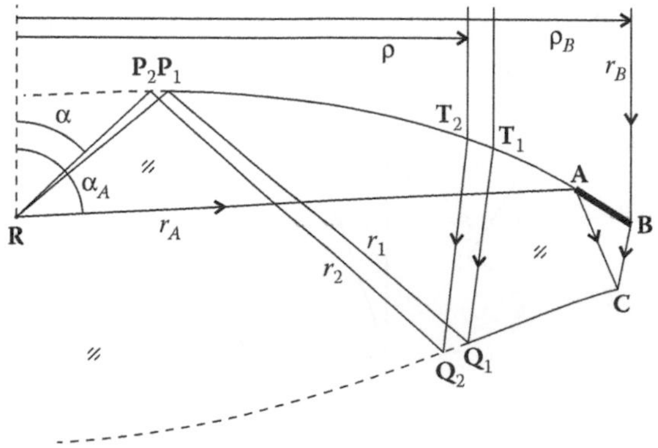

FIGURE 11.14
RXI aplanatic optic.

edge of the bottom surface. Suppose now that we already know portion $\mathbf{P}_1\mathbf{B}$ of the top surface and $\mathbf{Q}_1\mathbf{C}$ of the bottom surface. The path of ray r_1 is, therefore, known, and also the normals to the optical surfaces at points $\mathbf{Q}_1$ and $\mathbf{P}_1$ (not shown). Now, we may take a new vertical ray r_2 at a distance ρ from the axis, refract it on the top surface at a point $\mathbf{T}_2$, and intersect it with the tangent to the bottom surface at $\mathbf{Q}_1$. This defines the position of a new point $\mathbf{Q}_2$ on the bottom surface. Now we can calculate the value of angle α from $\rho = f \sin \alpha$. We launch a ray from **R** at angle α to the optical axis, and intersect with the tangent to the top surface at point $\mathbf{P}_1$, calculating the position of a new point $\mathbf{P}_2$ on the top surface. The path of ray r_2 is now fully known and the normals to the top and bottom surfaces may now be determined at points $\mathbf{P}_1$ and $\mathbf{Q}_1$. The process may now be repeated by decreasing ρ (by a small amount) and calculating new points to the left of $\mathbf{P}_2$ and $\mathbf{Q}_2$.

11.4 Fresnel Optics

It is also possible to design infinitesimal étendue optics with Fresnel (discontinuous) optical surfaces. Let us then start, for example, with an existing Fresnel reflector composed of helistats h and flat receiver R, as in Figure 11.15. Helistats h reflect incoming vertical light directly to receiver R. We may now add a new heliostat at the edge, starting at point $\mathbf{P}_0$. This new helistats m_1 is designed in conjunction with a secondary mirror m_S starting at point $\mathbf{S}_0$. Point $\mathbf{P}_0$ is at a horizontal distance ρ_0 from the axis of the system and point $\mathbf{S}_0$ makes an angle α_0 to that same axis (normal to R).

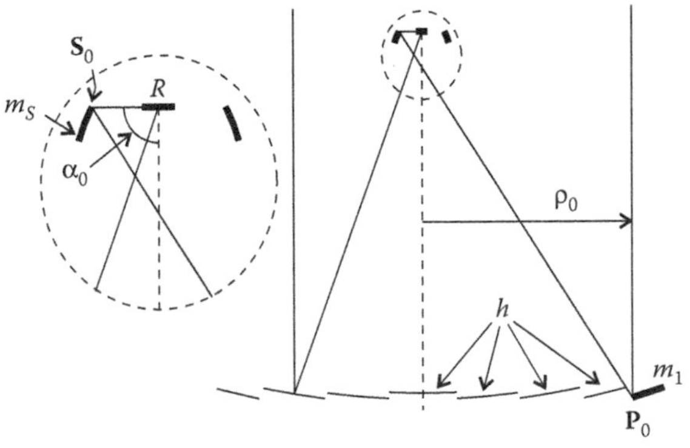

FIGURE 11.15
Fresnel concentrator with heliostats h and receiver R. Additional infinitesimal étendue optic composed of primary mirror m_1 and secondary mirror m_S.

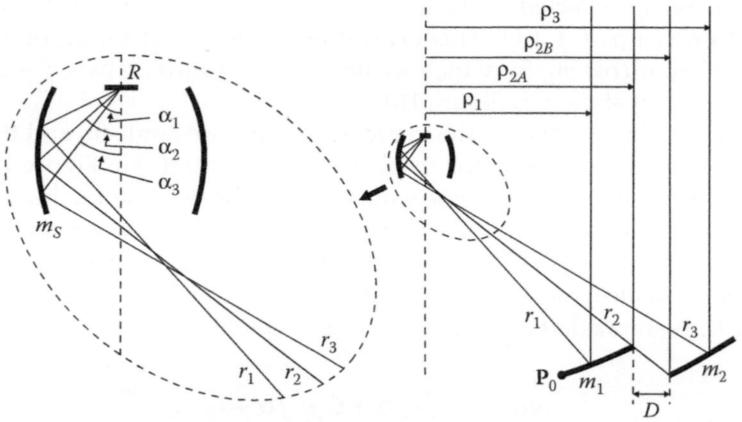

FIGURE 11.16
Infinitesimal étendue optic composed of primary mirrors m_1 and m_2 and secondary mirror m_S.

Figure 11.16 shows some ray paths for the portion of the concentrator to the right of $\mathbf{P}_0$. In this configuration, ρ increases as α decreases, and the expression to be used is then Equation 11.8 with a negative sign: $\rho = -f\sin\alpha + C$.

Mirror m_1 starts at point $\mathbf{P}_0$ for which the angle at the receiver is $\alpha_0 = \pi/2$, and the distance to the optical axis is ρ_0. Since the receiver is immersed in air, we have $n = 1$. From $\rho = -f\sin\alpha + C$ we, therefore, have for this starting position

$$\rho_0 = -\frac{R}{2\sin\theta}\sin\left(\frac{\pi}{2}\right) + C \tag{11.18}$$

or

$$C = \rho_0 + \frac{R}{2\sin\theta} \tag{11.19}$$

and the expression describing primary mirror m_1 and the corresponding section of the secondary is

$$\rho = -f\sin(\alpha) + \rho_0 + f \tag{11.20}$$

where $f = R/2\sin\theta$. For consistency, we may choose for θ the same half-acceptance angle of the Fresnel concentrator with heliostats h and receiver R.

For rays r_1 reflected on mirror m_1, we have $\rho_1 = -f\sin\alpha_1 + \rho_0 + f$ and the same is valid for specific ray r_2 reflected at the right edge of mirror m_1, for which we have $\rho_{2A} = -f\sin\alpha_2 + \rho_0 + f$. However, this same ray r_2 may also be seen as being reflected at the left edge of mirror m_2, for which we have $\rho_{2B} = -f\sin\alpha_2 + \rho_0 + f + D$. This expression is also valid for all other rays r_3 reflected at mirror m_2, for which we have $\rho_3 = -f\sin\alpha_3 + \rho_0 + f + D$. The complete optic is shown in Figure 11.17.

If given circular symmetry, the meridian acceptance angle θ_M and the sagittal acceptance angle θ_G will not match for the concentrator in Figure 11.17. In order for that to happen, the optic would have to be designed using expression (11.13) or its equivalent (11.14) and that is not the case for the central Fresnel mirror or the added heliostats $m_1, m_2, \ldots$ and secondary m_S.

It is also possible to design a mirror–mirror (XX) concentrator for a circular receiver R, as shown in Figure 11.18. In this example,

$$\rho(\alpha) = \frac{R}{2\sin\theta}\alpha + C = f\alpha + C \tag{11.21}$$

where $f = R/2\sin\theta$, and R is the diameter of the circular receiver. Expression (11.21) was obtained from (11.12) with a positive sign and $n = 1$ since the receiver is in air.

The first mirror m_1 can be calculated, for example, with the condition $\rho = 0$ for $\alpha = 0$, which makes $C = 0$, or

$$\rho_1(\alpha) = f\alpha_1 \tag{11.22}$$

The second mirror m_2 is then calculated with the expression

$$\rho_2(\alpha) = f\alpha_2 + D_1 \tag{11.23}$$

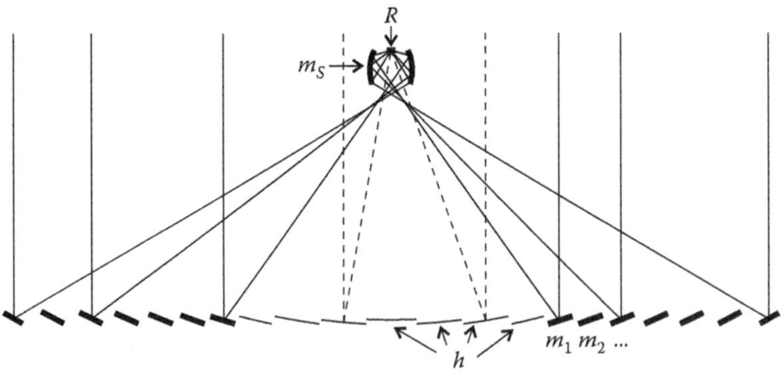

FIGURE 11.17
Fresnel concentrator made of heliostats h and receiver R and additional infinitesimal étendue optic made of heliostats $m_1, m_2, \ldots$ with corresponding secondary mirror m_S.

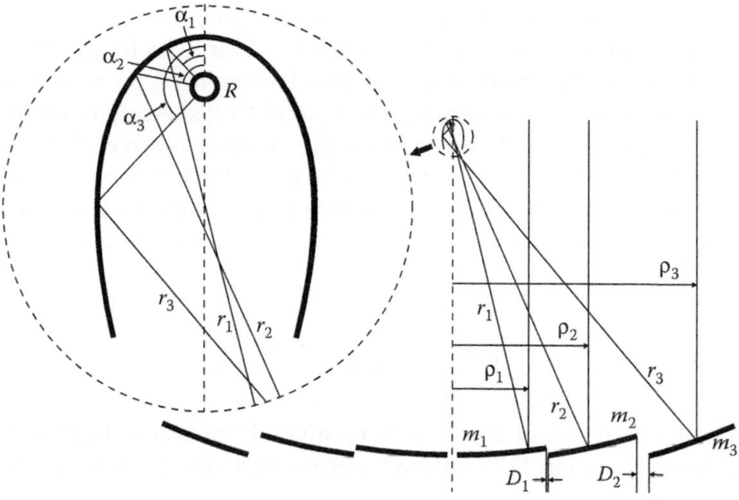

FIGURE 11.18
XX Fresnel concentrator for a circular receiver R.

The third mirror m_3 is then calculated with the expression

$$\rho_3(\alpha) = f\alpha_3 + D_1 + D_2 \tag{11.24}$$

and so on for all the heliostats.

A particular case of one of these designs is shown in Figure 11.19, which now has a primary with a single mirror m_1, and a corresponding secondary

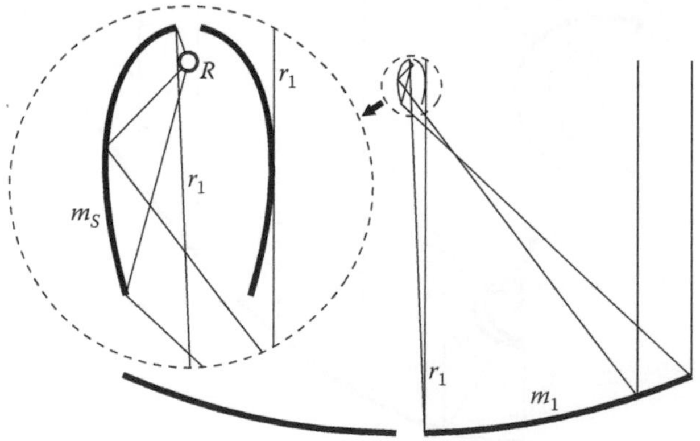

FIGURE 11.19
RR concentrator in which all rays undergo two reflections. On their way down, all rays miss the secondary mirror m_S and receiver R. Reflection on the primary mirror m_1, redirectes them toward secondary mirror m_S, from which they are again redirected toward the receiver R.

mirror m_S (and their symmetrical).[4] In this example, the whole optic is calculated by an expression similar to Equation 11.23 in which now D_1 is a constant calculated for the path of ray r_1. The left-most ray r_1 reflected by m_1 misses the secondary when coming down, avoiding shading of the primary by the secondary mirrors. It also misses the circular receiver R after reflection on the primary mirror m_1 and before being reflected by m_S toward R.

A similar method to that shown in Figure 11.10 can also be used to design a mirror–lens (XR) concentrator for a circular receiver R, as shown in Figure 11.20.[5] In this example,

$$\rho(\alpha) = \frac{nR}{2\sin\theta}\alpha = f\alpha \tag{11.25}$$

where R is the diameter of the circular receiver. Expression (11.25) was again obtained from Equation 11.12 with a positive sign, and $\rho = 0$ for $\alpha = 0$ which makes $C = 0$.

The inner section of the secondary is prescribed as a circular section c centered at the center of R, a straight line m along the flow-line pointing toward the center of R, and a straight line f (perpendicular to m).

Figure 11.21 shows a similar optic, but without the radial line m on the inner surface of the refractor. Incoming vertical light rays, after crossing the optic, are headed toward the center of R. These rays hit surface R perpendicularly and produce there a uniform illumination (Figure 11.21a). Incoming light rays at the acceptance angle θ are redirected by the optic in directions (approximately) tangent to R and, therefore, also produce a (quite) uniform illumination on R (Figure 11.21b). That is, parallel light

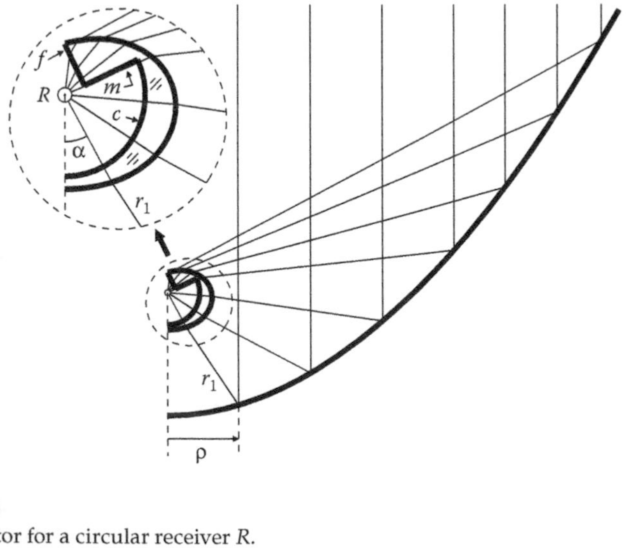

FIGURE 11.20
XR concentrator for a circular receiver R.

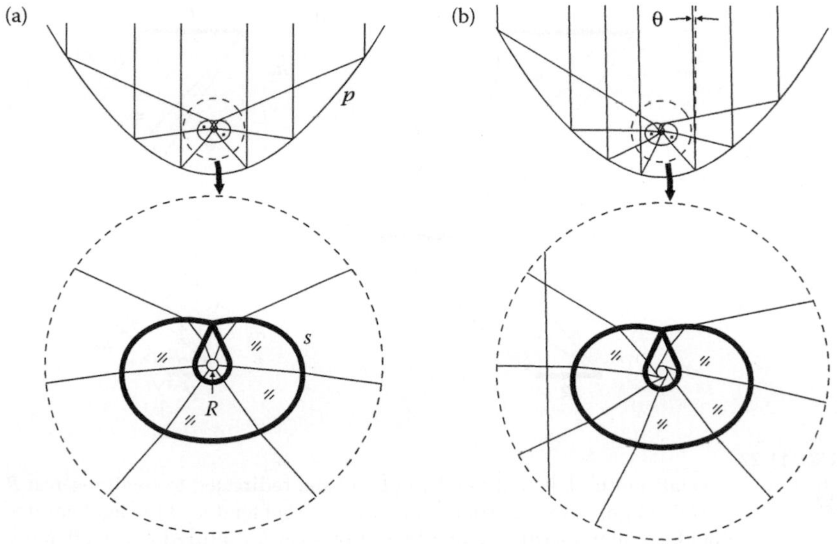

FIGURE 11.21
Circular receiver illumination uniformity for different incidence angles of sunlight: (a) vertical light and (b) light tilted by the acceptance angle θ of the concentrator.

rays inside the acceptance angle always produce a uniform illumination on receiver *R*.

Infinitesimal étendue designs may also be used as a basis design to be further improved by aplying the SMS design method.[6]

The optic in Figure 11.21 verifies Equation 11.25 where *f* is a constant. As seen from Figure 11.21a, incoming parallel vertical rays are directed to the center of receiver *R*. However, since *R* has a finite radius, these rays are absorbed by *R* as they impact all around the circular receiver. The same happens with edge rays tilted by an angle θ to the vertical. If these rays are coming from a very large and very far object (the sun, for example), no image of that object is formed on the tubular receiver. This is in contrast to what happens with, say, the optic in Figure 11.10, that verifies the condition of aplanatism (11.15), and does form an image of this large, far object onto a flat receiver.

11.5 Finite Distance Source

All the examples given above were for the case in which the light source is very far from the optic. This is useful in the designing of solar concentrators, since the sun is very far away from Earth (and, therefore, the concentrator). It is possible, however, to apply these concepts to a situation in which both the source (emitter) and receiver are close to the optic. Figure 11.22 shows a

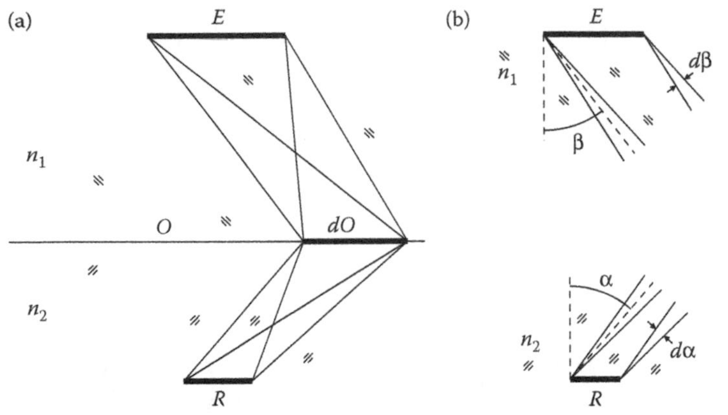

FIGURE 11.22
(a) The light of a small emitter *E* is captured by optic *O* and redirected to small receiver *R*. In particular, some of that light crosses portion *dO* of optic *O*. (b) Étendue of the light emitted by *E* in direction β with angular aperture *d*β and that of receiver *R* captured in direction α with angular aperture *d*α.

diagrammatic representation of that situation. An emitter *E* emits light that is captured by an optic *O* and redirected toward a receiver *R*.

Expression (11.2) may now be applied to both the emitter *E* and receiver *R*, in the limit case in which they are very small. Supose then that emitter *E* is immersed in a medium of refractive index n_1. The étendue of its emission at an angle β within angular aperture *d*β is given by

$$U_E = 2n_1 E\cos\beta\sin\left(\frac{d\beta}{2}\right) \tag{11.26}$$

This light is redirected by a portion *dO* of optic *O* toward receiver *R*, immersed in a medium of refractive index n_2. At the receiver, this étendue is

$$U_R = 2n_2 R\cos\alpha\sin\left(\frac{d\alpha}{2}\right) \tag{11.27}$$

Equating the étendues emitted by *E* and captured by *R*, we get

$$2n_1 E\cos\beta\sin\left(\frac{d\beta}{2}\right) = 2n_2 R\cos\alpha\sin\left(\frac{d\alpha}{2}\right) \tag{11.28}$$

or (when $\beta \to 0$ and $\alpha \to 0$)

$$n_1 E\cos\beta d\beta = n_2 R\cos\alpha d\alpha \tag{11.29}$$

and

$$n_1 E \sin\beta = n_2 R \sin\alpha + C \tag{11.30}$$

if $\beta = 0$ when $\alpha = 0$, we get $C = 0$ and[7]

$$\sin\beta = \frac{n_2 R}{n_1 E}\sin\alpha \tag{11.31}$$

This is the same expression as Equation 4.74 only now R and E are infinitesimal sizes while in Equation 4.74 a_1 and a_2 are finite sizes. Figure 11.23 shows one of these optics.

Here, angles α and β are related by expression (11.31), which can also be written as

$$\alpha(\beta) = \arcsin\left[\frac{n_1 E}{n_2 R}\sin\beta\right] \tag{11.32}$$

Let us now consider a different situation. A point emitter **E** coupled to an optic O_R with rotational symmetry around axis b produces a given output intensity pattern, as shown in Figure 11.24. Point **E** emits an intensity $I_S(\beta,\varphi)$ in direction (β,φ) within a solid angle $d\Omega_S$, resulting in a flux $d\Phi = I_S(\beta,\varphi)\, d\Omega_S$. The optic will redirect this light producing an output pattern with an intensity $I_0 I_P(\alpha,\varphi)$ in direction (α,φ) within a solid angle $d\Omega_P$, resulting in a flux $d\Phi = I_0 I_P(\alpha,\varphi)\, d\Omega_P$, where I_0 is a constant to be determined.

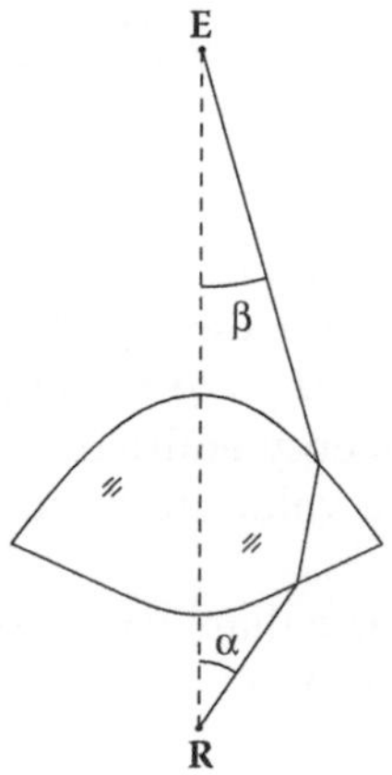

FIGURE 11.23
RR aplanatic optic with emitter **E** and receiver **R**.

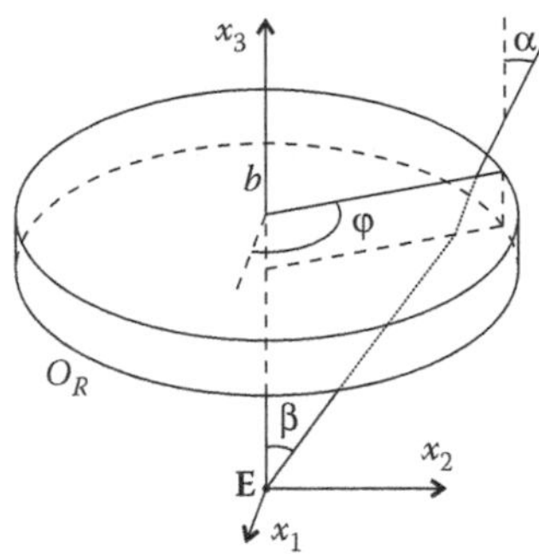

FIGURE 11.24
A ray emitted from point source **E** at an angle β to the optical axis crosses a rotational optic O_R and leaves making an angle α to the optical axis.

From conservation of energy, the flux emitted by point **E** must match that leaving the optic:

$$\begin{aligned} d\Phi &= I_S(\beta,\varphi)d\Omega_S = I_0 I_P(\alpha,\varphi)d\Omega_P \\ &= I_S(\beta)\sin\beta d\beta d\varphi = I_0 I_P(\alpha)\sin\alpha d\alpha d\varphi \\ &= I_S(\beta)\sin\beta d\beta = I_0 I_P(\alpha)\sin\alpha d\alpha \end{aligned} \tag{11.33}$$

since angle $d\varphi$ is the same, before and after the optic. This is a differential equation for $\alpha(\beta)$. As an initial condition, we assume that $\beta = \beta_0$ when $\alpha = \alpha_0$ and we may write:

$$\int_{\beta_0}^{\beta} I_S(\beta)\sin\beta\, d\beta = I_0 \int_{\alpha_0}^{\alpha} I_P(\alpha)\sin\alpha\, d\alpha \tag{11.34}$$

Point **E** emits light within an angle range $\beta_0 < \beta < \beta_M$ and the optic redistributes this light into an output pattern within an angle range $\alpha_0 < \alpha < \alpha_M$. When we replace $\beta = \beta_M$ and $\alpha = \alpha_M$ into Equation 11.34 we get an expression for the conservation of total energy emitted by **E** that allows us to calculate I_0. Now, we may give different values to β and calculate the corresponding values of α, giving us $\alpha(\beta)$.

We now consider the particular case in which $I_S(\beta) = \cos^N(\beta)$ and $I_P(\alpha) = \cos^M(\alpha)$. Equation 11.34 is now

$$\int_{\beta_0}^{\beta} \cos^N\beta \sin\beta\, d\beta = I_0 \int_{\alpha_0}^{\alpha} \cos^M\alpha \sin\alpha\, d\alpha \tag{11.35}$$

Now, we consider expression

$$\int \cos^B x \sin x \, dx = -\frac{\cos^{1+B} x}{1+B} \tag{11.36}$$

where B is a parameter. Replacing Equation 11.36 on both sides of Equation 11.35 and taking $\beta = \beta_M$ and $\alpha = \alpha_M$ we get

$$I_0 = \frac{\cos^{1+N}\beta_0 - \cos^{1+N}\beta_M}{1+N} \Bigg/ \frac{\cos^{1+M}\alpha_0 - \cos^{1+M}\alpha_M}{1+M} \tag{11.37}$$

Accordingly, replacing Equation 11.36 on both sides of Equation 11.35 we get

$$\frac{\cos^{1+N}\beta}{1+N} = I_0 \frac{\cos^{1+M}\alpha}{1+M} + C \tag{11.38}$$

where

$$C = \frac{\cos^{1+N}\beta_0}{1+N} - I_0 \frac{\cos^{1+M}\alpha_0}{1+M} \tag{11.39}$$

Equation 11.38 can be used to calculate $\alpha(\beta)$ as

$$\alpha(\beta) = \arccos\left(\left(\frac{1+M}{I_0} \left(\frac{\cos^{1+N}\beta}{1+N} - C \right) \right)^{\frac{1}{1+M}} \right) \tag{11.40}$$

As an example of application, we consider $N = 1$, resulting in $I_S(\beta) = \cos(\beta)$, so that **E** is a Lambertian emitter. We also make $M = -3$, resulting in $I_P(\alpha) = \cos^{-3}(\alpha)$. This output pattern will generate a uniform irradiance on a distant target (see Chapter 20). As initial condition we choose $\beta_0 = \alpha_0 = 0$ so that a ray emitted from **E** along the optical axis ($\beta = 0$) will leave the optic also along the optical axis ($\alpha = 0$). Finally, we choose $\beta_M = 90°$ (**E** emits up in all directions) and a value for α_M. With these conditions, we may calculate I_0 from Equation 11.37, constant C from Equation 11.39 and function $\alpha(\beta)$ from Equation 11.40.

A possible resulting optic is shown in Figure 11.25. Point **E** emits light into a dome with refractive index n. The design of the dome starts at point $\mathbf{P}_0$ and its normal $\mathbf{n}_0$. Suppose now that we already calculated the shape of the dome from $\mathbf{P}_0$ to $\mathbf{P}_1$. The path of ray r_1 is therefore known, as is the normal $\mathbf{n}_1$ to the dome at point $\mathbf{P}_1$. Now, launch a new ray r_2 from **E** at an angle $\beta_2 = \beta_1 + \Delta\beta$ where $\Delta\beta$ is a small value. We may intersect ray r_2 with the straight line through $\mathbf{P}_1$ and perpendicular to $\mathbf{n}_1$. This gives us the position of point $\mathbf{P}_2$. From Equation 11.40 we get $\alpha_2 = \alpha(\beta_2)$. We now have the directions of ray r_2

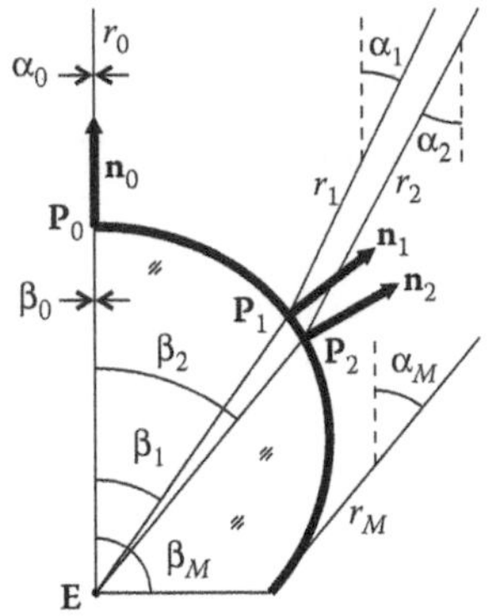

FIGURE 11.25
A refractive surface takes the light from a point emitter **E** and produces a prescribed intensity pattern.

before and after refraction at $\mathbf{P}_2$ and we may calculate the direction of normal $\mathbf{n}_2$ at $\mathbf{P}_2$. The process of calculating the shape of the dome continues by choosing a new value $\beta_3 = \beta_2 + \Delta\beta$ and calculating a new point $\mathbf{P}_3$ to the right of $\mathbf{P}_2$. Repeating this process with a very small value of $\Delta\beta$ gives us the shape of the dome. This process ends at ray r_M when β reaches value β_M and α reaches value α_M.

Figure 11.26 shows another example of an optic calculated for the same point emitter **E** and the same intensity pattern as in Figure 11.25. Now,

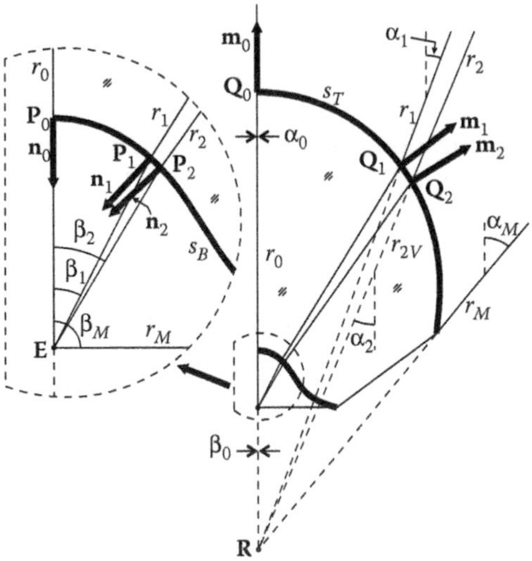

FIGURE 11.26
A lens with surfaces s_B and s_T takes the light from a point emitter **E** and produces a prescribed intensity pattern. The light emitted appears to come from point **R**.

however, the optic is a lens with two optical surfaces s_B and s_T and the emitted light appears to come from a point **R**.

The design of the lens starts at point $\mathbf{P}_0$ with its normal $\mathbf{n}_0$ on s_B and at point $\mathbf{Q}_0$ with its normal $\mathbf{m}_0$ on s_T. Suppose now that we already calculated the shape of the lens from $\mathbf{P}_0$ to $\mathbf{P}_1$ along s_B and from $\mathbf{Q}_0$ to $\mathbf{Q}_1$ along s_T. The path of ray r_1 is therefore known, as are the normals $\mathbf{n}_1$ at point $\mathbf{P}_1$ and $\mathbf{m}_1$ at point $\mathbf{Q}_1$. Now, launch a new ray r_2 from **E** at an angle $\beta_2 = \beta_1 + \Delta\beta$ where $\Delta\beta$ is a small value. We may intersect ray r_2 with a straight line through $\mathbf{P}_1$ and perpendicular to $\mathbf{n}_1$. This gives us the position of point $\mathbf{P}_2$. From Equation 11.40 we get $\alpha_2 = \alpha(\beta_2)$. Now, launch a new (virtual) ray r_{2V} from **R** at an angle α_2. We may intersect ray r_{2V} with the straight line through $\mathbf{Q}_1$ and perpendicular to $\mathbf{m}_1$. This gives us the position of point $\mathbf{Q}_2$. We now have the directions of ray r_2 before and after refraction at $\mathbf{P}_2$ and $\mathbf{Q}_2$ and we may calculate the directions of normals $\mathbf{n}_2$ at $\mathbf{P}_2$ and $\mathbf{m}_2$ at $\mathbf{Q}_2$. The process of calculating the shape of the lens continues by choosing a new value $\beta_3 = \beta_2 + \Delta\beta$ and calculating a new point $\mathbf{P}_3$ on s_B to the right of $\mathbf{P}_2$ and a new point $\mathbf{Q}_3$ on s_T to the right of $\mathbf{Q}_2$. Repeating this process with a very small value of $\Delta\beta$ gives us the shape of both surfaces s_B and s_T of the lens. This process ends at ray r_M when β reaches value β_M and α reaches value α_M.

Note that the design process of the lens in Figure 11.26 is the same used in the design of the lens in Figure 11.23, only now point **R** is virtual and function $\alpha(\beta)$ is given by Equation 11.40 instead of Equation 11.32.

Figure 11.27 shows another lens for the same emitter **E** and the same output pattern as the lens in Figure 11.26. The design process of the lens is also similar to that in Figure 11.26, only the path of each ray is now different.

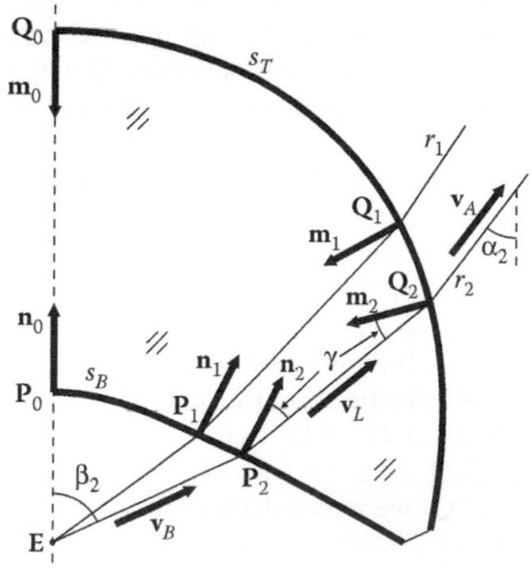

FIGURE 11.27
A lens with surfaces s_B and s_T takes the light from a point emitter **E** and produces a prescribed intensity pattern. The angle γ that a ray r_2 makes with the normal n_2 to s_B is the same it makes with the normal $\mathbf{m}_2$ to s_T.

Ray r_2 leaves **E** at an angle β_2 to the vertical in the direction of unit vector $\mathbf{v}_B = (\cos(\pi/2\text{-}\beta_2),\sin(\pi/2\text{-}\beta_2))$ and leaves the lens in the direction of unit vector $\mathbf{v}_A = (\cos(\pi/2\text{-}\alpha_2),\sin(\pi/2\text{-}\alpha_2))$ where $\alpha_2 = \alpha(\beta_2)$ is again given by Equation 11.40. We may define the direction $\mathbf{v}_L$ of the ray inside the lens as $\mathbf{v}_L = \text{nrm}(\mathbf{v}_B + \mathbf{v}_A)$ and $\mathbf{v}_L$ points in the direction of the bisector to $\mathbf{v}_B$ and $\mathbf{v}_A$. Launch ray r_2 from **E** with direction $\mathbf{v}_B$ and intersect it with the straight line through $\mathbf{P}_1$ and perpendicular to $\mathbf{n}_1$. This gives us the position of point $\mathbf{P}_2$. Now launch again ray r_2 from $\mathbf{P}_2$ in the direction of $\mathbf{v}_L$ and intersect it with a straight line through $\mathbf{Q}_1$ and perpendicular to $\mathbf{m}_1$. This gives us the position of point $\mathbf{Q}_2$. We now have the directions of ray r_2 before and after refraction at $\mathbf{P}_2$ and $\mathbf{Q}_2$ and we may calculate the directions of normals $\mathbf{n}_2$ at $\mathbf{P}_2$ and $\mathbf{m}_2$ at $\mathbf{Q}_2$. Like before, the complete design is obtained by proceeding in very small steps.

The rays crossing the lens make the same angle γ with the normals to both optical surfaces s_B and s_T. The bending of the light as it crosses the lens is then equally distributed between s_B and s_T.

11.6 Examples

The following examples use expressions for the curves and functions that are derived in Chapter 21.

EXAMPLE 11.1

Design an aplanatic (infinitesimal etendue) RX optic. The optic is designed for a refractive index $n = 1.5$, its point receiver is at position $\mathbf{R} = (0,0)$, the refractive surface starts at point $\mathbf{Q} = (0,2.5)$ with normal $\mathbf{m} = (0,1)$ and the mirror starts at point $\mathbf{P} = (0, -5)$ with normal $\mathbf{n} = (0, -1)$, as shown in Figure 11.28.

The magnification is calculated by specifying a receiver size $R = 1$ and a half-acceptance angle $\theta = 6°$. It may be obtained from Equation 11.16 as $f = 1.5 \times 1/(2 \sin 6°) = 7.17508$.

Choose a small step, say $\gamma = 12°$. Now take $\delta = \alpha_1 = \gamma = 12°$. Vector $\mathbf{v}_1$ may then be obtained as $\mathbf{v}_1 = (\cos(-\pi/2 + \delta),\ \sin(-\pi/2 + \delta)) = (0.207912, -0.978148)$. Point $\mathbf{P}_1$ is obtained from $\mathbf{P}_1 = \text{islp}(\mathbf{R},\mathbf{v}_1,\mathbf{P},\mathbf{n}) = (1.06278, -5)$. We choose the vertical coordinate of point $\mathbf{T}_1$ as, say, $x_2 = 5$ and calculate $\mathbf{T}_1 = (f \sin \delta, 5) = (1.49178, 5)$. Point $\mathbf{Q}_1$ may now be calculated as $\mathbf{Q}_1 = \text{islp}(\mathbf{T}_1,\mathbf{u},\mathbf{Q},\mathbf{m}) = (1.49178, 2.5)$, where $\mathbf{u} = (0, -1)$. Now that we have the positions of points $\mathbf{P}_1$ and $\mathbf{Q}_1$ we may calculate $\mathbf{n}_1 = \text{rfrnrm}(\text{nrm}(\mathbf{P}_1 - \mathbf{R}),\text{nrm}(\mathbf{Q}_1 - \mathbf{P}_1),n,n) = (0.0760773, -0.997102)$ and also $\mathbf{m}_1 = \text{rfrnrm}(\text{nrm}(\mathbf{Q}_1 - \mathbf{P}_1), -\mathbf{u},n,1) = (0.169667, 0.985501)$ where $-\mathbf{u} = (0,1)$.

Now take $\delta = \alpha_2 = 2\gamma = 24°$. Vector $\mathbf{v}_2$ may then be obtained as $\mathbf{v}_2 = (\cos(-\pi/2 + \delta),\ \sin(-\pi/2 + \delta)) = (0.406737, -0.913545)$. Point $\mathbf{P}_2$ is obtained from $\mathbf{P}_2 = \text{islp}(\mathbf{R},\mathbf{v}_2,\mathbf{P}_1,\mathbf{n}_1) = (2.18792, -4.91415)$. Point $\mathbf{T}_2$ is given by $\mathbf{T}_2 = (f \sin \delta, 5) = (2.91837, 5)$. Point $\mathbf{Q}_2$ may now be calculated as $\mathbf{Q}_2 = \text{i}$

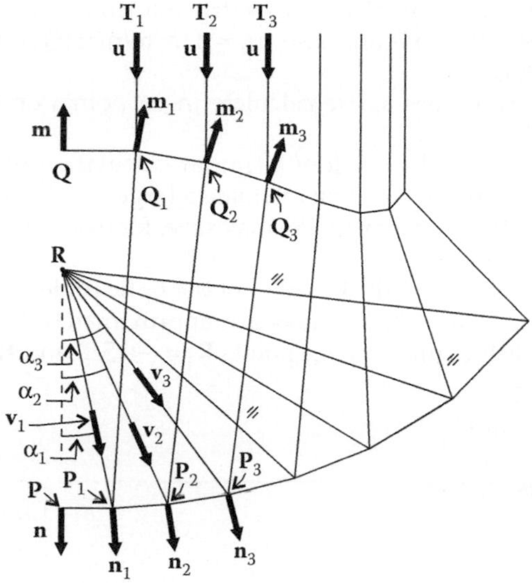

FIGURE 11.28
Calculation process of an aplanatic RX.

slp($\mathbf{T}_2$,(0, −1),$\mathbf{Q}_1$,$\mathbf{m}_1$) = (2.91837,2.2544). Now that we have the positions of points $\mathbf{P}_2$ and $\mathbf{Q}_2$, we may calculate $\mathbf{n}_2$ = rfrnrm(nrm($\mathbf{P}_2 - \mathbf{R}$),nrm($\mathbf{Q}_2 - \mathbf{P}_2$),$n$,$n$) = (0.158002, −0.987439) and also $\mathbf{m}_2$ = rfrnrm(nrm($\mathbf{Q}_2 - \mathbf{P}_2$),(0,1),$n$, 1) = (0.295128,0.955458).

Now take $\delta = \alpha_3 = 3\gamma = 36°$. Vector $\mathbf{v}_3$ may then be obtained as $\mathbf{v}_3 = (\cos(-\pi/2 + \delta),\ \sin(-\pi/2 + \delta))$ = (0.587785, −0.809017). Point $\mathbf{P}_3$ is obtained from $\mathbf{P}_3$ = islp($\mathbf{R}$,$\mathbf{v}_3$,$\mathbf{P}_2$,$\mathbf{n}_2$) = (3.42637, −4.71599). Point $\mathbf{T}_3$ is given by $\mathbf{T}_3 = (f \sin \delta, 5)$ = (4.21741,5). Point $\mathbf{Q}_3$ may now be calculated as $\mathbf{Q}_3$ = is lp($\mathbf{T}_3$,(0, −1),$\mathbf{Q}_2$,$\mathbf{m}_2$) = (4.21741,1.85314). Now that we have the positions of

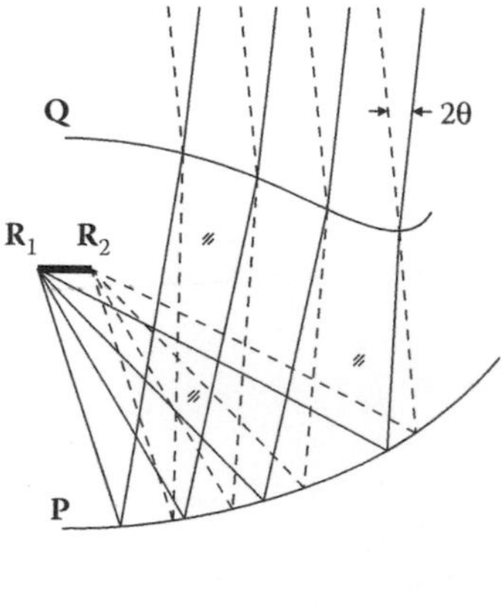

FIGURE 11.29
Aplanatic RX optic under radiation of angular aperture 2θ.

points $\mathbf{P}_3$ and $\mathbf{Q}_3$, we may calculate $\mathbf{n}_3$ = rfrnrm(nrm($\mathbf{P}_3 - \mathbf{R}$),nrm($\mathbf{Q}_3 - \mathbf{P}_3$), n,n) = (0.251509, −0.967855) and also $\mathbf{m}_3$ = rfrnrm(nrm($\mathbf{Q}_3 - \mathbf{P}_3$),(0,1),$n$,1) = (0.344158,0.938912).

The process continues as we calculate more points on the top and bottom optical surfaces.

Figure 11.29 shows the result of a similar calculation with a smaller step of $\gamma = 0.01°$. This process will calculate a large number of points, but we do not need to save them all. We may save, for example, one in every 10 points calculated.

Figure 11.29 also shows the behavior of the optic under radiation with angular aperture $2\theta = 12°$. The rays are approximately focused at the edges of a receiver bound by edge points $\mathbf{R}_1 = (-0.5,0)$ and $\mathbf{R}_2 = (0.5,0)$.

References

1. Koshel, R. J., *Illumination Engineering: Design with Nonimaging Optics*, Wiley IEEE Press, Piscataway, NJ, 2013.
2. Gordon, J. M. and Feuermann, D., Optical performance at the thermodynamic limit with tailored imaging designs, *Appl. Opt.*, 44(12), 2005.
3. Benitez, P. and Miñano, J. C., Ultrahigh-numerical-aperture imaging concentrator, *J. Opt. Soc. Am. A*, 14(8), 1997.
4. Canavarro, C. et al., New second-stage concentrators (XX SMS) for parabolic primaries; Comparison with conventional parabolic trough concentrators, *Sol. Energy*, 92, 98–105, 2013.
5. Benitez, P. and Miñano, J. C., Offence against the edge ray theorem? *SPIE Proceedings*, 5529, *Nonimaging Optics and Efficient Illumination Systems*, 2004.
6. Canavarro, D. et al., Infinitesimal etendue and Simultaneous Multiple Surface (SMS) concentrators for fixed receiver troughs, *Sol. Energy*, 97, 493–504, 2013.
7. Nakar, D. et al., Aplanatic near-field optics for efficient light transfer, *Opt. Eng.*, 45(3), 2006.

12

Köhler Optics and Color-Mixing

12.1 Introduction

Nonimaging optics transfer light from an emitter to a receiver. However, in many situations, the emitter is not uniform (may have hot spots, for example) and those nonuniformities are transferred by the optic from emitter to receiver, resulting in an undesirable irradiance pattern. In other situations, the emitter changes over time. This is the case, for example, of sunlight, as the sun moves across the sky; it is also the case of LED light sources composed of several emitters that may be on or off separately, and at different times. There are still other situations in which the position of the emitter (say an LED) may vary, for example, due to position tolerances in the assembly process, and this variation affects the light output. In these and other cases, it may be desirable that even if the emitter changes or has artifacts, the light output stays the same.

Köhler optics produce a stable output, independent of (limited) variations of the source.

12.2 Köhler Optics

Figure 12.1a shows an SMS optic **CD** with emitter **AB** and receiver **EF**. Rays coming from edge **A** of the emitter are redirected to the edge **F** of the receiver, and rays coming from edge **B** of the emitter are redirected to edge **E** of the receiver. If the whole emitter **AB** is lit, emitting light towards **CD**, then all points of the receiver **EF** will be illuminated, as in Figure 12.1b.

In real systems, when assembling the different components, there will be errors in their positions relative to the ideal placement. Suppose then that, due to assembly errors, an emitter may be placed anywhere inside region **AB**, as shown in Figure 12.2. If the emitter falls at position E_1, only region R_1 of the receiver will be illuminated. Also, if the emitter falls at position E_2 or E_3, only regions R_2 or R_3 will be illuminated. This leads to nonuniform illumination of receiver **EF**, as shown in Figure 12.2.

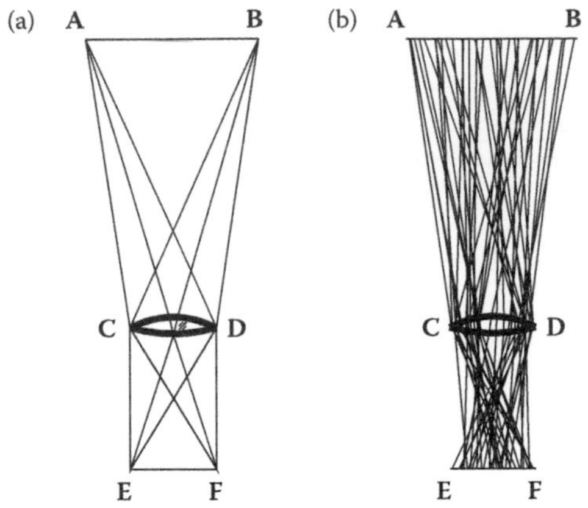

FIGURE 12.1
(a) SMS lens **CD** with emitter AB and receiver **EF**. (b) A uniform emitter **AB** fully illuminates receiver EF.

A different situation may be that position tolerances are tighter, and E_1, E_2, and E_3 may be placed more accurately, but these are emitters with different colors. If emitter E_1 is red, region R_1 of receiver **EF** will be illuminated with red light. Also, if emitters E_2 and E_3 are green and blue, respectively, regions R_2 and R_3 of receiver **EF** will be illuminated with green and blue

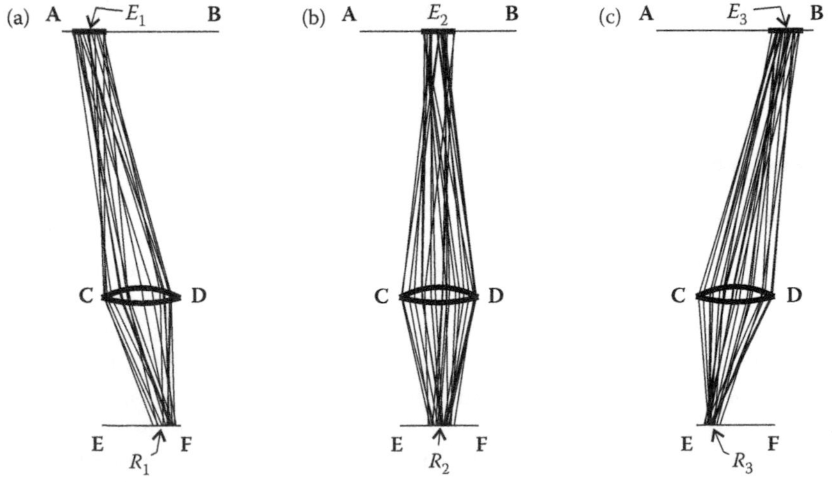

FIGURE 12.2
If an emitter E is placed inside **AB**, it will produce a nonuniform irradiance on receiver **EF**. (a), (b), and (c) respectively show the situations in which different emitters E_1, E_2, E_3 are placed inside **AB**. If these emitters E_1, E_2, E_3 represent different colors, different areas R_1, R_2, R_3 of the receiver **EF** will be illuminated with the light of different colors, leading to color separation on **EF**.

light, respectively. This leads to color separation: different regions of the receiver are illuminated by light of different colors.

This nonuniform illumination of the receiver, or color separation, may be undesirable in many applications. One way to avoid these effects is to combine two optics, **CD** and **EF**, in series, as shown in Figure 12.3a. Now, optic **CD** will redirect to **EF** the light it receives from **AB**. Also, optic **EF** will redirect to **GH** the light it receives from **CD**. This combination of optics **CD** and **EF** is called a Köhler integrator.[1–7] Region **AB** is called the integration area, and is where the light emitter(s) should be placed.

Figure 12.3b shows light rays r_1, r_2, and r_3 emitted from point **E**. Since these come from inside **AB**, optic **CD** redirects them to points inside **EF**. When seen from optic **EF**, edge ray r_1 comes from the edge **C** and, therefore, is redirected to edge **H**. Ray r_3 comes from the edge **D** and, therefore, is redirected to edge **G**. Ray r_2 comes from inside **CD**, and therefore, is redirected to a point inside **GH**. The result is that light coming from point **E** is spread out all over **GH**, fully illuminating it.

Figure 12.4 shows a Köhler integrator designed for an integration area **AB** and a receiver **GH**. An emitter E_1 inside **AB** fully illuminates receiver **GH**, as shown in Figure 12.4a. If this emitter moves to positions E_2 or E_3 inside **AB**, it will still fully illuminate **GH**, generating a similar illumination pattern. The integration area **AB** is, therefore, the position tolerance for emitter E_1, which may be positioned anywhere inside **AB**. This may be very important from a practical point of view since, when manufacturing an optic, there will be position tolerances of the emitter relative to the optic, and the illumination pattern (ideally) should not be affected by those tolerances.

A different situation occurs when E_1, E_2, and E_3 are emitters of different colors. In that case, all three emitters generate similar irradiance patterns

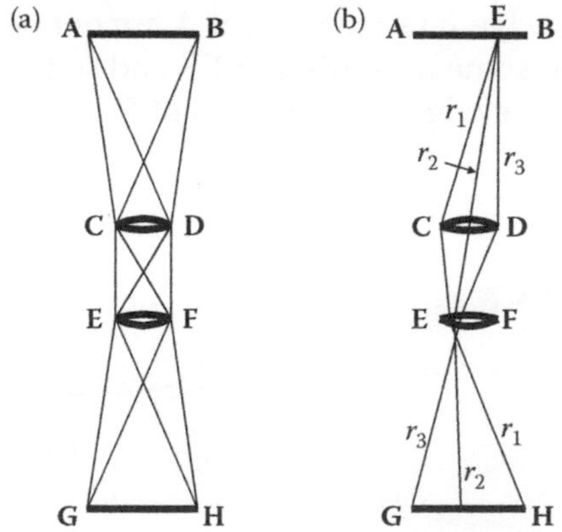

FIGURE 12.3
(a) Integrator: Optic **CD** redirects to **EF** the light it receives from **AB**, and optic **EF** redirects to **GH** the light it receives from **CD**. (b) Light coming from point **E** in region **AB** spreads over the whole receiver **GH**, fully illuminating it.

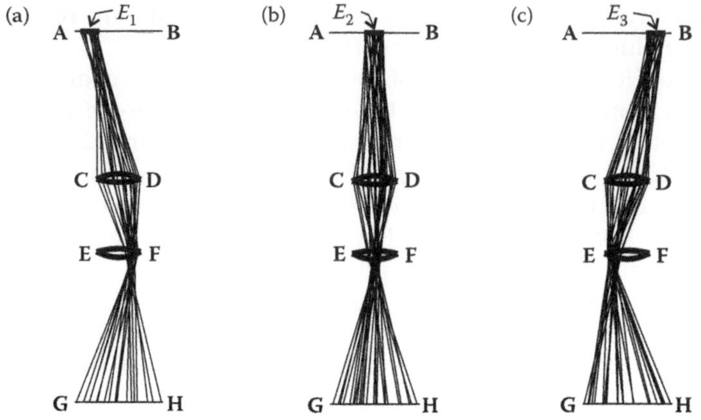

FIGURE 12.4
A source may be placed anywhere (shown are three different positions—$\mathbf{E}_1$, $\mathbf{E}_2$, and $\mathbf{E}_3$) inside integration area **AB**, and still produce the same irradiance pattern on receiver **GH**. Alternatively, $\mathbf{E}_1$, $\mathbf{E}_2$, and $\mathbf{E}_3$ may be sources of different colors, whose emissions are mixed at receiver **GH**. (a), (b), and (c) show the paths of the light rays through the optic when the source (emitter) is at three different positions E_1, E_2, and E_3 respectively.

on **GH**. Their light will, therefore, be mixed at **GH**, generating, for example, white light. This is important for color-mixing. Emitters E_1, E_2, and E_3 may be, for example, three LEDs of different colors inside a common package. Aside from color-mixing, the integrator optic also provides tolerance for the positions of colored emitters E_1, E_2, and E_3 inside the integration area **AB**.

The optic in Figure 12.1 may be simplified, as shown in Figure 12.5. The SMS lens **CD** in Figure 12.1 was replaced by a single refractive surface **CD** in Figure 12.5. Now, the edge rays from point **A** are not perfectly focused onto point **F**. This results in some rays missing **EF**, and some radiation is lost. The same also happens to the edge rays from point **B**.

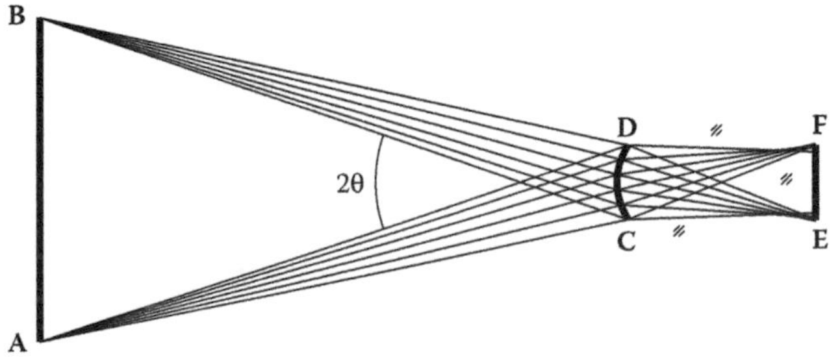

FIGURE 12.5
Refractive surface **CD** images (although not perfectly) **AB** onto **EF**.

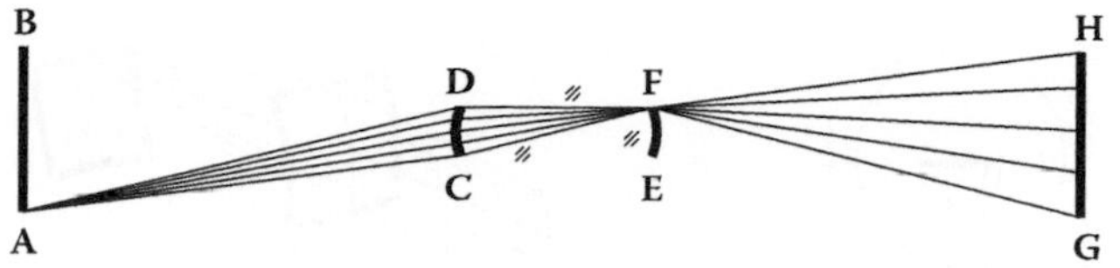

FIGURE 12.6
Two refractive surfaces, **CD** and **EF**, acting as an integrator.

If angle 2θ is small, and distance [**C**,**E**] is large compared to distance [**C**,**D**] (refractive surface **CD** has a long focal distance), the focusing may be good enough for practical applications. When this happens, surface **CD** images **AB** onto **EF**. That case is illustrated in Figure 12.6, which now shows an integrator composed of refractive surfaces **CD** and **EF**. These replace SMS lenses **CD** and **EF** in Figure 12.4.

Figure 12.7 shows a three-dimensional (3-D) view of the same device, where the two-dimensional (2-D) refractive surfaces **CD** and **EF** are now replaced by 3-D refractive surfaces L_1 and L_2. This integrator has an integration area E for the emitter and a receiver R.

First refractive surface L_1 images the light source inside integration area E onto the second refractive surface L_2, as shown in Figure 12.8.

Second refractive surface L_2 images the first refractive surface L_1 onto the receiver R, generating there the image I_1 of L_1, as shown in Figure 12.9.

Figure 12.10 shows the complete paths of light rays from light source to receiver R. Surface L_1 images the light source onto L_2 and surface L_2 images L_1 onto R. If L_1 is uniformly illuminated by the light source, image I_1 of L_1 will also be uniform, generating a uniform irradiance pattern on receiver R.

Let us now go back to the situation shown in Figure 12.1 and consider the case in which **AB** becomes very large and very distant from optic **CD**. This large source now spans an angle 2θ when seen from **CD**, as shown in Figure 12.11a. Optic **CD** focuses the edge rays from the source onto the edges **E** and

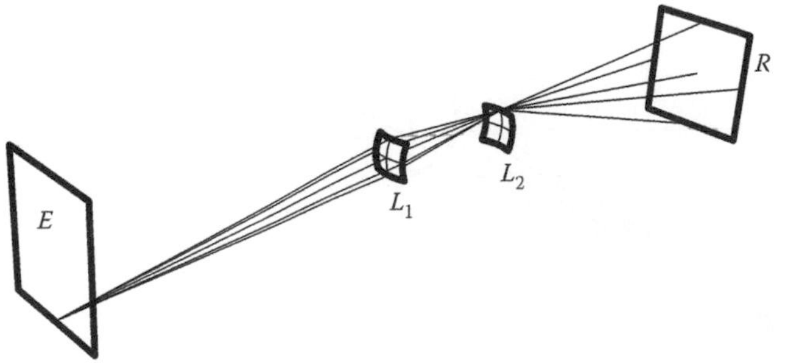

FIGURE 12.7
Three-dimensional view of two refractive surfaces, L_1 and L_2, acting as an integrator.

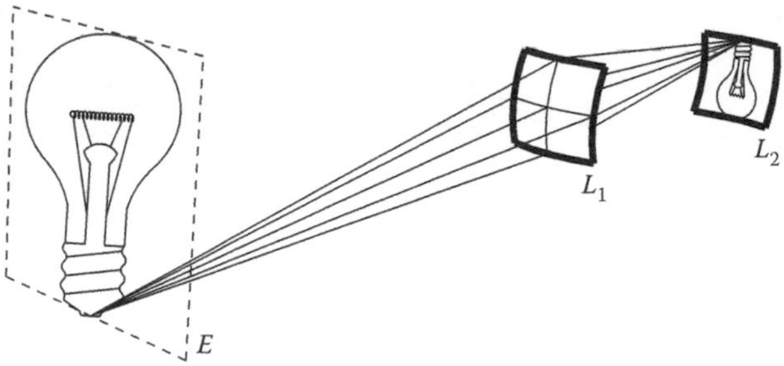

FIGURE 12.8
First optical surface L_1 images the light source onto the second optical surface L_2. Getting all the light from E into L_2 is important for efficiency.

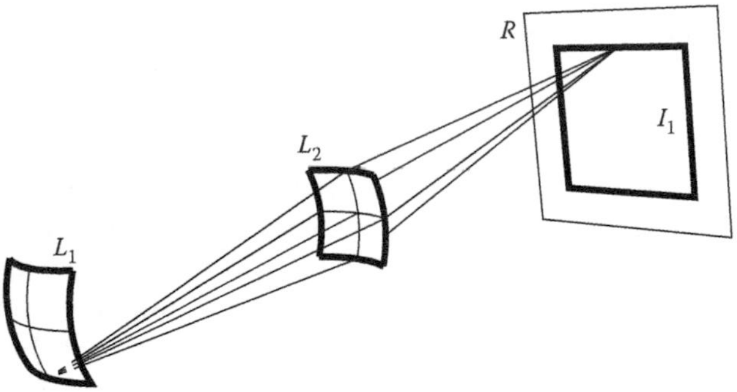

FIGURE 12.9
Second optical surface L_2 images L_1 onto receiver R creating there an image I_1 of L_1.

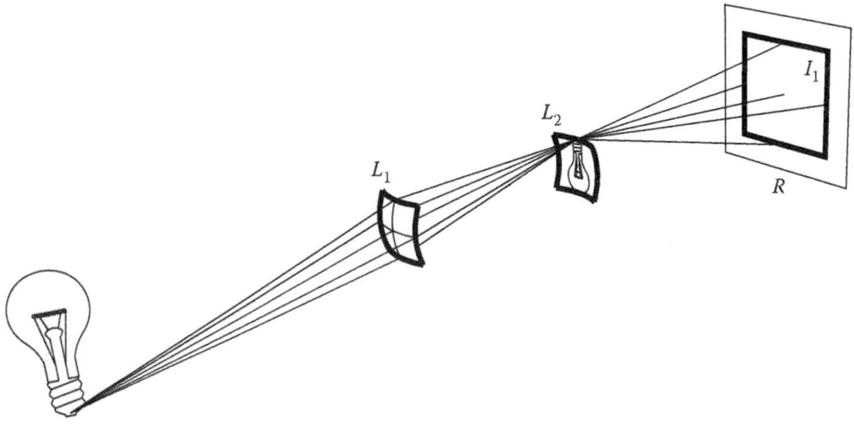

FIGURE 12.10
Path of light rays through an integrator from light source (left) to receiver (right).

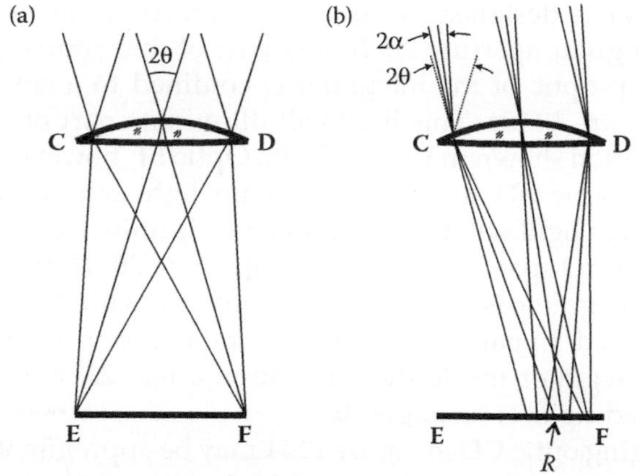

FIGURE 12.11
(a) Large source, very far from optic **CD**, spans an angular aperture 2θ. (b) A smaller emitter inside the source spans an angle 2α and illuminates a region *R* of receiver **EF**.

F of the receiver. An emitting region inside the source will span a smaller angle 2α, as in Figure 12.11b, and will illuminate a region *R* of receiver **EF**.

If the receiver **EF** is the same size as optic **CD**, two of these optics may be combined into a single device, as in Figure 12.12. Optic **EF** is symmetrical to **CD** with symmetry axis *s*.

The resulting device is an integrator that accepts light (coming from above) within a cone of angular aperture 2θ (integration angle), and emits light (going down) within the same angular aperture 2θ. Optic **CD** takes light inside acceptance angle 2θ and redirects it to **EF**. Optic **EF** (symmetric to optic **CD**) takes light coming from **CD** and emits it down confined to an angular aperture 2θ.

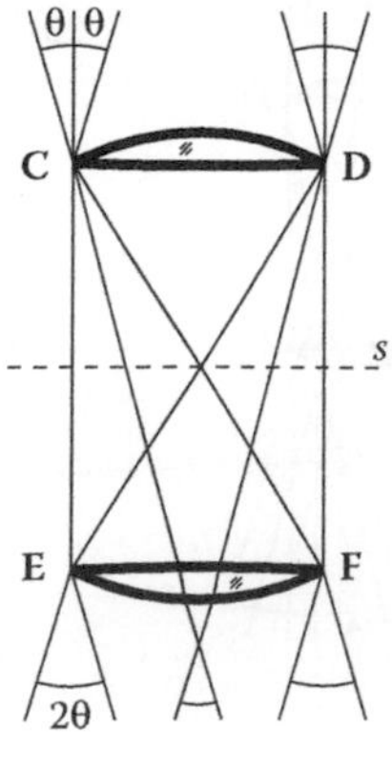

FIGURE 12.12
Köhler integrator made of two lenses (symmetric) **CD** and **EF** with entrance and exit angular aperture 2θ.

This integrator is designed for an infinite source at an infinite distance spanning an angular aperture 2θ. If only part of that source is lit, the light reaching the top optic of the integrator is confined to a smaller angle 2α, as shown in Figure 12.13a. This light will illuminate part of optic **EF** (similarly to the situation shown in Figure 12.11b. Optic **EF**, however, receives light from the whole optic **CD** and, therefore, emits light across the whole emission angle 2θ. This light will then illuminate an infinite receiver placed at an infinite distance and spanning an angular aperture 2θ. Therefore, even if the source of the integrator is not all lit (or is not uniform), the whole receiver will be illuminated. In particular, a set of parallel rays *r*, as in Figure 12.13b, reaching the integrator inside the integration angle 2θ, will also leave the integrator spreading over an angle 2θ.

The nonimaging optic **CD** in Figure 12.11 may be approximated by a focusing refractive surface **CD**, as in Figure 12.14. In general, the focusing by surface **CD** of the edge rays with angular aperture 2θ onto the edges of **EF** will not be perfect, as shown in Figure 12.14. However, this focusing may be good for practical applications when angle 2θ is small, and when distance [**C**,**E**] is large compared to distance [**C**,**D**], that is, lens **CD** has a large focal length.

In this case, the integrator in Figure 12.13 may be simplified, resulting in the configuration shown in Figure 12.15. Now the integrator is composed of two refractive surfaces, **CD** and **EF**. Light reaching the device at an angle θ is focused to the edge **F** of **EF**, leaving the device within a cone of angular aperture 2θ, as in Figure 12.15a. Light coming in vertically is focused to the center of **EF** and also leaves the device within the same cone of angular aperture 2θ, as in Figure 12.15b. In general, the light coming in from any direction inside the integration angle 2θ will leave the integrator spreading

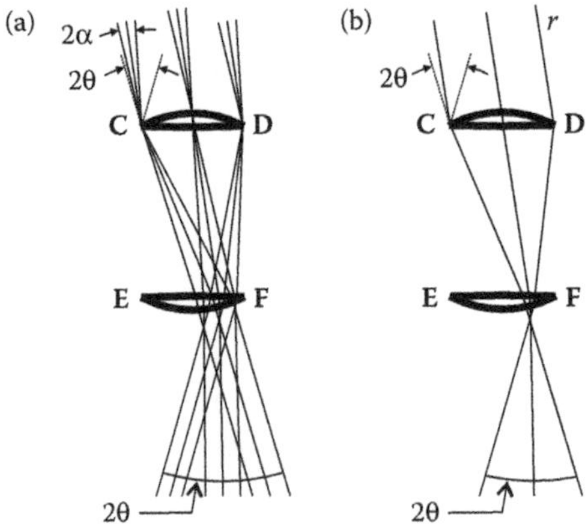

FIGURE 12.13
(a) Köhler optic designed for an integration angle 2θ is illuminated by light with a smaller angular aperture 2α. (b) Extreme case in which the incoming light is a set of parallel rays. In both situations (a) and (b) the exit angle at the bottom is 2θ.

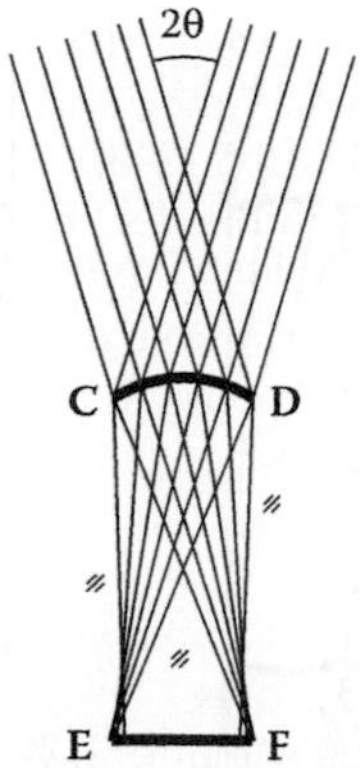

FIGURE 12.14
Refractive surface **CD** redirects to **EF** light inside an angular aperture 2θ.

over an angle 2θ. Figure 12.15c shows a three-dimensional (3-D) view of the geometry in Figure 12.15a, where two-dimensional (2-D) lenses **CD** and **EF** are now 3-D lenses L_1 and L_2, respectively.

Now, several of these elements may be placed side by side, as in Figure 12.16. The whole device is called a Köhler integrator, and each one of the elements is a Köhler channel.

Figure 12.17 shows the representation in phase space of the rays crossing a Köhler integrator.

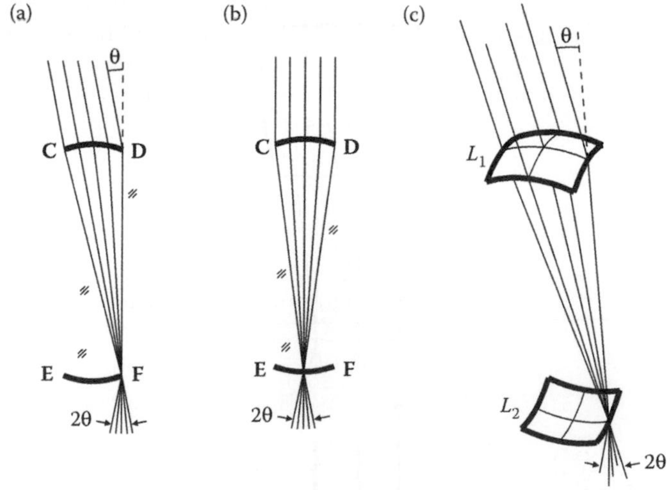

FIGURE 12.15
In (a) and (b), refractive surfaces **CD** and **EF** form an integrator with exit angle 2θ. (c) Three-dimensional view of same: light leaves the device inside a cone of angular aperture 2θ.

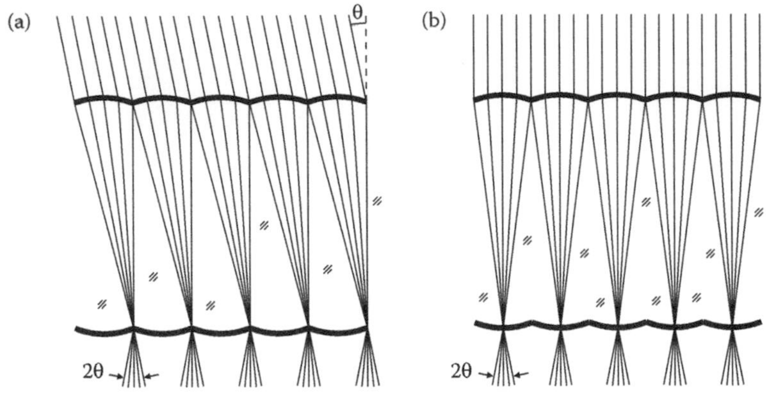

FIGURE 12.16
Array of Köhler integrator channels. (a) Paths of light rays for incoming light tilted by an angle θ to the vertical. (b) Paths of light rays for incoming vertical light.

The device in Figure 12.16 is designed for an infinite integration area and infinite receiver, both spanning and angular aperture 2θ. It is, however, possible to also design multichannel Köhler integrators for finite integration area and receiver. Figure 12.18 shows one such device designed for an integration area $\mathbf{E}_1\mathbf{E}_2$ and receiver $\mathbf{R}_1\mathbf{R}_2$.

Optical surface $\mathbf{P}_1\mathbf{P}_2$ must then image $\mathbf{E}_1\mathbf{E}_2$ onto $\mathbf{Q}_1\mathbf{Q}_2$, and optical surface $\mathbf{Q}_1\mathbf{Q}_2$ must image $\mathbf{P}_1\mathbf{P}_2$ onto $\mathbf{R}_1\mathbf{R}_2$. Optical surface $\mathbf{P}_1\mathbf{P}_2$ cannot guarantee

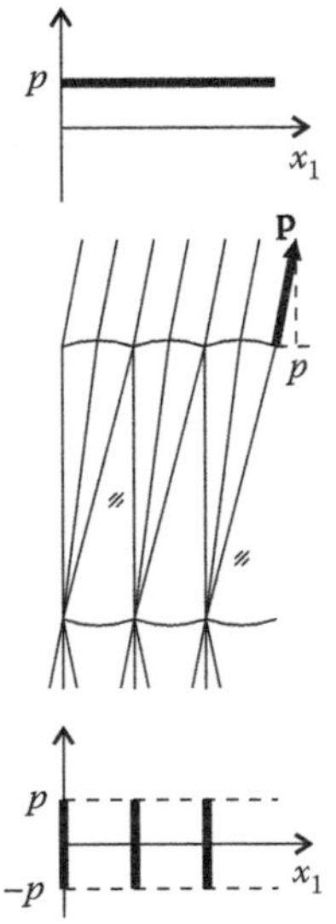

FIGURE 12.17
Rays of a Köhler integrator in phase space.

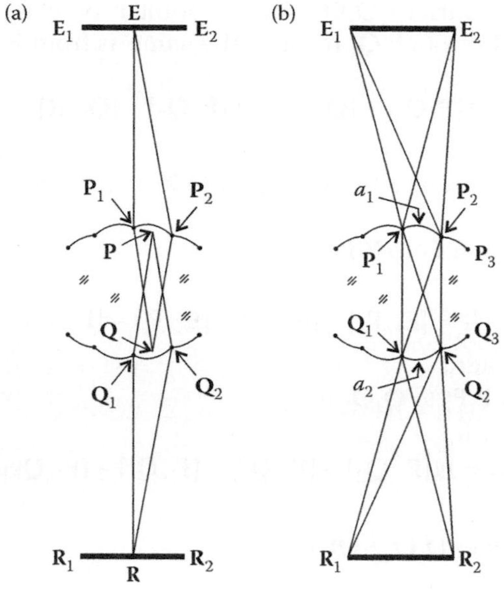

FIGURE 12.18
Multichannel Köhler integrator for a finite emitter $\mathbf{E}_1\mathbf{E}_2$, and receiver $\mathbf{R}_1\mathbf{R}_2$. (a) Light from **E** is focused at **Q** and from **P** is focused at **R**. (b) The etendue from $\mathbf{E}_1\mathbf{E}_2$ to $\mathbf{P}_1\mathbf{P}_2$ is the same as from $\mathbf{P}_1\mathbf{P}_2$ to $\mathbf{Q}_1\mathbf{Q}_2$ and the same from $\mathbf{Q}_1\mathbf{Q}_2$ to $\mathbf{R}_1\mathbf{R}_2$.

perfect imaging of $\mathbf{E}_1\mathbf{E}_2$ onto $\mathbf{Q}_1\mathbf{Q}_2$. However, it is possible to perfectly focus the midpoint **E** of $\mathbf{E}_1\mathbf{E}_2$ onto the midpoint **Q** of $\mathbf{Q}_1\mathbf{Q}_2$, using a Cartesian oval. Optical surface $\mathbf{P}_1\mathbf{P}_2$ is, then, a Cartesian oval with foci **E** and **Q**. Using the same reasoning, optical surface $\mathbf{Q}_1\mathbf{Q}_2$ is a Cartesian oval with foci **P** (the midpoint of $\mathbf{P}_1\mathbf{P}_2$) and **R** (the midpoint of $\mathbf{R}_1\mathbf{R}_2$).

Aside from the image formation, it is important to guarantee that no light is lost as it travels through the Köhler channel, from $\mathbf{E}_1\mathbf{E}_2$ through $\mathbf{P}_1\mathbf{P}_2$, $\mathbf{Q}_1\mathbf{Q}_2$, and $\mathbf{R}_1\mathbf{R}_2$. The étendue U_1 of the light from $\mathbf{E}_1\mathbf{E}_2$ to $\mathbf{P}_1\mathbf{P}_2$ must then be the same as the étendue U_2 of the light from $\mathbf{P}_1\mathbf{P}_2$ to $\mathbf{Q}_1\mathbf{Q}_2$, and the same as the étendue U_3 from $\mathbf{Q}_1\mathbf{Q}_2$ to $\mathbf{R}_1\mathbf{R}_2$.

Suppose now that the positions of $\mathbf{E}_1\mathbf{E}_2$ and $\mathbf{R}_1\mathbf{R}_2$ are known. Now choose the positions of $\mathbf{P}_1$ and $\mathbf{Q}_1$. Next, we must determine the positions of $\mathbf{P}_2 = (x_P, y_P)$ and $\mathbf{Q}_2 = (x_Q, y_Q)$. We have, therefore, four unknowns: x_P, y_P and x_Q, y_Q.

Since optical surface $\mathbf{P}_1\mathbf{P}_2$ images point **E** onto **Q**, the optical path length from **E** to **Q** through $\mathbf{P}_1$ must be the same as from **E** to **Q** through $\mathbf{P}_2$:

$$[\mathbf{E},\mathbf{P}_1] + n[\mathbf{P}_1,\mathbf{Q}] = [\mathbf{E},\mathbf{P}_2] + n[\mathbf{P}_2,\mathbf{Q}] \tag{12.1}$$

Also, since optical surface $\mathbf{Q}_1\mathbf{Q}_2$ images point **P** onto **R**, the optical path length from **P** to **R** through $\mathbf{Q}_1$ must be the same as from **P** to **R** through $\mathbf{Q}_2$,

$$n[\mathbf{P},\mathbf{Q}_1] + [\mathbf{Q}_1,\mathbf{R}] = n[\mathbf{P},\mathbf{Q}_2] + [\mathbf{Q}_2,\mathbf{R}] \tag{12.2}$$

where [**X**,**Y**] is the distance between points **X** and **Y** and n is the refractive index of the lens.

The étendue from $\mathbf{E}_1\mathbf{E}_2$ to $\mathbf{P}_1\mathbf{P}_2$ is

$$U_1 = [\mathbf{E}_1,\mathbf{P}_2] + [\mathbf{E}_2,\mathbf{P}_1] - [\mathbf{E}_1,\mathbf{P}_1] - [\mathbf{E}_2,\mathbf{P}_2] \tag{12.3}$$

The étendue from $\mathbf{P}_1\mathbf{P}_2$ to $\mathbf{Q}_1\mathbf{Q}_2$ is

$$U_2 = n([\mathbf{P}_1,\mathbf{Q}_2] + [\mathbf{P}_2,\mathbf{Q}_1] - [\mathbf{P}_1,\mathbf{Q}_1] - [\mathbf{P}_2,\mathbf{Q}_2]) \tag{12.4}$$

and the étendue from $\mathbf{Q}_1\mathbf{Q}_2$ to $\mathbf{R}_1\mathbf{R}_2$ is

$$U_3 = [\mathbf{Q}_1,\mathbf{R}_2] + [\mathbf{Q}_2,\mathbf{R}_1] - [\mathbf{Q}_1,\mathbf{R}_1] - [\mathbf{Q}_2,\mathbf{R}_2] \tag{12.5}$$

Now, conservation of étendue as light travels through the Köhler channel results in

$$\begin{aligned} U_1 &= U_2 \\ U_2 &= U_3 \end{aligned} \tag{12.6}$$

Equations 12.1, 12.2, and 12.6 are a system of four equations, which allows us to determine the four unknowns x_P, y_P, and x_Q, y_Q, giving us the positions of points $\mathbf{P}_2 = (x_P, y_P)$ and $\mathbf{Q}_2 = (x_Q, y_Q)$. Once the positions of $\mathbf{P}_2$ and $\mathbf{Q}_2$ are defined, Cartesian oval a_1 from $\mathbf{P}_1$ to $\mathbf{P}_2$, and Cartesian oval a_2 from $\mathbf{Q}_1$ to $\mathbf{Q}_2$ can be calculated. Curve a_1 focuses point **E** onto **Q** and curve a_2 focuses point **P** onto **R**.

Now that the positions of $\mathbf{P}_2$ and $\mathbf{Q}_2$ are known, applying the same procedure, we can calculate the positions of two new points $\mathbf{P}_3$ and $\mathbf{Q}_3$ at the edge of the next Köhler channel. The process is repeated for as many channels as needed.

Another possibility for calculating the Köhler channels is to start with an existing optic, as shown in Figure 12.19. Start, for example, with an SMS (or aplanatic) optic with emitter $\mathbf{E}_1\mathbf{E}_2$, receiver $\mathbf{R}_1\mathbf{R}_2$, and optical surfaces c_1 and c_2. Now, we may add Köhler channels to the exiting optic.

Suppose that we pick points $\mathbf{P}_1$ on c_1 and $\mathbf{Q}_1$ on c_2 as edges of a Köhler channel. The other edges of the same Köhler channel will be at points $\mathbf{P}_2$ on c_1 and $\mathbf{Q}_2$ on c_2 whose positions must be determined. Now there are only two degrees of freedom in the system: the parameter of point $\mathbf{P}_2$ on c_1 and the

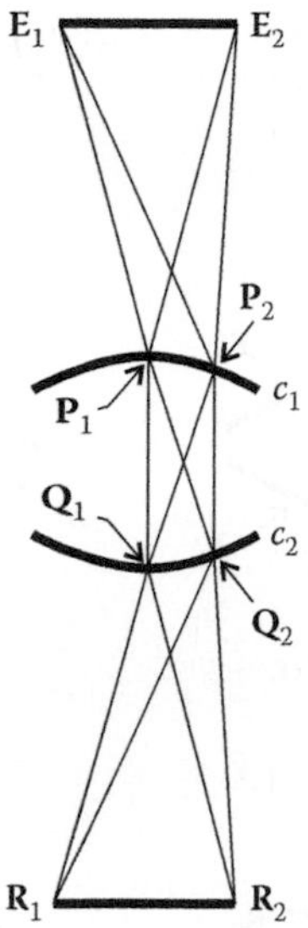

FIGURE 12.19
An existing optic may be modified to add Köhler channels.

parameter of $\mathbf{Q}_2$ on c_2. Therefore, it is not possible to impose the four conditions of Equations 12.1, 12.2, and 12.6, and we must choose two of these to determine the positions of $\mathbf{P}_2$ and $\mathbf{Q}_2$. We may then choose Equation 12.6 and ensure that the étendue from $\mathbf{E}_1\mathbf{E}_2$ to $\mathbf{P}_1\mathbf{P}_2$ matches the étendue from $\mathbf{P}_1\mathbf{P}_2$ to $\mathbf{Q}_1\mathbf{Q}_2$ and also that from $\mathbf{Q}_1\mathbf{Q}_2$ to $\mathbf{R}_1\mathbf{R}_2$.

When building Cartesian oval curves similar to those in Figure 12.18, there is no guarantee that a Cartesian oval curve a_1 starting at $\mathbf{P}_1$ on c_1, will end at point $\mathbf{P}_2$ or that the Cartesian oval a_2 starting at $\mathbf{Q}_1$ on c_2 will end at $\mathbf{Q}_2$. When building a complete Köhler optic with many channels, similar to that in Figure 12.18, there may be discontinuities at points $\mathbf{P}_1$, $\mathbf{P}_2$, … on top and $\mathbf{Q}_1$, $\mathbf{Q}_2$, … at the bottom.

Figure 12.20a shows another possibility to add Köhler channels to an existing SMS optic. Choose a point $\mathbf{P}_1$ on the top surface of the SMS optic and consider the flow-line f_P through it. This flow-line crosses the bottom surface at a point $\mathbf{Q}_1$. Take ray r_2 from $\mathbf{E}_2$ and calculate the normal $\mathbf{n}_P$ at point $\mathbf{P}_1$ that refracts this ray to point $\mathbf{Q}_1$. This is the normal that a Köhler refractive surface should have at point $\mathbf{P}_1$. Take ray r_1 coming from point $\mathbf{E}_1$ and refract it at point $\mathbf{P}_1$ with normal $\mathbf{n}_P$ intersecting it with the bottom surface of the SMS lens, determining point $\mathbf{Q}_2$. Flow-line f_Q through $\mathbf{Q}_2$ determines the position of point $\mathbf{P}_2$ on the top surface. Once the positions of points $\mathbf{P}_1$ and $\mathbf{P}_2$ on top, and $\mathbf{Q}_1$ and $\mathbf{Q}_2$ at the bottom surface are determined, the surfaces of a Köhler channel may be calculated (similar to a_1 and a_2 in Figure 12.18).

Since étendue is conserved between flow-lines (in this case f_P and f_Q), the amount of light transferred between $\mathbf{E}_1\mathbf{E}_2$ and $\mathbf{P}_1\mathbf{P}_2$ is the same transferred

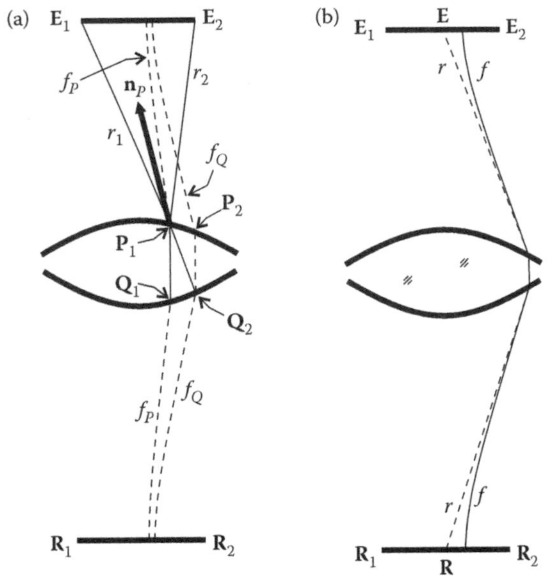

FIGURE 12.20
(a) Selecting two flow-lines f_P and f_Q that bound a Köhler channel to be added to an existing optic. (b) The path of a flow-line may be approximated by the middle ray r far from emitter $\mathbf{E}_1\mathbf{E}_2$ and receiver $\mathbf{R}_1\mathbf{R}_2$.

between $\mathbf{Q}_1\mathbf{Q}_2$ and $\mathbf{R}_1\mathbf{R}_2$. In Equation 12.6 this means $U_1 = U_3$. However, there is no guarantee that the equation $U_1 = U_2$ in Equation 12.6 is fulfilled, or that Equations 12.1 or 12.2 are fulfilled either. This method can, however, be extended to three-dimensional geometry.[1]

Flow-lines f from emitter $\mathbf{E}_1\mathbf{E}_2$ are hyperbolas with foci $\mathbf{E}_1$ and $\mathbf{E}_2$. As we move away from $\mathbf{E}_1\mathbf{E}_2$, these hyperbolas become closer and closer to their asymptote. For that reason, far from $\mathbf{E}_1\mathbf{E}_2$, flow-line f is well approximated by the path of ray r emitted from the center $\mathbf{E}$ of emitter $\mathbf{E}_1\mathbf{E}_2$. The same is true for the flow-line after it crosses the lens on its way to receiver $\mathbf{R}_1\mathbf{R}_2$. Far from $\mathbf{R}_1\mathbf{R}_2$ flow-line f is well approximated by the path of ray r through the center $\mathbf{R}$ of $\mathbf{R}_1\mathbf{R}_2$. Figure 12.20b shows the comparison of flow-line f and ray r through $\mathbf{E}$ and $\mathbf{R}$.

12.3 Solar Energy Concentration Based on Köhler Optics

Köhler optics may be used in solar energy concentration. Consider the aplanatic (infinitesimal étendue) lens in Figure 12.21a. Distance d to the optical axis of the incoming ray is related to the angle α to the vertical axis at focus

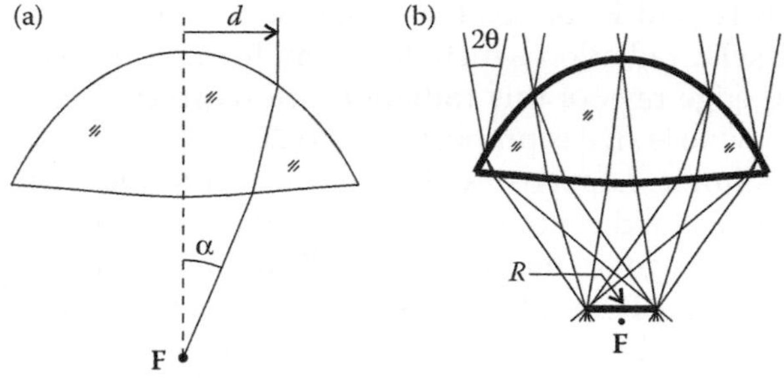

FIGURE 12.21
(a) Aplanatic lens with focus **F**. (b) Same optic used as a concentrator with acceptance angle 2θ and receiver *R*.

F by $d = f \sin \alpha$, where f is the magnification. This optic can also be used as concentrator for an acceptance angle 2θ and receiver **R**, as in Figure 12.21b. The focusing of the edge rays at the edges of **R** is not perfect (see Chapter 11). Magnification is, in this case, given by $f = nR/(2 \sin \theta)$ or approximately $f = R/(2\theta)$, since the receiver R is in air with refractive index $n = 1$ and, for small θ, we have $\sin \theta \approx \theta$.

Figure 12.22a shows a set of rays on meridian plane μ crossing a pinhole **P**. Rays with an angular aperture 2θ are redirected (approximately) to two

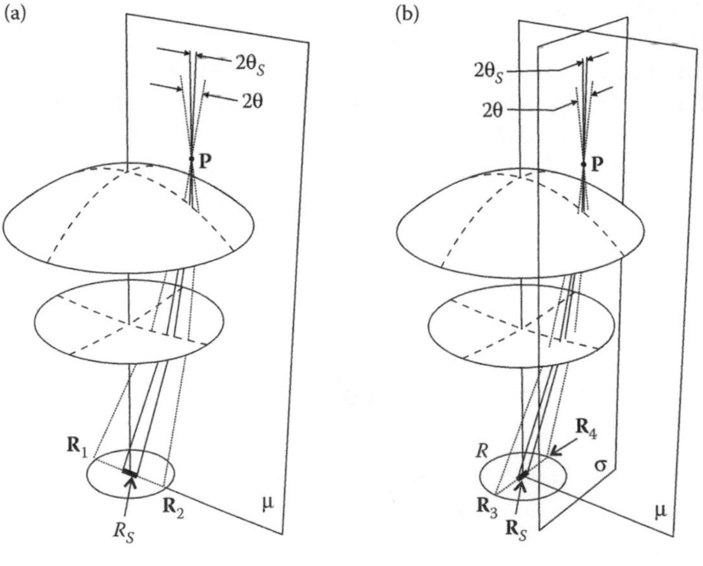

FIGURE 12.22
(a) Paths of meridian rays through an optic with circular symmetry. (b) Paths of sagittal rays through an optic with circular symmetry.

opposite points $\mathbf{R}_1$ and $\mathbf{R}_2$ on the boundary of receiver R. However, if used to concentrate solar radiation which has a smaller angular aperture $2\theta_S$, the corresponding edge rays of this radiation are redirected to the edges of a smaller line R_S, whose size is proportional to $2\theta_S$.

Something similar happens with the rays entering the concentrator through a pinhole $\mathbf{P}$ and contained in a sagittal plane σ, as in Figure 12.22b. Sagittal rays at the edges of the acceptance 2θ are redirected (approximately) to two opposite points $\mathbf{R}_3$ and $\mathbf{R}_4$ on the boundary of receiver R. However, for solar radiation, which has a smaller angular aperture $2\theta_S$, the corresponding edge rays are redirected to the edges of a smaller line R_S, whose size is proportional to $2\theta_S$.

The resulting irradiance pattern on the receiver is a bright spot at its center, as shown in Figure 12.23. Suppose now that the light flux crossing pinhole $\mathbf{P}$ is Φ_P. The irradiance dE_P on the illuminated area is proportional to $\Phi_P/(\pi\theta_S^2)$ or $dE_P \propto \Phi_P/\theta_S^2$. The average irradiance dE_R on the whole receiver is proportional to $\Phi_P/(\pi\,\theta^2)$, or $dE_R \propto \Phi_P/\theta^2$.

The ratio between the irradiance on the illuminated spot and that in the whole receiver is then $dE_P/dE_R \propto (\theta/\theta_S)^2$. This may lead to a high irradiance peak at the center of the receiver, which is undesirable in solar applications.

One option to lower the irradiance at the center of the receiver is to use a Köhler integrator. This is an example of an integrator based on an existing optic. Referring to Figure 12.24a, take points $\mathbf{P}_1$ and $\mathbf{Q}_1$ on the axis of symmetry of optical surfaces c_1 and c_2 (only the right halves of c_1 and c_2 are shown). Now calculate points $\mathbf{P}_2$ and $\mathbf{Q}_2$ (along curves c_1 and c_2) at the edge of a Köhler channel by conservation of étendue, as described above. Note that if point $\mathbf{P}_1 = (P_{11}, P_{12})$ and $\mathbf{P}_2 = (P_{21}, P_{22})$, the étendue of the incoming radiation at $\mathbf{P}_1\mathbf{P}_2$ is given by $U_1 = 2(P_{21} - P_{11}) \sin\theta$. The process may be repeated, calculating more pairs $\mathbf{P}_k$, $\mathbf{Q}_k$ on curves c_1 and c_2 limiting other Köhler channels. Optical surfaces c_1 and c_2 may now be replaced by two sets of Cartesian ovals, as in Figure 12.24b. These surfaces may have gaps (discontinuities) at point $\mathbf{P}_2$, $\mathbf{P}_3$, … on top and $\mathbf{Q}_2$, $\mathbf{Q}_3$, … at the bottom, as described above.

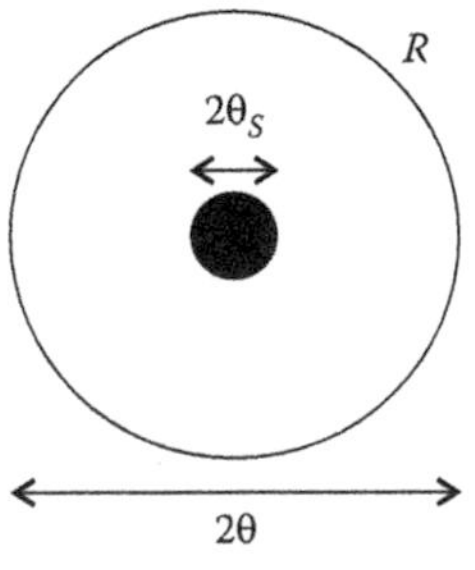

FIGURE 12.23
Irradiance hotspot (black circle at center) on a receiver R created by vertical sunlight of angular aperture $2\theta_S$ when focused by a concentrator with circular symmetry and acceptance angle 2θ.

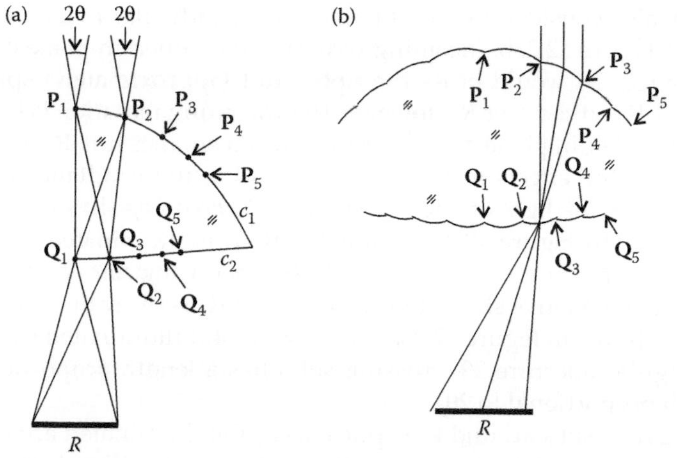

FIGURE 12.24
(a) Dividing an existing optic into edges of Köhler channels. (b) Original optic replaced by a set of Köhler channels.

Now, this design may be made into a three-dimensional optic by circular symmetry. Figure 12.25 shows one such option. Suppose we use the resulting device as a solar concentrator. The optic has an acceptance 2θ, but is used to concentrate sunlight with a smaller angular aperture $2\theta_S$. Consider now a meridional plane μ, as in Figure 12.25a, and a radial slit *s* on that plane covering the radial size of a lenticulation. Incoming radiation contained in plane μ and confined to angle $2\theta_S$ crosses the optic and spreads over diameter $\mathbf{R}_1\mathbf{R}_2$ on receiver *R*.

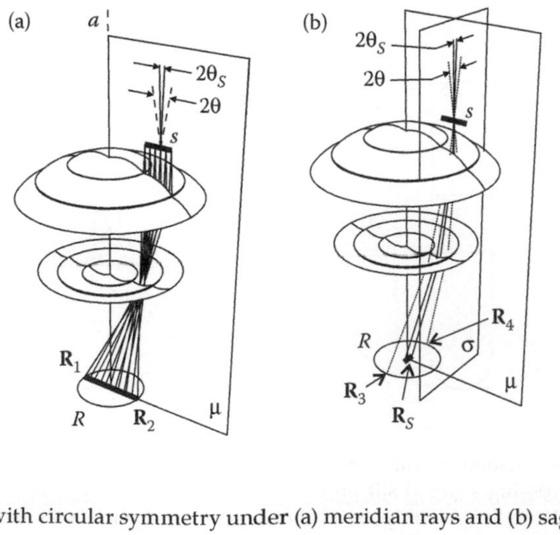

FIGURE 12.25
Köhler integrator with circular symmetry under (a) meridian rays and (b) sagittal rays.

We may also consider a sagittal plane σ perpendicular to μ and the same slit s, as in Figure 12.25b. Incoming radiation contained in plane σ and confined to angle 2θ would cross the optic and (approximately) spread over diameter $\mathbf{R}_3\mathbf{R}_4$ on receiver R. However, the incoming radiation is limited to a smaller angle $2\theta_S$ in plane σ, and spreads over a smaller line $\mathbf{R}_S$ on R.

Combining the effects in Figure 12.25a and b, the irradiance pattern on R, produced by the light of angular spread $2\theta_S$, crossing slit s is as shown in Figure 12.26a. In Figure 12.25b, diameter $\mathbf{R}_3\mathbf{R}_4$ is illuminated by radiation with angular aperture 2θ and $\mathbf{R}_S$ by radiation with angular aperture $2\theta_S$. For that reason, the relative size between $\mathbf{R}_3\mathbf{R}_4$ and $\mathbf{R}_S$ is the same as between 2θ and $2\theta_S$, as shown in Figure 12.26a. The region of R illuminated by the radiation of angular aperture $2\theta_S$ crossing slit s has a length proportional to 2θ, and width proportional to $2\theta_S$.

Rotating now slit s around the optical axis, the illuminated area shown in Figure 12.26a also rotates in receiver R, resulting in the illumination pattern shown in Figure 12.26b. It has a hotspot in the center and lower illumination at the periphery.

Suppose now that the light flux crossing slit s is Φ_S. The irradiance dE_S on the illuminated area shown in Figure 12.26a is proportional to $\Phi_S/(2\theta_S \times 2\theta) = \Phi_S/(4\theta\, \theta_S)$, or $dE_S \propto \Phi_S/(\theta\, \theta_S)$. The average irradiance on the whole receiver is proportional to $\Phi_S/(\pi\, \theta^2)$, or $dE_R \propto \Phi_S/\theta^2$. Ratio $dE_S/dE_R \propto \theta/\theta_S$ is the same for the whole shaded area in Figure 12.26a, and, in particular, for the central circle of diameter $2\theta_S$.

As the slit s rotates around axis a, the center region of diameter $2\theta_S$ (solid black circle in Figure 12.26b) is the only area of receiver R illuminated by all these positions of the rotating slit. For all these positions, the ratio of irradiances between the black circle and the whole receiver is proportional to θ/θ_S, and this is also the ratio of irradiances in a concentrator, or $dE_S/dE_R \propto \theta/\theta_S$.

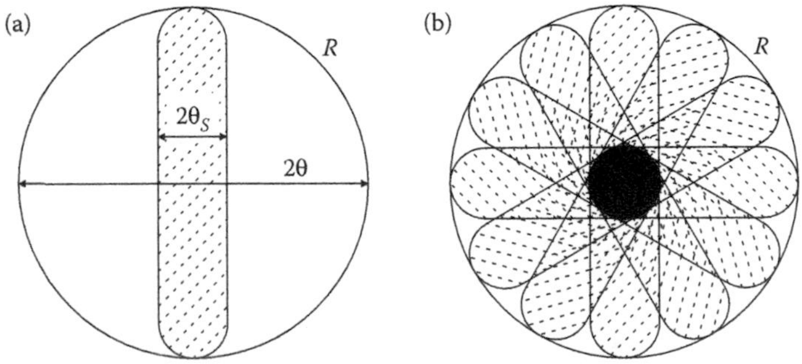

FIGURE 12.26
(a) Irradiance shape (shaded area at center) on a receiver R created by vertical sunlight of angular aperture $2\theta_S$ crossing a radial slit on a Köhler concentrator with circular symmetry and acceptance angle 2θ. (b) Irradiance pattern generated by the whole Köhler concentrator.

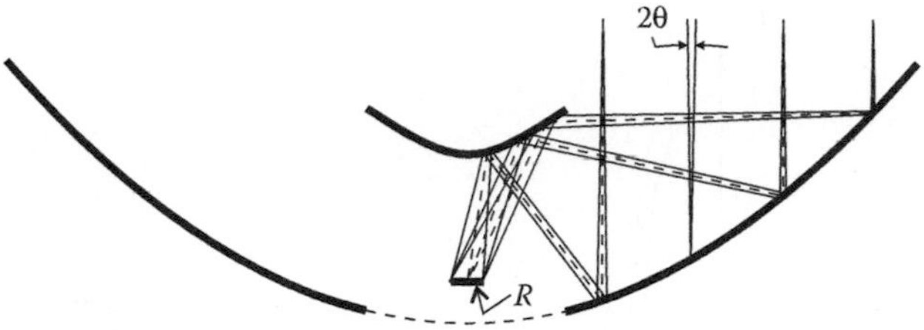

FIGURE 12.27
Two mirror concentrators (XX) with acceptance angle 2θ and receiver *R*.

The integrator with circular symmetry does not eliminate the hotspot at the center of the receiver, but it lowers it considerably from $(\theta/\theta_S)^2$ to θ/θ_S, relative to the average irradiance on the whole receiver.

Figure 12.27 shows a more practical configuration that can be used as a solar concentrator. Now, instead of two refractive surfaces, the optic has two reflective surfaces (mirrors).

Also here, the optical surfaces may be modified and replaced by several Köhler channels that act as an integrator, as shown in Figure 12.28.

If these designs are made into three-dimensional devices by rotational symmetry, the peak irradiance at the center of receiver *R* of the integrator (Figure 12.28) will be significantly lower than that of the optic in Figure 12.27. This is illustrated in Figure 12.29, which shows the relative irradiance curves I_R on the receiver, as a function of relative distance *r* to its center, for the configurations in Figure 12.27 (curve *b*) and Figure 12.28 (curve *a*).

The optics described above are designed in two dimensions, and then given rotational symmetry. However, it is also possible to design Köhler optics that do not have rotational (or linear) symmetry. Figure 12.30 shows one such possibility. This optic is composed of a reflective primary composed of four

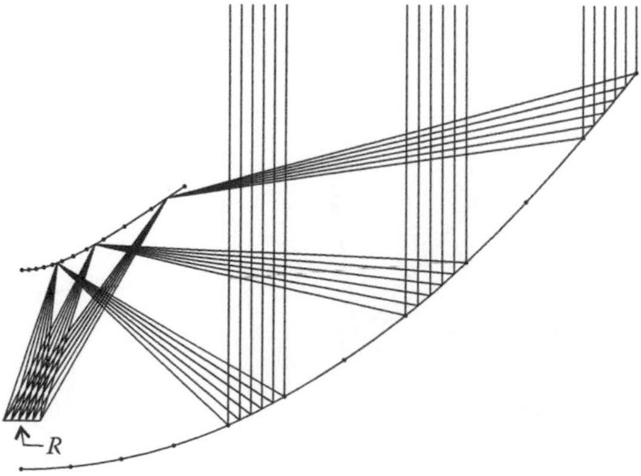

FIGURE 12.28
Replacing the mirrors of XX concentrators by Köhler channels.

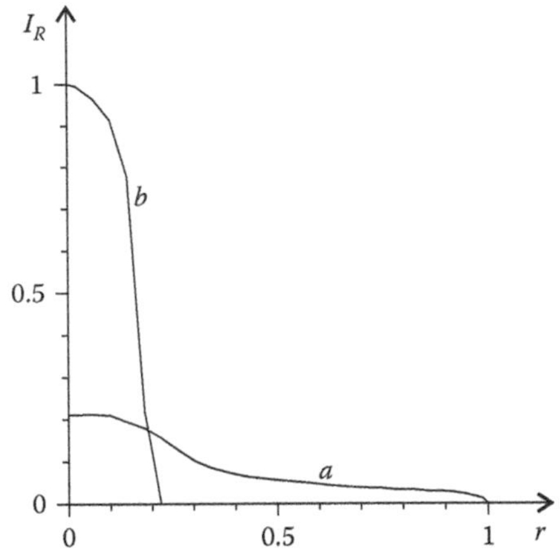

FIGURE 12.29
Comparison between the irradiance pattern on the receiver generated by a focusing optic with circular symmetry (curve *a*), and a similar optic with radial Köhler integration (curve *b*).

sections p_1, p_2, p_3, and p_4, a refractive secondary, also composed of four sections s_1, s_2, s_3, and s_4. All these sections are placed around a vertical central axis *b* through center point **C** of the primary mirror. Axis *b* also crosses the center of receiver *R*.

Section p_1 of the primary is a parabola with rotational symmetry around vertical axis a_1 through point $\mathbf{V}_1$, displaced relative to the center point **C**. The corresponding section s_1 of the secondary is a Cartesian oval with rotational symmetry around axis $\mathbf{P}_1\mathbf{R}_1$ that focuses point $\mathbf{P}_1$ onto $\mathbf{R}_1$, as shown in Figure 12.30a. Primary section p_1 focuses onto s_1 the vertical incoming light.

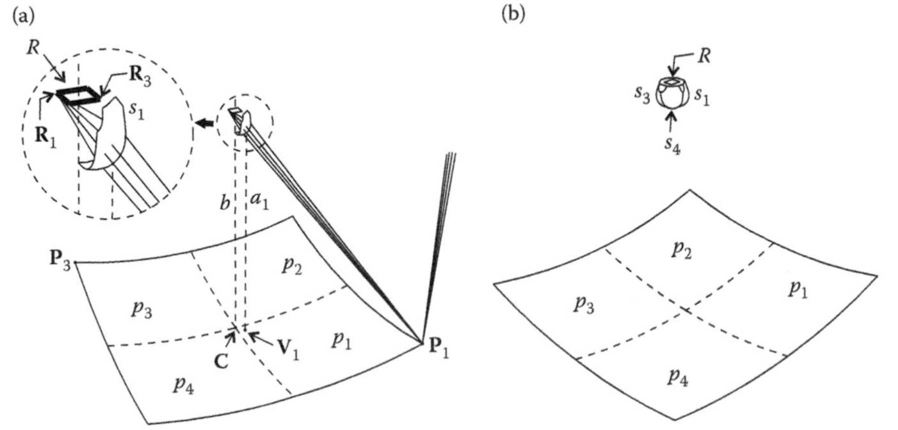

FIGURE 12.30
(a) Primary mirror p_1 and secondary refractive surface s_1 form a Köhler channel. (b) Köhler integrator composed of four separate channels.

Secondary section s_1 images primary section p_1 onto receiver R. The complete device is as shown in Figure 12.30b where primary sections p_1, p_2, p_3, and p_4 work with secondary sections s_1, s_2, s_3, and s_4, respectively.

The goal of this optic is to create a square illumination pattern on receiver R, fully and uniformly illuminating it. Section s_1 of the secondary sharply focuses $\mathbf{P}_1$ onto $\mathbf{R}_1$, and the primary corner at $\mathbf{P}_1$ onto an illumination corner at $\mathbf{R}_1$. But s_1 will not perfectly focus **C** onto $\mathbf{R}_3$ and, therefore, will not form a sharp illumination corner at $\mathbf{R}_3$. However, s_3 will focus sharply $\mathbf{P}_3$ onto $\mathbf{R}_3$, and the primary corner at $\mathbf{P}_3$ onto an illumination corner at $\mathbf{R}_3$. The result of superimposing on R the light coming from all the sections of the primary is a well-defined square and uniform illumination pattern.

Figure 12.31 shows a cross-section through axes b, a_1 and points $\mathbf{R}_1$, $\mathbf{R}_3$, and $\mathbf{P}_1$. Figure 12.31a shows how secondary section s_1 focuses edge point $\mathbf{P}_1$ of the primary onto edge point $\mathbf{R}_1$ of receiver R. Figure 12.31b shows how section p_1 of the primary focuses vertical incoming light onto focus **F** on the secondary s_1. During the design process, the position of focus **F** may have to be optimized. Although it is shown in this example as being on the surface of s_1, this may not be the optimum position, which may, for example, be inside s_1. Since this device has reflective (X) and refractive (R) optical surfaces, it is called an XR Köhler.

Figure 12.32a shows the irradiance pattern (in arbitrary units) on a receiver R (with size 1×1) for vertical sunlight. Figure 12.32b shows the same, but for sunlight making an angle 0.7° to the vertical (axis b of the concentrator). For vertical sunlight, the irradiance is very uniform, and the irradiance shape fits well with the square shape of receiver R. For sunlight incident at an angle, the irradiance pattern will vary, but still maintaining a high degree of uniformity. Depending on the design parameters, these results will vary.

A similar solar concentrator to that in Figure 12.30 may be obtained by replacing the primary mirror sections by Fresnel lens sections. That is the case of the optic shown in Figure 12.33a. Now, each section p_1, p_2, p_3, and p_4

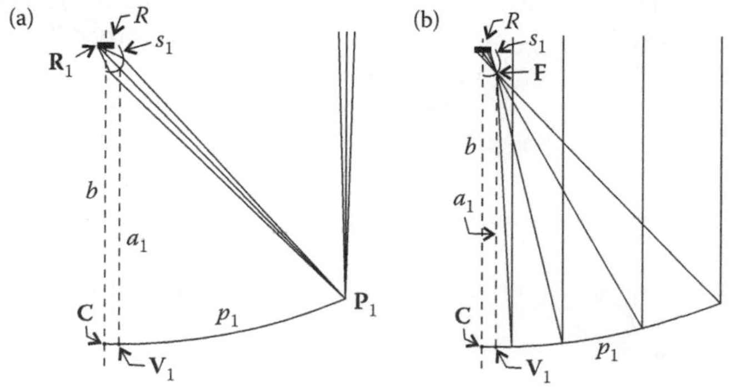

FIGURE 12.31
Vertical cross-section of the integrator in Figure 12.30. (a) Paths of the rays reflected at edge $\mathbf{P}_1$ of the primary mirror p_1. (b) Paths of the rays focused by p_1 onto s_1.

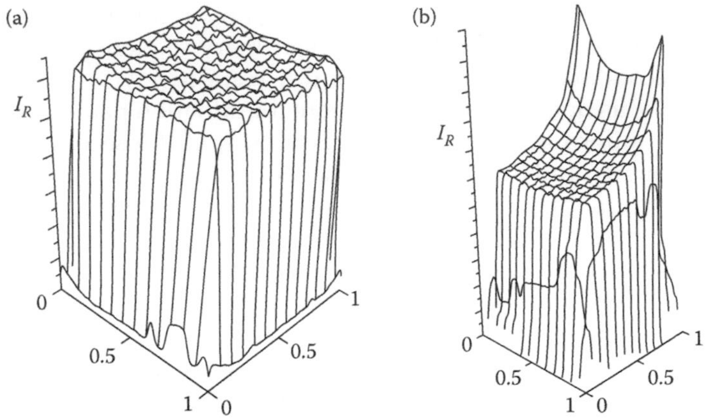

FIGURE 12.32
Irradiance pattern on the receiver of the concentrator in Figure 12.30 for (a) vertical sunlight, and (b) sunlight at an angle of 0.7° to the vertical.

of Fresnel lens focuses vertical light onto secondary section s_1, s_2, s_3, and s_4 respectively. The secondary sections focus the outer corner of each primary section onto the opposite corner of receiver R. Figure 12.33b shows a different view of the secondary optical element.

The geometry of this concentrator is shown in more detail in Figure 12.34, which is a vertical cross-section through the diagonal $\mathbf{P}_1\mathbf{P}_3$ of the primary optical element. Square section p_1 of the primary is a Fresnel lens with

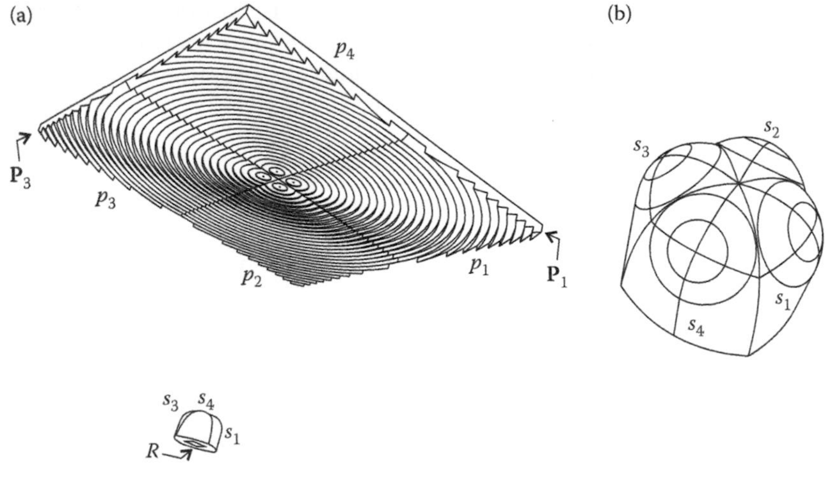

FIGURE 12.33
(a) Köhler concentrator composed of four sections of Fresnel lens (p_1, p_2, p_3, p_4) and corresponding sections of refractive secondary optic (s_1, s_2, s_3, s_4). (b) Detail of the secondary optical element.

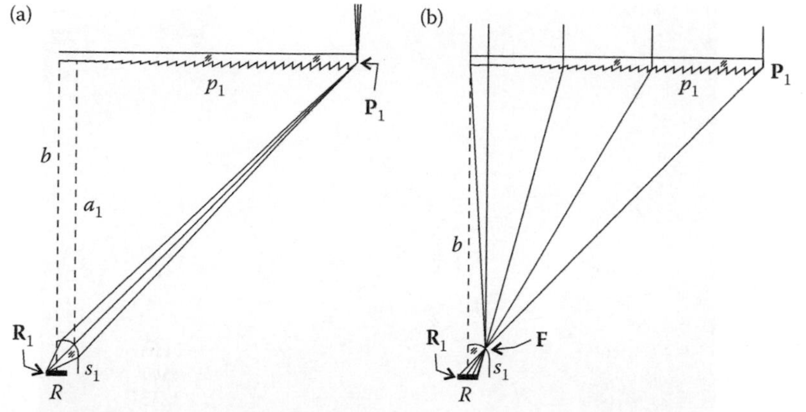

FIGURE 12.34
Vertical cross-section of the integrator in Figure 12.33. (a) Rays from $\mathbf{P}_1$ are focused to $\mathbf{R}$ by the secondary s_1. (b) Vertical incoming rays are focused to $\mathbf{F}$ by the Fresnel lens p_1.

circular symmetry around vertical axis a_1 and focus $\mathbf{F}$ on the secondary. Axis a_1 is displaced relative to axis b through the center of the receiver. Secondary section s_1 has circular symmetry around axis $\mathbf{R}_1\mathbf{P}_1$, and focuses corner $\mathbf{P}_1$ of the primary onto corner $\mathbf{R}_1$ of receiver R. As with the XR Köhler, also here the optimum position of focus $\mathbf{F}$ of the primary may not be on the surface of s_1, but somewhere else, for example, inside s_1.

Since this device has a Fresnel primary and refractive (R) secondary, it is called a Fresnel–R Köhler, or FK for short.

Figure 12.35a shows the central region of the Fresnel primary of an FK concentrator, and Figure 12.35b shows the corresponding secondary.

Figure 12.36 shows an array of several FK concentrators assembled together in a module.

12.4 Prescribed Irradiance Köhler Optics

It is possible to add Köhler channels to an exiting optic that generates a desired irradiance pattern on a receiver. We may start with an optic O_1 extending from $\mathbf{O}_1$ to $\mathbf{O}_2$ and designed for an emitter $\mathbf{E}_1\mathbf{E}_2$ and receiver $\mathbf{R}_1\mathbf{R}_2$, as shown in Figure 12.37 (see Chapter 10, Figure 10.25a).

Point $\mathbf{R}_6$ on receiver $\mathbf{R}_1\mathbf{R}_2$ is illuminated by a cone of rays bound by edge rays r_{6L} and r_{6R} perpendicular to wavefronts w_2 and w_1, respectively. If the radiance L of the light is the same in all directions (emitted from a Lambertian source), the irradiance at point $\mathbf{R}_6$ is defined by the cone bound by r_{6L} and r_{6R}.

It is possible to add Köhler channels to the optic in Figure 12.37, with optical surfaces c_1 and c_2, as shown in Figure 12.38. Suppose then that we already

(a)

(b)

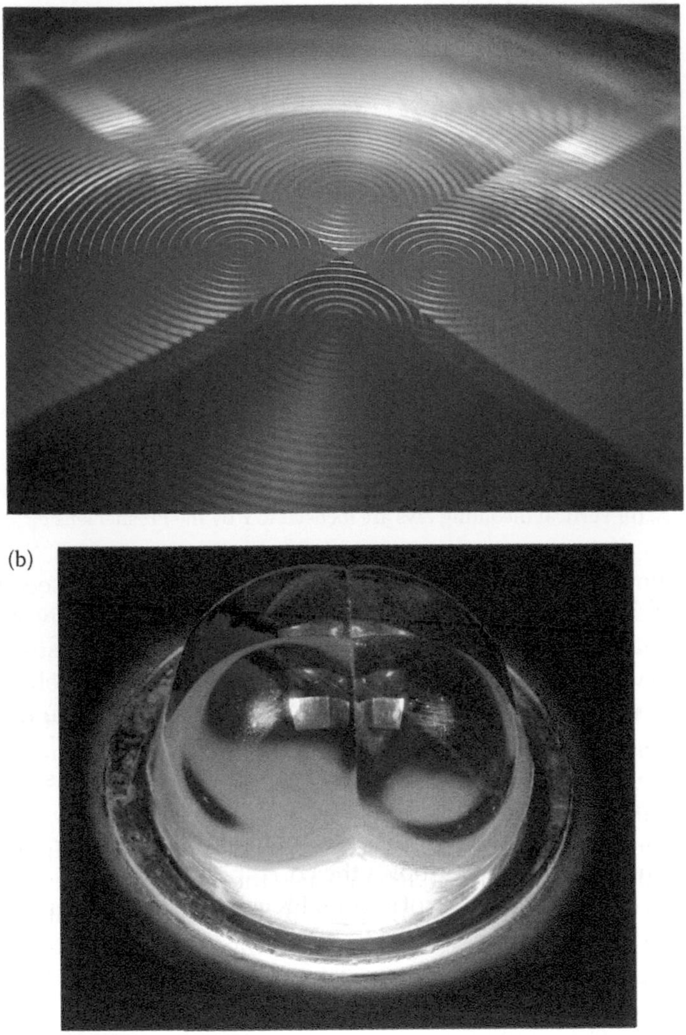

FIGURE 12.35
(a) Primary optical element (POE) and (b) secondary optical element (SOE) of a Fresnel–R Köhler concentrator. (Courtesy of Light Prescriptions Innovators.)

know the positions of points $\mathbf{P}_1$ on c_1 and $\mathbf{Q}_1$ on c_2. It is now possible to use conservation of étendue and calculate points $\mathbf{P}_2$ on c_1 and $\mathbf{Q}_2$ on c_2, and determine the edges of the Köhler channel.

The étendue U_1 from $\mathbf{E}_1\mathbf{E}_2$ to $\mathbf{P}_1\mathbf{P}_2$ is given by expression (12.3), the étendue U_2 from $\mathbf{P}_1\mathbf{P}_2$ to $\mathbf{Q}_1\mathbf{Q}_2$ is given by expression (12.4), and the étendue U_3 from $\mathbf{Q}_1\mathbf{Q}_2$ to wavefronts w_1 and w_2 is given by

$$U_3 = [\mathbf{Q}_1, \mathbf{W}_{12}] + [\mathbf{Q}_2, \mathbf{W}_{21}] - [\mathbf{Q}_1, \mathbf{W}_{11}] - [\mathbf{Q}_2, \mathbf{W}_{22}] \quad (12.7)$$

FIGURE 12.36
Fresnel–R Köhler module composed of several concentrators. (Courtesy of Light Prescriptions Innovators.)

as illustrated in Figure 12.38. Now, conservation of étendue as light travels through the Köhler channel results in $U_1 = U_2$ and $U_2 = U_3$. These two equations define the positions of points $\mathbf{P}_2$ and $\mathbf{Q}_2$ along curves c_1 and c_2, respectively, based on the known positions of $\mathbf{P}_1$ and $\mathbf{Q}_1$. The process may then continue, calculating the edges of more Köhler channels to the right of $\mathbf{P}_2$ and

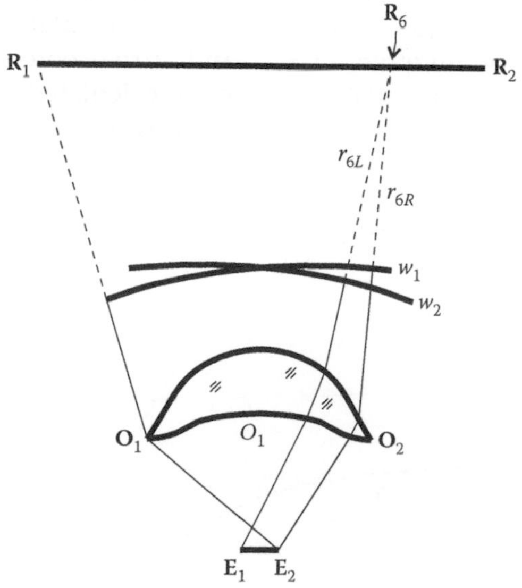

FIGURE 12.37
Optic O_1 generates a prescribed irradiance on receiver $\mathbf{R}_1\mathbf{R}_2$.

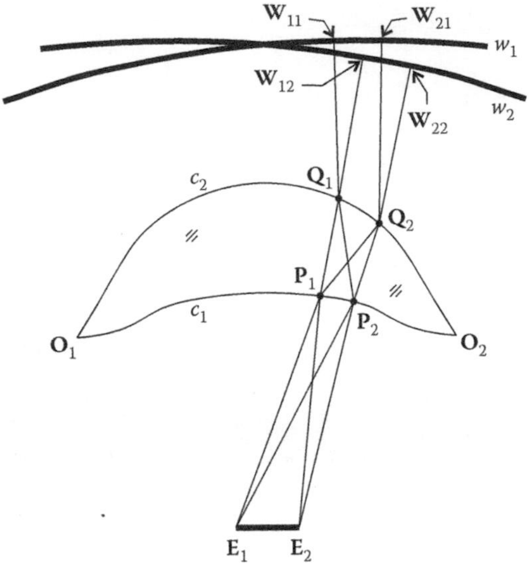

FIGURE 12.38
If two points $\mathbf{P}_1$ and $\mathbf{Q}_1$ at the edge of a Köhler channel are known, two new points $\mathbf{P}_2$ and $\mathbf{Q}_2$ (on curves c_1 and c_2) at the other edge can be calculated using conservation of étendue.

$\mathbf{Q}_2$, in a process similar to that shown in Figure 12.24. The result is shown in Figure 12.39.

Now the spaces between the Köhler edge points may be filled with lenticulations, as shown in Figure 12.40. The bottom lenticulation l_2 is a Cartesian oval with foci at the midpoint of $\mathbf{E}_1\mathbf{E}_2$ and at the center of opposing lenticulation l_1. The top lenticulation l_3 may be calculated, for example, as a Cartesian oval that focuses the rays from wavefront w_1 onto the edge of

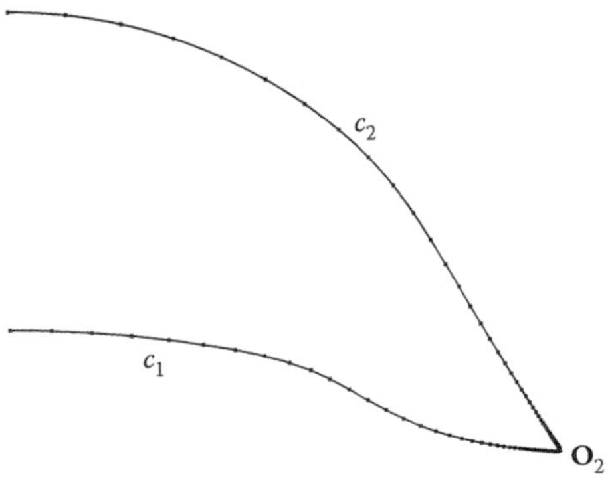

FIGURE 12.39
Points along curves c_1 and c_2 defining the edges of Köhler channels.

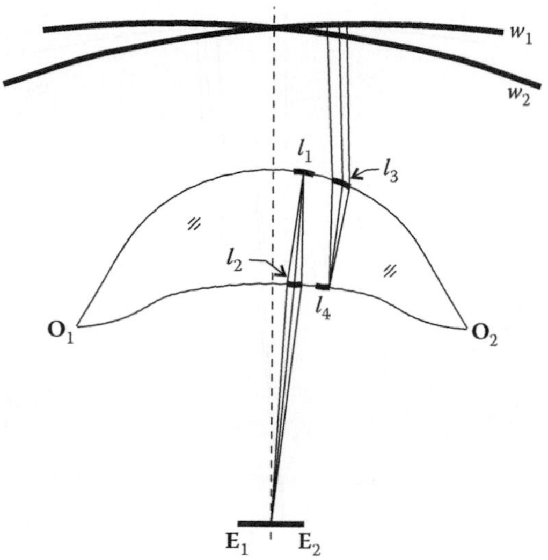

FIGURE 12.40
Bottom lenticulation l_2 focuses the center of $\mathbf{E}_1\mathbf{E}_2$ onto the center of l_1. Top lenticulation l_3 focuses the rays from w_1 onto the edge of l_4.

opposing lenticulation l_4. There will be small gaps between the edges of the lenticulations.

Another option for designing the top lenticulations is to build a bisector wavefront w_B whose rays bisect the edge rays (perpendicular to wavefronts w_1 and w_2), as shown in Figure 12.41.

Figure 12.42 shows a different option for calculating the top lenticulation l_3, now as a Cartesian oval that focuses the rays from bisector wavefront w_B onto the midpoint of the opposing lenticulation l_4.

In the Köhler optic, the edge rays exiting the lens are (approximately) perpendicular to wavefronts w_1 and w_2, as shown in Figure 12.43a. These exit

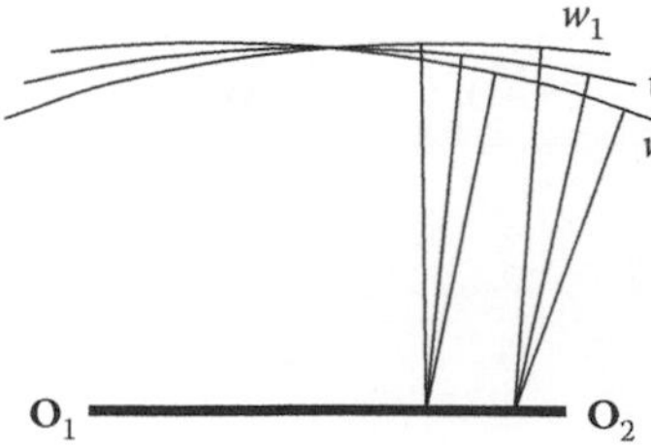

FIGURE 12.41
Rays perpendicular to bisector wavefront w_B bisect the edge rays perpendicular to wavefronts w_1 and w_2.

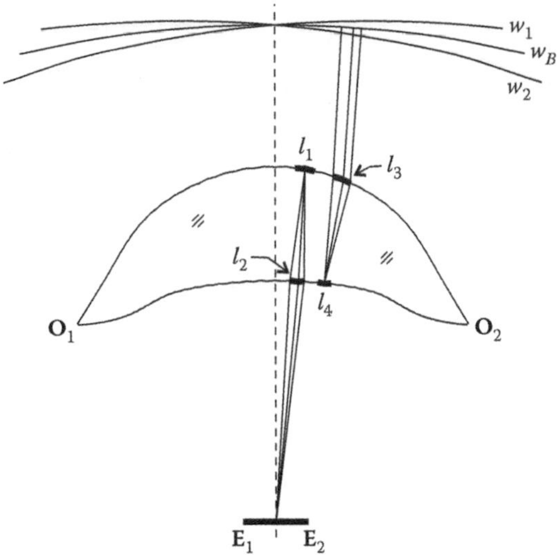

FIGURE 12.42
Bottom lenticulation l_2 focuses the center of E_1E_2 onto the center of l_1. Top lenticulation l_3 focuses the bisector rays from w_B onto the center of of l_4.

edge rays are defined by the geometry of the top and bottom lenticulations l_3 and l_4, and not by the emitter $\mathbf{E}_1\mathbf{E}_2$.

Figure 12.43b shows the same, but now for the smooth optic. Here, the edge rays exiting the optic are also perpendicular to wavefronts w_1 and w_2, but these exit rays are defined by the emitter $\mathbf{E}_1\mathbf{E}_2$.

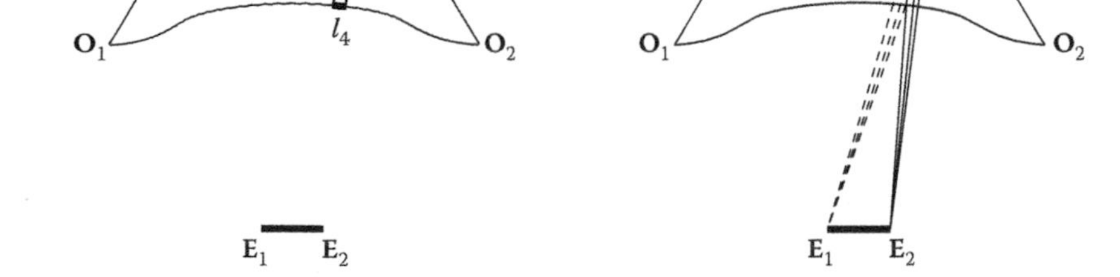

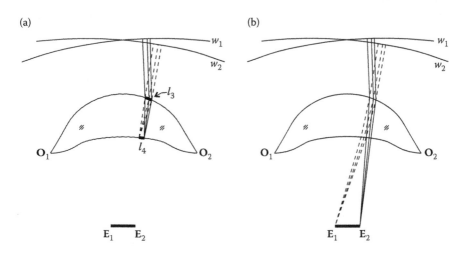

FIGURE 12.43
(a) The emission of the Köhler optic is defined by the geometry of lenticulations l_3 and l_4.
(b) The emission of the smooth optic is defined by the emitter $\mathbf{E}_1\mathbf{E}_2$.

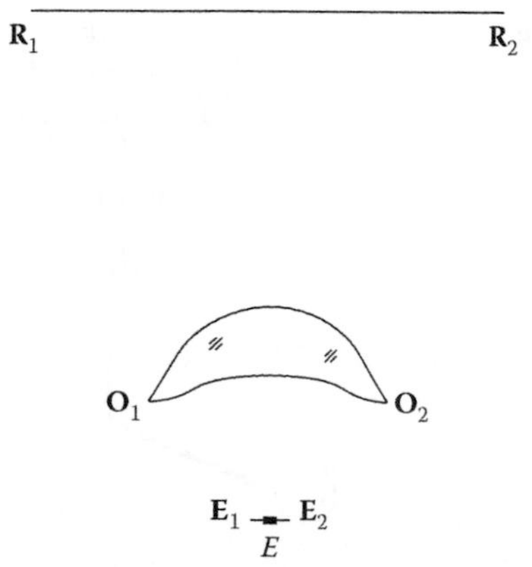

FIGURE 12.44
A large emitter $\mathbf{E_1E_2}$ will produce the desired irradiance pattern on $\mathbf{R_1R_2}$ through Köhler optic $\mathbf{O_1O_2}$. The same also holds for a smaller emitter E inside the integration area $\mathbf{E_1E_2}$.

In both cases, the exit edge rays are perpendicular to wavefronts w_1 and w_2 and, therefore, both optics produce the same emission pattern. However, in the case of the Köhler optic, this emission pattern is defined by the geometry of the lenticulations, while in the smooth optic it is defined by the emitter $\mathbf{E_1E_2}$. For that reason, the Köhler optic is tolerant to changes in the emitter $\mathbf{E_1E_2}$, while the smooth optic is not. However, different Köhler channels may contribute to different regions of the pattern and this may result in artifacts.

If the Köhler optic in Figure 12.44 is used with a large Lambertian source $\mathbf{E_1E_2}$, it will (approximately) produce the same irradiance pattern on $\mathbf{R_1R_2}$ as the smooth optic in Figure 12.37. Also, if the same Köhler optic is used with a smaller Lambertian emitter E inside the integration area $\mathbf{E_1E_2}$, the irradiance pattern on $\mathbf{R_1R_2}$ will still be (approximately) the same.

If the Köhler optic is combined with a CEC, with mirrors e_1 and e_2, that couples a Lambertian emitter $\mathbf{E_3E_4}$ onto $\mathbf{O_1O_2}$, the irradiance pattern on $\mathbf{R_1R_2}$ will still be the same. That configuration is shown in Figure 12.45.

However, if emitter $\mathbf{E_3E_4}$ is replaced by a smaller Lambertian emitter E, the irradiance pattern generated by the CEC on the bottom surface c_1 of the Köhler optic will be different, and this will result in a different irradiance pattern on receiver $\mathbf{R_1R_2}$.

12.5 Color-Mixing Based on Köhler Optics

As stated above, Köhler optics can be used to mix the light form different sources. An example of one such integrator optic is the shell mixer.[5] Here,

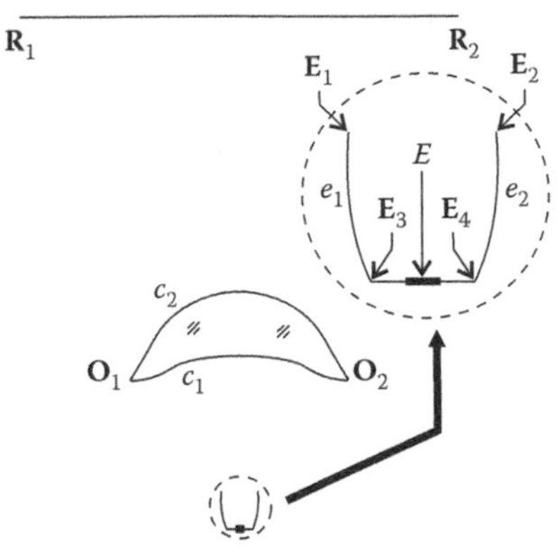

FIGURE 12.45
Köhler optic $\mathbf{O}_1\mathbf{O}_2$ may be combined with a CEC e_1, e_2. In that case, an emitter $\mathbf{E}_3\mathbf{E}_4$ will produce the desired irradiance pattern on $\mathbf{R}_1\mathbf{R}_2$. This, however, will not be the case if a smaller emitter E is used.

a nonuniform source is replaced by an apparent uniform virtual source by means of an optic. Figure 12.46 shows a Köhler channel of a shell mixer. First refractive surface $\mathbf{P}_0\mathbf{P}_1$ images emitter $\mathbf{E}_1\mathbf{E}_2$ onto second refractive surface $\mathbf{Q}_0\mathbf{Q}_1$, as shown in Figure 12.46a.

Refractive surface $\mathbf{Q}_0\mathbf{Q}_1$ creates a virtual image of $\mathbf{P}_0\mathbf{P}_1$ on top of emitter $\mathbf{E}_1\mathbf{E}_2$, as in Figure 12.46b. Rays coming from a point on $\mathbf{P}_0\mathbf{P}_1$, now appear to come from a point on $\mathbf{E}_1\mathbf{E}_2$. Even if emitter $\mathbf{E}_1\mathbf{E}_2$ is not uniform, surface $\mathbf{P}_0\mathbf{P}_1$ will be uniformly illuminated (since it is far from $\mathbf{E}_1\mathbf{E}_2$) and, therefore, the virtual image of $\mathbf{P}_0\mathbf{P}_1$ created on top of $\mathbf{E}_1\mathbf{E}_2$ will also be uniform. Nonuniform emitter $\mathbf{E}_1\mathbf{E}_2$ is then replaced by a uniform virtual emitter. Figure 12.46c shows how rays coming from a point $\mathbf{E}$ inside emitter $\mathbf{E}_1\mathbf{E}_2$ appear to come from the whole emitter after crossing the optic.

The calculation of points $\mathbf{P}_2$ and $\mathbf{Q}_2$ of a Köhler channel based on the known positions of $\mathbf{P}_1$ and $\mathbf{Q}_1$ is similar to that defined by expressions (12.1), (12.2), and (12.6). However, now "receiver" $\mathbf{R}_1\mathbf{R}_2$ is virtual and superimposed on emitter $\mathbf{E}_1\mathbf{E}_2$, that is, $\mathbf{R}_1 = \mathbf{E}_1$ and $\mathbf{R}_2 = \mathbf{E}_2$.

The equations for the optical path length are given by expression (12.1) for optical surface $\mathbf{P}_1\mathbf{P}_2$ (Figure 12.47a) and

$$n[\mathbf{P},\mathbf{Q}_1] - [\mathbf{Q}_1,\mathbf{E}] = n[\mathbf{P},\mathbf{Q}_2] - [\mathbf{Q}_2,\mathbf{E}] \tag{12.8}$$

for optical surface $\mathbf{Q}_1\mathbf{Q}_2$, since rays $\mathbf{Q}_1\mathbf{E}$ and $\mathbf{Q}_2\mathbf{E}$ are now virtual (see Figure 12.47b and Chapter 21).

The étendue from $\mathbf{E}_1\mathbf{E}_2$ to $\mathbf{P}_1\mathbf{P}_2$ and the étendue from $\mathbf{P}_1\mathbf{P}_2$ to $\mathbf{Q}_1\mathbf{Q}_2$ are given by expressions (12.3) and (12.4), respectively (Figure 12.47c). The étendue from $\mathbf{Q}_1\mathbf{Q}_2$ to $\mathbf{E}_1\mathbf{E}_2$ is

$$U_3 = [\mathbf{Q}_1,\mathbf{E}_2] + [\mathbf{Q}_2,\mathbf{E}_1] - [\mathbf{Q}_1,\mathbf{E}_1] - [\mathbf{Q}_2,\mathbf{E}_2] \tag{12.9}$$

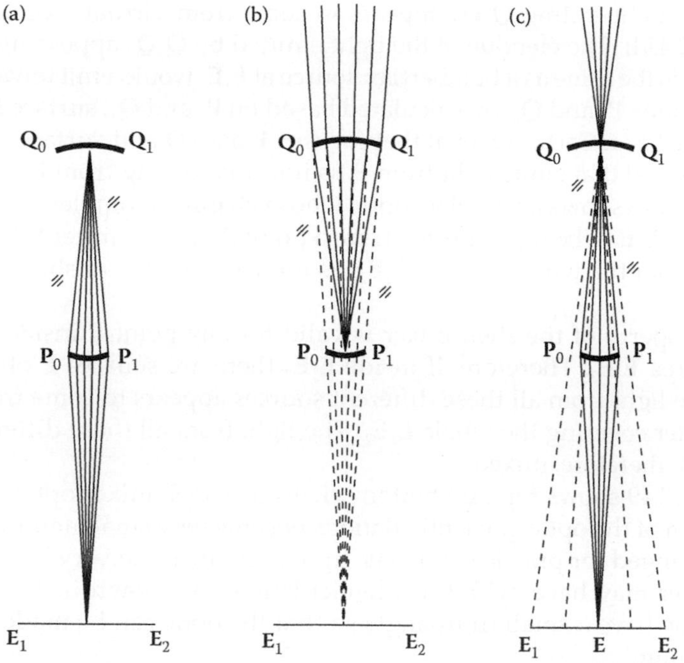

FIGURE 12.46
Shell mixer integrator Köhler channel. (a) Rays from the center of the emitter $\mathbf{E}_1\mathbf{E}_2$ are focused by the first refractive surface $\mathbf{P}_0\mathbf{P}_1$ onto the center $\mathbf{Q}_0\mathbf{Q}_1$ of the second refractive surface. (b) Rays from the center of $\mathbf{P}_0\mathbf{P}_1$ leave the optic as if coming from the center of $\mathbf{E}_1\mathbf{E}_2$. (c) Rays from a point **E** on emitter $\mathbf{E}_1\mathbf{E}_2$ leave the optic appearing to come from the entire emitter $\mathbf{E}_1\mathbf{E}_2$.

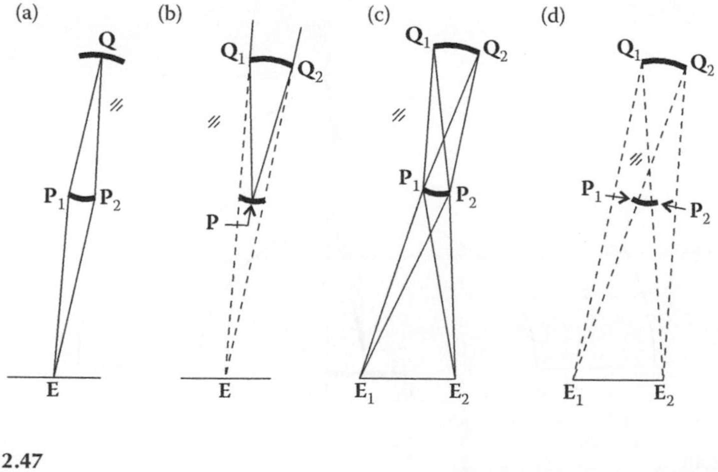

FIGURE 12.47
Conservation of optical path length, (a) and (b), and étendue, (c) and (d), in a Köhler channel of a shell mixer.

since now light exiting $\mathbf{Q}_1\mathbf{Q}_2$ appears to come from virtual "receiver" $\mathbf{E}_1\mathbf{E}_2$ (Figure 12.47d). The étendue of the light emitted by $\mathbf{Q}_1\mathbf{Q}_2$ appearing to come from $\mathbf{E}_1\mathbf{E}_2$ is the same as a Lambertian source at $\mathbf{E}_1\mathbf{E}_2$ would emit toward $\mathbf{Q}_1\mathbf{Q}_2$.

Once points $\mathbf{P}_2$ and $\mathbf{Q}_2$ are calculated based on $\mathbf{P}_1$ and $\mathbf{Q}_1$, surface $\mathbf{P}_1\mathbf{P}_2$ may be obtained as a Cartesian oval that focuses **E** onto **Q** and surface $\mathbf{Q}_1\mathbf{Q}_2$ as a Cartesian oval that emits light from **P**, as if it was coming from **E**.

Figure 12.48 shows a complete optic whose channels completely surround the emitter $\mathbf{E}_1\mathbf{E}_2$. The light emitted from a point **P** inside emitter $\mathbf{E}_1\mathbf{E}_2$ appears to come from the whole emitter $\mathbf{E}_1\mathbf{E}_2$, as indicated by the dashed lines (virtual rays).

This property of the shell mixer is valid for any point **P** inside the integration area $\mathbf{E}_1\mathbf{E}_2$. Therefore, if inside $\mathbf{E}_1\mathbf{E}_2$ there are sources **P** of different colors, the light from all these different sources appears to come from a virtual emitter covering the whole $\mathbf{E}_1\mathbf{E}_2$. The light from all these different light sources is, therefore, mixed.

Figure 12.49 shows top and bottom views of a shell mixer optic. Closer to the bottom of the optic, the lenticulations become very small and the tesselation is changed for practical reasons. In particular, at the very bottom *b*, the shell mixer may have cylindrical lenticulations, as shown in Figure 12.49, which may have a small draft angle so that the optic can be made by injection molding.

In 3-D, the Cartesian oval $\mathbf{P}_1\mathbf{P}_2$ in Figure 12.47a has rotational symmetry around the axis through its foci **E** and **Q**. The same holds for Cartesian oval $\mathbf{Q}_1\mathbf{Q}_2$ in Figure 12.47b, which, in 3-D, has rotational symmetry around the axis through its foci **E** and **P**. These surfaces make up the lenticulations of the shell mixer.

A possible approximation is to make the lenticulations spherical, which may even be calculated using paraxial optics.[8]

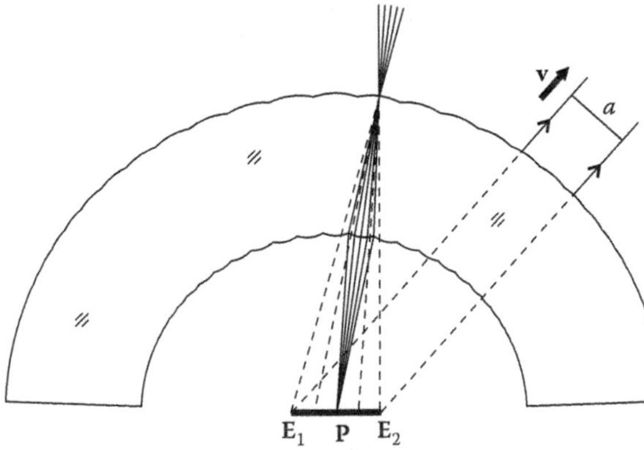

FIGURE 12.48
Shell mixer optic with channels surrounding the emitter $\mathbf{E}_1\mathbf{E}_2$. Flashed area *a* for the emission in direction **v** may encompass several Köhler channels.

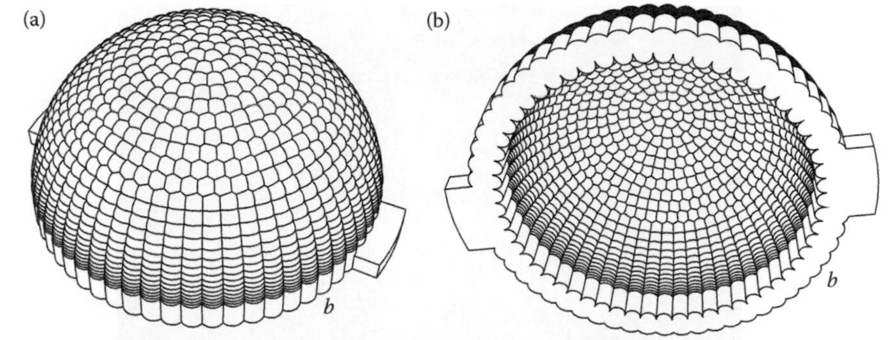

FIGURE 12.49
(a) Top view of a shell mixer optic. For practical reasons, at the bottom *b* the lenticulations are almost vertical and cylindrical. (b) Bottom view of the same.

Figure 12.50 shows a shell mixer optic.

Figure 12.51a and b shows details of the top and bottom lenticulations of a shell mixer.

12.6 SMS-Based Köhler Optics

Köhler optics may be extended to 3-D geometry, based on 3-D SMS designs.[1] These designs, however, make use of some concepts that are described below, before applying them to the SMS 3-D Köhler designs.

FIGURE 12.50
Shell mixer optic. (Courtesy of Light Prescriptions Innovators.)

(a)

(b)

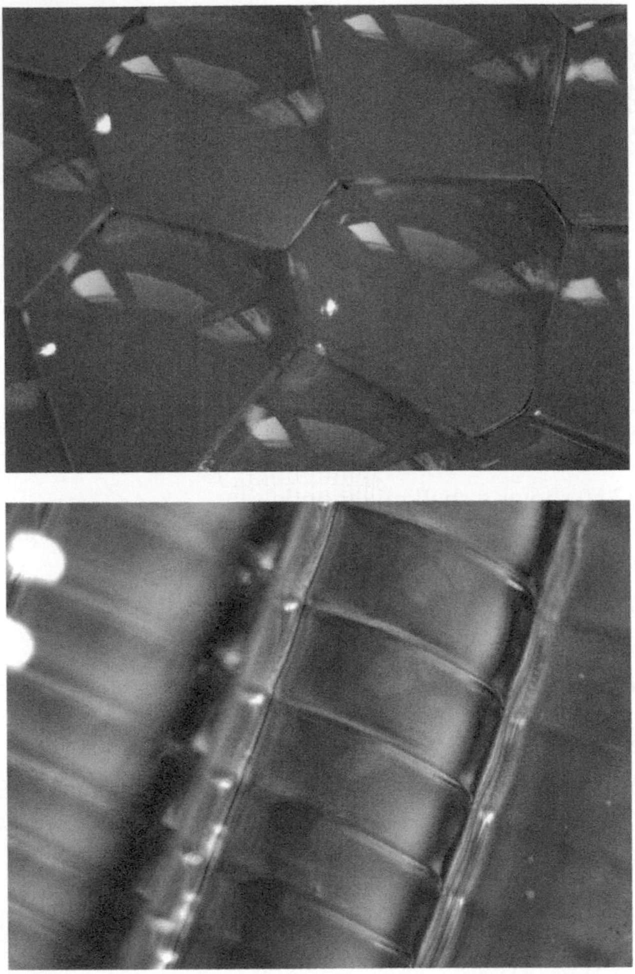

FIGURE 12.51
Detail of the top (a) and bottom (b) lenticulations of a shell mixer. (Courtesy of Light Prescriptions Innovators.)

A first concept is that of a constant p wavefront. Let us consider a curve f whose points are parameterized as $\mathbf{f}(\sigma)$ and whose tangent vectors (derivative) are given by $\mathbf{t}(\sigma)$, where σ is the curve parameter, as shown in Figure 12.52. From each point $\mathbf{f}(\sigma)$ on the curve, two light rays $r_1(\sigma)$ and $r_2(\sigma)$ radiate at an angle $\alpha(\sigma)$, to the tangent vector $\mathbf{t}(\sigma)$. Rays $r_1(\sigma)$ and $r_2(\sigma)$ are perpendicular to wavefronts w_A and w_B respectively.

Consider now the situation in Figure 12.53 showing a bundle of rays r making an angle α to a vector $\mathbf{t}$. All rays r make the same angle α to vector $\mathbf{t}$. These rays are on a cone with vertex at point $\mathbf{C}$ and may be parameterized by, say, an angle φ around vector $\mathbf{t}$. These rays are then given by $r(\varphi)$ and have

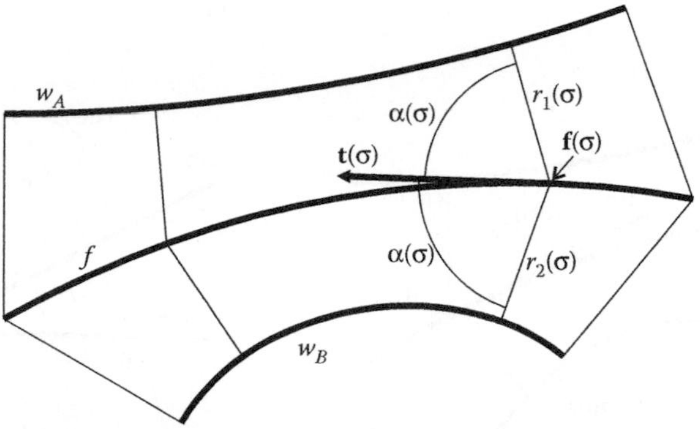

FIGURE 12.52
Rays $r_1(\sigma)$ and $r_2(\sigma)$ make an angle $\alpha(\sigma)$ to the tangent vector $\mathbf{t}(\sigma)$ to curve $\mathbf{f}(\sigma)$. These rays are perpendicular to wavefronts w_A and w_B.

momenta $\mathbf{p}(\varphi)$ pointing at a circle b perpendicular to $\mathbf{t}$. Since all rays make the same angle α to vector $\mathbf{t}$, their projection p onto $\mathbf{t}$ is contant for all φ:

$$p = \text{constant} \tag{12.10}$$

This is sometimes called a contant p bundle of rays.

The concept in Figure 12.52 may now be extended to 3-D geometry, as shown in Figure 12.54a. Here, rays r radiate from a point $\mathbf{f}(\sigma)$ on curve f, making an angle $\alpha(\sigma)$ to the tangent vector $\mathbf{t}(\sigma)$ to curve f at point $\mathbf{f}(\sigma)$. These rays have circular symmetry around vector $\mathbf{t}(\sigma)$ and, therefore, are on a cone that has a vertex $\mathbf{f}(\sigma)$ and whose base is defined by a circle $b(\sigma)$, perpendicular to $\mathbf{t}(\sigma)$. This is a similar situation to that in Figure 12.53. Figure 12.54b shows the

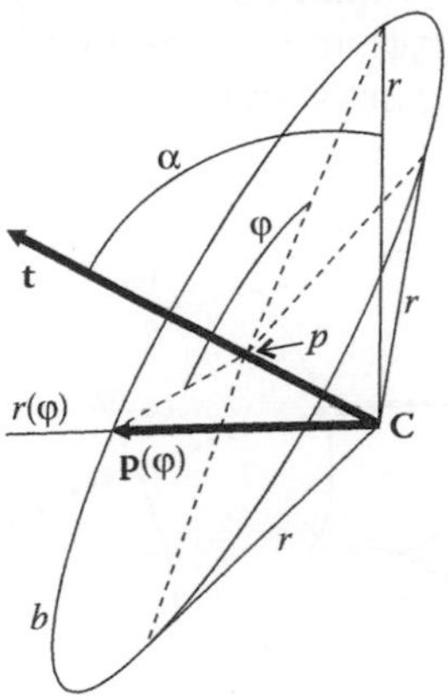

FIGURE 12.53
A bundle of rays r makes a contant angle α to a line defined by point $\mathbf{C}$ and vector $\mathbf{t}$. Each one of these rays may be parameterized by, say, an angle φ around vector $\mathbf{t}$.

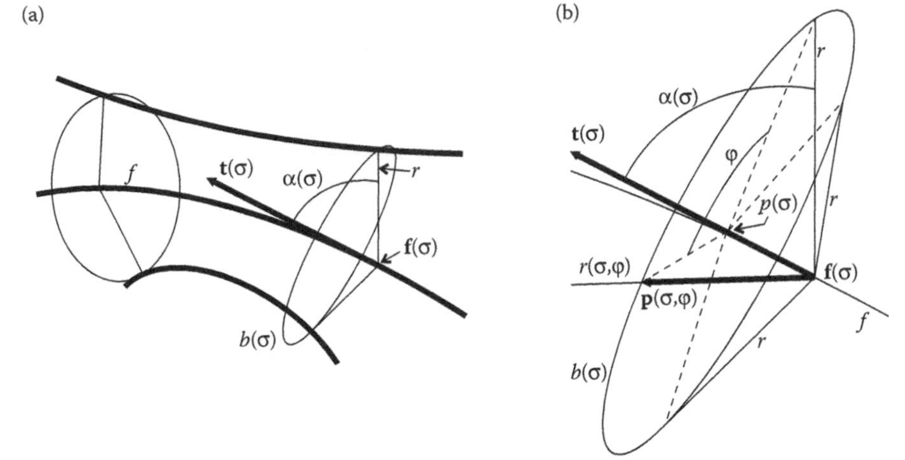

FIGURE 12.54
(a) Rays r form a cone with vertex $\mathbf{f}(\sigma)$ and angular aperture $\alpha(\sigma)$ around tangent vector $\mathbf{t}(\sigma)$ to curve f. (b) Rays r radiating from a point $\mathbf{f}(\sigma)$ on curve f have momenta $\mathbf{p}(\sigma,\varphi)$ with a constant component $p(\sigma)$ along tangent vector $\mathbf{t}(\sigma)$.

geometry of the rays r radiating from point $\mathbf{f}(\sigma)$ on curve f. To be uniquely identified, aside from parameter σ, the rays of this cone may be parameterized by, say, an angle φ around vector $\mathbf{t}(\sigma)$. This is, therefore, a two parameter bundle of rays $r(\sigma,\varphi)$. The momenta $\mathbf{p}(\sigma,\varphi)$ of all rays $r(\sigma,\varphi)$ point at a circle $b(\sigma)$ perpendicular to $\mathbf{t}(\sigma)$. Also, for a given value of σ, all momenta $\mathbf{p}(\sigma,\varphi)$ have the same component $p(\sigma)$ along direction $\mathbf{t}(\sigma)$. For a given parameter σ, the value p is constant for all rays of the cone and does not depend on φ. This is again similar to the situation in Figure 12.53.

The different cones of rays radiating from the points of curve f form a two parameter bundle of rays $r(\sigma,\varphi)$ and are perpendicular to a wavefront w_2, as shown in Figure 12.55. These are sometimes called constant p wavefronts.

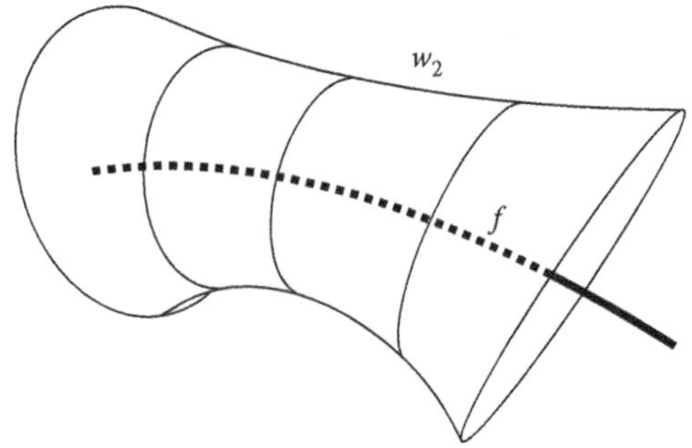

FIGURE 12.55
Rays radiating from curve f in cones of constant p are perpendicular to a wavefront w_2.

Now we consider a second concept: refraction (or reflection) on a curve (line). Consider then that a ray i with momentum $\mathbf{p}_1$ travelling in a medium of refractive index n_1 refracts at an interface with normal $\mathbf{n}$ into another medium of refractive index n_2, becoming ray r with momentum $\mathbf{p}_2$, as shown in Figure 12.56.

The momenta of the incoming and refracted rays are related by $\mathbf{p}_1 \cdot \mathbf{t} = \mathbf{p}_2 \cdot \mathbf{t}$, where $\mathbf{t}$ is a unit vector perpendicular to $\mathbf{n}$ and on the same plane as $\mathbf{n}$, $\mathbf{p}_1$, and $\mathbf{p}_2$. We may also write $\mathbf{p}_1 = n_1\mathbf{i}$ and $\mathbf{p}_2 = n_2\mathbf{r}$, resulting in $n_1\mathbf{i} \cdot \mathbf{t} = n_2\mathbf{r} \cdot \mathbf{t}$ where $\mathbf{i}$ and $\mathbf{r}$ are unit vectors in the directions of $\mathbf{p}_1$ and $\mathbf{p}_2$, respectively. Angle α that the refracted ray makes to tangent vector $\mathbf{t}$ may then be obtained from $n_1\mathbf{i} \cdot \mathbf{t} = n_2\cos\alpha$, or

$$\alpha = \arccos\left(\frac{n_1}{n_2}\mathbf{i} \cdot \mathbf{t}\right) = \arccos(\mathbf{p}_1 \cdot \mathbf{t}/n_2) \tag{12.11}$$

Figure 12.57 shows the same, but now in a 3-D view. Normal $\mathbf{n}$ is perpendicular to axis x_1 and is, therefore, given by $\mathbf{n} = (0,m_2,m_3)$, where m_2 and m_3 are its coordinates along axes x_2 and x_3 respectively. Vector $\mathbf{t}$ is perpendicular to $\mathbf{n}$ and lays on the same plane as $\mathbf{p}_1$ and $\mathbf{p}_2$.

The momenta $\mathbf{p}_1 = (p_1(n_1),p_2(n_1),p_3(n_1))$ and $\mathbf{p}_2 = (p_1(n_2),p_2(n_2),p_3(n_2))$ of the ray before and after refraction are related by $\mathbf{p}_2 = \mathbf{p}_1 + k\mathbf{n}$ where k is a scalar. Since the x_1 component of $\mathbf{n}$ is zero, we get $p_1(n_1) = p_1(n_2)$, that is

$$p_1 = \text{constant} \tag{12.12}$$

Angle α that the refracted ray makes to tangent vector $\mathbf{t}$ is again given by expression (12.11).

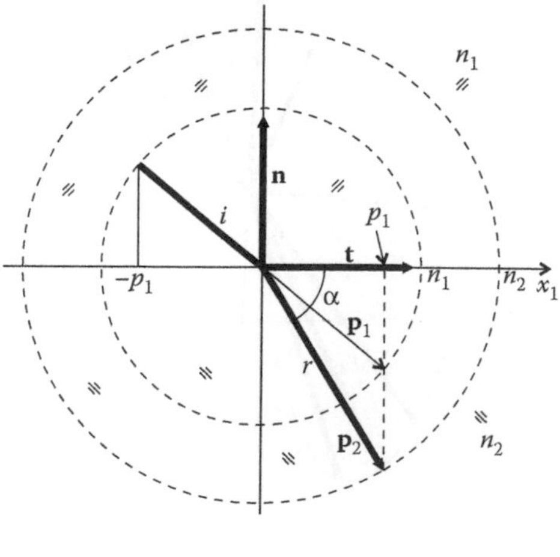

FIGURE 12.56
Refraction of a ray i from a medium of refractive index n_1 into another medium of refractive index n_2.

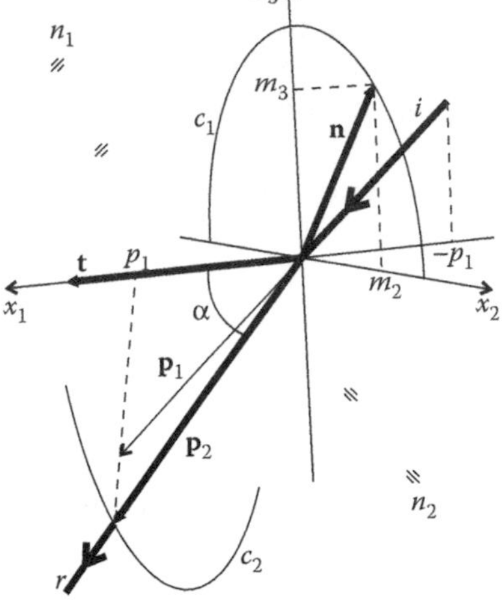

FIGURE 12.57
Refraction of a ray i from a medium of refractive index n_1 into another medium of refractrive index n_2. Normal $\mathbf{n}$ is perpendicular to axis x_1.

Now consider the situation in Figure 12.58 where the normal $\mathbf{n}$ rotates around axis x_1 (or vector $\mathbf{t}$) on a circle c_1 centered at the origin. Incident ray i does not change.

In all of its orientations $\mathbf{n}_1$, $\mathbf{n}_2$, $\mathbf{n}_3$, ... the normal is perpendicular to axis x_1 (or vector $\mathbf{t}$) and, therefore, is given by an expression $\mathbf{n}_k = (0, m_{k2}, m_{k3})$ with

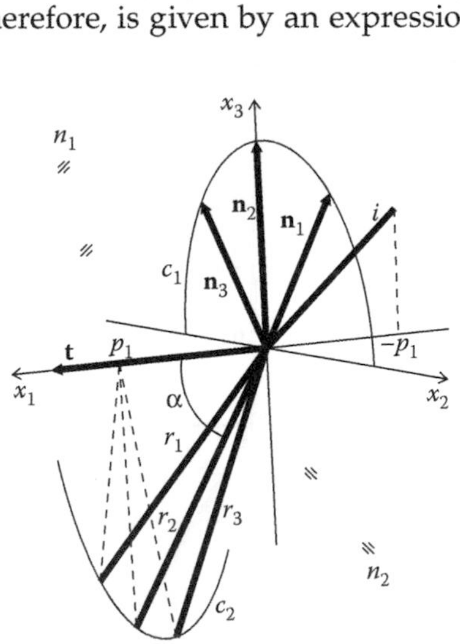

FIGURE 12.58
Refraction of a ray i from a medium of refractive index n_1 into another medium of refractive index n_2. Normal $\mathbf{n}$ is perpendicular to axis x_1 and rotates around axis x_1 to orientations $\mathbf{n}_1$, $\mathbf{n}_2$, $\mathbf{n}_3$, ...

$k = 1, 2, 3, \ldots$ In all cases expression (12.12) still holds and the momenta of the refracted rays $r_1, r_2, r_3, \ldots$ is then on a curve c_2 parallel to plane x_2x_3 and centered at point $(p_1,0,0)$. Rays $r_1, r_2, r_3, \ldots$ are then on a cone defined by circle c_2 and a vertex at the origin (where all refractions occur). All refracted rays $r_1, r_2, r_3, \ldots$ on that cone make the same angle α to tangent vector **t**. Angle α is again given by expression (12.11).

A similar result but with $n_1 = n_2$ is also valid in the case of reflection since $\mathbf{p}_2 = \mathbf{p}_1 + k\mathbf{n}$ still holds, where k is a scalar.

Now we consider a third concept: a Cartesian oval that takes the rays from a wavefront and deflects them in directions that cross a given curve. The shape of Cartesian oval surface is based on another given curve. The calculation of this Cartesian oval surface makes use of the other two concepts presented above: contant p wavefronts and refraction on a line.

Let us then consider a situation in which we want to calculate an optical surface that take rays from a wavefront w_1 (in this example, a point) and redirects them in directions that cross a given curve f. We further impose the condition that this optical surface must contain another given curve c.

We then start by taking a point **P** on curve c, as show in Figure 12.59. Vector **t** is tangent to c at **P**. The normal to the optical surface at point **P** will be perpendicular to tangent vector **t**. Consider then a unit radius circle c_1 centered at **P** and perpendicular to **t**. Consider also a range of possible normal vectors to the optical surface at **P**. These possible normal vectors point at curve c_1 and the corresponding refracted rays are contained in a cone of angular aperture α defined by vertex **P** and a circle c_2 perpendicular to **t**. This cone of possible refracted rays at **P** intersects curve f at point **X**. Path **X**, **P**, w_1 defines ray r through **P**.

In order to determine the intersection point of a cone and a curve we may, for example, use the configuration in Figure 12.60. The intersection point $\mathbf{X} = \mathbf{f}(\sigma_X)$ verifies

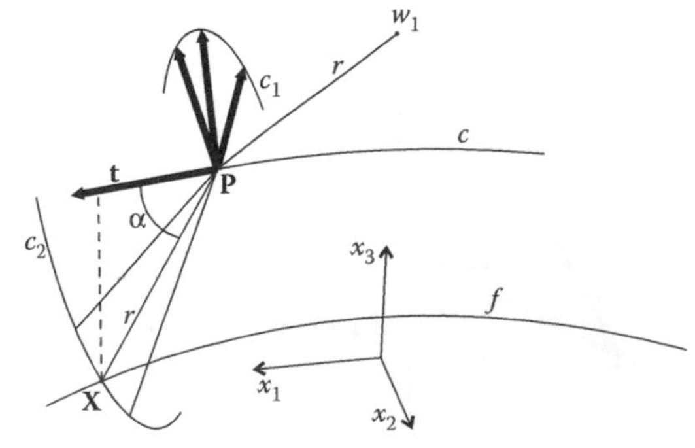

FIGURE 12.59
Possible rays from wavefront w_1 and refracted (or reflected) at point **P** on curve c are contained in a cone with anglular aperture α and axis **t** (tangent to c at **P**). This cone intersects curve f at point **X** defining the path of ray r.

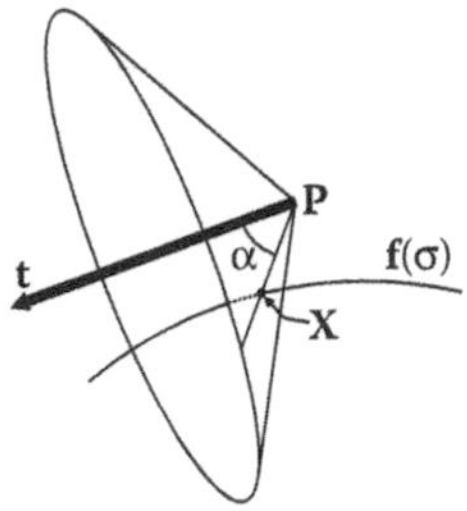

FIGURE 12.60
The intersection point **X** of a curve **f**(σ) with a cone defined by vertex **P**, axis **t** and angular aperture α is such that the angle of vector **X-P** with **t** is α.

$$(\mathbf{f}(\sigma_X) - \mathbf{p}) \cdot \mathbf{t} = \|\mathbf{f}(\sigma_X) - \mathbf{p}\| \cos\alpha \tag{12.13}$$

where α is the angle that the wall of the cone makes to the bisector unit vector **t** through its symmetry axis. We may now solve Equation 12.13 for σ_X determining the intersection point **X**.

Since we have the path w_1-**P-X** of ray r, we may determine the optical path length S_1 from w_1 to **X**. Now, reverse the direction of ray r, as shown in Figure 12.61. It now leaves point **X** in the direction of vector **u**, and is refracted at point **P** toward wavefront w_1. The unit tangent vector to curve f at point **X** is given by $\mathbf{t}_X$. Rotating vector **u** around vector $\mathbf{t}_X$ results in a cone of rays emitted from point **X**. Imposing a constant optical path length S_1 from **X** to w_1, it is possible to determine the paths of rays r_1, r_2, r_3, ... and new points $\mathbf{Q}_1$, $\mathbf{Q}_2$, $\mathbf{Q}_3$, ... on a curve g_1 on the optical surface.

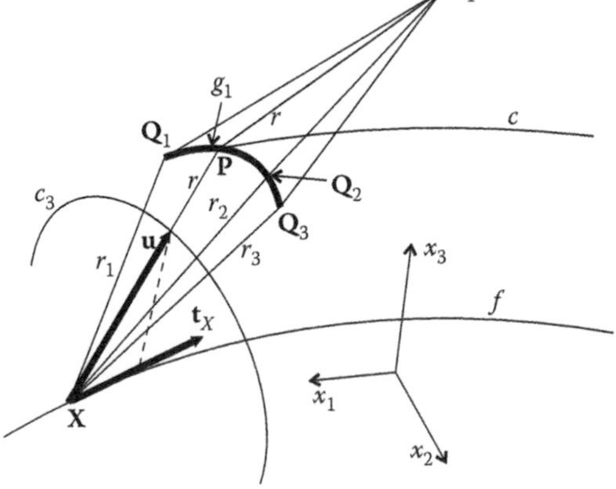

FIGURE 12.61
Given the path of ray r, with direction **u** at point **X**, a curve g_1 on the optical surface may be obtained by constant optical path length S_1 between **X** and w_1. The rays r_1, r_2, r_3, ... refracted at curve g_1 are defined by rotating **u** around axis $\mathbf{t}_X$ (tangent to curve f at point **X**).

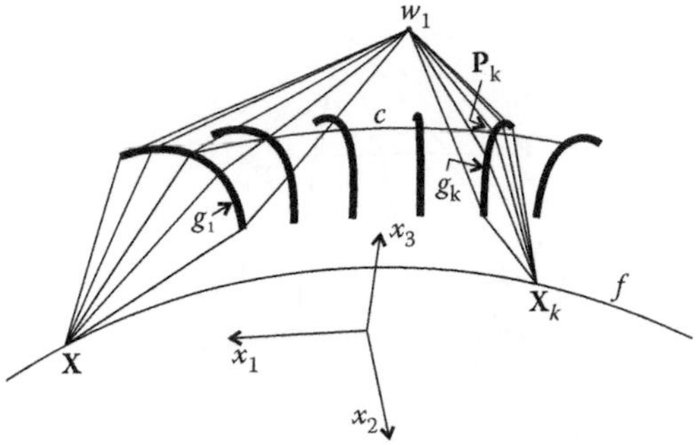

FIGURE 12.62
Different curves g_k start at points $\mathbf{P}_k$ on cuve c and are obtained by constant optical path length S_k between wavefront w_1 and point $\mathbf{X}_k$ on f. The set of all these curve g_k defines the Cartesian oval optical surface that contains curve c and redirects the rays from wavefront w_1 to points of curve f.

Moving point **P** along curve c and repeating the same process, one can calculate new curves $g_2, g_3, \ldots, gk, \ldots$ on the optical surface. The set of all these curves fully defines the Cartesian oval optical surface, as shown in Figure 12.62.

Please note that the optical path length S_1 is constant for the points of curve g_1, but will have a different value S_2 for the points of curve g_2 and so on, as point **P** moves along curve c. In particular, curve g_k is obtained by starting at point $\mathbf{P}_k$ on curve c and setting a constant optical path length S_k between w_1 and $\mathbf{X}_k$.

Rays converging to curve f are perpendicular to a constant p wavefront w_2 similar to that in Figure 12.55. However, using the method in Figure 12.62, it is possible to calculate the Cartesian oval surface (defined by curves g_k) without calculating the shape of the constant p wavefront around curve f.

The constant p wavefront around curve f may, however, be calculated using a similar method to that used to calculate the Cartesian oval surface. Consider then ray r in Figure 12.59, shown again in Figure 12.63. We may now choose a point **W** along its path. This defines an optical path length $[[w_1,\mathbf{P},\mathbf{W}]]$ from w_1 to **W**. We may now rotate point **W** around the tangent vector $\mathbf{t}_X$ to curve f at point **X**, resulting in circle c_3.

Using the same method illustrated in Figure 12.59 we may calculate the path of other rays r_k from wavefront w_1 to curve f, crossing curve c. Also, we may determine the positions of points $\mathbf{W}_k$ along these rays such that the optical path length $[[w_1,\mathbf{P}_k,\mathbf{W}_k]]$ equals the optical path length $[[w_1,\mathbf{P},\mathbf{W}]]$, that is, we calculate $\mathbf{W}_k$ such that $[[w_1,\mathbf{P}_k,\mathbf{W}_k]]= [[w_1,\mathbf{P},\mathbf{W}]]$. Again, we may now rotate point $\mathbf{W}_k$ around the tangent vector $\mathbf{t}_k$ to curve f at point $\mathbf{X}_k$, resulting in circle c_k. The set of all these circles c_k defines the shape of the constant p wavefront w_2 around curve f. Its shape is similar to that shown in Figure 12.55.

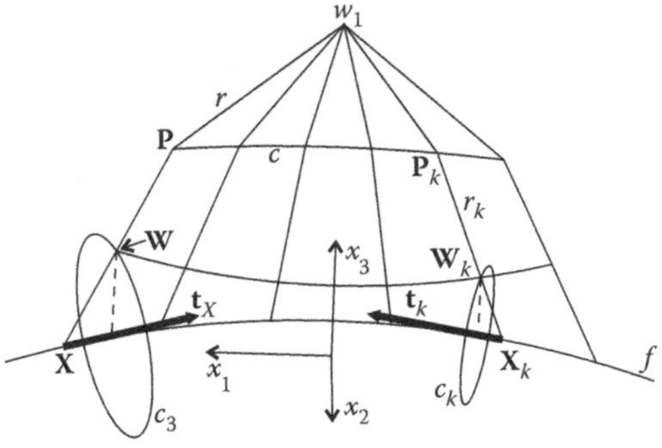

FIGURE 12.63
Point $\mathbf{W}_k$ is located along ray r_k and its position is such that $[[w_1,\mathbf{P}_k,\mathbf{W}_k]] = [[w_1,\mathbf{P},\mathbf{W}]]$. We may now rotate $\mathbf{W}_k$ around tangent $\mathbf{t}_k$ to curve f at point $\mathbf{X}_k$, resulting in circle c_k. The set of all these circles defines a contant p wavefront around curve f.

Now we consider a fourth concept: an SMS based Köhler optic in 2-D. Here, we calculate a Köhler optic in 2-D geometry based on the chains of an SMS design. Consider then an emitter $\mathbf{E}_1\mathbf{E}_2$, a receiver $\mathbf{R}_1\mathbf{R}_2$, and a starting point $\mathbf{P}_0$ and its normal. We may now calculate an SMS chain starting at $\mathbf{P}_0$ and defining two optical surfaces. The rays of this SMS chain are show as solid lines in Figure 12.64a. Now take point $\mathbf{Q}_0$ and its normal at the middle point of the bottom optical surface. Note that $\mathbf{P}_0$ and $\mathbf{Q}_0$ are on the same flow line that, in this case, is also the symmetry axis of the optic. Using the same optical path

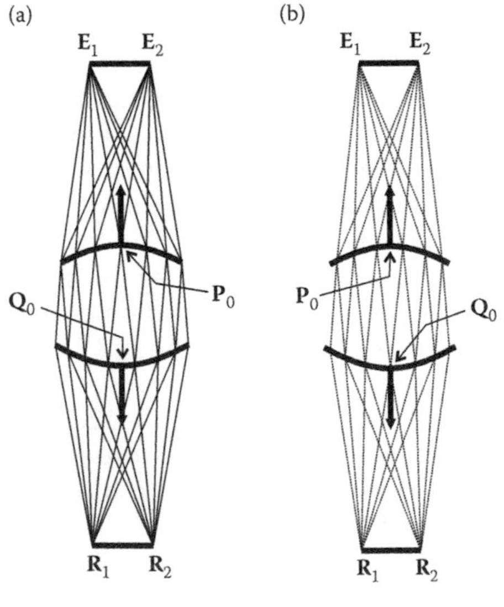

FIGURE 12.64
(a) SMS chain starting at point $\mathbf{P}_0$. (b) SMS chain starting at point $\mathbf{Q}_0$, the mid point of the bottom surface calculated in (a).

lengths as before, we may calculate a new SMS chain starting at $\mathbf{Q}_0$. The rays of this new SMS chain are show as dotted lines in Figure 12.64b.

Figure 12.65a shows these two SMS chains together. Now take the points where the solid lines rays intersect the top surface and the points where the dotted lines rays intersect the bottom surface. Figure 12.65b shows those points and some rays crossing through points $\mathbf{P}_0$ and $\mathbf{P}_1$ on the top surface and $\mathbf{Q}_0$ and $\mathbf{Q}_1$ on the bottom surface.

These points $\mathbf{P}_0$, $\mathbf{P}_1$ and $\mathbf{Q}_0$, $\mathbf{Q}_1$ may be taken for the edges of a Köhler channel. This method does not guarantee étendue conservation. The étendue of the radiation from $\mathbf{E}_1\mathbf{E}_2$ to $\mathbf{P}_0\mathbf{P}_1$ does not necessarily match that from $\mathbf{P}_0\mathbf{P}_1$ to $\mathbf{Q}_0\mathbf{Q}_1$ and that from $\mathbf{Q}_0\mathbf{Q}_1$ to $\mathbf{R}_1\mathbf{R}_2$. Although this method may not be very precise in some configurations, it intuitively illustrates the approximate relation between SMS chains and Köhler channels.

The dotted rays coming from the edges of $\mathbf{E}_1\mathbf{E}_2$ and crossing the center of $\mathbf{P}_0\mathbf{P}_1$ are redirected to edges $\mathbf{Q}_0$ and $\mathbf{Q}_1$ of the bottom lenticulation. Also, the solid rays coming from the edges of $\mathbf{R}_1\mathbf{R}_2$ and crossing the center of $\mathbf{Q}_0\mathbf{Q}_1$ are redirected to edges $\mathbf{P}_0$ and $\mathbf{P}_1$ of the top lenticulation. Figure 12.66 shows the resulting Köhler optic and some rays crossing it. The paths of rays $\mathbf{P}_1$-$\mathbf{R}_1$ and $\mathbf{P}_0$-$\mathbf{R}_2$ in Figure 12.65b and Figure 12.66 are very similar (ideally the paths would be the same).

We may now utilize the four concepts presented above in the design of an SMS 3-D Köhler optic. We start with a configuration as shown in Figure 12.67. It shows an emitter $\mathbf{E}_1\mathbf{E}_2$, a receiver $\mathbf{R}_1\mathbf{R}_2$, and a starting point $\mathbf{P}_0$ and its

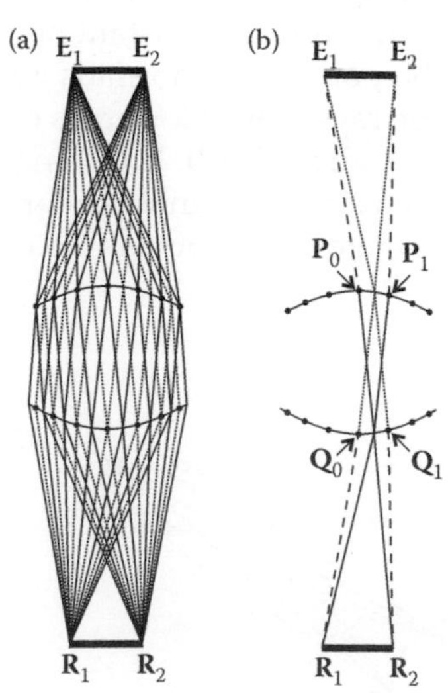

FIGURE 12.65
(a) The two SMS chains of Figure 12.64. We take the points where the solid lines intersect the top surface and those where the dotted lines intersect the bottom surface. (b) Same points with some rays passing through some of those points: $\mathbf{P}_0$, $\mathbf{P}_1$, $\mathbf{Q}_0$, and $\mathbf{Q}_1$.

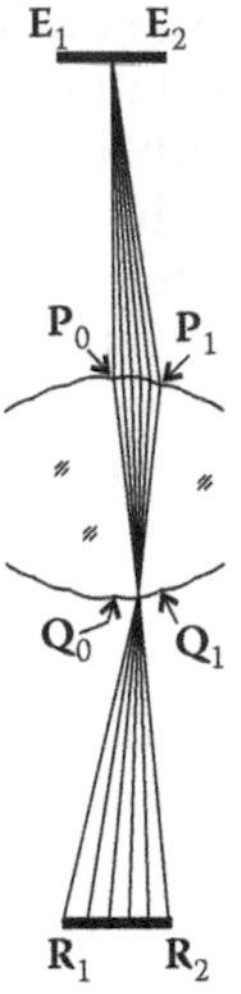

FIGURE 12.66
Köhler optic based on the SMS chains points in Figure 12.65. The paths of rays $\mathbf{P}_1$-$\mathbf{R}_1$ and $\mathbf{P}_0$-$\mathbf{R}_2$ in this figure and in Figure 12.65 are very similar.

normal. We may now calculate an SMS chain starting at $\mathbf{P}_0$ and defining two optical surfaces (curves).

We take curve c_0 as the initial curve of an SMS 3-D design, as in Figure 12.68a. Rays coming from point $\mathbf{E}_4$ are refracted at the points (and normals are not shown) of curve c_0. Imposing a constant optical path length from $\mathbf{E}_4$ to $\mathbf{R}_3$, we may calculate the points (and normals are not shown) of another curve c_1 that redirects these rays to $\mathbf{R}_3$. Now, rays coming from point $\mathbf{R}_4$ are refracted at curve c_1, as in Figure 12.68b. Imposing a constant optical path length from $\mathbf{R}_4$ to $\mathbf{E}_3$, we may calculate another curve c_2 that redirects these rays to $\mathbf{E}_3$. This process may now continue as we calculate more curves on both surfaces of the lens.

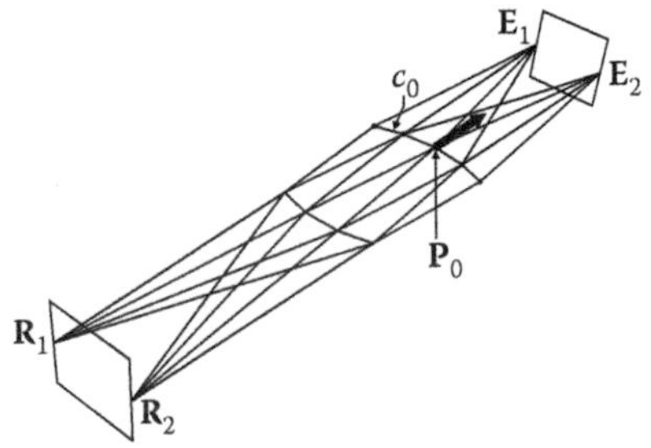

FIGURE 12.67
SMS 2-D used to determine curve c_0 that will be later used as a starting curve for an SMS 3-D design.

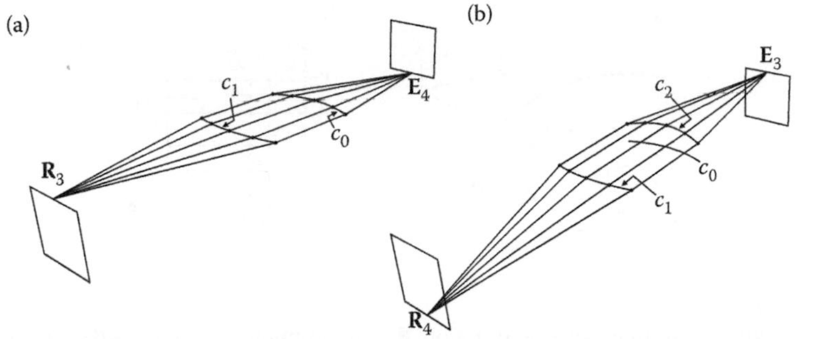

FIGURE 12.68
(a) Refracting at curve c_0 rays from $\mathbf{E}_4$ and imposing a constant optical path length to $\mathbf{R}_3$ determines curve c_1. (b) Refracting at curve c_1 rays from $\mathbf{R}_4$ and imposing a constant optical path length to $\mathbf{E}_3$ determines curve c_2.

The complete set of curves c_0, c_1, c_2, … is shown in Figure 12.69 as a set of solid curves. They define two optical surfaces s_1 and s_2. If needed, the optical surfaces may be better defined by calculating more points using a skinning method (see Chapter 9).

We now need to parameterize the optical surfaces. Consider then a set of points $\mathbf{P}_{ij}$. We may interpolate those points with a surface s, as in Figure 12.70a.[9] This surface may be parameterized as $s(u,v)$ with $u_A \le u \le u_B$ and $v_A \le u \le v_B$ where (u,v) is the parameter space shown in Figure 12.70b. We may choose the parameterization of the surface in such a way that the points defining a curve g_0 have a constant parameter v_0, that is, curve g_0 is parameterized as $g_0(u) = s(u,v_0)$ where v_0 is constant and $u_A \le u \le u_B$. Accordingly, the points defining a curve g_1 have a constant parameter u_1, that is, curve g_1 is parameterized as $g_1(v) = s(u_1,v)$ where u_1 is constant and $v_A \le v \le v_B$. Typically, the parameter range is chosen as $0 \le u \le 1$ and $0 \le v \le 1$, that is, $u_A = v_A = 0$ and $u_B = v_B = 1$.

The same parameterization may be applied to the interpolation of the points in the optical surfaces calculated by the SMS method. Figure 12.71 shows that for surface $s_1 = s_1(u,v)$ in Figure 12.69.

A point $\mathbf{P}$ on s_1 is given by $\mathbf{P} = s_1(u_P,v_P)$ and defines two isoparametric curves c_H and c_V on the surface. Curve c_H is parameterized by $c_H(u) = s_1(u,v_P)$ with

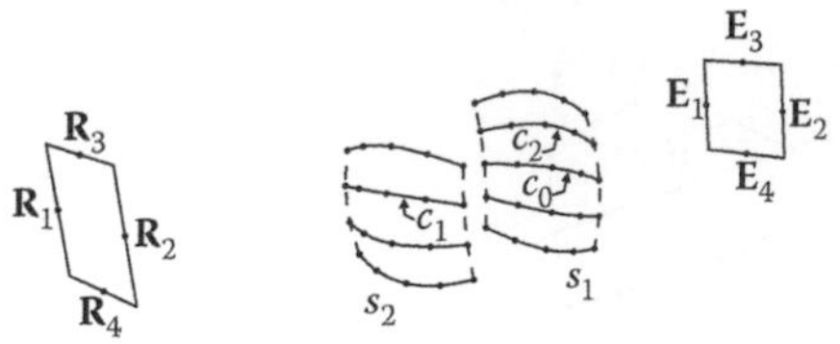

FIGURE 12.69
SMS 3-D curves of a lens that focuses $\mathbf{E}_3$ onto $\mathbf{R}_4$ and $\mathbf{E}_4$ onto $\mathbf{R}_3$, based on a starting curve c_0.

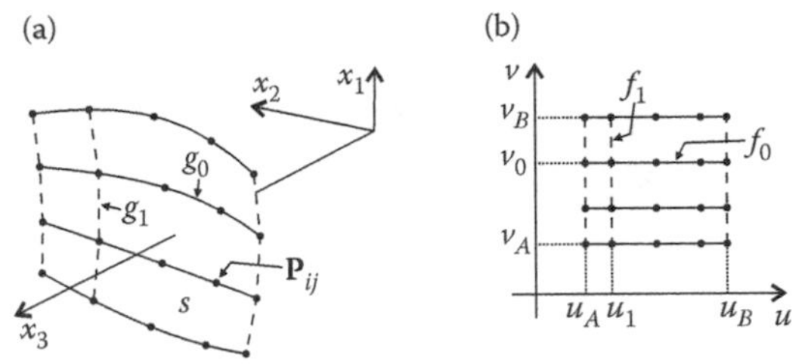

FIGURE 12.70
(a) Surface $s(u,v)$ is an interpolation of points $\mathbf{P}_{ij}$. (b) Parameter space of surface s. Curve g_0 is defined on a horizontal line $f_0 = (u,v_0)$ in parameter space where v_0 is constant and $u_A \le u \le u_B$. Curve g_1 is defined on a vertical line $f_1 = (u_1,v)$ in parameter space where u_1 is constant and $v_A \le v \le v_B$.

$u_A \le u \le u_B$ and curve c_V is parameterized by $c_V(v) = s_1(u_P,v)$ with $v_A \le v \le v_B$. Something similar is valid for the interpolation of surface s_2.

Going back to a configuration similar to that in Figure 12.69, take a flow line f from emitter to receiver. In the example in Figure 12.72a this flow line crosses the center **E** of the emitter and center **R** of the receiver. It intersects surface s_1 at point $\mathbf{P}_1$ and surface s_2 at point $\mathbf{Q}_1$.

Point $\mathbf{P}_1$ defines two curves on s_1: g_P in the vertical direction and b_1 in the horizontal direction. Also, point $\mathbf{Q}_1$ defines two curves on s_2: g_Q in the vertical direction and e_1 in the horizontal direction. In this example, curves g_P and g_Q are on a symmetry plane of the system through points $\mathbf{E}_3$, $\mathbf{E}_4$, $\mathbf{R}_3$, and $\mathbf{R}_4$. We can now move along curves g_P and g_Q to find points $\mathbf{P}_2$ and $\mathbf{Q}_2$ that equal the étendues from $\mathbf{E}_3\mathbf{E}_4$ to $\mathbf{P}_1\mathbf{P}_2$, from $\mathbf{P}_1\mathbf{P}_2$ to $\mathbf{Q}_1\mathbf{Q}_2$ and from $\mathbf{Q}_1\mathbf{Q}_2$ to $\mathbf{R}_3\mathbf{R}_4$. This is similar to what was done in the 2-D case in Figure 12.24. Point $\mathbf{P}_2$ defines a horizontal curve b_2 and point $\mathbf{Q}_2$ defines a horizontal curve e_2. We take curves b_1 and b_2 as the boundaries of a Köhler lenticulation on surface s_1 and curves e_1 and e_2 as the boundaries of the corresponding lenticulation on surface s_2.

An alternative way to calculate the approximate positions of points $\mathbf{P}_2$ and $\mathbf{Q}_2$ is to use a method similar to that in Figure 12.20, but now in 3-D. The wavefronts used in the calculation are now $\mathbf{E}_3$, $\mathbf{E}_4$ and $\mathbf{R}_3$, $\mathbf{R}_4$. Mid point **E** is

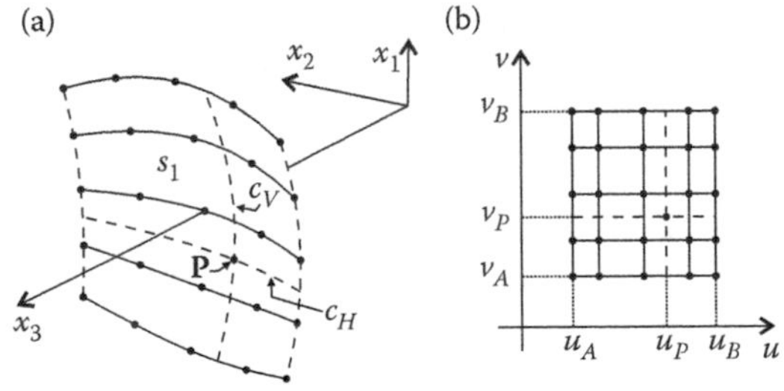

FIGURE 12.71
(a) Surface s_1. A point **P** defines a horizontal curve c_H and a vertical curve c_V on s_1. (b) Parameter space of surface s_1 parameterization.

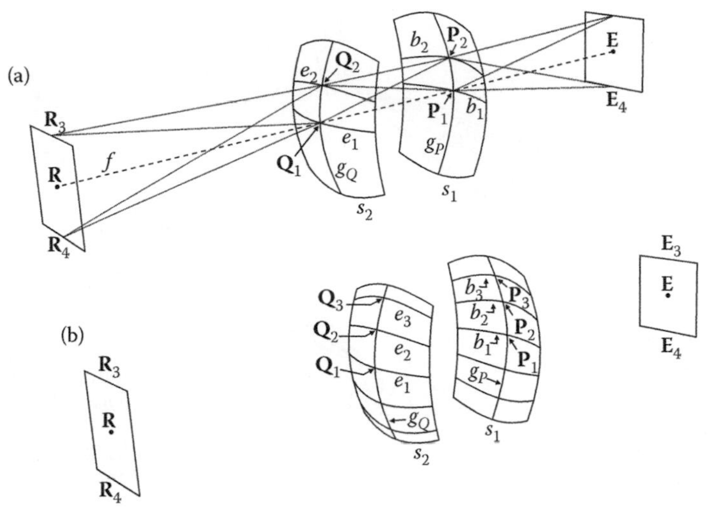

FIGURE 12.72
(a) A flow line f intersects surface s_1 at point $\mathbf{P}_1$ and surface s_2 at point $\mathbf{P}_2$. New points $\mathbf{P}_2$ and $\mathbf{Q}_2$ may be obtained by étendue conservation through $\mathbf{E}_3\mathbf{E}_4$, $\mathbf{P}_1\mathbf{P}_2$, $\mathbf{Q}_1\mathbf{Q}_2$, and $\mathbf{R}_3\mathbf{R}_4$. Curves b_1 through $\mathbf{P}_1$ and b_2 through $\mathbf{P}_2$ define the edges of a Köhler lenticulation on s_1. Curves e_1 through $\mathbf{Q}_1$ and e_2 through $\mathbf{Q}_2$ define the edges of a Köhler lenticulation on s_2. (b) The same process may be repeated calculating curves b_1, b_2, b_3, ... on s_1 and curves e_1, e_2, e_3, ... on s_2 defining the boundaries of other Köhler channels.

obtained as $\mathbf{E} = (\mathbf{E}_3 + \mathbf{E}_4)/2$ and mid point $\mathbf{R}$ is obtained as $\mathbf{R} = (\mathbf{R}_3 + \mathbf{R}_4)/2$. Referring then to Figure 12.72a, the flow lines may be approximated by the rays emitted from $\mathbf{E}$. Select a point $\mathbf{P}_1$ on s_1 determining curve b_1. Propagate a ray from $\mathbf{E}$ through point $\mathbf{P}_1$ of surface s_1. It impacts s_2 at point $\mathbf{Q}_1$ determining curve e_1. Consider a ray from $\mathbf{E}_3$ and refract it at point $\mathbf{P}_1$ of curve b_1. The normal $\mathbf{n}_P$ to the Köhler lenticulation at point $\mathbf{P}_1$ will be perpendicular to curve b_1. We rotate $\mathbf{n}_P$ around b_1 until the refracted ray from $\mathbf{E}_3$ intersects line e_1. This defines the orientation of $\mathbf{n}_P$. Now take a ray from $\mathbf{E}_4$ and refract it at $\mathbf{P}_1$ with normal $\mathbf{n}_P$. This ray impacts s_2 at point $\mathbf{Q}_2$, determining curve e_2. Take a ray coming from $\mathbf{R}$, refract it on s_2 at $\mathbf{Q}_2$ and intersect it with s_1, determining the position of $\mathbf{P}_2$ (in general, this will approximate the path of the flow line through $\mathbf{Q}_2$). Point $\mathbf{P}_2$ determines curve b_2.

We may now calculate the right lenticulation (on surface s_1) as a Cartesian oval surface that, for example, focuses point $\mathbf{E}_4$ onto curve e_2 and contains curve b_1. The optical surface thus calculated will not, in general, end at curve b_2 and there will be a small gap there. The result of this calculation is shown in Figure 12.73a.

First, take a point $\mathbf{P}$ on b_1 and calculate the path of ray $\mathbf{E}_4$–$\mathbf{P}$–$\mathbf{X}$, as in Figure 12.59. This defines a cone of rays at point $\mathbf{X}$ on curve e_2, as in Figure 12.61. This cone is shown diagrammatically in Figure 12.74. Curve b_2 intersects this cone at point $\mathbf{A}$. Points $\mathbf{P}$ and $\mathbf{X}$ define the direction of vector $\mathbf{u}$. Points $\mathbf{A}$ and $\mathbf{X}$

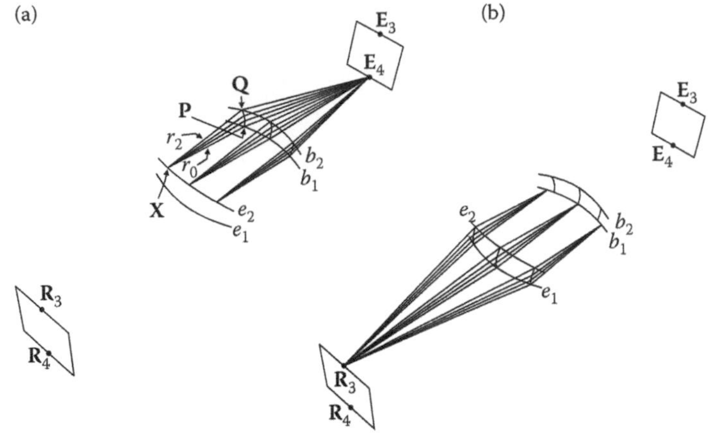

FIGURE 12.73
(a) The right lenticulation is calculated, for example, as a Cartesian oval surface that focuses $\mathbf{E}_4$ onto line e_2. The surface contains, for example, curve b_1. (b) The left lenticulation is calculated, for example, as a Cartesian oval surface that focuses $\mathbf{R}_3$ onto line b_1. The surface contains, for example, curve e_1.

define the direction of vector $\mathbf{u}_A$. Curve **P–Q** on the Cartesian oval in Figure 12.73a is calculated by emitting rays from **X** and imposing a constant optical path length from **X** to $\mathbf{E}_4$. These rays emitted from **X** start with direction **u**, then rotate around vector $\mathbf{t}_X$ (tangent to e_2 at **X**) to direction $\mathbf{u}_A$, spanning an angle γ around $\mathbf{t}_X$. Angle γ is that between vectors $\mathbf{v} = \mathbf{P} - \mathbf{T}$ and $\mathbf{v}_A = \mathbf{A} - \mathbf{T}_A$, where $\mathbf{T} = \mathbf{X} + (\mathbf{P} - \mathbf{X}) \cdot \mathbf{t}_X\mathbf{t}_X$ and $\mathbf{T}_A = \mathbf{X} + (\mathbf{A} - \mathbf{X}) \cdot \mathbf{t}_X\mathbf{t}_X$. Points **T** and $\mathbf{T}_A$ are the projections of points **P** and **A** onto the line defined by point **X** and unit vector

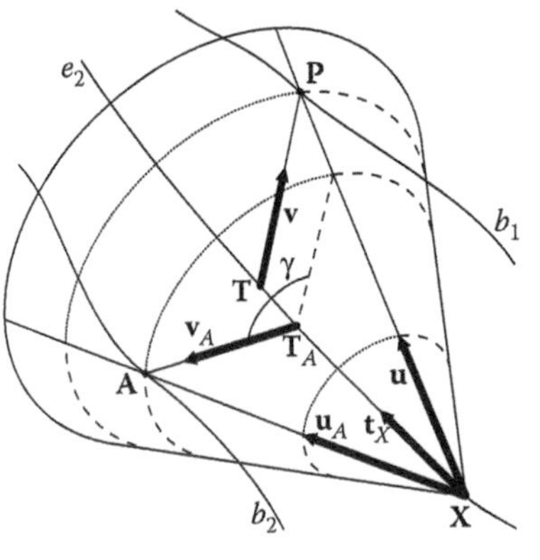

FIGURE 12.74
Curve **PQ** on the right lenticulation of Figure 12.73a is obtained by constant optical path length from **X** to $\mathbf{E}_4$. The points on curve PQ are obtained for rays emitted from **X**. The emission directions rotate from **u** to $\mathbf{u}_A$ around $\mathbf{t}_X$ and span an angle γ defined by curves b_1 and b_2.

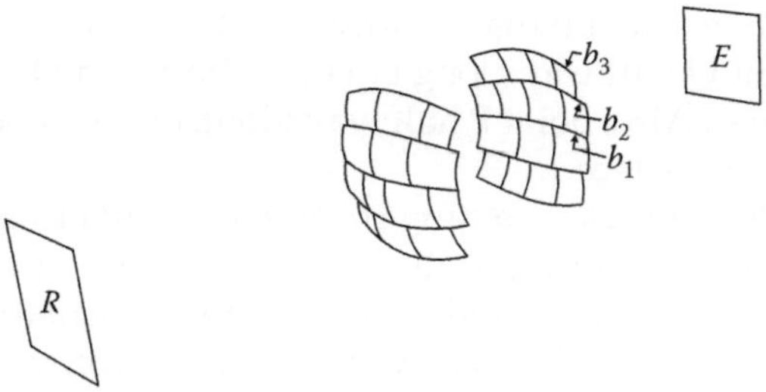

FIGURE 12.75
Köhler optic coupling emitter E with receiver R.

$\mathbf{t}_X$. Moving point **P** along curve b_1 and repeating the same process, we calculate new curves on the lenticulation.

Acting similarly, we may now calculate the left lenticulation (on surface s_2) as, for example, a Cartesian oval surface that focuses point $\mathbf{R}_3$ onto curve b_1 and contains curve e_1. The optical surface thus calculated will not, in general, end at curve e_2 and there will be a small gap there. The result of this calculation is shown in Figure 12.73b.

New lenticulations may now be calculated between edge curves b_2 and b_3 on s_1 and between edge curves e_2 and e_3 on s_2 shown in Figure 12.72b. Using the same procedure, new lenticulations may also be added below points $\mathbf{P}_1$ and $\mathbf{Q}_1$. The result is shown in Figure 12.75.

Figure 12.76 shows some illustrative rays emitted from a point $\mathbf{E}_4$ in the emitter. After crossing the optic, they reach the middle of receiver R, approximately along a vertical line.

If the emission point was to move vertically along line $\mathbf{E}_4\mathbf{E}_3$, the irradiance pattern on the receiver would remain (almost) unchanged. If, however, the emission point moved horizontally along direction $\mathbf{E}_1\mathbf{E}_2$, the illuminated dashed line on R would also move horizontally (in the opposite direction to the movement of the emission point). This is, therefore, a one-directional integrator, as it integrates only in one direction $\mathbf{E}_4\mathbf{E}_3$ and not in the other direction $\mathbf{E}_1\mathbf{E}_2$.

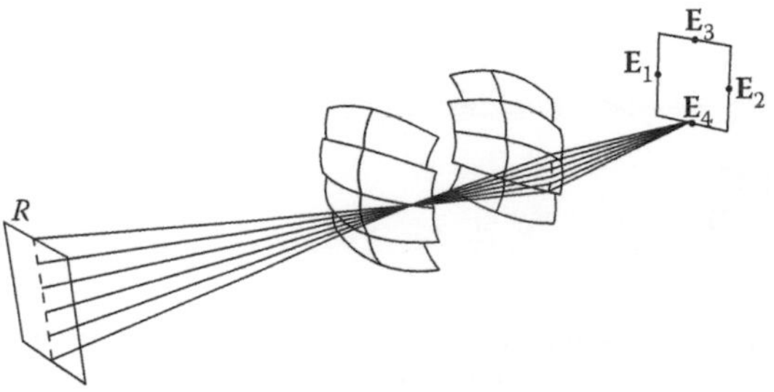

FIGURE 12.76
Raytracing of a Köhler integrator based on an SMS 3-D optic.

Figure 12.77a shows an alternative way to calculate the lenticulations of a Köhler optic. Point **Q** half way along curve g_Q between $\mathbf{Q}_1$ and $\mathbf{Q}_2$ defines a curve e on surface s_2. Also, point **P** half way along curve g_P between $\mathbf{P}_1$ and $\mathbf{P}_2$ defines a curve b on surface s_1.

Note that the dotted rays r_E coming from wavefronts $\mathbf{E}_3$ and $\mathbf{E}_4$ and crossing at the center curve b of the right lenticulation are approximately redirected by s_1 to edge curves e_1 and e_2 of the left lenticulation, as shown in Figure 12.77b. Also, the solid rays r_R coming from wavefronts $\mathbf{R}_3$ and $\mathbf{R}_4$ and crossing at the center curve e of the left lenticulation are approximately redirected by s_2 to edge curves b_1 and b_2 of the right lenticulation. This is similar to what is shown in Figure 12.65b, but now in 3-D geometry.

The lenticulations may now be calculated as focusing center **E** of the emitter onto curve e and center **R** of the receiver onto b, as in Figure 12.78.

Figure 12.79 shows an optic calculated with these lenticulations further extended, both horizontally and vertically.

Two-directional integrators are also possible. Examples include the shell mixer or the Fresnel–R Köhler (FK).

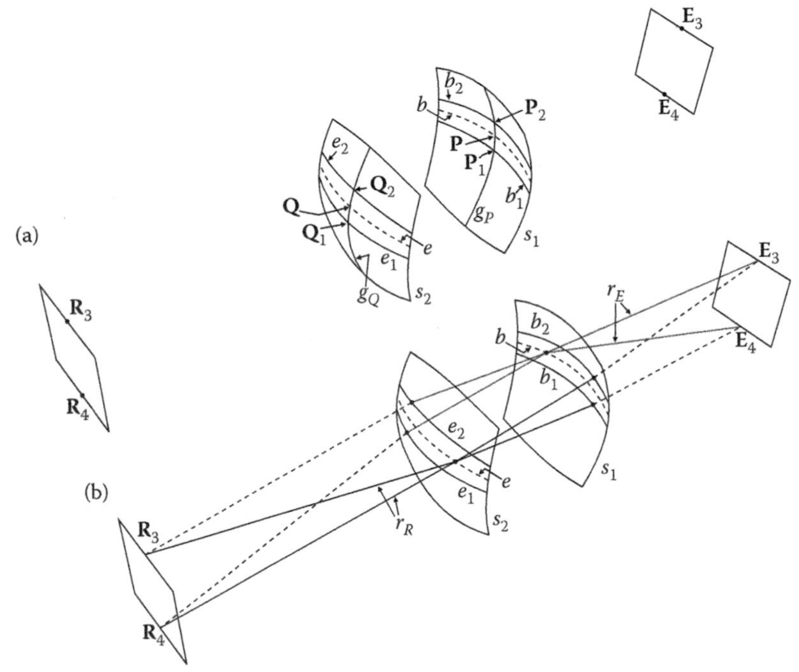

FIGURE 12.77
(a) Point **P** between $\mathbf{P}_1$ and $\mathbf{P}_2$ defines a curve b that may be used as the focus of the left lenticulation between curves e_1 and e_2. Point **Q** between $\mathbf{Q}_1$ and $\mathbf{Q}_2$ defines a curve e that may be used as the focus of the right lenticulation between curves b_1 and b_2. (b) Rays from $\mathbf{R}_3$ and $\mathbf{R}_4$ crossing mid curve e are approximately redirected to edge curves b_1 and b_2. Rays from $\mathbf{E}_3$ and $\mathbf{E}_4$ crossing mid curve b are approximately redirected to edge curves e_1 and e_2.

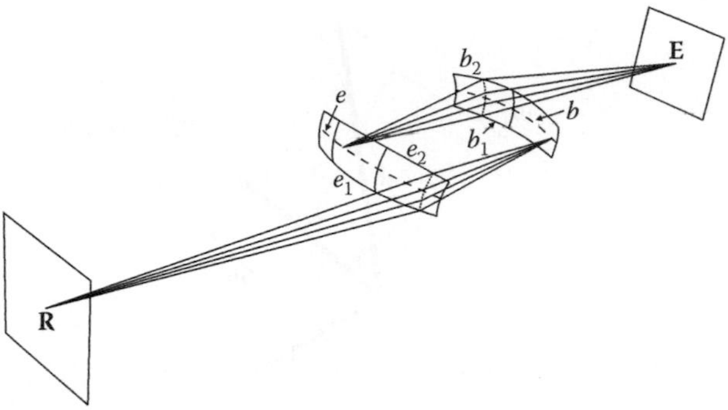

FIGURE 12.78
The right lenticulation is a Cartesian oval surface that focuses **E** onto *e*. The left lenticulation is a Cartesian oval surface that focuses **R** onto *b*.

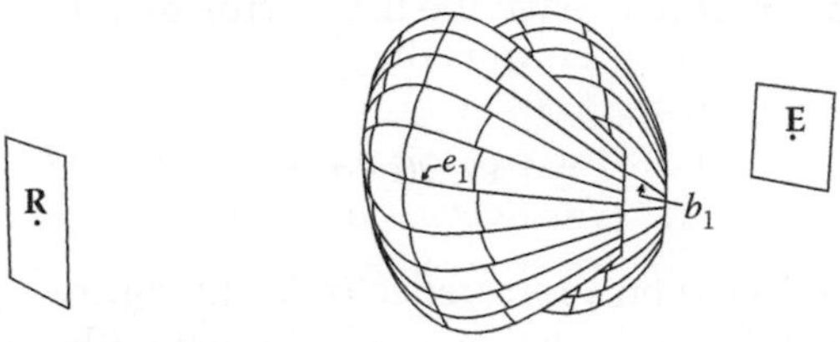

FIGURE 12.79
One-directional integrator (Köhler) optic.

12.7 Color-Mixing with Grooved Reflectors

The optics described in this section are not Köhler optics. However, they too are able to mix light from different light sources, similar to what Köhler optics can do.

A linear retroreflector is composed of two flat mirrors m_1 and m_2 at a right angle to each other, forming a V-groove reflector, as shown in Figure 12.80. The retroreflector is characterized by a local coordinate system with vectors **s**, **t**, and **n**. The edge, where mirrors m_1 and m_2 meet points in the direction of the unit, tangent vector **t**. Unit normal vector **n** is perpendicular to **t** and points in the direction of the bisector to the 90° angle that the two mirrors make with each other. Unit vector **s** is perpendicular to both **t** and **n**.

A ray r hitting the retroreflector will be twice reflected, first by mirror m_1 and then by mirror m_2, and its path is shown in top and side views in Figure 12.81.

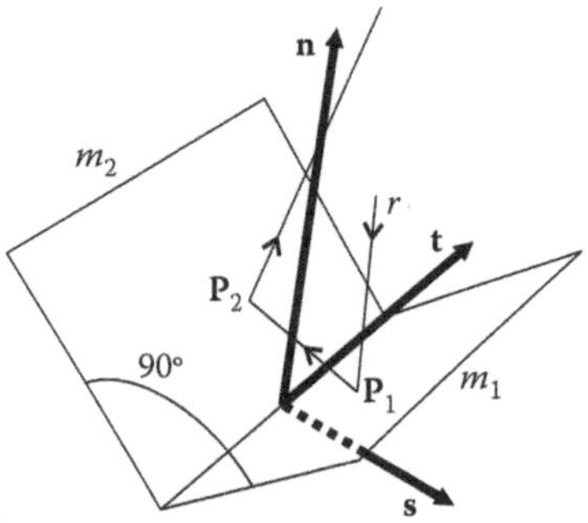

FIGURE 12.80
A retroreflector (V-groove reflector) is composed of two mirrors, m_1 and m_2, at a right angle to each other.

Let us now consider an incident ray with direction **i** reflected on a flat mirror with normal **n**. It will leave the flat mirror in a direction **r** given (see Chapter 16) by

$$\mathbf{r} = \mathbf{i} - 2(\mathbf{i} \cdot \mathbf{n})\mathbf{n} \tag{12.10}$$

This same ray, if reflected on a retroreflector (as in Figure 12.80) will leave in direction $\mathbf{r}_R$ symmetrical to **r** relative to the plane μ with normal **s**, as shown in Figure 12.82.

Therefore, the retroreflected ray $\mathbf{r}_R$ may be seen as twice reflected: once on the plane with normal **n** and another on the plane with normal **s**, resulting in[10]

$$\mathbf{r}_R = \mathbf{i} - 2(\mathbf{i} \cdot \mathbf{n})\mathbf{n} - 2(\mathbf{i} \cdot \mathbf{s})\mathbf{s} \tag{12.11}$$

The property of retroreflection shown in Figure 12.82 may be used to design collimator optics that perform color-mixing.

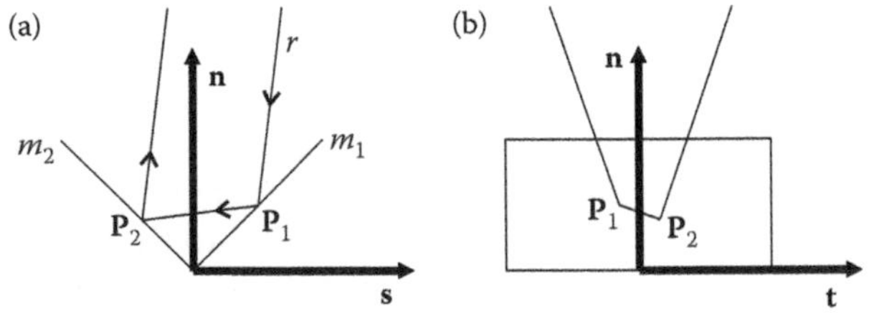

FIGURE 12.81
Top (a) and side (b) views of a ray *r* reflected on a retroreflector composed of two mirrors, m_1 and m_2, at a right angle to each other.

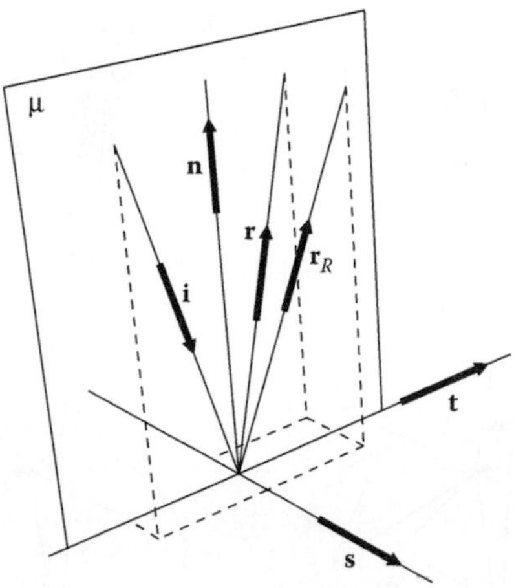

FIGURE 12.82
Incident ray direction **i**. If reflected on a flat mirror with normal **n**, it leaves in direction **r**. If retroreflected, it leaves in direction $\mathbf{r}_R$.

More complex V-groove reflectors may also be designed with free-form walls.[11,12]

Figure 12.83 shows a smooth elliptical mirror *m* with foci **R** and **F**. It focuses to point **F** all rays emitted from **R**, as in Figure 12.83a. Also, off-axis rays coming from a point **P** are reflected to a point **Q**, which is an image of **P**, as in Figure 12.83b.

This smooth mirror *m* may now be made into a grooved reflector by adding radial 90° V-shaped grooves to it, resulting in the reflector in Figure 12.84.

We now compare the behavior of the smooth reflector in Figure 12.83 with that of the grooved reflector in Figure 12.84. A ray emitted from off-axis point **P** towards a point **M** on the optic is reflected by a smooth mirror *m* to point **Q** (as in Figure 12.83b). However, a grooved reflector m_G will reflect this same ray at point **M** to another point **T**, symmetrical to **Q** relative to radial (meridian) plane μ through point **M**, as shown in Figure 12.85. Plane μ intersects mirror *m* at curve *c*. The V-groove reflector through point **M** follows radial curve *c*.

In general, when ray from **P** to **M** is reflected by a radial retroreflector, if point **M** and plane μ rotate by an angle φ to axis x_1 (around axis x_3), point **Q** rotates by an angle 2φ (around axis x_3) to point **T**. In particular, as point **M** rotates by an angle π around the optical axis of a grooved reflector m_G, point **Q** rotates by an angle 2π completing a circle around the optical axis, as shown in Figure 12.86a, forming a ring-spot.[13] Also, the light emitted by a radial line *R* generates an illuminated disk *D*, as shown in Figure 12.86b.

Suppose now that we use this grooved reflector optic with a light source composed of four emitters *R*, *G*, *G*, *B*, as in Figure 12.87. These four emitters may be of different colors, say one is red, two are green, and one is blue.

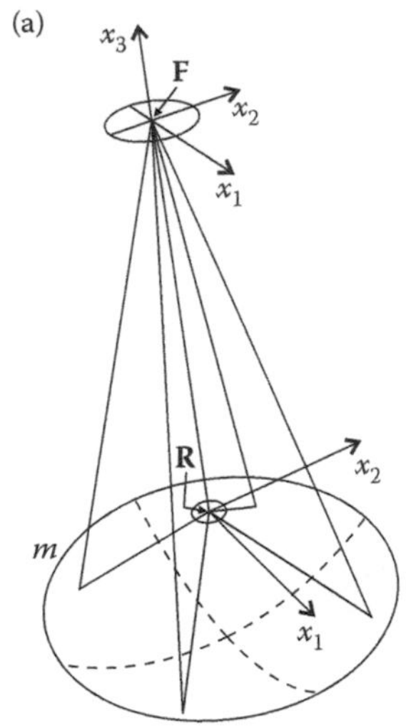

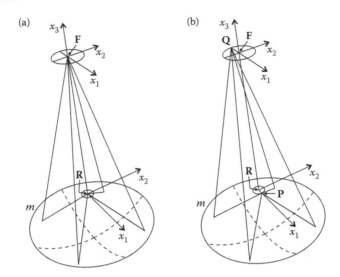

FIGURE 12.83
(a) Elliptical mirror m focuses point **R** onto point **F**. (b) An off-axis point **P** is focused by m onto point **Q**.

The R emitter generates a red disk D. Also, the G emitters generate green disks D and the B emitter generates a blue disk D. All these disks of different colors superimpose, generating white light, resulting in color-mixing.

This same principle may be used to design a more practical collimator, based on an SMS RXI optic.[14,15] Figure 12.88 shows a cross section of an RXI optic with a refractive central lens. This central lens spreads the light as it exits the optic, generating a halo around the central spot of angular aperture 2θ generated by the RXI. Also, the lens eliminates the need for a mirrored central portion on the top surface of the RXI (see Chapter 9).

Figure 12.89 shows the device in Figure 12.88, with a grooved mirror on the back.

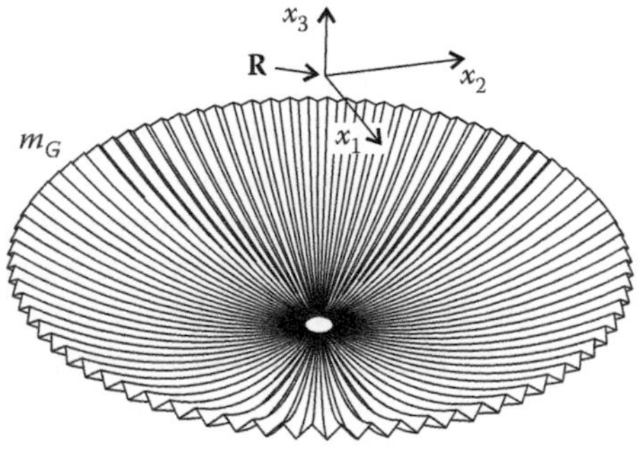

FIGURE 12.84
Grooved mirror m_G with radial retroreflector grooves.

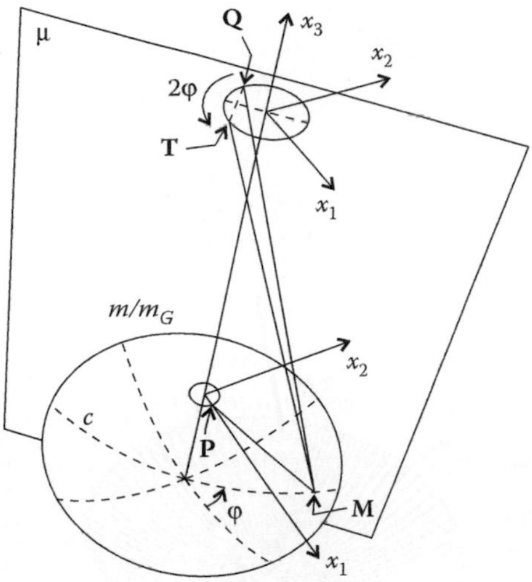

FIGURE 12.85
A ray emitted from **P** towards **M** is reflected by a smooth mirror m to point **Q**. However, a radial retroreflector groove along radial curve c (of a grooved mirror m_G) sends this same ray to point **T** symmetrical to **Q** relative to plane μ along the retroreflector groove.

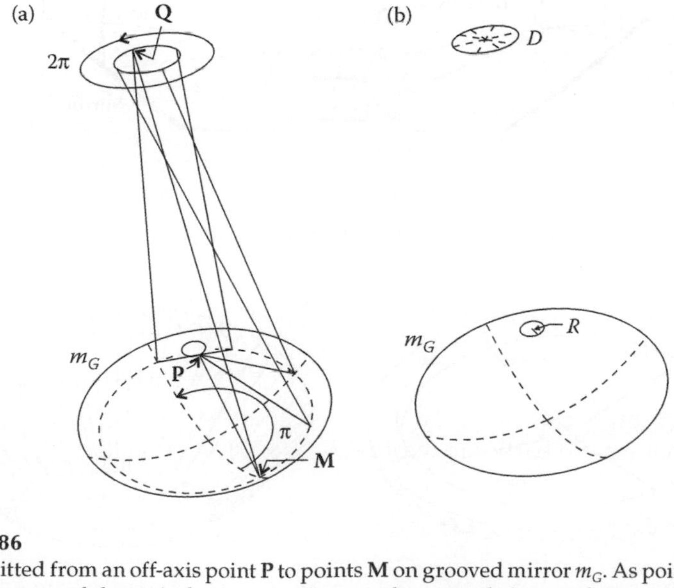

FIGURE 12.86
(a) Light emitted from an off-axis point **P** to points **M** on grooved mirror m_G. As point **M** rotates by an angle π around the optical axis, point **Q** rotates by an angle 2π completing a circle around the optical axis. (b) Light emitted by a radial line R generates a disk D.

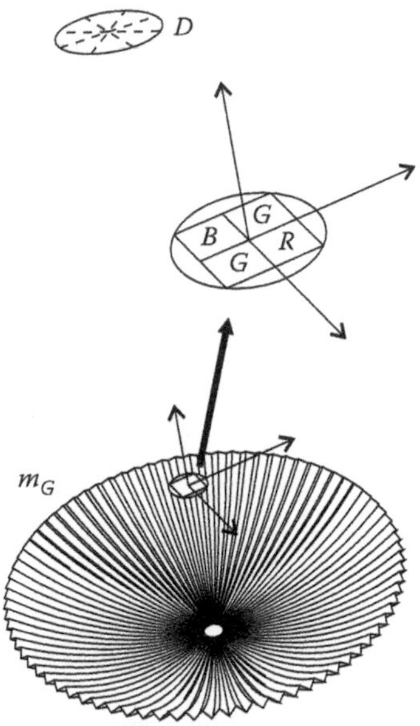

FIGURE 12.87
An emitter R generates a disk D. The same is true for emitters G and B. If these have different colors (say red, green, and blue), these (red, green, and blue) disks superimpose, generating white light.

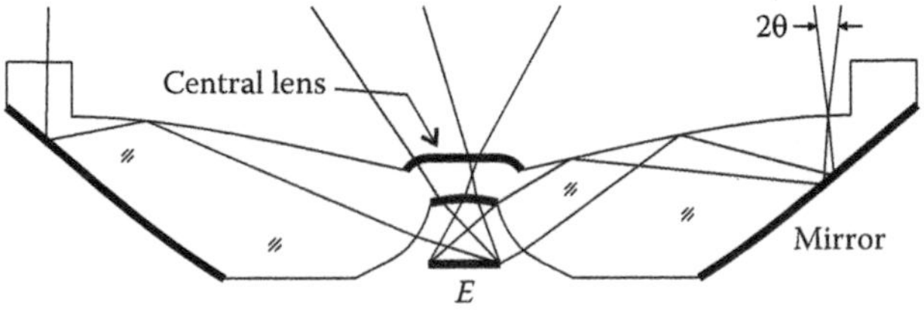

FIGURE 12.88
Cross-section of an RXI with a refractive central lens.

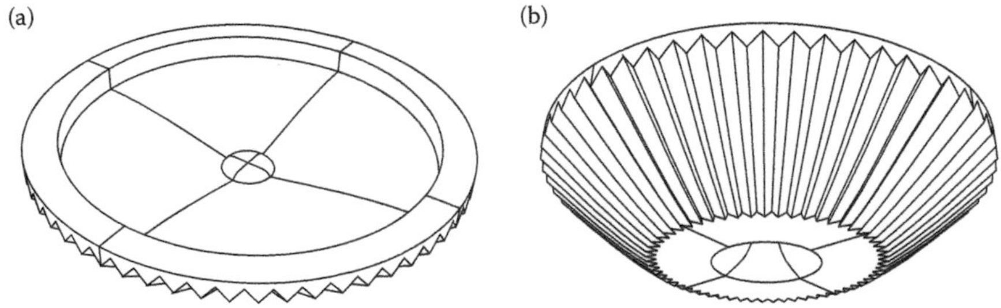

FIGURE 12.89
(a) Top and (b) bottom view of a grooved RXI.

(a)

(b)

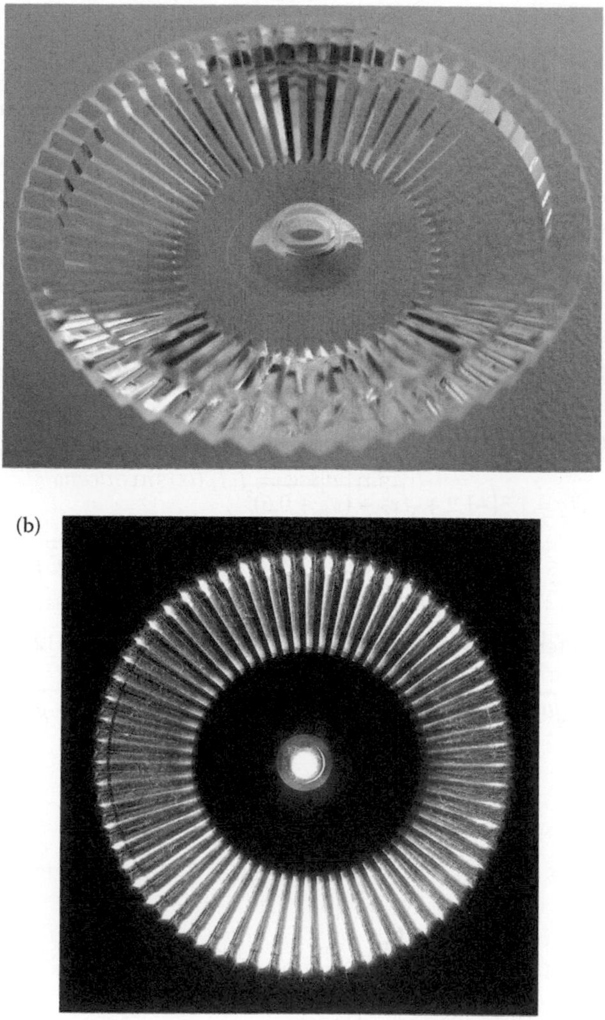

FIGURE 12.90
(a) Grooved RXI optic. (b) Lit grooved RXI optic. (Courtesy of Light Prescriptions Innovators.)

Figure 12.90a shows a grooved RXI optic. Figure 12.90b shows the lit RXI in a luminaire.

12.8 Examples

The following examples use expressions for the curves and functions that are derived in Chapter 21.

EXAMPLE 12.1

Design a Köhler integrator lens for an emitter with edge points $\mathbf{E}_1 = (-0.4,2)$, $\mathbf{E}_2 = (0.4,2)$, receiver edge points $\mathbf{R}_1 = (-0.5,-2.5)$, $\mathbf{R}_2 = (0.5,-2.5)$, and refractive index $n = 1.5$. The starting points for the Köhler channels are $\mathbf{P}_1 = (0,0.6)$ and $\mathbf{Q}_1 = (0,-0.6.)$, as shown in Figure 12.91.

We now calculate the positions of points $\mathbf{P}_2 = (x_P,y_P)$ and $\mathbf{Q}_2 = (x_Q,y_Q)$ at the other edges of the Köhler channel. We may now determine the positions of points $\mathbf{E} = (0,2)$ and $\mathbf{R} = (0,-2.5)$. Also, points $\mathbf{P}$ and $\mathbf{Q}$ may be obtained as functions of the unknowns x_P, y_P, x_Q, y_Q as $\mathbf{P} = 0.5(\mathbf{P}_1 + \mathbf{P}_2) = (0.5x_P,0.5(y_P + 0.6))$ and $\mathbf{Q} = (0.5x_Q,0.5(y_Q - 0.6))$.

The étendue from $\mathbf{E}_1\mathbf{E}_2$ to $\mathbf{P}_1\mathbf{P}_2$ is given by Equation 12.3,

$$U_1 = \sqrt{(-0.4 - x_P)^2 + (2 - y_P)^2} - \sqrt{(x_P - 0.4)^2 + (y_P - 2)^2} \tag{12.12}$$

The étendue from $\mathbf{P}_1\mathbf{P}_2$ to $\mathbf{Q}_1\mathbf{Q}_2$ is given by Equation 12.4,

$$\begin{aligned} U_2 = 1.5\Big(&-1.2 + \sqrt{x_P^2 + (y_P + 0.6)^2} \\ &+ \sqrt{x_Q^2 + (y_Q - 0.6)^2} - \sqrt{(x_Q - x_P)^2 + (y_Q - y_P)^2}\Big) \end{aligned} \tag{12.13}$$

and the étendue from $\mathbf{Q}_1\mathbf{Q}_2$ to $\mathbf{R}_1\mathbf{R}_2$ is given by Equation 12.5,

$$U_3 = -\sqrt{(0.5 - x_Q)^2 + (-2.5 - y_Q)^2} + \sqrt{(0.5 + x_Q)^2 + (2.5 + y_Q)^2} \tag{12.14}$$

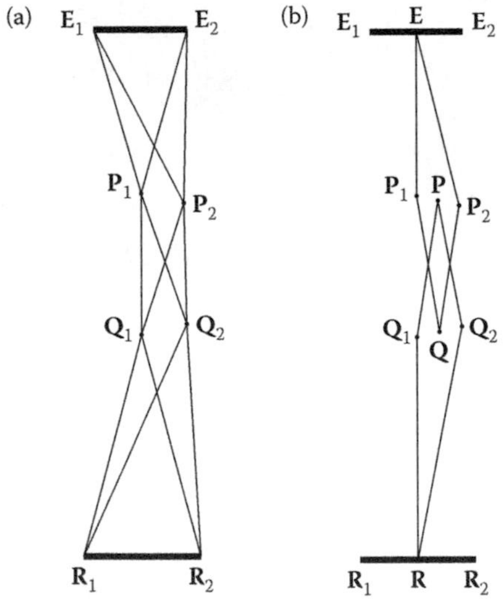

FIGURE 12.91
Given emitter edges $\mathbf{E}_1$ and $\mathbf{E}_2$, receiver edges $\mathbf{R}_1$ and $\mathbf{R}_2$, and starting points $\mathbf{P}_1$ and $\mathbf{Q}_1$, at the edges of a Köhler channel, calculate the other edges of the integrator channel using both (a) conservation of étendue and (b) constant optical path length.

Equations 12.1 and 12.2 for the optical path length are given by

$$\begin{aligned}1.4 + 1.5\sqrt{0.25x_Q^2 + 0.25(y_Q - 1.8)^2}\\ = \sqrt{x_P^2 + (y_P - 2)^2} + 1.5\sqrt{(x_P - 0.5x_Q)^2 + (0.3 + y_P - 0.5y_Q)^2}\end{aligned} \tag{12.15}$$

and

$$\begin{aligned}1.9 + 1.5\sqrt{0.25x_P^2 + 0.25(y_P + 1.8)^2}\\ = \sqrt{x_Q^2 + (y_Q + 2.5)^2} + 1.5\sqrt{(x_Q - 0.5x_P)^2 + (y_Q - 0.3 - 0.5y_P)^2}\end{aligned} \tag{12.16}$$

respectively. Solving Equation 12.6 with the expressions in (12.12), (12.13), and (12.14), together with Equations 12.15 and 12.16, we get the values of the four unknowns: $x_P = 0.365958$, $y_P = 0.518326$, $x_Q = 0.387795$, and $y_Q = -0.509867$. This gives us the positions of points $\mathbf{P}_2 = (0.365958, 0.518326)$ and $\mathbf{Q}_2 = (0.387795, -0.509867)$.

Points **P** and **Q** are now given by $\mathbf{P} = 0.5(\mathbf{P}_1 + \mathbf{P}_2) = (0.182979, 0.559163)$ and $\mathbf{Q} = 0.5(\mathbf{Q}_1 + \mathbf{Q}_2) = (0.193898, -0.554934)$.

We may now calculate the shape of the refractive surface (Cartesian oval) linking $\mathbf{P}_1$ and $\mathbf{P}_2$ as $a_1 = \text{cco}(\mathbf{Q}, n, \mathbf{E}, 1, \mathbf{P}_1)$ or

$$\begin{aligned}a_1 = (&-0.193898 + K_1 \cos(1.49505 + \phi),\\ &-0.554934 + K_1 \sin(1.49505 + \phi))\end{aligned} \tag{12.17}$$

with

$$\begin{aligned}K_1 = {} & 3.78797 - 2.04982 \cos\phi\\ & - 0.8\sqrt{21.4537 - 24.2646 \cos\phi + 3.28264 \cos(2\phi)}\end{aligned} \tag{12.18}$$

with $\phi_1 \le \phi \le \phi_2$ where $\phi_1 = \text{angpn}(\mathbf{P}_1 - \mathbf{Q}, \mathbf{E} - \mathbf{Q}) = -0.0905889$ and $\phi_2 = \text{angpn}(\mathbf{P}_2 - \mathbf{Q}, \mathbf{E} - \mathbf{Q}) = 0.234709$.

The shape of the refractive surface (Cartesian oval) linking $\mathbf{Q}_1$ and $\mathbf{Q}_2$ is given by $a_2 = \text{cco}(\mathbf{P}, n, \mathbf{R}, \mathbf{Q}_1)$ or

$$\begin{aligned}a_2 = (&-0.182979 + K_2 \cos(4.77213 + \phi),\\ &0.559163 + K_2 \sin(4.77213 + \phi))\end{aligned} \tag{12.19}$$

with

$$\begin{aligned}K_2 = {} & 4.39233 - 2.4517 \cos\phi\\ & - 0.8\sqrt{29.8335 - 33.6522 \cos\phi + 4.69598 \cos(2\phi)}\end{aligned} \tag{12.20}$$

with $\phi_1 \le \phi \le \phi_2$ where $\phi_1 = \text{angpn}(\mathbf{Q}_1 - \mathbf{P}, \mathbf{R} - \mathbf{P}) = 0.0968203$ and $\phi_2 = \text{angpn}(\mathbf{Q}_2 - \mathbf{P}, \mathbf{R} - \mathbf{P}) = -0.249039$.

The other Köhler channels are calculated using the same procedure. A complete device is similar to that shown in Figure 12.18.

References

1. Miñano, J. C., Free-form integrator array optics, *Proc. SPIE 5942, Nonimaging Optics and Efficient Illumination Systems II*, September 2, 2005.
2. Benitez, P. et al., High-concentration mirror-based Köhler integrating system for tandem solar cells, *Photovoltaic Energy Conversion, IEEE 4th World Conference*, Vol. 1, pp. 690–693, 2006.
3. Dross, O. et al., Köhler integrators embedded into illumination optics add functionality, *Proc. SPIE 7103, Illumination Optics*, September 24, 2008.
4. Hernandez, M. et al., 1000× shadow-free mirror-based Köhler concentrator, *Proc. SPIE 8468, High and Low Concentrator Systems for Solar Electric Applications VII*, October 10, 2012.
5. Cvetkovic, A. et al., Primary optics for efficient high-brightness LED colour mixing, *Proc. SPIE 8485, Nonimaging Optics: Efficient Design for Illumination and Solar Concentration IX*, October 11, 2012.
6. Hernandez, M. et al., CPV and illumination systems based on XR-Köhler devices, *Proc. SPIE 7785, Nonimaging Optics: Efficient Design for Illumination and Solar Concentration VII*, August 18, 2010.
7. Hernandez, M. et al., High-performance Köhler concentrators with uniform irradiance on solar cell, *Proc. SPIE 7059, Nonimaging Optics and Efficient Illumination Systems V*, September 2, 2008.
8. Hecht, E., *Optics*, Addison-Wesley, San Francisco, CA, 2002.
9. Piegel, L. and Tiller, W., *The NURBS book*, Springer, Germany, 1997.
10. Buljan, M., *Free-Form Optical Systems for Nonimaging Applications*, PhD thesis, Technical University of Madrid, 2014.
11. Grabovickic, D., Benitez, P. and Miñano, J. C., Aspheric V-groove reflector design with the SMS method in two dimensions, *Opt. Express*, 18(3), 2515–2521, 2010.
12. Grabovickic, D., Benitez, P. and Miñano, J. C., Free-form V-groove reflector design with the SMS method in three dimensions, *Opt. Express*, 19(S4), A747–A756, 2011.
13. Benitez, P. et al., On the analysis of rotational symmetric microstructured surfaces, *Opt. Express*, 15(5), 2219, 2007.
14. Grabovickic, D. et al., Metal-less V-groove RXI collimator, *Proc. SPIE 8170, Illumination Optics II*, September 21, 2011.
15. Grabovickic, D. et al., Design, manufacturing, and measurements of a metal-less V-groove RXI collimator, *Proc. SPIE 8124, Nonimaging Optics: Efficient Design for Illumination and Solar Concentration VIII*, September 21, 2011.

13

The Miñano Design Method Using Poisson Brackets

13.1 Introduction

Nonimaging optics are usually designed as 2-D profiles that are then extruded to form a trough-like optic, or rotated to generate a rotational optic, or crossed to form a square cross-section optic. These optics are usually not ideal in 3-D geometry. The Miñano design method using Poisson brackets utilizes an extra degree of freedom to design ideal 3-D optics: a variable refractive index inside the optic. This enables, for example, the design of an ideal 3-D concentrator with flat entrance and exit apertures, acceptance angle θ, and maximum concentration.

13.2 Design of Two-Dimensional Concentrators for Inhomogeneous Media

Consider the calculation of the refractive index distribution that makes a given set of light rays on a plane possible. If we could calculate the optical path length along the rays, we could use the eikonal equation,

$$n^2 = \|\nabla S\|^2 \tag{13.1}$$

to calculate the refractive index n, except that to calculate S, we need to know n. Although we cannot calculate S, we can define a set of curves on the plane which are perpendicular to the light rays. These are given by $i(x_1, x_2) = C$, where C is a constant. For different values of C, different perpendicular lines to the rays are obtained. We then have $S(x_1, x_2) = S(i(x_1, x_2))$ or $S = S(i)$ and

$$\nabla S = \frac{dS}{di}\nabla i \tag{13.2}$$

∇i is known because the curves $i(x_1, x_2) = C$ were defined. Giving the function dS/di can be calculated using the eikonal equation. Substituting Equation 13.2 in Equation 13.1 we get

$$n^2 = \left(\frac{dS}{di}\right)^2 \|\nabla i\|^2 \tag{13.3}$$

Consider two sets of rays perpendicular to two wavefronts S_1 = constant and S_2 = constant propagating in a given medium of refractive index $n(x_1, x_2)$. For these two wavefronts, we can write

$$\begin{aligned} \mathbf{p}_1 &= \nabla S_1 \\ \mathbf{p}_2 &= \nabla S_2 \end{aligned} \tag{13.4}$$

or

$$\begin{aligned} \mathbf{p}_1 + \mathbf{p}_2 &= \nabla(S_1 + S_2) \\ \mathbf{p}_1 - \mathbf{p}_2 &= \nabla(S_1 - S_2) \end{aligned} \tag{13.5}$$

If the refractive index is the same for both wavefronts, then $\|\mathbf{p}_1\| = \|\mathbf{p}_2\| = n$ and the vector $\mathbf{p}_1 + \mathbf{p}_2$ points in the direction of the bisector of the two sets of rays.

We now have

$$\|\mathbf{p}_1 + \mathbf{p}_2\|^2 = (\mathbf{p}_1 + \mathbf{p}_2) \cdot (\mathbf{p}_1 + \mathbf{p}_2) = \|\mathbf{p}_1\|^2 + \|\mathbf{p}_2\|^2 + 2\mathbf{p}_1 \cdot \mathbf{p}_2 \tag{13.6}$$

or

$$2n^2 + 2\mathbf{p}_1 \cdot \mathbf{p}_2 = \|\nabla(S_1 + S_2)\|^2 \tag{13.7}$$

and

$$\|\mathbf{p}_1 - \mathbf{p}_2\|^2 = (\mathbf{p}_1 - \mathbf{p}_2) \cdot (\mathbf{p}_1 - \mathbf{p}_2) = \|\mathbf{p}_1\|^2 + \|\mathbf{p}_2\|^2 - 2\mathbf{p}_1 \cdot \mathbf{p}_2 \tag{13.8}$$

or

$$2n^2 - 2\mathbf{p}_1 \cdot \mathbf{p}_2 = \|\nabla(S_1 - S_2)\|^2 \tag{13.9}$$

Summing the two preceding equations gives

$$4n^2 = \|\nabla(S_1 + S_2)\|^2 + \|\nabla(S_1 - S_2)\|^2 \tag{13.10}$$

or

$$n^2 = \left\| \nabla\left(\frac{S_1 + S_2}{2}\right) \right\|^2 + \left\| \nabla\left(\frac{S_1 - S_2}{2}\right) \right\|^2 \tag{13.11}$$

We can also write

$$\nabla(S_1 + S_2) \cdot \nabla(S_1 - S_2) = (\mathbf{p}_1 + \mathbf{p}_2) \cdot (\mathbf{p}_1 - \mathbf{p}_2) = \|\mathbf{p}_1\|^2 - \|\mathbf{p}_2\|^2 = 0 \tag{13.12}$$

It can then be concluded that the light rays perpendicular to two given wavefronts can coexist in a medium of refractive index n if Equation 13.12 is fulfilled. In this case, the refractive index $n(x_1, x_2)$ can be calculated using Equation 13.11.

Consider that the two sets of light rays are the edge rays of radiation propagating in the medium. In this case, all that light propagates between the two given sets of rays. This is the same assumption used in the design of nonimaging optics by the flow-line method.

Now, having in consideration Equation 13.4, it can be seen that $\mathbf{p}_1 + \mathbf{p}_2$ points in the direction of the bisector of the edge rays, and therefore, in the direction of the vector flux **J** also, as seen in Figure 13.1. Consider the following functions:

$$\begin{aligned} G &= (S_1 - S_2)/2 \\ F &= (S_1 + S_2)/2 \end{aligned} \tag{13.13}$$

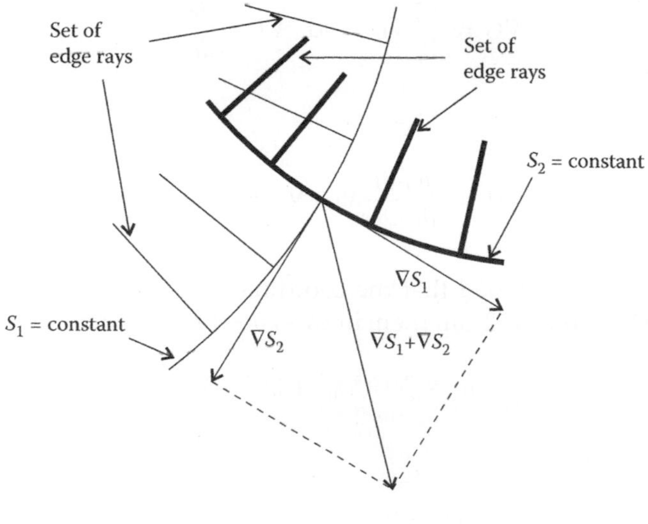

FIGURE 13.1
Vector $\mathbf{p}_1 + \mathbf{p}_2$ points in the direction of the bisector of the two sets of rays. Note that $\mathbf{p}_1 + \mathbf{p}_2$ points in the direction of $\nabla(S_1 + S_2) = \nabla S_1 + \nabla S_2$ and that $\|\nabla S_1\| = \|\nabla S_2\|$.

In this case, Equations 13.11 and 13.12 can be written as

$$n^2 = \|\nabla F\|^2 + \|\nabla G\|^2 \tag{13.14}$$

and

$$\nabla F \cdot \nabla G = 0 \tag{13.15}$$

Equation 13.15 tells us that the lines of constant G are tangent to the vector ∇F. Also, ∇F points in the direction of $\mathbf{p}_1 + \mathbf{p}_2$ and therefore, in the direction of the vector flux $\mathbf{J}$. We have seen (see Chapter 4) that the étendue is conserved between any two lines of constant G.

If we place mirrors along the flow-lines, the resulting optical system conserves étendue, and the existing set of light rays is unaltered. This is true because such a mirror transforms rays of one wavefront into rays of the another wavefront. Note that the vector flux lines bisect the rays of the two wavefronts.

Because $\nabla F \cdot \nabla G = 0$, the lines of $F = \text{constant}$ and $G = \text{constant}$ are perpendicular to each other. We can then define a new coordinate system $(i_1(x_1, x_2), i_2(x_1, x_2))$ on the plane, such that the lines $i_1 = \text{constant}$ coincide with the lines $G = \text{constant}$ and the lines $i_2 = \text{constant}$ coincide with the lines $F = \text{constant}$. In this case, we can write $G = G(i_1)$ and $F = F(i_2)$. Note that $G(x_1, x_2) = G(i_1(x_1, x_2))$ and $F(x_1, x_2) = F(i_2(x_1, x_2))$. From $G = G(i_1)$ we can see that $i_1 = \text{constant} \Rightarrow G = \text{constant}$ and from $F = F(i_2)$ we can see that $i_2 = \text{constant} \Rightarrow F = \text{constant}$. We also have

$$\nabla G = \frac{dG}{di_1}\nabla i_1 \quad \text{and} \quad \nabla F = \frac{dF}{di_2}\nabla i_2 \tag{13.16}$$

so that

$$\nabla G \cdot \nabla F = 0 \Leftrightarrow \frac{dG}{di_1}\frac{dF}{di_2}\nabla i_1 \cdot \nabla i_2 = 0 \Leftrightarrow \nabla i_1 \cdot \nabla i_2 = 0 \tag{13.17}$$

It can then be concluded that the coordinate system (i_1, i_2) is orthogonal. Equations 13.11 and 13.12 can then, in this coordinate system, be reduced to

$$n^2 = \|\nabla G(i_1)\|^2 + \|\nabla F(i_2)\|^2 \tag{13.18}$$

or

$$n^2 = \left(\frac{dG}{di_1}\right)^2 \|\nabla i_1\|^2 + \left(\frac{dF}{di_2}\right)^2 \|\nabla i_2\|^2 \tag{13.19}$$

since $\nabla F \cdot \nabla G = 0$ is already implicitly contained in the fact that we have $\nabla i_1 \cdot \nabla i_2 = 0$, that is, the coordinate system is orthogonal. Making $\alpha(i_1) = dG/di_1$ and $\beta(i_2) = dF/di_2$, and making $a_1 = \|\nabla i_1\|$ and $a_2 = \|\nabla i_2\|$ gives

$$n^2 = \alpha(i_1)^2 a_1{}^2 + \beta(i_2)^2 a_2{}^2 \tag{13.20}$$

An example of an ideal 2-D concentrator will be given in Section 13.7, which is based on this equation.

Note that when the two sets of edge rays are the same, we have $S_1 = S_2 = S$ and, therefore, $G = 0$. Accordingly, $F = (S_1 + S_2)/2 = S$. Therefore, Equation 13.19 simplifies to Equation 13.3 for the case of a single wavefront.

Note that given the functions F and G, expression (13.13) can be inverted to give S_1 and S_2.

$$\begin{aligned} S_1 &= F + G \\ S_2 &= F - G \end{aligned} \tag{13.21}$$

We now define the i_1-lines as those for which $i_2 =$ constant (in these lines only i_1 varies). Accordingly, we define the i_2-lines as those for which $i_1 =$ constant (in these lines only i_2 varies). The lines $G =$ constant (i_2-lines and $i_1 =$ constant) are the vector flux lines and bisect the edge rays. The lines $F =$ constant (i_1-lines and $i_2 =$ constant) are perpendicular to the vector flux lines.

13.3 Edge Rays as a Tubular Surface in Phase Space

Three-dimensional optical systems are described by the canonical Hamiltonian equations in which H is the Hamiltonian (see Chapter 14).

$$\begin{aligned} \frac{dx_1}{dx_3} &= \frac{\partial H}{\partial p_1} \qquad \frac{dp_1}{dx_3} = -\frac{\partial H}{\partial x_1} \\ \frac{dx_2}{dx_3} &= \frac{\partial H}{\partial p_2} \qquad \frac{dp_2}{dx_3} = -\frac{\partial H}{\partial x_2} \\ H &= -\sqrt{n^2 - p_1^2 - p_2^2} \end{aligned} \tag{13.22}$$

These equations can also be written in another form for the 3-D systems in which P is a new Hamiltonian for the system.

$$\begin{aligned}
&\frac{dx_1}{d\sigma} = \frac{\partial P}{\partial p_1} \quad \frac{dp_1}{d\sigma} = -\frac{\partial P}{\partial x_1} \\
&\frac{dx_2}{d\sigma} = \frac{\partial P}{\partial p_2} \quad \frac{dp_2}{d\sigma} = -\frac{\partial P}{\partial x_2} \\
&\frac{dx_3}{d\sigma} = \frac{\partial P}{\partial p_3} \quad \frac{dp_2}{d\sigma} = -\frac{\partial P}{\partial x_2} \\
&P = p_1^2 + p_2^2 + p_3^2 - n^2 = 0
\end{aligned} \tag{13.23}$$

Two-dimensional optical systems have one less dimension (along x_3) and are described by the canonical Hamiltonian equations in which H is the Hamiltonian.

$$\begin{aligned}
&\frac{dx_1}{dx_2} = \frac{\partial H}{\partial p_1} \quad \frac{dp_1}{dx_2} = -\frac{\partial H}{\partial x_1} \\
&H = -\sqrt{n^2 - p_1^2}
\end{aligned} \tag{13.24}$$

These equations can also be written in another form for the 2-D systems in which P is a new Hamiltonian for the system.

$$\begin{aligned}
&\frac{dx_1}{d\sigma} = \frac{\partial P}{\partial p_1} \quad \frac{dp_1}{d\sigma} = -\frac{\partial P}{\partial x_1} \\
&\frac{dx_2}{d\sigma} = \frac{\partial P}{\partial p_2} \quad \frac{dp_2}{d\sigma} = -\frac{\partial P}{\partial x_2} \\
&P = p_1^2 + p_2^2 - n^2 = 0
\end{aligned} \tag{13.25}$$

To understand the essential difference between imaging and nonimaging optics, it is instructive to analyze 2-D systems, from which the conclusions can then be extended to 3-D systems. For this presentation, the formalism defined by Equation 13.24 is more appropriate than that defined by Equation 13.25. The ray trajectories are then defined by the variables (x_1, x_2, p_1).

Suppose that we have an optical system with entrance aperture a_1 and exit aperture a_2, as presented in Figure 13.2. Further, suppose that the entrance aperture receives radiation with variable angular aperture from point to point.

In an imaging device, the objective is to transfer all the rays of light coming from a point with horizontal (x_1) coordinate X (object) independent of the angle of incidence to a point with horizontal (x_1) coordinate x (image). In an x_1p_1 plot, this objective is translated into transforming the vertical lines L_1 at the entrance aperture into vertical lines L_2 at the exit aperture.[1] Note that in an x_1p_1 plot, a vertical line $x_1 = x$, represents all the possible directions of

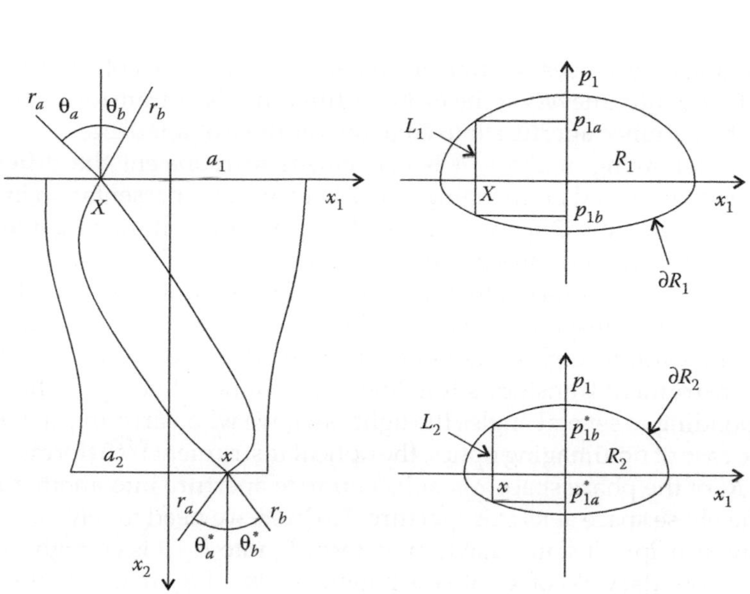

FIGURE 13.2
In an imaging optical system, the light coming from a point with horizontal (x_1) coordinate X at its entrance aperture a_1 must be concentrated onto a point with horizontal coordinate x at the exit aperture a_2. An imaging optical system transforms vertical lines in phase space at the entrance aperture of the device into vertical lines in phase space at the exit. A nonimaging system transforms line ∂R_1 at the entrance aperture into line ∂R_2 at the exit, that is, transforms the edge rays at the entrance into edge rays at the exit.

incidence of light rays at x. Mathematically, the relation between X and x can be written for an imaging system as

$$x = MX \tag{13.26}$$

where M is the magnification of the optical system as it tells us how many times the image is larger than the object. As can be seen, the angle of light (momentum $\mathbf{p}$) at x does not appear in this equation. This is because the incidence direction is not important. Only the size relations between image and object are important.

In the case of nonimaging optics, the approach is completely different. If the entrance aperture a_1 is illuminated by radiation with a given angular distribution, the optic must transfer this radiation to the exit a_2, and cause it to be emitted therefrom with a different angular distribution.[2]

At the entrance aperture of the device, line ∂R_1 in the x_1p_1 space represents the set of its edge rays, because the line corresponds to the extreme values of p_1 for each value of x_1. For example, for a point with horizontal coordinate X at the entrance aperture, the edge rays are represented by phase space points (X, p_{1a}) and (X, p_{1b}), which are on the line ∂R_1. At the exit aperture of the device, the edge rays define a line ∂R_2 in the phase space x_1p_1. What a nonimaging device does is it transforms the line ∂R_1 at the entrance aperture of the device into line ∂R_2 at the exit aperture, that is, it transforms the edge rays at the entrance aperture into edge rays at the exit aperture.[3]

For the following analysis, it is convenient to represent the differences between imaging and nonimaging optics in terms of representation in phase space, that is, in the 3-D space (x_1, x_2, p_1). Imaging and nonimaging optics can be represented in phase space in the form presented in Figure 13.3.[1]

In the imaging approach, to each point with horizontal (x_1) coordinate X of the object, a corresponding vertical line L_1 is in the plane x_1p_1, which represents the various angles that light can have when exiting X. An imaging optical instrument transforms this line L_1 into a new line L_2 on the image, corresponding to several angles the light can have when arriving at x from X.

In the case of nonimaging optics, the optical instrument transforms a given region R_1 of the phase space x_1p_1 at its entrance aperture into another region R_2 of the phase space at its exit aperture. To do so, we need to rely only on the edge ray principle. It states that to transform R_1 into R_2, it is enough to transform the boundary ∂R_1 of R_1 into the boundary ∂R_2 of R_2, that is, to transform the edge rays of R_1 into the edge rays of R_2. If the light rays from ∂R_1 are transformed into ∂R_2, then all the rays coming from R_1 will go through R_2.[4]

Transforming one boundary into the other can be made by joining them with a surface of the form

$$\omega(x_1, x_2, p_1) = 0 \tag{13.27}$$

as presented in Figure 13.4.[1,3,4] For example, a surface of the form $x_1^2 + p_1^2 = R^2$ would be a tube of radius R along axis x_2.

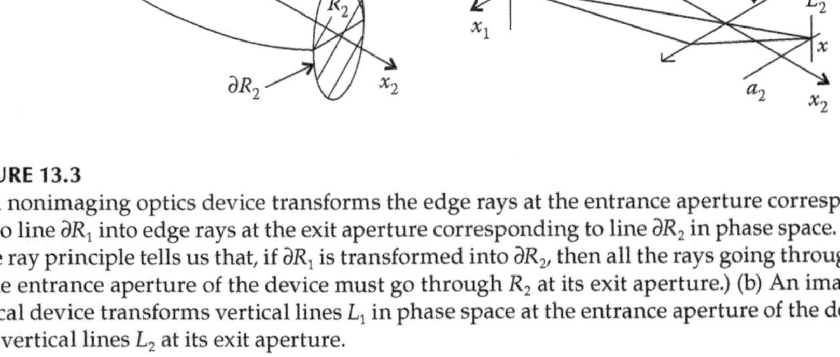

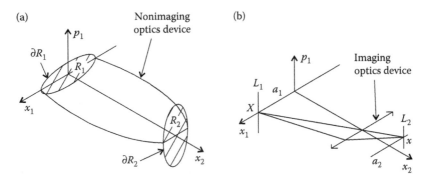

FIGURE 13.3
(a) A nonimaging optics device transforms the edge rays at the entrance aperture corresponding to line ∂R_1 into edge rays at the exit aperture corresponding to line ∂R_2 in phase space. (The edge ray principle tells us that, if ∂R_1 is transformed into ∂R_2, then all the rays going through R_1 at the entrance aperture of the device must go through R_2 at its exit aperture.) (b) An imaging optical device transforms vertical lines L_1 in phase space at the entrance aperture of the device into vertical lines L_2 at its exit aperture.

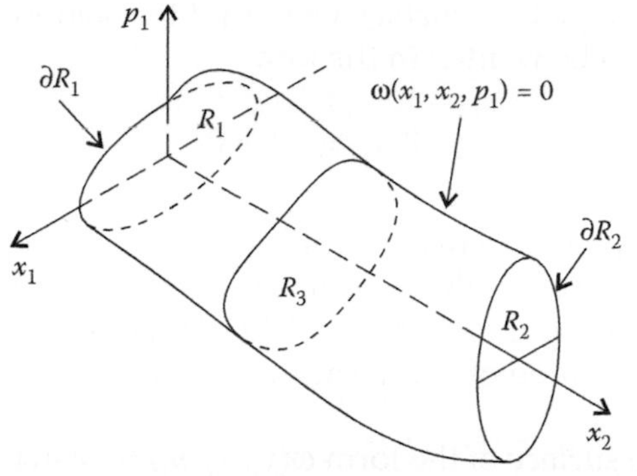

FIGURE 13.4
To guarantee that ∂R_1 is transformed into ∂R_2, the two lines can be connected by a surface of the form $\omega(x_1, x_2, p_1) = 0$. The trajectories of light on this surface in phase space correspond to trajectories of the edge rays inside the device. To guarantee that the étendue is conserved, it is necessary to guarantee that the area of any cut parallel to the plane x_1p_1 has an area R_3, which is constant and equal to R_1 and R_2.

Transformation of the edge rays at the entrance aperture into edge rays at the exit aperture guarantees that all rays of R_1 are transported to R_2. In fact, for a ray of R_1 not to pass through R_2, it would have to "escape" through the surface $\omega = 0$. Nonetheless, before it "escapes," it becomes a ray of $\omega = 0$. But the rays of $\omega = 0$ are transferred into the boundary of R_2 and, therefore, $\omega = 0$ does not allow the leakage of light rays from its interior.[5]

This cannot, however, be any surface. The area of region R_1 at the entrance aperture of the optical device equals the étendue there.[3,4] The same happens with the area of R_2 at the exit. By cutting between the entrance and the exit, the surface $\omega(x_1, x_2, p_1) = 0$ by planes parallel to the plane x_1p_1, we obtain the regions R_3, whose area correspond to the étendue along the device. To guarantee the conservation of étendue, it is necessary that areas of R_1, R_3, and R_2 are equal to one another.

The surface $\omega(x_1, x_2, p_1) = 0$ is made of the trajectories of the edge rays in phase space. Note that in the 2-D case analyzed here, we have

$$p_1 = n\cos\theta_1 = n\frac{dx_1}{\sqrt{dx_1^2 + dx_2^2}} = n\frac{x_1'}{\sqrt{1 + x_1'^2}} \tag{13.28}$$

with x_1' given by $x_1' = dx_1/dx_2$, and θ_1 is the angle the optical momentum makes to axis x_1, and (dx_1,dx_2) is an infinitesimal displacement along the path of the light ray. Expression $\omega(x_1, x_2, p_1) = 0$ can then be written as

$$\omega\left(x_1, x_2, \frac{dx_1}{dx_2}\right) = 0 \tag{13.29}$$

This differential equation enables us to find the solutions for the trajectories $x_1(x_2)$. These can be written in the form[6]

$$\Psi(x_1, x_2, c) = 0 \tag{13.30}$$

where c is the integration constant. For each value of c, a possible trajectory is obtained. The set of all of them forms the surface $\omega(x_1, x_2, p_1) = 0$. Equation 13.30 represents a one-parameter manifold of rays, where c is the parameter of the family. Each value of c determines a trajectory on the x_1x_2 plane of one ray.

An example of a surface of the form $\omega(x_1, x_2, p_1) = 0$ can be made with sinusoidal trajectories for the rays in the plane x_1x_2. This kind of trajectory occurs within optical fibers with a parabolic profile of refractive index.[1,3,6] These trajectories on the plane correspond to trajectories shaped as helices in phase space, as shown in Figure 13.5. Two trajectories on the plane describe two different lines in phase space. The set of all the possible sinusoidal trajectories of light in the 2-D optical fiber forms a cylindrical tube in phase space corresponding to the surface $\omega(x_1, x_2, p_1) = 0$, as presented in Figure 13.5.

In this case, we must have $\omega(x_1, x_2, p_1) = x_1^2 + p_1^2 - R^2 = 0$, where R is the radius of the cylinder, as well as the amplitude of the sinusoids drawn by the rays of light in their path in the plane x_1x_2.

Consider the case in which the optical system contains mirrors as in Figure 13.6. Here light is reflected between two flat parallel mirrors. The corresponding trajectories in phase space for two rays are presented in Figure 13.7.

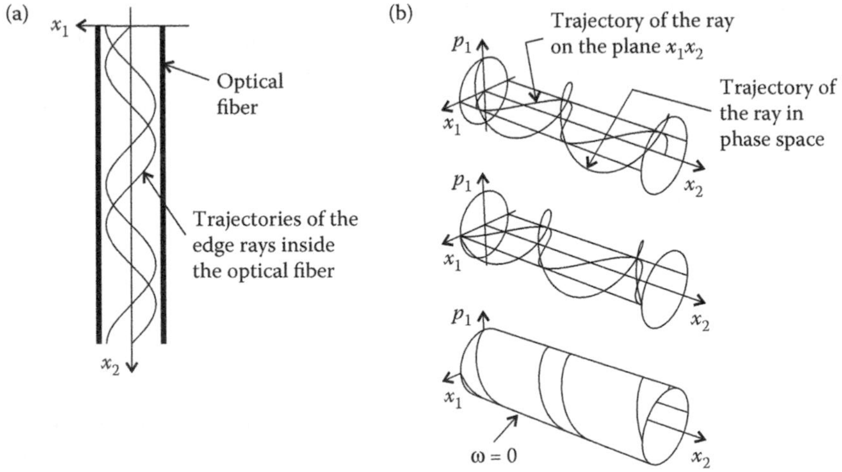

FIGURE 13.5
(a) Sinusoidal trajectories of light inside an optical fiber. (b) Corresponding helicoidal trajectories in phase space. The set of all the possible sinusoidal trajectories with the same amplitude forms a tube in phase space.

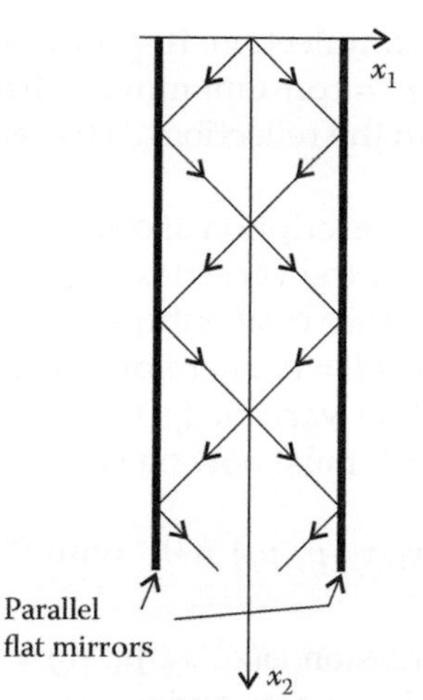

FIGURE 13.6
Trajectory of two light rays between two flat parallel mirrors in a medium with constant refractive index.

The rays of light in phase space now move onto either a top surface or a bottom surface, being "reflected" from one to the other by vertical walls.

The surface $\omega(x_1, x_2, p_1) = 0$ is now made up of top and bottom surfaces connected by two vertical lateral walls representing the mirrors.[6] Note that a surface $\omega = 0$, as presented in Figure 13.7, has edges. It can be seen, however, as a limit case of a surface with no edges, as the one presented in Figure 13.5.

In the path between reflections, the angle of the ray with the x_1 axis is not altered, and the ray moves in phase space on a plane p_1 = constant, that is, parallel to the plane x_1x_2. When a reflection occurs on one of the mirrors, the angle of the ray with the x_1 axis now becomes its symmetric version, and p_1 changes sign. In phase space, the ray now moves on a plane

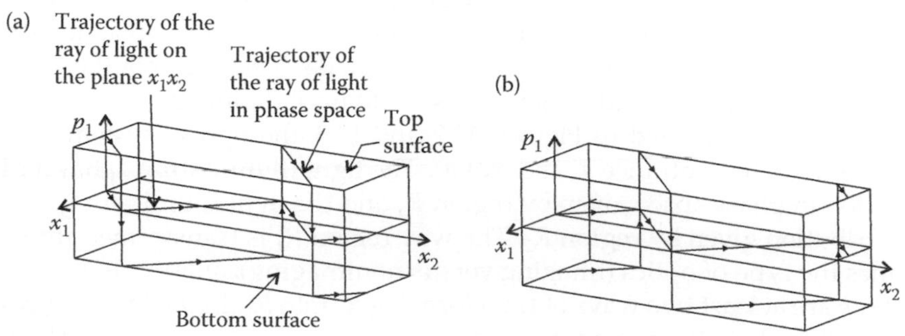

FIGURE 13.7
Trajectories in phase space of the rays presented in Figure 13.6. (a) Ray starting at $x_2 = 0$ with $p_1 > 0$. (b) Ray starting at $x_2 = 0$ with $p_1 < 0$.

$-p_1$ = constant until the next reflection. In phase space, the rays move on the planes p_1 = constant and $-p_1$ = constant moving from one to the other by vertical lines corresponding to the reflections. The set of all rays forms a rectangular tube-shaped surface.

Let us now go back to the description of optics in the formalism of Equation 13.25. In this case, instead of the variables (x_1, x_2, p_1), we will have (x_1, x_2, p_1, p_2). We have, therefore, another momentum but also another equation relating these variables, which is $P = 0$. Therefore, when changing to this formalism, we add not only another variable (p_2) but also another equation ($P = 0$). The condition $\omega(x_1, x_2, p_1) = 0$ must now be written as follows:

$$\omega(x_1, x_2, p_1, p_2) = 0 \quad \text{with } P = 0 \tag{13.31}$$

Note that this new expression $\omega(x_1, x_2, p_1, p_2) = 0$ is not different from the earlier one, $\omega(x_1, x_2, p_1) = 0$, because p_1 and p_2 are related by $p_1^2 + p_2^2 = n^2(x_1, x_2)$, that is, $P = 0$, and, therefore, we can obtain $p_2 = p_2(x_1, x_2, p_1)$, which, when replaced in Equation 13.31, results in $\omega(x_1, x_2, p_1) = 0$.

13.4 Poisson Brackets

The way light is transferred in phase space from the entrance to the exit apertures of an optical system defines its characteristic type of optics (imaging or nonimaging). This transfer must, nonetheless, always obey the general laws of optics, that is, the expressions (13.23), because it describes every optical system. A system obeying these equations must also obey the conservation of étendue law, because this law results from these equations (see Chapter 18).

The transfer of the light in phase space from the entrance aperture to the exit aperture must be characterized in mathematical terms by an equation (or possibly several), which must now be added to Equation 13.23 to form a set that describes this particular type of optics. This transfer must, nonetheless, obey the conservation of étendue, otherwise it would "conflict" with Equation 13.23, and it would not be possible to find a solution.

In the cases presented in Figures 13.2 and 13.3, the optical system under consideration has entrance a_1 and exit a_2. The light illuminating a_1 has a distribution in phase space given by region R_1 and the light illuminating a_2 has a distribution given by region R_2. The way region R_1 is transformed into R_2, defines the type of optics (imaging versus nonimaging).

There are at least two ways of transforming R_1 into R_2. In the imaging case, vertical lines L_1 of R_1 are transformed into vertical lines L_2 of R_2, and the spatial coordinates are related by expression (13.26), that is, $x = MX$.

In the other case, nonimaging or anidolic optics, the boundary of R_1 (called ∂R_1) is transformed into the boundary of R_2 (called ∂R_2). This transformation is achieved in mathematical terms by connecting ∂R_1 with ∂R_2 by Equation 13.27, that is, $\omega(x_1, x_2, p_1) = 0$ To guarantee that this surface ensures the conservation of étendue (and therefore, that the optical system is possible), it is necessary only to guarantee constancy of the areas corresponding to cuts parallel to the plane x_1p_1.

Consider now the case of a general 3-D nonimaging optical system. Its entrance and exit apertures are now connected by a surface of the form

$$\omega(x_1, x_2, x_3, p_1, p_2, p_3) = 0 \tag{13.32}$$

Equation 13.32, characteristic of nonimaging optics, plus the general equations of optics (Equation 13.23) define a set of equations defining nonimaging optical systems in mathematical terms. From all the possible optical systems described by the general equations (Equation 13.23), we are thus interested solely in those systems that satisfy Equation 13.32, that is, nonimaging optical systems.

$$\begin{aligned}
&\frac{dx_1}{d\sigma} = \frac{\partial P}{\partial p_1} \quad \frac{dp_1}{d\sigma} = -\frac{\partial P}{\partial x_1}\\
&\frac{dx_2}{d\sigma} = \frac{\partial P}{\partial p_2} \quad \frac{dp_2}{d\sigma} = -\frac{\partial P}{\partial x_2}\\
&\frac{dx_3}{d\sigma} = \frac{\partial P}{\partial p_3} \quad \frac{dp_3}{d\sigma} = -\frac{\partial P}{\partial x_3}\\
&P = p_1^2 + p_2^2 + p_3^2 - n(x_1, x_2, x_3) = 0\\
&\omega(x_1, x_2, x_3, p_1, p_2, p_3) = 0
\end{aligned} \tag{13.33}$$

This system of equations can, fortunately, be simplified. Since $\omega = 0$ represents a surface in phase space where edge rays move, $d\omega/d\sigma$ represents the variation of ω as the edge rays progress in the system. But, as seen in Equation 13.33, ω is constant and equal to zero, implying $d\omega/d\sigma = 0$.[6]

$$\frac{d\omega}{d\sigma} = \sum_{j=1}^{3}\left(\frac{\partial\omega}{\partial x_j}\frac{dx_j}{d\sigma} + \frac{\partial\omega}{\partial p_j}\frac{dp_j}{d\sigma}\right) \tag{13.34}$$

The term $\partial\omega/\partial\sigma$ does not appear because we are only considering surfaces ω not depending explicitly on σ.[7] Note that Equation 13.32 is $\omega(x_1(\sigma), x_2(\sigma), x_3(\sigma), p_1(\sigma), p_2(\sigma), p_3(\sigma)) = 0$ and does not depend explicitly on σ. Using the first set of equations in Equation 13.33, we can now write

$$\frac{d\omega}{d\sigma} = \sum_{j=1}^{3}\left(\frac{\partial\omega}{\partial x_j}\frac{\partial P}{\partial p_j} - \frac{\partial\omega}{\partial p_j}\frac{\partial P}{\partial x_j}\right) = \{\omega, P\} = 0 \tag{13.35}$$

where $\{\omega, P\}$ is defined by expression (13.35) and is called a Poisson bracket.[8–10] Expression (13.35) already "contains" the differential equations of Equation 13.33 for $dx_i/d\sigma$ and $dp_i/d\sigma$ because these have already been used in its derivation. We can then conclude that the trajectories obeying equation $\omega = 0$ and restricted by conditions $\{\omega, P\} = 0$ and $P = 0$ obey all the equations of Equation 13.33 and, therefore, represent the rays of light in the nonimaging optical system. The final system of equations is then given by

$$\begin{aligned} \{\omega, P\} &= 0 \\ P &= 0 \\ \omega &= 0 \end{aligned} \tag{13.36}$$

Equation 13.36 can also be written as

$$\begin{aligned} &\sum_{j=1}^{3}\left(\frac{\partial\omega}{\partial x_j}\frac{\partial P}{\partial p_j} - \frac{\partial\omega}{\partial p_j}\frac{\partial P}{\partial x_j}\right) = 0 \\ &P = p_1^2 + p_2^2 + p_3^2 - n^2(x_1, x_2, x_3) = 0 \\ &\omega(x_1, x_2, x_3, p_1, p_2, p_3) = 0 \end{aligned} \tag{13.37}$$

The Hamiltonian Equation 13.23 has the same form in the coordinate systems (x_1, x_2, x_3) and in another generalized coordinate system

$$(i_1(x_1, x_2, x_3), i_2(x_1, x_2, x_3), i_3(x_1, x_2, x_3)) \tag{13.38}$$

(see Chapter 14), thus Equation 13.36 is still valid in this new coordinate system.[1,3,7] To the new coordinates (i_1, i_2, i_3) correspond the new moments (u_1, u_2, u_3), and $\{\omega, P\} = 0$ can be written as

$$\{\omega, P\} = \sum_{j=1}^{3}\left(\frac{\partial\omega}{\partial i_j}\frac{\partial P}{\partial u_j} - \frac{\partial\omega}{\partial u_j}\frac{\partial P}{\partial i_j}\right) = 0 \tag{13.39}$$

And in this case $P = 0$ is given by the expression

$$P = u_1^2 a_1^2(i_1, i_2, i_3) + u_2^2 a_2^2(i_1, i_2, i_3) + u_3^2 a_3^2(i_1, i_2, i_3) - n^2 = 0 \tag{13.40}$$

where $a_k = \|\nabla i_k\|$ with $k = 1,2$, and 3. Because $i_k = i_k(x_1, x_2, x_3)$, we have $a_k = a_k(x_1, x_2, x_3)$ or, writing x_1, x_2, and x_3 as functions of i_1, i_2, and i_3, $a_k = a_k(i_1, i_2, i_3)$. Equation 13.37 can now be written as

$$\sum_{j=1}^{3}\left(\frac{\partial\omega}{\partial i_j}\frac{\partial P}{\partial u_j}-\frac{\partial\omega}{\partial u_j}\frac{\partial P}{\partial i_j}\right)=0$$
$$P=u_1^2a_1^2(i_1,i_2,i_3)+u_2^2a_2^2(i_1,i_2,i_3)+u_3^2a_3^2(i_1,i_2,i_3)-n^2(i_1,i_2,i_3)=0 \tag{13.41}$$
$$\omega(i_1,i_2,i_3,u_1,u_2,u_3)=0$$

This is the general system of equations describing a 3-D nonimaging optical system.

13.5 Curvilinear Coordinate System

As mentioned earlier, the surface $\omega = 0$ is made of the trajectories of the edge rays. The calculation of these trajectories is simplified in a coordinate system, where the components p_1 and p_2 of the optical momentum $\mathbf{p}$ have simple expressions.

Let us consider, for example, the crossing of two edge rays in the interior of a compound parabolic concentrator (CPC), as presented in Figure 13.8.

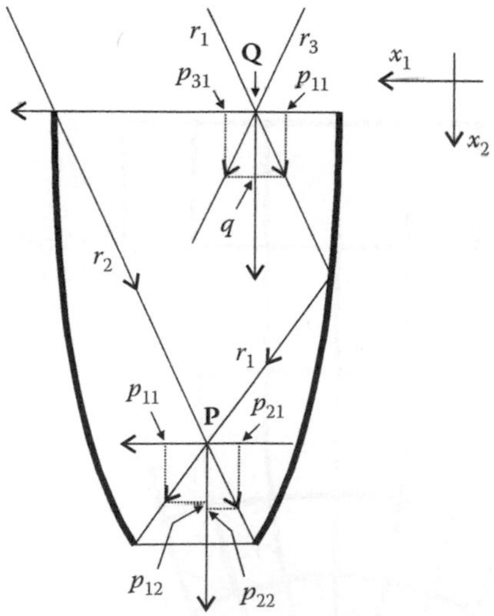

FIGURE 13.8
At a point **P** in a CPC filled with air ($n = 1$), the optical momenta for the edge rays are $\mathbf{p}_1 = (p_{11}, p_{12})$ and $\mathbf{p}_2 = (p_{21}, p_{22})$ with $p_{11} \neq p_{12} \neq p_{21} \neq p_{22}$. However, at another point **Q**, at the entrance aperture where the coordinate axis x_2 bisects the edge rays, their optical momenta are $\mathbf{p}_1 = (p_{11}, q)$ and $\mathbf{p}_3 = (p_{31}, q)$ with $p_{11} = -p_{31}$ and are, therefore, defined by fewer variables.

Two rays r_1 and r_2 go through a point **P** in the interior of the CPC filled with air ($n = 1$). The projections in the directions of the coordinate axes x_1 and x_2 of the unit vectors in the directions of the edge rays are $\mathbf{p}_1 = (p_{11}, p_{12})$ and $\mathbf{p}_2 = (p_{21}, p_{22})$, and enable us to conclude that the p_1 components p_{11} and p_{21} and the p_2 components p_{12} and p_{22} are different from each other and different for the two edge rays. At point **Q** at the entrance aperture, however, the situation is different. Here the coordinate axis x_2 bisects the edge rays. The projections, in the directions of the coordinate axes x_1 and x_2, of the unit vectors in the directions of the edge rays are $\mathbf{p}_1 = (p_{11}, q)$ and $\mathbf{p}_3 = (p_{31}, q)$ and are, therefore, symmetric. Thus, at point **Q**, p_1 has symmetric values p_{31} and $p_{11} = -p_{31}$ for the two edge rays, whereas, p_2 has the same value q for both.

To simplify the expressions for p_1 and p_2, it is advantageous to consider a curvilinear coordinate system (i_1, i_2). The lines i_1 = constant and i_2 = constant must be orthogonal and one of them must bisect the edge rays at each point. In the case of a CPC, these lines, which bisect the edge rays at each point, have the shape presented in Figure 13.9.

As mentioned earlier, the lines of vector flux **J** bisect the trajectories of the edge rays. The curvilinear coordinates to be considered must then coincide with the lines of the vector flux. Note also that the lines G = constant referred to earlier, are along the lines of vector flux **J**. Therefore, the lines G = constant and F = constant must coincide with i_1 = constant and i_2 = constant, that is, with the curvilinear coordinates to be considered. In fact, we had already considered that we would have $G = G(i_1)$ and $F = F(i_2)$, and the lines i_1 = constant would bisect the edge rays.

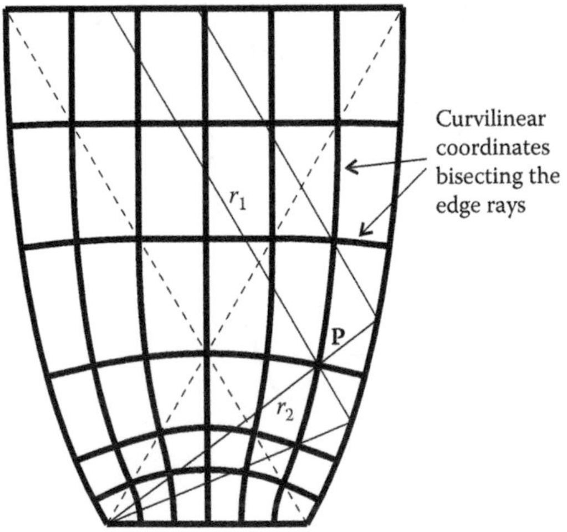

FIGURE 13.9
Curvilinear coordinate system which points, at each point **P**, in the direction of the bisector to the edge rays r_1 and r_2 inside a CPC.

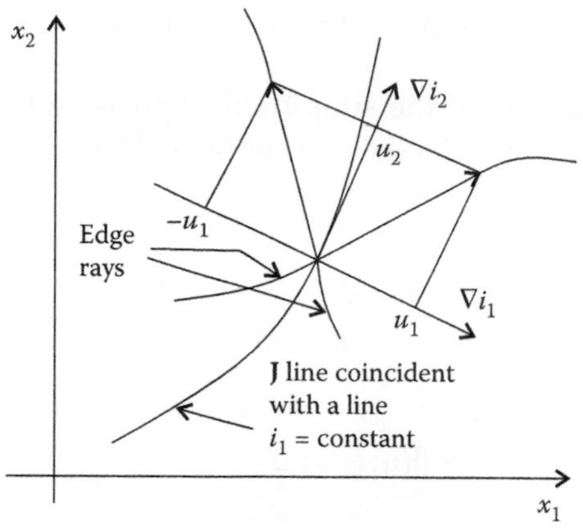

FIGURE 13.10
In the curvilinear coordinate system i_1, i_2, the line i_1 = constant points in the same direction as the bisector to edge rays. In this case, the magnitude of the u_2 component of the moment is the same for both edge rays and the u_1 component is symmetrical. The edge rays form a V-shape around the line i_1 = constant.

We must now define these new lines i_1 = constant and i_2 = constant. In fact, it suffices to define one of these sets (e.g., i_1 = constant), because the second set is made of lines perpendicular to the first ones. Figure 13.10 presents two edge rays crossing a given point on the plane.

As can be seen, the component u_2 in the direction of ∇i_2 of both edge rays is the same. The components u_1 in the direction of ∇i_1 have the same magnitude, but different signs. Therefore, u_2 is the same for both edge rays, and depends only on the coordinates of the point under consideration. We have $u_2 = \beta(i_1, i_2)$ and, therefore, we can write, in this case, Equation 13.27 as[3]

$$\omega(i_1, i_2, u_2) = 0 \Leftrightarrow u_2 - \beta(i_1, i_2) = 0 \tag{13.42}$$

which is a simple form of the equation $\omega = 0$.

13.6 Design of Two-Dimensional Concentrators

In case of 2-D geometry, Equation 13.41 can be written as

$$\begin{aligned} &\sum_{j=1}^{2}\left(\frac{\partial \omega}{\partial i_j}\frac{\partial P}{\partial u_j} - \frac{\partial \omega}{\partial u_j}\frac{\partial P}{\partial i_j}\right) = 0 \\ &P = u_1^2 a_1^2(i_1, i_2) + u_2^2 a_2^2(i_1, i_2) - n^2(i_1, i_2) = 0 \\ &\omega(i_1, i_2, u_1, u_2) = 0 \end{aligned} \tag{13.43}$$

This is the general system of equations describing a 2-D nonimaging optical system.

We can now make use of the simplified expression (13.42) for $\omega = 0$. The system of equations to be solved can be obtained from Equation 13.43 and is, in this case,

$$\{\omega, P\} = \sum_{j=1}^{2}\left(\frac{\partial\omega}{\partial i_j}\frac{\partial P}{\partial u_j} - \frac{\partial\omega}{\partial u_j}\frac{\partial P}{\partial i_j}\right) = 0$$
$$P = u_1^2 a_1^2(i_1, i_2) + u_2^2 a_2^2(i_1, i_2) - n^2(i_1, i_2) = 0 \tag{13.44}$$
$$\omega = u_2 - \beta(i_1, i_2) = 0$$

From the first of these expressions, we get

$$\{\omega, P\} = \frac{\partial\omega}{\partial i_1}\frac{\partial P}{\partial u_1} - \frac{\partial\omega}{\partial u_1}\frac{\partial P}{\partial i_1} + \frac{\partial\omega}{\partial i_2}\frac{\partial P}{\partial u_2} - \frac{\partial\omega}{\partial u_2}\frac{\partial P}{\partial i_2} = 0 \tag{13.45}$$

From the second expression, we get

$$\begin{aligned} \frac{\partial P}{\partial i_1} &= 2a_1\frac{\partial a_1}{\partial i_1}u_1^2 + 2a_2\frac{\partial a_2}{\partial i_1}u_2^2 - 2n\frac{\partial n}{\partial i_1} \qquad \frac{\partial P}{\partial u_1} = 2a_1^2 u_1 \\ \frac{\partial P}{\partial i_2} &= 2a_1\frac{\partial a_1}{\partial i_2}u_1^2 + 2a_2\frac{\partial a_2}{\partial i_2}u_2^2 - 2n\frac{\partial n}{\partial i_2} \qquad \frac{\partial P}{\partial u_2} = 2a_2^2 u_2 \end{aligned} \tag{13.46}$$

and from the third expression, we get

$$\begin{aligned} \frac{\partial\omega}{\partial i_1} &= -\frac{\partial\beta}{\partial i_1} \qquad \frac{\partial\omega}{\partial u_1} = 0 \\ \frac{\partial\omega}{\partial i_2} &= -\frac{\partial\beta}{\partial i_2} \qquad \frac{\partial\omega}{\partial u_2} = 1 \end{aligned} \tag{13.47}$$

making $u_2 = \beta$ the Poisson bracket of ω and P can be calculated as

$$a_1^2 u_1\frac{\partial\beta}{\partial i_1} + a_2^2\beta\frac{\partial\beta}{\partial i_2} - n\frac{\partial n}{\partial i_2} + a_1\frac{\partial a}{\partial i_2}u_1^2 + a_2\frac{\partial a_2}{\partial i_2}\beta^2 = 0 \tag{13.48}$$

From the condition $P = 0$ we can obtain

$$u_1^2 = \frac{n^2 - a_2^2\beta^2}{a_1^2} \Leftrightarrow u_1 = \pm\frac{\sqrt{n^2 - a_2^2\beta^2}}{a_1} \tag{13.49}$$

where $u_2 = \beta$. Equation 13.49 gives the two values of u_1 for a given value of u_2, as represented in Figure 13.10. Substituting for u_1^2 from Equation 13.49 into Equation 13.48 gives

$$\left(a_1{}^2\frac{\partial\beta}{\partial i_1}\right)u_1 + \left(a_2{}^2\beta\frac{\partial\beta}{\partial i_2} - n\frac{\partial n}{\partial i_2} + \frac{n^2 - a_2^2\beta^2}{a_1}\frac{\partial a_1}{\partial i_2} + a_2\frac{\partial a_2}{\partial i_2}\beta^2\right) = 0 \quad (13.50)$$

This has the form $Au_1 + B = 0$. This expression must be fulfilled for both possible values of u_1, thus $Au_1 + B = 0$, and also $-Au_1 + B = 0$ and, therefore, $A = B = 0$. Then $a_1 = \|\nabla i_1\| \neq 0$ gives

$$\begin{aligned}&\frac{\partial\beta}{\partial i_1} = 0\\ &a_2^2\beta\frac{\partial\beta}{\partial i_2} - n\frac{\partial n}{\partial i_2} + \frac{n^2 - a_2^2\beta^2}{a_1}\frac{\partial a_1}{\partial i_2} + a_2\frac{\partial a_2}{\partial i_2}\beta^2 = 0\end{aligned} \quad (13.51)$$

from $\partial\beta/\partial i_1 = 0$, we obtain

$$\beta = \beta(i_2) \quad (13.52)$$

The second condition of Equation 13.51 can be written as

$$\left[\left(2n\frac{\partial n}{\partial i_2} - 2\beta\frac{\partial\beta}{\partial i_2}a_2^2 - 2a_2\frac{\partial a_2}{\partial i_2}\beta^2\right)a_1^2 - 2a_1\frac{\partial a_1}{\partial i_2}\left(n^2 - \beta^2 a_2^2\right)\right]\Big/ a_1^4 = 0 \quad (13.53)$$

that is,

$$\frac{\partial}{\partial i_2}\left(\frac{n^2 - \beta^2 a_2^2}{a_1^2}\right) = 0 \Leftrightarrow \frac{n^2 - \beta^2 a_2^2}{a_1^2} = \alpha(i_1)^2 \Leftrightarrow n^2 = \alpha(i_1)^2 a_1^2 + \beta(i_2)^2 a_2^2 \quad (13.54)$$

Note that $\beta(i_2)a_2 = u_2 a_2$ is the component of $\mathbf{p}$ in the direction of vector ∇i_2, and, therefore, $\alpha(i_1)a_1$ must be the component of $\mathbf{p}$ in the direction of ∇i_1. This equation corresponds to Equation 13.20 obtained earlier.

13.7 An Example of an Ideal Two-Dimensional Concentrator

An example of application of Miñano's design method in two dimensions is presented. It applies the earlier ideas in the design of a concentrator with maximum concentration and flat entrance and exit apertures. The entrance aperture will be at the x_1 axis ($x_2 = 0$) and the exit aperture will be at the line $x_2 = 1$. We choose the shape of the i_1-lines. The i_2-lines are perpendicular to the i_1-lines and can be obtained from them. The refractive index can then be calculated using Equation 13.20 or Equation 13.54, and the boundary conditions for the problem.

Figure 13.11 shows the shape of an i_1-line (i_2 = constant). It is a circumference centered at the x_2 axis. The reason for choosing this shape for the i_1-lines becomes apparent in Section 13.9 when we apply these results to the design of a 3-D concentrator.

If this circumference was centered at $(x_1, x_2) = (0, 0)$, it would be defined by an equation of the form $x_1^2 + x_2^2 = R^2$. However, displacing its center along the axis x_2 to a position $R + i_2$, its equation is now $x_1^2 + [(R + i_2) - x_2]^2 = R^2$ that is,

$$x_2 = R(i_2) + i_2 - \sqrt{R(i_2)^2 - x_1^2} \tag{13.55}$$

Later, an expression for $R(i_2)$ will be given. This is the equation for the i_1-lines (i_2 = constant). For each value of i_2, a value for $R(i_2)$ is defined, and a circumference is obtained. It can be seen that these circumferences cross the x_2 axis at $x_2 = i_2$. Therefore, the line $i_2 = 0$ crosses the x_2 axis at $x_2 = 0$ (the x_1 axis) and the line $i_2 = 1$ crosses the x_2 axis at $x_2 = 1$. Lines $i_2 = 0$ and $i_2 = 1$ correspond to the entrance and exit apertures of the concentrator, so that they are chosen to be flat. Therefore, for $x_2 = i_2 = 0$ and $x_2 = i_2 = 1$ we must have $R \to \infty$.

The i_2-lines are perpendicular to the i_1-lines. These i_2-lines can be defined so that they cross the x_1 axis with $i_1 = x_1$. The i_1- and i_2-lines are presented in Figure 13.12.

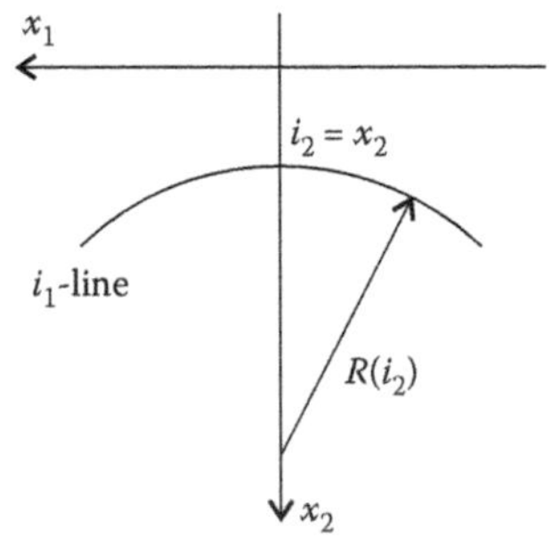

FIGURE 13.11
The i_1-lines (i_2 = constant) are chosen to be circumferences centered at the x_2 axis having radius $R(i_2)$.

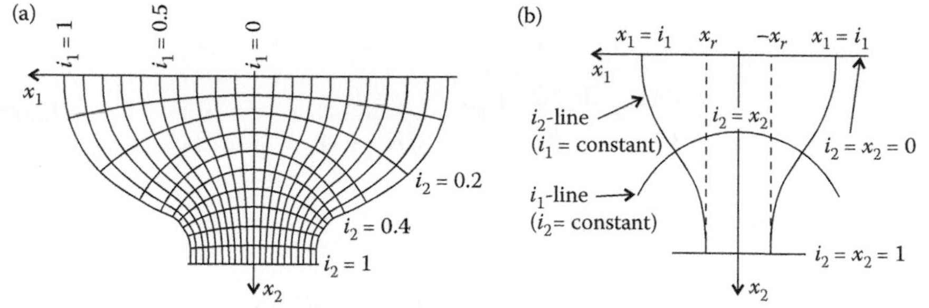

FIGURE 13.12
(a) A set of i_1- and i_2-lines for the concentrator being designed. (b) The i_1-lines cross the x_2 axis (optical axis) with $i_2 = x_2$ and the i_2-lines cross the x_1 axis (entrance aperture) for $i_1 = x_1$ and the exit aperture ($x_2 = 1$) for $x_1 = x_r$.

The expression for i_1 is given by

$$i_1 = \frac{2Rx_1}{R + \sqrt{R^2 - x_1^2}} \exp\left(\int_0^{i_2} \frac{1}{R} di_2 \right) \tag{13.56}$$

where R is a function $R(i_2)$ of i_2 and obeys Equation 13.55. We are not going to derive Equation 9.56, but instead, we prove that the i_1- and i_2-lines just defined fulfill the conditions mentioned earlier.[7] Define a function $M(i_1)$ by

$$M(i_1) = \ln\left(\frac{i_1}{2}\right) = \ln\left[\frac{R}{R/x_1 + \sqrt{R^2/x_1^2 - 1}} \exp\left(\int_0^{i_2} \frac{1}{R} di_2 \right) \right] \tag{13.57}$$

or

$$M(i_1) = \ln R - \ln\left(\frac{R}{x_1} + \sqrt{\frac{R^2}{x_1^2} - 1} \right) + \int_0^{i_2} \frac{1}{R} di_2 \tag{13.58}$$

Calculating the x_1 and x_2 derivatives of $M(i_1)$ gives

$$\frac{\partial M}{\partial x_1} = \frac{\partial i_2/\partial x_1}{R} \left(R' - \frac{R'R}{\sqrt{R^2 - x_1^2}} + 1 \right) + \frac{R}{x_1\sqrt{R^2 - x_1^2}} \tag{13.59}$$

and

$$\frac{\partial M}{\partial x_2}=\frac{\partial i_2/\partial x_2}{R}\left(R'-\frac{R'R}{\sqrt{R^2-x_1^2}}+1\right) \tag{13.60}$$

where $R' = dR(i_2)/di_2$. Note that

$$\int_0^{i_2}\frac{1}{R}di_2 = F(i_2) \quad \text{and} \quad \frac{\partial F(i_2)}{\partial x_2}=\frac{dF}{di_2}\frac{\partial i_2}{\partial x_2} \tag{13.61}$$

where $F(i_2)$ is a function of i_2, and, therefore,

$$\frac{\partial}{\partial x_1}\left(\int_0^{i_2}\frac{1}{R}di_2\right)=\frac{1}{R}\frac{\partial i_2}{\partial x_1} \quad \text{and} \quad \frac{\partial}{\partial x_2}\left(\int_0^{i_2}\frac{1}{R}di_2\right)=\frac{1}{R}\frac{\partial i_2}{\partial x_2} \tag{13.62}$$

The derivatives $\partial i_2/\partial x_1$ and $\partial i_2/\partial x_2$ can be obtained by calculating the x_1 and x_2 derivatives of Equation 13.55 and solving for $\partial i_2/\partial x_1$ and $\partial i_2/\partial x_2$. (Note that $dx_2/dx_2 = 1$ and that $dx_2/dx_1 = 0$.) These partial derivatives are

$$\frac{\partial i_2}{\partial x_1}=\frac{-x_1\big/\sqrt{R^2-x_1^2}}{1+R'-\left(R'R\big/\sqrt{R^2-x_1^2}\right)} \tag{13.63}$$

and

$$\frac{\partial i_2}{\partial x_2}=\frac{1}{1+R'-\left(R'R\big/\sqrt{R^2-x_1^2}\right)} \tag{13.64}$$

or

$$\nabla i_2=\left(1+R'-\frac{R'R}{\sqrt{R^2-x_1^2}}\right)^{-1}\left(-x_1\frac{1}{\sqrt{R^2-x_1^2}},1\right) \tag{13.65}$$

Replacing these expressions in Equations 13.59 and 13.60 gives

$$\left(\frac{\partial M}{\partial x_1}\right)^2=\frac{1}{x_1^2}-\frac{1}{R^2} \quad \text{and} \quad \left(\frac{\partial M}{\partial x_2}\right)^2=\frac{1}{R^2} \tag{13.66}$$

Expression (13.66) can also be written as

$$\nabla M = \left(\sqrt{\frac{1}{x_1^2} - \frac{1}{R^2}}, \frac{1}{R} \right) = \frac{1}{R}\left(\frac{1}{x_1}\sqrt{R^2 - x_1^2}, 1 \right) \tag{13.67}$$

and then we get

$$\nabla i_2 \cdot \nabla M = 0 \tag{13.68}$$

Since $M = M(i_1)$ as given by Equation 13.57, we have

$$\nabla M = \frac{dM}{di_1}\nabla i_1 \Leftrightarrow \nabla M = \frac{1}{i_1}\nabla i_1 \tag{13.69}$$

and because $dM/di_1 \neq 0$, we have

$$\nabla i_1 \cdot \nabla i_2 = 0 \tag{13.70}$$

As mentioned earlier, the i_2-lines were defined in such a way that they cross the x_1 axis ($x_2 = 0$) with $i_1 = x_1$. This can be seen in expression (13.56),

$$i_1 = \frac{2x_1}{1 + \sqrt{1 - x_1^2/R^2}} \exp\left(\int_0^{i_2} \frac{1}{R} di_2 \right) \tag{13.71}$$

When $i_2 \to 0$ and $R \to \infty$ then $i_1 \to x_1$, because for $i_2 = 0$, we have $\exp\left(\int_0^{i2}(1/R)di_2\right) = \exp\left(\int_0^0(1/R)di_2\right) = \exp 0 = 1$. At the entrance aperture we then have $i_1 = x_1$.

The i_2-lines intercept the exit aperture (receiver) for values of x_1 such that $x_1 = x_r$, where x_r can be obtained from expression (13.71) making $i_2 \to 1$ and $R \to \infty$.

$$i_1 = x_r \exp\left(\int_0^1 \frac{1}{R} di_2 \right) \tag{13.72}$$

Note that $R(i_2) \to \infty$ when $i_2 \to 0$ and $i_2 \to 1$, but $R(i_2)$ has finite values for $0 < i_2 < 1$. Now consider that the i_2-lines are vector flux lines. The final device will then be limited by two of these lines, converted to mirrors. The points where these two lines cross the entrance aperture ($i_2 = 0$) will then define the entrance aperture of the final device, and the points where these two lines cross the exit aperture will define the exit aperture of the final device. Since each one of these lines crosses the entrance aperture at $x_1 = i_1$, and the exit aperture at $x_1 = x_r$, the ratio between the dimensions for the entrance and exit apertures will be $C_{2D} = i_1/x_r$. The geometrical concentration for a symmetrical concentrator will then be $C_g = i_1/x_r$. From expression (13.72) we then obtain

$$C_g = \exp\left(\int_0^1 \frac{1}{R} di_2\right) \tag{13.73}$$

We still have not given an expression for $R(i_2)$. As stated earlier, this function must be such that $R \to \infty$ when $i_2 \to 0$ and $i_2 \to 1$. A possibility is to choose the following function,

$$R(i_2) = \frac{m}{i_2^2(1 - i_2^2)} \tag{13.74}$$

where m is a constant. To obtain the value of m, we replace this expression for $R(i_2)$ in expression (13.73) for C_g, giving

$$\ln C_g = \frac{1}{m}\int_0^1 i_2^2(1 - i_2^2)di_2 \Leftrightarrow m = \frac{2}{15 \ln C_g} \tag{13.75}$$

We now have a completely defined set of i_1-lines and the corresponding i_2-lines.

We must next find the 2-D refractive index distribution that transforms the i_2-lines into vector flux lines. The refractive index distribution can be found from Equation 13.20 or Equation 13.54, that is, $n^2 = \alpha(i_1)^2 a_1^2 + \beta(i_2)^2 a_2^2$. First, we note that $a_1^2 = \|\nabla i_1\|^2$ and $a_2^2 = \|\nabla i_2\|^2$. We can then write (see Equations 13.63 and 13.64):

$$a_2^2 = \left(\frac{\partial i_2}{\partial x_1}\right)^2 + \left(\frac{\partial i_2}{\partial x_2}\right)^2 = \left[(1 + R')\sqrt{1 - \frac{x_1^2}{R^2}} - R'\right]^{-2} \tag{13.76}$$

From expression (13.66) we can see that

$$\|\nabla M\|^2 = \frac{1}{x_1^2} \tag{13.77}$$

and from Equation 13.69 we get

$$\|\nabla i_1\|^2 = \frac{1}{x_1^2} i_1^2 \tag{13.78}$$

and, therefore,

$$n^2 = \frac{1}{x_1^2}\alpha^*(i_1)^2 + \left[(1+R')\sqrt{1-\frac{x_1^2}{R^2}} - R'\right]^{-2} \beta(i_2)^2 \tag{13.79}$$

where $\alpha^*(i_1) = i_1\alpha(i_1)$ and from Equation 13.74:

$$R' = \frac{dR(i_2)}{di_2} = \frac{2m(2i_2^2 - 1)}{i_2^3(i_2^2 - 1)^2} \tag{13.80}$$

If we want maximum concentration at the exit aperture, the edge rays must reach it making angles $\pi/2$ with the i_2-lines. This implies that the component a_2u_2 of the optical momentum must be zero. Since $a_2 = \|\nabla i_2\|$, this component of the optical momentum is given by $\|\nabla i_2\|\beta(i_2)$. If we have $\nabla i_2 = 0$, it would not be possible to define a local system of coordinates because one of the unit vectors of this local coordinate system is $\mathbf{e}_2 = \nabla i_2/\|\nabla i_2\|$. For this component of the optical momentum to be zero at the exit aperture, we must then have $\beta(i_2) = 0$ for $i_2 = 1$. In this case, expression (13.79) can be written as

$$n^2(i_2 = 1) = \frac{1}{x_r^2}\alpha^*(i_1)^2 \tag{13.81}$$

For the receiver, we have $x_1 = x_r$, given by expression (13.72). Combining this expression with expression (13.73), we get $i_1 = x_r C_g$. If we want the refractive index to be $n = n_r$ at the receiver, we get from Equation 13.81

$$\alpha^*(i_1) = \frac{i_1 n_r}{C_g} \tag{13.82}$$

Expression (13.79) can now be written as

$$n^2 = \frac{1}{x_1^2}\frac{i_1^2 n_r^2}{C_g^2} + \left[(1+R')\sqrt{1-\frac{x_1^2}{R^2}} - R'\right]^{-2} \beta(i_2)^2 \tag{13.83}$$

or by using expression (13.71) for i_1 as

$$n^2 = \left[\frac{2}{1+\sqrt{1-x_1^2/R^2}}\exp\left(\int_0^{i_2}\frac{1}{R}di_2\right)\right]^2 \frac{n_r^2}{C_g^2} + \left[(1+R')\sqrt{1-\frac{x_1^2}{R^2}} - R'\right]^{-2} \beta(i_1)^2 \quad (13.84)$$

At the optical axis, that is, for $x_1 = 0$, we have

$$n^2(x_1 = 0) = \beta(i_2)^2 + \left[\exp\left(\int_0^{i_2}\frac{1}{R}di_2\right)\right]^2 \frac{n_r^2}{C_g^2} \quad (13.85)$$

If we now make for the optical axis $n = n_r$, we get

$$\beta(i_2) = n_r\sqrt{1 - \frac{1}{C_g^2}\left[\exp\left(\int_0^{i_2}\frac{1}{R}di_2\right)\right]^2} \quad (13.86)$$

As we progress along the optical axis, the component of the optical momentum relative to the i_2-lines must obey $a_2u_2 = a_2\beta(i_2) > 0$, and, therefore, $\beta(i_2) > 0$ between the entrance and exit apertures, that is, for $0 < i_2 < 1$. (Note that $a_2 = \|\nabla i_2\| > 0$ and $\beta(i_2) = 0$ for $i_2 = 1$, because maximum concentration is required.) The expression obtained for $\beta(i_2)$ fulfills these conditions, as seen from expressions (13.86) and (13.73). The expression for the refractive index can now be written as

$$n^2 = \frac{1}{x_1^2}\frac{i_1^2 n_r^2}{C_g^2} + \left[(1+R')\sqrt{1-\frac{x_1^2}{R^2}} - R'\right]^{-2} n_r^2\left[1 - \frac{\exp\left(2\int_0^{i_2}(1/R)di_2\right)}{C_g^2}\right] \quad (13.87)$$

For the entrance aperture, that is, for $i_2 = 0$, we have $R \to \infty$ and $i_1 = x_1$. This expression then simplifies to

$$n^2(i_2 = 0) = \frac{n_r^2}{C_g^2} + n_r^2\left[1 - \frac{1}{C_g^2}\right] = n_r^2 \quad (13.88)$$

We can thus conclude that for the points of the entrance aperture, we have a constant refractive index, $n = n_r$.

From Equations 13.84 and 13.87, the refractive index is

$$n^2 = \frac{4R^2 \dfrac{n_r^2}{C_g^2} \exp\left(2\displaystyle\int_0^{i_2} \frac{1}{R} di_2\right)}{\left(R + \sqrt{R^2 - x_1^2}\right)^2} + \frac{n_r^2 - \dfrac{n_r^2}{C_g^2} \exp\left(2\displaystyle\int_0^{i_2} \frac{1}{R} di_2\right)}{\left[(1 + R')\sqrt{1 - \left(x_1^2/R^2\right)} - R'\right]^2} \tag{13.89}$$

We still have to relate the geometrical concentration C_g to the acceptance angle φ of the concentrator. At the entrance aperture, the refractive index n_r will cause the rays to refract and, therefore, to be angularly confined within a cone with half-angle φ_1 given by the law of refraction.

$$n_r \sin \varphi_1 = \sin \varphi \tag{13.90}$$

Since the concentrator has maximum concentration and the exit aperture has a refractive index of n_r, inside the device, we have a concentration (which equals the geometrical concentration) $C_g = 1/\sin \varphi_1$ so that

$$\sin \varphi = \frac{n_r}{C_g} \tag{13.91}$$

We can now write

$$n^2 = \frac{n_r^2 - \sin^2 \varphi \exp\left(2\displaystyle\int_0^{i_2} \frac{1}{R} di_2\right)}{\left[(1 + R')\sqrt{1 - \dfrac{x_1^2}{R^2}} - R'\right]^2} + \frac{4R^2 \sin^2 \varphi \exp\left(2\displaystyle\int_0^{i_2} (1/R) di_2\right)}{\left(R + \sqrt{R^2 - x_1^2}\right)^2} \tag{13.92}$$

Let us, for example, presume that $n_r = 1.5$ and $C_g = 3$. We have, for the acceptance angle,

$$\varphi = \arcsin\left(\frac{1.5}{3}\right) = 30° \tag{13.93}$$

Giving values to i_2, we can calculate R and R' using expressions (13.74) and (13.80). Giving values to x_1 also, it is possible to calculate $n(x_1, i_2)$ using expression (13.92). Using the values for x_1 and i_2, it is possible to obtain $x_2(x_1, i_2)$ using Equation 13.55 and, therefore, $n(x_1, x_2)$. The resulting refractive index distribution is presented in Figure 13.13.

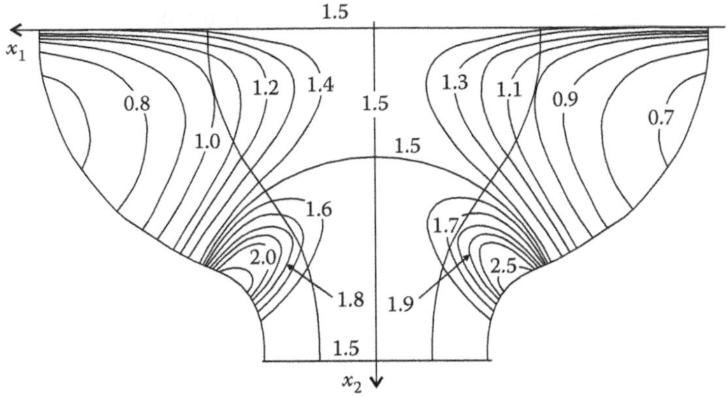

FIGURE 13.13
Refractive index distribution inside the concentrator, manifested by contours of constant refractive index.

Using expression (13.74) for R in Equation 13.55, we get for x_2

$$x_2 = i_2 + \frac{m}{i_2^2 - i_2^4} - \sqrt{\frac{m^2}{i_2^4(i_2^2-1)^2} - x_1} \tag{13.94}$$

We can also write

$$\int_0^{i_2} \frac{1}{R}\,di_2 = \int_0^{i_2} \frac{i_2^2(1-i_2^2)}{m}\,di_2 = \left[\frac{5i_2^3 - 3i_2^5}{15m}\right]_0^{i_2} = \frac{5i_2^3 - 3i_2^5}{15m} \tag{13.95}$$

Also expression (13.56) for i_1 can now be written as follows

$$i_1 = \frac{2m\,x_1 \exp((5i_2^3 - 3i_2^5)/15m)}{m - i_2^2(i_2^2-1)\sqrt{m^2/(i_2^4(i_2^2-1)^2) - x_1^2}} \tag{13.96}$$

This expression can be solved for x_1 to give

$$x_1 = \frac{4\,i_1\,m^2 \exp(i_2^3(5+3i_2^2)/15m)}{i_1^2 i_2^4(i_2^2-1)^2 \exp(2i_2^5/5m) + 4m^2\exp(2i_2^3/3m)} \tag{13.97}$$

Now, maintaining a fixed value for i_1 and varying i_2, different values for x_1 can be obtained. The corresponding value for x_2 can be calculated using expression (13.94) and the i_2-lines obtained, as presented in Figure 13.12.

Replacing expressions (13.95) and (13.80) for R' and expression (13.74) for R in expression (13.92), we get

$$n^2 = 4m^2A\left(m - i_2^2(i_2^2-1)\sqrt{\frac{m^2}{i_2^4(i_2^2-1)^2} - x_1^2}\right)^{-2}$$

$$+ (n_r^2 - A)\left(1 + \frac{2m(2i_2^2-1)}{i_2^3(i_2^2-1)^2}\right)^{-2}\left(1 + \frac{m(2-4i_2^2)}{i_2^3(i_2^2-1)^2} - \frac{i_2^4x_1^2(i_2^2-1)^2}{m^2}\right)^{-1} \quad (13.98)$$

with $A = \sin^2\varphi \exp[2i_2^3(5-3i_2^2)/(15m)]$. Giving values to i_2 and x_1, n can be obtained. For the same values of i_2 and x_1, we can also obtain x_2, using expression (13.94), finally giving $n(x_1, x_2)$.

Two vector flux lines (i_2-lines) can now be chosen as mirrors, completing this design for an ideal 2-D concentrator.

13.8 Design of Three-Dimensional Concentrators

The design method for 3-D concentrators is similar to that presented earlier for the 2-D case. Here, only systems with rotational symmetry are analyzed.[3,7] For solving this problem, we must choose a coordinate system that is nothing more than an extension of the coordinate system used in the 2-D case. We now have a coordinate system (i_1, θ, i_3) of space (x_1, x_2, x_3), where θ measures the angle around the axis of symmetry (it therefore corresponds to the angular coordinate of cylindrical coordinates). For θ = constant, we obtain a plane containing the optical axis. On this plane, the coordinates i_1 and i_3 are two curvilinear coordinates similar to the ones (i_1, i_2) considered earlier for the 2-D case. These coordinates are represented in Figure 13.14.

The orthogonality of the new curvilinear coordinates is now ensured by

$$\nabla i_1 \cdot \nabla\theta = \nabla\theta \cdot \nabla i_3 = \nabla i_1 \cdot \nabla i_3 = 0 \quad (13.99)$$

We can now make

$$a_1 = \|\nabla i_1\|,\ b = 1/\rho = \|\nabla\theta\|,\ a_3 = \|\nabla i_3\| \quad (13.100)$$

Let u_1, h, and u_3 be the momenta corresponding to these coordinates. Similar to what happened in the 2-D case, we can obtain the components of the optical momentum a_1u_1, bh, and a_3u_3 relative to ∇i_1, $\nabla\theta$, and ∇i_3 respectively. Since the refractive index of the system does not depend on θ, the quantity h is a

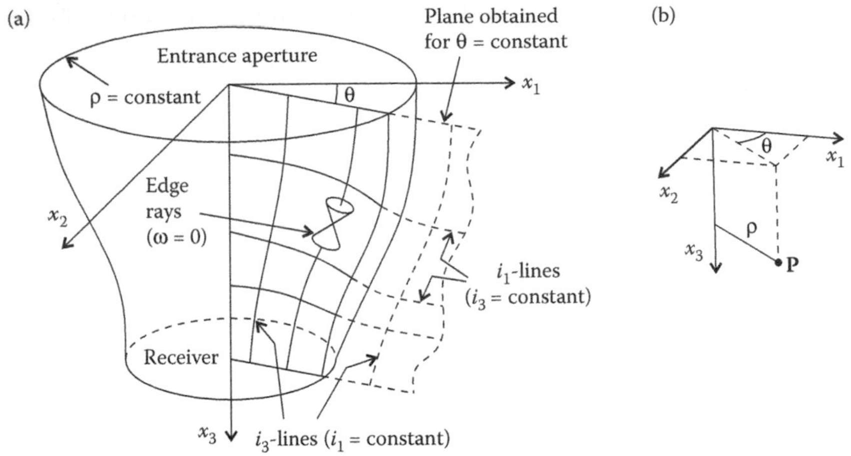

FIGURE 13.14
(a) In a system with circular symmetry, the angle θ around the optical axis can be chosen as a coordinate. On each plane obtained with θ = constant, a curvilinear coordinate system i_1, i_3 is used. (In this case, the edge rays form a cone around the i_3-lines, that is, these lines point in the direction of the bisector to the edge rays.) (b) A cylindrical coordinate system. Axis x_3 points in the direction of the optical axis and ρ is the distance from a point **P** to axis x_3.

constant called the skew invariant or skewness (see Chapter 17). Note that ∇i_1 is obtained with θ = constant and i_3 = constant, $\nabla\theta$ is obtained with i_1 = constant and i_3 = constant, and ∇i_3 is obtained with i_1 = constant and θ = constant, that is, in the direction of ∇i_1 only i_1 varies; in the direction of $\nabla\theta$ only θ varies; and in the direction of ∇i_3 only i_3 varies. The lines in which only i_1 varies are called i_1-lines, those in which only θ varies are called θ-lines, and those in which only i_3 varies are called i_3-lines.

In this coordinate system, the expression for P is given by

$$P = a_1^2 u_1^2 + b^2 h^2 + a_3^2 u_3^2 - n^2 \tag{13.101}$$

where

$$bh = n\cos\phi = h/\rho \tag{13.102}$$

in which h is again the skew invariant (a constant), $b = 1/\rho$ where ρ is the distance to the axis of symmetry and ϕ is the angle between the tangent to the light ray and vector $\mathbf{e}_\theta = \nabla\theta/\|\nabla\theta\|$. The expression for P can now be written for this coordinate system in the following form:

$$P = a_1^2(i_1, i_3)u_1^2 + b^2(i_1, i_3)h^2 + a_3^2(i_1, i_3)u_3^2 - n^2(i_1, i_3) \tag{13.103}$$

The fact that a_1, b, a_3, and n do not depend on θ is a consequence of the rotational symmetry.

The Poisson bracket of ω and P must now be zero for $\omega = 0$ and $P = 0$:

$$\{\omega,P\} = \frac{\partial\omega}{\partial i_1}\frac{\partial P}{\partial u_1} - \frac{\partial\omega}{\partial u_1}\frac{\partial P}{\partial i_1} + \frac{\partial\omega}{\partial\theta}\frac{\partial P}{\partial h} - \frac{\partial\omega}{\partial h}\frac{\partial P}{\partial\theta} + \frac{\partial\omega}{\partial i_3}\frac{\partial P}{\partial u_3} - \frac{\partial\omega}{\partial u_3}\frac{\partial P}{\partial i_3} = 0 \quad (13.104)$$

The system of equations to be solved is then as follows:

$$\begin{aligned}
&\{\omega,P\} = \frac{\partial\omega}{\partial i}\frac{\partial P}{\partial u_1} - \frac{\partial\omega}{\partial u_1}\frac{\partial P}{\partial i_1} + \frac{\partial\omega}{\partial\theta}\frac{\partial P}{\partial h} - \frac{\partial\omega}{\partial h}\frac{\partial P}{\partial\theta} + \frac{\partial\omega}{\partial i_3}\frac{\partial P}{\partial u_3} - \frac{\partial\omega}{\partial u_3}\frac{\partial P}{\partial i_3} = 0\\
&\quad P = a_1^2(i_1,i_3)u_1^2 + b^2(i_1,i_3)h^2 + a_3^2(i_1,i_3)u_3^2 - n^2(i_1,i_3) = 0 \qquad (13.105)\\
&\omega(i_1,\theta,i_3,u_1,h,u_3) = 0
\end{aligned}$$

A simple form of the equation $\omega = 0$ can also be used in this case. Here, expression (13.42) can be written in the following form,[3,7]

$$\omega = u_3 - \beta(i_1,i_3) = 0 \qquad (13.106)$$

where β depends only on i_1 and i_3 in the same way a_1, b, a_3, and n do. There is, at this point, no guarantee that an expression such as condition (13.106) for the edge rays is valid. Only if a solution for the optical system can be found, it is possible to verify that this relation is true. Condition (13.106) requires that all the edge rays at the point under consideration have a momentum **p** with the same value for the component u_3. This means that the edge rays must form a circular cone around the i_3-line. In the 2-D cases presented earlier, this circular cone was just a V-shape around the i_2-line (i_1 = constant line). If the i_3-lines point in the direction of the bisector to the edge rays, they must also coincide with the lines of vector flux **J**.

Since $\omega = \omega(i_1, i_3, u_3)$, the system of equations to be solved is then in this case as follows

$$\begin{aligned}
&\{\omega,P\} = \frac{\partial\omega}{\partial i_1}\frac{\partial P}{\partial u_1} + \frac{\partial\omega}{\partial i_3}\frac{\partial P}{\partial u_3} - \frac{\partial\omega}{\partial u_3}\frac{\partial P}{\partial i_3} = 0\\
&P = a_1^2(i_1,i_3)u_1^2 + b^2(i_1,i_3)h^2 + a_3^2(i_1,i_3)u_3^2 - n^2(i_1,i_3) = 0 \qquad (13.107)\\
&\omega = u_3 - \beta(i_i,i_3) = 0
\end{aligned}$$

To solve this system of equations, we can start by substituting expression (13.103) for P and condition (13.106) for ω into expression (13.104) for $\{\omega, P\}$, that is, substituting the second and third equations of Equation 13.107 in the first equation. Then we get

$$\frac{\partial\beta}{\partial i_1}a_1^2u_1+\frac{\partial\beta}{\partial i_3}a_3^2u_3-\left(n\frac{\partial n}{\partial i_3}-a_1\frac{\partial a_1}{\partial i_3}u_1^2-b\frac{\partial b}{\partial i_3}h^2-a_3\frac{\partial a_3}{\partial i_3}u_3^2\right)=0 \quad (13.108)$$

We can now introduce the condition $P=0$ to eliminate h^2 and $\omega=0$ to replace u_3 by β. We get

$$u_1^2\left(a_1\frac{\partial a_1}{\partial i_3}-\frac{\partial b/\partial i_3}{b}a_1^2\right)+u_1\left(\frac{\partial\beta}{\partial i_1}a_1^2\right)$$
$$+\left(-n\frac{\partial n}{\partial i_3}+\frac{\partial b/\partial i_3}{b}n^2+a_3\frac{\partial a_3}{\partial i_3}\beta^2-\frac{\partial b/\partial i_3}{b}a_3^2\beta^2+\beta\frac{\partial\beta}{\partial i_3}a_3^2\right)=0 \quad (13.109)$$

Note that these expressions could also be written in terms of $(\partial\rho/\partial i_3)/\rho$ instead of $(\partial b/\partial i_3)/b$ because $b=1/\rho$; calculating the i_3 derivative

$$\frac{\partial b}{\partial i_3}=-\frac{\partial\rho/\partial i_3}{\rho^2}\Leftrightarrow\frac{\partial b/\partial i_3}{b}=-\frac{\partial\rho/\partial i_3}{\rho} \quad (13.110)$$

Since expression (13.109) must be zero for any value of u_1, we must have

$$\begin{aligned}\partial\beta/\partial i_1&=0\\ a_1\frac{\partial a_1}{\partial i_3}-\frac{\partial b/\partial i_3}{b}a_1^2&=0\\ -n\frac{\partial n}{\partial i_3}+\frac{\partial b/\partial i_3}{b}n^2+a_3\frac{\partial a_3}{\partial i_3}\beta^2-\frac{\partial b/\partial i_3}{b}a_3^2\beta^2+\beta\frac{\partial\beta}{\partial i_3}a_3^2&=0\end{aligned} \quad (13.111)$$

from the first of these equations

$$\partial\beta/\partial i_1=0\Leftrightarrow\beta=\beta(i_3) \quad (13.112)$$

The second equation can be written as

$$\frac{b}{a_1^3}\left(a_1\frac{\partial a_1}{\partial i_3}-\frac{\partial b/\partial i_3}{b}a_1^2\right)=0\Leftrightarrow\frac{(\partial a_1/\partial i_3)b-(\partial b/\partial i_3)a_1}{a_1^2}=0$$
$$\Leftrightarrow\frac{\partial(a_1/b)/\partial i_3}{a_1/b}=0\Leftrightarrow\frac{\partial}{\partial i_3}\ln\left(\frac{a_1}{b}\right)=0 \quad (13.113)$$

which can now be integrated, resulting in

$$\ln\left(\frac{a_1}{b}\right) = F_1(i_1) \Leftrightarrow \left|\frac{a_1}{b}\right| = F_2(i_1) \Leftrightarrow \frac{b^2}{a_1^2} = F_3(i_1)^2 \tag{13.114}$$

Making now $F_3(i_1) = dM(i_1)/di_1$, we get

$$\frac{b^2}{a_1^2} = \left(\frac{dM(i_1)}{di_1}\right)^2 \tag{13.115}$$

Since M is a function of $i_1(x_1, x_2, x_3)$, we can write

$$\nabla M = \left(\frac{\partial M}{\partial x_1}, \frac{\partial M}{\partial x_2}, \frac{\partial M}{\partial x_3}\right) = \left(\frac{dM}{di_1}\frac{\partial i_1}{\partial x_1}, \frac{dM}{di_1}\frac{\partial i_1}{\partial x_2}, \frac{dM}{di_1}\frac{\partial i_1}{\partial x_3}\right) = \frac{dM}{di_1}\nabla i_1 \tag{13.116}$$

Now considering the definition of a_1, we can write

$$\|\nabla M\|^2 = \left(\frac{dM}{di_1}\right)^2 a_1^2 \tag{13.117}$$

Inserting Equation 13.117 into Equation 13.115 gives

$$b^2 = \|\nabla M\|^2 \Leftrightarrow \|\nabla M(i_1)\|^2 = 1/\rho^2 \tag{13.118}$$

To integrate the third equation of Equation 13.111, we can write

$$\begin{aligned}&\left[\left(2n\frac{\partial n}{\partial i_3} - 2a_3\frac{\partial a_3}{\partial i_3}\beta^2 - 2\beta\frac{\partial \beta}{\partial i_3}a_3^2\right)b^2 - 2b\frac{\partial b}{\partial i_3}(n^2 - a_3^2\beta^2)\right]b^{-4} = 0\\ &\Leftrightarrow \frac{\partial}{\partial i_3}\left(\frac{n^2 - a_3^2\beta^2}{b^2}\right) = 0\end{aligned} \tag{13.119}$$

Considering expression (13.103) for P and that $P = 0$, we can conclude that $n^2 - a_3^2\beta^2 = a_1^2u_1^2 + b^2h^2 \geq 0$; this expression can now be integrated, resulting in

$$\frac{n^2 - a_3^2\beta^2}{b^2} = \eta_1(i_1)^2 \tag{13.120}$$

We can then make

$$n^2 = b^2\eta(i_1)^2 + a_3^2\beta(i_3)^2 \tag{13.121}$$

Making use of Equation 13.114, we can write

$$n^2 = a_1^2\alpha^2(i_1) + a_3^2\beta^2(i_3) \tag{13.122}$$

where $\alpha^2(i_1) = (b^2/a_1^2)\eta^2(i_1) = F_3^2(i_1)\eta^2(i_1)$. We then obtain the following two equations:

$$\begin{aligned} \|\nabla M(i_1)\|^2 &= 1/\rho^2 \\ n^2 &= a_1^2\alpha^2(i_1) + a_3^2\beta^2(i_3) \end{aligned} \tag{13.123}$$

These are the equations used in the example given in Section 13.9. The first step is to solve the first equation of Equation 13.123 $\|\nabla M(i_1)\|^2 = 1/\rho^2$ using the boundary conditions. This enables us to obtain the shape of the i_1 and i_3 lines. Then the second equation $n^2 = a_1^2\alpha^2(i_1) + a_3^2\beta^2(i_3)$ is used to obtain the refractive index distribution.

Equation 13.123 can be given other forms. One of them can be obtained by making use of Equation 13.114 to get $a_1^2F_3^2(i_1) = 1/\rho^2$. We can then write

$$\begin{aligned} \|\nabla i_1\|^2 F_3^2(i_1) &= \frac{1}{\rho^2} \\ \|\nabla i_1\|^2 \alpha^2(i_1) + \|\nabla i_3\|^2 \beta^2(i_3) &= n^2 \end{aligned} \tag{13.124}$$

Another possible form of Equation 13.123 can be obtained by defining two functions $A(i_1)$ and $C(i_3)$ such that $dA/di_1 = \alpha$ and $dC/di_3 = \beta$. We can write

$$\begin{aligned} \|\nabla A\|^2 &= \left(\frac{dA}{di_1}\right)^2 \|\nabla i_1\|^2 = \alpha^2 a_1{}^2 \\ \|\nabla C\|^2 &= \left(\frac{dC}{di_3}\right)^2 \|\nabla i_3\|^2 = \beta^2 a_3{}^2 \end{aligned} \tag{13.125}$$

and the second equation of Equation 13.123 can be written as[3]

$$n^2 = \|\nabla A(i_1)\|^2 + \|\nabla C(i_3)\|^2 \tag{13.126}$$

This expression corresponds to expression (13.18) of the 2-D case. So, from the initial system of Equation 13.107, we get two equations.

$$\begin{aligned} \|\nabla M(i_1)\|^2 &= 1/\rho^2 \\ n^2 &= \|\nabla A(i_1)\|^2 + \|\nabla C(i_3)\|^2 \end{aligned} \tag{13.127}$$

The systems of two equations, Equation 13.123 and Equation 13.124, or Equation 13.127 are equivalent. Either one of these systems of two equations can be used to design ideal 3-D concentrators with a variable refractive index.[7]

13.9 An Example of an Ideal Three-Dimensional Concentrator

We now present an example of application of Miñano's design method in three dimensions. We use Equation 13.123, which is rewritten here:

$$\begin{aligned} \|\nabla M(i_1)\|^2 &= 1/\rho^2 \\ n^2 &= a_1^2\alpha^2(i_1) + a_3^2\beta^2(i_3) \end{aligned} \tag{13.128}$$

The example described here can be found in the relevant literature.[3,7,11,12] Begin with the first equation, which is similar to the eikonal equation $\|\nabla S\|^2 = n^2$ where S is the optical path length. In this case, S = constant represents a wavefront, with the rays of light perpendicular to these surfaces. The situation is similar to the first of equations in Equation 13.128 if M is seen as optical path length S and $1/\rho$ is seen as refractive index n. In this case, the surfaces M = constant are the wavefronts, and the lines perpendicular to these surfaces are the rays of light. However, $M = M(i_1)$ and, therefore, M = constant implies that i_1 = constant. Thus, the surfaces i_1 = constant correspond to wavefronts and the lines perpendicular to these surfaces correspond to light rays. Lines perpendicular to surfaces i_1 = constant are the i_1-lines, which must then have the same shape as light rays propagating in a medium of refractive index $n = 1/\rho$. Given the symmetry of the problem, i_3- and i_1-lines are on planes θ = constant, as seen in Figure 13.14. Also, i_3-lines are perpendicular to i_1-lines. Therefore, the i_3-lines must be shaped as the wavefronts of light rays traveling in a medium having a refractive index $n = 1/\rho$.

In a medium of refractive index $n = 1/\rho$, light rays are shaped as circles centered at the axis of symmetry of the optical system, that is, at $\rho = 0$, or, what is the same, the x_3 axis.[7,13] These light rays are solutions of the first of equations in Equation 13.128: $\|\nabla M(i_1)\|^2 = 1/\rho^2$. Therefore, the i_1-lines must be circles centered at the x_3 axis. Rotating the i_1-lines around this axis gives the i_3 = constant surfaces. Two of these surfaces will be used as entrance and exit apertures. Because these surfaces are obtained by rotating circles around the x_3 axis; they are spherical surfaces centered on this axis, the radius of each being a function of i_3 and, if the entrance and exit apertures are flat,

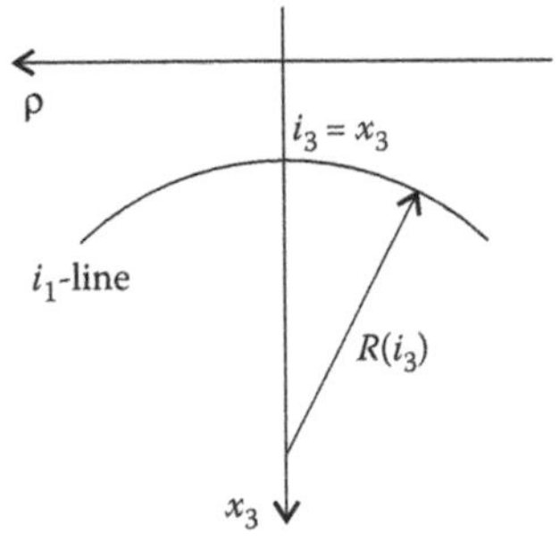

FIGURE 13.15
The i_1-lines (i_3 = constant, θ = constant) are circles centered at the x_3 axis with radius $R(i_3)$.

their radii must be infinite for both these apertures. Also choose the entrance aperture as the surface defined by $i_3 = 0$, and the exit aperture as the surface defined by $i_3 = 1$. We must then have $R(i_3)$ infinite for $i_3 = 0$ and $i_3 = 1$. Figure 13.15 shows the shape of an i_1-line, a circle centered on the x_3 axis.

The equation of this circle is

$$x_3 = R(i_3) + i_3 - \sqrt{R^2(i_3) - \rho^2} \tag{13.129}$$

This is the equation for the i_1-lines (i_3 = constant, θ = constant). For each value of i_3, a value for $R(i_3)$ is obtained, and a circumference defined. Angle θ defines its angle around the x_3 axis. It can be seen that these circumferences cross the x_3 axis ($\rho = 0$) at $x_3 = i_3$. Therefore, the surface $i_3 = 0$ coincides with the surface $x_3 = 0$ and the surface $i_3 = 1$ coincides with the surface $x_3 = 1$. Remember that surfaces $i_3 = 0$ and $i_3 = 1$ correspond to the entrance and exit apertures of the concentrator, and that they were chosen to be flat.

On the planes θ = constant, the i_3-lines are perpendicular to the i_1-lines. They correspond to the wavefronts that are solutions of the first of equations in Equation 13.128, and can be defined so that they cross the plane $x_3 = 0$ with $i_1 = \rho$. The i_1- and i_3-lines are shown in Figure 13.16.

On a plane θ = constant we have, therefore, the same shapes for the i_1- and i_3-lines as we chose earlier for the i_1- and i_2-lines in the 2-D example. The reason why we chose these (apparently strange) shapes in the 2-D case is they work in the design of a 3-D device.

The planes ρ-x_3 in the 3-D case have now the same geometry of the planes x_1-x_2 in the 2-D case. We can use the results of the 2-D case, or 2-D → 3-D, which comprises $x_1 \rightarrow \rho$, $x_2 \rightarrow x_3$, and $i_2 \rightarrow i_3$. Using expression (13.71) for the 2-D case gives i_1 in the 3-D case as

$$i_1 = \frac{2R\rho}{R + \sqrt{R^2 - \rho^2}} \exp\left(\int_0^{i_3} \frac{1}{R}\, di_3 \right) \tag{13.130}$$

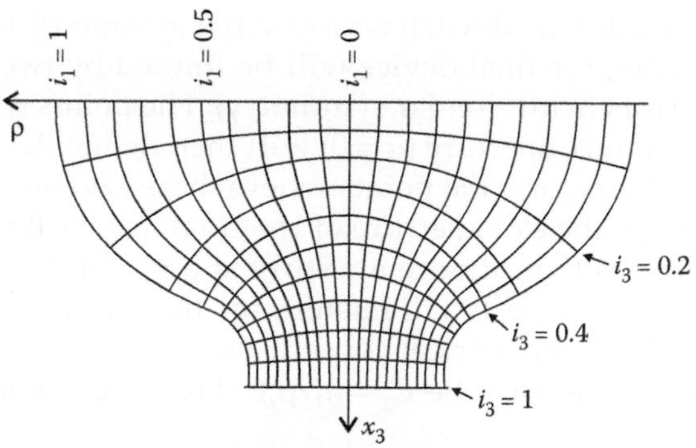

FIGURE 13.16
A set of i_1- and i_3-lines for the concentrator being designed.

and we choose function $M(i_1)$ as

$$M(i_1) = \ln\left(\frac{i_1}{2}\right) = \ln\left[\frac{R}{R/\rho + \sqrt{R^2/\rho^2 - 1}}\exp\left(\int_0^{i_3}\frac{1}{R}di_3\right)\right] \tag{13.131}$$

which is similar to the 2-D case in Equation 13.57. We can rewrite the expression for $M(i_1)$ as

$$M(i_1) = \ln R - \ln\left(\frac{R}{\rho} + \sqrt{\frac{R^2}{\rho^2} - 1}\right) + \int_0^{i_3}\frac{1}{R}di_3 \tag{13.132}$$

It can be verified that $M(i_1)$ given by this equation with the i_1-lines defined by Equation 13.130 satisfies the first equation of Equation 13.128. Calculating the x_3 and ρ derivatives of $M(i_1)$, and using the expressions for $\partial i_3/\partial x_3$ and $\partial i_3/\partial \rho$ obtained from expression (13.129), gives

$$\left(\frac{\partial M}{\partial \rho}\right)^2 = \frac{1}{\rho^2} - \frac{1}{R^2} \quad \text{and} \quad \left(\frac{\partial M}{\partial x_3}\right)^2 = \frac{1}{R^2} \tag{13.133}$$

as we did in expression (13.66) for the 2-D case. It can, therefore, be seen that the first of equations in Equation 13.128 is satisfied.

As mentioned earlier, the i_3-lines were defined in such a way that they cross the plane $x_3 = 0$ with $i_1 = \rho$. The i_3-lines intercept the exit aperture (receiver) for values of ρ such that $\rho = \rho_r$, where ρ_r is given by (see expression (13.72))

$$i_1 = \rho_r \exp\left(\int_0^1 \frac{1}{R}di_3\right) \tag{13.134}$$

The device being designed will have a circular symmetry. Because i_3-lines are vector flux lines, the final device will be limited by two of these lines converted to mirrors (with circular symmetry). The points where these two lines cross the entrance aperture ($i_3 = 0$) will then define the entrance aperture of the final device, and the points where these two lines cross the exit aperture will define the exit aperture of the final device. Because each one of these lines crosses the entrance aperture at $\rho = i_1$, and the exit aperture at $\rho = \rho_r$, the ratio between the diameters for the entrance and exit aperture will be i_1/ρ_r. The geometrical concentration for the concentrator with circular symmetry will then be $C_g = (i_1/\rho_r)^2$. From expression (13.134), we then obtain

$$C_g = \left[\exp\left(\int_0^1 \frac{1}{R}\,di_3\right)\right]^2 \tag{13.135}$$

We still have not given an expression for $R(i_3)$. As stated earlier, this function must be such that $R \to \infty$ when $i_3 \to 0$ and $i_3 \to 1$. One possibility is to choose a function similar to that of the 2-D case earlier (Equation 13.74).

$$R(i_3) = \frac{m}{i_3^2(1 - i_3^2)} \tag{13.136}$$

where m is a constant. To obtain the value of m, we replace this expression for $R(i_3)$ in the expression (13.135) for C_g. We get

$$\ln\left(\sqrt{C_g}\right) = \frac{1}{m}\int_0^1 i_3^2(1 - i_3^2)\,di_3 \Leftrightarrow m = \frac{4}{15\ln C_g} \tag{13.137}$$

We now have a completely defined set of i_1-lines and the corresponding i_3-lines. Because these lines are contained in θ = constant planes, the following step is similar to solving a 2-D problem for these lines, that is, we must now find the 2-D refractive index distribution that transforms the i_3-lines into vector flux lines. The refractive index distribution can be found using the second expression of Equations 13.128, that is, $n^2 = a_1^2\alpha^2(i_1) + a_3^2\beta^2(i_3)$.

This problem can be solved in a manner similar to the previous 2-D problem. Again, the concentrator is designed for maximum concentration. We get the following for x_3 from expression (13.129) and Equation 13.136 for R:

$$x_3 = i_3 + \frac{m}{i_3^2 - i_3^4} - \sqrt{\frac{m^2}{i_3^4(i_3^2 - 1)^2} - \rho} \tag{13.138}$$

This expression is similar to expression (13.94) obtained earlier for the 2-D case. The refractive index is given by

$$n^2 = 4m^2A\left(m - i_3^2(i_3^2 - 1)\sqrt{\frac{m^2}{i_3^4(i_3^2 - 1)^2} - \rho^2}\right)^{-2}$$
$$+ (n_r^{\,2} - A)\left(1 + \frac{2m(2i_3^2 - 1)}{i_3^3(i_3^2 - 1)^2}\right)^{-2}\left(1 + \frac{m(2 - 4i_3^2)}{i_3^3(i_3^2 - 1)^2} - \frac{i_2^4\rho^2(i_3^2 - 1)^2}{m^2}\right)^{-1} \quad (13.139)$$

with $A = \sin^2\varphi \exp[2i_3^3(5 - 3i_3^2)/(15m)]$ and where φ is the acceptance angle of the optic. Giving values to i_3 and ρ enables to obtain n. For the same values of i_3 and ρ we can also obtain x_3 using expression (13.138). Therefore, $n(\rho, x_3)$ can be obtained.

References

1. Miñano, J.C and Benitez, P., Poisson bracket design method review. Application to the elliptic bundles, *SPIE Conference on Nonimaging Optics: Maximum Efficiency Light Transfer V*, SPIE Vol. 3781, 2, 1999.
2. Welford, W.T. and Winston, R., *High Collection Nonimaging Optics*, Academic Press, San Diego, 1989.
3. Miñano, J.C., Poisson brackets method of design of nonimaging concentrators: A review, *SPIE Conference on Nonimaging Optics: Maximum Efficiency Light Transfer II*, SPIE Vol. 2016, 98, 1993.
4. Miñano, J.C, Two-dimensional nonimaging concentrators with inhomogeneous media: A new look, *J. Opt. Soc. Am. A*, 2, 11, 1826, 1985.
5. Gordon, J., *Solar Energy—The State of the Art, ISES Position Papers*, James & James Science Publishers Ltd, London, 2001.
6. Miñano, J.C., Refractive-index distribution in two-dimensional geometry for a given one-parameter manifold of rays, *J. Opt. Soc. Am. A*, 2, 11, 1821, 1985.
7. Miñano, J.C., Design of three-dimensional nonimaging concentrators with inhomogeneous media, *J. Opt. Soc. Am. A*, 3, 9, 1345, 1986.
8. Goldstein, H., *Classical Mechanics*, Addison-Wesley Publishing Company, Reading, 1980.
9. Leech, J.W., *Classical Mechanics*, Chapman & Hall, London, 1965.
10. Honerkamp, J. and Römer, H., *Theoretical Physics, A Classical Approach*, Springer-Verlag, Berlin, 1993.
11. Luque, A., *Solar Cells and Optics for Photovoltaic Concentration*, Adam Hilger, Bristol and Philadelphia, 1989.
12. Winston, R., Miñano, J.C., Benitez, P., with contributions by Shatz, N., Bortz, J.C., *Nonimaging Optics*, Elsevier Academic Press, Amsterdam, 2005.
13. Stavroudis, O.N., *The Optics of Rays, Wave Fronts, and Caustics*, Academic Press, New York, 1972.

Section II

Geometrical Optics

14

Lagrangian and Hamiltonian Geometrical Optics

14.1 Fermat's Principle

All geometrical optics can be derived from Fermat's Principle, which states that, given a light ray between two points and its travel time between them, any adjacent path close to it should have the same travel time. Being a principle, it is not demonstrated, but accepted as being true and used to derive the entire mathematical framework of geometrical optics. It is, however, possible to infer why it describes the behavior of light by analyzing reflection and refraction.

The law of reflection has long been known. Back in the Hellenistic Age, the Hero of Alexandria stated that light travels along the shortest path in a homogeneous medium.[1] His reasoning is illustrated in Figure 14.1a. A light ray is emitted from point $\mathbf{P}_1$, reflects off mirror M at point **A**, and is thereby redirected to point $\mathbf{P}_2$. The distance between the points $\mathbf{P}_1$ and $\mathbf{P}_2$ is the same as that from **Q** to $\mathbf{P}_2$, where **Q** is the mirror image of point $\mathbf{P}_1$. If light would follow a path $\mathbf{P}_1\mathbf{B}\mathbf{P}_2$, which equals $\mathbf{QBP}_2$, or $\mathbf{P}_1\mathbf{C}\mathbf{P}_2$, which equals $\mathbf{QCP}_2$, it would be traveling a longer distance. This principle explains why the angle α between the incident ray and the normal to the surface equals the angle between the normal and the reflected ray.

In mathematical terms, the distance S between the two points $\mathbf{P}_1$ and $\mathbf{P}_2$, as represented in Figure 14.1b, is

$$S = \sqrt{a^2 + x^2} + \sqrt{b^2 + (d - x)^2} \tag{14.1}$$

and, therefore,

$$\frac{dS}{dx} = \frac{1}{2}\frac{2x}{\sqrt{a^2 + x^2}} - \frac{1}{2}\frac{2(d - x)}{\sqrt{b^2 + (d - x)^2}} = \sin\alpha_1 - \sin\alpha_2 \tag{14.2}$$

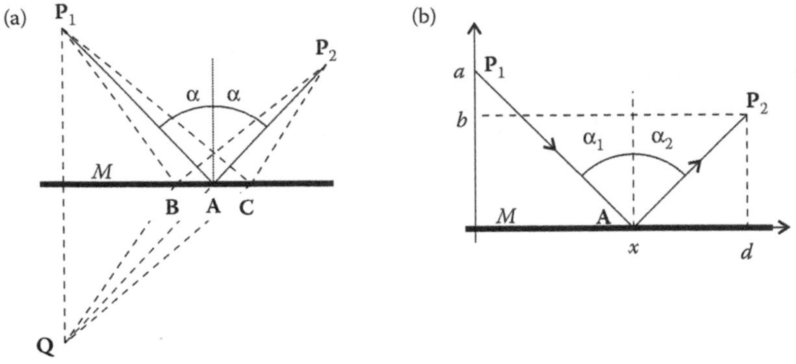

FIGURE 14.1
(a) On reflection, light follows the shortest path between the emitting point $\mathbf{P}_1$ and the end point $\mathbf{P}_2$. (b) The angles α_1 and α_2 that the incident and reflected rays make to the surface normal are equal to each other.

We are looking for the position x of point **A** for which the distance between $\mathbf{P}_1$ and $\mathbf{P}_2$ is minimal. The value of x that minimizes S is obtained by making $dS/dx = 0$, and, therefore,

$$\sin\alpha_1 = \sin\alpha_2 \Leftrightarrow \alpha_1 = \alpha_2 \tag{14.3}$$

which is again the law of reflection.

The principle that light travels between two points along the shortest possible distance does not explain refraction. This is apparent from Figure 14.2a,

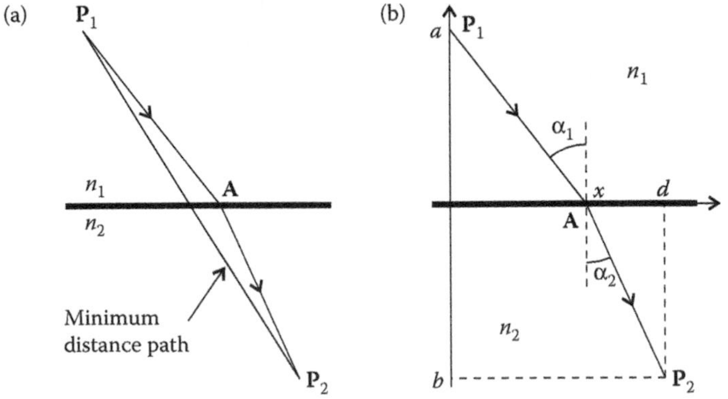

FIGURE 14.2
If the light would follow the shortest path between two points $\mathbf{P}_1$ and $\mathbf{P}_2$, then refraction would not occur, as shown by the minimum distance path in (a). Light, instead, follows the path of minimal time (b).

where we have two media of different refractive indices (e.g., air and water). If light would follow the shortest path, it would go straight from $\mathbf{P}_1$ to $\mathbf{P}_2$ with no refraction. It was Pierre de Fermat who first proposed that light has different speeds in different media, and that it is time that is minimized when light travels between two points $\mathbf{P}_1$ and $\mathbf{P}_2$.[1] Reflection would be a particular case of this principle, in which light always travels in the same medium (always with the same speed), as illustrated in Figure 14.1, and therefore minimizing time is equivalent to minimizing distance. In the case of refraction, the time T taken by light to go from point $\mathbf{P}_1$ to point $\mathbf{P}_2$ (as shown in Figure 14.2b) is given by

$$T = \frac{\sqrt{a^2 + x^2}}{v_1} + \frac{\sqrt{b^2 + (d - x)^2}}{v_2} \tag{14.4}$$

where v_1 is the speed of light in the medium where $\mathbf{P}_1$ is located, and v_2 the speed of light where $\mathbf{P}_2$ is located. We, therefore, have

$$\begin{aligned} \frac{dT}{dx} &= \frac{x}{v_1\sqrt{a^2 + x^2}} - \frac{d - x}{v_2\sqrt{b^2 + (d - x)^2}} = \frac{\sin\alpha_1}{v_1} - \frac{\sin\alpha_2}{v_2} \\ &= (n_1 \sin\alpha_1 - n_2 \sin a_2)/c \end{aligned} \tag{14.5}$$

where $n = c/v$ is the refractive index of the material, and c the speed of light *in vacuum*. We, therefore, have $n_1 = c/v_1$ and $n_2 = c/v_2$. We are looking for the position x of point $\mathbf{A}$ that minimizes the time taken by the light to travel between the points $\mathbf{P}_1$ and $\mathbf{P}_2$. The value of x that minimizes T is obtained by making $dT/dx = 0$, and, therefore,

$$n_1 \sin\alpha_1 = n_2 \sin\alpha_2 \tag{14.6}$$

which is Snell's Law of refraction. Note that when $n_1 = n_2$, we get the law of reflection.

When minimizing T, we used the expression $dT/dx = 0$. An alternative way of thinking about this minimization problem is to consider that, for a small variation dx, we must have $dT = 0$. Expression (14.5) may then be rewritten as

$$dT = \frac{1}{c}(n_1 \sin\alpha_1 - n_2 \sin\alpha_2)dx \tag{14.7}$$

and, therefore,

$$dT = 0 \Leftrightarrow n_1 \sin\alpha_1 - n_2 \sin\alpha_2 = 0 \Leftrightarrow n_1 \sin\alpha_1 = n_2 \sin\alpha_2 \tag{14.8}$$

Expression $dT = 0$ then gives us the laws of refraction and reflection (particularly the case in which $n_1 = n_2$). Expression (14.4) may also be written as

$$T = \frac{s_1}{v_1} + \frac{s_2}{v_2} = \frac{1}{c}(n_1 s_1 + n_2 s_2) = \frac{1}{c} S \tag{14.9}$$

where s_1 is the distance between points $\mathbf{P}_1$ and **A**, and s_2 the distance between points **A** and $\mathbf{P}_2$. Now defining a new quantity called optical path length S, that is the product of the refractive index and distance, we can see that $S = n_1 s_1 + n_2 s_2$ is the optical path length between the points $\mathbf{P}_1$ and $\mathbf{P}_2$. Minimizing T is equivalent to minimizing S, since the latter can be obtained from the former by multiplying with the same constant c. Also, $dT = 0 \Leftrightarrow dS = 0$ since $dS = cT$. We then conclude that $dS = 0$ will yield the laws of reflection and refraction.

The case of reflection is shown in Figure 14.3, where we have a light ray $\mathbf{P}_1\mathbf{A}\mathbf{P}_2$ (an actual path of a light ray). The normal to the surface at **A** is **n**. We also consider a varied path $\mathbf{P}_1\mathbf{B}\mathbf{P}_2$ (not an actual light ray) obtained by displacing point **A** by a distance dx. The refractive index in this example is considered constant with value $n = 1$, and, therefore, light will travel in straight lines. The optical path length for the light ray is $S_1 = [\mathbf{P}_1, \mathbf{A}] + [\mathbf{A}, \mathbf{P}_2]$, where $[\mathbf{X}, \mathbf{Y}]$ is the distance between the points **X** and **Y**. The varied path has an optical path length $S_2 = [\mathbf{P}_1, \mathbf{B}] + [\mathbf{B}, \mathbf{P}_2] = [\mathbf{P}_1, \mathbf{A}] + ds_1 + [\mathbf{A}, \mathbf{P}_2] - ds_2$. But $ds_1 = ds_2$ and, therefore,

$$dS = S_2 - S_1 = 0 \tag{14.10}$$

for the reflection at point **A**.

The principle that light travels along the path for which the optical path length S is minimal still does not explain all the situations. An example that escapes this principle is the reflection by an elliptical mirror, as depicted in

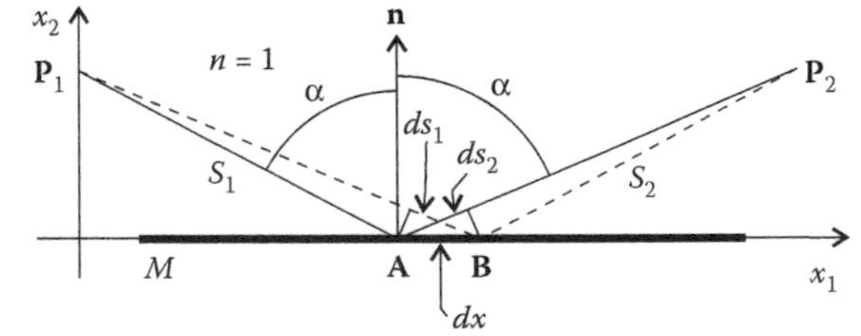

FIGURE 14.3
Path $\mathbf{P}_1\mathbf{A}\mathbf{P}_2$ corresponds to a light ray and path $\mathbf{P}_1\mathbf{B}\mathbf{P}_2$ is a varied path that deviates only infinitesimally from that light ray. If S_1 is the optical path length for $\mathbf{P}_1\mathbf{A}\mathbf{P}_2$ and S_2 for $\mathbf{P}_1\mathbf{B}\mathbf{P}_2$, then $dS = S_2 - S_1 = 0$.

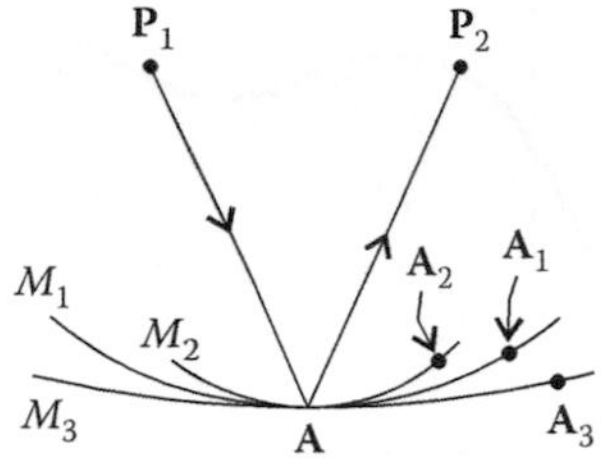

FIGURE 14.4
In general, the path that light follows is not necessarily the one that minimizes travel time. Mirror M_1 is an ellipse with foci $\mathbf{P}_1$ and $\mathbf{P}_2$ and the light travel time (and distance) is the same for all points $\mathbf{A}_1$ on the mirror. Mirror M_2, however, is inside the ellipse and the reflection at point **A** corresponds to a maximum in travel time and distance. Finally, mirror M_3 is exterior to the ellipse and the reflection at point **A** corresponds to a minimum in travel time and distance.

Figure 14.4, where a light ray is emitted from point $\mathbf{P}_1$, reflected off a point **A**, and redirected to point $\mathbf{P}_2$. If we consider that light is reflected by mirror M_1, which is an ellipse with foci $\mathbf{P}_1$ and $\mathbf{P}_2$, the optical path length is constant for all the points on the mirror. If, for example, the reflection was at point $\mathbf{A}_1$, the optical path length would still be the same, since from the definition of an ellipse, $[\mathbf{P}_1, \mathbf{A}] + [\mathbf{A}, \mathbf{P}_2] = [\mathbf{P}_1, \mathbf{A}_1] + [\mathbf{A}_1, \mathbf{P}_2]$ for any point $\mathbf{A}_1$.

If, however, the light is reflected by mirror M_2, which lies inside the ellipse, reflection at any other point $\mathbf{A}_2$ would mean a smaller optical path length and, therefore, for mirror M_2, light travels on a path that maximizes the optical path length.

Finally, if the light is reflected by mirror M_3, which lies outside the ellipse, reflection at any other point $\mathbf{A}_3$ would mean a larger optical path length and, therefore, for mirror M_3, light travels on a path that minimizes the optical path length. In all the three cases, however, $dS = 0$, as can be seen by a similar reasoning to the one Figure 14.3.

In all the situations presented so far, we considered either a constant index of refraction n or two media with indices n_1 and n_2 separated by one surface. In general, however, light will travel in some path in a medium with a refractive index varying from point to point and is therefore given by $n(x_1, x_2, x_3)$. The definition of optical path length between the two points $\mathbf{P}_1$ and $\mathbf{P}_2$ is generalized in this case to

$$S = \int_{\mathbf{P}_1}^{\mathbf{P}_2} n\,ds = \int_{\mathbf{P}_1}^{\mathbf{P}_2} n(x_1, x_2, x_3)\,ds \tag{14.11}$$

Figure 14.5 illustrates this more general situation, showing an arbitrary ray of light going from one point $\mathbf{P}_1$ to another point $\mathbf{P}_2$ in a medium of variable refractive index.

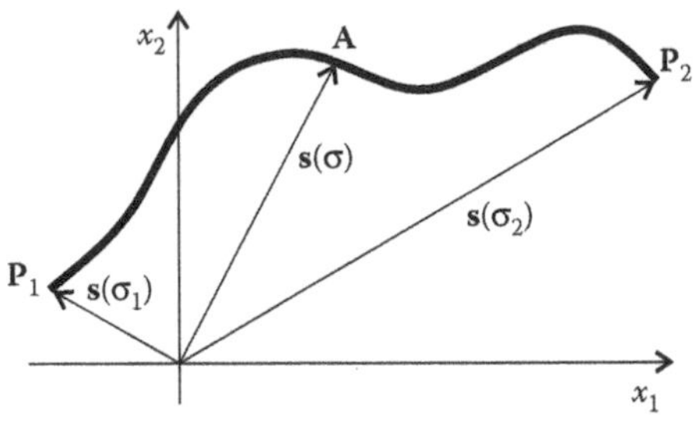

FIGURE 14.5
In general, light will travel in a medium wherein the refractive index changes from point to point and light rays will take the shape of general curves.

Here, any trajectory of light is given by a curve $\mathbf{s}(\sigma) = (x_1(\sigma), x_2(\sigma), x_3(\sigma))$. For each value of the parameter σ there exists a point in space, so that when σ varies continuously between the values σ_1 and σ_2, there exists a space curve that mathematically represents the light ray. We then have $\mathbf{P}_1 = \mathbf{s}(\sigma_1) = (x_1(\sigma_1), x_2(\sigma_1), x_3(\sigma_1))$ and $\mathbf{P}_2 = \mathbf{s}(\sigma_2) = (x_1(\sigma_2), x_2(\sigma_2), x_3(\sigma_2))$. The infinitesimal curve length ds is given by

$$ds = \sqrt{dx_1^2 + dx_2^2 + dx_3^2} = \sqrt{(dx_1/d\sigma)^2 + (dx_2/d\sigma)^2 + (dx_3/d\sigma)^2}\,d\sigma \qquad (14.12)$$

and, therefore, the optical path length along the curve $\mathbf{s}$ from point $\mathbf{P}_1$ to point $\mathbf{P}_2$ is given by

$$S = \int_{\sigma_1}^{\sigma_2} n(x_1, x_2, x_3)\sqrt{(dx_1/d\sigma)^2 + (dx_2/d\sigma)^2 + (dx_3/d\sigma)^2}\,d\sigma \qquad (14.13)$$

In the preceding examples, condition $dS = 0$ was used to describe the light rays in the case in which the varied path depended only on a single parameter x. In general, however, the varied path will have some complex shape that is no longer a function of only one parameter x, but is rather another independent curve that is very close to the first one. The variation to be considered in this more general case is then $\delta S = 0$, where $\delta S = S_2 - S_1$ and S_1 is the optical path length for the light ray, and S_2 the optical path length for the varied path. This more general situation is depicted in Figure 14.6.

This condition ($\delta S = 0$) means that the optical path length is stationary along a light ray. In mathematical terms, we have

$$\delta S = \delta \int_{\mathbf{P}_1}^{\mathbf{P}_2} n\,ds = 0 \qquad (14.14)$$

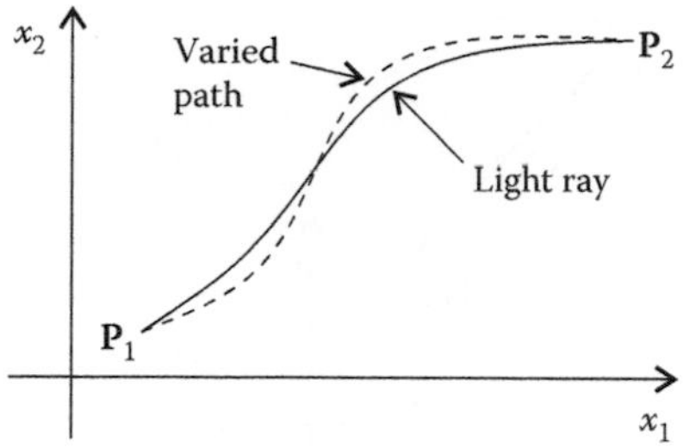

FIGURE 14.6
If a light ray between points $\mathbf{P}_1$ and $\mathbf{P}_2$ has an optical path length S_1 and a varied path that is very close to it has an optical path length S_2, then the variation $\delta S = S_2 - S_1 = 0$.

From Equation 14.13, we can see that expression (14.14) can also be written as

$$\delta \int_{\sigma_1}^{\sigma_2} n(x_1, x_2, x_3)\sqrt{x_1'^2 + x_2'^2 + x_3'^2}\, d\sigma = 0$$

$$\Leftrightarrow \delta \int_{\sigma_1}^{\sigma_2} L(x_1, x_2, x_3, x_1', x_2', x_3')\, d\sigma = 0 \tag{14.15}$$

with $x_k' = dx_k/d\sigma$, where $k = 1,2,3$ and function L is defined by this expression. An example of a geometrical interpretation of the principle defined by expression is represented in Figure 14.7 for the case of refraction on a surface.[2] A light ray coming from a point $\mathbf{P}_1$ refracts at a point $\mathbf{A}$ on the surface, and is redirected toward another point $\mathbf{P}_2$. The optical path length for this ray is

$$S_1 = \int_{\mathbf{P}_1}^{\mathbf{P}_2} n\, ds = n_1 s_1 + n_2 s_2 \tag{14.16}$$

where s_1 and s_2 are, respectively, the distances between $\mathbf{P}_1$ and $\mathbf{A}$ and $\mathbf{A}$ and $\mathbf{P}_2$, and n_1 and n_2 are the refractive indices above and below the surface, respectively.

The optical path length for path $\mathbf{P}_1\mathbf{A}\mathbf{P}_2$ is S_1. Now if point $\mathbf{A}$ is moved slightly in the direction $\delta\mathbf{s}$ tangent to the surface, another (varied) path $\mathbf{P}_1\mathbf{B}\mathbf{P}_2$ is obtained. This path, in general, is not a possible light ray, but its optical path length S_2 is such that $\delta S = S_2 - S_1 = 0$. Path $\mathbf{P}_1\mathbf{A}\mathbf{P}_2$ will only be a possible light ray if this condition is met.

For the varied path $\mathbf{P}_1\mathbf{B}\mathbf{P}_2$, the variation in optical path length is $\delta s_1 = \mathbf{i} \cdot \delta\mathbf{s}$ before refraction and $\delta s_2 = -\mathbf{r} \cdot \delta\mathbf{s}$ after refraction. Unit vector $\mathbf{i}$ points in the

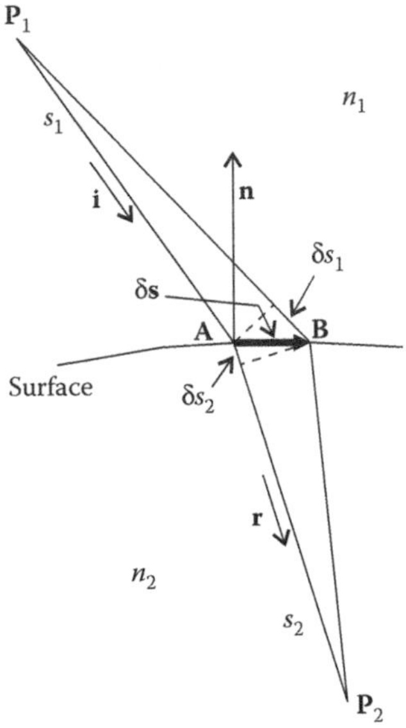

FIGURE 14.7
Light ray path $\mathbf{P}_1\mathbf{AP}_2$ and varied path $\mathbf{P}_1\mathbf{BP}_2$ for the case of refraction on a surface.

direction of the incident ray $\mathbf{P}_1\mathbf{A}$, and unit vector $\mathbf{r}$ points in the direction of the refracted ray $\mathbf{AP}_2$. The variation in the optical path length from path $\mathbf{P}_1\mathbf{AP}_2$ to path $\mathbf{P}_1\mathbf{BP}_2$ is then

$$\delta S = n_1 \delta s_1 + n_2 \delta s_2 = n_1 \mathbf{i} \cdot \boldsymbol{\delta}\mathbf{s} - n_2 \mathbf{r} \cdot \boldsymbol{\delta}\mathbf{s} \tag{14.17}$$

where $\boldsymbol{\delta}\mathbf{s}$ is an infinitesimal vector tangent to the surface. It is also perpendicular to the surface (unit vector) normal $\mathbf{n}$. We can, therefore, write $\boldsymbol{\delta}\mathbf{s} \cdot k\mathbf{n} = 0$ where k is a constant. Since we also have $\delta S = 0$, we obtain

$$(n_1 \mathbf{i} - n_2 \mathbf{r}) \cdot \boldsymbol{\delta}\mathbf{s} = k\mathbf{n} \cdot \boldsymbol{\delta}\mathbf{s} \Leftrightarrow n_1 \mathbf{i} - n_2 \mathbf{r} = k\mathbf{n} \tag{14.18}$$

and, therefore, **i**, **r**, and **n**, that is, the incident ray, refracted ray, and surface normal, are coplanar. We can, therefore, obtain the direction of the refracted ray **r** as a linear combination of the direction of the incident ray **i** and the normal to the surface **n** as

$$\mathbf{r} = \lambda \mathbf{i} + \mu \mathbf{n} \tag{14.19}$$

where λ and μ are scalars. Taking the cross-product of both terms in Equation 14.18, and noting that $\mathbf{n} \times \mathbf{n} = 0$ we obtain Snell's Law,

$$n_1\mathbf{i}\times\mathbf{n}-n_2\mathbf{r}\times\mathbf{n}=k\mathbf{n}\times\mathbf{n}\Leftrightarrow n_1\mathbf{i}\times\mathbf{n}=n_2\mathbf{r}\times\mathbf{n}\Leftrightarrow n_1\sin\alpha_1=n_2\sin\alpha_2 \tag{14.20}$$

where α_1 and α_2 are the angles that the incident and refracted rays make to the normal to the surface. Although the geometry shown in Figure 14.7 is 2-D, the same calculations and results still hold in 3-Ds. In the case of reflection, Equation 14.19 is also valid and Equation 14.20 would be $\alpha_1=\alpha_2$.

14.2 Lagrangian and Hamiltonian Formulations

We start with a mathematical construction and later apply the results to the particular case of optics. We define

$$S=\int_{\mathbf{P}_1}^{\mathbf{P}_2} L(x_1,x_2,\sigma,x_1',x_2')\,d\sigma \tag{14.21}$$

as the integral of a given function $L(x_1,x_2,\sigma,x_1',x_2')$ along a path between the points $\mathbf{P}_1$ and $\mathbf{P}_2$. In this expression, $x_k'=dx_k/d\sigma$. We want to find the path for which

$$\delta S=0\Leftrightarrow\delta\int_{\mathbf{P}_1}^{\mathbf{P}_2} L(x_1,x_2,\sigma,x_1',x_2')\,d\sigma=0 \tag{14.22}$$

Suppose that we have a given path on the plane x_1x_2 parameterized by $\mathbf{c}_1(\sigma)=(x_1(\sigma),x_2(\sigma))$, where σ is some parameter with $\sigma_1\le\sigma\le\sigma_2$. Let us further consider that this path starts at point $\mathbf{P}_1=\mathbf{c}_1(\sigma_1)$ and ends at point $\mathbf{P}_2=\mathbf{c}_1(\sigma_2)$. We define S_1 as the integral of function L along this path $\mathbf{c}_1$ as

$$S_1=\int_{\sigma_1}^{\sigma_2} L(x_1,x_2,\sigma,x_1',x_2')\,d\sigma \tag{14.23}$$

where $x_1(\sigma)$ and $x_2(\sigma)$ now define the curve $\mathbf{c}_1$. For calculating the variation δS we consider a different path $\mathbf{c}_2$ that deviates slightly from $\mathbf{c}_1$ but also starts at $\mathbf{P}_1$ and ends at $\mathbf{P}_2$ as shown in Figure 14.8, a new path is given by $\mathbf{c}_2(\sigma)=(\overset{*}{x}_1(\sigma),\overset{*}{x}_2(\sigma))$ also with $\sigma_1\le\sigma\le\sigma_2$. These two paths are related by

$$\mathbf{c}_2(\sigma)=\mathbf{c}_1(\sigma)+(\delta x_1(\sigma),\delta x_2(\sigma)) \tag{14.24}$$

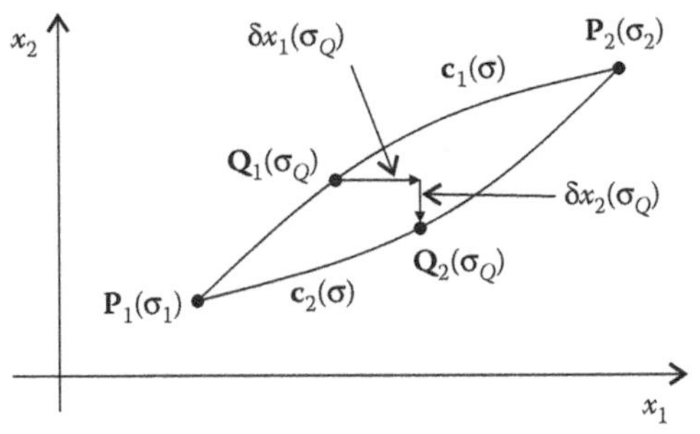

FIGURE 14.8
Path $\mathbf{c}_1$ for a light ray between two points $\mathbf{P}_1$ and $\mathbf{P}_2$ and a separate path $\mathbf{c}_2$ between the same two points. Path $\mathbf{c}_2$ can be related to $\mathbf{c}_1$ by $\mathbf{c}_2 = \mathbf{c}_1 + \eta\ \delta\alpha$ in which $\eta = (\eta_1(\sigma), \eta_2(\sigma))$ with $\eta(\sigma_1) = \eta(\sigma_2) = (0,0)$ and $\delta\alpha$ is an infinitesimal constant.

For example, point $\mathbf{Q}_1 = \mathbf{c}_1(\sigma_Q)$ on the curve $\mathbf{c}_1$ corresponds to point $\mathbf{Q}_2 = \mathbf{c}_2(\sigma_Q)$ on the curve $\mathbf{c}_2$ with $\mathbf{Q}_2 = \mathbf{Q}_1 + (\delta x_1(\sigma_Q), \delta x_2(\sigma_Q))$. We now write

$$\begin{aligned} \delta x_1(\sigma) &= \eta_1(\sigma)\delta\alpha \\ \delta x_2(\sigma) &= \eta_2(\sigma)\delta\alpha \end{aligned} \tag{14.25}$$

in which $\delta\alpha$ is an infinitesimal constant and $\eta_1(\sigma)$ and $\eta_2(\sigma)$ are any two functions of σ. To ensure that $\mathbf{c}_2$ starts at $\mathbf{P}_1$, we must make $\eta_1(\sigma_1) = \eta_2(\sigma_1) = 0$. To assure that $\mathbf{c}_2$ ends at $\mathbf{P}_2$, we must have $\eta_1(\sigma_2) = \eta_2(\sigma_2) = 0$.

The σ derivative of expression (14.25), that is, $\delta x_1 = \eta_1\delta\alpha$ and $\delta x_2 = \eta_2\delta\alpha$ can be written in the following form:

$$\begin{aligned} \delta x_1' &= \eta_1'\delta\alpha \\ \delta x_2' &= \eta_2'\delta\alpha \end{aligned} \tag{14.26}$$

The integral S of Equation 14.21 for the path $\mathbf{c}_2$ is given by

$$S_2 = \int_{\sigma_1}^{\sigma_2} \left(L + \frac{\partial L}{\partial x_1}\eta_1\delta\alpha + \frac{\partial L}{\partial x_2}\eta_2\delta\alpha + \frac{\partial L}{\partial x_1'}\eta_1'\delta\alpha + \frac{\partial L}{\partial x_2'}\eta_2'\delta\alpha \right) d\sigma \tag{14.27}$$

Therefore, the variation of S is given by

$$\delta S = S_2 - S_1 = \delta\alpha \int_{\sigma_1}^{\sigma_2} \left(\frac{\partial L}{\partial x_1}\eta_1 + \frac{\partial L}{\partial x_1'}\eta_1' + \frac{\partial L}{\partial x_2}\eta_2 + \frac{\partial L}{\partial x_2'}\eta_2' \right) d\sigma \tag{14.28}$$

We can now write ($k = 1,2$)

$$\int_{\sigma_1}^{\sigma_2} \left(\frac{\partial L}{\partial x'_k} \eta'_k \right) d\sigma = \int_{\sigma_1}^{\sigma_2} \left[\frac{d}{d\sigma} \left(\frac{\partial L}{\partial x'_k} \eta_k \right) - \frac{d}{d\sigma} \left(\frac{\partial L}{\partial x'_k} \right) \eta_k \right] d\sigma$$
$$= -\int_{\sigma_1}^{\sigma_2} \left[\frac{d}{d\sigma} \left(\frac{\partial L}{\partial x'_k} \right) \eta_k \right] d\sigma \tag{14.29}$$

since $\eta_k(\sigma_1) = \eta_k(\sigma_2) = 0$. The expression for δS can then be written as

$$\delta S = \delta\alpha \int_{\sigma_1}^{\sigma_2} \left(\left[\frac{\partial L}{\partial x_1} - \frac{d}{d\sigma} \left(\frac{\partial L}{\partial x'_1} \right) \right] \eta_1 + \left[\frac{\partial L}{\partial x_2} - \frac{d}{d\sigma} \left(\frac{\partial L}{\partial x'_2} \right) \right] \eta_2 \right) d\sigma \tag{14.30}$$

Considering Equation 14.22 that $\delta S = 0$ and that it is true for any η_1 and η_2, we can write

$$\begin{aligned} \frac{d}{d\sigma} \left(\frac{\partial L}{\partial x'_1} \right) &= \frac{\partial L}{\partial x_1} \\ \frac{d}{d\sigma} \left(\frac{\partial L}{\partial x'_2} \right) &= \frac{\partial L}{\partial x_2} \end{aligned} \tag{14.31}$$

These are the Euler equations for the path $(x_1(\sigma), x_2(\sigma))$.[3–7]

The approach described to solve Equation 14.22 is called the Lagrangian formulation. We now describe an alternative formulation of the problem, called the Hamiltonian formulation. We first define

$$\begin{aligned} p_1 &\equiv \frac{\partial L}{\partial x'_1} \\ p_2 &\equiv \frac{\partial L}{\partial x'_2} \end{aligned} \tag{14.32}$$

Each of these quantities is called momentum. The Euler equation (14.31) can now be written as

$$\begin{aligned} \frac{dp_1}{d\sigma} &= \frac{\partial L}{\partial x_1} \\ \frac{dp_2}{d\sigma} &= \frac{\partial L}{\partial x_2} \end{aligned} \tag{14.33}$$

We now define a new function H by

$$H \equiv x_1' p_1 + x_2' p_2 - L \tag{14.34}$$

Since $L = L(x_1, x_2, \sigma, x_1', x_2')$, from expression (14.32) we can obtain

$$\begin{aligned} x_1' &= x_1'(x_1, x_2, p_1, p_2, \sigma) \\ x_2' &= x_2'(x_1, x_2, p_1, p_2, \sigma) \end{aligned} \tag{14.35}$$

And, therefore, from expression (14.34) we have $H = H(x_1, x_2, p_1, p_2, \sigma)$. We can now obtain for the differential of H as follows[3]

$$dH = \sum_k \frac{\partial H}{\partial x_k} dx_k + \frac{\partial H}{\partial p_k} dp_k + \frac{\partial H}{\partial \sigma} d\sigma \quad (k = 1, 2) \tag{14.36}$$

However, from the definition of H in expression (14.34), we can also obtain for the differential of H as follows:

$$dH = \sum_k x_k' dp_k + p_k dx_k' - \frac{\partial L}{dx_k'} dx_k' - \frac{\partial L}{dx_k} dx_k - \frac{\partial L}{\partial \sigma} d\sigma \quad (k = 1, 2) \tag{14.37}$$

Now considering expression (14.32) and the Euler equation (14.33), we obtain

$$dH = \sum_k x_k' dp_k - p_k' dx_k - \frac{\partial L}{\partial \sigma} d\sigma \quad (k = 1, 2) \tag{14.38}$$

where $p_k' = dp_k / d\sigma$. Comparing Equation 14.36 with Equation 14.38, we obtain

$$\begin{aligned} x_k' &= \frac{\partial H}{\partial p_k} \quad p_k' = -\frac{\partial H}{\partial x_k} \quad (k = 1, 2) \\ \frac{\partial H}{\partial \sigma} &= -\frac{\partial L}{\partial \sigma} \end{aligned} \tag{14.39}$$

The differential equations for $dx_1/d\sigma$, $dx_2/d\sigma$, $dp_1/d\sigma$ and $dp_2/d\sigma$ are called canonical Hamilton equations, in which H is the Hamiltonian.

In the Lagrangian formulation, paths were calculated in space (x_1, x_2), called the configuration space, defined by two second-order differential equations called the Euler equations.

In the Hamiltonian formulation, we have two more variables p_1 and p_2. Paths are now calculated in the space (x_1,x_2,p_1,p_2) called a phase space and described by four first-order differential equations called canonical Hamilton equations. Each variable p_1 and p_2 is called momentum.

An alternative way of deriving the canonical Hamilton equations is from a modified version of Equation 14.22. We replace L by using expression (14.34) to obtain

$$\delta\int_{P_1}^{P_2} (x_1' p_1 + x_2' p_2 - H)\, d\sigma = 0 \tag{14.40}$$

Since $H = H(x_1, x_2, p_1, p_2, \sigma)$, Equation 14.40 is a particular case of a more general equation of the form,[3]

$$\delta\int_{P_1}^{P_2} f(x_1, x_2, p_1, p_2, \sigma, x_1', x_2', p_1', p_2')\, d\sigma = 0 \tag{14.41}$$

which is the same form as Equation 14.22, only with more variables. The corresponding Euler equations are now

$$\begin{aligned} \frac{d}{d\sigma}\left(\frac{\partial f}{\partial x_k'}\right) &= \frac{\partial f}{\partial x_k} \qquad k = 1,2 \\ \frac{d}{d\sigma}\left(\frac{\partial f}{\partial p_k'}\right) &= \frac{\partial f}{\partial p_k} \qquad k = 1,2 \end{aligned} \tag{14.42}$$

In our case, we have $f = x_1' p_1 + x_2' p_2 - H$ and therefore $\partial f/\partial x_k' = p_k$ and $\partial f/\partial x_k = -\partial H/\partial x_k$ and we can write the first group of equations as

$$p_k' = -\frac{\partial H}{\partial x_k} \qquad k = 1,2 \tag{14.43}$$

However, f does not depend explicitly on p_k' and therefore $\partial f/\partial p_k' = 0$ so that the second group of equations reduces to $\partial f/\partial p_k = 0$ or

$$x_k' = \frac{\partial H}{\partial p_k} \qquad k = 1,2 \tag{14.44}$$

These two sets of equations correspond to the canonical Hamilton equation 14.39.

14.3 Optical Lagrangian and Hamiltonian

We now apply the mathematical results obtained earlier to the particular case of optics. Consider a ray of light traveling between points $\mathbf{P}_1$ and $\mathbf{P}_2$. The time T taken by the light ray to travel from point $\mathbf{P}_1$ to point $\mathbf{P}_2$, is given by

$$T = \int_{\mathbf{P}_1}^{\mathbf{P}_2} dt = T_2 - T_1 \tag{14.45}$$

that is, the time of arrival, (T_2), minus the time of departure, (T_1). A different situation occurs when, instead of knowing the time of departure, T_1, and the time of arrival, T_2, we know the path of the light ray and the speed of light at each point. We can now write $ds = vdt$ in which ds is an infinitesimal displacement, v the speed of light in the medium in which it is propagating, and dt an infinitesimal interval of time. Expression (14.45) can then be written as

$$T = \frac{1}{c}\int_{\mathbf{P}_1}^{\mathbf{P}_2} \frac{c}{v}\frac{ds}{dt}dt = \frac{1}{c}\int_{\mathbf{P}_1}^{\mathbf{P}_2} n\,ds \tag{14.46}$$

where $n = c/v$ is the refractive index and c the speed of light in vacuum. We are now calculating T by accumulating (integrating) the infinitesimal times dt taken by the light to cover the infinitesimal distances ds between the points $\mathbf{P}_1$ and $\mathbf{P}_2$.

The optical path length S of a light ray traveling between the two points $\mathbf{P}_1$ and $\mathbf{P}_2$ is given by the integral

$$S = \int_{\mathbf{P}_1}^{\mathbf{P}_2} n\,ds \tag{14.47}$$

in which $n(x_1, x_2, x_3)$ is the refractive index and ds an infinitesimal displacement along the path of the light ray. From the definitions of S and T, it can be seen that they are related by the expression $S = cT$; therefore, if one of them is known, the other can be obtained. The expression for S is, however, totally geometrical since time does not appear in it.

Assume that light propagates in a direction in which x_3 increases, so that we can take x_3 as a parameter and make $x_1 = x_1(x_3)$ and $x_2 = x_2(x_3)$. The trajectories for the light rays can be written as $\mathbf{s} = (x_1(x_3), x_2(x_3), x_3)$. The optical path length can then be written as

$$S = \int n\,ds = \int n\frac{ds}{dx_3}dx_3 = \int L\,dx_3 \tag{14.48}$$

where L is given by

$$L = n\frac{ds}{dx_3} = n\frac{\sqrt{dx_1^2 + dx_2^2 + dx_3^2}}{dx_3} = n\sqrt{1 + x_1'^2 + x_2'^2} \tag{14.49}$$

where $x_1' = dx_1/dx_3$ and $x_2' = dx_2/dx_3$. Considering that $n = n(x_1, x_2, x_3)$, it can be seen that $L = L(x_1, x_2, x_3, x_1', x_2')$. This function L is known as the Lagrangian of the optical system. The laws of geometrical optics can be obtained from Fermat's Principle, which states that,

$$\delta S = 0 \Leftrightarrow \delta\int_{P_1}^{P_2} n\, ds = 0 \tag{14.50}$$

or

$$\delta\int_{P_1}^{P_2} L(x_1, x_2, x_3, x_1', x_2')\, dx_3 = 0 \tag{14.51}$$

As we can see, Equation 14.51 has the same form as Equation 14.23. The main difference is that, instead of parameter σ, we now have coordinate x_3 and, therefore, the paths defined by $(x_1(x_3), x_2(x_3), x_3)$ are light rays in 3-D space. The Euler equation 14.31 now becomes

$$\begin{aligned} \frac{d}{dx_3}\left(\frac{\partial L}{\partial x_1'}\right) &= \frac{\partial L}{\partial x_1} \\ \frac{d}{dx_3}\left(\frac{\partial L}{\partial x_2'}\right) &= \frac{\partial L}{\partial x_2} \end{aligned} \tag{14.52}$$

This is the Lagrangian formulation of geometrical optics.

We now consider the Hamiltonian formulation. Since $dx_3/dx_3 = x_3' = 1$, we can also write $L = n\sqrt{x_1'^2 + x_2'^2 + x_3'^2}$. From the definition of p_k in expression (14.32) and also defining $p_3 = \partial L/\partial x_3'$ we can write

$$\begin{aligned} p_k &= \frac{\partial L}{\partial x_k'} = n\frac{x_k'}{\sqrt{x_1'^2 + x_2'^2 + x_3'^2}} = n\frac{dx_k}{\sqrt{dx_1^2 + dx_2^2 + dx_3^2}} \\ &= n\frac{dx_k}{ds} \quad (k = 1,2,3) \end{aligned} \tag{14.53}$$

or in the vector form $\mathbf{p} = n d\mathbf{s}/ds$. To interpret the physical meaning of vector $\mathbf{p}$, we consider an infinitesimal displacement $d\mathbf{s}$ along a light ray. It can be

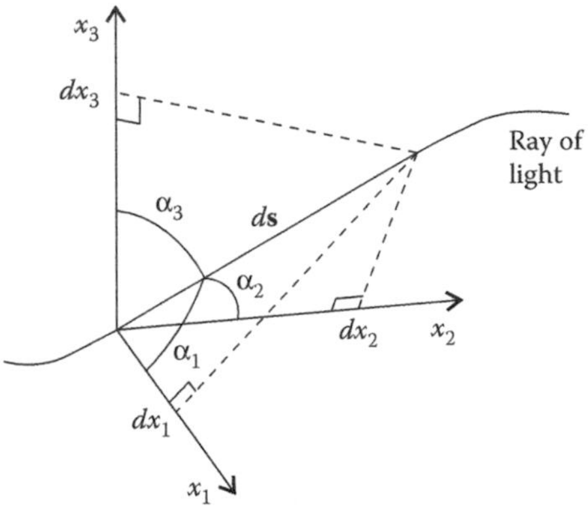

FIGURE 14.9
A displacement $d\mathbf{s}$ along a light ray can be written as $d\mathbf{s} = (dx_1, dx_2, dx_3) = (ds \cos \alpha_1, ds \cos \alpha_2, ds \cos \alpha_3)$.

written in terms of its components along the x_1, x_2, and x_3 axes, as presented in Figure 14.9:

$$d\mathbf{s} = (dx_1, dx_2, dx_3) = (ds \cos\alpha_1, ds \cos\alpha_2, ds \cos\alpha_3) \tag{14.54}$$

with $ds = \|d\mathbf{s}\|$ and α_1, α_2, and α_3 being the angles that displacement $d\mathbf{s}$ makes with the axes x_1, x_2, and x_3, respectively.

Multiplying by the refractive index gives

$$n\frac{d\mathbf{s}}{ds} = (n\cos\alpha_1, n\cos\alpha_2, n\cos\alpha_3) \tag{14.55}$$

From expression (14.53), the optical momentum vector $\mathbf{p}$ is also given by

$$\mathbf{p} = n\frac{d\mathbf{s}}{ds} = \left(n\frac{dx_1}{ds}, n\frac{dx_2}{ds}, n\frac{dx_3}{ds}\right) = (p_1, p_2, p_3) \tag{14.56}$$

This vector is such that $\|\mathbf{p}\| = n$, or

$$p_1^2 + p_2^2 + p_3^2 = n^2 \tag{14.57}$$

and it points along the direction of propagation of the light. It is tangent to the light ray at each point. From Equations 14.55 and 14.56, we can also write $\mathbf{p} = n(\cos \alpha_1, \cos \alpha_2, \cos \alpha_3) = n\mathbf{t}$ with $\|\mathbf{t}\| = 1$. Note that a unit vector projected onto the x_1, x_2, and x_3 axes has coordinates $(\cos\alpha_1, \cos\alpha_2, \cos\alpha_3)$, and, therefore, $\cos^2\alpha_1 + \cos^2\alpha_2 + \cos^2\alpha_3 = 1$.

The Lagrangian L defined by Equation 14.49 can be rewritten as

$$L = n\sqrt{1 + x_1'^2 + x_2'^2} = x_1' \frac{n x_1'}{\sqrt{1 + x_1'^2 + x_2'^2}} + x_2' \frac{n x_2'}{\sqrt{1 + x_1'^2 + x_2'^2}} + n\frac{1}{\sqrt{1 + x_1'^2 + x_2'^2}} \tag{14.58}$$

or, having in consideration expression (14.53)

$$L = x_1' p_1 + x_2' p_2 + p_3 \tag{14.59}$$

Comparing expression (14.59) with expression (14.34) we can see that

$$H = -p_3 = -n\frac{1}{\sqrt{1 + x_1'^2 + x_2'^2}} \tag{14.60}$$

We have $n > 0$ and also $p_3 > 0$ and therefore $H < 0$. From expression $p_3 > 0$ it can be seen that we are considering rays of light with $\cos\alpha_3 > 0$. From Figure 14.9 it can be seen that we must have $0 \le \alpha_k \le \pi$, and, therefore, $\cos\alpha_3 > 0$ implies that $0 \le \alpha_3 < \pi/2$, which confirms the preceding assumption that the rays of light are propagating through the system in the direction of increasing x_3.

We have seen earlier (expression (14.57)) that $p_1^2 + p_2^2 + p_3^2 = n^2$ and, therefore, $p_3 = \sqrt{n^2 - p_1^2 - p_2^2}$ since $p_3 > 0$. Since $H = -p_3$, we can write

$$H = -\sqrt{n^2 - p_1^2 - p_2^2} \tag{14.61}$$

We can finally write, combining Equation 14.61 with Equation 14.39,[8,9]

$$\begin{aligned} \frac{dx_1}{dx_3} &= \frac{\partial H}{\partial p_1} \quad \frac{dp_1}{dx_3} = -\frac{\partial H}{\partial x_1} \\ \frac{dx_2}{dx_3} &= \frac{\partial H}{\partial p_2} \quad \frac{dp_2}{dx_3} = -\frac{\partial H}{\partial x_2} \\ H &= -\sqrt{n^2 - p_1^2 - p_2^2} \end{aligned} \tag{14.62}$$

Since $n = n(x_1, x_2, x_3)$, we have $H = H(x_1, x_2, x_3, p_1, p_2)$ and H depends explicitly on parameter x_3.

14.4 Another Form for the Hamiltonian Formulation

Let us now consider a more general situation than the one described by Equation 14.21 and add one more dimension x_3 to the problem so that we now have

$$S = \int_{P_1}^{P_2} L(x_1, x_2, x_3, \sigma, x_1', x_2', x_3')\, d\sigma \tag{14.63}$$

Equation 14.22 now becomes

$$\delta \int_{P_1}^{P_2} L(x_1, x_2, x_3, \sigma, x_1', x_2', x_3')\, d\sigma = 0 \tag{14.64}$$

where the path between points $\mathbf{P}_1$ and $\mathbf{P}_2$ is now parameterized as $\mathbf{c}(\sigma) = (x_1(\sigma), x_2(\sigma), x_3(\sigma))$. The Euler equation (14.31) of the Lagrangian formulation becomes

$$\begin{aligned} \frac{d}{d\sigma}\left(\frac{\partial L}{\partial x_1'}\right) &= \frac{\partial L}{\partial x_1} \\ \frac{d}{d\sigma}\left(\frac{\partial L}{\partial x_2'}\right) &= \frac{\partial L}{\partial x_2} \\ \frac{d}{d\sigma}\left(\frac{\partial L}{\partial x_3'}\right) &= \frac{\partial L}{\partial x_3} \end{aligned} \tag{14.65}$$

For the Hamiltonian formulation, we define the optical momentum similarly to what we did earlier in expression (14.32) as

$$p_k \equiv \frac{\partial L}{\partial x_k'} \quad (k = 1, 2, 3) \tag{14.66}$$

And if we also define a new Hamiltonian P similar to what we did earlier in expression (14.34) as

$$P \equiv x_1' p_1 + x_2' p_2 + x_3' p_3 - L \tag{14.67}$$

The canonical Hamilton equation 14.39 now becomes

$$\begin{aligned}
\frac{dx_1}{d\sigma} &= \frac{\partial P}{\partial p_1} \quad \frac{dp_1}{d\sigma} = -\frac{\partial P}{\partial x_1} \\
\frac{dx_2}{d\sigma} &= \frac{\partial P}{\partial p_2} \quad \frac{dp_2}{d\sigma} = -\frac{\partial P}{\partial x_2} \\
\frac{dx_3}{d\sigma} &= \frac{\partial P}{\partial p_3} \quad \frac{dp_3}{d\sigma} = -\frac{\partial P}{\partial x_3} \\
\frac{\partial P}{\partial \sigma} &= -\frac{\partial L}{\partial \sigma}
\end{aligned} \tag{14.68}$$

where $P = P(x_1, x_2, x_3, \sigma, p_1, p_2, p_3)$. The differential equations for $dx_k/d\sigma$, and $dp_k/d\sigma$ are the canonical Hamilton equations, in which P is the Hamiltonian. These equations can also be obtained by replacing Equation 14.67 with Equation 14.64 to obtain

$$\delta \int_{P_1}^{P_2} (x_1' p_1 + x_2' p_2 + x_3' p_3 - P)\, d\sigma = 0 \tag{14.69}$$

which has the same form as Equation 14.40 with a new spatial variable x_3 and a new momentum p_3.

We may now apply this result to optics. Now, instead of considering coordinate x_3 as the parameter for the path of the light rays, we consider a generic parameter σ. We then have

$$L = n\frac{ds}{d\sigma} = n\frac{\sqrt{dx_1^2 + dx_2^2 + dx_3^2}}{d\sigma} = n(x_1, x_2, x_3)\sqrt{x_1'^2 + x_2'^2 + x_3'^2} \tag{14.70}$$

where $x_k = x_k(\sigma)$ and $x_k' = dx_k/d\sigma$ Therefore, we obtain a Lagrangian

$$L(x_1, x_2, x_3, x_1', x_2', x_3') \tag{14.71}$$

This is a particular case of Equation 14.64, in which the Lagrangian L does not depend explicitly on parameter σ. We then have $\partial L/\partial\sigma = 0$ and, therefore, from the last equation of Equation 14.68, we get $\partial P/\partial\sigma = 0$ and, therefore, P also does not depend explicitly on σ and we have

$$P = P(x_1, x_2, x_3, p_1, p_2, p_3) \tag{14.72}$$

Now, from Equation 14.70 we can write

$$p_k = \frac{\partial L}{\partial x_k'} = n\frac{x_k'}{\sqrt{x_1'^2 + x_2'^2 + x_3'^2}} \quad (k = 1, 2, 3) \tag{14.73}$$

and, therefore,

$$L = x_1' \frac{n x_1'}{\sqrt{x_1'^2 + x_2'^2 + x_3'^2}} + x_2' \frac{n x_2'}{\sqrt{x_1'^2 + x_2'^2 + x_3'^2}} + x_3' \frac{n x_3'}{\sqrt{x_1'^2 + x_2'^2 + x_3'^2}} \tag{14.74}$$

or

$$L = x_1' p_1 + x_2' p_2 + x_3' p_3 \tag{14.75}$$

From Equation 14.67, we then get $P = 0$ which, together with Equation 14.72, becomes

$$P(x_1, x_2, x_3, p_1, p_2, p_3) = 0 \tag{14.76}$$

From Equation 14.73, we can see that

$$p_1^2 + p_2^2 + p_3^2 - n^2(x_1, x_2, x_3) = 0 \tag{14.77}$$

The optical Hamiltonian is chosen as

$$P = p_1^2 + p_2^2 + p_3^2 - n^2(x_1, x_2, x_3) = 0 \tag{14.78}$$

Expression (14.78) for P together with all except the last equation of (14.68) forms a set of equations that describe the light rays:[5,10–12]

$$\begin{aligned} \frac{dx_1}{d\sigma} &= \frac{\partial P}{\partial p_1} \quad \frac{dp_1}{d\sigma} = -\frac{\partial P}{\partial x_1} \\ \frac{dx_2}{d\sigma} &= \frac{\partial P}{\partial p_2} \quad \frac{dp_2}{d\sigma} = -\frac{\partial P}{\partial x_2} \\ \frac{dx_3}{d\sigma} &= \frac{\partial P}{\partial p_3} \quad \frac{dp_3}{d\sigma} = -\frac{\partial P}{\partial x_3} \\ P &= p_1^2 + p_2^2 + p_3^2 - n^2(x_1, x_2, x_3) = 0 \end{aligned} \tag{14.79}$$

where P is a new Hamiltonian for the system. In this case σ is a parameter along the trajectories of the light rays.

Note that the choice of P is not unique.[13] For example, we have a function $f(x)$ such that $f(x) = 0$ only if $x = 0$ and that $f'(x) \neq 0$ with $f' = df/dx$. We may now choose as a new Hamiltonian $f(P)$. Replacing this in expression (14.79), we obtain

$$\frac{dx_k}{d\sigma} = \frac{\partial(f(P))}{\partial p_k} = f'(P) \frac{\partial P}{\partial p_k} \tag{14.80}$$

Since for the light rays we must have $P = 0$, we can write

$$\frac{dx_k}{d\sigma} = f'(0)\frac{\partial P}{\partial p_k} \tag{14.81}$$

We can now change coordinates to get

$$\frac{dx_k}{d\tau}\frac{d\tau}{d\sigma} = f'(0)\frac{\partial P}{\partial p_k} \tag{14.82}$$

And if we make $d\tau/d\sigma = f'(0)$, we obtain

$$\frac{dx_k}{d\tau} = \frac{\partial P}{\partial p_k} \tag{14.83}$$

which are the same as the original equations, just with a different parameterization. The same can be done for the equations for $dp_k/d\sigma$. For this new Hamiltonian $f(P)$ we also have $f(P) = 0$ since $f(0) = 0$ and $P = 0$.

To verify that the systems of equations (14.62) and (14.79) are equivalent, the equation $P = 0$ can be used to eliminate two of the other equations. We then have

$$P = p_3^2 - (n^2 - p_1^2 - p_2^2) = p_3^2 - H^2 = 0 \Leftrightarrow p_3 = \pm H \tag{14.84}$$

where $H^2 = n^2 - p_1^2 - p_2^2$. Of the two possible solutions of $p_3 = \pm H$, we choose (the reason for which will be presented after the derivation of Equation 14.94)

$$p_3 = -H \tag{14.85}$$

Suppose that light travels along the x_3 axis in the direction of increasing x_3, equivalent to $p_3 > 0$. We should then have $H < 0$, so that

$$H = -\sqrt{n^2(x_1, x_2, x_3) - p_1^2 - p_2^2} \tag{14.86}$$

H is called the Hamiltonian and $H = H(x_1, x_2, x_3, p_1, p_2)$. From expression (14.79), we then have

$$\frac{dx_3}{d\sigma} = \frac{\partial P}{\partial p_3} = \frac{\partial(p_3^2 - H^2)}{\partial p_3} = 2p_3 = -2H \tag{14.87}$$

Expression (14.87) can now be used in the calculation of

$$\frac{dx_2}{d\sigma} = \frac{\partial(p_3^2 - H^2)}{\partial p_2} = -2H\frac{\partial H}{\partial p_2} \Leftrightarrow \frac{dx_2}{d\sigma} = \frac{dx_3}{d\sigma}\frac{\partial H}{\partial p_2} \tag{14.88}$$

so that

$$\frac{dx_2}{dx_3} = \frac{\partial H}{\partial p_2} \tag{14.89}$$

Similarly, we obtain

$$\frac{dx_1}{dx_3} = \frac{\partial H}{\partial p_1} \tag{14.90}$$

We can now make similar calculations for the remaining expressions of (14.79). We then have

$$\frac{dp_3}{d\sigma} = -\frac{\partial P}{\partial x_3} = 2H\frac{\partial H}{\partial x_3} = -\frac{dx_3}{d\sigma}\frac{\partial H}{\partial x_3} \tag{14.91}$$

so that

$$\frac{dp_3}{dx_3} = -\frac{\partial H}{\partial x_3} \Leftrightarrow \frac{dH}{dx_3} = \frac{\partial H}{\partial x_3} \tag{14.92}$$

Similarly, we can further calculate

$$\frac{dp_2}{d\sigma} = -\frac{\partial P}{\partial x_2} = 2H\frac{\partial H}{\partial x_2} = -\frac{dx_3}{d\sigma}\frac{\partial H}{\partial x_2} \Leftrightarrow \frac{dp_2}{dx_3} = -\frac{\partial H}{\partial x_2} \tag{14.93}$$

and also

$$\frac{dp_1}{dx_3} = -\frac{\partial H}{\partial x_1} \tag{14.94}$$

Equations 14.86, 14.89, 14.90, 14.93, and 14.94 can now be put together as the system of equation (14.62). It can be noted that if, in Equation 14.85, we had chosen $p_3 = H$ instead of $p_3 = -H$, the equations for dx_k/dx_3 would be of the form $dx_k/dx_3 = -\partial H/dp_k$ and those for dp_k/dx_3 would be of the form $dp_k/dx_3 = \partial H/dx_k$ instead of the form of Equation 14.62.

From expression (14.86) for H, we have $H = H(x_1(x_3), x_2(x_3), x_3, p_1(x_3), p_2(x_3))$. The total x_3 derivative of H is then given as

$$\frac{dH}{dx_3} = \frac{\partial H}{\partial x_1}\frac{dx_1}{dx_3} + \frac{\partial H}{\partial x_2}\frac{dx_2}{dx_3} + \frac{\partial H}{\partial p_1}\frac{dp_1}{dx_3} + \frac{\partial H}{\partial p_2}\frac{dp_2}{dx_3} + \frac{\partial H}{\partial x_3} \tag{14.95}$$

By using Equation 14.62, we obtain

$$\frac{dH}{dx_3} = \frac{\partial H}{\partial x_1}\frac{dx_1}{dx_3} + \frac{\partial H}{\partial x_2}\frac{dx_2}{dx_3} + \frac{\partial H}{\partial p_1}\frac{dp_1}{dx_3} + \frac{\partial H}{\partial p_2}\frac{dp_2}{dx_3} + \frac{\partial H}{\partial x_3} \tag{14.96}$$

Equation 14.96 is similar to Equation 14.92. It can be seen that Equation 14.96 is implicitly contained in Equation 14.62 and, thus, it need not be included in this set.

14.5 Change of Coordinate System in the Hamilton Equations

The general equation

$$\delta\int L\,d\sigma = 0 \Leftrightarrow \delta\int_{P_1}^{P_2} (x_1' p_1 + x_2' p_2 + x_3' p_3 - P)\,d\sigma = 0 \tag{14.97}$$

in the particular case of optics, corresponds to Fermat's Principle and results in the canonical Hamilton equation 14.79,

$$\begin{aligned} \frac{dx_1}{d\sigma} &= \frac{\partial P}{\partial p_1} & \frac{dp_1}{d\sigma} &= -\frac{\partial P}{\partial x_1} \\ \frac{dx_2}{d\sigma} &= \frac{\partial P}{\partial p_2} & \frac{dp_2}{d\sigma} &= -\frac{\partial P}{\partial x_2} \\ \frac{dx_3}{d\sigma} &= \frac{\partial P}{\partial p_3} & \frac{dp_3}{d\sigma} &= -\frac{\partial P}{\partial x_3} \end{aligned} \tag{14.98}$$

subject to the condition that

$$P = p_1^2 + p_2^2 + p_3^2 - n^2(x_1, x_2, x_3) = 0 \tag{14.99}$$

Now we consider a change of coordinates in the Hamiltonian formulation defined by Equations 14.98 and 14.99. Since we have six independent variables: $x_1(\sigma)$, $x_2(\sigma)$, $x_3(\sigma)$, $p_1(\sigma)$, $p_2(\sigma)$, and $p_3(\sigma)$, the change of coordinates will, in general, be given by[4]

$$\begin{aligned}
i_1 &= i_1(x_1, x_2, x_3, p_1, p_2, p_3, \sigma) \\
i_2 &= i_2(x_1, x_2, x_3, p_1, p_2, p_3, \sigma) \\
i_3 &= i_3(x_1, x_2, x_3, p_1, p_2, p_3, \sigma) \\
u_1 &= u_1(x_1, x_2, x_3, p_1, p_2, p_3, \sigma) \\
u_2 &= u_2(x_1, x_2, x_3, p_1, p_2, p_3, \sigma) \\
u_3 &= u_3(x_1, x_2, x_3, p_1, p_2, p_3, \sigma)
\end{aligned} \tag{14.100}$$

where i_1, i_2, and i_3 are the new spatial coordinates, and u_1, u_2, and u_3 the new momenta. We want the equations of the light rays in these new coordinates to have the same form as Equation 14.98, so the new coordinates defined by Equation 14.100 must also verify

$$\delta \int_{P_1}^{P_2} (i_1' u_1 + i_2' u_2 + i_3' u_3 - Q)\, d\sigma = 0 \tag{14.101}$$

where $i_k' = di_k/d\sigma$ and Q is the new Hamiltonian for these new variables. Equation 14.97 results in Equations 14.98 and 14.99. Also, Equation 14.101, that has the same form as Equation 14.97, merely with new variables, will result in equations of the same form as Equations 14.98 and 14.99 but with the new variables: $x_1 \to i_1$, $x_2 \to i_2$, $x_3 \to i_3$, $p_1 \to u_1$, $p_2 \to u_2$, $p_3 \to u_3$ and $P \to Q$.

Equations 14.97 and 14.101 may be combined to give[3,4]

$$\delta \int_{P_1}^{P_2} [(x_1' p_1 + x_2' p_2 + x_3' p_3 - P) - (i_1' u_1 + i_2' u_2 + i_3' u_3 - Q)]\, d\sigma = 0 \tag{14.102}$$

The condition $\delta \int g\, d\sigma = 0$ is, in general, satisfied by $g = dG/d\sigma$ where G is an arbitrary function.[3,4] Applying this result to Equation 14.102 gives

$$(x_1' p_1 + x_2' p_2 + x_3' p_3 - P) - (i_1' u_1 + i_2' u_2 + i_3' u_3 - Q) = \frac{dG}{d\sigma} \tag{14.103}$$

The transformations of coordinates fulfilling Equation 14.103 are called canonical, and the equations of the light rays in these new coordinates i_k and u_k ($k = 1, 2, 3$) have the same form as Equation 14.98.

In general, the left-hand-side of Equation 14.103 is a function of x_k, p_k, i_k, u_k, and σ, where $k = 1,2,3$. Function G would, therefore, in general, be a function of all the 13 of these variables. They are, however, related by the six equations in Equation 14.100, so that we can reduce the number of independent variables to seven.[4] Depending on whether we choose the old or the new spatial coordinates, and the old and the new momenta as the parameters of G, we obtain different types of generating functions.

Consider a particular case of function G given by

$$G = G_2(x_1, x_2, x_3, u_1, u_2, u_3) - \sum_{k=1}^{3} u_k\, i_k \qquad (14.104)$$

Function G_2 is called a generating function of type 2. Inserting it into Equation 14.103 gives

$$\left(\sum_{k=1}^{3} x_k' p_k - P\right) - \left(\sum_{k=1}^{3} i_k' u_k - Q\right) = \frac{d}{d\sigma}\left(G_2(x_1, x_2, x_3, u_1, u_2, u_3) - \sum_{k=1}^{3} u_k\, i_k\right) \qquad (14.105)$$

and

$$\left(\sum_{k=1}^{3} p_k - \frac{\partial G_2}{\partial x_k}\right) dx_k + \left(\sum_{k=1}^{3} i_k - \frac{\partial G_2}{\partial u_k}\right) du_k + (Q - P)\, d\sigma = 0 \qquad (14.106)$$

Since x_k, u_k, and σ are independent variables, we have

$$\begin{aligned} p_k &= \frac{\partial G_2}{\partial x_k} \\ i_k &= \frac{\partial G_2}{\partial u_k} \\ Q &= P \end{aligned} \qquad (14.107)$$

with $k = 1, 2, 3$. We now choose a specific function G_2 given by

$$\begin{aligned} G_2 &= G_2(x_1, x_2, x_3, u_1, u_2, u_3) \\ &= u_1 i_1(x_1, x_2, x_3) + u_2 i_2(x_1, x_2, x_3) + u_3 i_3(x_1, x_2, x_3) \end{aligned} \qquad (14.108)$$

and from Equation 14.107 we obtain

$$p_k = \frac{\partial G_2}{\partial x_k} = u_1 \frac{\partial i_1}{\partial x_k} + u_2 \frac{\partial i_2}{\partial x_k} + u_3 \frac{\partial i_3}{\partial x_k} \qquad (14.109)$$

or in the vector form

$$\mathbf{p} = u_1\nabla i_1 + u_2\nabla i_2 + u_3\nabla i_3 \tag{14.110}$$

Also, since

$$G_2 = \sum_{k=1}^{3} u_k i_k \tag{14.111}$$

we have $i_k = \partial G_2/\partial u_k$, with $k = 1, 2, 3$ or

$$\begin{aligned} i_1 &= i_1(x_1, x_2, x_3) \\ i_2 &= i_2(x_1, x_2, x_3) \\ i_3 &= i_3(x_1, x_2, x_3) \end{aligned} \tag{14.112}$$

and this set of equations gives us the transformation from the old coordinates (x_1,x_2,x_3) to the new coordinates (i_1,i_2,i_3), called a point transformation, because it involves only the spatial coordinates and not the momenta.

The transformations between G_2 and G given by Equation 14.104, and that between the Lagrangian and the Hamiltonian in Equation 14.67, are called Legendre transformations.

Since the transformations of coordinates between x_1, x_2, x_3, p_1, p_2, p_3, and i_1, i_2, i_3, u_1, u_2, u_3 is canonical, it preserves the form of Equation 14.98. In the same way as from Equation 14.97, we get Equations 14.98 and 14.99, from Equation 14.101 we get, since $Q = P$,

$$\begin{aligned} \frac{di_1}{d\sigma} &= \frac{\partial P}{\partial u_1} \quad \frac{du_1}{d\sigma} = -\frac{\partial P}{\partial i_1} \\ \frac{di_2}{d\sigma} &= \frac{\partial P}{\partial u_2} \quad \frac{du_2}{d\sigma} = -\frac{\partial P}{\partial i_2} \\ \frac{di_3}{d\sigma} &= \frac{\partial P}{\partial u_3} \quad \frac{du_3}{d\sigma} = -\frac{\partial P}{\partial i_3} \\ P &= \mathbf{p}\cdot\mathbf{p} - n^2 = 0 \end{aligned} \tag{14.113}$$

We can now rewrite the expression for the optical momentum $\mathbf{p}$ as

$$\mathbf{p} = u_1 \,\|\nabla i_1\|\, \frac{\nabla i_1}{\|\nabla i_1\|} + u_2 \,\|\nabla i_2\|\, \frac{\nabla i_2}{\|\nabla i_2\|} + u_3 \,\|\nabla i_3\|\, \frac{\nabla i_3}{\|\nabla i_3\|} \tag{14.114}$$

or

$$\mathbf{p} = u_1 a_1 \mathbf{e}_1 + u_2 a_2 \mathbf{e}_2 + u_3 a_3 \mathbf{e}_3 \tag{14.115}$$

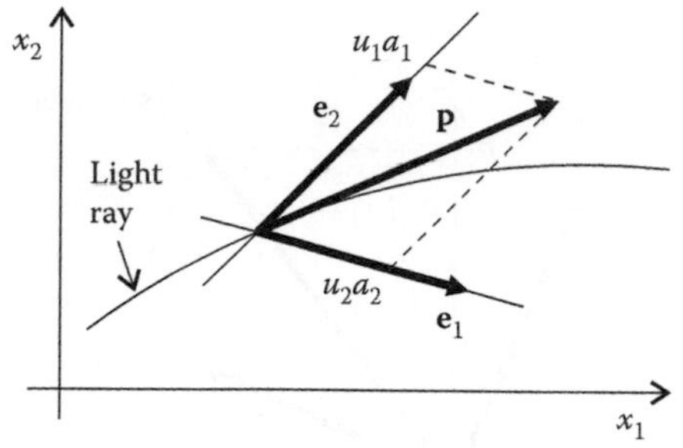

FIGURE 14.10
Components of the optical momentum $\mathbf{p}$ in a basis defined by unit vectors $\mathbf{e}_1$ and $\mathbf{e}_2$ in a 2-D geometry.

with $a_k = \|\nabla i_k\|$ and $\mathbf{e}_k = \nabla i_k / \|\nabla i_k\|$. Vectors $\mathbf{e}_1$, $\mathbf{e}_2$, and $\mathbf{e}_3$ form a basis of unit vectors. This geometry is shown in Figure 14.10 for the 2-D case. It should be noted that, since $i_k = i_k(x_1, x_2, x_3)$, and $a_k = \|\nabla i_k\|$, we have $a_k = a_k(x_1, x_2, x_3)$ or, writing x_1, x_2, and x_3 as functions of i_1, i_2, and i_3 by using Equation 14.112, we have $a_k = a_k(i_1, i_2, i_3)$.

We now further restrict the transformation of coordinates in Equation 14.112 to the case in which vectors $\mathbf{e}_1$, $\mathbf{e}_2$, and $\mathbf{e}_3$ are orthogonal, that is,

$$\nabla i_1 \cdot \nabla i_2 = \nabla i_2 \cdot \nabla i_3 = \nabla i_1 \cdot \nabla i_3 = 0 \tag{14.116}$$

Vector ∇i_1 is perpendicular to the surface i_1 = constant, as shown in Figure 14.11. If, instead of having i_1 as constant, both i_2 and i_3 are kept constant simultaneously, we obtain a line along which only i_1 varies, called an i_1-line.

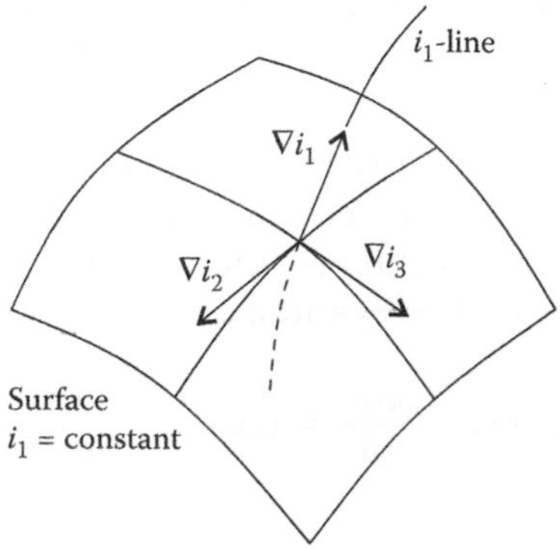

FIGURE 14.11
A system of three curvilinear orthogonal coordinates, i_1, i_2, and i_3. Also shown is the i_1-line and the surface i_1 = constant.

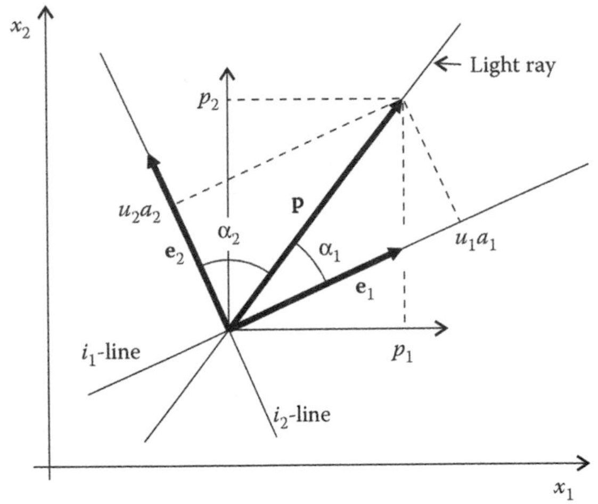

FIGURE 14.12
Momentum **p** in a 2-D coordinate system i_1, i_2 is given as $\mathbf{p} = u_1a_1\mathbf{e}_1 + u_2a_2\mathbf{e}_2$ in which $a_k = \|\nabla i_k\|$ and the unit vectors $\mathbf{e}_k$ are given as $\mathbf{e}_k = \nabla i_k/\|\nabla i_k\|$. Since $\|\mathbf{p}\| = n$, one can conclude that $u_k a_k = n \cos \alpha_k$, which can now be extended to 3-D systems.

Such a line is perpendicular to the surface i_1 = constant. Vector ∇i_1 is tangent to the i_1-line. Similar conclusions can be drawn for i_2 and i_3.

This case, in which vectors $\mathbf{e}_1$, $\mathbf{e}_2$, and $\mathbf{e}_3$ form an orthogonal unit basis, is presented in Figure 14.12 for 2-D geometry. Note that, in general, the i_1- and i_2-lines, and the light rays will be curved.

From the last expression of (14.113) we have $\|\mathbf{p}\| = n$. Multiplying expression (14.115) by $1/n$ gives

$$\frac{\mathbf{p}}{\|\mathbf{p}\|} = \frac{u_1a_1}{n}\mathbf{e}_1 + \frac{u_2a_2}{n}\mathbf{e}_2 + \frac{u_3a_3}{n}\mathbf{e}_3 \tag{14.117}$$

where $\mathbf{p}/\|\mathbf{p}\|$ is a unit vector. Thus, we can conclude that u_1a_1/n is the direction cosine of the angle α_1 that the vector **p** makes with the vector $\mathbf{e}_1$. Similarly, u_2a_2/n and u_3a_3/n are the direction cosines of the angles α_2 and α_3 that the vector **p** makes with the vectors $\mathbf{e}_2$ and $\mathbf{e}_3$:

$$\frac{u_1a_1}{n} = \cos\alpha_1, \quad \frac{u_2a_2}{n} = \cos\alpha_2, \quad \frac{u_3a_3}{n} = \cos\alpha_3 \tag{14.118}$$

We then obtain

$$\mathbf{p} = n\cos\alpha_1\mathbf{e}_1 + n\cos\alpha_2\mathbf{e}_2 + n\cos\alpha_3\mathbf{e}_3 \tag{14.119}$$

With expression (14.115), expression (14.113) can be finally written as

$$\begin{aligned} &\frac{di_1}{d\sigma} = \frac{\partial P}{\partial u_1} \quad \frac{du_1}{d\sigma} = -\frac{\partial P}{\partial i_1} \\ &\frac{di_2}{d\sigma} = \frac{\partial P}{\partial u_2} \quad \frac{du_2}{d\sigma} = -\frac{\partial P}{\partial i_2} \\ &\frac{di_3}{d\sigma} = \frac{\partial P}{\partial u_3} \quad \frac{du_3}{d\sigma} = -\frac{\partial P}{\partial i_3} \\ &P = u_1^2 a_1^2 + u_2^2 a_2^2 + u_3^2 a_3^2 - n^2 = 0 \end{aligned} \tag{14.120}$$

These are the canonical Hamilton equations for the generalized coordinates.

14.6 Integral Invariants

Let us take S given by Equation 14.63, but now along a path of the system from point $\mathbf{P}_A = \mathbf{P}_A(\sigma_A) = (x_1(\sigma_A),x_2(\sigma_A),x_3(\sigma_A)) = (x_{1A},x_{2A},x_{3A})$ to point $\mathbf{P}_B = \mathbf{P}_B(\sigma_B) = (x_1(\sigma_B),x_2(\sigma_B),x_3(\sigma_B)) = (x_{1B},x_{2B},x_{3B})$. If this path is known, then S is given by[14]

$$\begin{aligned} S &= S(\mathbf{P}_A,\sigma_A,\mathbf{P}_B,\sigma_B) = S(x_{1A},x_{2A},x_{3A},\sigma_A,x_{1B},x_{2B},x_{3B},\sigma_B) \\ &= \int_{\sigma_A}^{\sigma_B} L(x_1,x_2,x_3,\sigma,x_1',x_2',x_3')\, d\sigma \end{aligned} \tag{14.121}$$

Let us now consider a trajectory t_P in configuration space (x_1,x_2,x_3). It obeys the Hamilton equations of motion from point $\mathbf{P}_A(x_{1A},x_{2A},x_{3A},\sigma_A)$ to a point $\mathbf{P}_B(x_{1B},x_{2B},x_{3B},\sigma_B)$ as in Figure 14.13. For this trajectory, function S has a value $S_P = S(x_{1A},x_{2A},x_{3A},\sigma_A,\ x_{1B},x_{2B},x_{3B},\sigma_B)$. Now, consider another trajectory t_Q an infinitesimal distance away from t_P. Trajectory t_Q extends from a point $\mathbf{Q}_A(x_{1A} + dx_{1A},\ x_{2A} + dx_{2A},\ x_{3A} + dx_{3A},\ \sigma_A + d\sigma_A)$ to a point $\mathbf{Q}_B(x_{1B} + dx_{1B},\ x_{2B} + dx_{2B},\ x_{3B} + dx_{3B},\ \sigma_B + d\sigma_B)$. For this trajectory t_Q, function S has another value $S_Q = S(x_{1A} + dx_{1A},\ x_{2A} + dx_{2A},\ x_{3A} + dx_{3A},\ \sigma_A + d\sigma_A,\ x_{1B} + dx_{1B},\ x_{2B} + dx_{2B},\ x_{3B} + dx_{3B},\ \sigma_B + d\sigma_B)$. The variation of S from t_P to t_Q is given by $dS = S_Q - S_P$.

Now consider a trajectory t_R a finite distance away from t_P. Trajectory t_R extends from a point $\mathbf{R}_A(x_{1C},x_{2C},x_{3C},\sigma_C)$ to a point $\mathbf{R}_B(x_{1D},x_{2D},x_{3D},\sigma_D)$. For this trajectory, function S has a value $S_R = S(x_{1C},x_{2C},x_{3C},\sigma_C,\ x_{1D},x_{2D},x_{3D},\sigma_D)$. The total variation of S from t_P to t_R is $S_R - S_P = \int dS$.

The variation dS of function S relative to its parameters is given by

$$\begin{aligned} dS = &\frac{\partial S}{\partial x_{1A}} dx_{1A} + \frac{\partial S}{\partial x_{2A}} dx_{2A} + \frac{\partial S}{\partial x_{3A}} dx_{3A} + \frac{\partial S}{\partial \sigma_A} d\sigma_A \\ &+ \frac{\partial S}{\partial x_{1B}} dx_{1B} + \frac{\partial S}{\partial x_{2B}} dx_{2B} + \frac{\partial S}{\partial x_{3B}} dx_{3B} + \frac{\partial S}{\partial \sigma_B} d\sigma_B \end{aligned} \tag{14.122}$$

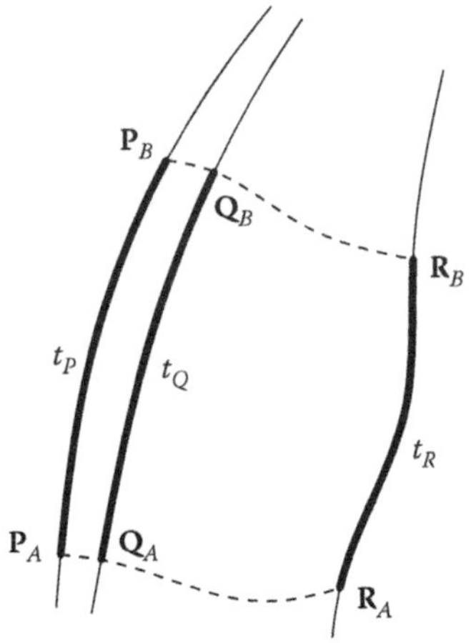

FIGURE 14.13
Trajectory t_P in configuration space from point $\mathbf{P}_A$ to point $\mathbf{P}_B$. Also shown are trajectories t_Q at an infinitesimal distance from t_P and trajectory t_R at a finite distance from t_P.

In order to obtain an equation for dS, we need to calculate the partial derivatives in this expression.

We first consider variations of the end coordinates x_{1A}, x_{2A}, x_{3A} (constant σ_A), and x_{1B}, x_{2B}, x_{3B} (constant σ_B), which will allow us to obtain $\partial S/\partial x_{1A}$, $\partial S/\partial x_{2A}$, $\partial S/\partial x_{3A}$, and $\partial S/\partial x_{1B}$, $\partial S/\partial x_{2B}$, $\partial S/\partial x_{3B}$ in expression (14.122). The variation of S is in this case given by

$$\delta S = \int_{\sigma_A}^{\sigma_B} \sum_{k=1}^{3} \left(\frac{\partial L}{\partial x_k} \delta x_k + \frac{\partial L}{\partial x'_k} \delta x'_k \right) d\sigma \tag{14.123}$$

Using the Euler equation 14.65 we get

$$\begin{aligned} \delta S &= \int_{\sigma_A}^{\sigma_B} \sum_{k=1}^{3} \left(\frac{d}{d\sigma}\left(\frac{\partial L}{\partial x'_k} \right) \delta x_k + \frac{\partial L}{\partial x'_k} \delta x'_k \right) d\sigma \\ &= \int_{\sigma_A}^{\sigma_B} \sum_{k=1}^{3} \frac{d}{d\sigma}\left(\frac{\partial L}{\partial x'_k} \delta x_k \right) d\sigma \end{aligned} \tag{14.124}$$

Using Equation 14.66 this can also be written as

$$\begin{aligned} \delta S &= \int_{\sigma_A}^{\sigma_B} \frac{d}{d\sigma}(p_1 \delta x_1 + p_2 \delta x_2 + p_3 \delta x_3) d\sigma = (p_1 \delta x_1 + p_2 \delta x_2 + p_3 \delta x_3)\Big|_{\sigma_A}^{\sigma_B} \\ &= (p_{1B}\delta x_{1B} + p_{2B}\delta x_{2B} + p_{3B}\delta x_{3B}) - (p_{1A}\delta x_{1A} + p_{2A}\delta x_{2A} + p_{3A}\delta x_{3A}) \end{aligned} \tag{14.125}$$

where $p_{kA} = p_k(\sigma_A)$, $p_{kB} = p_k(\sigma_B)$, $\delta x_{kA} = \delta x_k(\sigma_A)$, and $\delta x_{kB} = \delta x_k(\sigma_B)$, $k = 1,2,3$. However, from expression (14.121), we also have

$$\delta S = \frac{\partial S}{\partial x_{1A}}\delta x_{1A} + \frac{\partial S}{\partial x_{2A}}\delta x_{2A} + \frac{\partial S}{\partial x_{3A}}\delta x_{3A} + \frac{\partial S}{\partial x_{1B}}\delta x_{1B} + \frac{\partial S}{\partial x_{2B}}\delta x_{2B} + \frac{\partial S}{\partial x_{3B}}\delta x_{3B} \tag{14.126}$$

From Equations 14.125 and 14.126 we get

$$\begin{aligned} p_{1A} &= -\frac{\partial S}{\partial x_{1A}} \quad & p_{2A} &= -\frac{\partial S}{\partial x_{2A}} \quad & p_{3A} &= -\frac{\partial S}{\partial x_{3A}} \\ p_{1B} &= \frac{\partial S}{\partial x_{1B}} \quad & p_{2B} &= \frac{\partial S}{\partial x_{2B}} \quad & p_{3B} &= \frac{\partial S}{\partial x_{3B}} \end{aligned} \tag{14.127}$$

and in particular, at $\mathbf{P}_B$, the end of the path, we get

$$\mathbf{p} = \nabla S \tag{14.128}$$

where $\mathbf{p} = (p_1, p_2, p_3)$ and $\nabla S = (\partial S/\partial x_1, \partial S/\partial x_2, \partial S/\partial x_3)$.

Equation 14.127 is a set of six first-order differential equations. Note that if function S and six initial conditions $q_{1A}, q_{2A}, q_{3A}, p_{1A}, p_{2A}, p_{3A}$ are known for $\sigma = \sigma_A$, Equation 14.127 may be used to solve for $q_{1B}, q_{2B}, q_{3B}, p_{1B}, p_{2B}, p_{3B}$ at another parameter $\sigma = \sigma_B$.[14] Initial conditions q_{1A}, q_{2A}, q_{3A} define the initial position of a light ray and p_{1A}, p_{2A}, p_{3A} its initial direction and refractive index. Now we consider variations in σ_A and σ_B, which will allow us to obtain $\partial S/\partial\sigma_A$ and $\partial S/\partial\sigma_B$. For variations of parameter value $\sigma = \sigma_B$ from expression (14.121) and the fundamental theorem of calculus,[15] we have

$$\frac{dS}{d\sigma_B} = L_B \tag{14.129}$$

where $dS/d\sigma_B$ is $dS/d\sigma$ calculated for $\sigma = \sigma_B$ and $L_B = L(\sigma_B)$. Using Equation 14.127 we can rewrite this expression as

$$\begin{aligned} L_B &= \frac{\partial S}{\partial \sigma_B} + \frac{\partial S}{\partial x_{1B}}\frac{\partial x_{1B}}{\partial \sigma_B} + \frac{\partial S}{\partial x_{2B}}\frac{\partial x_{2B}}{\partial \sigma_B} + \frac{\partial S}{\partial x_{3B}}\frac{\partial x_{3B}}{\partial \sigma_B} \\ &= \frac{\partial S}{\partial \sigma_B} + p_{1B}x'_{1B} + p_{2B}x'_{2B} + p_{3B}x'_{3B} \end{aligned} \tag{14.130}$$

and from Equation 14.67 we get

$$\frac{\partial S}{\partial \sigma_B} = -P_B \tag{14.131}$$

where P_B is the Hamiltonian at $\sigma = \sigma_B$.

Now, for variations of parameter value $\sigma = \sigma_A$ from (14.121) and the fundamental theorem of calculus, we have

$$\frac{dS}{d\sigma_A} = -L_A \tag{14.132}$$

and from Equation 14.127 we get

$$\begin{aligned} -L_A &= \frac{\partial S}{\partial \sigma_A} + \frac{\partial S}{\partial x_{1A}}\frac{\partial x_{1A}}{\partial \sigma_A} + \frac{\partial S}{\partial x_{2A}}\frac{\partial x_{2A}}{\partial \sigma_A} + \frac{\partial S}{\partial x_{3A}}\frac{\partial x_{3A}}{\partial \sigma_A} \\ &= \frac{\partial S}{\partial \sigma_A} - p_{1A}x'_{1A} - p_{2A}x'_{2A} - p_{3A}x'_{3A} \end{aligned} \tag{14.133}$$

and from Equation 14.67 we get

$$\frac{\partial S}{\partial \sigma_A} = P_A \tag{14.134}$$

where P_A is the Hamiltonian at $\sigma = \sigma_A$.

If we now allow variations of all arguments in expression (14.121), we get from Equation 14.122 and from Equations 14.127, 14.131, and 14.134,

$$\begin{aligned} dS &= (p_{1B}dx_{1B} + p_{2B}dx_{2B} + p_{3B}dx_{3B} - P_B d\sigma_B) \\ &\quad - (p_{1A}dx_{1A} + p_{2A}dx_{2A} + p_{3A}dx_{3A} - P_A d\sigma_A) \end{aligned} \tag{14.135}$$

which can also be written as

$$\begin{aligned} dS &= (p_1dx_1 + p_2dx_2 + p_3dx_3 - Pd\sigma)_{\sigma=\sigma_B} \\ &\quad - (p_1dx_1 + p_2dx_2 + p_3dx_3 - Pd\sigma)_{\sigma=\sigma_A} \end{aligned} \tag{14.136}$$

Now consider a closed curve c_A in configuration space (x_1,x_2,x_3), as shown in Figure 14.14. Each one of its points is crossed by a trajectory that obeys the Hamilton equations of motion (none of these trajectories are tangent to c_A). As these trajectories evolve in configuration space, each one of these trajectories crosses a point of another closed curve c_B (none of these trajectories are tangent to c_B). Each one of these trajectories t_P then crosses one point $\mathbf{P}_A(x_{1A},x_{2A},x_{3A},\sigma_A)$ of c_A and one point $\mathbf{P}_B(x_{1B},x_{2B},x_{3B},\sigma_B)$ of c_B and there is, therefore, a one-to-one correspondence between the points of c_A and c_B. Now, as point $\mathbf{P}_A(x_{1A},x_{2A},x_{3A},\sigma_A)$ moves to a neighboring point $\mathbf{Q}_A(x_{1A} + dx_{1A}, x_{2A} + dx_{2A}, x_{3A} + dx_{3A}, \sigma_A + d\sigma_A)$ on curve c_A, point $\mathbf{P}_B(x_{1B},x_{2B},x_{3B},\sigma_B)$ moves to a neighboring point $\mathbf{Q}_B(x_{1B} + dx_{1B}, x_{2B} + dx_{2B}, x_{3B} + dx_{3B}, \sigma_B + d\sigma_B)$ on curve c_B. The difference

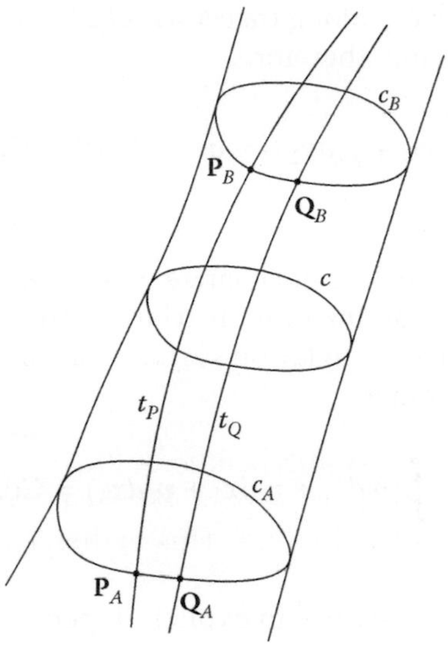

FIGURE 14.14
Trajectories t_P, t_Q, … crossing closed curve c_A in configuration space evolve and cross another closed curve c_B.

dS in the value of S when calculated along trajectory t_P from $\mathbf{P}_A$ to $\mathbf{P}_B$ and calculated along trajectory t_Q from $\mathbf{Q}_A$ to $\mathbf{Q}_B$ is given by expression (14.122), or its equivalent (14.135). Here $p_{1B}dx_{1B} + p_{2B}dx_{2B} + p_{3B}dx_{3B} - P_B d\sigma_B$ corresponds to the displacement from $\mathbf{P}_B$ to $\mathbf{Q}_B$ and $p_{1A}dx_{1A} + p_{2A}dx_{2A} + p_{3A}dx_{3A} - P_A d\sigma_A$ corresponds to the displacement from $\mathbf{P}_A$ to $\mathbf{Q}_A$.

If now point $\mathbf{P}_A$ goes all around curve c_A returning to the starting point $\mathbf{P}_A$ and corresponding point $\mathbf{P}_B$ goes all around curve c_B returning to the starting point $\mathbf{P}_B$, the total variation of S will be zero, $\oint dS = 0$. This corresponds, in Figure 14.13, to point $\mathbf{R}_A$ going back to point $\mathbf{P}_A$ and point $\mathbf{R}_B$ going back to point $\mathbf{P}_B$. From expression (14.136) we get

$$\oint_{c_A} (p_1 dx_1 + p_2 dx_2 + p_3 dx_3 - P d\sigma) - \oint_{c_B} (p_1 dx_1 + p_2 dx_2 + p_3 dx_3 - P d\sigma) = 0 \tag{14.137}$$

or

$$\oint_{c_A} (p_1 dx_1 + p_2 dx_2 + p_3 dx_3 - P d\sigma) = \oint_{c_B} (p_1 dx_1 + p_2 dx_2 + p_3 dx_3 - P d\sigma) \tag{14.138}$$

If we move closed curve c_B along trajectories t_P, t_Q, … to another curve c, we still get the same result and, therefore,

$$I = \oint_c (p_1 dx_1 + p_2 dx_2 + p_3 dx_3 - P d\sigma) = \text{Constant} \tag{14.139}$$

An integral whose value does not change when taken over a given domain (such as curve c) is called an integral invariant of the system.[14,16–21]

Now, if we confine curves c to lay on surfaces of constant σ, we have $d\sigma = 0$ and Equation 14.139 becomes

$$I = I_1 = \oint_c (p_1 dx_1 + p_2 dx_2 + p_3 dx_3) = \text{Constant} \tag{14.140}$$

Note that if we did not assume an explicit dependence of S on σ from the beginning, then Equation 14.121 was

$$\begin{aligned} S &= S(\mathbf{P}_A, \mathbf{P}_B) = S(x_{1A}, x_{2A}, x_{3A}, x_{1B}, x_{2B}, x_{3B}) \\ &= \int_{\sigma_A}^{\sigma_B} L(x_1, x_2, x_3, x_1', x_2', x_3')\, d\sigma \end{aligned} \tag{14.141}$$

where $\mathbf{P}_A$ and $\mathbf{P}_B$ lie on surfaces of constant σ. In that case, Equation 14.122 was

$$\begin{aligned} dS = {} & \frac{\partial S}{\partial x_{1A}} dx_{1A} + \frac{\partial S}{\partial x_{2A}} dx_{2A} + \frac{\partial S}{\partial x_{3A}} dx_{3A} \\ & + \frac{\partial S}{\partial x_{1B}} dx_{1B} + \frac{\partial S}{\partial x_{2B}} dx_{2B} + \frac{\partial S}{\partial x_{3B}} dx_{3B} \end{aligned} \tag{14.142}$$

Equation 14.127 would be derived the same way and Equation 14.135 was

$$dS = (p_{1B} dx_{1B} + p_{2B} dx_{2B} + p_{3B} dx_{3B}) - (p_{1A} dx_{1A} + p_{2A} dx_{2A} + p_{3A} dx_{3A}) \tag{14.143}$$

In Figure 14.14 curves c_A, c_B, and c would be on surfaces of constant σ and we would get an expression (14.140).

Now, going back to Equation 14.140, we may consider that curve c encloses a surface A in configuration space (x_1,x_2,x_3). This surface may be parameterized by two parameters (u,v), and its points are defined by $x_1 = x_1(u,v)$, $x_2 = x_2(u,v)$ and $x_3 = x_3(u,v)$. If each point of surface A is crossed by one trajectory, then A defines a two-parameter bundle of trajectories: for each (u,v) there is a trajectory.

In general, closed curve c is given by $(x_1(\tau),x_2(\tau),x_3(\tau))$ where τ is the parameter of the curve. Each point of the curve is crossed by one trajectory and, therefore, curve c defines a one-parameter manifold of trajectories: for each value of parameter τ there is a trajectory. This one-parameter set of trajectories defines the boundary a two-parameter set of trajectories, each one of those defined by parameters (u,v). Curve c then defines the boundary of a two-parameter (u,v) set of trajectories. A displacement (dx_1,dx_2,dx_3) along curve c is then given by

$$dx_k = \frac{\partial x_k}{\partial u}du + \frac{\partial x_k}{\partial v}dv \tag{14.144}$$

where $k = 1, 2, 3$. Substituting these expressions in Equation 14.140, we get

$$I_1 = \oint_b \sum_{k=1}^{3} p_k dx_k = \oint_b \sum_{k=1}^{3} p_k \frac{\partial x_k}{\partial u}du + p_k \frac{\partial x_k}{\partial v}dv \tag{14.145}$$

Note that line integral (14.140) is calculated along a curve c in configuration space x_1, x_2, x_3. However, after changing coordinates from x_1,x_2,x_3 to u,v line integral (14.145) is now calculated in parameter plane u,v along a curve b. Curve b is then an image in plane u,v of curve c in configuration space. Equation 14.145 may be rewritten as

$$I_1 = \sum_{k=1}^{3} \oint_b p_k \frac{\partial x_k}{\partial u}du + p_k \frac{\partial x_k}{\partial v}dv \tag{14.146}$$

Green's theorem states that

$$\oint F du + G dv = \iint \left(\frac{\partial G}{\partial u} - \frac{\partial F}{\partial v} \right) du\, dv \tag{14.147}$$

Curve b in parameter plane u,v encloses an area B in that same plane u,v. Applying Green's Theorem to each term in summation (14.146), the line integral over curve b may be transformed into a surface integral over B

$$\begin{aligned} I_1 &= \sum_{k=1}^{3} \iint_B \left(\frac{\partial}{\partial u}\left(p_k \frac{\partial x_k}{\partial v} \right) - \frac{\partial}{\partial v}\left(p_k \frac{\partial x_k}{\partial u} \right) \right) du\, dv \\ &= \iint_B \sum_{k=1}^{3} \left(\frac{\partial}{\partial u}\left(p_k \frac{\partial x_k}{\partial v} \right) - \frac{\partial}{\partial v}\left(p_k \frac{\partial x_k}{\partial u} \right) \right) du\, dv \end{aligned} \tag{14.148}$$

Now ($k = 1, 2, 3$)

$$\frac{\partial}{\partial u}\left(p_k\frac{\partial x_k}{\partial v}\right)-\frac{\partial}{\partial v}\left(p_k\frac{\partial x_k}{\partial u}\right)=\frac{\partial p_k}{\partial u}\frac{\partial x_k}{\partial v}+p_k\frac{\partial}{\partial u}\frac{\partial x_k}{\partial v}-\frac{\partial p_k}{\partial v}\frac{\partial x_k}{\partial u}-p_k\frac{\partial}{\partial v}\frac{\partial x_k}{\partial u}$$

$$=\frac{\partial p_k}{\partial u}\frac{\partial x_k}{\partial v}-\frac{\partial x_k}{\partial u}\frac{\partial p_k}{\partial v}=\frac{\partial(p_k,x_k)}{\partial(u,v)} \tag{14.149}$$

is the Jacobian determinant of the transformation between canonical variables x_k, p_k, and parameters u, v. We then get from expression (14.148)

$$I_1=\iint_B\left(\sum_{k=1}^{3}\frac{\partial(p_k,x_k)}{\partial(u,v)}\right)du\,dv=J_2=\iint_A\sum_{k=1}^{3}dp_k dx_k \tag{14.150}$$

where, from expression (14.140), $I_1 = J_2$ is a constant. Integral J_2 is now calculated in phase space x_1,x_2,x_3,p_1,p_2,p_3.

From Equation 14.140 and Equation 14.150, we then have an invariant[22]

$$I_1=\oint_b\sum_{k=1}^{3}p_k dx_k=\oint_c(p_1dx_1+p_2dx_2+p_3dx_3)=\oint_c\mathbf{p}\cdot d\mathbf{x}$$

$$=J_2=\iint_A\sum_{k=1}^{3}dp_k dx_k=\iint_A dp_1dx_1+dp_2dx_2+dp_3dx_3=\iint_A d\mathbf{p}\cdot d\mathbf{x} \tag{14.151}$$

where $\mathbf{p} = (p_1,p_2,p_3)$, $d\mathbf{x} = (dx_1,dx_2,dx_3)$, and $d\mathbf{p} = (dp_1,dp_2,dp_3)$. By applying Stokes' Theorem, we get[22]

$$\oint_c\mathbf{p}\cdot d\mathbf{x}=\iint_A\nabla\times\mathbf{p}\cdot\mathbf{n}da=\iint_A\nabla\times\mathbf{p}\cdot d\mathbf{a} \tag{14.152}$$

with $d\mathbf{a} = da\,\mathbf{n}$ where $\mathbf{n}$ is a unit normal vector to surface A bound by curve c and da is an infinitesimal area in A.

From Equations 14.150 and 14.149 we get

$$I_1=\iint_B\left(\sum_{k=1}^{3}\frac{\partial p_k}{\partial u}\frac{\partial x_k}{\partial v}-\sum_{k=1}^{3}\frac{\partial x_k}{\partial u}\frac{\partial p_k}{\partial v}\right)du\,dv \tag{14.153}$$

or

$$I_1=\iint_B(\mathbf{p}_u\cdot\mathbf{x}_v-\mathbf{p}_v\cdot\mathbf{x}_u)\,du\,dv \tag{14.154}$$

where

$$\begin{aligned} \mathbf{x}_u &= (\partial x_1/\partial u, \partial x_2/\partial u, \partial x_3/\partial u) \\ \mathbf{x}_v &= (\partial x_1/\partial v, \partial x_2/\partial v, \partial x_3/\partial v) \\ \mathbf{p}_u &= (\partial p_1/\partial u, \partial p_2/\partial u, \partial p_3/\partial u) \\ \mathbf{p}_v &= (\partial p_1/\partial v, \partial p_2/\partial v, \partial p_3/\partial v) \end{aligned} \tag{14.155}$$

The results of expressions (14.151), (14.152), and (14.154) may now be summarized as

$$\begin{aligned} I_1 &= \oint_C \mathbf{p}\cdot d\mathbf{x} = J_2 = \iint_A d\mathbf{p}\cdot d\mathbf{x} = \iint_A \nabla\times\mathbf{p}\cdot d\mathbf{a} \\ &= \iint_B (\mathbf{p}_u\cdot\mathbf{x}_v - \mathbf{p}_v\cdot\mathbf{x}_u)\,du\,dv \end{aligned} \tag{14.156}$$

There are also integral invariants of higher order[14,18-21]

$$J_4^* = \iiiint \sum_{i=1}^{3}\sum_{j=1}^{3} dx_i dx_j dp_i dp_j \tag{14.157}$$

where J_4^* is constant. This expression may be rewritten making $x_k = x_k(s,t,u,v)$, $p_k = p_k(s,t,u,v)$ with $k = 1, 2, 3$, where s, t, u, v are parameters of the 4-parameter space of integral J_4^*. This results in

$$J_4^* = \iiiint \sum_{i=1}^{3}\sum_{j=1}^{3} \frac{\partial(dx_i dx_j dp_i dp_j)}{\partial(s,t,u,v)}\,ds\,dt\,du\,dv \tag{14.158}$$

Now, the determinant of a matrix is zero when two lines (or columns) are equal, and we get

$$dx_k dx_k dp_i dp_j = dx_i dx_j dp_k dp_k = 0 \tag{14.159}$$

Also, changing the order of two lines (or columns) changes the sign of a determinant, and we get

$$dx_i dx_j dp_k dp_l = -dx_j dx_i dp_k dp_l = dx_j dx_i dp_l dp_k \tag{14.160}$$

where i, j, k, l may take values 1, 2, 3. This property is denoted as $dx_1 \wedge dx_2 = -dx_2 \wedge dx_1$ in exterior algebra, from which $dx_1 \wedge dx_1 = -dx_1 \wedge dx_1 = 0$.

Exterior algebra may also be used in the derivation of integral invariants.[23] Instead of the J_4^* in expression (14.157), we may then use[24,25]

$$J_4 = \iiiint \sum_{i \neq j} dx_i dx_j dp_i dp_j \tag{14.161}$$

or

$$J_4 = \iiiint dx_1 dx_2 dp_1 dp_2 + dx_1 dx_3 dp_1 dp_3 + dx_2 dx_3 dp_2 dp_3 \tag{14.162}$$

which is constant. An integral invariant of order 6 also exists,

$$J_6 = \iiiiiint dx_1 dx_2 dx_3 dp_1 dp_2 dp_3 \tag{14.163}$$

where J_2, J_4, and J_6 are integral invariants of orders 2, 4, and 6 respectively. Integral invariants are called conservation of étendue in optics, and are further discussed in Chapter 18.

14.7 Movements of the System as Canonical Transformations

Equation 14.135 has the same form as Equation 14.103. Therefore, the movement of the system according to the equations of motion from coordinates (x_{1A}, x_{2A}, x_{3A}) to coordinates (x_{1B}, x_{2B}, x_{3B}) may be seen as a canonical transformation with $-S$ as the generating function of the transformation.[24] Now, the change of coordinates

$$(x_1, x_2, x_3, p_1, p_2, p_3) \rightarrow (i_1, i_2, i_3, u_1, u_2, u_3) \tag{14.164}$$

becomes

$$(x_{1A}, x_{2A}, x_{3A}, p_{1A}, p_{2A}, p_{3A}) \rightarrow (x_{1B}, x_{2B}, x_{3B}, p_{1B}, p_{2B}, p_{3B}) \tag{14.165}$$

with $P \rightarrow P_A$ and $Q \rightarrow P_B$. Generating function G_2 in (14.104) is now

$$G_2(x_{1A}, x_{2A}, x_{3A}, p_{1B}, p_{2B}, p_{3B}) \tag{14.166}$$

and Equation 14.107 becomes

$$
\begin{aligned}
p_{kA} &= \frac{\partial G_2}{\partial x_{kA}} \\
x_{kB} &= \frac{\partial G_2}{\partial p_{kB}} \\
P_B &= P_A
\end{aligned}
\tag{14.167}
$$

with $k = 1, 2, 3$. These results may now be used to show that[26]

$$
\begin{aligned}
J_6 &= \iiiint\!\!\iint dx_{1A}dx_{2A}dx_{3A}dp_{1A}dp_{2A}dp_{3A} \\
&= \iiiint\!\!\iint dx_{1B}dx_{2B}dx_{3B}dp_{1B}dp_{2B}dp_{3B}
\end{aligned}
\tag{14.168}
$$

or that volume J_6 in phase space is constant as the system evolves according to the equations of motion.

We may now use the canonical transformation defined by generating function G_2 in (14.166). Before the transformation

$$
dx_{1A}dx_{2A}dx_{3A}dp_{1A}dp_{2A}dp_{3A} = \left|\frac{\partial(x_{kA}, p_{kA})}{\partial(x_{kA}, p_{kB})}\right| dx_{1A}dx_{2A}dx_{3A}dp_{1B}dp_{2B}dp_{3B} \tag{14.169}
$$

with

$$
\left|\frac{\partial(x_{kA}, p_{kA})}{\partial(x_{kA}, p_{kB})}\right| = \frac{\partial(x_{1A}, x_{2A}, x_{3A}, p_{1A}, p_{2A}, p_{3A})}{\partial(x_{1A}, x_{2A}, x_{3A}, p_{1B}, p_{2B}, p_{3B})} \tag{14.170}
$$

and after the transformation

$$
dx_{1B}dx_{2B}dx_{3B}dp_{1B}dp_{2B}dp_{3B} = \left|\frac{\partial(x_{kB}, p_{kB})}{\partial(x_{kA}, p_{kB})}\right| dx_{1A}dx_{2A}dx_{3A}dp_{1B}dp_{2B}dp_{3B} \tag{14.171}
$$

with

$$
\left|\frac{\partial(x_{kB}, p_{kB})}{\partial(x_{kA}, p_{kB})}\right| = \frac{\partial(x_{1B}, x_{2B}, x_{3B}, p_{1B}, p_{2B}, p_{3B})}{\partial(x_{1A}, x_{2A}, x_{3A}, p_{1B}, p_{2B}, p_{3B})} \tag{14.172}
$$

Now we may write

$$\left|\frac{\partial(x_{kA},p_{kA})}{\partial(x_{kA},p_{kB})}\right| = \left|\begin{matrix}\left[\dfrac{\partial x_{iA}}{\partial x_{jA}}\right] & \left[\dfrac{\partial x_{iA}}{\partial p_{jB}}\right] \\ \left[\dfrac{\partial p_{iA}}{\partial x_{jA}}\right] & \left[\dfrac{\partial p_{iA}}{\partial p_{jB}}\right]\end{matrix}\right| = \left|\begin{matrix}\mathbf{I} & \left[\dfrac{\partial x_{iA}}{\partial p_{jB}}\right] \\ \mathbf{0} & \left[\dfrac{\partial p_{iA}}{\partial p_{jB}}\right]\end{matrix}\right| = \left|\frac{\partial p_{iA}}{\partial p_{jB}}\right| = \left|\frac{\partial G_2}{\partial p_{jB}\partial x_{iA}}\right| \tag{14.173}$$

where **I** is the identity matrix and **0** a zero matrix. Also, p_{iA} and x_{jA} are independent variables and therefore

$$\begin{aligned}\partial x_{iA}/\partial x_{jA} &= \delta_{ij} \\ \partial p_{iA}/\partial x_{jA} &= 0\end{aligned} \tag{14.174}$$

where $\delta_{ij} = 0$ if $i \neq j$ and $\delta_{ij} = 1$ if $i = j$. Now we may also write

$$\left|\frac{\partial(x_{kB},p_{kB})}{\partial(x_{kA},p_{kB})}\right| = \left|\begin{matrix}\left[\dfrac{\partial x_{iB}}{\partial x_{jA}}\right] & \left[\dfrac{\partial x_{iB}}{\partial p_{jB}}\right] \\ \left[\dfrac{\partial p_{iB}}{\partial x_{jA}}\right] & \left[\dfrac{\partial p_{iB}}{\partial p_{jB}}\right]\end{matrix}\right| = \left|\begin{matrix}\left[\dfrac{\partial x_{iB}}{\partial x_{jA}}\right] & \mathbf{0} \\ \left[\dfrac{\partial p_{iB}}{\partial x_{jA}}\right] & \mathbf{I}\end{matrix}\right| = \left|\frac{\partial x_{iB}}{\partial x_{jA}}\right| = \left|\frac{\partial G_2}{\partial x_{jA}\partial p_{iB}}\right| \tag{14.175}$$

where p_{iB} and x_{jB} are independent variables and, therefore,

$$\begin{aligned}\partial x_{iB}/\partial x_{jB} &= \delta_{ij} \\ \partial p_{iB}/\partial x_{jB} &= 0\end{aligned} \tag{14.176}$$

and therefore from Equations 14.175 and 14.173

$$\left|\frac{\partial(x_{kA},p_{kA})}{\partial(x_{kA},p_{kB})}\right| = \left|\frac{\partial(x_{kB},p_{kB})}{\partial(x_{kA},p_{kB})}\right| \tag{14.177}$$

Combining now this result with Equations 14.171 and 14.169 we get

$$dx_{1A}dx_{2A}dx_{3A}dp_{1A}dp_{2A}dp_{3A} = dx_{1B}dx_{2B}dx_{3B}dp_{1B}dp_{2B}dp_{3B} \tag{14.178}$$

or

$$J_6 = \int dx_1 dx_2 dx_3 dp_1 dp_2 dp_3 \tag{14.179}$$

is an integral invariant that remains constant as the trajectories of the system progress according to the equations of motion.

References

1. Pedrotti, L. S. and Pedrotti, F. L., *Optics and Vision*, Prentice-Hall, Upper Saddle River, New Jersey, 1998.
2. Buchdahl, H. A., *An Introduction to Hamiltonian Optics*, Dover Publications, Inc, New York, 1970.
3. Goldstein, H., *Classical Mechanics*, Addison-Wesley Publishing Company, Reading, Massachusetts, 1980.
4. Leech, J. W., *Classical Mechanics*, Chapman & Hall, London, 1965.
5. Arnaud, J. A., *Beam and Fiber Optics*, Academic Press, New York, 1976.
6. Guenther, R. D., *Modern Optics*, John Wiley & Sons, New York, 1990.
7. Stavroudis, O. N., *The Optics of Rays, Wavefronts, and Caustics*, Academic Press, New York, 1972.
8. Marcuse, D., *Light Transmission Optics*, Van Nostrand Reinhold Company, New York, 1972.
9. Luneburg, R. K., *Mathematical Theory of Optics*, University of California Press, Berkeley and Los Angeles, 1964, 90.
10. Miñano, J. C., Design of three-dimensional nonimaging concentrators with inhomogeneous media, *J. Opt. Soc. Am. A*, 3, 9, 1345, 1986.
11. Miñano, J. C., Poisson brackets method of design of nonimaging concentrators: A review, *SPIE Conference on Nonimaging Optics: Maximum Efficiency Light Transfer II*, SPIE Vol. 2016, 98, San Diego, California, USA, 1993.
12. Miñano, J. C. and Benitez, P., Poisson bracket design method review. Application to the elliptic bundles, *SPIE Conference on Nonimaging Optics: Maximum Efficiency Light Transfer V*, SPIE Vol. 3781, 2, 1999.
13. Winston, R. et al., *Nonimaging Optics*, Elsevier Academic Press, Amsterdam, 2005.
14. Meirovitch, L., *Methods of Analytical Dynamics*, McGraw-Hill, USA, 1970.
15. Apostol, T. M., *Calculus, Volume 1, One-Variable Calculus, with an Introduction to Linear Algebra*, 2nd edition, John Wiley & Sons, 1967.
16. Poincaré, H., *Les Méthodes Nouvelles De Mécanique Céleste, Tome III*, Gauthier-Villars, Paris, 1899.
17. Cartan, E., *Leçons Sur Les Invariants Intégraux*, Hermann, Paris, 1922.
18. Gantmacher, F., *Lectures in Analytical Mechanics*, Mir Publishers, Moscow, 1975.
19. Rund, H., *The Hamilton-Jacobi Theory in the Calculus of Variations: Its Role in Mathematics and Physics*, D. Van Nostrand Company LTD, London, 1966.
20. Tiwari, R. N. and Thakur, B. S., *Classical Mechanics: Analytical Dynamics*, Prentice-Hall, New Delhi, 2007.
21. Pars, L. A., *A Treatise on Analytical Dynamics*, Heinemann, London, 1965.
22. Miñano, J. C. and Benitez, P., Fermat's principle and conservation of 2D etendue, *Proc. SPIE 5529, Nonimaging Optics and Efficient Illumination Systems*, Denver, Colorado, USA, 2004.
23. Flanders, H., *Differential Forms with Applications to the Physical Sciences*, Dover Publications, Inc., New York, 1963.
24. Landau, L. D. and Lifshitz, E. M., *Mechanics*, 3rd edition, Butterworth-Heinemann, 2000.

25. Whittaker, E. T., *A Treatise on the Analytical Dynamics of Particles and Rigid Bodies*, 2nd edition, Cambridge University Press, London, Edinburgh, New York, Bombay, Toronto, Tokyo, 1917.
26. Teodorescu, P. P., *Mechanical Systems, Classical Models, Volume III: Analytical Mechanics*, Springer, Bucharest, 2002.

15

Rays and Wavefronts

15.1 Optical Momentum

The optical momentum vector **p** at a point **Q** on a light ray is tangent to the light ray at that point **Q**. We will now see that **p** is also perpendicular to the surfaces S = constant and, therefore, these surfaces are perpendicular to the light rays.

For the Lagrangian $L = L(x_1, x_2, x_3, x_1', x_2')$ used in Equation 14.51 we can write

$$\frac{dL}{dx_3} = \sum_{k=1}^{2} \frac{\partial L}{\partial x_k} x_k' + \frac{\partial L}{\partial x_k'} \frac{dx_k'}{dx_3} + \frac{\partial L}{\partial x_3} \tag{15.1}$$

Considering the Euler Equation 14.52, we get[1]

$$\frac{dL}{dx_3} = \sum_{k=1}^{2} \frac{d}{dx_3}\left(\frac{\partial L}{\partial x_k'}\right) x_k' + \frac{\partial L}{\partial x_k'} \frac{dx_k'}{dx_3} + \frac{\partial L}{\partial x_3} = \frac{d}{dx_3}\left(\sum_{k=1}^{2} \frac{\partial L}{\partial x_k'} x_k'\right) + \frac{\partial L}{\partial x_3} \tag{15.2}$$

and therefore,

$$\frac{d}{dx_3}\left(L - \sum_{k=1}^{2} \frac{\partial L}{\partial x_k'} x_k'\right) = \frac{\partial L}{\partial x_3} \tag{15.3}$$

Considering Equation 14.49 and $dx_3/dx_3 = x_3' = 1$, we can now write

$$L - \sum_{k=1}^{2} \frac{\partial L}{\partial x_k'} x_k' = n\sqrt{1 + x_1'^2 + x_2'^2} - \sum_{k=1}^{2} n \frac{x_k'^2}{\sqrt{1 + x_1'^2 + x_2'^2}} = n \frac{1}{\sqrt{1 + x_1'^2 + x_2'^2}} \tag{15.4}$$

This relation can be written as

$$L - \sum_{k=1}^{2} \frac{\partial L}{\partial x'_k} x'_k = n \frac{x'_3}{\sqrt{x_1'^2 + x_2'^2 + x_3'^2}} = \frac{\partial L}{\partial x'_3} \tag{15.5}$$

Replacing it into Equation 15.3, we can finally write

$$\frac{d}{dx_3}\left(\frac{\partial L}{\partial x'_3}\right) = \frac{\partial L}{\partial x_3} \tag{15.6}$$

Combining this equation with the Euler equations (Equation 14.52), we get

$$\frac{d}{dx_3}\left(\frac{\partial L}{\partial x'_k}\right) = \frac{\partial L}{\partial x_k} \quad (k = 1,2,3) \tag{15.7}$$

Since $p_k = \partial L/\partial x'_k$, we can also write

$$\frac{dp_k}{dx_3} = \frac{\partial L}{\partial x_k} \quad (k = 1,2,3) \tag{15.8}$$

In the context of Lagrangian optics, considering Equation 14.48, and the Euler equations in the form just mentioned, we can write

$$\frac{\partial S}{\partial x_k} = \frac{\partial}{\partial x_k}\int L dx_3 = \int \frac{\partial L}{\partial x_k} dx_3 = \int \frac{dp_k}{dx_3} dx_3 = p_k \tag{15.9}$$

or $p_k = \partial S/\partial x_k$, which can be written as

$$\mathbf{p} = \nabla S \tag{15.10}$$

From this expression, it can be concluded that vector **p** is perpendicular to the surfaces S = constant, S being the optical path length. Since **p** is tangent to the rays of light, it can be concluded that the surfaces S = constant are perpendicular to the rays of light. Such surfaces are called wavefronts.

The Lagrangian and Hamiltonian formulations are just two alternative formulations of optics. As an example of this, Equation 15.10 relating the rays of light with the wavefronts can also be obtained in the context of Hamiltonian optics. Considering expression (14.34) and Equation 14.49, we can write

$$n\frac{ds}{dx_3} = x'_1 p_1 + x'_2 p_2 - H \tag{15.11}$$

and therefore,

$$S = \int n\,ds = \int n\frac{ds}{dx_3}dx_3 = \int (x_1'p_1 + x_2'p_2 - H)dx_3 \tag{15.12}$$

From which we can obtain

$$\frac{\partial S}{\partial x_1} = \frac{\partial}{\partial x_1}\int (x_1'p_1 + x_2'p_2 - H)dx_3 = \int \frac{\partial}{\partial x_1}(x_1'p_1 + x_2'p_2 - H)dx_3 \tag{15.13}$$

since we have $x_1 = x_1(x_3)\, x_1' = x_1'(x_3)$ and, thus, $\partial x_1'/\partial x_1 = 0$. Accordingly, $\partial x_2'/\partial x_2 = 0$. Considering the Hamilton Equation 14.62, we get

$$\frac{\partial S}{\partial x_1} = \int -\frac{\partial H}{\partial x_1}dx_3 = \int \frac{dp_1}{dx_3}dx_3 = p_1 \tag{15.14}$$

The same way

$$\frac{\partial S}{\partial x_2} = p_2 \tag{15.15}$$

From expression (14.39) with parameter σ now given by coordinate x_3, we have $\partial L/\partial x_3 = -\partial H/\partial x_3$ and therefore,

$$\frac{\partial}{\partial x_3}[x_1'p_1 + x_2'p_2 - H] = -\frac{\partial H}{\partial x_3} \tag{15.16}$$

and, considering that $H = -p_3$,

$$\frac{\partial S}{\partial x_3} = \frac{\partial}{\partial x_3}\int (x_1'p_1 + x_2'p_2 - H)dx_3 = \int -\frac{\partial H}{\partial x_3}dx_3 = \int \frac{dp_3}{dx_3}dx_3 = p_3 \tag{15.17}$$

Equations 15.14, 15.15, and 15.17 can be combined in the following equation

$$\mathbf{p} = \nabla S \tag{15.18}$$

which is the same as Equation 15.10.

Optical momentum **p** points in the direction of the light rays at each point. Since it is perpendicular to the wavefronts defined by S = constant, the light rays are also perpendicular to the wavefronts, as shown in Figure 15.1.

In a material with continuously varying refractive index, the light rays are curved, and the optical momentum vector **p** at a point **Q** on the light ray is

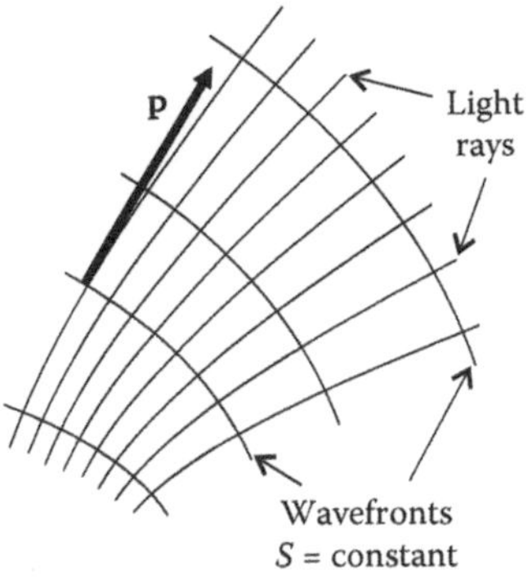

FIGURE 15.1
Light rays are perpendicular to the wavefronts defined by S = constant. Optical momentum **p** is also perpendicular to the wavefronts.

tangent to the light ray and has a magnitude equal to the refractive index at **Q**, that is, $\|\mathbf{p}(\mathbf{Q})\| = n(\mathbf{Q})$, as shown in Figure 15.2a. If the refractive index is constant and equal to n, the light ray is a straight line, and **p** is parallel to the light ray, with $\|\mathbf{p}\| = n$, as shown in Figure 15.2b.

For vector $\mathbf{p} = (p_1, p_2)$, its p_1 component along the x_1 axis can then be obtained as the product of its magnitude and the cosine of the angle it makes to the x_1 axis: $p_1 = \|\mathbf{p}\| \cos \alpha_1 = \|\mathbf{p}\| \cos \beta_1$. The same is true for the x_2 component: $p_2 = \|\mathbf{p}\| \cos \alpha_2 = \|\mathbf{p}\| \cos \beta_2$, as shown in Figure 15.3.

In 3-D systems, something similar happens, for vector $\mathbf{p} = (p_1, p_2, p_3)$. Also in this case, $p_1 = \|\mathbf{p}\| \cos \alpha_1 = \|\mathbf{p}\| \cos \beta_1$, where α_1 or β_1 are the angles that the vector **p** makes with axis x_1. This can be seen to be the same as the situation of Figure 15.3a if we consider the plane γ containing the x_1 axis and vector **p**, as shown in Figure 15.4. Similar conclusions can be drawn for the x_2 and x_3 components of **p**.

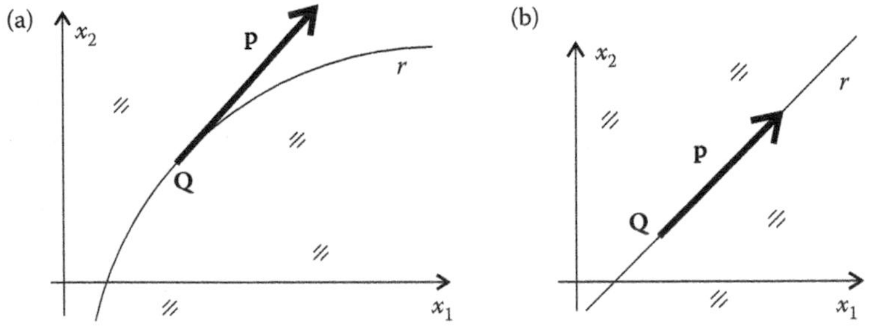

FIGURE 15.2
(a) In a medium of continuously varying refractive index, the light rays are curved and the optical momentum is tangent to the light rays. (b) When the refractive index is constant, the light ray is a straight line, and the optical momentum has the same direction as the light ray.

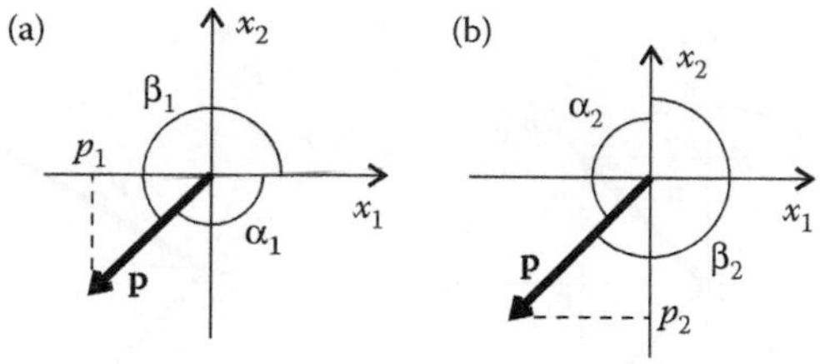

FIGURE 15.3
The p_1 component of a vector **p** is given by $v_1 = ||\mathbf{p}||\cos\alpha_1 = ||\mathbf{p}||\cos\beta_1$. Also, the value of x_2 is given by $p_2 = ||\mathbf{p}||\cos\alpha_2 = ||\mathbf{p}||\cos\beta_2$.

In 2-D systems, when we specify the value of p_1, there are two possible rays r_A and r_B that have the same value as p_1 (Figure 15.5a). The two possible p_2 values for these two light rays can be obtained in 2-D systems from $p_1^2 + p_2^2 = n^2$ as

$$p_2 = \pm\sqrt{n^2 - p_1^2} \tag{15.19}$$

Something similar happens in 3-D systems, as shown in Figure 15.5b. Specifying the values of p_1 and p_2, we can obtain the value of p_3 as

$$p_3 = \pm\sqrt{n^2 - p_1^2 - p_2^2} \tag{15.20}$$

In most optical systems, however, rays propagate in a given direction. Let us say that, in the 2-D case, rays propagate in the direction in which x_2 increases, as shown in Figure 15.6.

In this case, we would have $p_2 > 0$ and, therefore, by specifying the value of p_1, we would know the corresponding value of p_2 and could completely determine the ray of light. It would be ray r_A in Figure 15.5a for the 2-D case.

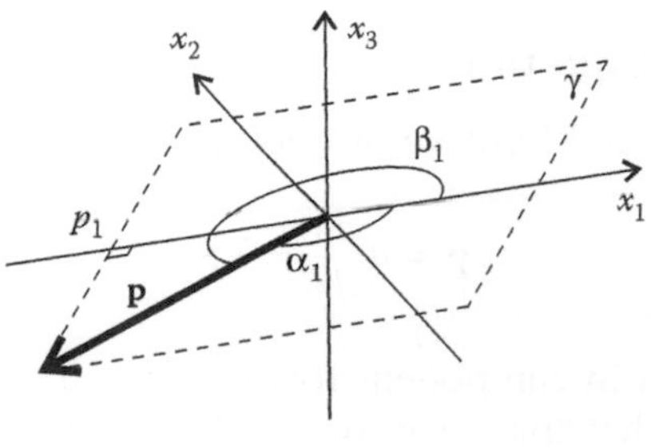

FIGURE 15.4
Considering the plane γ defined by vector **p** and the x_1 axis, it can be seen that its p_1 component is given by $p_1 = ||\mathbf{p}||\cos\alpha_1 = ||\mathbf{p}||\cos\beta_1$, where α_1 and β_1 are the angles the vector makes to the x_1 axis.

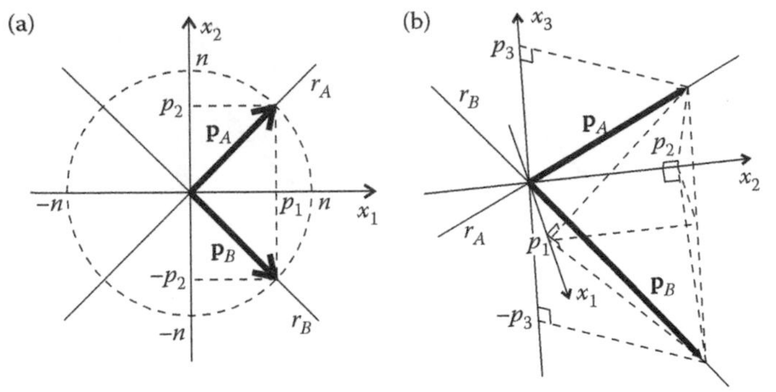

FIGURE 15.5
(a) Specifying the value of p_1 for a light ray in a 2-D system is not enough to define a light ray, since there are two light rays that have the same p_1 value. (b) Something similar happens when specifying p_1 and p_2 in a 3-D system.

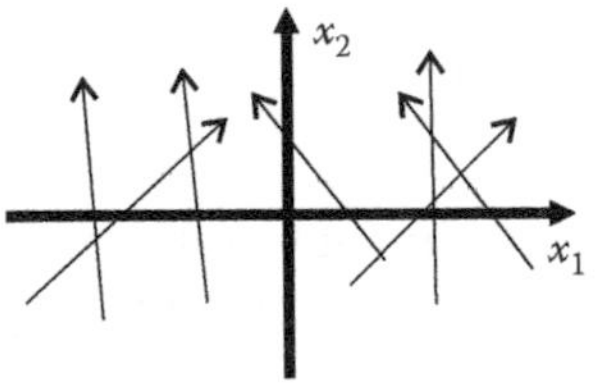

FIGURE 15.6
In most optical systems light travels in a given direction along the optical axis. In this case, this axis is x_2.

For the 3-D case, if rays propagate in a direction of increasing x_3, then $p_3 > 0$, and specifying p_1 and p_2 will define the ray. It would be ray r_A in Figure 15.5b that has $p_3 > 0$.

15.2 The Eikonal Equation

From Equations 14.56 and 15.10 we can obtain

$$\mathbf{p} = n\frac{d\mathbf{s}}{ds} = \nabla S \tag{15.21}$$

which can be written in component form as $p_i = \partial S/\partial x_i$. From expression (14.57) it can be seen that $\|\mathbf{p}\| = n$. Equation 15.21 then enables us to write

$$\left(\frac{\partial S}{\partial x_1}\right)^2 + \left(\frac{\partial S}{\partial x_2}\right)^2 + \left(\frac{\partial S}{\partial x_3}\right)^2 = n^2 \Leftrightarrow \| \nabla S \|^2 = n^2 \tag{15.22}$$

This is the eikonal equation.[2–5] It should be noted that, considering Equation 15.21, the eikonal equation reduces itself to the equation $P = 0$ of the equation system (14.79).

Another possible way to derive the eikonal equation from Equation 15.21 is by considering the third component of this expression, which is $\partial S/\partial x_3 - p_3 = 0$. Considering expression (14.60), $p_3 = -H$, we get $\partial S/\partial x_3 + H = 0$. As can be seen in expression (14.86) $H = H(x_1, x_2, x_3, p_1, p_2)$. Considering the first two expressions of Equation 15.21, we can finally write

$$\frac{\partial S}{\partial x_3} + H\left(x_1, x_2, x_3, \frac{\partial S}{\partial x_1}, \frac{\partial S}{\partial x_2}\right) = 0 \tag{15.23}$$

This is a differential equation for S, called the Hamilton–Jacobi equation.[3,4] The eikonal equation (Equation 15.22) can now be obtained from this equation. Introducing the expression for H given by Equation 14.61 into expression (15.23) gives

$$\frac{\partial S}{\partial x_3} - \sqrt{n^2 - \left(\frac{\partial S}{\partial x_1}\right)^2 - \left(\frac{\partial S}{\partial x_2}\right)^2} = 0 \tag{15.24}$$

which corresponds to the eikonal Equation 15.22.

15.3 The Ray Equation

Let us consider Equation 15.21 again, which, in its components, can be written as

$$n\frac{dx_k}{ds} = \frac{\partial S}{\partial x_k} \tag{15.25}$$

The derivative of this expression with respect to path length s is

$$\frac{d}{ds}\left(n\frac{dx_k}{ds}\right) = \frac{d}{ds}\frac{\partial S}{\partial x_k} \tag{15.26}$$

Considering that $S = \int n\,ds$, we get

$$\frac{d}{ds}\left(n\frac{dx_k}{ds}\right) = \frac{d}{ds}\frac{\partial}{\partial x_k}\int n\,ds = \frac{d}{ds}\int \frac{\partial n}{\partial x_k}ds = \frac{\partial n}{\partial x_k} \tag{15.27}$$

or

$$\frac{d}{ds}\left(n\frac{d\mathbf{s}}{ds}\right) = \nabla n \tag{15.28}$$

This expression is called the ray equation. Considering Equation 15.21, the ray equation can also be written as

$$\frac{d\mathbf{p}}{ds} = \nabla n \tag{15.29}$$

which, in its components, can be written as $dp_k/ds = \partial n/\partial x_k$, where $k = 1, 2, 3$.

To facilitate the solution of this equation, it is assumed in several applications that the light rays travel almost parallel to the optical axis of the system (paraxial). We first note that, with $x' = dx_k/dx_3$ and $x_3' = 1$, we have

$$\frac{d}{ds} = \left(\frac{1}{\sqrt{x_1'^2 + x_2'^2 + x_3'^2}}\right)\frac{d}{dx_3} = \left(\frac{1}{\sqrt{1 + x_1'^2 + x_2'^2}}\right)\frac{d}{dx_3} \tag{15.30}$$

In the paraxial approximation, it is assumed that light rays have paths through the optical system such that the trajectories keep almost parallel to the optical axis, and that the changes in direction are small. This does not mean that light rays cannot spatially propagate far from the optical axis; only the angles with them are small. Light rays can become distant from the optical axis by traveling along a long path. Since x_1' and x_2' describe the slope of the light ray relative to the optical axis x_3, these quantities should be small. We then have $x_1' \ll 1$ and $x_2' \ll 1$. From Equation 15.30, we get the following approximation:

$$\frac{d}{ds} \approx \frac{d}{dx_3} \tag{15.31}$$

If the optical axis is considered to be along the x_3 axis, we can then make $ds \approx dx_3$. The ray equation can then be written as

$$\frac{d}{dx_3}\left(n\frac{d\mathbf{s}}{dx_3}\right) = \nabla n \tag{15.32}$$

This is the paraxial ray equation.[2,4]

Given then the refractive index $n(x_1, x_2, x_3)$, Equation 15.32 can be written in component form, given that $d\mathbf{s} = (dx_1, dx_2, dx_3)$,

$$\frac{d}{dx_3}\left(n(x_1,x_2,x_3)\frac{dx_1}{dx_3}\right) = \frac{\partial n(x_1,x_2,x_3)}{\partial x_1}$$
$$\frac{d}{dx_3}\left(n(x_1,x_2,x_3)\frac{dx_2}{dx_3}\right) = \frac{\partial n(x_1,x_2,x_3)}{\partial x_2} \tag{15.33}$$

This is a system of differential equations for $x_1(x_3)$ and $x_2(x_3)$ so that, given the refractive index $n(x_1, x_2, x_3)$, we can calculate the paths of the light rays, which are given by $(x_1(x_3), x_2(x_3), x_3)$. Note that the third component of Equation 15.32 reduces itself to $dn/dx_3 = \partial n/\partial x_3$ since $dx_3/dx_3 = 1$ and, therefore, is not included in Equation 15.33.

If a function $S(x_1, x_2, x_3)$ is given such that $S = \text{constant}$ defines the wavefronts for the light rays, the refractive index $n(x_1, x_2, x_3)$ that makes those wavefronts possible can be obtained by the following eikonal equation:

$$n = \|\nabla S\| \tag{15.34}$$

Also, the ray equation enables us to obtain the corresponding paths for the light rays.

It should be noted, however, that in nonimaging optics, the angles of light rays to the optical axis are often large, and therefore, the paraxial approximation does not apply. In such a case, the ray equation as given by Equation 15.28 must be used to determine the ray trajectories for a given refractive index $n(x_1, x_2, x_3)$.

The ray equation can also be obtained from the Lagrangian optics. In this case, making $k = 1$ in Equation 15.7, and considering Equation 14.49 and expression (14.53), we get

$$\frac{d}{dx_3}\left(\frac{\partial L}{\partial x_1'}\right) = \frac{\partial L}{\partial x_1} \Leftrightarrow \frac{d}{dx_3}\left(\frac{n x_1'}{\sqrt{x_1'^2 + x_2'^2 + x_3'^2}}\right) = \sqrt{x_1'^2 + x_2'^2 + x_3'^2}\,\frac{\partial n}{\partial x_1} \tag{15.35}$$

and making use of Equation 15.30, we get

$$\frac{d}{ds}\left(n\frac{dx_1}{ds}\right) = \frac{\partial n}{\partial x_1} \tag{15.36}$$

Writing the equations for the other components x_2 and x_3, we get

$$\frac{d}{ds}\left(n\frac{d\mathbf{s}}{ds}\right) = \nabla n \tag{15.37}$$

This is again the ray equation.[2,4] It specifies the light ray direction in terms of the refractive index. If the parameterization now $(x_1(\sigma), x_2(\sigma), x_3(\sigma))$ the ray trajectory will be a continuous curve, $\mathbf{s}(\sigma)$.[6]

As mentioned earlier, the Lagrangian and Hamiltonian formulations are just two alternative formulations of optics. Another example of this can be given by deriving the ray equation from the Hamiltonian formulation. From Equation 14.62 and calculating the derivatives $\partial H/\partial p_i$ and $\partial H/\partial x_i$ we obtain

$$x_1' = \frac{p_1}{\sqrt{n^2 - p_1^2 - p_2^2}}, \quad x_2' = \frac{p_2}{\sqrt{n^2 - p_1^2 - p_2^2}} \tag{15.38}$$

and

$$p_1' = \frac{n}{\sqrt{n^2 - p_1^2 - p_2^2}} \frac{\partial n}{\partial x_1}, \quad p_2' = \frac{n}{\sqrt{n^2 - p_1^2 - p_2^2}} \frac{\partial n}{\partial x_2} \tag{15.39}$$

From Equation 14.92 we can also obtain

$$p_3' = \frac{n}{\sqrt{n^2 - p_1^2 - p_2^2}} \frac{\partial n}{\partial x_3} \tag{15.40}$$

From expression (15.38) we can obtain

$$1 + x_1'^2 + x_2'^2 = \frac{n^2}{n^2 - p_1^2 - p_2^2} \quad \text{or} \quad \sqrt{1 + x_1'^2 + x_2'^2} = \frac{n}{\sqrt{n^2 - p_1^2 - p_2^2}} \tag{15.41}$$

From Equation 14.49, we can see that $ds/dx_3 = \sqrt{1 + x_1'^2 + x_2'^2}$. Combining this with Equation 15.41, we get

$$\frac{ds}{dx_3} = \frac{n}{\sqrt{n^2 - p_1^2 - p_2^2}} \tag{15.42}$$

Equations 15.39 and 15.40 can now be written in a simplified form:

$$\frac{dp_1}{ds} = \frac{\partial n}{\partial x_1}, \quad \frac{dp_2}{ds} = \frac{\partial n}{\partial x_2}, \quad \frac{dp_3}{ds} = \frac{\partial n}{\partial x_3} \tag{15.43}$$

These are the components of the ray equation (Equation 15.37), which can also be written as

$$\frac{d\mathbf{p}}{ds} = \nabla n \tag{15.44}$$

15.4 Optical Path Length between Two Wavefronts

The second fundamental theorem of calculus for line integrals can be written as[7]

$$\int_{\mathbf{P}_1}^{\mathbf{P}_2} \nabla\varphi \cdot d\mathbf{r} = \varphi(\mathbf{P}_2) - \varphi(\mathbf{P}_1) \tag{15.45}$$

in which the integral is calculated along a curve $\mathbf{r}(\xi)$ in space with $\xi_1 \leq \xi \leq \xi_2$ and $\mathbf{P}_1 = \mathbf{r}(\xi_1)$ and $\mathbf{P}_2 = \mathbf{r}(\xi_2)$. A consequence the line integral of a gradient field is independent of the path chosen, depending only on the initial and final points of the path of integration. The line integral of $\nabla\varphi$ along a line connecting points $\mathbf{P}_1$ and $\mathbf{P}_2$ depends only on $\mathbf{P}_1$ and $\mathbf{P}_2$, that is, on $\varphi(\mathbf{P}_1)$ and $\varphi(\mathbf{P}_2)$, and not on the path between $\mathbf{P}_1$ and $\mathbf{P}_2$. In case if point $\mathbf{P}_1$ coincides with point $\mathbf{P}_2$, then

$$\oint \nabla\varphi \cdot d\mathbf{r} = 0 \tag{15.46}$$

This result can now be applied to optics. Replacing the function φ by the optical path length S in expression (15.46), gives

$$\oint \nabla S \cdot d\mathbf{r} = 0 \tag{15.47}$$

and in expression (15.45),

$$\int_{\mathbf{P}_1}^{\mathbf{P}_2} \nabla S \cdot d\mathbf{r} = S(\mathbf{P}_2) - S(\mathbf{P}_1) \tag{15.48}$$

Equation 15.21 can be written as

$$n\frac{d\mathbf{s}}{ds} = \nabla S \tag{15.49}$$

Inserting, accordingly, in expression (15.47), gives

$$\oint n\frac{d\mathbf{s}}{ds} \cdot d\mathbf{r} = 0 \tag{15.50}$$

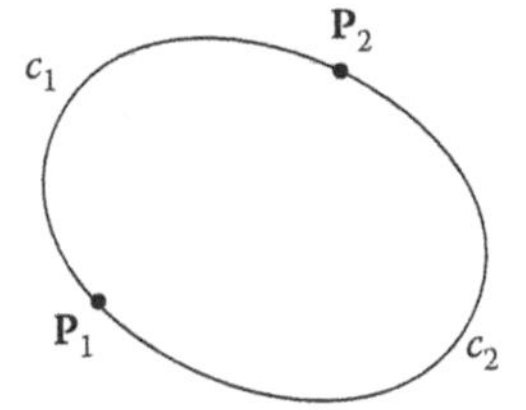

FIGURE 15.7
Two points $\mathbf{P}_1$ and $\mathbf{P}_2$ divide a closed curve into two curves c_1 and c_2.

Consider two points $\mathbf{P}_1$ and $\mathbf{P}_2$ on the closed curve used for the integration. These two points divide the closed curve into two parts (curves c_1 and c_2), both starting at $\mathbf{P}_1$ and ending at $\mathbf{P}_2$, as presented in Figure 15.7.

This integral can then be written as[4]

$$\int_{\mathrm{P}_1}^{\mathrm{P}_2} n\frac{d\mathbf{s}_1}{ds_1}\cdot d\mathbf{r}_1 = \int_{\mathrm{P}_1}^{\mathrm{P}_2} n\frac{d\mathbf{s}_2}{ds_2}\cdot d\mathbf{r}_2 \tag{15.51}$$

in which the integrals are calculated along the curves c_1 and c_2 resulting from the division of the closed curve by points $\mathbf{P}_1$ and $\mathbf{P}_2$. Vector $d\mathbf{s}/ds$ is a unit vector pointing in the direction of $\mathbf{p}$, that is, in the direction of the light ray. Making $\mathbf{t} = d\mathbf{s}/ds$ gives

$$\int_{\mathrm{P}_1}^{\mathrm{P}_2} n\mathbf{t}_1\cdot d\mathbf{r}_1 = \int_{\mathrm{P}_1}^{\mathrm{P}_2} n\mathbf{t}_2\cdot d\mathbf{r}_2 \Leftrightarrow \int_{\mathrm{P}_1}^{\mathrm{P}_2} \mathbf{p}_1\cdot d\mathbf{r}_1 = \int_{\mathrm{P}_1}^{\mathrm{P}_2} \mathbf{p}_2\cdot d\mathbf{r}_2 \tag{15.52}$$

The integrand $n\mathbf{t}\cdot d\mathbf{r}$ is the projection of vector $n\mathbf{t}$, that is, the momentum $\mathbf{p}$, in the direction of the curve. This result shows that the integration of $n\mathbf{t}$ along curve c_1 is the same as the integral of this quantity along curve c_2, both beginning at $\mathbf{P}_1$ and ending at $\mathbf{P}_2$. It can then be concluded that the integral of this quantity along a curve does not depend on the shape of the curve, but only on the initial and final points $\mathbf{P}_1$ and $\mathbf{P}_2$.

Consider two wavefronts $S = s_1$ and $S = s_2$, as well as two points $\mathbf{P}_1$ and $\mathbf{P}_2$ on these wavefronts, as shown in Figure 15.8. The integral $\int_{\mathbf{P}_1}^{\mathbf{P}_2} n\mathbf{t}\cdot d\mathbf{r}$ taken along a curve between points $\mathbf{P}_1$ and $\mathbf{P}_2$ does not depend on the integration path, so that we can choose the portion of the wavefront $S = s_1$ between $\mathbf{P}_1$ and $\mathbf{Q}_1$ for the integration path and then the portion of the light ray between $\mathbf{Q}_1$ and $\mathbf{P}_2$. The integral between $\mathbf{P}_1$ and $\mathbf{Q}_1$ is zero because light rays are perpendicular to the wavefronts. The integral between $\mathbf{Q}_1$ and $\mathbf{P}_2$ equals the optical path length between $\mathbf{Q}_1$ and $\mathbf{P}_2$. We could, nonetheless, have chosen

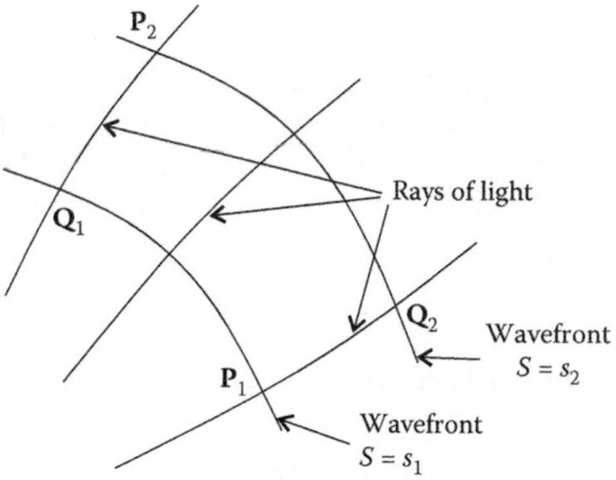

FIGURE 15.8
The optical path length between the two points $\mathbf{P}_1$ and $\mathbf{P}_2$ belonging to the two wavefronts $S = s_1$ and $S = s_2$ equals $s_2 - s_1$ and does not depend on the particular path between $\mathbf{P}_1$ and $\mathbf{P}_2$.

another path, for example, from $\mathbf{P}_1$ to $\mathbf{Q}_2$ along light ray and then from $\mathbf{Q}_2$ to $\mathbf{P}_2$ along the wavefront $S = s_2$, to generate the optical path between $\mathbf{P}_1$ and $\mathbf{Q}_2$. It can then be concluded that the optical path lengths between $\mathbf{Q}_1$ and $\mathbf{P}_2$ and between $\mathbf{P}_1$ and $\mathbf{Q}_2$ must be the same. This proves that the optical path length between two wavefronts must be the same along any light ray connecting them. We can then write

$$\int_{P_1}^{P_2} n\mathbf{t} \cdot d\mathbf{r} = s_2 - s_1 \tag{15.53}$$

A particular case can be considered when the two wavefronts involve two points. In this case, the optical path length between these two points must be a constant for the rays connecting them.[8]

This result can also be obtained directly from Equation 15.51. Consider that the curves connecting the points $\mathbf{P}_1$ and $\mathbf{P}_2$ are two light rays. In this case, the displacement $d\mathbf{r}$ along the curve coincides with a displacement $d\mathbf{s}$ along the light ray. Therefore, $d\mathbf{r} = d\mathbf{s}$, and so we can write

$$S = \int_{P_1}^{P_2} n\, ds = \int_{P_1}^{P_2} n\frac{d\mathbf{s}}{ds} \cdot d\mathbf{s} = \int_{P_1}^{P_2} \mathbf{p} \cdot d\mathbf{s} = \int_{P_1}^{P_2} \nabla S \cdot d\mathbf{s} = S(\mathbf{P}_2) - S(\mathbf{P}_1) \tag{15.54}$$

Since we are considering integrations along the rays of light, $\mathbf{p}$ has the same direction as $d\mathbf{s}$. We can also write $\nabla \times \nabla S = 0$, where $\nabla \times$ is the rotational operator (curl).[7] From this, we can obtain $\nabla \times \mathbf{p} = 0$.

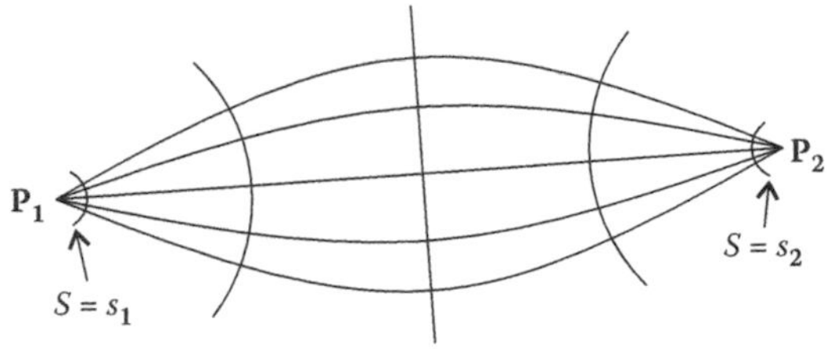

FIGURE 15.9
The optical path length for the light rays connecting the points $\mathbf{P}_1$ and $\mathbf{P}_2$ must be the same for all of them.

While we are considering integrations along the light rays, **t** and $d\mathbf{r}$ are parallel, such that their scalar product is simply an element ds of the light ray. Then from Equation 15.52, we obtain

$$\int_{P_1}^{P_2} n\, ds_1 = \int_{P_1}^{P_2} n\, ds_2 \tag{15.55}$$

It can now be seen from this expression that if several light rays leaving $\mathbf{P}_1$ meet at $\mathbf{P}_2$, their optical path lengths must be equal. Figure 15.9 shows one such situation.

This result is, nonetheless, valid only if the luminous field is continuous, so that infinite light rays smoothly fill the space between any two rays considered. This result will not be valid if, for example, a light ray travels from $\mathbf{P}_1$ to $\mathbf{P}_2$ directly and the other one reaches $\mathbf{P}_2$ after reflecting off a mirror.[4]

References

1. Leech, J.W., *Classical Mechanics*, Chapman & Hall, London, 1965.
2. Guenther, R.D., *Modern Optics*, John Wiley and Sons, New York, 1990.
3. Stavroudis, O.N., *The Optics of Rays, Wavefronts, and Caustics*, Academic Press, New York, 1972.
4. Marcuse, D., *Light Transmission Optics*, Van Nostrand Reinhold Company, New York, 1972.
5. Born, M. and Wolf, E., *Principles of Optics*, Pergamon Press, Oxford, 1980.
6. Gomez-Reino, C., Perez, M.V. and Bao, C., *GRIN Gradient-Index Optics, Fundamentals and Applications*, Springer, Berlin, 2002.
7. Apostol, T.M., *Calculus –Volume II*, 2nd ed., John Wiley and Sons, New York, 1969.
8. Luneburg, R.K., *Mathematical Theory of Optics*, University of California Press, Berkeley and Los Angeles, 1964, 130.

16

Reflection and Refraction

16.1 Reflected and Refracted Rays

A ray i traveling in a medium of refractive index n_1 incident on a surface A with normal $\mathbf{n}$ is refracted thereupon into a medium of refractive index n_2. The angle α_1 that the ray makes with the normal before refraction is related to the angle it makes with the normal after refraction, by Snell's law of refraction:

$$n_1 \sin\alpha_1 = n_2 \sin\alpha_{2R} \tag{16.1}$$

Note that expression (16.1) for a refracted ray can also be written as

$$\sin\alpha_1 = \frac{n_2}{n_1}\sin\alpha_2 \tag{16.2}$$

which is equivalent to a refraction from a medium of refractive index $n_A = 1$ into another medium of refractive index $n_B = n_2/n_1$. In that case, expression (16.2) can be written as $n_A \sin \alpha_1 = n_B \sin \alpha_2$ or its equivalent $\sin \alpha_1 = n_B \sin \alpha_2$.

Going back to expression (16.1), if surface A was a mirror, the ray would be reflected, and it would continue traveling in the medium of refractive index n_1. In this case, expression (16.1) still holds if we make $n_1 = n_2$ and replace α_{2R} for refraction with α_{2X} for reflection, obtaining (Figure 16.1).

$$\sin\alpha_1 = \sin\alpha_{2X} \Leftrightarrow \alpha_1 = \alpha_{2X} \tag{16.3}$$

The incident ray i is traveling in a direction defined by unit vector $\mathbf{i}$, the normal to the surface is given by unit vector $\mathbf{n}$, and the refracted ray r_R travels in a direction defined by unit vector $\mathbf{r}_R$. In the case of reflection, the reflected ray r_X travels in a direction defined by unit vector $\mathbf{r}_X$. As seen in Chapter 14, in the case of refraction, unit vectors $\mathbf{i}$, $\mathbf{n}$, and $\mathbf{r}_R$ are all in the same plane. In the case of reflection too, unit vectors $\mathbf{i}$, $\mathbf{n}$, and $\mathbf{r}_X$ are all in the same plane. This means that the direction of the refracted or reflected rays can be obtained by

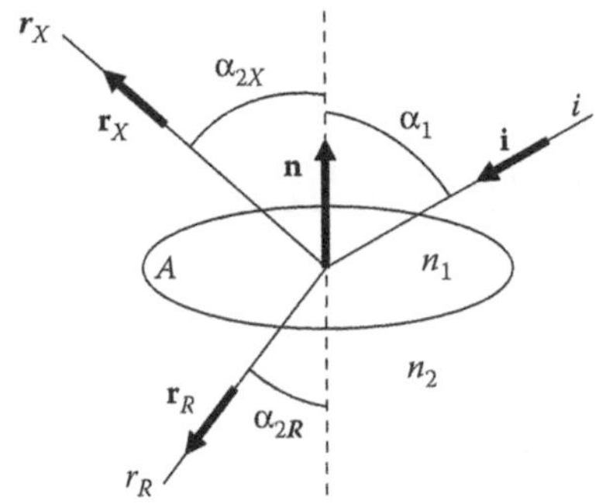

FIGURE 16.1
Refracted and reflected rays from an incident ray i. In the case of refraction, $n_1 \sin \alpha_1 = n_2 \sin \alpha_{2R}$. In the case of reflection $\alpha_1 = \alpha_{2X}$.

a linear combination of the incident direction **i** and normal **n** to the surface as (in Equation 14.19)

$$\mathbf{r} = \lambda \mathbf{i} + \mu \mathbf{n} \tag{16.4}$$

where $\mathbf{r} = \mathbf{r}_R$ in the case of refraction, and $\mathbf{r} = \mathbf{r}_X$ in the case of reflection. Coefficients λ and μ are also different in both cases.

We can now derive the expressions for the direction of the refracted or reflected ray as a function of the direction of the incident ray.[1,2]

If **n** is a unit vector normal to the surface, the sine of the angle between **n** and **i** can be obtained from the magnitude of the cross-product of these two unit vectors. The same is true for the vectors **n** and **r**. Therefore, expressions (16.1) and (16.3) can be written as

$$n_1 \mathbf{i} \times \mathbf{n} = n_2 \mathbf{r} \times \mathbf{n} \tag{16.5}$$

This is because **i**, **n**, and **r** are all contained in the same plane (expression (16.4)). As mentioned earlier, in the case of reflection, $n_1 = n_2$. Expression (16.5) can be written as

$$\mathbf{p}_1 \times \mathbf{n} = \mathbf{p}_2 \times \mathbf{n} \tag{16.6}$$

where $\mathbf{p}_1 = n_1 \mathbf{i}$ and $\mathbf{p}_2 = n_2 \mathbf{r}$ are the optical momenta of the ray before and after refraction or reflection, and where $\|\mathbf{i}\| = \|\mathbf{r}\| = 1$.

Finding the external product of both the sides of expression (16.4) by **n**, and considering expression (16.5) and $\mathbf{n} \times \mathbf{n} - 0$ gives

$$\mathbf{r} \times \mathbf{n} = \lambda \mathbf{i} \times \mathbf{n} \Rightarrow \lambda = n_1/n_2 \Rightarrow \mathbf{r} = n_1/n_2\, \mathbf{i} + \mu \mathbf{n} \tag{16.7}$$

Squaring both the sides of Equation 16.7 and considering that $(\mathbf{a}+\mathbf{b}) \cdot (\mathbf{a}+\mathbf{b}) = \mathbf{a} \cdot \mathbf{a} + \mathbf{b} \cdot \mathbf{b} + 2\mathbf{a} \cdot \mathbf{b}$ and these are unit vectors, so that $\mathbf{r} \cdot \mathbf{r} = \mathbf{n} \cdot \mathbf{n} = 1$ gives

$$1 = \left(\frac{n_1}{n_2}\right)^2 + \mu^2 + 2\mu \frac{n_1}{n_2}\mathbf{i} \cdot \mathbf{n} \Leftrightarrow \mu^2 + \left(2q\frac{n_1}{n_2}\right)\mu + \left[\left(\frac{n_1}{n_2}\right)^2 - 1\right] = 0 \quad (16.8)$$

with $q = \mathbf{i} \cdot \mathbf{n}$. Equation 16.8 of second degree can be solved for μ, resulting in

$$\mu = -q\frac{n_1}{n_2} \pm \sqrt{\left(q\frac{n_1}{n_2}\right)^2 - \left[\left(\frac{n_1}{n_2}\right)^2 - 1\right]} = -q\frac{n_1}{n_2} \pm \sqrt{1 - \left(\frac{n_1}{n_2}\right)^2 (1 - q^2)}. \therefore \quad (16.9)$$

As can be verified, we have two possible solutions. Introducing them into Equation 16.7 and replacing q gives

$$\mathbf{r} = \frac{n_1}{n_2}\mathbf{i} + \left(-(\mathbf{i} \cdot \mathbf{n})\frac{n_1}{n_2} \pm \sqrt{1 - \left(\frac{n_1}{n_2}\right)^2 \left[1 - (\mathbf{i} \cdot \mathbf{n})^2\right]}\right)\mathbf{n} \quad (16.10)$$

Choosing the solution with a positive sign gives

$$\mathbf{r} = \frac{n_1}{n_2}\mathbf{i} + \left(-(\mathbf{i} \cdot \mathbf{n})\frac{n_1}{n_2} + \sqrt{1 - \left(\frac{n_1}{n_2}\right)^2 \left[1 - (\mathbf{i} \cdot \mathbf{n})^2\right]}\right)\mathbf{n} \quad (16.11)$$

which is the direction of the refracted ray as a function of the direction of the incoming ray, and of the normal to the surface. Remember that in expression (16.11) we have $\|\mathbf{i}\| = \|\mathbf{n}\| = \|\mathbf{r}\| = 1$ (i.e., all are unit vectors). Expression (16.11) can also be written as

$$n_2\mathbf{r} = n_1\mathbf{i} - (n_1\mathbf{i} \cdot \mathbf{n})\mathbf{n} + \sqrt{n_2^2 - n_1^2 + (n_1\mathbf{i} \cdot \mathbf{n})^2}\,\mathbf{n} \quad (16.12)$$

or

$$\mathbf{p}_2 = \mathbf{p}_1 - (\mathbf{p}_1 \cdot \mathbf{n})\mathbf{n} + \sqrt{n_2^2 - n_1^2 + (\mathbf{p}_1 \cdot \mathbf{n})^2}\,\mathbf{n} \quad (16.13)$$

To understand the meaning of the solution with a negative sign, in Equation 16.10 we must now consider that, also in the case of reflection, the reflected vector can be obtained as a linear combination of the incident vector and the normal to the surface, that is, expression (16.4) is also valid in this case. Besides, making $n_1 = n_2$, expression (16.5) can also be applied to reflection. In this case, as referred, we have $n_1 = n_2$, so that, by choosing the negative sign solution, in Equation 16.10 we obtain

$$\mathbf{r} = \mathbf{i} + \left(-(\mathbf{i} \cdot \mathbf{n}) - \sqrt{1 - [1 - (\mathbf{i} \cdot \mathbf{n})^2]}\right)\mathbf{n} \tag{16.14}$$

Expression (16.14) can be rewritten as

$$\mathbf{r} = \mathbf{i} - 2(\mathbf{i} \cdot \mathbf{n})\mathbf{n} \tag{16.15}$$

Expression (16.15) gives us the reflected ray as a function of the incident ray and the normal to the surface.

Given a surface, its normal can be given in two opposite directions, as presented in Figure 16.2, in which $\mathbf{n}_2 = -\mathbf{n}_1$. In the resulting expression for the reflected ray, the direction of the normal is actually not important, since $\mathbf{r} = \mathbf{i} - 2(\mathbf{i} \cdot \mathbf{n})\mathbf{n} = \mathbf{r} = \mathbf{i} - 2(\mathbf{i} \cdot (-\mathbf{n}))(-\mathbf{n})$. We can, therefore, choose either normal direction $\mathbf{n}_1$ or $\mathbf{n}_2$, when using expression (16.15).

The same, however, does not apply to expression (16.11) for refraction. In this case, the angle between the normal to the surface and the incident ray must be smaller than, or equal to, $\pi/2$. Therefore, given two unit vectors $\mathbf{i}$ and $\mathbf{n}$ corresponding to the incident ray, and the normal to the surface, if $\mathbf{i} \cdot \mathbf{n} \geq 0$, the refracted ray can be obtained using the normal $\mathbf{n}$. If $\mathbf{i} \cdot \mathbf{n} < 0$ the normal $-\mathbf{n}$ should be used in the calculation. In the case of Figure 16.2, refraction should then be calculated with normal $\mathbf{n}_2$ to the surface A.

Expression (16.15) can be interpreted geometrically.[3] Figure 16.3 represents a reflection showing the incident ray, the reflected ray, and the normal to the surface, which are all unit vectors, that is, $\|\mathbf{i}\| = \|\mathbf{r}\| = \|\mathbf{n}\|$, all contained in the

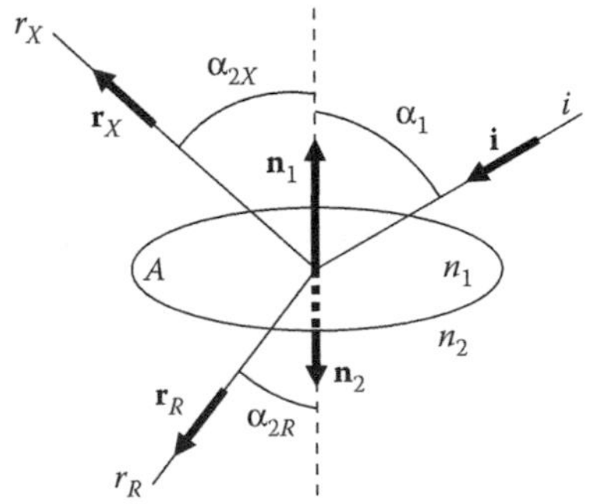

FIGURE 16.2
A surface A has two normal directions $\mathbf{n}_1$ and $\mathbf{n}_2 = -\mathbf{n}_1$. In the expression for the reflected ray, we can either use $\mathbf{n}_1$ or $\mathbf{n}_2$, but in the expression for the refracted ray we must use the direction $\mathbf{n} = \mathbf{n}_2$ that fulfills $\mathbf{i} \cdot \mathbf{n} > 0$.

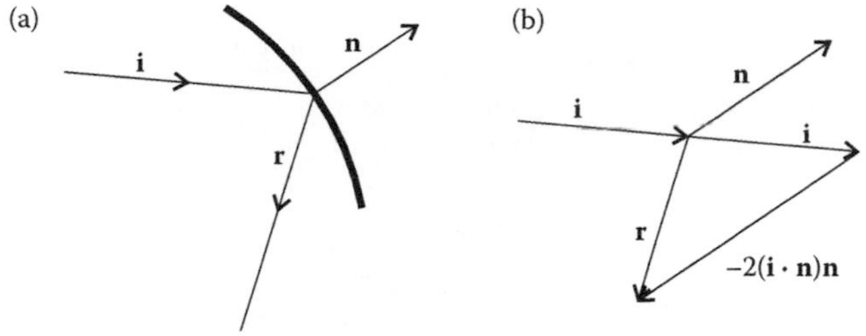

FIGURE 16.3
(a) A reflection showing the incident ray, the reflected ray, and the normal to the surface, which are all unit vectors; (b) The relationship among these three vectors.

same plane. Figure 16.3 shows that the reflected ray is related to the incident ray and the normal to the surface by expression (16.15).

From the earlier discussion, it can be concluded that, although Equation 16.10 can be applied to both reflection and refraction, only the solution with a positive sign has meaning for refraction, and only the solution with a negative sign has meaning for reflection.

Expression (16.11) for refraction is only valid for

$$1-\left(\frac{n_1}{n_2}\right)^2[1-(\mathbf{i}\cdot\mathbf{n})^2]\geq 0 \Rightarrow \frac{n_2}{n_1}\geq\sqrt{1-(\mathbf{i}\cdot\mathbf{n})^2}\Rightarrow n_2\geq n_1\sqrt{1-\cos^2\alpha_1} \tag{16.16}$$

Note that expression (16.16) can also be obtained from expression (16.13), making $n_2^2 - n_1^2 + (\mathbf{p}_1 \cdot \mathbf{n})^2 > 0$. We can now write

$$n_2 \geq n_1 \sin\alpha_1 \Leftrightarrow n_2 \sin(\pi/2) \geq n_1 \sin\alpha_1 \tag{16.17}$$

In the case of equality, we have $n_2 = n_1 \sin \alpha_1$. Since $\sin \alpha_1 < 1$, in this particular case of quality, it can be verified that we must have $n_2 < n_1$. The angle α for which the equality holds is called the critical angle, and can be obtained from Equation 16.17 as

$$\alpha_C = \arcsin(n_2/n_1) \tag{16.18}$$

Consider the refraction of a ray of light propagating from within a medium of refractive index n_1 into a medium of refractive index n_2, with $n_2 < n_1$, as presented in Figure 16.4.

An incident ray i_1 is refracted at the surface becoming refracted ray r_1. As the angle α_{1A} increases, the angle α_2 also increases until the limiting case, where the incident ray i_C makes an angle α_C with the normal to the surface, and the refracted ray r_C makes an angle $\pi/2$ with the normal to the surface,

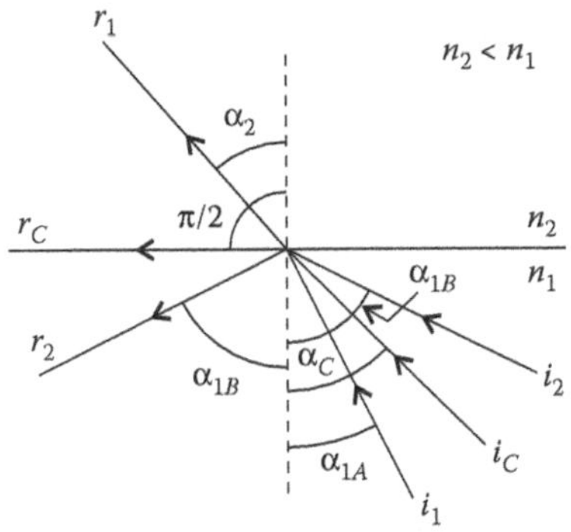

FIGURE 16.4
Three rays of light propagating in a medium of refractive index n_1. Ray i_1 makes an angle $\alpha_{1A} < \alpha_C$ with the normal to the surface. This ray is refracted by the surface, becoming ray r_1 propagating through the medium of refractive index $n_2 < n_1$. Ray i_C, making the angle α_C with the normal, called the critical angle, is refracted becoming ray r_C tangent to the surface of separation between the two media. Ray i_2, making with the normal an angle larger than α_C, is reflected by TIR to become ray r_2.

that is, the refracted ray is tangent to the surface. This case occurs when $\alpha_2 = \pi/2$, so that α_C fulfills Equation 16.18. For incidence angles α_{1B} larger than α_C, the ray of light is no longer refracted, but is reflected by the surface. This phenomenon is called TIR. Therefore, the incident ray i_2 is reflected to become ray r_2.

Given the incident and refracted or reflected rays, it is also possible to find the normal to a surface that transforms one into the other.

Consider a refraction with the incoming ray having momentum $\mathbf{p}_1$ and the refracted ray having momentum $\mathbf{p}_2$, as presented in Figure 16.5a. In this case, the law of refraction can be written $\|\mathbf{p}_2\| \sin \alpha_2 = \|\mathbf{p}_1\| \sin \alpha_1$. The normal to the surface has the direction of $\mathbf{p}_1 - \mathbf{p}_2$, so that

$$\mathbf{n} = \frac{\mathbf{p}_1 - \mathbf{p}_2}{\| \mathbf{p}_1 - \mathbf{p}_2 \|} \tag{16.19}$$

Expression (16.19) enables us to obtain the normal to the surface from the incident and refracted rays.

In the case of reflection, the normal has the direction of $\mathbf{p}_2 - \mathbf{p}_1$, as can be seen from Figure 16.5b, so that it can be written as $\mathbf{n} = (\mathbf{p}_2 - \mathbf{p}_1)/\|\mathbf{p}_2 - \mathbf{p}_1\|$.

Since, for reflection, we can choose for the normal to the surface either $\mathbf{n}$ or $-\mathbf{n}$, in this case we can also write

$$\mathbf{n} = \frac{\mathbf{p}_1 - \mathbf{p}_2}{\| \mathbf{p}_1 - \mathbf{p}_2 \|} \tag{16.20}$$

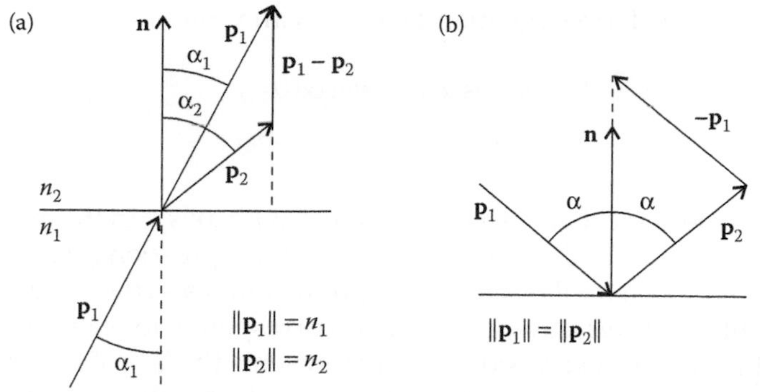

FIGURE 16.5
The plane defined by the incident and refracted rays (a) or reflected rays (b) and the normal to the surface. In (a), the incident and refracted rays are represented by the momentum vectors $\mathbf{p}_1$ and $\mathbf{p}_2$ that have magnitude $\|\mathbf{p}_1\| = n_1$ and $\|\mathbf{p}_2\| = n_2$, respectively, and in (b) $\|\mathbf{p}_1\| = \|\mathbf{p}_2\|$. In both cases, the normal vector points in the direction of the subtraction of these two vectors.

Expression (16.20) is the same as that presented earlier in the case of refraction. In the case of reflection, the refractive index is the same for the incident and refracted rays. Therefore, we have $\|\mathbf{p}_1\| = \|\mathbf{p}_2\| = n$. We can then write expression (16.20) as

$$\mathbf{n} = \frac{\mathbf{i} - \mathbf{r}}{\|\mathbf{i} - \mathbf{r}\|} \tag{16.21}$$

where $\mathbf{i} = \mathbf{p}_1/n$ and $\mathbf{r} = \mathbf{p}_2/n$ and therefore $\|\mathbf{i}\| = \|\mathbf{r}\| = 1$.

16.2 The Laws of Reflection and Refraction

The expressions to derive the directions of the refracted and reflected rays were based on expressions (16.1) through (16.4), which we now derive.

Suppose that a plane x_1x_2 separates two media of different refractive indices n_1 and n_2. The refractive index does not vary along the plane x_1x_2 but only in its perpendicular direction. We then have, in this case,

$$\frac{\partial n}{\partial x_1} = \frac{\partial n}{\partial x_2} = 0 \tag{16.22}$$

From the ray equation, that is, from expression (15.29), it can be concluded that p_1 and p_2 must be constant, that is,

$$p_1(n_1) = p_1(n_2) \quad \text{and} \quad p_2(n_1) = p_2(n_2) \tag{16.23}$$

Therefore, considering Equation 14.56, we can write

$$\begin{aligned} n_1 \cos\alpha_{1n1} &= n_2 \cos\alpha_{1n2} \\ n_1 \cos\alpha_{2n1} &= n_2 \cos\alpha_{2n2} \end{aligned} \tag{16.24}$$

where α_{inj} is the angle that the light ray makes with axis x_i in the medium with a refractive index n_j. In the case in which $n_1 \neq n_2$, expression (16.24) represents the law of refraction for the passage of light through a surface separating two media of different refractive indices, and corresponds to Snell's Law.[4] In the case in which $n_1 = n_2$, expression (16.24) represents the law of reflection.

The angles α_{inj} are represented in Figure 16.6a for the case of refraction. Figure 16.6b presents the angles that the incident and refracted rays make with the surface normal. Note that for the incident ray *i*, the momentum (and the direction of the ray) points in the direction from **O** to **A**, and in the case of the refracted ray *r*, the momentum (and the direction of the ray) points in the direction from **O** to **B**.

In the case of reflection, the situation is similar, and is presented in Figure 16.7. Figure 16.7b shows the angles that the incident and reflected rays make with the normal to the surface. Angles α_{i1} and α_{i2} are those that the ray of light makes with axis x_i before and after the reflection, respectively.

For a ray of light before and after the refraction ($n_1 \neq n_2$) or reflection ($n_1 = n_2$), we can write:

$$\begin{aligned} p_1^2(n_1) + p_2^2(n_1) + p_3^2(n_1) &= n_1^2 \\ p_1^2(n_2) + p_2^2(n_2) + p_3^2(n_2) &= n_2^2 \end{aligned} \tag{16.25}$$

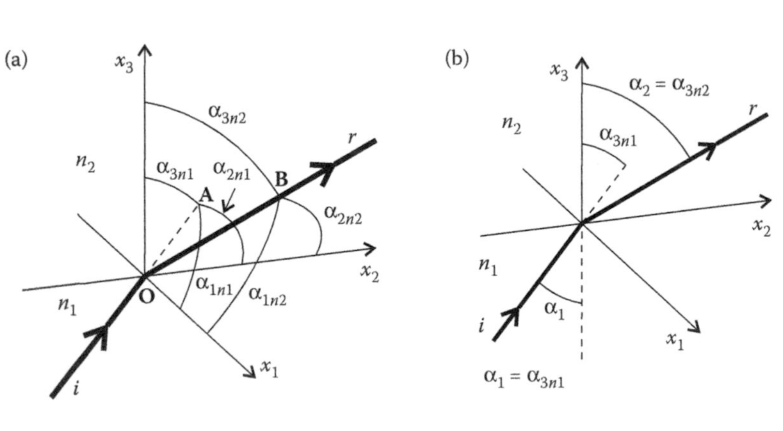

FIGURE 16.6
Plane x_1x_2 separates two media of different refractive indices n_1 and n_2. A ray of light *i* traveling in the medium of refractive index n_1 is refracted at x_1x_2 and transformed into *r*. (a) The angles that *i* and *r* make with the coordinate axes x_1, x_2, and x_3. (b) The angles that *i* and *r* make with the normal to the surface, that is, the angles with the positive portion of axis x_3. The rays *i* and *r* and the x_3 axis are in the same plane.

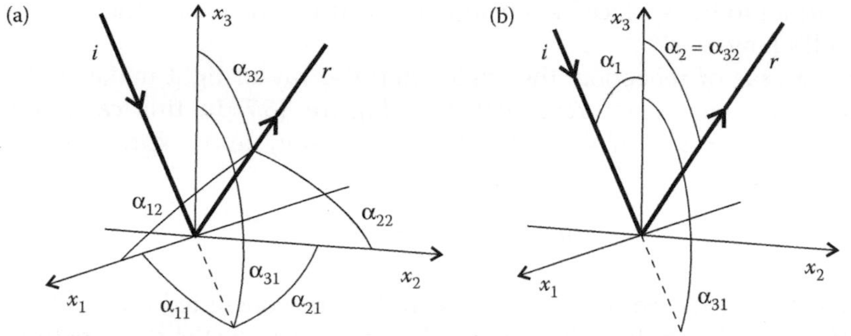

FIGURE 16.7
Plane x_1x_2 represents a mirror. A ray of light i is reflected at x_1x_2 and transformed into r. (a) The angles that i and r make with the axes x_1, x_2, and x_3. (b) The angles that i and r make with the normal to the surface, that is, the angles with the positive portion of axis x_3. The rays i and r and axis x_3 are in the same plane.

Considering that $p_1(n_1) = p_1(n_2)$ and $p_2(n_1) = p_2(n_2)$ and subtracting Equation 16.25, we obtain

$$n_1^2 - p_3^2(n_1) = n_2^2 - p_3^2(n_2) \tag{16.26}$$

This expression can also be written as

$$n_1^2 - n_1^2 \cos\alpha_{3n1} = n_2^2 - n_2^2 \cos\alpha_{3n2} \tag{16.27}$$

that is,

$$n_1^2 \sin^2\alpha_{3n1} = n_2^2 \sin^2\alpha_{3n2} \tag{16.28}$$

In the case of refraction, making $\alpha_1 = \alpha_{3n1}$ and $\alpha_2 = \alpha_{3n2}$, as presented in Figure 16.6b, gives

$$n_1^2 \sin^2\alpha_1 = n_2^2 \sin^2\alpha_2 \tag{16.29}$$

Since $0 \le \alpha_1 \le \pi/2$ and $0 \le \alpha_2 \le \pi/2$, we have $\sin\alpha_1 \ge 0$ and $\sin\alpha_2 \ge 0$, and we can obtain

$$n_1 \sin\alpha_1 = n_2 \sin\alpha_2 \tag{16.30}$$

The surface separating two media of different indices n_1 and n_2 is the plane x_1x_2, therefore, having as normal the axis x_3, with unit vector $\mathbf{e}_3 = (0,0,1)$. It can then be seen that α_1 and α_2 are the angles that the ray of light makes with the normal to the surface before and after refraction. This is the usual form of Snell's Law.

In the case of reflection, the angles that the ray of light makes with the axes x_1, x_2, and x_3 are represented in Figure 16.7. In this case, $\alpha_2 = \alpha_{32}$ and $\alpha_1 + \alpha_{31} = \pi \Leftrightarrow \alpha_1 = \pi - \alpha_{31}$. Besides, $n_1 = n_2$ and expression (16.28) simplifies to

$$\sin\alpha_1 = \sin\alpha_2 \Leftrightarrow \alpha_1 = \alpha_2 \tag{16.31}$$

since $0 \le \alpha_1 \le \pi/2$ and $0 \le \alpha_2 \le \pi/2$. It can then be concluded that, in the case of reflection, the angle with the normal to the surface is the same, before and after reflection.

In the preceding derivation of the law of refraction, we assumed that the refracted ray must be in the plane defined by the incident ray, and by the normal to the surface $\mathbf{e}_3 = (0,0,1)$ separating the two media of refractive indices n_1 and n_2. We assumed the same for the case of reflection in which $n_1 = n_2$. We can now verify this by showing that the refracted or reflected rays can be obtained by a linear combination of the incident ray and the normal to the surface. From expression (16.26), we obtain

$$p_3(n_2) = \pm\sqrt{p_3^2(n_1) + n_2^2 - n_1^2} \tag{16.32}$$

The incident ray has the direction of

$$\mathbf{p}_I = (p_1(n_1), p_2(n_1), p_3(n_1)) \tag{16.33}$$

The refracted ray $\mathbf{p}_R = (p_1(n_2), p_2(n_2)p_3(n_2))$, and, therefore, from expressions (16.23) and (16.32) we can obtain

$$\mathbf{p}_R = \left(p_1(n_1), p_2(n_1), \pm\sqrt{p_3^2(n_1) + n_2^2 - n_1^2}\right) \tag{16.34}$$

To see that $\mathbf{p}_R$ is in the plane defined by $\mathbf{p}_I$ and $\mathbf{e}_3$, we can verify that $\mathbf{p}_R$ can be obtained as a linear combination of $\mathbf{p}_I$ and $\mathbf{e}_3$, that is,

$$\mathbf{p}_R = a\mathbf{p}_I + b\mathbf{e}_3 \tag{16.35}$$

which corresponds to a system of three equations and two unknowns (a and b):

$$\begin{aligned} p_1(n_1) &= a\, p_1(n_1) \Rightarrow a = 1 \\ p_2(n_1) &= a\, p_2(n_1) \Rightarrow a = 1 \\ \pm\sqrt{p_3^2(n_1) + n_2^2 - n_1^2} &= a p_3(n_1) + b \Rightarrow b = -p_3(n_1) \pm \sqrt{p_3^2(n_1) + n_2^2 - n_1^2} \end{aligned} \tag{16.36}$$

The system of equations has a solution, and, therefore, $\mathbf{p}_R$, $\mathbf{p}_I$, and $\mathbf{e}_3$ are vectors in the same plane. Given that these expressions can be applied either to reflection or to refraction, it can be concluded that the reflected or refracted ray is in the plane defined by the incident ray and the normal to the surface (which can be a mirror or a surface separating two media having different refractive index).

Expression (16.30) is general and can be applied to either refraction or reflection by any surface. To verify this, again consider a surface separating two media having different refractive indices (refraction) or a mirror (reflection). Further, consider a ray of light arriving at point **P**. We can make the x_3 axis coincident with the normal to the surface at **P**. In this case, the plane x_1x_2 is tangent to the surface. In the neighborhood of **P**, the tangent plane well approximates the surface. That way, the refracted ray at **P** should refract the same way, as if it was refracted in the plane tangent to the surface, and expression (16.30) is, therefore, still applicable. In this case, α_1 and α_2 are the angles that the incident and refracted rays make with the normal to the surface. Also in this case, the incident and refracted rays and the normal to the surface are contained in the same plane.

References

1. Stavroudis, O.N., *The Optics of Rays, Wavefronts, and Caustics*, Academic Press, New York, 1972.
2. Kush, O., *Computer-Aided Optical Design of Illuminating and Irradiating Devices*, ASLAN Publishing House, Moscow, 1993.
3. Welford, W.T. and Winston, R., *High Collection Nonimaging Optics*, Academic Press, San Diego, 1989.
4. Jenkins, F.A. and White, H.E., *Fundamentals of Optics*, 3rd ed, McGraw-Hill Book Company, New York, 1957.

17

Symmetry

17.1 Conservation of Momentum and Apparent Refractive Index

If we refract a light ray with momentum $\mathbf{p}_1$ at a surface with normal $\mathbf{n}$, and it comes out as a ray with momentum $\mathbf{p}_2$ after refraction, $\mathbf{p}_2$ and $\mathbf{p}_1$ are related (see Chapter 16) by

$$\mathbf{p}_2 = \mathbf{p}_1 - \left[(\mathbf{p}_1 \cdot \mathbf{n}) + \sqrt{n_2^2 - n_1^2 + (\mathbf{p}_1 \cdot \mathbf{n})^2}\right]\mathbf{n} \tag{17.1}$$

where n_1 and n_2 are the refractive indices before and after refraction. If we had a reflection, instead of refraction, the reflected ray would have momentum

$$\mathbf{p}_2 = \mathbf{p}_1 - 2(\mathbf{p}_1 \cdot \mathbf{n})\mathbf{n} \tag{17.2}$$

where n is the refractive index of the material in which the reflection occurs. In any case, we can write

$$\mathbf{p}_2 = \mathbf{p}_1 + \sigma\mathbf{n} \tag{17.3}$$

where σ is a scalar.

We now consider a different situation in which we have the general coordinate axes $i_1 i_2 i_3$ and obtain the mathematical relations between the angles α_1, α_2, and α_3, which a ray of light makes with the coordinate axes i_1, i_2, and i_3, as well as the angles β_1 and β_2 that its projection onto the plane $i_1 i_2$ makes with the axes i_1 and i_2, as shown in Figure 17.1.

Consider a ray of light propagating in the direction of the unit vector $\mathbf{v}$, making angles α_1, α_2, and α_3 with the axes i_1, i_2, and i_3, respectively. From Figure 17.1 we can see that

$$\cos\alpha_2 = \sin\alpha_3 \cos\beta_2 \tag{17.4}$$

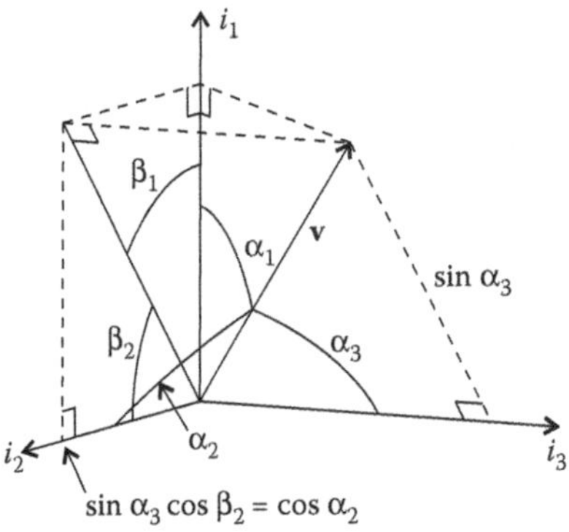

FIGURE 17.1
Projections of a unit vector **v**, that is, $||\mathbf{v}|| = 1$, onto the coordinate axes i_1, i_2, and i_3 and onto the plane i_1i_2.

and squaring both sides of Equation 17.4 yields

$$\sin^2\alpha_3\cos^2\beta_2 = \cos^2\alpha_2 \Leftrightarrow (1-\cos^2\alpha_3)\cos^2\beta_2 = \cos^2\alpha_2 \tag{17.5}$$

Multiplying both sides of Equation 17.5 by the refractive index n gives

$$(n^2 - n^2\cos^2\alpha_3)\cos^2\beta_2 = n^2\cos^2\alpha_2 \tag{17.6}$$

that is,

$$(n^2 - p_3^2)\cos^2\beta_2 = p_2^2 \tag{17.7}$$

which will prove useful next.

Consider the refraction (or reflection) at a point on a surface, and orient a set of coordinates $i_1i_2i_3$ such that the normal to the surface at that point, and in these local coordinates, point in direction i_1. The normal to the surface is then $\mathbf{n} = (m_1, 0, 0)$, and from Equation 17.3 we can see that the i_2 and i_3 components of the momentum do not change. If $\mathbf{p}_1 = (p_1(n_1), p_2(n_1), p_3(n_1))$ and $\mathbf{p}_2 = (p_1(n_2), p_2(n_2), p_3(n_2))$ we have $p_2(n_1) = p_2(n_2)$ and $p_3(n_1) = p_3(n_2)$, that is,

$$\begin{aligned} p_2 &= \text{constant} \\ p_3 &= \text{constant} \end{aligned} \tag{17.8}$$

Applying expression (17.7) to the media with refractive indices n_1 and n_2 yields

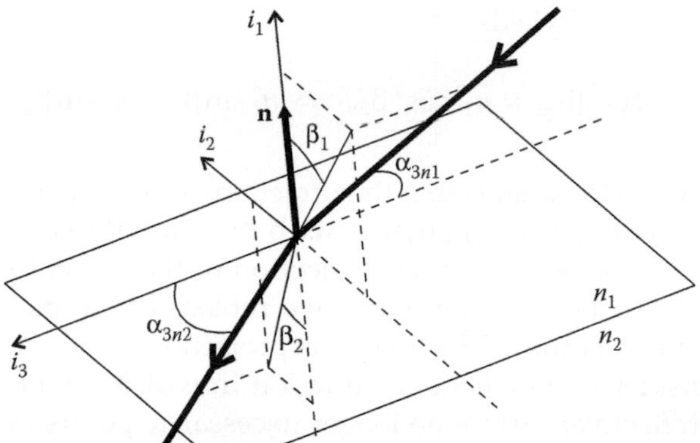

FIGURE 17.2
Refraction of a ray at a surface separating two media of refractive indices n_1 and n_2, also showing normal **n** and the projection of the ray trajectory onto plane i_1i_2.

$$\begin{aligned}\left[n_1^2 - p_3^2(n_1)\right]\cos^2\beta_{2n1} &= p_2^2(n_1)\\ \left[n_2^2 - p_3^2(n_2)\right]\cos^2\beta_{2n2} &= p_2^2(n_2)\end{aligned} \tag{17.9}$$

where β_{2nj} is the angle that the projection of the ray of light onto the plane i_1i_2 makes with the axis i_2 in the medium having a refractive index n_j, as presented in Figures 17.2 and 17.3. Since p_2 and p_3 are constant, that is, $p_2(n_1) = p_2(n_2)$ and $p_3(n_1) = p_3(n_2)$, we obtain

$$(n_1^2 - p_3^2)\cos^2\beta_{2n1} = (n_2^2 - p_3^2)\cos^2\beta_{2n2} \tag{17.10}$$

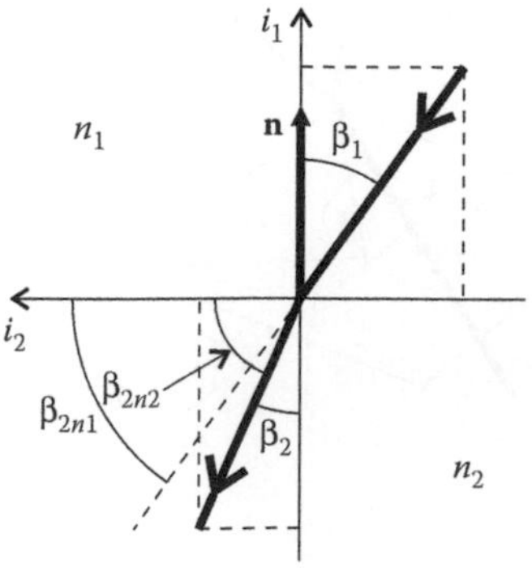

FIGURE 17.3
Projection of a 3-D ray onto plane i_1i_2.

Making $n_k^{*2} = n_k^2 - p_3^2$ yields

$$n_1^{*2}\cos^2\beta_{2n1} = n_2^{*2}\cos^2\beta_{2n2} \Rightarrow n_1^*\sin\beta_1 = n_2^*\sin\beta_2 \tag{17.11}$$

where β_1 and β_2 are the angles that the projection of the ray of light onto the plane i_1i_2 makes with the axis i_1, that is, with the normal to the surface.

It can then be concluded that the projection of the ray onto the plane i_1i_2 also fulfills the law of refraction when we replace the refractive indices n_1 and n_2 by n_1^* and n_2^*. Figure 17.3 shows this projection.

We now consider a more general situation in which the normal **n** to the refractive (or reflective) surface no longer necessarily points in the direction of i_1, but **n** is still contained in the plane i_1i_2 (plane ν) perpendicular to axis i_3, as shown in Figure 17.4. In this case, $\mathbf{n} = (m_1, m_2, 0)$ and p_3 is conserved on refraction (or reflection), as seen from Equation 17.3. Equation 17.11 is still valid in this case where angles β_1 and β_2 are the angles that the projection of the light ray onto the plane ν makes with the normal **n**, before and after refraction (or reflection). We can then write

$$\sqrt{n_1^2 - p_3^2}\sin\beta_1 = \sqrt{n_2^2 - p_3^2}\sin\beta_2 \tag{17.12}$$

The i_3 component of the optical momentum is constant and this means that

$$p_3 = n_1\cos\alpha_{3n1} = n_2\cos\alpha_{3n2} \tag{17.13}$$

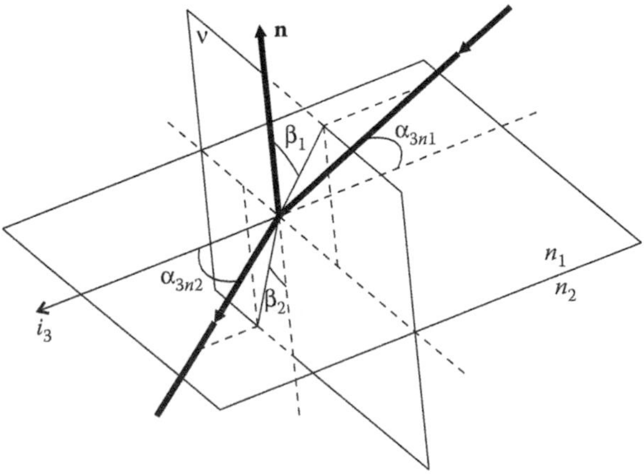

FIGURE 17.4
Axis i_3 is tangent to a surface separating two media of refractive indices n_1 and n_2. The normal **n** to the surface is perpendicular to i_3 and contained in plane ν. The projection of the ray trajectory onto plane ν follows the law of refraction with a modified refractive index.

If medium n_1 is air, then $n_1 = 1$ and from expression (17.12) we get the particular case in which

$$\sqrt{1-p_3^2}\sin\beta_1 = \sqrt{n_2^2 - p_3^2}\sin\beta_2 \tag{17.14}$$

Expression (17.14) may also be rewritten as

$$\sin\beta_1 = \frac{\sqrt{n_2^2 - p_3^2}}{\sqrt{1-p_3^2}}\sin\beta_2 \tag{17.15}$$

which is equivalent to a refraction from a medium of refractive index $n_A = 1$ into another medium of refractive index $n_B = \sqrt{n_2^2 - p_3^2}/\sqrt{1-p_3^2}$. In that case, expression (17.15) can be written as $n_A \sin\beta_1 = n_B \sin\beta_2$ or its equivalent $\sin\beta_1 = n_B \sin\beta_2$.

17.2 Linear Symmetry

We can now apply the result obtained earlier to the case of a system with linear symmetry along the axis x_3, as shown in Figure 17.5. We align axis i_3 along x_3. The normals to the optical surfaces are perpendicular to x_3 and, therefore, contained in the plane x_1x_2 (plane v, perpendicular to x_3). These

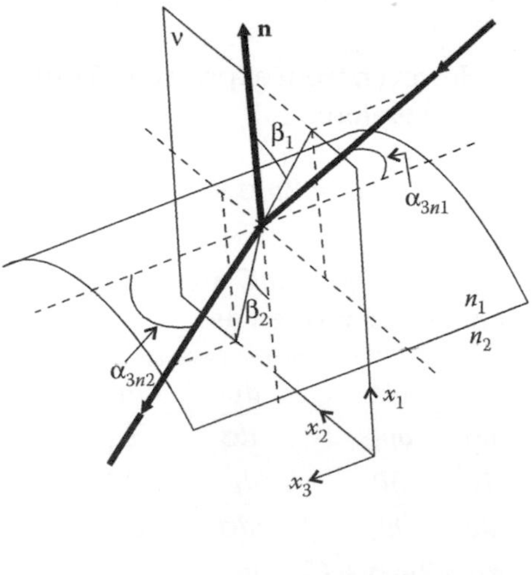

FIGURE 17.5
A surface having linear symmetry along axis x_3 and normal **n**, separates two media of refractive indices n_1 and n_2. The projected trajectory of a light ray onto plane v perpendicular to x_3 follows the law of refraction with a modified refractive index.

normals have x_3 component as zero and, from expression (17.3), we can then conclude that the p_3 component of the momentum (along x_3) is conserved by reflections and refractions. It is also conserved as the ray travels in a homogeneous medium of refractive index n between reflections or refractions, because the angle the ray makes with the axis x_3 is constant, as it travels in a straight line. Component p_3 of the optical momentum is then always conserved in a linear system extruded along the axis x_3.

The result in expression (17.12) also applies to the linear system in Figure 17.5. We can then study the linear system as a 2-D system, in which the refractive index is replaced by

$$n^* = \sqrt{n^2 - p_3^2} \tag{17.16}$$

and expression (17.12) is, in this case,

$$n_1^* \sin\beta_1 = n_2^* \sin\beta_2 \tag{17.17}$$

Then the projections of the light rays of a linear system onto plane x_1x_2 behave as a 2-D system with refractive index given by Equation 17.16.

Consider next a more general way to derive these results. A linear optical system aligned along axis x_3 may be described by the coordinate system (x_1, x_2, x_3) in which the refractive index does not depend on x_3 and $n = n(x_1, x_2)$. From expression (14.79), it can be concluded that in this case $\partial P/\partial x_3 = \partial n/\partial x_3 = 0$; therefore, $dp_3/d\sigma = 0$ and

$$p_3 = C_3 \tag{17.18}$$

where C_3 is a constant. However, from expression (14.79) we can also see that $\partial P/\partial p_3 = 2p_3 = dx_3/d\sigma$, and therefore,

$$x_3 = 2p_3\sigma + C \tag{17.19}$$

that is, $x_3 = C_1\sigma + C$, where $C_1 = 2p_3$ and C are constants.

Expression (14.79) can then be written as

$$\begin{aligned}
&\frac{dx_1}{d\sigma} = \frac{\partial P}{\partial p_1} \qquad \frac{dp_1}{d\sigma} = -\frac{\partial P}{\partial x_1} \\
&\frac{dx_2}{d\sigma} = \frac{\partial P}{\partial p_2} \qquad \frac{dp_2}{d\sigma} = -\frac{\partial P}{\partial x_2} \\
&x_3 = 2p_3\,\sigma + C \qquad p_3 = C_3 \\
&P = p_1^2 + p_2^2 - \left[n^2(x_1, x_2) - p_3^2\right] = 0
\end{aligned} \tag{17.20}$$

where C and C_3 are constants. The behavior of the optical system along the axis x_3 is then known. The analysis of a 3-D system with linear symmetry can then be reduced to the analysis of the 2-D system described by the following system of equations:

$$\begin{aligned} \frac{dx_1}{d\sigma} &= \frac{\partial P}{\partial p_1} \quad \frac{dp_1}{d\sigma} = -\frac{\partial P}{\partial x_1} \\ \frac{dx_2}{d\sigma} &= \frac{\partial P}{\partial p_2} \quad \frac{dp_2}{d\sigma} = -\frac{\partial P}{\partial x_2} \\ P &= p_1^2 + p_2^2 - n^{*2} = 0 \end{aligned} \tag{17.21}$$

With $n^{*2} = n^2(x_1, x_2) - p_3^2$, that is,

$$n^* = \sqrt{n^2(x_1, x_2) - p_3^2} \tag{17.22}$$

where p_3 is conserved (is constant).

This is a particular result of a general case in which the Hamiltonian does not depend on one of the coordinates x_k, that is, $\partial P/\partial x_k = 0 \Rightarrow p_k = \text{constant}$. This coordinate x_k is called cyclic, and the corresponding momentum is constant. The system can then be described with one less independent coordinate (one less dimension).[1] Further examples of such systems are those with circular symmetry, as described in Section 17.3.

Since $\partial P/\partial x_k = -\partial L/\partial x_k$, where L is the Lagrangian, we have $\partial L/\partial x_k = 0$ for the cyclic coordinate x_k. The corresponding Euler equation is then

$$\frac{d}{d\sigma}\left(\frac{\partial L}{\partial x_k'}\right) = 0 \tag{17.23}$$

which can be integrated once resulting in $\partial L/\partial x_k' = \text{Constant}$. Expression (17.23) is called a first integral of the second-order Euler equation.[2] Note that this first integral can also be written as $p_k = \text{Constant}$.

17.3 Circular Symmetry and Skew Invariant

Often, optical instruments have circular or axial symmetry, that is, they are symmetric around an axis of rotation, assumed here to be x_3. Further, postulate that light rays progress along the axis x_3. Here it is convenient to choose a cylindrical coordinate system. Each point **P** in space is then defined by coordinate x_3, by distance ρ to the x_3 axis, and by an angle θ around that axis, as presented in Figure 17.6. These coordinates define a local coordinate system

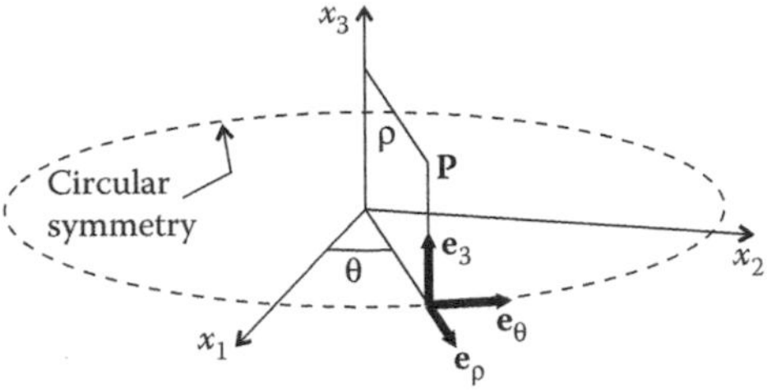

FIGURE 17.6
Cylindrical coordinate system used in 3-D optical systems with circular symmetry.

$\mathbf{e}_\rho$, $\mathbf{e}_\theta$, $\mathbf{e}_3$. A system with circular symmetry is characterized by the fact that a cut with any vertical plane containing axis x_3, yields the same cross-section. In this case, the refractive index will be a function of ρ and x_3 because the optical surfaces do not change with θ.

A light ray is traveling in a medium of refractive index n_1 and refracts into a medium of refractive index n_2. If $n_1 = n_2$, then the ray is reflected. Before refraction (or reflection), the ray has an optical momentum $\mathbf{p}_1$, and after refraction (or reflection), it has an optical momentum $\mathbf{p}_2$. If the system has circular symmetry, the normals to the optical surfaces at each point have no component along $\mathbf{e}_\theta$, that is, $\mathbf{n} = (m_\rho, 0, m_3)$. The component of $\mathbf{p}_1$ along $\mathbf{e}_\theta$ will, therefore, remain unchanged by refraction or reflection. This component is given by

$$p_\theta = n\cos\phi \qquad (17.24)$$

where ϕ is the angle the light ray makes with $\mathbf{e}_\theta$. We then have

$$n_1\cos\phi_1 = n_2\cos\phi_2 \qquad (17.25)$$

as represented in Figure 17.7.

Note that although p_θ is conserved in a refraction (or reflection), it is not conserved as the ray propagates straight through the system between refractions or reflections. As we can see from Figure 17.8, going from point **A** to point **B** along the path of a ray causes angle ϕ to change and, therefore, p_θ also changes because the refractive index n is constant between refractions or reflections.

We consider again the ray as it propagates through the system between refractions or reflections. We have a situation, as shown in Figure 17.9, in which Figure 17.9b is the top view of Figure 17.9a.

The projection of ray r onto plane x_1x_2, parallel to the plane defined by $\mathbf{e}_\rho$ and $\mathbf{e}_\theta$, is line r_P. For this projection (and for ray r), we have

$$\rho\sin\alpha = M \qquad (17.26)$$

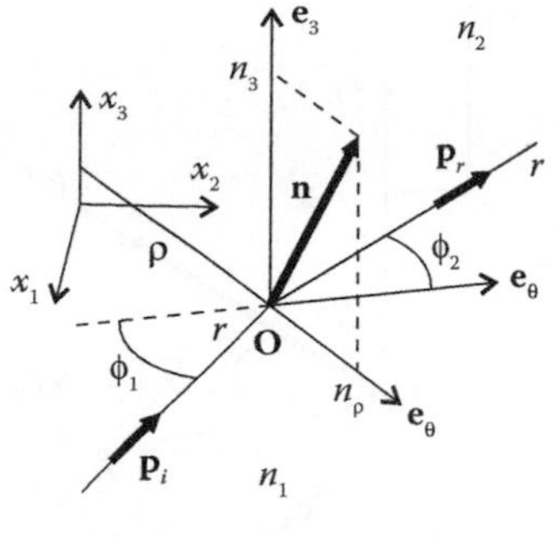

FIGURE 17.7
A light ray refracts at a point **O** of an optical system with circular symmetry. The θ component p_θ of the optical momentum does not change with refraction.

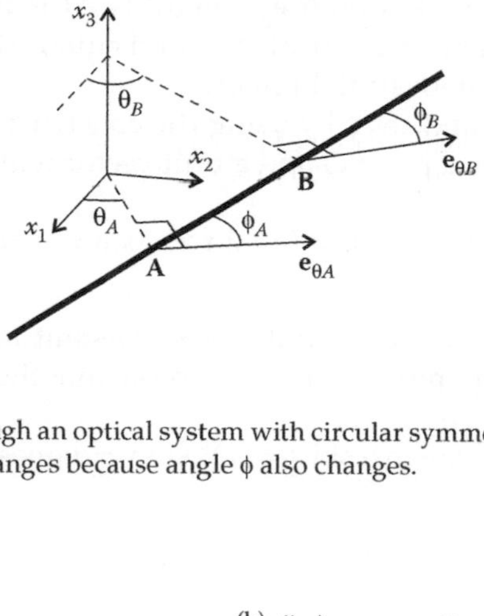

FIGURE 17.8
As a light ray travels through an optical system with circular symmetry, the θ component p_θ of the optical momentum changes because angle ϕ also changes.

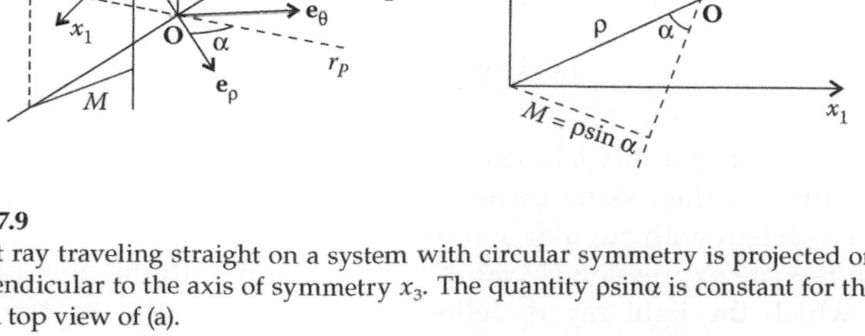

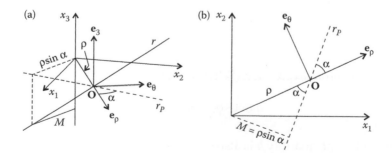

FIGURE 17.9
(a) A light ray traveling straight on a system with circular symmetry is projected onto plane x_1x_2 perpendicular to the axis of symmetry x_3. The quantity ρsinα is constant for this projection. (b) A top view of (a).

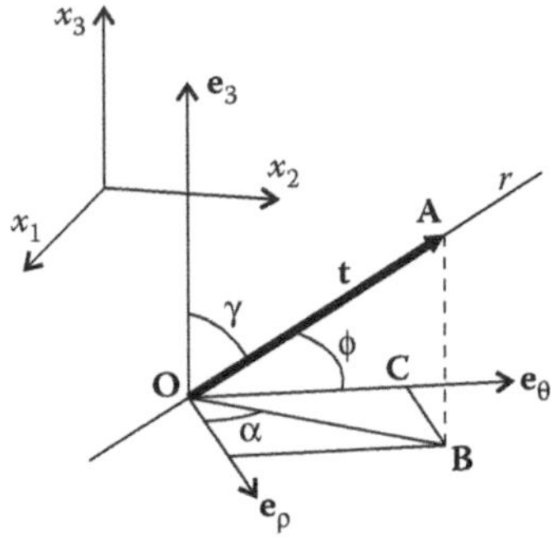

FIGURE 17.10
Relating angle ϕ that a light ray makes with $\mathbf{e}_\theta$ with the angle γ it makes with $\mathbf{e}_3$.

where M is a constant, ρ is the distance from a point **O** on the ray to axis x_3, and α the angle r_p makes with $\mathbf{e}_\rho$. Quantity M is the minimal distance between the light ray and the optical axis, and equals the length of the common perpendicular to both straight lines.

We now relate sin α and cos ϕ by using the construction in Figure 17.10.

For a ray r defined by a point **O** and a unit vector with direction **t**, we have

$$\mathbf{OC} = \cos\phi = \mathbf{OB}\sin\alpha = \sin\gamma\sin\alpha \tag{17.27}$$

Since sin γ and the refractive index are constant as the ray propagates between reflections and refractions, we can say that the quantity

$$h = n\rho\sin\alpha\sin\gamma = nM\sin\gamma = n\rho\cos\phi \tag{17.28}$$

is conserved. We can also write

$$p_\theta = n\cos\phi = h\frac{1}{\rho} = bh \tag{17.29}$$

where $b = 1/\rho$. From expression (17.25) we have

$$n_1\rho\cos\phi_1 = n_2\rho\cos\phi_2 \tag{17.30}$$

and, therefore, the quantity h is also conserved in refractions and reflections. This quantity is called skew invariant or skewness, and we see it is conserved in a system with circular symmetry around axis x_3.

We can rewrite expression (17.28) in another form with the help of Figure 17.11, in which the light ray is defined by a point **O** and by the optical

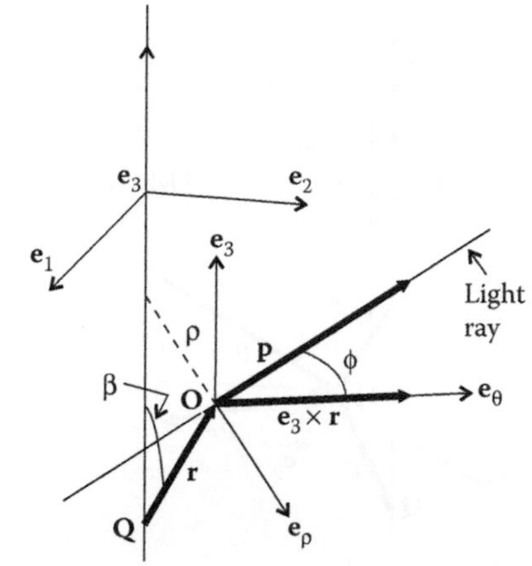

FIGURE 17.11
Construction for writing the expression for the skew invariant h as a scalar triple product of $\mathbf{p}$, $\mathbf{e}_3$, and $\mathbf{r}$, basically the volume of the parallelepiped they form.

momentum $\mathbf{p}$, which is given by $\mathbf{p} = n\mathbf{t}$ for $\mathbf{t}$ being a unit vector in the direction of the propagation of the ray. Then $\|\mathbf{p}\| = n$ and

$$\rho = \| \mathbf{r} \| \sin\beta = \| \mathbf{e}_3 \times \mathbf{r} \| \tag{17.31}$$

where $\mathbf{r}$ is a vector from a point $\mathbf{Q}$ on the axis of symmetry to a point $\mathbf{O}$ on the ray. We can write

$$h = \| \mathbf{e}_3 \times \mathbf{r} \| \|\mathbf{p}\| \cos\phi = \mathbf{p} \cdot (\mathbf{e}_3 \times \mathbf{r}) \tag{17.32}$$

obtaining h as a scalar triple product of $\mathbf{p}$, $\mathbf{e}_3$, and $\mathbf{r}$ (note that vector $\mathbf{e}_3 \times \mathbf{r}$ points in the direction of $\mathbf{e}_\theta$).

Yet another way to write the expression for h is to consider the geometry of Figure 17.12. The light ray is again defined by a point $\mathbf{O}$ and the optical momentum $\mathbf{p}$. Vector $\mathbf{p}_P$ is the projection of $\mathbf{p}$ in the plane x_1x_2 (parallel to the plane defined by $\mathbf{e}_\rho$ and $\mathbf{e}_\theta$) and has magnitude $\| \mathbf{p}_P \| = n \sin\gamma$. Also, $\mathbf{r}_P$ is the projection of $\mathbf{r}$ onto the x_1x_2 plane with $\|\mathbf{r}_P\| = \rho$ and we can write expression (17.28) as

$$h = \| \mathbf{p}_P | \| \, |\mathbf{r}_P \| \, \sin\alpha = \|\mathbf{r}_P \times \mathbf{p}_P \| \tag{17.33}$$

Therefore, if we have $\mathbf{r} = (x_1, x_2, x_3)$ and $\mathbf{p} = (p_1, p_3, p_3)$ we can rewrite expression (17.33) as

$$h = \|(x_1, x_2, 0) \times (p_1, p_2, 0)\| \tag{17.34}$$

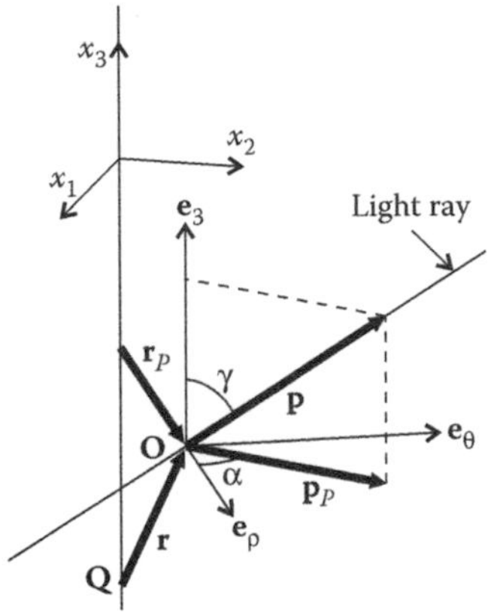

FIGURE 17.12
The skew invariant can be calculated as a function of the projection onto the plane x_1x_2 of vectors **r** and **p**.

Just as in the case of the linear system, also in the case of the circular optics, we can use expression (17.12). Now we take coordinate i_3 as coordinate θ, and we get the system in Figure 17.13. Now plane ν is defined as a surface θ = constant and is perpendicular to $\mathbf{e}_\theta$.

The normals to the optical surfaces are perpendicular to $\mathbf{e}_\theta$ and, therefore, contained in the planes ν defined by $\mathbf{e}_\rho$ and $\mathbf{e}_3$ (perpendicular to $\mathbf{e}_\theta$). The p_θ component of the optical momentum is unchanged by refraction and is given by

$$p_\theta = n_1 \cos\alpha_{\theta n1} = n_2 \cos\alpha_{\theta n2} \tag{17.35}$$

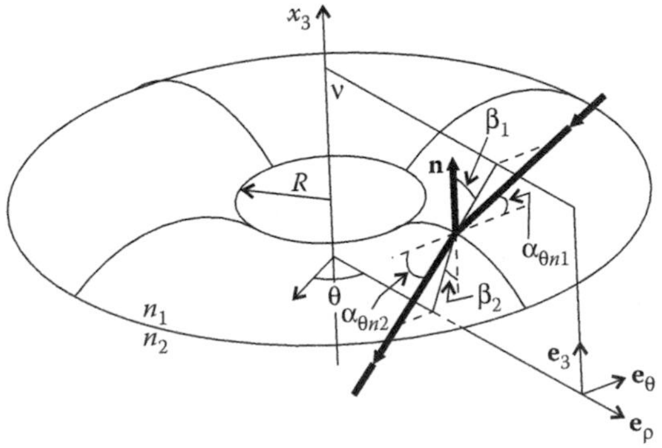

FIGURE 17.13
Projection of a light ray traveling in an optical system with circular symmetry onto a plane ν defined by θ = constant. The projected trajectory of the light ray onto plane ν follows the law of refraction with a modified refractive index.

Note, however, that p_θ is not constant while the ray propagates after refraction, rather what is constant is the skewness h. If the trajectory of the light rays is projected onto plane v, then expresssion (17.12) now becomes

$$\sqrt{n_1^2 - p_\theta^2}\,\sin\beta_1 = \sqrt{n_2^2 - p_\theta^2}\,\sin\beta_2 \tag{17.36}$$

because we are considering that i_3 is now θ. From expressions (17.24) and (17.28) we obtain $p_\theta = h/\rho$ and we can write

$$\sqrt{n_1^2 - (h/\rho)^2}\,\sin\beta_1 = \sqrt{n_2^2 - (h/\rho)^2}\,\sin\beta_2 \tag{17.37}$$

And therefore, when projected onto the plane v, the refraction appears to happen with a refractive index

$$n^* = \sqrt{n^2 - (h/\rho)^2} \tag{17.38}$$

Note that as the radius R of the circular system tends to infinity, its behavior tends to become that of a linear system.[3]

An alternative way to derive the conservation of h is from the Hamilton Equation 14.120 making $i_1 \to \rho$, $i_2 \to \theta$, $i_3 \to x_3$. Expression (14.119) for the vector $\mathbf{p}$ can be written in this case as

$$\mathbf{p} = n\cos\varphi\,\mathbf{e}_\rho + n\cos\phi\,\mathbf{e}_\theta + n\cos\gamma\,\mathbf{e}_3 \tag{17.39}$$

where φ, ϕ, and γ are, respectively, the angles which the direction of the light ray makes with unit vectors $\mathbf{e}_\rho$, $\mathbf{e}_\theta$, and $\mathbf{e}_3$, as shown in Figure 17.14a.

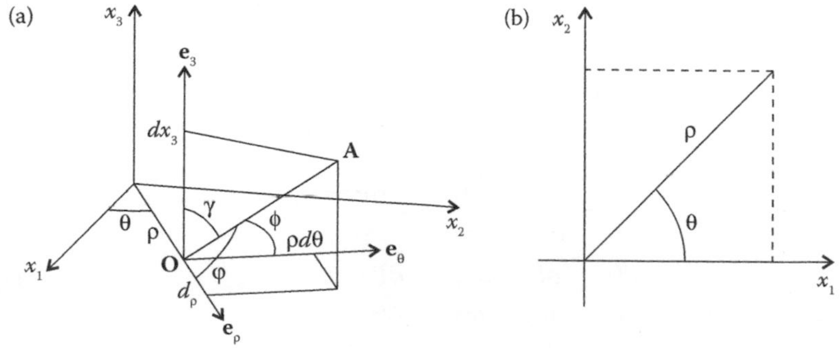

FIGURE 17.14
(a) A light ray through points **O** and **A** makes angles φ, ϕ, and γ with unit vectors $\mathbf{e}_\rho$, $\mathbf{e}_\theta$, and $\mathbf{e}_3$, respectively. (b) The angle θ is a function of x_1 and x_2.

Expression (14.115), however, states that

$$\mathbf{p} = u_1 a_1 \mathbf{e}_1 + u_2 a_2 \mathbf{e}_2 + u_3 a_3 \mathbf{e}_3 \tag{17.40}$$

with $a_k = ||\nabla i_k||$ and $\mathbf{e}_k = \nabla i_k / ||\nabla i_k||$. In this case, with $i_1 \to \rho$, $i_2 \to \theta$, $i_3 \to x_3$

$$\mathbf{p} = u_\rho \, ||\nabla \rho|| \, \mathbf{e}_\rho + u_\theta \, ||\nabla \theta|| \, \mathbf{e}_\theta + u_3 \, ||\nabla x_3|| \, \mathbf{e}_3 \tag{17.41}$$

From Figure 17.14b we can see that

$$\theta = \arccos\left(x_1 / \sqrt{x_1^2 + x_2^2} \right) = \arccos(x_1/\rho) \tag{17.42}$$

And we can calculate the gradient of θ as

$$\nabla\theta = \left(\frac{\partial\theta}{\partial x_1}, \frac{\partial\theta}{\partial x_2}, \frac{\partial\theta}{\partial x_3} \right) = \left(-\frac{\sqrt{x_2^2}}{\rho^2}, \frac{x_1 x_2}{\rho^2 \sqrt{x_2^2}}, 0 \right) \tag{17.43}$$

And therefore,

$$||\, \nabla\theta \,|| = 1/\rho = b \tag{17.44}$$

where b is defined by expression (17.44). Also, from $\rho = \sqrt{x_1^2 + x_2^2}$ we obtain $||\nabla\rho|| = 1$. We also have $||\nabla x_3|| = 1$. This enables us to write Equation 17.41 as

$$\mathbf{p} = u_\rho \, \mathbf{e}_\rho + u_\theta b \, \mathbf{e}_\theta + u_3 \, \mathbf{e}_3 \tag{17.45}$$

From Equation 14.118 we can see that $u_\theta = h$ since

$$u_\theta b = n \cos\phi \Leftrightarrow u_\theta = n\rho\cos\phi \tag{17.46}$$

The system of differential equation (14.120) can now be rewritten, making $i_1 \to \rho$, $i_2 \to \theta$, $i_3 \to x_3$ and renaming $u_\rho = p_\rho$ and $u_3 = p_3$ to obtain

$$\begin{aligned}
&\frac{d\rho}{d\sigma} = \frac{\partial P}{\partial p_\rho} \quad \frac{dp_\rho}{d\sigma} = -\frac{\partial P}{\partial \rho} \\
&\frac{d\theta}{d\sigma} = \frac{\partial P}{\partial u_\theta} \quad \frac{du_\theta}{d\sigma} = \frac{\partial P}{\partial \theta} \\
&\frac{dx_3}{d\sigma} = \frac{\partial P}{\partial p_3} \quad \frac{dp_3}{d\sigma} = -\frac{\partial P}{\partial x_3} \\
&P = p_\rho^2 + u_\theta^2 b^2 + p_3^2 - n^2(\rho, x_3) = 0
\end{aligned} \tag{17.47}$$

Since P does not depend on θ we have $\partial P/\partial\theta = 0$ and

$$u_\theta = h \tag{17.48}$$

where h is a constant. The quantity h is the skew invariant or skewness. Also, from the equation for $d\theta/d\sigma$ we have

$$\frac{d\theta}{\sigma} = 2hb^2 = \frac{2h}{\rho^2(\sigma)} \tag{17.49}$$

or

$$\theta = \int \frac{2h}{\rho^2(\sigma)} d\sigma + C_\theta = F(\sigma) + C_\theta \tag{17.50}$$

where C_θ is also a constant. We, therefore, have

$$\begin{aligned}
&\frac{d\rho}{d\sigma} = \frac{\partial P}{\partial p_\rho} \quad \frac{dp_\rho}{d\sigma} = -\frac{\partial P}{\partial \rho} \\
&\theta = F(\sigma) + C_\theta \quad u_\theta = h \\
&\frac{dx_3}{d\sigma} = \frac{\partial P}{\partial p_3} \quad \frac{dp_3}{d\sigma} = -\frac{\partial P}{\partial x_3} \\
&P = p_\rho^2 + p_3^2 - \left[n^2(\rho, x_3) - h^2b^2\right] = 0
\end{aligned} \tag{17.51}$$

The analysis of a 3-D system with circular symmetry can then be reduced to the analysis of the 2-D system described by the following system of equations:

$$\begin{aligned}
&\frac{d\rho}{d\sigma} = \frac{\partial P}{\partial p_\rho} \quad \frac{dp_\rho}{d\sigma} = -\frac{\partial P}{\partial \rho} \\
&\frac{dx_3}{d\sigma} = \frac{\partial P}{\partial p_3} \quad \frac{dp_3}{d\sigma} = -\frac{\partial P}{\partial x_3} \\
&P = p_\rho^2 + p_3^2 - n^{*2} = 0
\end{aligned} \tag{17.52}$$

where $n^{*2} = n^2(\rho, x_3) - h^2b^2$, that is,

$$n^* = \sqrt{n^2(\rho, x_3) - \frac{h^2}{\rho^2}} \tag{17.53}$$

Skew rays can then be described as rays in the plane if the refractive index n is replaced by n^*.[4]

These equations enable us to obtain $\rho(\sigma)$ and $x_3(\sigma)$. By using $\rho(\sigma)$, function $\theta(\sigma)$ can be obtained. Constant C_θ can be obtained from the initial conditions: $\sigma = \sigma_1 \Rightarrow \theta = \theta_1$. We then have

$$C_\theta = \theta_1 - F(\sigma_1) \tag{17.54}$$

so that the value of C_θ can be calculated. The optical momentum is now given by Equation 17.45 as

$$\mathbf{p} = u_\rho\,\mathbf{e}_\rho + hb\,\mathbf{e}_\theta + u_3\,\mathbf{e}_3 = u_\rho\,\mathbf{e}_\rho + h\frac{1}{\rho}\mathbf{e}_\theta + u_3\,\mathbf{e}_3 \tag{17.55}$$

Yet another way to derive the conservation of h is directly from Fermat's principle.[4] In this case of circular symmetry, the refractive index will be a function of ρ and x_3 with $\rho = \sqrt{x_1^2 + x_2^2}$. Fermat's principle, written in the form of Equation 14.50, with the Lagrangian given by Equation 14.49, can now be written as

$$\delta\int n(\rho, x_3)\sqrt{1 + x_1'^2 + x_2'^2}\,dx_3 = 0 \tag{17.56}$$

Introducing

$$\begin{aligned} x_1 &= \rho\cos\theta \\ x_2 &= \rho\sin\theta \end{aligned} \tag{17.57}$$

we can write for Equation 17.56:[4]

$$\delta\int n(\rho, x_3)\sqrt{1 + \rho'^2 + \rho^2\theta'^2}\,dx_3 = 0 \tag{17.58}$$

where $\rho' = d\rho/dx_3$ and $\theta' = d\theta/dx_3$. Equation 17.58 can be written as

$$\delta\int F(\rho, x_3, \rho', \theta')\,dx_3 = 0 \tag{17.59}$$

with $F(\rho, x_3, \rho', \theta') = n(\rho, x_3)\sqrt{1 + \rho'^2 + \rho^2\theta'^2}$. The euler equation in θ (see Equation 14.31) can now be written as

$$\frac{\partial F}{\partial \theta} - \frac{d}{dx_3}\left(\frac{\partial F}{\partial \theta'}\right) = 0 \tag{17.60}$$

Since F does not explicitly depend on θ, we have $\partial F/\partial\theta = 0$, so that

$$\frac{d}{dx_3}\left(\frac{\partial F}{\partial\theta'}\right) = 0 \Leftrightarrow \frac{d}{dx_3}\left(\frac{n\rho^2\theta'}{\sqrt{1+\rho'^2+\rho^2\theta'^2}}\right) = 0 \tag{17.61}$$

therefore,

$$\frac{n\rho^2\theta'}{\sqrt{1+\rho'^2+\rho^2\theta'^2}} = h \tag{17.62}$$

where h is a constant.[4] The quantity h is again the skew invariant or skewness. Now, let $\mathbf{e}_3$, $\mathbf{e}_\theta$. and $\mathbf{e}_\rho$ be the mutually orthogonal unit vectors, respectively, tangent to the lines in which only x_3, θ, and ρ vary. We then have $\mathbf{e}_k = \nabla i_k / \|\nabla i_k\|$, that is: $\mathbf{e}_3 = \nabla i_3 / \|\nabla i_3\|$, $\mathbf{e}_\theta = \nabla i_\theta / \|\nabla i_\theta\|$, $\mathbf{e}_\rho = \nabla i_\rho / \|\nabla i_\rho\|$. From expression (17.62) we see that

$$h = n\rho\frac{\rho d\theta}{\sqrt{dx_3{}^2 + d\rho^2 + \rho^2 d\theta^2}} = \rho n\cos\phi \tag{17.63}$$

where, as seen from Figure 17.14a, ϕ corresponds to the angle that the light ray passing through point **O** and **A** makes with the vector $\mathbf{e}_\theta$ and $\rho = \sqrt{x_1^2 + x_2^2}$.

References

1. Arnold, V. I., *Mathematical Methods of Classical Mechanics*, Mir, Moscow, 1989.
2. Boas, M. L., *Mathematical Methods in the Physical Sciences*, John Wiley & Sons, New York, 1966.
3. Miñano, J. C., Cylindrical concentrators as a limit case of toroidal concentrators, *Appl. Opt.*, 23, 2017, 1984.
4. Luneburg, R. K., *Mathematical Theory of Optics*, University of California Press, Berkeley and Los Angeles, 1964, 191.

18

Étendue in Phase Space

18.1 Étendue and the Point Characteristic Function

Here we derive the conservation of étendue from optical first principles, utilizing a reference wavefront from which we can calculate the optical path length to a given point $\mathbf{P} = (x_1, x_2, x_3)$. It is then possible to define a function $S(\mathbf{P}) = S(x_1, x_2, x_3)$ that gives the optical path length between the reference wavefront and any given point. The momentum or a light ray at point $\mathbf{P}$ is given by $\mathbf{p} = \nabla S$, where $\nabla = (\partial/\partial x_1, \partial/\partial x_2, \partial/\partial x_3)$. Accordingly, if we now consider another point $\mathbf{P}^* = (x_1^*, x_2^*, x_3^*)$, we have $\mathbf{p}^* = \nabla^* S$ where $\nabla^* = (\partial/\partial x_1^*, \partial/\partial x_2^*, \partial/\partial x_3^*)$.

Based on the definition of function $S(\mathbf{P})$, we can now define the point characteristic function, $V(\mathbf{P}, \mathbf{P}^*) = V(x_1, x_2, x_3, x_1^*, x_2^*, x_3^*)$ which gives the optical path length between two given points $\mathbf{P}$ and $\mathbf{P}^*$ in the medium.[1,2] It is given by

$$V(x_1, x_2, x_3, x_1^*, x_2^*, x_3^*) = \int_{\mathbf{P}}^{\mathbf{P}^*} n\, ds = S(\mathbf{P}^*) - S(\mathbf{P}) = S(x_1^*, x_2^*, x_3^*) - S(x_1, x_2, x_3) \tag{18.1}$$

We have $\nabla V = -\nabla S$ and $\nabla^* V = \nabla^* S$. And also, as we have seen, $\mathbf{p} = \nabla S$ and $\mathbf{p}^* = \nabla^* S$. Then

$$\mathbf{p} = -\nabla V \quad \text{and} \quad \mathbf{p}^* = \nabla^* V \tag{18.2}$$

or, in terms of its components,

$$\begin{aligned} (p_1, p_2, p_3) &= (-V_{x1}, -V_{x2}, -V_{x3}) \\ (p_1^*, p_2^*, p_3^*) &= (V_{x1^*}, V_{x2^*}, V_{x3^*}) \end{aligned} \tag{18.3}$$

where $V_i = \partial V/\partial i$.

Let **P** and **P*** be defined, respectively, at the entrance and exit apertures of an optical system.[3] Let us further consider that **P** is located in the plane x_1x_2 of its coordinate system, as shown in Figure 18.1.

Considering differentials dx_1 and dx_2 for the position of **P** and dx_1^* and dx_2^* for the position of **P***, we can write for the corresponding momentum variations by using $V_{ij} = \partial(\partial V/\partial i)/\partial j$:

$$\begin{aligned} dp_1 &= -V_{x1x1}dx_1 - V_{x1x2}dx_2 - V_{x1x1^*}dx_1^* - V_{x1x2^*}dx_2^* \\ dp_2 &= -V_{x2x1}dx_1 - V_{x2x2}dx_2 - V_{x2x1^*}dx_1^* - V_{x2x2^*}dx_2^* \\ dp_1^* &= V_{x1^*x1}dx_1 + V_{x1^*x2}dx_2 + V_{x1^*x1^*}dx_1^* + V_{x1^*x2^*}dx_2^* \\ dp_2^* &= V_{x2^*x1}dx_1 + V_{x2^*x2}dx_2 + V_{x2^*x1^*}dx_1^* + V_{x2^*x2^*}dx_2^* \end{aligned} \tag{18.4}$$

Equation 18.4 can be rearranged in the following matrix form:

$$\begin{pmatrix} V_{x1x1^*} & V_{x1x2^*} & 0 & 0 \\ V_{x2x1^*} & V_{x2x2^*} & 0 & 0 \\ V_{x1^*x1^*} & V_{x1^*x2^*} & -1 & 0 \\ V_{x2^*x1^*} & V_{x2^*x2^*} & 0 & -1 \end{pmatrix} \begin{pmatrix} dx_1^* \\ dx_2^* \\ dp_1^* \\ dp_2^* \end{pmatrix} = \begin{pmatrix} -V_{x1x1} & -V_{x1x2} & -1 & 0 \\ -V_{x2x1} & -V_{x2x2} & 0 & -1 \\ -V_{x1^*x1} & -V_{x1^*x2} & 0 & 0 \\ -V_{x2^*x1} & -V_{x2^*x2} & 0 & 0 \end{pmatrix} \begin{pmatrix} dx_1 \\ dx_2 \\ dp_1 \\ dp_2 \end{pmatrix}$$
$$B \cdot M^* = A \cdot M \tag{18.5}$$

The determinant of matrix B is given by

$$\det B = V_{x1x1^*}V_{x2x2^*} - V_{x1x2^*}V_{x2x1^*} \tag{18.6}$$

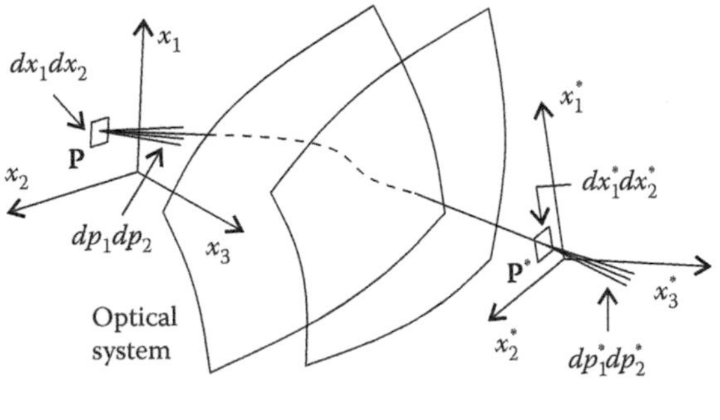

FIGURE 18.1
Plane x_1x_2 is at the entrance aperture of the optical system and the plane $x_1^*x_2^*$ is at its exit aperture. The rays of a bundle passing through dx_1 dx_2 in plane x_1x_2 have different directions, such that p_1 varies by dp_1, and p_2 varies by dp_2, for these rays. On another plane $x_1^*x_2^*$, these rays pass through an elemental area $dx_1^*dx_2^*$, and have different directions, such that p_1^* varies by dp_1^*, and p_2^* varies by dp_2^*, for these rays. The conservation of étendue is expressed as $dx_1dx_2dp_1dp_2 = dx_1^*dx_2^*dp_1^*dp_2^*$.

The determinant of matrix A can also be obtained as

$$\det A = V_{x1^*x1}V_{x2^*x2} - V_{x2^*x1}V_{x1^*x2} \tag{18.7}$$

Making $V_{x1^*x1} = V_{x1x1^*}$, $V_{x1^*x2} = V_{x2x1^*}$, $V_{x2^*x1} = V_{x1x2^*}$, and $V_{x2^*x2} = V_{x2x2^*}$, we can write

$$\det A = \det B \tag{18.8}$$

Now, noting that the determinant of the product of two matrices is the product of the determinants of the matrices, that is, if C and D are two matrices, we have $\det(C \cdot D) = \det C \det D$, we can write $\det(B^{-1} \cdot B) = \det B^{-1} \det B$. Since $\det(B^{-1} \cdot B) = 1$, we obtain $\det B^{-1} = 1/\det B$. Or considering Equation 18.8

$$\det B^{-1} = 1/\det A \tag{18.9}$$

Multiplying the left-hand-side of Equation 18.5 by matrix B^{-1}, we obtain

$$M^* = (B^{-1} \cdot A) \cdot M \tag{18.10}$$

Considering Equation 18.9, we can write

$$\det(B^{-1} \cdot A) = \det B^{-1} \det A = 1 \tag{18.11}$$

But Equation 18.10 can also be written as

$$\begin{pmatrix} dx_1^* \\ dx_2^* \\ dp_1^* \\ dp_2^* \end{pmatrix} = \begin{pmatrix} \partial x_1^*/\partial x_1 & \partial x_1^*/\partial x_2 & \partial x_1^*/\partial p_1 & \partial x_1^*/\partial p_2 \\ \partial x_2^*/\partial x_1 & \partial x_2^*/\partial x_2 & \partial x_2^*/\partial p_1 & \partial x_2^*/\partial p_2 \\ \partial p_1^*/\partial x_1 & \partial p_1^*/\partial x_2 & \partial p_1^*/\partial p_1 & \partial p_1^*/\partial p_2 \\ \partial p_2^*/\partial x_1 & \partial p_2^*/\partial x_2 & \partial p_2^*/\partial p_1 & \partial p_2^*/\partial p_2 \end{pmatrix} \begin{pmatrix} dx_1 \\ dx_2 \\ dp_1 \\ dp_2 \end{pmatrix} \tag{18.12}$$

$$M^* = C \cdot M$$

And, therefore, $C = B^{-1} \cdot A$. But, we can also write

$$dx_1^* dx_2^* dp_1^* dp_2^* = \frac{\partial(x_1^*, x_2^*, p_1^*, p_2^*)}{\partial(x_1, x_2, p_1, p_2)} dx_1 dx_2 dp_2 dp_2 \tag{18.13}$$

where

$$\frac{\partial(x_1^*, x_2^*, p_1^*, p_2^*)}{\partial(x_1, x_2, p_1, p_2)} = \det C \tag{18.14}$$

since det $C = 1$,

$$dx_1^* dx_2^* dp_1^* dp_2^* = dx_1 dx_2 dp_2 dp_2 \tag{18.15}$$

which means that the quantity

$$dU = dx_1 dx_2 dp_1 dp_2 \tag{18.16}$$

is then conserved as light travels within optical systems. The coordinate system (x_1, x_2, P_1, P_2) of these special coordinates and momenta is called a phase space. A point **R** in phase space has coordinates (x_{1R}, x_{2R}, p_{1R}, p_{2R}), corresponding to a point (x_{1R}, x_{2R}) in the x_1x_2 plane and a direction (p_{1R}, p_{2R}). This point in phase space then uniquely defines both a spatial point and a direction, and, therefore, a ray. A continuous set of points in phase space defines a region within which each point represents a ray of light, so that the region defines a bundle of rays.

The elemental region in phase space has a volume $dU = dx_1 dx_2 dp_1 dp_2$ (x_1, x_2, p_1, p_2) that is called étendue, and Equation 18.15 defines the conservation of étendue in an optical system: $dU^* = dU$.

Consider an elemental area dx_1dx_2 and a set of rays leaving it in different directions. For these rays, p_1 varies by a value dp_1, and p_2 varies by a value dp_2. These rays will travel through an optical system and pass through another area $dx_1^* dx_2^*$. Now, the directions of these rays are such that p_1^* varies by a value dp_1^*, and p_2^* varies by a value dp_2^* as presented in Figure 18.1. The conservation of étendue is expressed as $dx_1^* dx_2^* dp_1^* dp_2^* = dx_1 dx_2 dp_2 dp_2$. These rays passing through dx_1dx_2 and $dx_1^* dx_2^*$ are called a light beam, so that étendue is conserved for light beams.

Note that $dx_1dx_2dp_2dp_2$ is an elementary region in phase space (x_1, x_2, p_1, p_2). Therefore, the conservation of étendue states that, if a given set of rays occupies a given region of elemental volume $dx_1dx_2dp_2dp_2$ in phase space at a given point in an optical system, then, after traveling through that optical system, these rays will still occupy a region of elemental volume $dx_1^* dx_2^* dp_1^* dp_2^* = dx_1 dx_2 dp_2 dp_2$, that is, the same volume as earlier, although the new region may have a different shape. Therefore, the volume in phase space occupied by a set of light rays is constant, as they travel through the optical system.

For 2-D systems, we have one less dimension, and a situation similar to that shown in Figure 18.2.

In this case, the quantity that is conserved is $dU_{2D} = dx_1 dp_1$.

18.2 Étendue in Hamiltonian Optics

Here we present the conservation of étendue from the point of view of Hamiltonian optics. Consider a volume V moving in space, as presented in

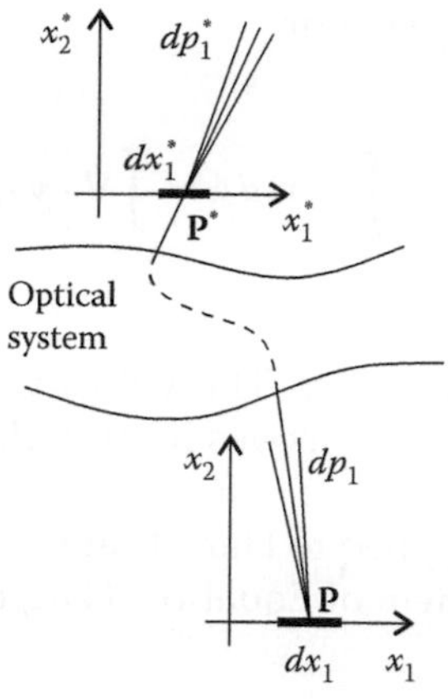

FIGURE 18.2
Conservation of étendue for 2-D optical systems.

Figure 18.3. Let dA be an element of its surface and $\mathbf{n}$ be a unit vector perpendicular to dA. Further, consider that the element of area (dA) is moving with velocity $\mathbf{v} = \dot{\mathbf{x}} = d\mathbf{x}/dt$, as shown in Figure 18.3a. During a period of time dt, this element of area moves $\mathbf{v}dt$, producing an increase in volume. In the 2-D case, we have $dV = da\,v dt \cos\gamma = da(\mathbf{v}\cdot\mathbf{n})dt$, as shown in Figure 18.3b, where v is the magnitude of $\mathbf{v}$. In the 3-D case, we have $dV = dA(\mathbf{v}\cdot\mathbf{n})dt$. Integrating on the whole surface A delimiting V, the total volume variation is

$$\frac{dV}{dt} = \int_A \mathbf{v}\cdot\mathbf{n}\,dA \tag{18.17}$$

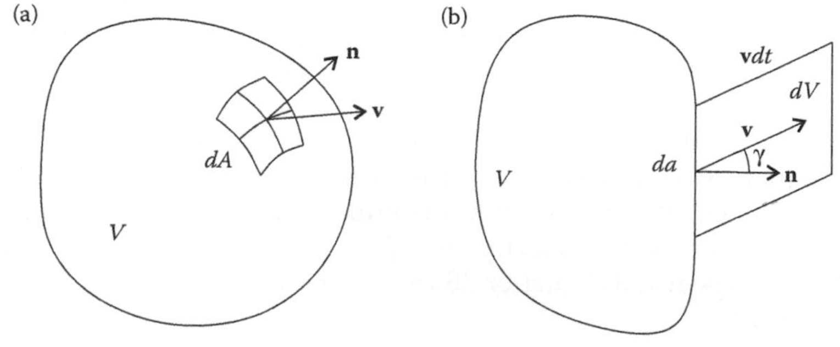

FIGURE 18.3
(a) A volume V moves in space. A small portion dA of its surface moves with it. (b) The 2-D case: if da moves with a velocity $\mathbf{v}$, the increase in volume due to the movement of da during a time period dt is given by $dV = da\ vdt\ \cos\gamma = da(\mathbf{v}\cdot\mathbf{n})dt$, $\mathbf{n}$ being a unit vector perpendicular to da. In the general 3-D case, this relation is still valid, and we have $dV = dA\ (\mathbf{v}\cdot\mathbf{n})dt$.

Using Gauss's theorem[4] we obtain[5]

$$\frac{dV}{dt} = \int_A \mathbf{v} \cdot \mathbf{n}\, dA = \int_V \nabla \cdot \mathbf{v}\, dV \tag{18.18}$$

where $\nabla \cdot \mathbf{v}$ is the divergence of $\mathbf{v}$. The velocity is given by $\mathbf{v} = \dot{\mathbf{x}}$. In case the volume under consideration moves in an n-dimensional space, we have $\mathbf{v} = (\dot{x}_1, \ldots, \dot{x}_n)$.

This result can now be applied to Hamiltonian optical systems. A 3-D system is described by the system of Equation 14.62, or

$$\begin{aligned} x_1' &= \frac{\partial H}{\partial p_1} \qquad p_1' = -\frac{\partial H}{\partial x_1} \\ x_2' &= \frac{\partial H}{\partial p_2} \qquad p_2' = -\frac{\partial H}{\partial x_2} \\ H &= -\sqrt{n^2 - p_1^2 - p_2^2} \end{aligned} \tag{18.19}$$

where the primes represent x_3 derivatives, since now x_3 takes the role of time. That is, instead of $\dot{x} = dx/dt$, we have geometrical derivatives $x_k' = dx_k/dx_3$, since we now have $x_1 = x_1(x_3)$ and $x_2 = x_2(x_3)$. Coordinate x_3 corresponds to the system optical axis, that is, the axis along which light propagates in the optical system in the direction of increasing x_3, so that $p_3 > 0$. Each light ray can be defined, for each value of x_3, by (x_1, x_2, p_1, p_2). A point in space (x_1, x_2, p_1, p_2) defines the position (x_1, x_2) of the light ray and its direction of propagation (p_1, p_2). Note that p_1 and p_2 enable the definition of the direction of the light ray in space x_1, x_2, x_3 since, from $p_1^2 + p_2^2 + p_3^2 = n^2$, p_3 can be obtained if p_1 and p_2 are given. As the light ray propagates along the axis x_3, coordinates x_1, x_2, p_1, and p_2 vary, and the light ray moves in a 4-D phase space (x_1, x_2, p_1, p_2). Let us now consider a large number of light rays propagating in the system, each occupying a point in phase space, and the entire set of rays constituting a region in that space. In particular, if the rays are continuously distributed in the optical system, they will occupy a volume V in phase space. Each point of this volume moves with "velocity" $\mathbf{v} = (x_1', x_2', p_1', p_2')$ with $x_k' = dx_k/dx_3$ and $p_k' = dp_k/dx_3$. The result (Equation 18.18) obtained earlier can now be applied to phase space:

$$\frac{dV}{dx_3} = \int_V \nabla \cdot \mathbf{v}\, dV \tag{18.20}$$

Here, as stated earlier, axis x_3 now takes the role of time. The expression for $\nabla \cdot \mathbf{v}$ can now be calculated from expression (18.19)

$$\begin{aligned}\nabla \cdot \mathbf{v} &= \left(\frac{\partial x_1'}{\partial x_1} + \frac{\partial x_2'}{\partial x_2} + \frac{\partial p_1'}{\partial p_1} + \frac{\partial p_2'}{\partial p_2}\right) \\ &= \left(\frac{\partial}{\partial x_1}\frac{\partial H}{\partial p_1} + \frac{\partial}{\partial x_2}\frac{\partial H}{\partial p_2} - \frac{\partial}{\partial p_1}\frac{\partial H}{\partial x_1} - \frac{\partial}{\partial p_2}\frac{\partial H}{\partial x_2}\right) = 0 \end{aligned} \qquad (18.21)$$

so that

$$dV/dx_3 = 0 \qquad (18.22)$$

This result is called Liouville's Theorem[5–7] and applies to any Hamiltonian system.

Equation 18.22 enables us to conclude that the elemental region has a constant volume $dV = dx_1dx_2dp_1dp_2$ as the light propagates in the optical system, that is, along axis x_3. In the case of Hamiltonian optics, this volume dV in phase space is called étendue, and is represented by dU. We can then write $dU =$ constant or[3]

$$U = \int dx_1dx_2dp_1dp_2 = \text{constant} \qquad (18.23)$$

Let us suppose, for example, that $x_3 = 0$ corresponds to the rectangular entrance aperture of a 3-D optical device. This entrance extends between $x_{1A} < x_1 < x_{1B}$ and $x_{2A} < x_2 < x_{2B}$. Let us further consider that the light entering the device makes angles with the coordinate axes such that $p_{1A} < p_1 < p_{1B}$ and $p_{2A} < p_2 < p_{2B}$. We then see that these conditions define a region in the 4-D space (x_1, x_2, p_1, p_2) at the entrance of the device. As the light progresses through the optical system, the coordinates, and the angles that the light rays make with the axes, change. The phase space volume, however, remains constant.

18.3 Integral Invariants and Étendue

Integral invariants are called conservation of étendue in optics. For a 3-D system there are three invariants, J_2, J_4, and J_6 of orders 2, 4, and 6, respectively.

Order 2 integral invariant $U_{2\text{-D}} = J_2$ is called étendue 2-D (two dimensions)[8–11] in optics or Lagrange invariant.[8] It may be written in different forms, as per expression (14.156):

$$\begin{aligned} U_{2-D} = I_1 &= \oint_c \mathbf{p} \cdot d\mathbf{x} = \oint_c (p_1 dx_1 + p_2 dx_2 + p_3 dx_3) \\ &= J_2 = \iint_A d\mathbf{p} \cdot d\mathbf{x} = \iint_A dp_1 dx_1 + dp_2 dx_2 + dp_3 dx_3 \\ &= \iint_A \nabla \times \mathbf{p} \cdot d\mathbf{a} \\ &= \iint_B (\mathbf{p}_u \cdot \mathbf{x}_v - \mathbf{p}_v \cdot \mathbf{x}_u) du dv \end{aligned} \tag{18.24}$$

Here, c is a closed curve in configuration space (x_1, x_2, x_3) parameterized as $(x_1(\tau), x_2(\tau), x_3(\tau))$. Curve c encloses an area A also in configuration space parameterized as $(x_1(u, v), x_2(u, v), x_3 = x_3(u, v))$. In parameter plane u, v the integral is now taken over an area B. A bundle of rays crosses area A in configuration space.

Order 4 integral invariant $U_{3\text{-D}} = J_4$ given by expression (14.162) is called étendue 3-D (three dimensions) in optics, or simply étendue U and is given by

$$U = J_4 = \int dx_1 dx_2 dp_1 dp_2 + dx_1 dx_3 dp_1 dp_3 + dx_2 dx_3 dp_2 dp_3 \tag{18.25}$$

(the four integral signs were replaced by just one for notation simplicity). Note that if we calculate U at a planes $x_3 = \text{constant}$, we get $dx_3 = 0$ for those planes and, therefore,

$$U = \int dx_1 dx_2 dp_1 dp_2 \tag{18.26}$$

is conserved. This is the same as expression (18.23) for the case in which light propagates along axis x_3.[12]

Order 6 integral invariant J_6 is given by expression (14.163)

$$J_6 = \int dx_1 dx_2 dx_3 dp_1 dp_2 dp_3 \tag{18.27}$$

(the six integral signs were replaced by just one for notation simplicity). However, in optics, only the solutions of expression (14.79), contained in the subspace $P = 0$, are light ray trajectories.[9] At a position (x_1, x_2, x_3), the momentum components are such that $p_1^2 + p_2^2 + p_3^2 = n^2$ and, therefore, the solutions describing light rays are contained on the spherical surface $p_1^2 + p_2^2 + p_3^2 = n^2$

inside space (p_1, p_2, p_3). This surface has zero volume and, therefore, $dp_1dp_2dp_3 = 0$ resulting in

$$J_6 = \int dx_1dx_2dx_3dp_1dp_2dp_3 = 0 \tag{18.28}$$

For that reason, integral invariant J_6 in expression (18.27) is not used in optics.

18.4 Refraction, Reflection, and Étendue 2-D

The étendue 2-D invariant may also be obtained directly from the laws of refraction and reflection.[13] Consider a surface A parameterized by $\mathbf{x}(u,v)$ in three-dimensional (3-D) space, that is, $\mathbf{x}(u, v) = (x_1(u, v), x_2(u, v), x_3(u, v))$, as shown in Figure 18.4. Its normal at each point is given by $\mathbf{n}(u,v)$ and $\mathbf{n}$ is perpendicular to both partial devivatives (tangent vectors) $\mathbf{x}_u = \partial\mathbf{x}/\partial u$ and $\mathbf{x}_v = \partial\mathbf{x}/\partial v$ resulting in

$$\begin{aligned}\mathbf{n}\cdot\mathbf{x}_u &= 0\\ \mathbf{n}\cdot\mathbf{x}_v &= 0\end{aligned} \tag{18.29}$$

Calculating the partial derivative of $\mathbf{n}\cdot\mathbf{x}_u = 0$ relative to v, we get $\mathbf{n}_v\cdot\mathbf{x}_u + \mathbf{n}\cdot\mathbf{x}_{uv} = 0$ or $\mathbf{n}_v\cdot\mathbf{x}_u = -\mathbf{n}\cdot\mathbf{x}_{uv}$. Also, calculating the partial derivative of $\mathbf{n}\cdot\mathbf{x}_v = 0$ relative to u, we get $\mathbf{n}_u\cdot\mathbf{x}_v + \mathbf{n}\cdot\mathbf{x}_{vu} = 0$ or $\mathbf{n}_u\cdot\mathbf{x}_v = -\mathbf{n}\cdot\mathbf{x}_{uv}$. We then get

$$\mathbf{n}_v\cdot\mathbf{x}_u = \mathbf{n}_u\cdot\mathbf{x}_v \tag{18.30}$$

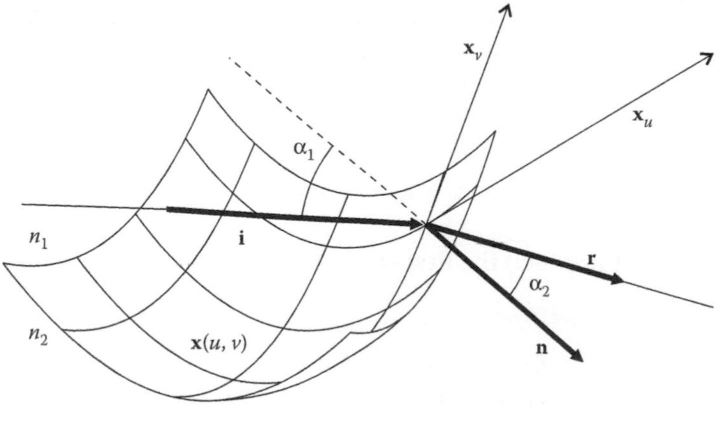

FIGURE 18.4
Surface $\mathbf{x}(u, v)$ with normal $\mathbf{n}$, perpendicular to partial derivatives $\mathbf{x}_u$ and $\mathbf{x}_v$. An incident ray with direction $\mathbf{i}$ is deflected at the surface and leaves in direction $\mathbf{r}$.

Let us now consider that surface $\mathbf{x}(u, v)$ is crossed by a two-parameter set of rays, one ray per point on the surface. Let us also suppose that the surface separates two media of refractive indices n_1 and n_2. The incident rays on the surface have directions $\mathbf{i}(u, v)$, and the refracted (or reflected if $n_1 = n_2$) rays have directions $\mathbf{r}(u, v)$, where $\mathbf{i}$ and $\mathbf{r}$ are unit vectors. Since vectors $\mathbf{i}$, $\mathbf{r}$ and the normal $\mathbf{n}$ to the surface are all on the same plane, we may write $n_1\mathbf{i} - n_2\mathbf{r} = k\mathbf{n}$, or $n_1\mathbf{i}(u, v) - n_2\mathbf{r}(u, v) = k(u, v)\mathbf{n}(u, v)$ where k is a scalar.

Now, we calculate the u derivative of $n_1\mathbf{i} - n_2\mathbf{r} = k\mathbf{n}$ to get $n_1\mathbf{i}_u - n_2\mathbf{r}_u = k_u\mathbf{n} + k\mathbf{n}_u$. Multiplying (dot product) by $\mathbf{x}_v$, resuts in

$$n_1\mathbf{i}_u \cdot \mathbf{x}_v - n_2\mathbf{r}_u \cdot \mathbf{x}_v = k_u\mathbf{n} \cdot \mathbf{x}_v + k\mathbf{n}_u \cdot \mathbf{x}_v \tag{18.31}$$

Accordingly, calculating the v derivative of $n_1\mathbf{i} - n_2\mathbf{r} = k\mathbf{n}$ we get $n_1\mathbf{i}_v - n_2\mathbf{r}_v = k_v\mathbf{n} + k\mathbf{n}_v$ and multiplying by $\mathbf{x}_u$, results in

$$n_1\mathbf{i}_v \cdot \mathbf{x}_u - n_2\mathbf{r}_v \cdot \mathbf{x}_u = k_v\mathbf{n} \cdot \mathbf{x}_u + k\mathbf{n}_v \cdot \mathbf{x}_u \tag{18.32}$$

From expressions (18.29) and (18.30) we get $n_1\mathbf{i}_u \cdot \mathbf{x}_v - n_2\mathbf{r}_u \cdot \mathbf{x}_v = n_1\mathbf{i}_v \cdot \mathbf{x}_u - n_2\mathbf{r}_v \cdot \mathbf{x}_u$, or

$$n_1(\mathbf{i}_u \cdot \mathbf{x}_v - \mathbf{i}_v \cdot \mathbf{x}_u) = n_2(\mathbf{r}_u \cdot \mathbf{x}_v - \mathbf{r}_v \cdot \mathbf{x}_u) \tag{18.33}$$

which means that

$$n(\mathbf{i}_u \cdot \mathbf{x}_v - \mathbf{i}_v \cdot \mathbf{x}_u) = \mathbf{p}_u \cdot \mathbf{x}_v - \mathbf{p}_v \cdot \mathbf{x}_u = \text{constant} \tag{18.34}$$

with $\mathbf{p} = n\mathbf{i}$. The quantity in expression (18.34) is constant, and conserved on deflection (refraction or reflection).

This result is still valid if the rays cross another surface $\mathbf{z}(u, v)$. Suppose then that we write $\mathbf{z} = \mathbf{x} + \lambda\mathbf{i}$, which is the same as $\mathbf{z}(u, v) = \mathbf{x}(u, v) + \lambda(u, v)\mathbf{i}(u, v)$. Calculating the u and v derivatives we get

$$\begin{aligned} \mathbf{z}_u &= \mathbf{x}_u + \lambda_u\mathbf{i} + \lambda\mathbf{i}_u \\ \mathbf{z}_v &= \mathbf{x}_v + \lambda_v\mathbf{i} + \lambda\mathbf{i}_v \end{aligned} \tag{18.35}$$

multiplying (dot product) the first expression by $\mathbf{i}_v$ and the second by $\mathbf{i}_u$ we get

$$\begin{aligned} \mathbf{z}_u \cdot \mathbf{i}_v &= \mathbf{x}_u \cdot \mathbf{i}_v + \lambda_u\mathbf{i} \cdot \mathbf{i}_v + \lambda\mathbf{i}_u \cdot \mathbf{i}_v \\ \mathbf{z}_v \cdot \mathbf{i}_u &= \mathbf{x}_v \cdot \mathbf{i}_u + \lambda_v\mathbf{i} \cdot \mathbf{i}_u + \lambda\mathbf{i}_v \cdot \mathbf{i}_u \end{aligned} \tag{18.36}$$

Now, $\mathbf{i}$ is a unit vector so that $\mathbf{i} \cdot \mathbf{i} = 1$. Calculating the u derivative, we get $\mathbf{i}_u \cdot \mathbf{i} + \mathbf{i} \cdot \mathbf{i}_u = 0$ since 1 (one) is a constant and does not depend on u. We then get $\mathbf{i} \cdot \mathbf{i}_u = 0$. Also for the v derivative $\mathbf{i} \cdot \mathbf{i}_v = 0$.[14] Having in consideration that $\mathbf{i}_u \cdot \mathbf{i}_v = \mathbf{i}_v \cdot \mathbf{i}_u$ we then get, by subtracting the two expressions in (18.36),

$$\mathbf{i}_u \cdot \mathbf{x}_v - \mathbf{i}_v \cdot \mathbf{x}_u = \mathbf{i}_u \cdot \mathbf{z}_v - \mathbf{i}_v \cdot \mathbf{z}_u \tag{18.37}$$

Combining this with expression (18.34), one may conclude that (18.34) is an invariant, independent of how the bundle of rays is intersected.[13] Note that the rays intersect surface $\mathbf{z}(u,v)$ but are not refracted or reflected there, and the refractive index n is the same before and after surface $\mathbf{z}(u,v)$. Integrating over the whole surface $\mathbf{x}(u, v)$, we get that

$$U_{2\text{-D}} = \iint (\mathbf{p}_u \cdot \mathbf{x}_v - \mathbf{p}_v \cdot \mathbf{x}_u) du dv \tag{18.38}$$

is an invariant, as in (18.24). This expression may also be written as

$$\begin{aligned} U_{2-D} &= \iint \left(\sum_{k=1}^{3} \frac{\partial p_k}{\partial u}\frac{\partial x_k}{\partial v} - \frac{\partial x_k}{\partial u}\frac{\partial p_k}{\partial v} \right) du dv = \iint \left(\sum_{k=1}^{3} \frac{\partial(p_k, x_k)}{\partial(u,v)} \right) du dv \\ &= \iint \sum_{k=1}^{3} dp_k dx_k = \iint d\mathbf{p} \cdot d\mathbf{x} \end{aligned} \tag{18.39}$$

Now, having consideration expressions (14.149) with $k = 1,2,3$, we get

$$\begin{aligned} U_{2-D} &= \sum_{k=1}^{3} \iint dp_k dx_k = \sum_{k=1}^{3} \iint \frac{\partial(p_k, x_k)}{\partial(u,v)} du dv \\ &= \sum_{k=1}^{3} \iint \left(\frac{\partial}{\partial u}\left(p_k \frac{\partial x_k}{\partial v} \right) - \frac{\partial}{\partial v}\left(p_k \frac{\partial x_k}{\partial u} \right) \right) du dv \end{aligned} \tag{18.40}$$

Applying now Green's Theorem (14.147), to each term in the summation, this surface integral may be changed to a curve integral as

$$\begin{aligned} U_{2-D} &= \sum_{k=1}^{3} \oint p_k \frac{\partial x_k}{\partial u} du + p_k \frac{\partial x_k}{\partial v} dv = \sum_{k=1}^{3} \oint p_k \left(\frac{\partial x_k}{\partial u} du + \frac{\partial x_k}{\partial v} dv \right) \\ &= \sum_{k=1}^{3} \oint p_k dx_k = \oint \sum_{k=1}^{3} p_k dx_k = \oint \mathbf{p} \cdot d\mathbf{x} \end{aligned} \tag{18.41}$$

which is the same expression as (18.24). By applying Stokes' Theorem, we get expression (14.152), also in (18.24).

18.5 Étendue 2-D Examples

We now consider a few examples to illustrate the physical meaning of etendue 2-D given by $U_{2\text{-D}}$.

Figure 18.5 shows a set of rays perpendicular to a wavefront A, enclosed by curve c. This wavefront is a surface parameterized by two parameters, and there is a ray crossing each point of A. This is then a two-parameter bundle of rays. For those rays, Equation 15.10 applies

$$\mathbf{p} = \nabla S \tag{18.42}$$

where S is the optical path length. Replacing this in the third line expression of (18.24) we get

$$U_{2\text{-D}} = \iint_A (\nabla \times (\nabla S)) \cdot d\mathbf{a} = 0 \tag{18.43}$$

since $\nabla \times (\nabla S) = 0$.

Since the light rays are perpendicular to wavefront A, they are also perpendicular to curve c and, therefore, $\mathbf{p} \cdot d\mathbf{x} = 0$ along curve c, since momentum $\mathbf{p}$ is perpendicular to $d\mathbf{x}$, which is tangent to curve c. We then get

$$U_{2\text{-D}} = \oint_c \mathbf{p} \cdot d\mathbf{x} = 0 \tag{18.44}$$

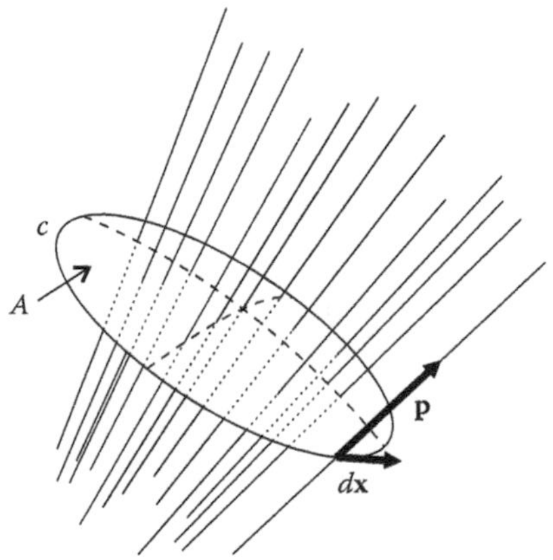

FIGURE 18.5
A set of rays perpendicular to a wavefront A has $U_{2D} = 0$.

Consider now a different bundle of rays crossing a surface A, as shown in Figure 18.6a. All rays are perpendicular to curves l, which are horizontal and parallel to each other (dashed lines). The directions of the rays rotate from direction $\mathbf{p}_L$ to direction $\mathbf{p}_R$ as we move along these curves. Each point on surface A is parameterized by two parameters, and is crossed by a single ray. This is, therefore, a two-parameter bundle of rays.

The étendue 2-D of this bundle may be obtained from Equation 18.24 by a line integral along curve c, the boundary of area A, as shown in Figure 18.6b,

$$U_{2\text{-D}} = \oint_c \mathbf{p} \cdot d\mathbf{x} = \int_{\mathbf{A}}^{\mathbf{B}} \mathbf{p}_R \cdot d\mathbf{x} + \int_{\mathbf{B}}^{\mathbf{C}} \mathbf{p} \cdot d\mathbf{x} + \int_{\mathbf{C}}^{\mathbf{D}} \mathbf{p}_L \cdot d\mathbf{x} + \int_{\mathbf{D}}^{\mathbf{A}} \mathbf{p} \cdot d\mathbf{x} \tag{18.45}$$

Between **B** and **C**, and also between **D** and **A**, the light rays are perpencicular to curve c, and, therefore, $\mathbf{p} \cdot d\mathbf{x} = 0$. The expression for $U_{2\text{-D}}$ becomes

$$U_{2\text{-D}} = \oint_c \mathbf{p} \cdot d\mathbf{x} = \int_{\mathbf{A}}^{\mathbf{B}} \mathbf{p}_R \cdot d\mathbf{x} + \int_{\mathbf{C}}^{\mathbf{D}} \mathbf{p}_L \cdot d\mathbf{x} \tag{18.46}$$

Let us also suppose that $\mathbf{p}_L$ and $\mathbf{p}_R$ make an angle θ to the vertical, as shown in Figure 18.7.

If the rays are immersed in a medium of refractive index n, and $[\mathbf{A},\mathbf{B}] = [\mathbf{C},\mathbf{D}]$ we then get

$$U_{2\text{-D}} = n[\mathbf{A},\mathbf{B}]\sin\theta + n[\mathbf{C},\mathbf{D}]\sin\theta = 2n[\mathbf{A},\mathbf{B}]\sin\theta \tag{18.47}$$

Since $U_{2\text{-D}} \neq 0$, this bundle of rays is not perpendicular to a wavefront and Equation 18.42 does not apply.

In general, we do not have to consider that each point of the configuration space is crossed by a single trajectory.[9] Consider now the limit case in which distances $[\mathbf{A},\mathbf{D}]$ and $[\mathbf{B},\mathbf{C}]$ go to zero. Curve c collapses onto plane μ crossed

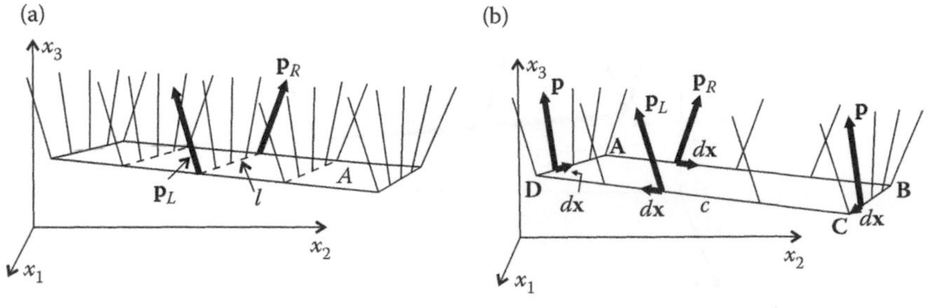

FIGURE 18.6
(a) A two-parameter bundle of rays crosses a surface A, one ray per each point of A. (b) The étendue 2-D may be calculated as a line integral along curve c, the boundary of A.

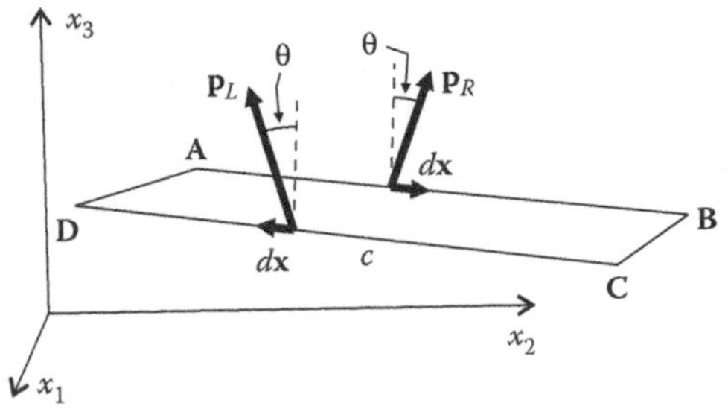

FIGURE 18.7
Rays crossing segments **AB** and **CD** of curve c make an angle θ to the vertical.

by a set of rays also in plane μ spanning an angular aperture 2θ, constant across curve c, as shown in Figure 18.8.

We may now calculate U_{2D} considering that curve c is closed going from **A** to **B** and back to **A** along the same path. Each point of this curve is crossed by one ray with momentum $\mathbf{p}_R$ in segment **AB** and momentum $\mathbf{p}_L$ in segment **BA**, as shown in Figure 18.9.

We then get from one of the forms in Equation 18.24,

$$U_{2\text{-D}} = \oint_c \mathbf{p} \cdot d\mathbf{x} = \int_{\mathbf{A}}^{\mathbf{B}} \mathbf{p}_R \cdot d\mathbf{x} + \int_{\mathbf{B}}^{\mathbf{A}} \mathbf{p}_L \cdot d\mathbf{x} = 2n[\mathbf{A},\mathbf{B}]\sin\theta \tag{18.48}$$

Now, the rays shown in Figure 18.9 are the boundary of a two-parameter set of rays crossing curve c, as shown in Figure 18.8. That two-parameter set of rays may be parameterized, for example, with parameters (τ, α) where τ is

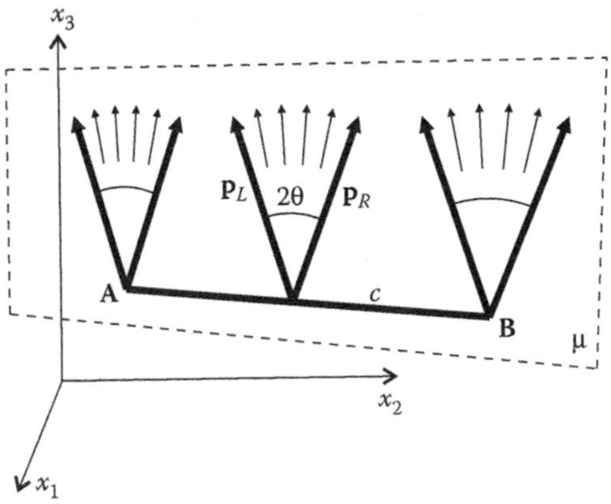

FIGURE 18.8
Two-parameter set of rays crossing curve c in plane μ and spanning an angular aperture 2θ at each point in c.

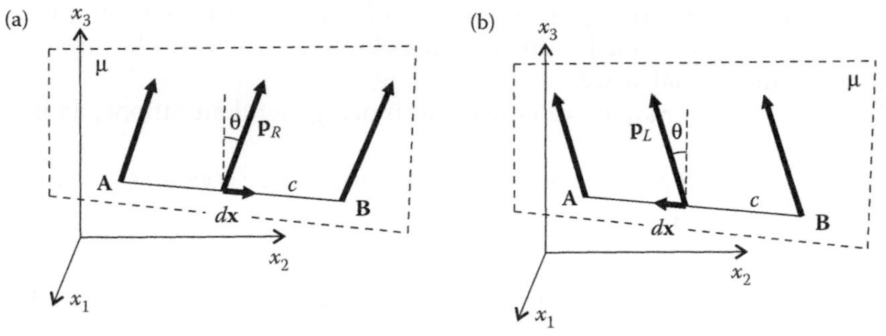

FIGURE 18.9
Curve c seen as a closed curve from **A** to **B** and then back to **A** along the same path c. (a) From **A** to **B** the curve is considered to be crossed by rays with momentum $\mathbf{p}_R$. (b) From **B** to **A** the curve is considered to be crossed by rays with momentum $\mathbf{p}_L$.

the parameter of c at point $c(\tau)$ where the ray crosses c and α is the angle of the ray with the normal to c in plane μ, as shown in Figure 18.10.

Another way to calculate $U_{2\text{-D}}$ is by using another form of Equation 18.24,

$$U_{2\text{-D}} = \iint d\mathbf{p} \cdot d\mathbf{x} = \int_{\mathbf{A}}^{\mathbf{B}} \int_{\mathbf{p}_L}^{\mathbf{p}_R} d\mathbf{p} \cdot d\mathbf{x} \tag{18.49}$$

Since momentum $\mathbf{p}$ does not depend on the position along curve c, one may write[15]

$$U_{2\text{-D}} = \int_{\mathbf{A}}^{\mathbf{B}} (\mathbf{p}_R - \mathbf{p}_L) \cdot d\mathbf{x} = 2n \sin\theta \int_{\mathbf{A}}^{\mathbf{B}} d\mathbf{x} = 2n[\mathbf{A}, \mathbf{B}] \sin\theta \tag{18.50}$$

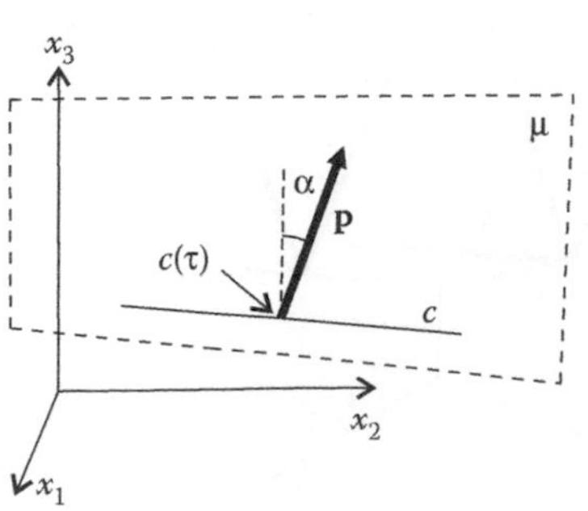

FIGURE 18.10
The two-parameter set of rays in Figure 18.8 may be parameterized, for example, by parameter σ on curve c and angle α the ray makes to the normal to c on plane μ.

Although the rays in Figure 18.8 are in 3-D space, all rays are in plane μ, and the value of $U_{2\text{-D}}$ matches the étendue obtained when similar rays travel in a two-dimensional space.

The étendue 2-D may be calculated in more general situations, as in the case of Figure 18.11.

Also in this example, we consider that $\mathbf{p}_R$ and $\mathbf{p}_L$ are constant along curve c_1 and Equation 18.49 may be rewritten as

$$U_{2\text{-D}} = \int_{\mathbf{A}}^{\mathbf{B}} (\mathbf{p}_R - \mathbf{p}_L) \cdot d\mathbf{x} = 2nL\sin\theta \tag{18.51}$$

where the integral sums the infinitesimal projections of $d\mathbf{x}$ in the direction of $\Delta\mathbf{p} = \mathbf{p}_R - \mathbf{p}_L$ adding up to length L, parallel to $\Delta\mathbf{p}$.

Consider now the situation in Figure 18.12, where an emitter $\mathbf{R}_L\mathbf{R}_R$ emits light to a curve c_2 stretching from points **A** to **B**. Each point **P** on c_2 receives light from the whole emitter $\mathbf{R}_L\mathbf{R}_R$ and, therefore, this is a two-parameter bundle of rays. The two parameters being, for example, a parameter along curve c_2 and a parameter along $\mathbf{R}_L\mathbf{R}_R$.

We may again use Equation 18.49 to calculate $U_{2\text{-D}}$ from $\mathbf{R}_L\mathbf{R}_R$ to c_2. Now, $\mathbf{p}_R = \nabla S_R$, where function $S_R(\mathbf{P})$ is the optical path length from $\mathbf{R}_R$ to a point **P** in space. Also, $\mathbf{p}_L = \nabla S_L$, where function $S_L(\mathbf{P})$ is the optical length from $\mathbf{R}_L$ to a point **P** in space. Replacing these expressions into Equation 18.49 we get for curve c_2 stretching from points **A** to **B**,

$$U_{2\text{-D}} = \int_{\mathbf{A}}^{\mathbf{B}} (\nabla S_R - \nabla S_L) \cdot d\mathbf{x} \tag{18.52}$$

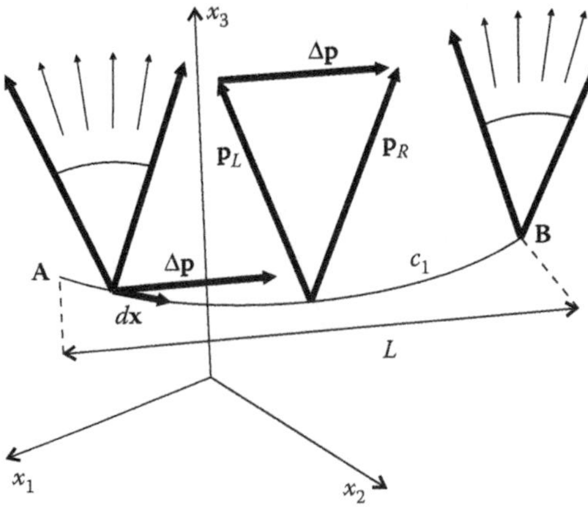

FIGURE 18.11
Two-parameter set of rays crossing a curve c_1.

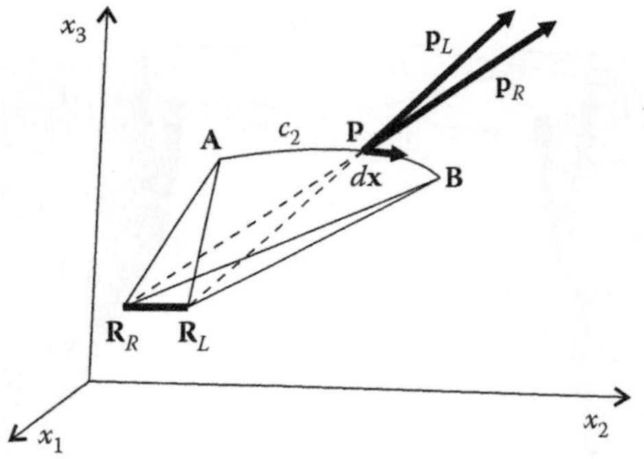

FIGURE 18.12
Two-parameter set of rays crossing a curve c_2 and coming from $\mathbf{R}_L\mathbf{R}_R$.

or

$$\begin{aligned} U_{2\text{-D}} &= (S_R(\mathbf{B}) - S_R(\mathbf{A})) - (S_L(\mathbf{B}) - S_L(\mathbf{A})) \\ &= S_R(\mathbf{B}) + S_L(\mathbf{A}) - S_R(\mathbf{A}) - S_L(\mathbf{B}) \end{aligned} \tag{18.53}$$

Now, if light is propagating in a medium of refractive index n, then $S_R(\mathbf{B}) = n[\mathbf{R}_R, \mathbf{B}]$, also $S_L(\mathbf{A}) = n[\mathbf{R}_L, \mathbf{A}]$, $S_R(\mathbf{A}) = n[\mathbf{R}_R, \mathbf{A}]$ and $S_L(\mathbf{B}) = n[\mathbf{R}_L, \mathbf{B}]$. The expression for $U_{2\text{-D}}$ then becomes

$$U_{2\text{-D}} = n([\mathbf{R}_R, \mathbf{B}] + [\mathbf{R}_L, \mathbf{A}] - [\mathbf{R}_R, \mathbf{A}] - [\mathbf{R}_L, \mathbf{B}]) \tag{18.54}$$

which is the same as the Hottel string formula used in two-dimensional optics.

The SMS 3-D optic in Figure 18.13 illustrates the combination of the situations shown in Figures 18.11 and 18.12. Here, a lens with thin edge e_H focuses to points $\mathbf{R}_L$ and $\mathbf{R}_R$ the incoming rays perpendicular to flat wavefronts w_1 and w_2. The étendue 2-D of the light incident on curve c_1 is given by expression (18.51), where, in this case, $L = [\mathbf{A}, \mathbf{B}]$ is the distance between points $\mathbf{A}$ and $\mathbf{B}$ (measured along the direction $\mathbf{p}_R - \mathbf{p}_L$, where $\mathbf{p}_L$ is perpendicular to w_1 and $\mathbf{p}_R$ is perpendicular to w_2). The étendue 2-D of the light crossing curve c_2 toward $\mathbf{R}_R\mathbf{R}_L$ is given by Equation 18.54. This 2-D "sheet of light" crossing curves c_1 and c_2 then conserves étendue 2-D as it propagates through air and optic. Curves c_1 and c_2 are free-form.

Figure 18.14 shows a particular case in which all rays are contained in plane γ. This situation is 2-D and the étendue 2-D is the same as that calculated for a two-dimensional optic. Curve c_3, where the rays enter the optic, is also contained in plane γ.

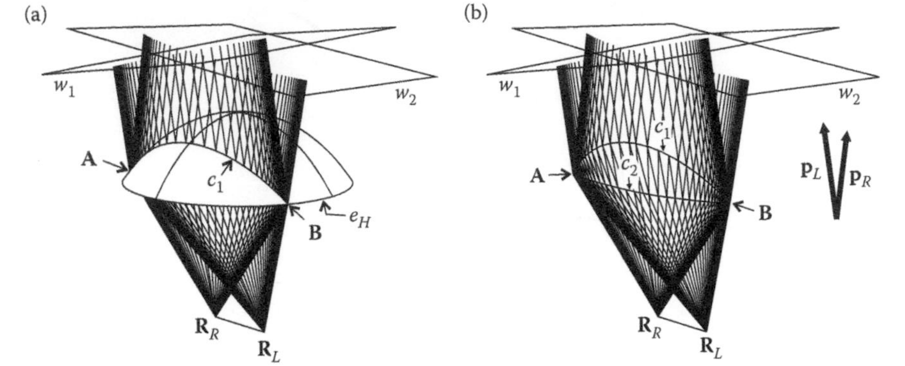

FIGURE 18.13
(a) A "sheet of light" crossing curves c_1 and c_2 on the top and bottom surfaces of an RR SMS 3-D lens conserves étendue 2-D. (b) Same light rays, but without showing the optic so that their complete paths inside the lens can be seen.

Figure 18.15 shows the same optic as Figures 18.13 and 18.14. Also shown is a bundle of rays perpendicular to flat wavefront w_1. As these rays cross the optic, they converge to point $\mathbf{R}_2$ and, therefore, are now perpendicular to spherical wavefron w_4 centered at $\mathbf{R}_2$.

The optic then couples wavefronts w_1 and w_4. Before entering the otpic, this bundle of rays is perpendicular to wavefront w_1 and, therefore, has zero etendue 2-D, that is, $U_{2\text{-D}} = 0$ as per Equation 18.43 or Equation 18.44. After

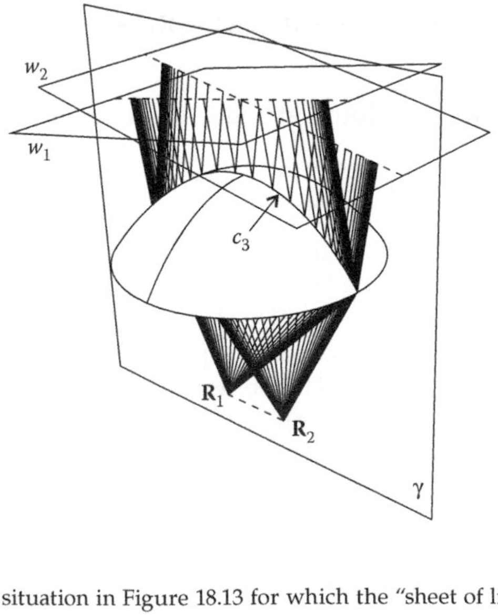

FIGURE 18.14
Particular case of the situation in Figure 18.13 for which the "sheet of light" is contained in a plane γ.

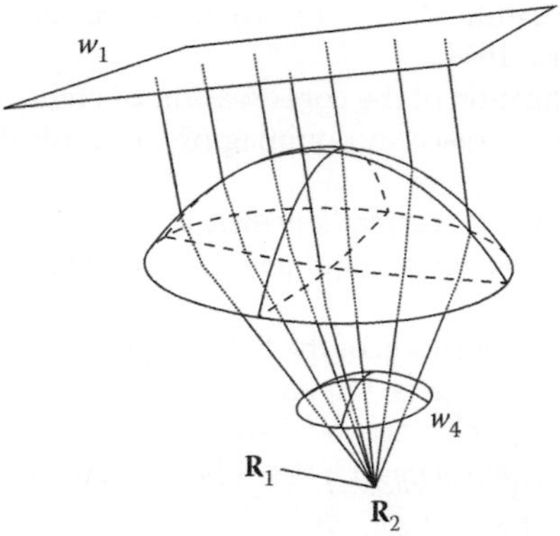

FIGURE 18.15
Etendue 2-D is conserved as the light from an input wavefront w_1 is compled by an optic onto an output wavefront w_4.

crossing the optic, these rays are perpendicular to wavefront w_4 and also have $U_{2\text{-D}} = 0$. Étendue 2-D is then conserved for this bundle of rays as it crosses the optic.

The SMS designs may then be seen as a result of imposing conservation of etendue-2D for two bundles of rays (perpendicular to two input wavefronts). Also, the 2-D flow line designs conserve etendue 2-D. Conservation of etendue 2-D deals with two parameter bundles of rays and is then used in the design of nonimaging optics. Conservation of etendue U deals with four parameter bundle of rays and is typically used to evaluate a design once it is done: calculate the etendue of the light at the entrance and exit apertures and determine if the optic is performing close to the theoretical limit of étendue conservation.

References

1. Born, M. and Wolf, E., *Principles of Optics*, Pergamon Press, Oxford, 1980.
2. Mahajan, V. N., *Optical Imaging and Aberrations, Part I, Ray Geometrical Optics*, SPIE Optical Engineering Press, Bellingham, 1998.
3. Welford, W. T. and Winston, R., *High Collection Nonimaging Optics*, Academic Press, San Diego, 1989.
4. Apostol, T. M., *Calculus—Volume II*, 2nd ed., John Wiley and Sons, New York, 1969.
5. Symon, K. R., *Mechanics*, 3rd Ed., Addison-Wesley Publishing Co. Inc., Reading, 1982.
6. Goldstein, H., *Classical Mechanics*, Addison-Wesley Publishing Company, Reading, 1980.

7. Synge, J. L. and Griffith, B. A., *Principles of Mechanics*, McGraw-Hill Book Company, New York, 1959.
8. Miñano, J. C., Application of the conservation of etendue theorem for 2-D subdomains of the phase space in nonimaging concentrators, *Appl. Opt.*, 23(12), 1984.
9. Miñano, J. C., Benitez, P., Fermat's principle and conservation of 2D etendue, *Proc. SPIE 5529, Nonimaging Optics and Efficient Illumination Systems*, Denver, Colorado, USA, 2004.
10. Miñano, J. C. et al., Application of the 2D etendue conservation to the design of achromatic aplanatic doublets, *Proceedings of SPIE Vol. 4446, Nonimaging Optics: Maximum Efficiency Light Transfer VI*, San Diego, California, USA, 2001.
11. Winston, R. et al., *Nonimaging Optics*, Elsevier Academic Press, Amsterdam, 2005.
12. Benitez, P., *Advanced Concepts of Non-Imaging Optics: Design and Manufacture*, PhD thesis, Polytechnic University of Madrid, 1998.
13. Herzberger, M., On the fundamental optical invariant, the optical tetrality principle, and on the new development of Gaussian optics based on this law, *JOSA*, 25, 295, September 1935.
14. Stavroudis, O. N., *The Mathematics of Geometrical and Physical Optics—The k-Function and Its Ramifications*, Wiley-VCH Verlag GmbH & Co. KGaA, Weinheim, Germany, 2006.
15. Benitez, P. et al., SMS freeforms for illumination, *Adv. Opt. Techn.*, 2(4), 323–329, 2013.

19

Classical Mechanics and Geometrical Optics

19.1 Fermat's Principle and Maupertuis' Principle

Equation 14.51 has the form of Hamilton's principle of classical mechanics. We could then be tempted to conclude that Hamilton's principle of mechanics corresponds to Fermat's principle of optics. In fact, this is not quite true. The correct variational principle of mechanics with which Fermat's principle can be related, is Maupertuis' principle, or the principle of least action.[1–5]

Hamilton's principle of classical mechanics is written as

$$\delta\int_{t_1}^{t_2} L(x_1, x_2, x_3, \dot{x}_1, \dot{x}_2, \dot{x}_3, t)dt \tag{19.1}$$

where L is the Lagrangian, x_i are the generalized coordinates, t the time, and $\dot{x}_k = dx_k/dt$. The Euler equations are

$$\frac{d}{dt}\left(\frac{\partial L}{\partial \dot{x}_1}\right) = \frac{\partial L}{\partial x_1}, \quad \frac{d}{dt}\left(\frac{\partial L}{\partial \dot{x}_2}\right) = \frac{\partial L}{\partial x_2}, \quad \frac{d}{dt}\left(\frac{\partial L}{\partial \dot{x}_3}\right) = \frac{\partial L}{\partial x_3} \tag{19.2}$$

The total derivative of L with respect to time t is

$$\frac{dL}{dt} = \frac{\partial L}{\partial t} + \sum_{k=1}^{3} \frac{\partial L}{\partial \dot{x}_k}\frac{d\dot{x}_k}{dt} + \frac{\partial L}{\partial x_k}\dot{x}_k \tag{19.3}$$

Let us now assume that L does not depend explicitly on time t,

$$L = L(x_1, x_2, x_3, \dot{x}_1, \dot{x}_2, \dot{x}_3) \tag{19.4}$$

We then have $\partial L/\partial t = 0$. We can now replace $\partial L/\partial x_k$ from the Euler equations to give

$$\frac{dL}{dt} = \sum_{k=1}^{3} \frac{\partial L}{\partial \dot{x}_k} \frac{d\dot{x}_k}{dt} + \frac{d}{dt}\left(\frac{\partial L}{\partial \dot{x}_k} \right) \dot{x}_k = \frac{d}{dt}\left(\sum_{k=1}^{3} \frac{\partial L}{\partial \dot{x}_k} \dot{x}_k \right) \tag{19.5}$$

or

$$\frac{d}{dt}\left(\sum_{k=1}^{3} \frac{\partial L}{\partial \dot{x}_k} \dot{x}_k - L \right) = 0 \tag{19.6}$$

Now, considering that

$$p_k = \frac{\partial L}{\partial \dot{x}_k} \tag{19.7}$$

we have

$$\frac{d}{dt}\left(\sum_{i=1}^{3} p_k \dot{x}_k - L \right) = 0 \tag{19.8}$$

The Hamiltonian H is defined by

$$H = \sum_{k} p_k \dot{x}_k - L \tag{19.9}$$

We can then conclude that the Hamiltonian does not depend on time and, therefore, is constant. Equation 19.8 expresses the law of conservation of energy, since the Hamiltonian corresponds to the energy E of the system.[3,4] We then have

$$H = \sum_{k=1}^{3} p_k \dot{x}_k - L = E \tag{19.10}$$

where E is a constant. In this case, it is possible to reduce the number of dimensions from four to three by eliminating time.[1] Consider x_1 and x_2 as functions of x_3, that is, $x_1 = x_1(x_3)$ and $x_2 = x_2(x_3)$, so that

$$\dot{x}_1 = \frac{dx_1}{dx_3} \dot{x}_3, \quad \dot{x}_2 = \frac{dx_2}{dx_3} \dot{x}_3 \tag{19.11}$$

and thus,

$$L(x_1,x_2,x_3,\dot{x}_1,\dot{x}_2,\dot{x}_3) = L\left(x_1,x_2,x_3,\frac{dx_1}{dx_3},\frac{dx_2}{dx_3},\dot{x}_3\right) \tag{19.12}$$

Accordingly,

$$\frac{\partial L}{\partial \dot{x}_k} = f_k\left(x_1,x_2,x_3,\frac{dx_1}{dx_3},\frac{dx_2}{dx_3},\dot{x}_3\right) \tag{19.13}$$

and considering Equation 19.7, Equation 19.10 can be written as

$$\sum_{k=1}^{3} f_k\left(x_1,x_2,x_3,\frac{dx_1}{dx_3},\frac{dx_2}{dx_3},\dot{x}_3\right)\dot{x}_k - L\left(x_1,x_2,x_3,\frac{dx_1}{dx_3},\frac{dx_2}{dx_3},\dot{x}_3\right) = E \tag{19.14}$$

Equation 19.14 can now be solved for $\dot{x}_3$ to give

$$\dot{x}_3 = \Phi\left(x_1,x_2,x_3,\frac{dx_1}{dx_3},\frac{dx_2}{dx_3},E\right) \tag{19.15}$$

From Equation 19.11, we also have

$$\begin{aligned} \dot{x}_1 &= \frac{dx_1}{dx_3}\Phi\left(x_1,x_2,x_3,\frac{dx_1}{dx_3},\frac{dx_2}{dx_3},E\right) \\ \dot{x}_2 &= \frac{dx_2}{dx_3}\Phi\left(x_1,x_2,x_3,\frac{dx_1}{dx_3},\frac{dx_2}{dx_3},E\right) \end{aligned} \tag{19.16}$$

Consider all the paths for which the system has some given constant energy E. We then compare only varied paths of the same energy as the real path. From Equation 19.10, we can write Equation 19.1 as[1]

$$\delta\int_{t_1}^{t_2}\left(\sum_{k=1}^{3} p_k\dot{x}_k - E\right)dt = \delta\int_{t_1}^{t_2}\sum_{k=1}^{3} p_k\dot{x}_k dt - \delta\int_{t_1}^{t_2} E\,dt = 0 \tag{19.17}$$

or

$$\delta\int_{t_1}^{t_2}\sum_{k=1}^{3} p_k\dot{x}_k\, dt = 0 \tag{19.18}$$

since $\delta\int E dt = 0$ because E is a constant. Considering Equation 19.7, the integral in Equation 19.18 can be rewritten as

$$\int_{t_1}^{t_2}\sum_{k=1}^{3}\frac{\partial L}{\partial \dot{x}_k}\dot{x}_k dt = \int_{x_{31}}^{x_{32}}\left(\frac{\partial L}{\partial \dot{x}_1}\frac{dx_1}{dx_3}+\frac{\partial L}{\partial \dot{x}_2}\frac{dx_2}{dx_3}+\frac{\partial L}{\partial \dot{x}_3}\right)dx_3 \tag{19.19}$$

since we are now considering $x_k = x_k(x_3)$, and therefore,

$$\dot{x}_k dt = \frac{dx_k}{dt}dt = \frac{dx_k}{dx_3}\frac{dx_3}{dt}dt = \frac{dx_k}{dx_3}dx_3 \tag{19.20}$$

Now making

$$F(x_1,x_2,x_3,\dot{x}_1,\dot{x}_2,\dot{x}_3) = \frac{\partial L}{\partial \dot{x}_1}\frac{dx_1}{dx_3}+\frac{\partial L}{\partial \dot{x}_2}\frac{dx_2}{dx_3}+\frac{\partial L}{\partial \dot{x}_3} \tag{19.21}$$

and replacing for $\dot{x}_1$, $\dot{x}_2$, and $\dot{x}_3$ from Equations 19.15 and 19.16, we have[6]

$$\delta\int_{x_{31}}^{x_{32}} F\left(x_1,x_2,x_3,\frac{dx_1}{dx_3},\frac{dx_2}{dx_3},E\right)dx_3 = 0 \tag{19.22}$$

or since the energy E is given to be constant,

$$\delta\int_{x_{31}}^{x_{32}} F(x_1,x_2,x_3,x_1',x_2')dx_3 = 0 \tag{19.23}$$

This is the Maupertuis principle of least action. Equation 19.10 enables the elimination of time derivatives $\dot{x}_k = dx_k/dt$, which can now be expressed as geometrical derivatives, given by $x_k' = dx_k/dx_3$, $(k = 1,2)$ where x_3 is now an independent variable. Equation 19.23 is purely geometrical and describes the orbits, not the evolution of the system in time. The latter can be found from the canonical equations,[1] which are obtained directly from Equations 19.1, 19.7, and 19.9 as

$$\frac{dx_k}{dt} = \frac{\partial H}{\partial p_k},\quad \frac{dp_k}{dt} = -\frac{\partial H}{\partial x_k}\quad (k = 1,2,3) \tag{19.24}$$

The differential equations derived from Equation 19.23 have the form of the Euler equations

$$\frac{d}{dx_3}\left(\frac{\partial F}{\partial x_1'}\right) = \frac{\partial F}{\partial x_1}, \quad \frac{d}{dx_3}\left(\frac{\partial F}{\partial x_2'}\right) = \frac{\partial F}{\partial x_2} \tag{19.25}$$

The integral in Equation 19.18 can also be written by using Equation 19.10 as

$$\delta\int_{t_1}^{t_2}(L+E)dt = \delta\int_{t_1}^{t_2}\sum_k p_k\dot{x}_k\,dt = \delta\int_{\mathrm{P}_1}^{\mathrm{P}_2}\sum_k p_k\,dx_k = \delta\int_{\mathrm{P}_1}^{\mathrm{P}_2}\mathbf{p}\cdot d\mathbf{s} \tag{19.26}$$

If T is the kinetic energy of the system and V the potential energy, then the total energy is $E = T + V$. Also, $L = T - V$.[3,4] Replacing this in Equation 19.10, we have

$$\sum_{k=1}^{3} p_k\dot{x}_k = 2T \tag{19.27}$$

Replacing this in expression (19.26) we have[7,8]

$$\delta\int_{t_1}^{t_2} 2T\,dt = 0 \Leftrightarrow \delta\int_{t_1}^{t_2} T\,dt = 0 \tag{19.28}$$

Maupertuis' principle of least action is also presented in classical mechanics textbooks as[3,6]

$$\Delta\int\sum_k p_k\dot{x}_k\,dt = 0 \tag{19.29}$$

The variation appearing in the expression for the principle of least action is the Δ variation. The δ variation corresponds to displacements in which the time is held fixed and the coordinates of the system are varied subject to the constraints imposed on the system. In contrast, the Δ variation deals with displacements in which, not only the coordinates of the system are varied, but it also involves a change in time. We are only considering cases, however, in which the energy is constant. In this case, we have seen that the dependence on time can be eliminated from the integral of expression (19.29). In this case, the Δ and δ variations can be made identical, and, therefore, expression (19.29) can be written as expression (19.26).[6]

It can be seen that Equation 19.23 is the same as Fermat's principle:

$$\delta\int_{P_1}^{P_2} n\,ds = \delta\int_{x_{31}}^{x_{32}} n(x_1,x_2,x_3)\sqrt{1+x_1'^2+x_2'^2}\,dx_3$$

$$= \delta\int_{x_{31}}^{x_{32}} L(x_1,x_2,x_3,x_1',x_2')\,dx_3 = 0 \qquad (19.30)$$

Euler equations (Equation 19.25) are also the same as those found in optics. Maupertuis' principle is, therefore, the equivalent in mechanics of Fermat's principle of optics.

It should be noted, however, that, if we consider a mechanical system with one less dimension, Equation 19.1 can be written as

$$\delta\int_{t_1}^{t_2} L(x_1,x_2,\dot{x}_1,\dot{x}_2,t)\,dt = 0 \qquad (19.31)$$

which is also mathematically similar to the form of Fermat's principle of optics (Equation 19.30) if time in Equation 19.31 is replaced by the independent variable x_3.[9] It should be noted, however, that the physical interpretations of Equations 19.31 and 19.30 are different. Equation 19.30 enables the determination of the paths of the system in 3-D space (x_1, x_2, x_3) but not of their evolution in time. However, Equation 19.31 enables the determination of the evolution in time of a 2-D system in space (x_1, x_2). This analogy is used in the following section.

19.2 Skew Invariant and Conservation of Angular Momentum

There is a relation between the conservation of angular momentum in mechanics, and the skew invariant in optics.

The constant h defined by expression (17.62) corresponds to the angular momentum in mechanics. To make this parallel clearer, we can give expression (17.62) a different form. Using expression (14.53), we obtain

$$x_1p_2 - x_2p_1 = n\frac{x_1x_2' - x_2x_1'}{\sqrt{1+x_1'^2+x_2'^2}} \qquad (19.32)$$

which, considering expressions (17.57) and (17.62), can be written as[10]

$$x_1p_2 - x_2p_1 = \frac{n\rho^2\theta'}{\sqrt{1+\rho'^2+\rho^2\theta'^2}} = h \tag{19.33}$$

It can now be noted that $x_1p_2 - x_2p_1$ is the magnitude of vector $\| (x_1, x_2, 0) \times (p_1, p_2, 0) \| = \| \mathbf{r} \times \mathbf{p} \| = \| \mathbf{L} \|$, where **L** is the angular momentum of a 2-D system. This analogy can also be derived from Equation 17.34. In the optics–mechanics analogy between Equations 19.31 and 19.30, the coordinate x_3 takes the role of time. If so, the analysis of a 3-D optical system in space (x_1, x_2, x_3) corresponds to the analysis of a mechanical system in space (x_1, x_2, t), t being the time. The trajectories in mechanics will then be 2-D in space (x_1, x_2), whereas in optics the corresponding light rays progress in a 3-D space (x_1, x_2, x_3).

19.3 Potential in Mechanics and Refractive Index in Optics

The ray equation (15.29) is

$$\frac{d\mathbf{p}}{ds} = \nabla n \tag{19.34}$$

Consider an optical system with its optical axis along x_3. Let us further suppose that light rays make small angles with the optical axis, so that we can make $ds \approx dx_3$. As discussed in Chapter 15, this approximation is called the paraxial approximation. Since we are considering the angles of light rays with the optical axis to remain small, the variations of the refractive index must also be small (otherwise, light rays could undergo large curvatures). This being the case, we can write $n = n_0 - \Delta n$, where n_0 is a constant and Δn a small variation. In this case, expression (19.34) can be written as

$$\frac{d\mathbf{p}}{dx_3} = \nabla(n_0 - \Delta n) \Leftrightarrow \frac{d\mathbf{p}}{dx_3} = -\nabla(\Delta n) \tag{19.35}$$

If coordinate x_3 is replaced by time t and the refractive index n by the potential V, it can be verified that Equation 19.35 is similar to the equation in mechanics for the movement of a particle in a potential field:

$$\frac{d\mathbf{p}}{dt} = -\nabla V \tag{19.36}$$

We then verify that, in the paraxial approximation, a refractive index distribution in optics takes the role of a potential in mechanics,[11] the momentum in optics takes the role of a mass's momentum in mechanics, and coordinate x_3 takes the role of time.

References

1. Born, M. and Wolf, E., *Principles of Optics*, Pergamon Press, Oxford, 1980.
2. Miñano, J. C and Benitez, P., Poisson bracket design method review. Application to the elliptic bundles, *SPIE Conference on Nonimaging Optics: Maximum Efficiency Light Transfer V*, SPIE Vol. 3781, 2, 1999.
3. Leech, J. W., *Classical Mechanics*, Chapman & Hall, London, 1965.
4. Goldstein, H., *Classical Mechanics*, Addison-Wesley Publishing Company, Reading, 1980.
5. Landau, L. and Lifshitz, E., *Mechanics*, Mir, Moscow, 1981.
6. Goldstein, H., *Classical Mechanics*, Addison-Wesley Press, Inc., Cambridge, 1951.
7. Sommerfeld, A., *Mechanics, Lectures on Theoretical Physics,* Vol. 1, Academic Press, New York, 1952.
8. Chetaev, N. G., *Theoretical Mechanics*, Mir Publishers, Moscow, Springer-Verlag, Berlin, 1989.
9. Welford, W. T. and Winston, R., *High Collection Nonimaging Optics*, Academic Press, San Diego, 1989.
10. Luneburg, R. K., *Mathematical Theory of Optics*, University of California Press, Berkeley and Los Angeles, 1964, 189.
11. Marcuse, D., *Light Transmission Optics*, Van Nostrand Reinhold Company, New York, 1972.

20

Radiometry, Photometry, and Radiation Heat Transfer

Radiometry deals with radiant quantities and applies to the entire electromagnetic spectrum. Photometry is a subdivision of radiometry that deals only with the part of the spectrum perceived by the human eye as light. In radiometry, it is possible to study nonvisible radiation, but in photometry only the visible part of the spectrum is considered. Radiation heat transfer, as the name suggests, deals with heat exchange by exchange of radiation, with bodies absorbing its heat much as the eye absorbs light.

Some concepts of radiometry, photometry, and radiation heat transfer are presented here briefly.

20.1 Definitions

The following definitions usually appear in books on radiometry, photometry, or optics, sometimes with entire chapters dedicated to these topics.

The central concept in radiometry is the radiation flux. It is the quantity of energy that is emitted, transmitted, or received per unit time,

$$\Phi = \frac{dQ}{dt} \tag{20.1}$$

where Q is the energy and t the time.

The human eye has differing sensitivity to different wavelengths (colors) of light. We must, therefore, distinguish two concepts. Radiant flux is the power of the radiation, measured in watts. Luminous flux is the measure of the perceived power of light by the human eye, measured in lumens. These quantities are related by the luminous efficacy function shown in Figure 20.1.[1] This function tells us how many lumens are there for each 1 W of power at a given wavelength. It has a maximum of 683 lm/W at 555 nm wavelength. For example, for 1 W power of radiation with a wavelength of 555 nm, we have 683 lm of visual sensation. For 1 W power of radiation of other wavelengths, the corresponding visual sensation in lumens is given by

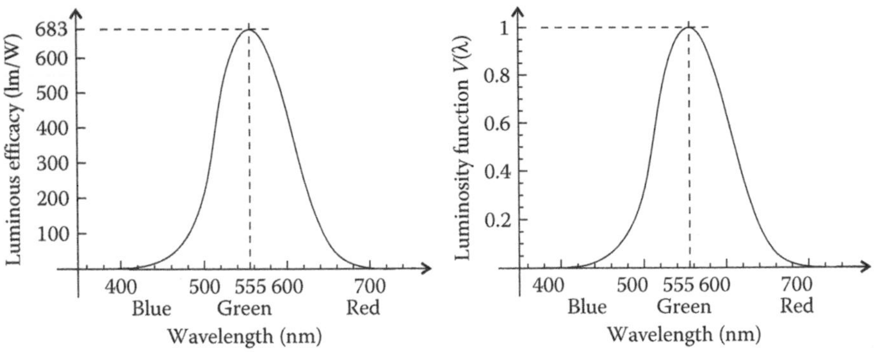

FIGURE 20.1
The human eye's sensitivity as a function of the wavelength of the light.

the luminous efficacy function. For example, radiation of wavelength 900 nm (infrared) will not be visible, so its luminous efficacy is zero.

We may now define the luminosity function $V(\lambda)$ (or photopic luminous efficiency function) the same way as the luminous efficacy, but normalized to its maximum value of 683, which occurs at 555 nm. We then have $V(555) = 1$. The luminous efficacy function can then be given by $683V(\lambda)$ where $V(\lambda)$ is the luminosity function.[2] Note that $V(\lambda)$ is dimensionless, but is multiplied by 683 lm/W to give the luminous efficacy.

Actually, the eye's sensitivity varies with the overall light level. We call photopic vision as the vision of the eye under well-lit conditions (normal lighting conditions during the day) and call scotopic vision as the vision of the eye in dim light (low-light conditions). The curve in Figure 20.1 refers to the eye's photopic sensitivity.

If a light source emits multiwavelength light with some kind of spectrum, there will be a power distribution as a function of wavelength. The strength of the corresponding total visual sensation can be calculated by

$$\Phi_V = 683\int_0^{\infty} \Phi(\lambda)V(\lambda)d\lambda \tag{20.2}$$

where $\Phi(\lambda)$ is the power in watts per unit wavelength, and Φ_V the total luminous flux in lumens. The integration limits, in practice, do not need to exceed the range of appreciable values of $V(\lambda)$, for example, 380–760 nm, rather than zero to infinity.

We now define some more quantities. The radiation flux emitted per unit surface is called emittance and is defined as

$$M = \frac{d\Phi}{dA} \tag{20.3}$$

where dA is an infinitesimal area emitting radiation. The radiation flux falling on a surface is called irradiance (W/m^2) and is defined by

$$E = \frac{d\Phi}{dA} \tag{20.4}$$

where dA is an infinitesimal area receiving radiation. The corresponding photometric quantity is called illuminance and is measured in lux (1 lx = 1 Lm/m^2).

The intensity of the radiation is defined as the flux per unit solid angle,

$$I = \frac{d\Phi}{d\Omega} \tag{20.5}$$

and, again, we distinguish between the radiometric quantity, which is given in watts per steradian (W/sr), and the photometric quantity, which is given in candelas, where 1 cd = 1 lm/sr.

The radiation flux per unit projected area and per unit solid angle is given by

$$L = \frac{d\Phi}{dA\cos\theta d\Omega} \tag{20.6}$$

where θ is the angle that normal **n** to area dA makes with the direction of the solid angle $d\Omega$, as shown in Figure 20.2. This quantity is called radiance and is measured in watts per steradian per square meter (W/sr/m^2). The corresponding photometric quantity is the luminance, also defined as

$$L_V = \frac{d\Phi_V}{dA\cos\theta d\Omega} \tag{20.7}$$

The quantity L_V is measured in candelas per square meter (cd/m^2). The flux used to define it is the "visual" flux Φ_V.

Instead of the notation $d\Phi$, the notation $d^2\Phi$ is customarily used to stress the fact that the flux in the definition of the radiance is proportional to the

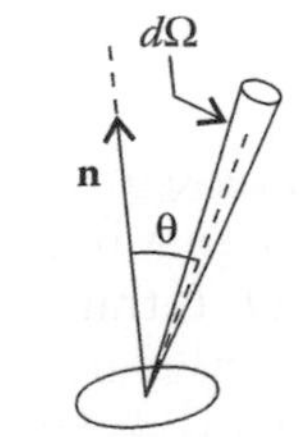

FIGURE 20.2
Radiation emitted by a solid angle $d\Omega$ in a direction making an angle θ with the normal **n** to area dA.

product of the two differentials dA and $d\Omega$ and, thus, is a second-order differential. Here, nonetheless, the notation $d\Phi$ is used instead of $d^2\Phi$.

Luminance or radiance may be a function of wavelength. In that case, if $L_\lambda(\lambda)$ is the spectral radiance, defined as the radiance per unit wavelength interval, the radiance is

$$L = \int_0^\infty L_\lambda(\lambda)d\lambda \tag{20.8}$$

Also, if $L_{V\lambda}(\lambda)$ is the spectral luminance, defined as the luminance per unit wavelength interval, the luminance is

$$L_V = \int_0^\infty L_{V\lambda}(\lambda)d\lambda \tag{20.9}$$

Luminance can be obtained from the spectral radiance as[3,4]

$$L_V = 683\int_0^\infty L(\lambda)V(\lambda)d\lambda \tag{20.10}$$

The radiation intensity emitted by an area dA is given as

$$I_{dA} = \frac{d\Phi}{d\Omega} = L\cos\theta dA \tag{20.11}$$

A similar expression could be written for the photometric quantity. Consider the particular case in which the radiance L (or luminance L_V) of the emitted radiation is uniform over a finite area A. The total intensity in direction θ can be obtained by

$$I(\theta) = L\cos\theta\int_A dA = L(\theta)A\cos\theta \tag{20.12}$$

Further, consider the particular case in which the radiance L (or luminance L_V) is independent of the direction, that is, $L(\theta) = L$, where L is a constant. For $\theta = 0$, the intensity is given by $I_0 = LA$, thus this expression can be written as

$$I = I_0\cos\theta \tag{20.13}$$

This is the Lambert's cosine law. A surface is called Lambertian if it emits, or if it intercepts, radiation with an intensity pattern following this cosine law solely due to variation of projected area.[3]

20.2 Conservation of Radiance in Homogeneous Media

Étendue conservation can be derived in many contexts. Here we present its conservation in the context of radiometry. Consider an infinitesimal area dA_1 emitting radiation in the direction of dA_2, as presented in Figure 20.3. These two areas are separated by a distance r. Note that r is a finite quantity, but dA_1 and dA_2 are infinitesimal. Areas dA_1 and dA_2 have normals $\mathbf{n}_1$ and $\mathbf{n}_2$ that make angles θ_1 and θ_2 in the direction of r.

By definition, an elementary light beam is composed of a central ray and all the rays passing through both dA_1 and dA_2, as shown in Figure 20.4.[3,5–7]

The solid angle $d\Omega_1$ is that defined at area dA_1 by area dA_2, and is given as

$$d\Omega_1 = \frac{dA_2 \cos\theta_2}{r^2} \tag{20.14}$$

In an equal manner, the solid angle defined by dA_1 in dA_2 is given by

$$d\Omega_2 = \frac{dA_1 \cos\theta_1}{r^2} \tag{20.15}$$

Multiplying $d\Omega_2$ by $dA_2 \cos \theta_2$ and $d\Omega_1$ by $dA_1 \cos \theta_1$, we can write

$$\begin{aligned} dA_1 \cos\theta_1 d\Omega_1 &= \frac{dA_1 dA_2 \cos\theta_1 \cos\theta_2}{r^2} \\ dA_2 \cos\theta_2 d\Omega_2 &= \frac{dA_1 dA_2 \cos\theta_1 \cos\theta_2}{r^2} \end{aligned} \tag{20.16}$$

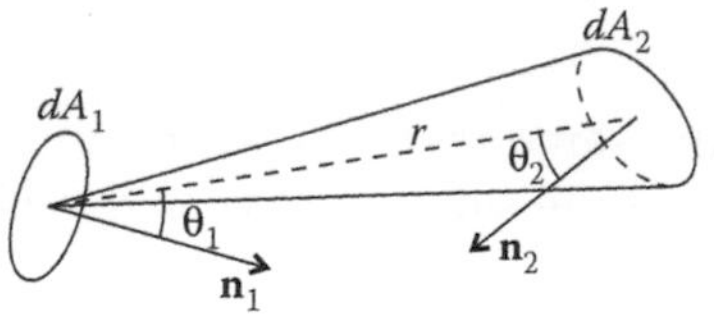

FIGURE 20.3
Radiation heat transfer between two surfaces dA_1 and dA_2.

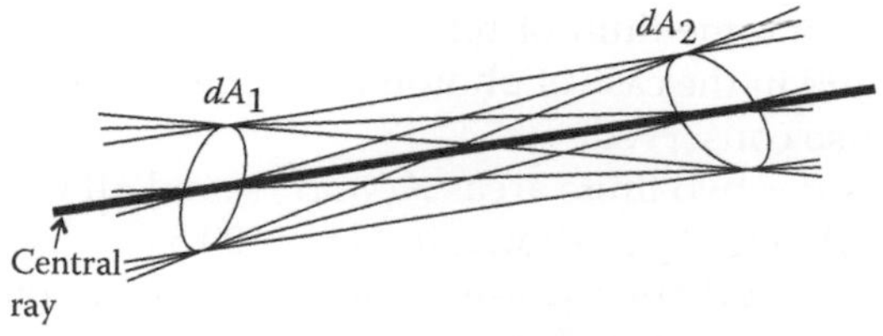

FIGURE 20.4
Illustration of an elementary beam of radiation.

Let us now consider the quantity

$$dU = dA\cos\theta d\Omega \tag{20.17}$$

We, therefore, have

$$\begin{aligned} dU_1 &= dA_1\cos\theta_1 d\Omega_1 \\ dU_2 &= dA_2\cos\theta_2 d\Omega_2 \end{aligned} \tag{20.18}$$

For an elementary light beam, all the light passing through dA_1 (see Figure 20.3) is that passing through dA_2. Therefore, dU, as defined in Equation 20.17, is, in this case, given by the first equation of expression (20.18). However, for the same elementary light beam, all the light passing through dA_2 is that coming from dA_1. Therefore, dU, as defined by Equation 20.17, is, in this case, given by the second equation of expression (20.18). For an elementary light beam, dU is then conserved since, as seen from Equation 20.16,

$$dU_1 = dU_2 \tag{20.19}$$

This quantity U is called étendue, throughput, or geometrical extent.[3,7–9] The étendue of the light beam as it crosses dA_1 is the same as when it crosses dA_2.

From the definition of radiance in Equation 20.6, we can see that it is related to the étendue by

$$d\Phi = L\,dU \tag{20.20}$$

and, for a light beam, the same rays pass through dA_1 and dA_2, so the flux through these two areas is the same, that is, $d\Phi_1 = d\Phi_2$. Since the étendue is also conserved, we obtain

$$L_1 = L_2 \tag{20.21}$$

and radiance is also conserved.[3,5–7,10] Note that relation (20.21) is also valid for light traveling in a medium of refractive index n. The arguments used earlier are also valid in the case of photometric quantities and, therefore, the luminance L_V is also conserved.

Let us now consider two finite areas A_1 and A_2 and all the light rays passing through both A_1 and A_2, as shown in Figure 20.5.

The étendue of the radiation passing through A_1 and going toward A_2 is given by the integration of the first equation of expression (20.18) in dA_1 and dA_2. However, the étendue passing through A_2 for the light coming from A_1 is given by the integration of the second equation of expression (20.18) in dA_1

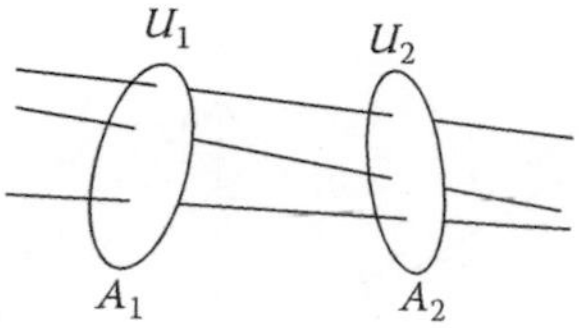

FIGURE 20.5
The étendue is conserved for the light passing through both areas A_1 and A_2.

and dA_2. But $dU_1 = dU_2$, and therefore, these integrals are also the same, and the étendue is conserved from A_1 to A_2.

20.3 Conservation of Basic Radiance in (Specular) Reflections and Refractions

Consider a light beam reflected by a mirror, as shown in Figure 20.6. This light beam is composed of a central ray and all the rays passing through dA_1 and dA_2. Mirror M creates an image dA_1^* of area dA_1. But our previous result on étendue conservation can be applied to elementary areas dA_1^* and dA_2, establishing that the étendue is conserved for the light beam passing through dA_2 and the mirror image dA_1^*. Therefore, the étendue is conserved for the light beam passing through dA_1 and dA_2, and it can be concluded that étendue is conserved during reflection.

Consider a light beam falling on the surface dA_1 separating two media (mediums 1 and 2) with refractive indices n_1 and n_2, respectively, as shown in Figure 20.7a. The flux coming from medium 1 and falling on dA_1 is given by

$$d\Phi_1 = L_1 dA_1 \cos\theta_1 d\Omega_1 \tag{20.22}$$

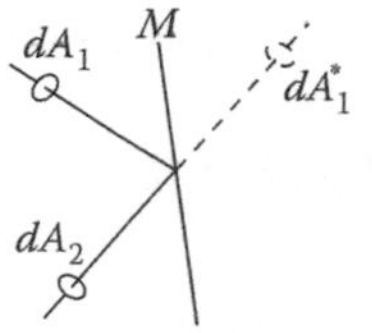

FIGURE 20.6
The étendue is conserved between area dA_2 and the mirror image dA_1^*. Therefore, it is also conserved between areas dA_1 and dA_2. Étendue is then conserved during reflection.

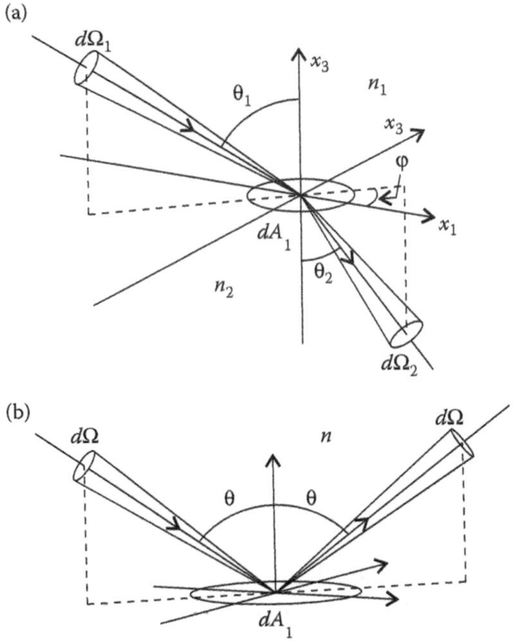

FIGURE 20.7
(a) The refraction of light from a medium with refractive index n_1 to another medium with refractive index n_2. The solid angle occupied by the radiation varies, but the quantity L/n^2 is constant. (b) A reflection, where the solid angle is constant. Since n does not vary either, the quantity L/n^2 is also invariant.

In Figure 20.8, surface dA_2 defines a solid angle $d\Omega_1$, which, in spherical coordinates, is given by

$$d\Omega_1 = dA_2/r^2 = \sin\theta_1 d\theta_1 d\varphi \tag{20.23}$$

And therefore,

$$d\Phi_1 = L_1 dA_1 \cos\theta_1 \sin\theta_1 d\theta_1 d\varphi \tag{20.24}$$

However, the flux propagating into medium 2 is given by

$$d\Phi_2 = L_2 dA_1 \cos\theta_2 d\Omega_2 = L_2 dA_1 \cos\theta_2 \sin\theta_2 d\theta_2 d\varphi \tag{20.25}$$

Note that angle φ is the same for both incident and refracted rays, since a refracted ray is contained in the plane defined by the incident ray and the normal to the surface. Assume that the normal to the surface points in the direction of axis x_3.

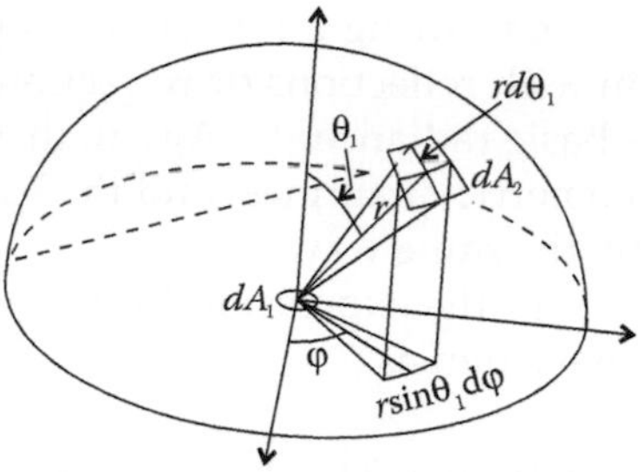

FIGURE 20.8
Radiation transfer between the surface dA_1 and the hemisphere above it, through all of which dA_1 radiates. The solid angle $d\Omega_1$ defined at area dA_1 by area dA_2, is given as $d\Omega_1 = \sin\theta_1 d\theta_1 d\varphi$.

Assuming there are no losses at the surface (i.e., Fresnel reflections are neglected, or have been suppressed by antireflection coatings), we can write $d\Phi_1 = d\Phi_2$ and, therefore,

$$\frac{L_1 \cos\theta_1 \sin\theta_1 d\theta_1}{L_2 \cos\theta_2 \sin\theta_2 d\theta_2} = 1 \qquad (20.26)$$

Snell's Law is

$$n_1 \sin\theta_1 = n_2 \sin\theta_2 \qquad (20.27)$$

Calculating the derivatives of both sides of the equation,

$$n_1 \cos\theta_1 d\theta_1 = n_2 \cos\theta_2 d\theta_2 \qquad (20.28)$$

therefore,

$$\frac{\sin\theta_1}{\sin\theta_2} = \frac{\cos\theta_1 d\theta_1}{\cos\theta_2 d\theta_2} = \frac{n_2}{n_1} \qquad (20.29)$$

Inserting this into expression (20.26) gives

$$\frac{L_1}{n_1^2} = \frac{L_2}{n_2^2} \Leftrightarrow L_1^* = L_2^* \qquad (20.30)$$

It can then be concluded that the quantity $L^* = L/n^2$ is conserved in refraction, and, therefore, also conserved in optical systems containing surfaces separating two media with different indices of refraction.[2,3,5–7,10] A similar calculation with $n_1 = n_2$, as in Figure 20.7b, enables us to conclude that L, and therefore L/n^2, are conserved during a reflection, as previously concluded. Thus, any optical system with reflections or refractions conserves the quantity $L^* = L/n^2$, known as basic radiance.[3,7,10] Again, these arguments are also valid in the case of photometric quantities, and the basic luminance $L_V^* = L_V/n^2$ is conserved in exactly the same way.

In terms of this quantity L^*, the expression for the energy flux through an elemental area dA can be written as

$$d\Phi = L dA \cos\theta d\Omega = L^*(n^2 dA \cos\theta d\Omega) \tag{20.31}$$

Assuming loss-less reflections or refractions (i.e., no scattering or Fresnel reflection), flux $d\Phi$ must be conserved. Since L^* is conserved, it can be concluded that $n^2 dA \cos\theta d\Omega$ is conserved as well. Generalizing, now, the definition of étendue to

$$dU = n^2 dA \cos\theta d\Omega \tag{20.32}$$

we verify that the étendue is conserved in an optical system with reflections or refractions. Note that the previous definition given by Equation 20.17 is still valid in the particular case where $n = 1$.

As an example, let us consider the optical system presented in Figure 20.9. This optical system has entrance aperture A_1, exit aperture A_2, and consists of two parallel flat mirrors M_1 and M_2. Consider two elementary light beams passing through dA_1 and dA_2 at the entrance and exit apertures, respectively. The light beam b_1 is composed of a central ray and all the rays passing through dA_1 and dA_2, and the light beam b_2 is composed of a central ray and all the rays passing through dA_1, being reflected at the mirror M_2 and then passing through dA_2. We have seen that for beam b_1, étendue is conserved. We have also seen that reflection conserves étendue, and therefore, it is also

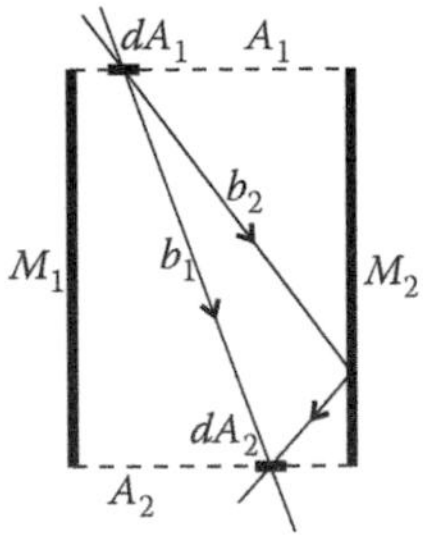

FIGURE 20.9
Étendue is conserved between areas dA_1 and dA_2, either for light beams that cross them directly, as is the case of beam b_1, or for light beams reflected by the mirrors, as is the case of beam b_2. If étendue is conserved for any of the areas of dA_1 and dA_2 of A_1 and A_2, then it is conserved between A_1 and A_2.

conserved for the elementary light beam b_2 passing through dA_1 and dA_2. Integrating, now, over areas A_1 and A_2, we can conclude that étendue is conserved for the radiation passing through A_1 and A_2. The same conclusion could be drawn if the optical system would have refractions, since refraction also conserves étendue.

The expression for the energy flux through an area dA can be written as

$$d\Phi = L^{*}dU \tag{20.33}$$

We then verify that, if the basic radiance is multiplied by the étendue, we obtain the energy flux.

Note that, in the definition of étendue in expression (20.32), n is a dimension-less quantity, since it is the ratio of two light speeds (*in vacuo* and *in medio*). The same happens with solid angle $d\Omega$, which is the ratio of any elemental area on a sphere, and the square of the sphere's radius. The solid angle, therefore, has the units of area divided by area and is a-dimensional. Something similar happens with angle θ, which has dimensions of length divided by length, and is therefore dimensionless, but nonetheless, not unitless, being expressed in degrees or radians. Therefore, dU has the units of dA, that is, units of area.

In the real world, reflections and refractions do not have the perfect characteristics as in geometric optics, because actual engineered surfaces always exhibit some scattering that adds to étendue, as well as reduces basic radiance and luminance (see Chapter 4).

20.4 Étendue and the Shape Factor

The energy flux per unit time that dA_1 emits in the direction of dA_2 or that passes through dA_1 in the direction of dA_2, as shown in Figure 20.3, is given by expression (20.20) as[3,8,10]

$$d\Phi_{12} = L_1 dA_1 \cos\theta_1 \, d\Omega_1 \tag{20.34}$$

Replacing the solid angle $d\Omega_1$ from expression (20.14), the flux emitted by dA_1 in the direction of dA_2 is given by

$$d\Phi_{12} = L_1 \frac{dA_1 dA_2 \cos\theta_1 \cos\theta_2}{r^2} \tag{20.35}$$

Consider the total flux emitted by surface dA_1, emitted into a hemisphere centered at dA_1 and covering it, as shown in Figure 20.8. For the light emitted by dA_1 we can write an expression similar to Equation 20.24 in spherical coordinates, and integrating over the entire hemisphere gives

$$d\Phi_{hem} = L_1 dA_1 \int_0^{2\pi}\int_0^{\pi/2} \cos\theta_1 \sin\theta_1 \, d\theta_1 d\varphi = \pi L_1 dA_1 \tag{20.36}$$

where $d\Phi_{hem}$ is the flux radiated by dA_1 into the hemisphere above it. In the situation shown in Figure 20.8, and in the particular case in which dA_1 is a blackbody surface, the radiation flux emitted toward the hemisphere above it is given as

$$d\Phi_{hem1} = \sigma T_1^4 dA_1 \tag{20.37}$$

where σ is the Stephan–Boltzmann constant ($\sigma = 5.670 \times 10^{-8}$ Wm^{-2}K^{-4}) and T_1 the temperature of the body containing area dA_1.[3,11] Comparing expressions (20.36) and (20.37) it can be verified that, in this case,

$$L_1 = \frac{\sigma T_1^4}{\pi} \tag{20.38}$$

which is the expression for the radiance of a blackbody at a temperature T_1.[3]

We can now consider again the situation presented in Figure 20.3 and calculate the ratio between the radiation emitted by dA_1 that arrives at dA_2 and all the radiation emitted by dA_1. This ratio is given by[3,7,8]

$$\begin{aligned} dF_{dA1-dA2} &= \frac{d\Phi_{12}}{d\Phi_{hem1}} = \frac{1}{\pi L_1 dA_1}\left(L_1 \frac{dA_1 dA_2 \cos\theta_1 \cos\theta_2}{r^2}\right) \\ &= \frac{dA_2 \cos\theta_1 \cos\theta_2}{\pi r^2} \end{aligned} \tag{20.39}$$

Note that $dF_{dA1-dA2}$ is a differential because it is proportional to an infinitesimal area dA_2. The quantity $dF_{dA1-dA2}$ is variously called the shape factor, angle factor, or configuration factor, and is used in radiation heat transfer[11,12] to designate the fraction of radiation leaving dA_1 that arrives at dA_2. With this definition, the flux emitted by dA_1 and received by dA_2 can now be rewritten from expression (20.35) as

$$d\Phi_{12} = \pi L_1 dA_1 \frac{dA_2 \cos\theta_1 \cos\theta_2}{\pi r^2} = \pi L_1 dA_1 dF_{dA1-dA2} \tag{20.40}$$

We now consider dA_1 and dA_2 as two blackbody emitters. The flux emitted by dA_1 toward dA_2 is given by Equation 20.40, which can be rewritten with Equation 20.38 as

$$d\Phi_{12} = \sigma T_1^4 dA_1 dF_{dA1-dA2} \tag{20.41}$$

However, the flux $d\Phi_2$ emitted by dA_2 and arriving at dA_1 is given by

$$d\Phi_{21} = \sigma T_2^4 dA_2 dF_{dA2-dA1} \tag{20.42}$$

In thermal equilibrium, the temperatures T_1 and T_2 of dA_1 and dA_2 are the same ($T_1 = T_2$) and give $d\Phi_{12} = d\Phi_{21}$, which can be written as

$$dA_1 dF_{dA1-dA2} = dA_2 dF_{dA2-dA1} \tag{20.43}$$

In radiation transfer, this expression is called the reciprocity relation[11,12] and it tells us that, in thermal equilibrium, the radiation $d\Phi_{12}$ emitted from dA_1 toward dA_2 equals $d\Phi_{21}$ emitted from dA_2 toward dA_1.

The reciprocity relation can also be written as

$$dA_1 \frac{dA_2 \cos\theta_1 \cos\theta_2}{\pi r^2} = dA_2 \frac{dA_1 \cos\theta_1 \cos\theta_2}{\pi r^2} \Leftrightarrow dU_1 = dU_2 \tag{20.44}$$

and, therefore, corresponds to the conservation of étendue as previously obtained. The étendue and the shape factor are, thus, related quantities.

To obtain this relation, from $dU = d\Phi/L$ and expression (20.36), we obtain $dU_{hem1} = d\Phi_{hem1}/L_1 = \pi dA_1$, and, therefore, expression (20.39) can be written in terms of étendue as

$$dF_{dA1-dA2} = \frac{d\Phi_{12}}{d\Phi_{hem1}} = \frac{d\Phi_{12}/L_1}{d\Phi_{hem1}/L_1} = \frac{dU_{12}}{dU_{hem1}} = \frac{dU_{12}}{\pi dA_1} \tag{20.45}$$

where U_{12} is the étendue of the light emitted from dA_1 toward dA_2. This expression enables us to use the shape factors to calculate the étendue. Shape factors are sometimes listed for different geometries in textbooks of radiation transfer.[11]

Another possible way to calculate the étendue of the light emitted from the area A_1 to the area A_2 is by using the Monte Carlo Method of randomized computer ray tracing.[3] For this, we may also relate the étendue of the light

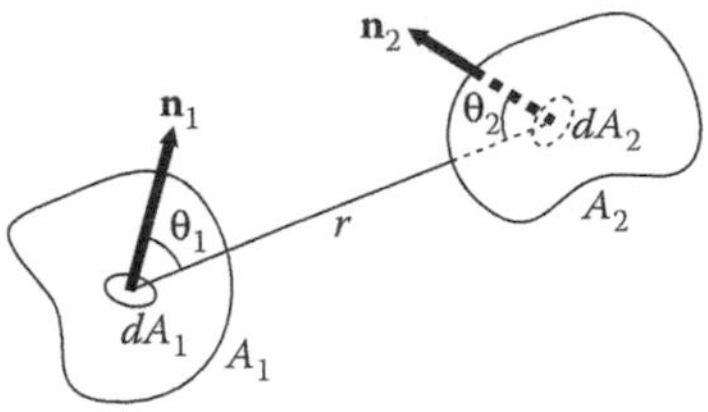

FIGURE 20.10
The flux emitted from A_1 toward A_2 is obtained by integration of dA_1 and dA_2 on A_1 and A_2.

emitted from the area A_1 toward the area A_2 with the fraction of light emitted by A_1 that reaches A_2. From Equation 20.45, we have

$$dU_{12} = \pi dA_1 \frac{d\Phi_{12}}{d\Phi_{hem1}} \tag{20.46}$$

If we now consider dA_1 as part of a uniform Lambertian emitter A_1, and dA_2 as part of an area A_2, as shown in Figure 20.10, we get

$$U_{12} = \pi A_1 \frac{\Phi_{12}}{\Phi_{hem1}} \tag{20.47}$$

where U_{12} is the étendue of the light emitted from A_1 toward A_2, Φ_{12} is the portion of the flux emitted by A_1 that is captured by A_2, and Φ_{hem1} the total flux emitted by A_1.

The étendue U_{12} of the radiation emitted from A_1 toward A_2 can then be obtained using a ray-tracing package (Monte Carlo Method). If A_1 emits unit flux, so that $\Phi_{hem1} = 1$, and if we assume that A_2 is a perfect absorber, the étendue of the light emitted from A_1 to A_2 is given by $U_{12} = \pi A_1\Phi_{12}$, where Φ_{12} is the flux absorbed by A_2.

20.5 Two-Dimensional Systems

Consider a 2-D system, as presented in Figure 20.11.

In a 3-D system, the flux from A_1 to A_2 was given by Equation 20.34. In the 2-D geometry, the flux (energy per unit time) passing through da_2 coming from da_1 is given as

$$d\Phi_1 = L_1 da_1 \cos\theta_1 d\theta_1 \tag{20.48}$$

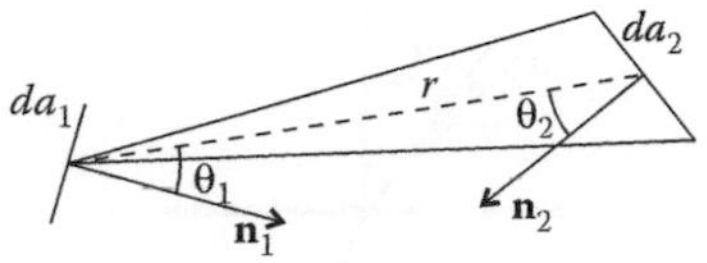

FIGURE 20.11
Radiation transfer between two lines da_1 and da_2 in a 2-D system.

with

$$d\theta_1 = \frac{da_2 \cos\theta_2}{r} \tag{20.49}$$

Note that da_1 and da_2 are no longer areas in a 3-D space, but lines (infinitesimal lengths) in a plane in a 2-D space.

The flux radiated by da_1 through the entire semicircumference above it, as shown in Figure 20.12, is given by

$$d\Phi_{hem1} = L_1 da_1 \int_{-\pi/2}^{\pi/2} \cos\theta_1 \, d\theta_1 = 2L_1 da_1 \tag{20.50}$$

The 2-D shape factor from da_1 to da_2 is now given by

$$dF_{da1-da2} = \frac{d\Phi_1}{d\Phi_{hem1}} = \cos\theta_1 \frac{da_2 \cos\theta_2}{2r} = \frac{1}{2}\cos\theta_1 \, d\theta_1 = \frac{1}{2} d(\sin\theta_1) \tag{20.51}$$

This result coincides with the shape factor obtained for a 3-D system where, in Figure 20.11, lines da_1 and da_2 extend to infinity in the direction perpendicular to the plane of the text, forming two parallel surfaces.[11,12]

The étendue given in 3-D by Equation 20.17 is now given in 2-D by

$$dU_{2\text{-}D} = da \cos\theta \, d\theta = da \, d(\sin\theta) \tag{20.52}$$

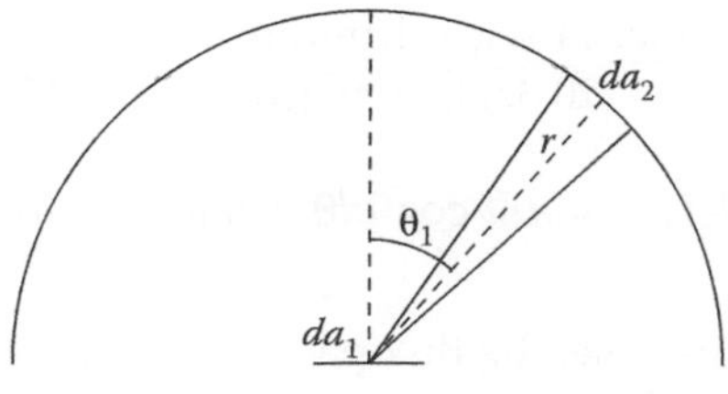

FIGURE 20.12
Radiation transfer between a line da_1 and a semicircumference above it. All the light emitted by da_1 crosses this semicircumference.

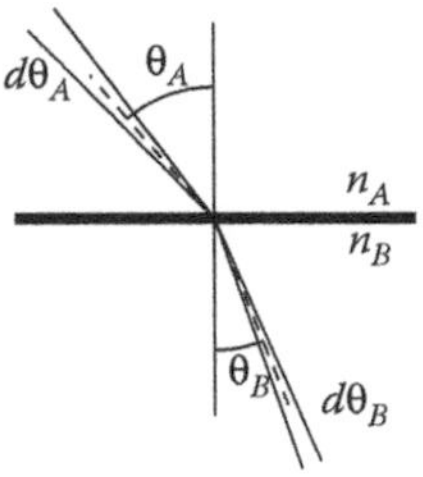

FIGURE 20.13
Refraction by a 2-D optic.

This must be conserved in the passage through an optical system. If we consider a 2-D system and the refraction on a line in the plane, we again have Snell's Law of refraction:

$$n_A \sin\theta_A = n_B \sin\theta_B \tag{20.53}$$

where n_A and θ_A are the refractive index and angle to the normal to the line before refraction, and n_B and θ_B the refractive index and angle to the normal to the curve after refraction, as shown in Figure 20.13.

Calculating the differentials of both sides gives

$$n_A \cos\theta_A d\theta_A = n_B \cos\theta_B d\theta_B \Leftrightarrow dU_A = dU_B \tag{20.54}$$

and, therefore, the 2-D étendue is conserved in 2-D refractions. In the case in which $n_A = n_B$, we obtain the conservation of étendue by reflections.

Also in this case, we have $dU = d\Phi/L$; thus, from Equation 20.50, we can obtain $dU_{hem1} = d\Phi_{hem1}/L_1 = 2da_1$, and the equivalent of Equation 20.45 for 2-D geometry becomes

$$dF_{da1-da2} = \frac{dU_1}{2da_1} \tag{20.55}$$

If light passes through media with different refractive indices, the étendue given in 3-D by expression (20.32) is now given in 2-D by

$$dU_{2\text{-}D} = n\,da\cos\theta\,d\theta = nda\,d(\sin\theta) \tag{20.56}$$

and is conserved. In this case, the flux passing through a line can be written in similar to Equation 20.33 as

$$d\Phi = L^* dU_{2\text{-}D} \tag{20.57}$$

where $L^* = L/n$ is the basic radiance. Again, a similar result is obtained for the basic luminance if photometric quantities are used.

20.6 Illumination of a Plane

Consider the cases in which we want to illuminate a plane using an infinitesimal flat light source parallel to the plane, as shown in Figure 20.14.

Let dA_2 be an elemental area of the plane to be illuminated, and dA_1 an infinitesimal source parallel to the plane and placed at a distance D, as shown in Figure 20.14.

In this case, dA_1 and dA_2 are parallel, and we have $\theta_1 = \theta_2 = \theta$. In addition, distance r between dA_1 and dA_2 can be related to distance D between dA_1 and the plane of dA_2 by

$$r = D/\cos\theta \tag{20.58}$$

Expression (20.35) can be written in this case as

$$d\Phi = L\frac{dA_1 dA_2}{D^2}\cos^4\theta \Leftrightarrow E = \frac{d\Phi}{dA_2} = L\frac{dA_1}{D^2}\cos^4\theta \tag{20.59}$$

where E is the irradiance at area dA_2 and is defined by $E = d\Phi/dA_2$. The irradiance E is, therefore, the energy flux per unit area passing through dA_2. For $\theta = 0$, expression (20.59) can be written as

$$E_0 = L_0\frac{dA_1}{D^2} \tag{20.60}$$

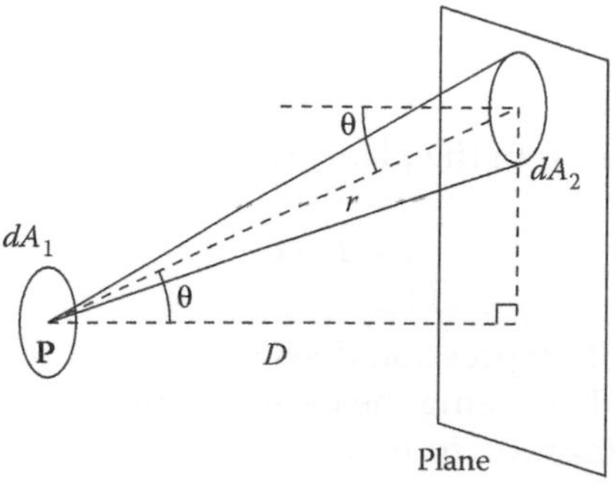

FIGURE 20.14
Illumination of a plane by an infinitesimal light source dA_1.

Dividing Equation 20.59 by Equation 20.60 gives

$$\frac{E}{E_0} = \frac{L}{L_0}\cos^4\theta \tag{20.61}$$

In the case where L is independent of the direction (i.e., in the case in which dA_1 is a Lambertian source) L in direction θ equals L_0 in the direction perpendicular to elemental area dA_1, giving $L = L_0$, so that[13,14]

$$E = E_0\cos^4\theta \tag{20.62}$$

from which we can conclude that the irradiance produced by the light emitted by an infinitesimal area on a parallel plane is proportional to $\cos^4\theta$. This steep fall is well known in, for example, the design of image projectors.

Now, we determine the (nonuniform) intensity pattern that dA_1 must emit to produce a uniform illumination of the plane. We no longer require dA_1 to be parallel to the plane. In this case, we can write

$$d\Phi = I d\Omega \tag{20.63}$$

where I is the intensity of the radiation coming from dA_1. Now, the quantity I characterizes the source only in terms of the direction in which the radiation is emitted. Considering expression (20.14) for $d\Omega_1$ and expression (20.58), we get

$$d\Phi = I\frac{dA_2\cos^3\theta}{D^2} \tag{20.64}$$

so that

$$\frac{E}{E_0} = \frac{I}{I_0}\cos^3\theta \tag{20.65}$$

For constant irradiance E in the plane, we must have $E = E_0$, so that

$$I = I_0/\cos^3\theta \tag{20.66}$$

As can be verified, this expression does not depend on the distance D from point **P** to the plane. This being the case, the distance D can have any value and can even go to infinity. In this case, expression (20.66) determines the angular intensity distribution of the radiation that a source must have to produce a constant irradiance on the plane placed at infinity. This result is still valid, even if the source now has a finite dimension.

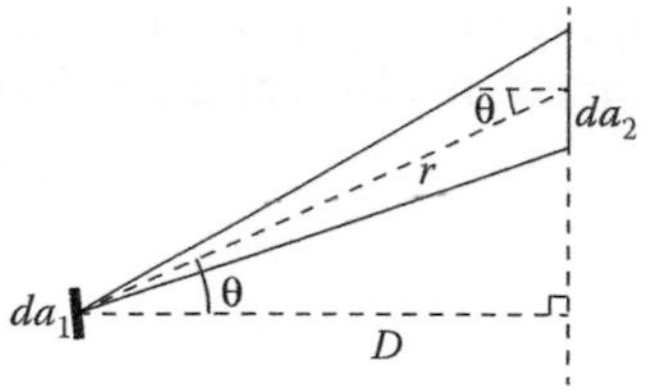

FIGURE 20.15
A 2-D system where an infinitesimal source da_1 illuminates a straight line.

Consider next the 2-D case in which an infinitesimal source da_1 illuminates a straight line, as shown in Figure 20.15. Expression (20.63) can be written in this case as

$$d\Phi = I d\theta \tag{20.67}$$

From Figure 20.15 we can see that

$$d\theta = \frac{da_2 \cos\theta}{r} \tag{20.68}$$

therefore,

$$d\Phi = I \frac{da_2 \cos^2\theta}{D} \tag{20.69}$$

and

$$\frac{E}{E_0} = \frac{I}{I_0} \cos^2\theta \tag{20.70}$$

from which it can be concluded that, in a 2-D system, for an infinitesimal source to produce a constant irradiance on a line, we must have

$$I = I_0/\cos^2\theta \tag{20.71}$$

As before, this expression does not depend on the distance D from the source to the line, so this angular distribution also enables a uniform radiation distribution produced on a line placed at an infinite distance. In this case, the dimension of the source can be taken as finite. It can then be concluded that, to have a finite-sized source producing a constant irradiance on a plane placed at an infinite distance, its angular distribution of radiation must fulfill Equation 20.71.

These arguments also apply to the case of photometric quantities, and the results obtained for irradiance are also valid for illuminance.

References

1. Henderson, S. T. and Marsden, A. M., *Lamps and Lighting: A Manual of Lamps and Lighting*, 2nd ed., Prepared by members of Thorn Lighting Ltd, Edward Arnold, London, 1972.
2. Klein, M. V. and Furtak, T. E., *Optics*, John Wiley & Sons, New York, 1986.
3. McCluney, W.R., *Introduction to Radiometry and Photometry*, Artech House, Boston, 1994.
4. Malacara, D., *Color Vision and Colorimetry, Theory and Applications*, SPIE Press, Bellingham, Washington, 2002.
5. Nicodemus, F. E., Radiance, *Am. J. Phys.*, 31, 368, 1963.
6. Spiro, I. J. and Thompson, B.J., *Selected Papers on Radiometry*, SPIE Milestone Series, Vol. MS 14, SPIE Optical Engineering Press, Bellingham, 1990.
7. Grum, F. and Becherer, R. J., *Optical Radiation Measurements, Volume I—Radiometry*, Academic Press, New York, 1979.
8. Wyatt, C. C., *Radiometric System Design*, Macmillan Publishing Company, New York, 1987.
9. Steel, W. H., Luminosity, Throughput or Étendue? Further Comments, *Appl. Opt.*, 14, 252, 1975.
10. Boyd, R. W., *Radiometry and the Detection of Optical Radiation*, John Wiley & Sons, New York, 1983.
11. Sparrow, E. M. and Cess, R. D., *Radiation Heat Transfer—Augmented Edition*, Hemisphere Publishing Corporation, Washington, London; McGraw-Hill Book Company, New York, 1978.
12. Siegel, R. and Howell, J. R., *Thermal Radiation Heat Transfer*, McGraw-Hill Book Company, New York, 1972.
13. Meyer-Arendt, J. R., *Introduction to Classical and Modern Optics*, 3rd. ed., Prentice-Hall, Englewood Cliffs, New Jersey, 1989.
14. Begunov, B. N. and Zakaznov, N. P., *Teoria de sistemas opticos*, Editorial Mir, Moscu, 1976. (Spanish translation of the book in Russian *Theory of Optical Systems*.)

21

Plane Curves

21.1 General Considerations

This chapter presents some plane curves[1] that are useful in designing non-imaging optics.

The magnitude of a vector **v** is given by

$$\| \mathbf{v} \| = \sqrt{\mathbf{v} \cdot \mathbf{v}} \tag{21.1}$$

A normalized vector with the same direction as **v**, but with unit magnitude, can be obtained from

$$\text{nrm}\,\mathbf{v} = \frac{\mathbf{v}}{\| \mathbf{v} \|} = \frac{\mathbf{v}}{\sqrt{\mathbf{v} \cdot \mathbf{v}}} \tag{21.2}$$

The distance **[A,B]** between two points **A** and **B** is given by the magnitude of the vector **B** – **A**, that is, ||**B** – **A**|| or

$$[\mathbf{A}, \mathbf{B}] = \sqrt{(\mathbf{B} - \mathbf{A}) \cdot (\mathbf{B} - \mathbf{A})} \tag{21.3}$$

The angle between two vectors **u** and **v** is given by

$$\text{ang}(\mathbf{v}, \mathbf{u}) = \theta = \arccos\left(\frac{\mathbf{v} \cdot \mathbf{u}}{\| \mathbf{v} \| \| \mathbf{u} \|} \right) = \arccos\left(\frac{\mathbf{v} \cdot \mathbf{u}}{\sqrt{\mathbf{v} \cdot \mathbf{v}}\sqrt{\mathbf{u} \cdot \mathbf{u}}} \right) \tag{21.4}$$

This angle, however, is $0 \le \theta \le \pi$. Consider then that vectors **u** and **v** are 2-D and that $\mathbf{u} = (u_1, u_2)$ and $\mathbf{v} = (v_1, v_2)$. We can define the vectors $\mathbf{U} = (u_1, u_2, 0)$ and $\mathbf{V} = (v_1, v_2, 0)$ in three dimensions. The cross product of **U** and **V** is

$$\mathbf{U} \times \mathbf{V} = (0, 0, u_1 v_2 - u_2 v_1) \tag{21.5}$$

If the third component of **U** × **V** is positive, then **v** is in the counterclockwise direction from **u**. Also, if the third component of **U** × **V** is negative, then

v is in the clockwise direction from **u**. We now define the angle in the positive direction from **u** to **v** as

$$\begin{aligned} &\mathrm{angp}(\mathbf{v},\mathbf{u}) = \mathrm{ang}(\mathbf{v},\mathbf{u}) && \text{if } u_1 v_2 - u_2 v_1 \geq 0 \\ &\mathrm{angp}(\mathbf{v},\mathbf{u}) = 2\pi - \mathrm{ang}(\mathbf{v},\mathbf{u}) && \text{if } u_1 v_2 - u_2 v_1 < 0 \end{aligned} \tag{21.6}$$

This is the angle that **v** makes relative to **u** in the positive direction and in the range $0 \leq \varphi < 2\pi$.

In the particular case where $\mathbf{u} = (1, 0)$ then angp(**v**,**u**) is the angle that vector $\mathbf{v} = (v_1, v_2)$ makes with axis x_1 and we define

$$\begin{aligned} &\mathrm{angh}\,\mathbf{v} = \arccos\left(\frac{v_1}{\sqrt{\mathbf{v}\cdot\mathbf{v}}}\right) && \text{if } v_2 \geq 0 \\ &\mathrm{angh}\,\mathbf{v} = 2\pi - \arccos\left(\frac{v_1}{\sqrt{\mathbf{v}\cdot\mathbf{v}}}\right) && \text{if } v_2 < 0 \end{aligned} \tag{21.7}$$

This is the angle vector $\mathbf{v} = (v_1, v_2)$ makes with the horizontal axis x_1. These functions are represented in Figure 21.1, where φ = angp(**v**,**u**) and β = angh(**v**)

Now consider the rotation of a point around the origin. A point $\mathbf{P} = (P_1, P_2)$ can be rotated by an angle α around the origin by applying a rotation matrix to it

$$R(\alpha) = \begin{pmatrix} \cos\alpha & -\sin\alpha \\ \sin\alpha & \cos\alpha \end{pmatrix} \tag{21.8}$$

thus, the rotated point is given by

$$R(\alpha)\cdot\mathbf{P} = \begin{pmatrix} \cos\alpha & -\sin\alpha \\ \sin\alpha & \cos\alpha \end{pmatrix}\cdot\begin{pmatrix} P_1 \\ P_2 \end{pmatrix} \tag{21.9}$$

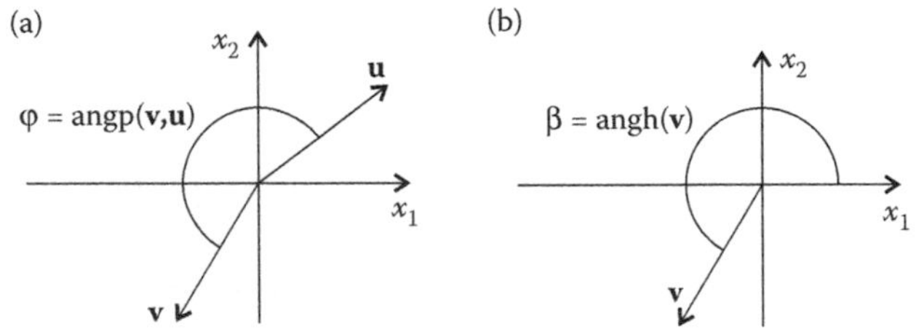

FIGURE 21.1
(a) Function angp(**v**,**u**) takes two vectors as parameters and gives the angle that the first makes relative to the second. (b) In the particular case where the second vector is in the x_1 direction, this function gives the angle of a vector to the x_1 axis.

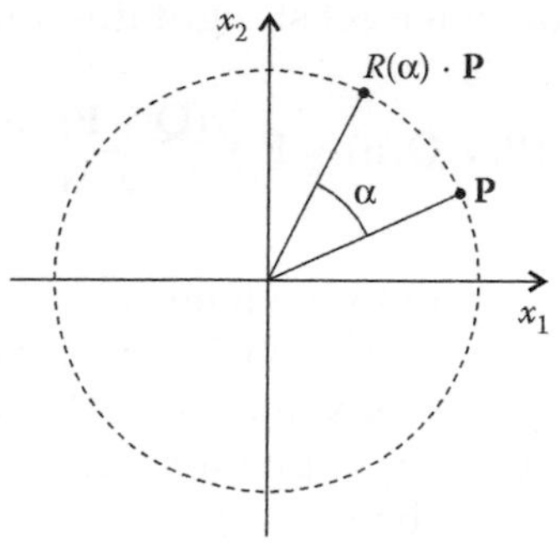

FIGURE 21.2
A rotation of a point **P** by an angle α around the origin accomplished by multiplying it on the left by a rotation matrix $R(\alpha)$.

The effect of the rotation matrix is represented in Figure 21.2.

The intersection point between a straight line and a plane can be obtained from the definitions of a straight line and a plane in terms of points and vectors. A point **P** and a vector **v**, as shown in Figure 21.3a, define a straight line. Accordingly, another point **Q** and a normal vector **n** define a plane μ.

The intersection point **X** fulfills the following

$$\begin{aligned} \mathbf{n} \cdot (\mathbf{X} - \mathbf{Q}) &= 0 \\ \mathbf{X} &= \mathbf{P} + d\mathbf{v} \end{aligned} \tag{21.10}$$

Replacing **X** in the first expression of Equation 21.10, solving for d, and introducing the result into the second expression gives

$$\mathbf{X} = \mathbf{P} + \frac{(\mathbf{Q} - \mathbf{P}) \cdot \mathbf{n}}{\mathbf{v} \cdot \mathbf{n}} \mathbf{v} \tag{21.11}$$

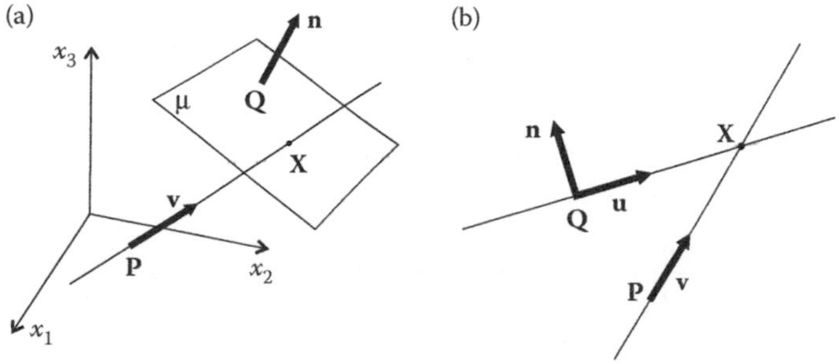

FIGURE 21.3
(a) Intersection between a straight line defined by point **P** and vector **v** and a plane μ defined by a point **Q** and normal **n**. (b) Intersection between two coplanar straight lines.

We can define a function (intersect straight line and plane) as follows

$$\text{islp}(\mathbf{P},\mathbf{v},\mathbf{Q},\mathbf{n}) = \mathbf{P} + \frac{(\mathbf{Q}-\mathbf{P})\cdot\mathbf{n}}{\mathbf{v}\cdot\mathbf{n}}\mathbf{v} \tag{21.12}$$

We may also apply this result to the intersection of two straight lines on the same plane, as shown in Figure 21.3b. The two coplanar straight lines are now defined by point **P** and vector **v** and point **Q** and vector **u**, respectively. The intersection is again given by expression (21.11) where **n** is a vector perpendicular to **u**. If $\mathbf{u} = (u_1, u_2)$, then

$$\mathbf{n} = R(\pi/2)\cdot\mathbf{u} = (-u_2, u_1) \tag{21.13}$$

We can also define a function (intersect straight lines) as follows:

$$\text{isl}(\mathbf{P},\mathbf{v},\mathbf{Q},\mathbf{u}) = \mathbf{P} + \frac{(\mathbf{Q}-\mathbf{P})\cdot\mathbf{n}}{\mathbf{v}\cdot\mathbf{n}}\mathbf{v} \tag{21.14}$$

Now consider a further situation, wherein a circle of radius r is centered at point **F** and a point **P** is exterior to the circle, that is, the distance from **F** to **P** is greater than r, with [F,P] > r, as presented in Figure 21.4. Also consider point **T*** on the tangent to the circle that contains point **P**. Direction **T*P** is in the counterclockwise direction from **FP**.

Now calculate the distance t_P from **T*** to **P** and angle ϕ_P that line **PT*** makes to a vector **u** tilted by an angle α to the horizontal.

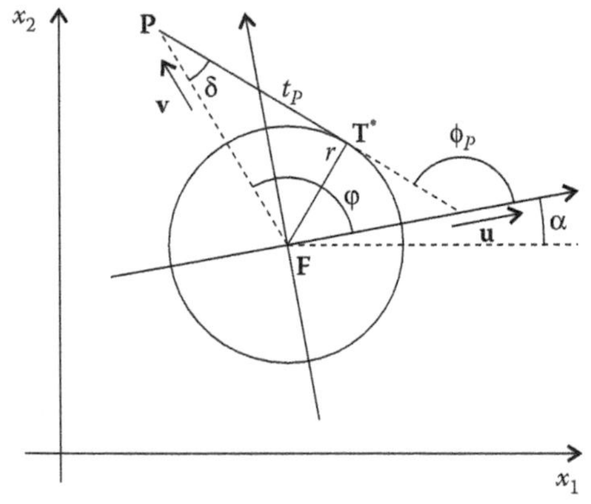

FIGURE 21.4
A circle with center **F** and radius r and an exterior point **P**. Distance t_P from **P** to the tangent point **T*** to the circle and the angle ϕ_P that **T*P** makes to a direction tilted by an angle α to the horizontal. Direction **T*P** is in counterclockwise from **FP**.

Distance t_P from $\mathbf{T}^*$ to $\mathbf{P}$ is given by

$$t_P = \sqrt{(\mathbf{P}-\mathbf{F})\cdot(\mathbf{P}-\mathbf{F}) - r^2} \tag{21.15}$$

Angle δ is given by

$$\delta = \arcsin\left(\frac{r}{\sqrt{(\mathbf{P}-\mathbf{F})\cdot(\mathbf{P}-\mathbf{F})}}\right) \tag{21.16}$$

Vector $\mathbf{u}$ is given by $\mathbf{u} = (\cos\alpha, \sin\alpha)$ and angle φ is given by

$$\varphi = \text{angp}(\mathbf{P}-\mathbf{F}, \mathbf{u}) \tag{21.17}$$

Angle ϕ_P can now be obtained from

$$\phi_P = \varphi + \delta \tag{21.18}$$

Now consider a similar situation, but point **T** is on the other side of the circle, as shown in Figure 21.5, that is, now **TP** is in the clockwise direction from **FP**.

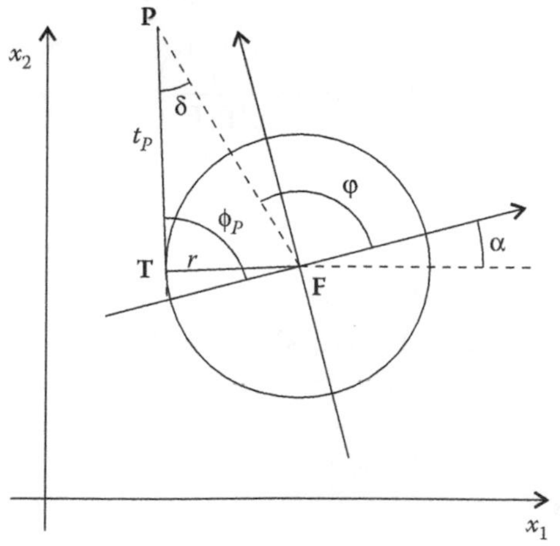

FIGURE 21.5
A circle with center **F** and radius r and an exterior point **P**. Distance t_P from **P** to the tangent point $\mathbf{T}^*$ to the circle and the angle ϕ_P that **TP** makes to a direction tilted by an angle α to the horizontal. Direction **TP** is in the clockwise direction from **FP**.

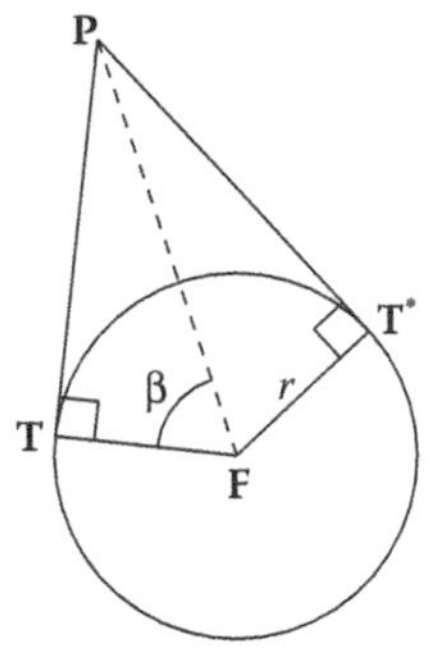

FIGURE 21.6
Circle with center **F** and radius r and an exterior point **P**. Tangent points **T** and **T*** for the lines that pass through **P** and are tangent to the circle.

The value for t_P is still given by expression (21.15). Also, angles φ and δ are still given by expressions (21.16) and (21.17) respectively. We can now obtain angle ϕ_P from φ and δ. To ensure that $0 \le \varphi_P < 2\pi$, we have $0 \le \phi_P < 2\pi$

$$\begin{aligned} \phi_P &= \varphi - \delta & \text{if } \varphi - \delta \ge 0 \\ \phi_P &= 2\pi + \varphi - \delta & \text{if } \varphi - \delta \ge 0 \end{aligned} \tag{21.19}$$

We now calculate the positions of points **T*** and **T** from the positions of point **P** and center **F** and radius r of the circle, as shown in Figure 21.6.

Angle β can be obtained from $r = [\mathbf{F},\mathbf{P}] \cos \beta$ as

$$\beta = \arccos\left(\frac{r}{[\mathbf{P},\mathbf{F}]}\right) \tag{21.20}$$

and

$$\begin{aligned} \mathbf{T} &= \mathbf{F} + rR(\beta) \cdot \text{nrm}(\mathbf{P} - \mathbf{F}) \\ \mathbf{T}^* &= \mathbf{F} + rR(-\beta) \cdot \text{nrm}(\mathbf{P} - \mathbf{F}) \end{aligned} \tag{21.21}$$

21.2 Parabola

For a parabola, we have $t + s = K$ where K is a constant and t and s are defined as shown in Figure 21.7. But $s = -t \cos \phi$ and therefore $t - t \cos \phi = K$ or

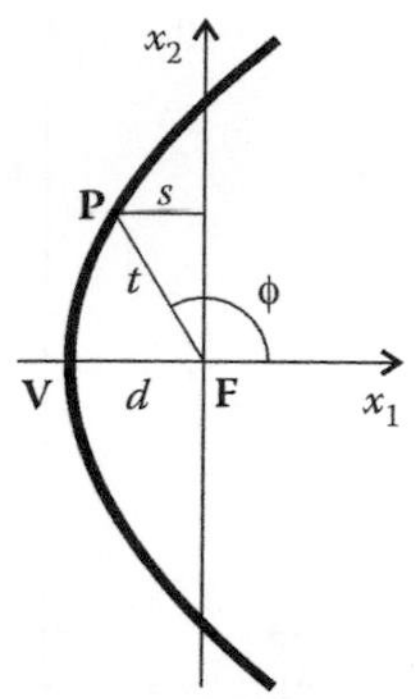

FIGURE 21.7
In a parabola with a horizontal axis, horizontal incoming rays are concentrated at the focus. This curve also fulfills $s + t = K$, where K is a constant.

$$t(\phi) = \frac{K}{1 - \cos\phi} \tag{21.22}$$

To find the value of K, we note that, when $\phi = \pi$ we have $t = d$, where d is the distance between the focus **F** of the parabola and its vertex **V**. Then $d - d\cos\pi = K$ or $K = 2d$.

The equation for a parabola in polar coordinates is then

$$t(\phi)(\cos\phi, \sin\phi) \tag{21.23}$$

with

$$t(\phi) = \frac{2d}{1 - \cos\phi} \tag{21.24}$$

A parabola rotated by an angle α has the equation

$$t(\phi)(\cos(\phi + \alpha), \sin(\phi + \alpha)) \tag{21.25}$$

A parabola rotated by an angle α and with focus at a point $\mathbf{F} = (F_1, F_2)$ has the equation

$$t(\phi)(\cos(\phi + \alpha), \sin(\phi + \alpha)) + (F_1, F_2) \tag{21.26}$$

Now consider that we want to determine the equation of a parabola having a given focus **F**, tilted by a given angle α to the horizontal, and passing

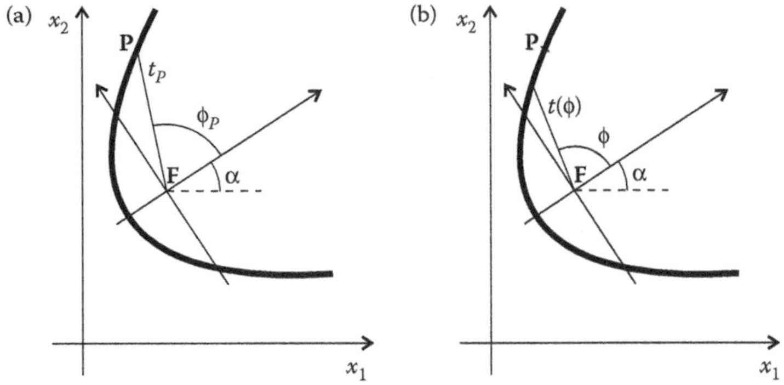

FIGURE 21.8
(a) A parabola can be completely determined by the position of its focus **F**, the angle α that its axis makes to the horizontal and a point **P**. (b) The parabola can be parameterized as a function of angle ϕ to its axis.

through a given point **P**, as shown in Figure 21.8. In this case, for $\phi = \phi_P$ we have $t = t_P = [\mathbf{F},\mathbf{P}]$ and, therefore, the constant K in the equation for the parabola is given by $[\mathbf{F},\mathbf{P}] - [\mathbf{F},\mathbf{P}] \cos \phi_P = K$. The equation for $t(\phi)$ for this parabola is then

$$\frac{\sqrt{(\mathbf{P}-\mathbf{F})\cdot(\mathbf{P}-\mathbf{F})} - (\mathbf{P}-\mathbf{F})\cdot(\cos\alpha, \sin\alpha)}{1-\cos\phi}(\cos(\phi+\alpha), \sin(\phi+\alpha)) + (F_1, F_2) \tag{21.27}$$

Note that instead of giving angle α to define the direction of the parabola's axis, we may alternatively give two points: focus **F** and another point **G** on the axis in the direction where the parabola opens, as shown in Figure 21.9.

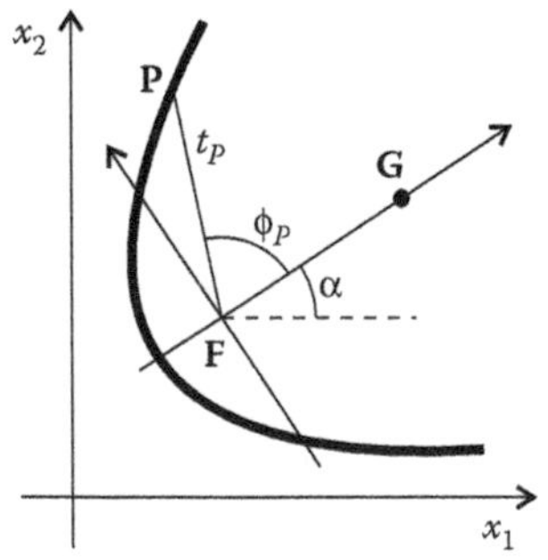

FIGURE 21.9
The direction of the axis of a parabola can be given by its focus **F** and another point **G** on the axis.

In this case, the angle α can be calculated by

$$\alpha = \text{angh}\,\mathbf{v} \quad \text{with } \mathbf{v} = (v_1, v_2) = \mathbf{G} - \mathbf{F} \tag{21.28}$$

with angh as defined in Equation 21.7. The parabola can then be obtained by expression (21.27). We can also obtain d by $d = K/2$ as

$$d = \frac{t_P - t_P \cos\phi_P}{2} = \frac{\sqrt{(\mathbf{P}-\mathbf{F})\cdot(\mathbf{P}-\mathbf{F})} - (\mathbf{P}-\mathbf{F})\cdot(\cos\alpha, \sin\alpha)}{2} \tag{21.29}$$

21.3 Ellipse

Let us now consider an ellipse, as in Figure 21.10, with foci $\mathbf{F} = (0,0)$ and $\mathbf{G} = (f, 0)$.

Point $\mathbf{P}$ has coordinates $\mathbf{P} = (P_1, P_2) = t(\cos\phi, \sin\phi)$ where t is the distance from $\mathbf{F}$ to $\mathbf{P}$. The distance s from point $\mathbf{P}$ to $\mathbf{G}$ is given by

$$s = \sqrt{(\mathbf{G}-\mathbf{P})\cdot(\mathbf{G}-\mathbf{P})} = \sqrt{f^2 + t^2 - 2ft\cos\phi} \tag{21.30}$$

Ellipses fulfill $t + s = K$, where K is a constant, that is, the distance between the vertices of the ellipse, as shown in Figure 21.10. This can also be written as

$$s^2 = (K-t)^2 \Leftrightarrow f^2 + t^2 - 2ft\cos\phi = (K-t)^2 \tag{21.31}$$

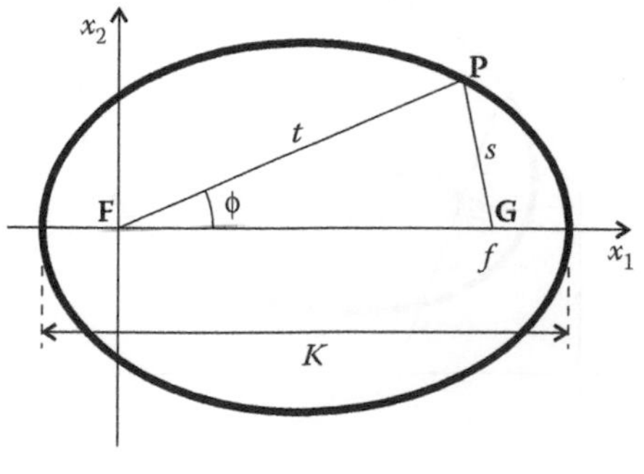

FIGURE 21.10
An ellipse reflects all the light emitted by a point source $\mathbf{F}$ to a point $\mathbf{G}$. Points $\mathbf{F}$ and $\mathbf{G}$ are the foci of the ellipse. This curve fulfills $t + s = K$, where K is a constant.

or

$$t(\phi) = \frac{K^2 - f^2}{2K - 2f\cos\phi} \tag{21.32}$$

and the ellipse is given by

$$\frac{K^2 - f^2}{2K - 2f\cos\phi}(\cos\phi, \sin\phi) \tag{21.33}$$

with $0 \leq \phi < 2\pi$ and $K > f$.

We can now write the equation for the general case of an ellipse, given foci **F** and **G**, that passes through a given point **P**, as presented in Figure 21.11.

From the positions of **F**, **G**, and **P**, we have

$$\begin{aligned} K &= t_P + s = [\mathbf{F},\mathbf{P}] + [\mathbf{P},\mathbf{G}] \\ f &= [\mathbf{F},\mathbf{G}] \\ \alpha &= \text{angh}\,\mathbf{v} \quad \text{with } \mathbf{v} = (v_1, v_2) = \mathbf{G} - \mathbf{F} \end{aligned} \tag{21.34}$$

where angh is defined in Equations 21.7. The ellipse is then given by

$$\frac{K^2 - f^2}{2K - 2f\cos\phi}(\cos(\phi + \alpha), \sin(\phi + \alpha)) + \mathbf{F} \tag{21.35}$$

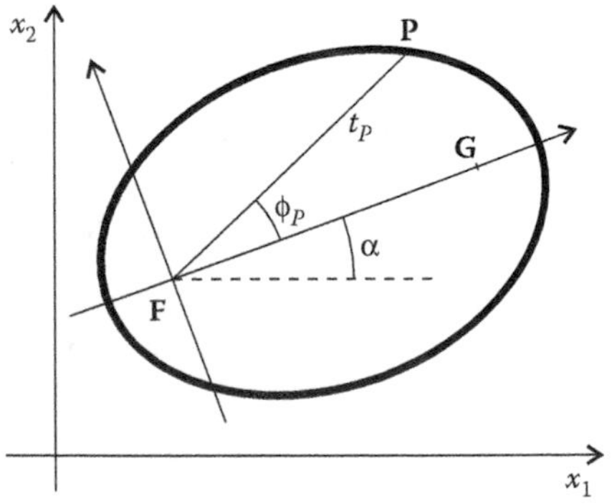

FIGURE 21.11
A general ellipse can be defined by the position of its two foci **F** and **G** and a point **P**.

21.4 Hyperbola

Let us now consider a hyperbola, as in Figure 21.12, with foci $\mathbf{F} = (0, 0)$ and $\mathbf{G} = (f, 0)$.

It is also described by an expression of the form $t(\phi) = (\cos\phi, \sin\phi)$. For angle ϕ corresponding to point **P** on the right-hand-side hyperbola, we have $t > 0$ and the points on this curve are defined by the condition $t - s = K$, where K is a constant. For angle ϕ^* corresponding to point **Q** on the left-hand-side hyperbola, we have $t < 0$ and the points on this curve are defined by the condition $s - |t| = K$ or $s + t = K$, where K is the constant same as that for the right-hand-side curve. In both the cases, we can write

$$s^2 = (K - t)^2 \tag{21.36}$$

where K is the distance between the two vertices of the hyperbola, as shown in Figure 21.12. The hyperbola is then given by the equation same as that of the ellipse:

$$\frac{K^2 - f^2}{2K - 2f\cos\phi}(\cos\phi, \sin\phi) \tag{21.37}$$

with $0 \leq \phi < 2\pi$ and $K < f$.

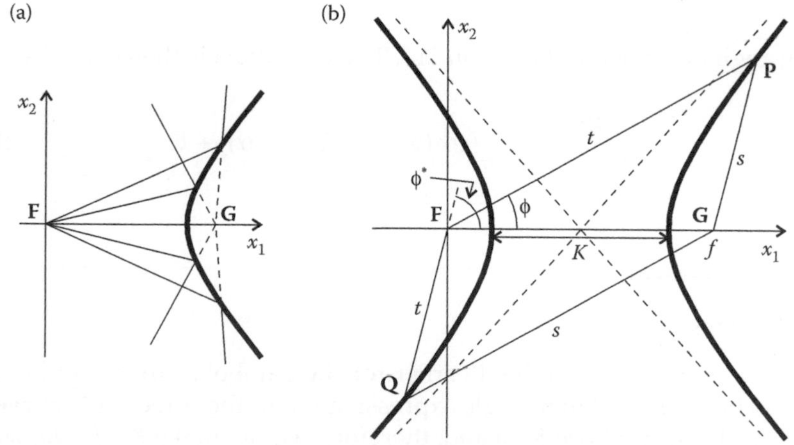

FIGURE 21.12
(a) If **F** is a point source, the hyperbola reflects its light as if it is diverging from another point **G**. Similarly, light directed toward **G** is reflected to **F**. Points **F** and **G** are its foci. The hyperbola has two branches with similar optical characteristics. (b) For points **P** on the right branch, we have $t - s = K$ (with $t > 0$) and for the left one $s + t = K$ (with $t < 0$), where K is a constant.

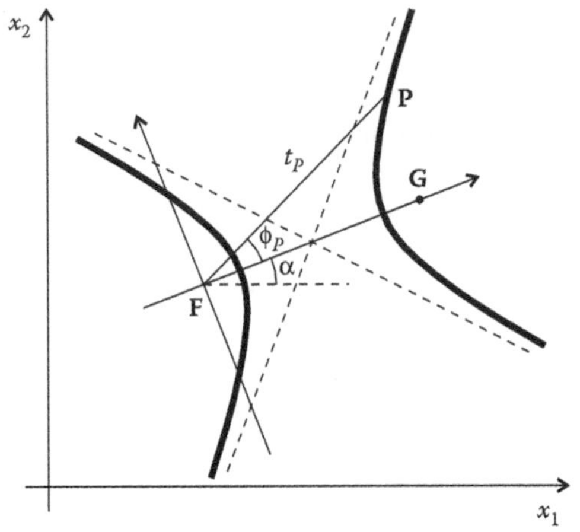

FIGURE 21.13
A general hyperbola can be defined by the position of its two foci **F** and **G** and a point **P**.

We can now write the equation for the general case of a hyperbola, given foci **F** and **G**, that passes through a given point **P**, as presented in Figure 21.13.

From the positions of **F**, **G**, and **P**, we have

$$\begin{aligned} K &= \|t - s| = |[\mathbf{F},\mathbf{P}] - [\mathbf{P},\mathbf{G}]| \\ f &= [\mathbf{F},\mathbf{G}] \\ \alpha &= \text{angh}\,\mathbf{v} \quad \text{with } \mathbf{v} = (v_1, v_2) = \mathbf{G} - \mathbf{F} \end{aligned} \tag{21.38}$$

where angh is defined in Equation 21.7. The hyperbola is then given by

$$\frac{K^2 - f^2}{2K - 2f\cos\phi}(\cos(\phi+\alpha), \sin(\phi+\alpha)) + \mathbf{F} \tag{21.39}$$

21.5 Conics

The three expressions obtained earlier for the parabola, ellipse, and hyperbola, can be combined in a single expression for all the three conic curves.

For the ellipse, we have $K > f$ and, therefore, we can make $K - f = 2d$, where $d > 0$. We then have $K = 2d + f$ and we can replace this in the expression for $t(\phi)$ for the ellipse

$$t(\phi) = \frac{K^2 - f^2}{2K - 2f\cos\phi} = \frac{2d^2 + 2fd}{f + 2d - f\cos\phi} \tag{21.40}$$

and, making $g = 1/f$, the ellipse is described by

$$\frac{2d^2g + 2d}{1 + 2dg - \cos\phi}(\cos\phi, \sin\phi) \tag{21.41}$$

Now consider a hyperbola rotated by an angle π. Its parametric representation is

$$\frac{K^2 - f^2}{2K - 2f\cos\phi}(\cos(\phi + \pi), \sin(\phi + \pi)) = \frac{K^2 - f^2}{-2K + 2f\cos\phi}(\cos\phi, \sin\phi) \tag{21.42}$$

For the hyperbola, we have $K < f$ and, therefore, we can make $f - K = 2d$ where $d > 0$. Then making $K = f - 2d$ and $g = 1/f$, we can write for the hyperbola rotated by an angle π as

$$\frac{-2d^2g + 2d}{1 - 2dg - \cos\phi}(\cos\phi, \sin\phi) \tag{21.43}$$

When $f \to \infty$, $g \to 0$ and both expressions, for the ellipse and the hyperbola rotated by an angle π, converge to the same parabola. We can then write as

$$\frac{2d^2\delta g + 2d}{1 + 2d\delta g - \cos\phi}(\cos\phi, \sin\phi) \tag{21.44}$$

If $\delta = 1$ we have an ellipse; if $\delta = 0$ we have a parabola; and if $\delta = -1$ we have a hyperbola. Figure 21.14 shows the three curves for the same value of d.

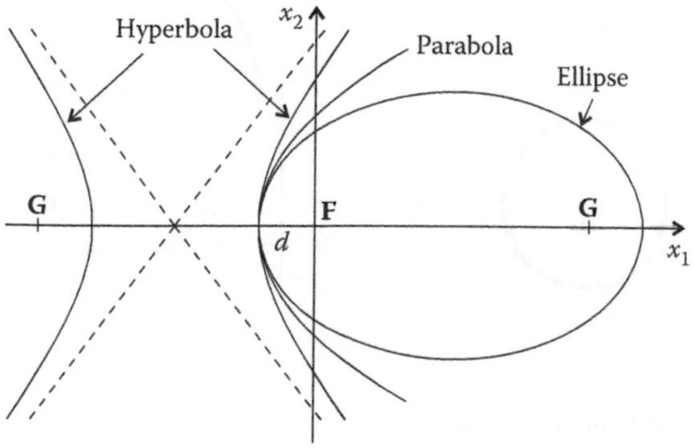

FIGURE 21.14
An ellipse has two foci **F** and **G**. As focus **G** tends to $+\infty$ the ellipse becomes a parabola. As the focus **G** now approaches **F**, but from $-\infty$, the parabola becomes a hyperbola.

21.6 Involute

Figure 21.15 shows an involute to a circle. It reflects rays tangent to the circle back to the circle again, still tangent (as indicated by the double arrow in the line connecting **T** and **P**).

This curve can be obtained by a string of constant length attached at point **A**, wrapping around the circle up to tangent point **T**, and then going straight to point **P** on the curve. The total length of the string **A-T-P** is given by $t + r\gamma$, which is a constant value K^* for the points on the curve. Since $\phi = \gamma + \pi/2$ we have

$$t + r\gamma = K^* \Leftrightarrow t(\phi) = K^* - r\gamma = K^* - r(\phi - \pi/2) = K - r\phi \quad (21.45)$$

where K is a constant. The value of K may be obtained if a point **P** on the curve is given, as shown in Figure 21.15b. For this point, we have

$$K = t_P + r\phi_P \quad (21.46)$$

The values of t_P and ϕ_P can be obtained from expressions (21.15) and (21.18). The curve is, therefore, given by

$$r(\cos(\phi - \pi/2), \sin(\phi - \pi/2)) + t(\phi)(\cos\phi, \sin\phi) \quad (21.47)$$

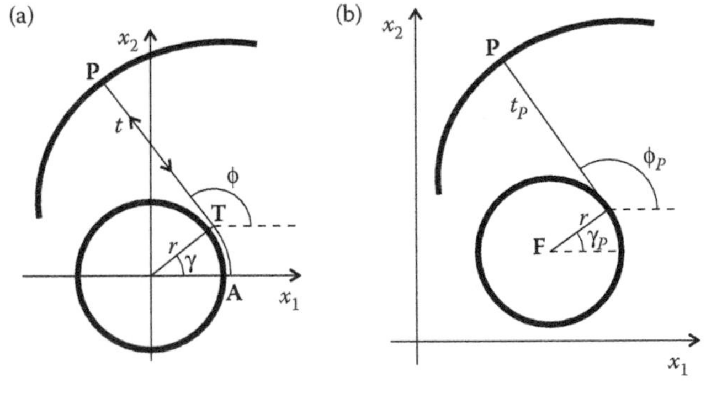

FIGURE 21.15
(a) An involute can be defined by a string of constant length **A–T–P** and, as angle γ increases, t decreases. Also, the string winds around the circle and, therefore, this curve is called a winding involute. (b) An involute can be defined by the center **F** of the circle, its radius r, and a point **P** on the curve.

where $t(\phi)$ is given by Equation 21.45, with K given by Equation 21.46. If the center of the circle is at a point $\mathbf{F} = (F_1, F_2)$ on the plane, the curve is given by

$$r(\sin\phi, -\cos\phi) + t(\phi)(\cos\phi, \sin\phi) + (F_1, F_2) \tag{21.48}$$

As angle ϕ increases, the string winds around the circle, thus this curve is called the winding involute through point **P**.

Another possibility is as angle ϕ increases, the string unwinds, as shown in Figure 21.16, and the curve is now called the unwinding involute through point **P**.

In this case, the string **A-T-P** has a constant length, given by $r\beta + t = K^*$, and is constant for all points **P** on the curve. Angle β is given by $\beta = 2\pi - (\phi + \pi/2) = 3\pi/2 - \phi$, giving

$$t(\phi) = K^* - r(3\pi/2 - \phi) = K + r\phi \tag{21.49}$$

If the center of the circle is at a point **F**, the curve is given by

$$r(-\sin\phi, \cos\phi) + t(\phi)(\cos\phi, \sin\phi) + \mathbf{F} \tag{21.50}$$

where $t(\phi)$ is given by Equation 21.49, and K is obtained from the position of a point **P** on the curve as

$$K = t_P - r\phi_P \tag{21.51}$$

where t_P and ϕ_P are given by expressions (21.15) and (21.19), respectively.

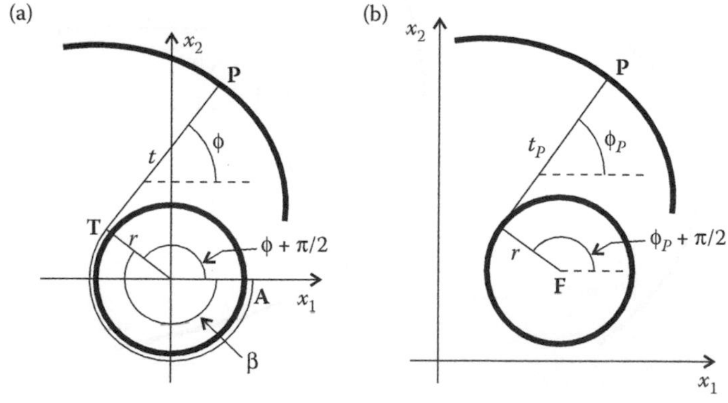

FIGURE 21.16
(a) As angle ϕ increases, the string unwinds around the circle and the curve is called an unwinding involute. (b) An involute can be defined by the center **F** of the circle, its radius r, and a point **P** on the curve.

Given a circle and an external point **P**, there are, therefore, two possible involutes passing through point **P** (winding and unwinding). The concepts of winding and unwinding will also be used in the definition of the macrofocal parabolas and ellipses in the following sections.

21.7 Winding Macrofocal Parabola

In a parabola, the sum of the distance t between its focus **F** and a point **P** and the distance s between the point and a line v_L perpendicular to the parabola's axis x_1 is a constant, that is, $t+s=K$ where K is a constant and t and s are defined as shown in Figure 21.17a. The points **P** on the parabola can then be generated by a string of constant length with one tip at the focus **F** and the other at line v_L in such a way that **PQ** is perpendicular to v_L.

In a macrofocal parabola,[2] the focus is replaced by a circle of radius r, as shown in Figure 21.17b. In this case, the string wraps around the macrofocus in such a way that one of its tips is at point **A** and the other tip at a point **Q** on line v_L. The length of the string in this case is, $r(\phi-\pi/2)+t+s$ and it is constant for the points **P** on the curve. As angle ϕ increases, the string winds around the macrofocus and, therefore, this curve is called a winding macrofocal parabola.

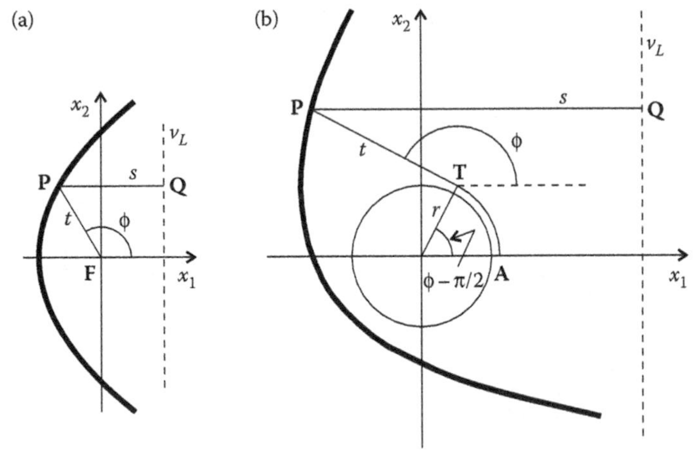

FIGURE 21.17
In the same way a parabola reflects parallel rays to a point (a), a macrofocal parabola reflects parallel rays tangent to a circular macrofocus of radius r (b). Just as with the parabola, the macrofocal parabola can be defined by a string of constant length **Q-P-T-A**. In this case, as angle ϕ increases, the string winds around the macrofocus, so this is called a winding macrofocal parabola.

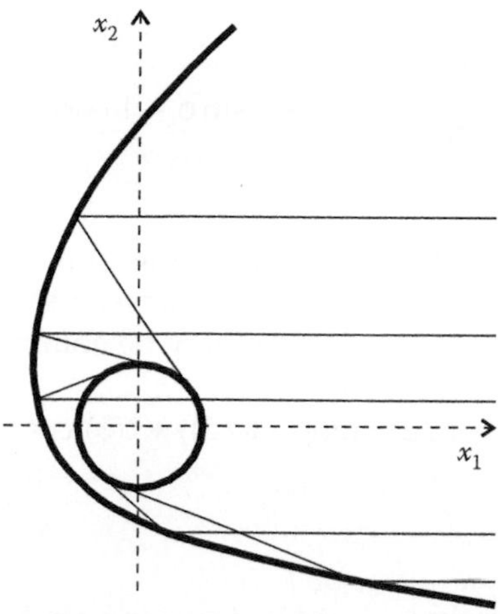

FIGURE 21.18
A macrofocal parabola reflects parallel rays to become tangent to a circular macrofocus. These parallel rays are also parallel to its axis, which, in this case, is the horizontal axis x_1.

If used as a mirror, this curve reflects parallel horizontal rays so they become tangent to a circular receiver, as shown in Figure 21.18.

Applying constant string length to this curve, we have the geometry presented in Figure 21.19, where M is the mirror.

In this case, we have $s_2 + s_1 + t + r(\phi - \pi/2) = K$, where K is a constant. But $s_1 = -t \cos \phi$ and $s_2 = r - r \cos(\phi - \pi/2)$. We can then write

$$r - r\cos(\phi - \pi/2) - t\cos\phi + t + r\phi = K_W \tag{21.52}$$

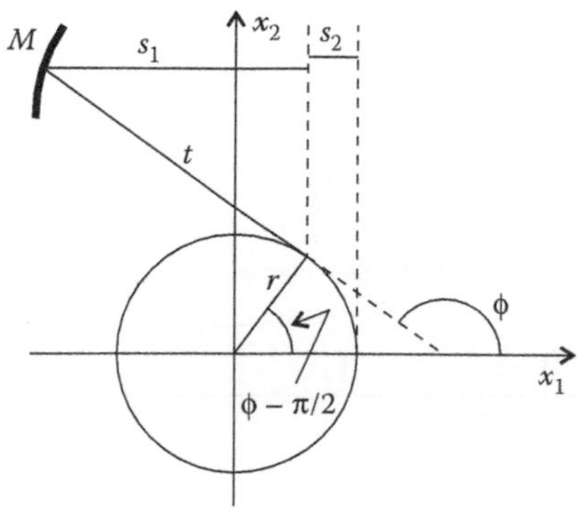

FIGURE 21.19
The points on the macrofocal parabola fulfill the condition $s_2 + s_1 + t + r(\phi - \pi/2) = K$, where K is a constant.

or

$$t(\phi) = \frac{K_W + r(\sin\phi - 1 - \phi)}{1 - \cos\phi} \qquad (21.53)$$

Note that for $r = 0$ we obtain an equation for a parabola. Constant K_W can now be determined from a point on the curve, just as with the case of the parabola. The macrofocal parabola can now be obtained from

$$r(\cos(\phi - \pi/2), \sin(\phi - \pi/2)) + t(\phi)(\cos\phi, \sin\phi) \qquad (21.54)$$

or

$$r(\sin\phi, -\cos\phi) + t(\phi)(\cos\phi, \sin\phi) \qquad (21.55)$$

A macrofocal parabola rotated by an angle α around the origin and the center of the macrofocus at a position $\mathbf{F} = (F_1, F_2)$ is given by

$$r(\sin(\phi + \alpha), -\cos(\phi + \alpha)) + t(\phi)(\cos(\phi + \alpha), \sin(\phi + \alpha)) + (F_1, F_2) \qquad (21.56)$$

This situation is presented in Figure 21.20 for the case in which the macrofocal parabola passes through a given point **P**.

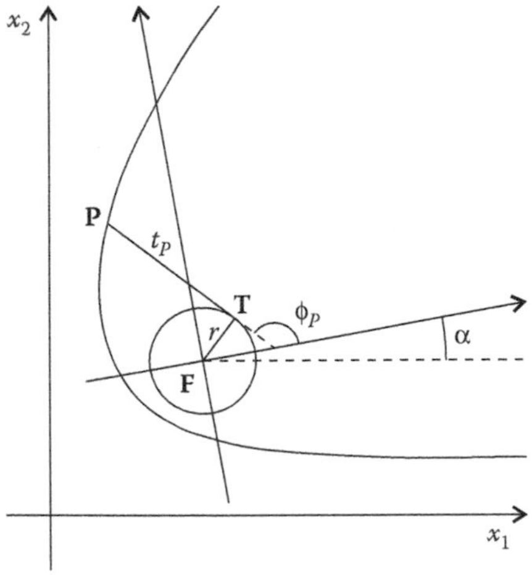

FIGURE 21.20
A general macrofocal parabola can be defined by the center **F** of its macrofocus and its radius, the angle that its axis makes to the horizontal, and a point **P** on the curve.

Constant K_W can now be obtained by solving Equation 21.53 with respect to K_W, noting that for point **P**, we have $\phi = \phi_P$ and $t(\phi) = t_P$. Given the distance t_P and angle ϕ_P, K_W is given by

$$K_W = t_P(1 - \cos\phi_P) + r(1 + \phi_P - \sin\phi_P) \tag{21.57}$$

The values of t_P and ϕ_P can be calculated from the position of point **P** by expressions (21.15) and (21.18).

21.8 Unwinding Macrofocal Parabola

Figure 21.21 shows another example of a macrofocal parabola. This curve can also be generated by a string of constant length. In this case, the string wraps around the macrofocus in such a way that one of its tips is at point **A** and the other tip at a point **Q** on line v_L. In this case, however, the string starts at point **A** and goes under the macrofocus instead of going over it, as in the case of the winding macrofocal parabola. The length of the string is, in this case, $r\beta + t + s$ and it is constant for the points **P** on the curve. As angle ϕ increases, the string unwinds around the macrofocus and, therefore, this curve is called an unwinding macrofocal parabola. Angle β is given by $\beta = 2\pi - (\phi + \pi/2) = 3\pi/2 - \phi$.

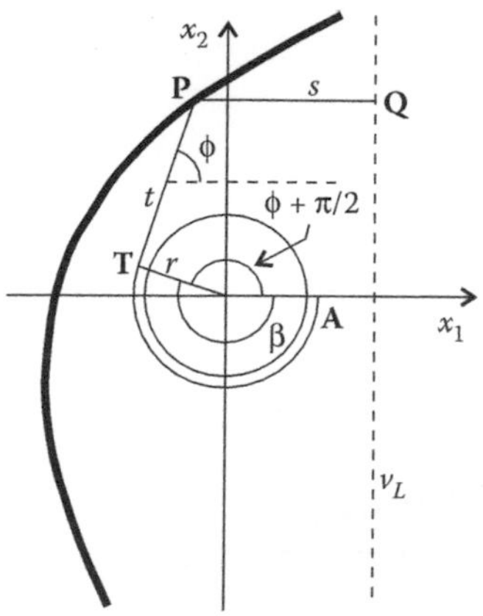

FIGURE 21.21
Macrofocal parabola generated by a string **Q-P-T-A** of constant length. As angle ϕ increases the string unwinds around the macrofocus, so that this curve is called a unwinding macrofocal parabola.

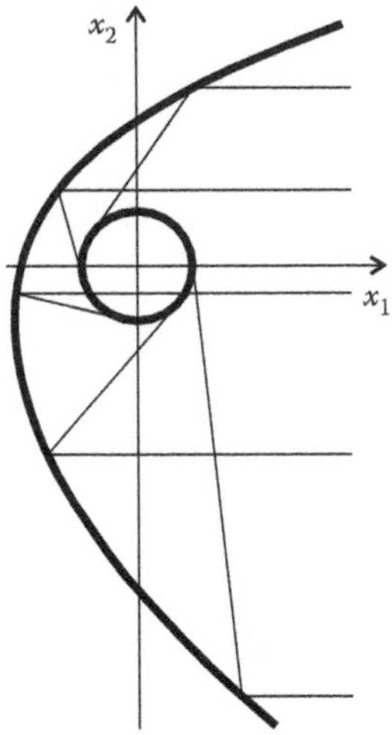

FIGURE 21.22
A macrofocal parabola reflects parallel rays to become tangent to the circular macrofocus. These parallel rays are also parallel to its axis that, in this case, is horizontal axis x_1.

If used as a mirror, this curve reflects parallel horizontal rays to become tangent to a circular receiver of radius, as shown in Figure 21.22.

This parabola is symmetrical with respect to the x_1 axis relative to a winding macrofocal parabola, as shown in Figure 21.23.

Point **P** on the winding macrofocal parabola is symmetrical to point **Q** on the unwinding macrofocal parabola. For these two points, we have $t_P = t_Q$

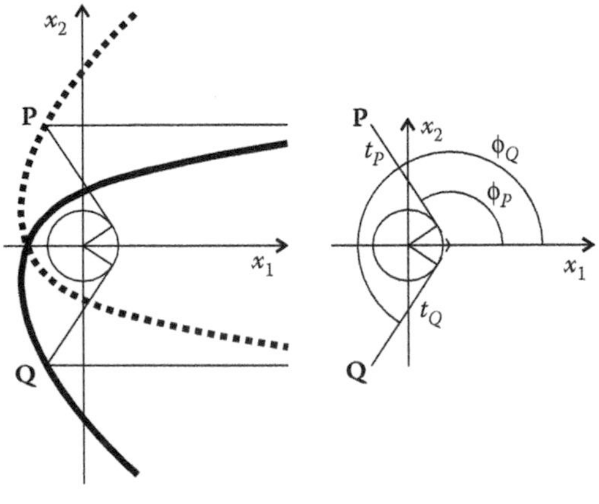

FIGURE 21.23
The winding and unwinding macrofocal parabolas with horizontal axis are the same curve, but reflected about the horizontal axis.

and $\phi_Q = 2\pi - \phi_P$. Replacing in the expression for $t(\phi)$ for the winding parabola ϕ by $2\pi - \phi$ we get

$$t(\phi) = \frac{K_W - r - r((2\pi - \phi) - \sin(2\pi - \phi))}{1 - \cos(2\pi - \phi)} = \frac{K_U + r(\phi - 1 - \sin\phi)}{1 - \cos\phi} \tag{21.58}$$

where K_U is a constant and the unwinding macrofocal parabola is given by

$$r(\cos(\phi + \pi/2), \sin(\phi + \pi/2)) + t(\phi)(\cos\phi, \sin\phi) \tag{21.59}$$

or

$$r(-\sin\phi, \cos\phi) + t(\phi)(\cos\phi, \sin\phi) \tag{21.60}$$

where angle ϕ and t are defined as shown in Figure 21.24, where M is the mirror.

Equation 21.60 can also be derived from a constant string length, as shown in Figure 21.25.

We have $r\beta + t + s = K$, where K is a constant. Since $\beta = 2\pi - (\phi + \pi/2) = 3\pi/2 - \phi$ and $s = r - s_1 = r - (t \cos\phi - r \sin\phi)$, we have $t(\phi)$ given by expression (21.58).

Constant K_U can be obtained from a point **P** on the curve. If this point is defined by t_P and ϕ_P, we have

$$K_U = t_P(1 - \cos\phi_P) + r(1 - \phi_P + \sin\phi_P) \tag{21.61}$$

The values of t_P and ϕ_P can be calculated from the position of point **P** by using expressions (21.15) and (21.19).

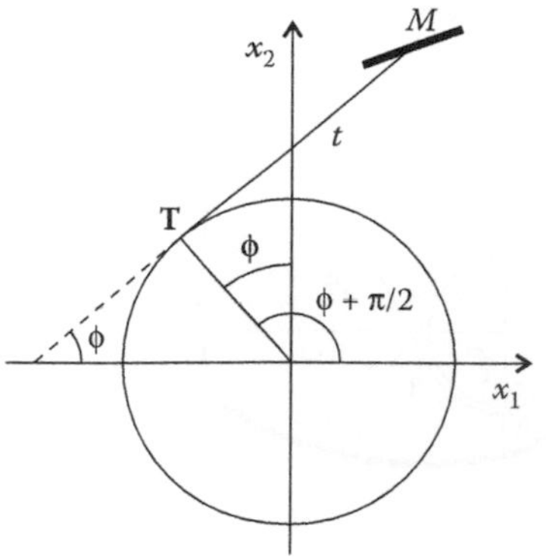

FIGURE 21.24
The unwinding macrofocal parabola defined by angle ϕ and distance t from the tangent point **T** to the mirror M.

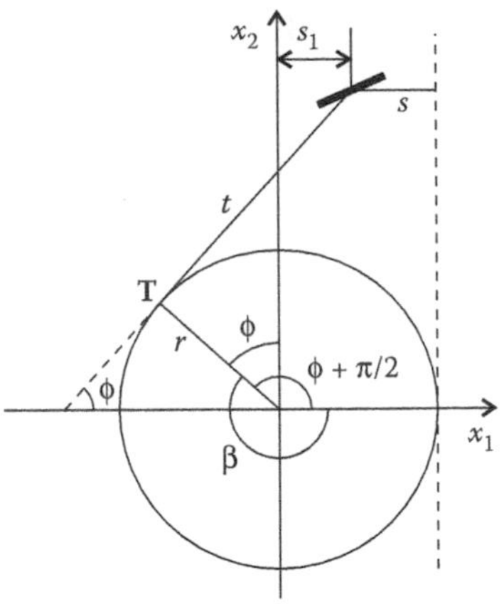

FIGURE 21.25
The unwinding macrofocal parabola can be obtained by a string of constant length $s + t + r\beta$.

21.9 Winding Macrofocal Ellipse

In an ellipse, the sum of the distances between a point and the foci is constant for all points. Therefore, attaching a string of constant length to both foci, keeping it stretched with a marker, and moving the marker, draws an ellipse.

In a macrofocal ellipse,[2] a circle of radius r, as shown in Figure 21.26, replaces one of the foci. In this case, the string wraps around the macrofocus in such a way that one of its tips is at point **A** and the other tip at the focus **G**,

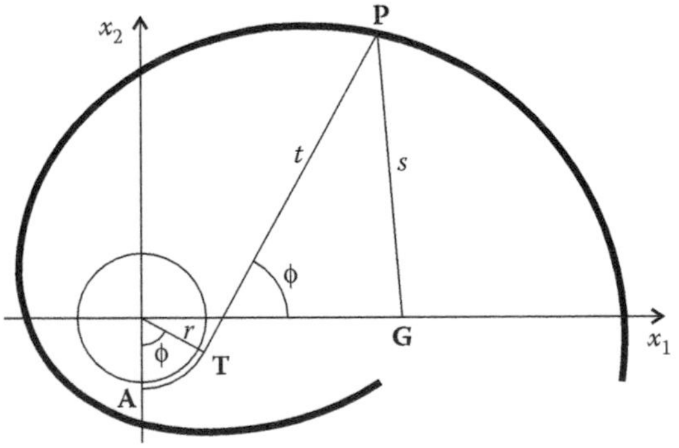

FIGURE 21.26
A macrofocal ellipse reflects the light rays emitted by a point source **G** tangent to a circular macrofocus. It can be generated by a string of constant length **G-P-T-A**. In this case, as angle ϕ increases, the string winds around the macrofocus and therefore this is a winding macrofocal ellipse.

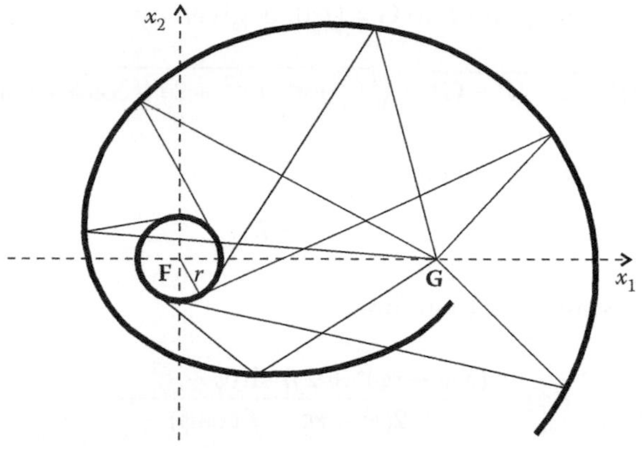

FIGURE 21.27
The light rays emitted by a point source **G** are reflected tangent to the macrofocus of center **F** and radius r. The macrofocal ellipse is actually a spiral curve.

which is still a point focus. The length of the string is, in this case, $r\phi + t + s$, and it is constant for the points **P** on the curve. As angle ϕ increases, the string winds around the macrofocus and this curve is accordingly called a winding macrofocal ellipse. The macrofocal ellipse is actually a spiral curve.

If used as a mirror, this curve reflects rays emitted from a point source **G** tangent to a circular receiver of radius r, as shown in Figure 21.27.

Applying constant string length to this curve, we have the geometry presented in Figure 21.28, where $\mathbf{F} = (0, 0)$ and $\mathbf{G} = (f, 0)$.

Point **P** is defined by

$$\begin{aligned} \mathbf{P} &= r(\cos(\phi - \pi/2), \sin(\phi - \pi/2)) + t(\cos\phi, \sin\phi) \\ &= (t\cos\phi + r\sin\phi, t\sin\phi - r\cos\phi) \end{aligned} \tag{21.62}$$

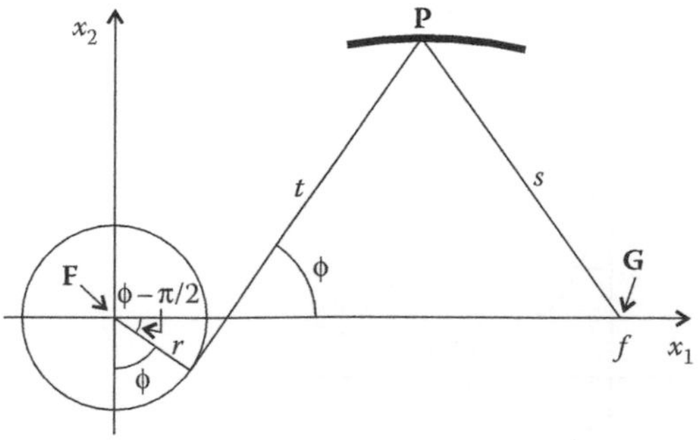

FIGURE 21.28
Geometry for determining the parameterization of a macrofocal ellipse. Its points are defined by $s + t + r\phi = K$, where K is a constant.

The distance s from point **P** to **G** = (f, 0) is given by

$$s = \sqrt{(\mathbf{P}-\mathbf{G})\cdot(\mathbf{P}-\mathbf{G})} = \sqrt{f^2 + r^2 + t^2 - 2f(t\cos\phi + r\sin\phi)} \quad (21.63)$$

From Figure 21.28 we now have

$$r\phi + t + s = K_W \quad (21.64)$$

where K is a constant and, therefore,

$$t(\phi) = \frac{(K_W - r\phi)^2 + 2fr\sin\phi - f^2 - r^2}{2(K - r\phi - f\cos\phi)} \quad (21.65)$$

The constant K_W can now be determined from any point on the curve. The macrofocal ellipse can now be obtained from

$$r(\cos(\phi - \pi/2), \sin(\phi - \pi/2)) + t(\phi(\cos\phi, \sin\phi)) \quad (21.66)$$

or

$$r(\sin\phi, -\cos\phi) + t(\phi)(\cos\phi, \sin\phi) \quad (21.67)$$

A winding macrofocal ellipse rotated by an angle α around the origin and with the center of the macrofocus at a position $\mathbf{F} = (F_1, F_2)$ is given by

$$r(\sin(\phi+\alpha), -\cos(\phi+\alpha)) + t(\phi)(\cos(\phi+\alpha), \sin(\phi+\alpha)) + (F_1, F_2) \quad (21.68)$$

This case is shown in Figure 21.29 where the macrofocal ellipse passes through a given point **P**.

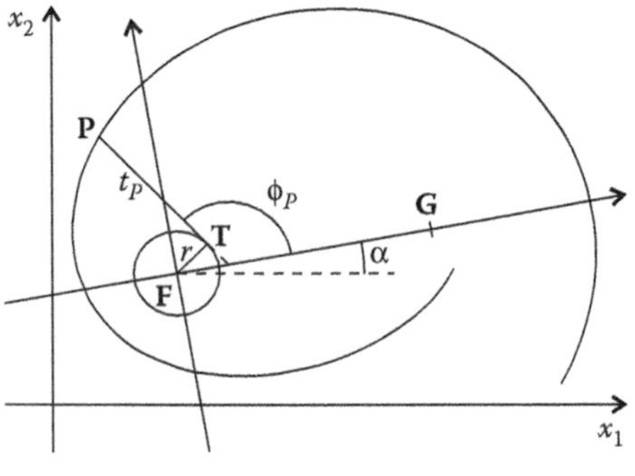

FIGURE 21.29
A general macrofocal ellipse can be defined by the radius r of its macrofocus and its center **F**, the position of the other focus **G**, and a point **P**.

Constant K can now be obtained by solving expression (21.65) with respect to K_W, noting that, for point **P**, we have $\phi = \phi_P$ and $t(\phi) = t_P$. Given distance t_P and angle ϕ_P, we have two possible values for constant K_W given by

$$K_W = t_P + r\phi_P \pm \sqrt{f^2 + r^2 + t_P^2 - 2f(t_P\cos\phi_P + r\sin\phi_P)} \tag{21.69}$$

We choose the following positive sign solution:

$$K_W = t_P + r\phi_P + \sqrt{f^2 + r^2 + t_P^2 - 2f(t_P\cos\phi_P + r\sin\phi_P)} \tag{21.70}$$

From the positions of **F** and **G**, we can calculate f, the distance between **F** and **G** as

$$f = [\mathbf{F},\mathbf{G}] = \sqrt{(\mathbf{F}-\mathbf{G})\cdot(\mathbf{F}-\mathbf{G})} \tag{21.71}$$

From the positions of **F** and **G**, we can also calculate angle α as

$$\alpha = \text{angh}\,\mathbf{v} \quad \text{with } \mathbf{v} = (v_1, v_2) = \mathbf{G} - \mathbf{F} \tag{21.72}$$

The values of t_P and ϕ_P can be calculated from the position of point **P** by using expressions (21.15) and (21.18). Note that from these expressions we have $0 \le \phi_P < 2\pi$. If point **P** should have a different parameter value (e.g., between 2π and 4π), we must change the curve's parameterization.

21.10 Unwinding Macrofocal Ellipse

Figure 21.30 shows another example of a macrofocal ellipse. As seen earlier, this curve can also be generated by a string, but the string wraps around the macrofocus, as shown in Figure 21.30, in such a way that one of its tips is at point **A** and the other tip at a point focus **G**. As angle ϕ increases, the string unwinds around the macrofocus, and thus this curve is called an unwinding macrofocal ellipse.

If used as a mirror, this curve reflects rays emitted from a point source at **G** tangent to a circular receiver of radius r, as shown in Figure 21.31.

We can derive the unwinding macrofocal ellipse equations from the winding case in the same way as with the macrofocal parabola. Accordingly, replacing ϕ by $2\pi - \phi$ in expression (21.65) gives

$$t(\phi) = \frac{(K_U + r\phi)^2 - 2fr\sin\phi - f^2 - r^2}{2(K_U + r\phi - f\cos\phi)} \tag{21.73}$$

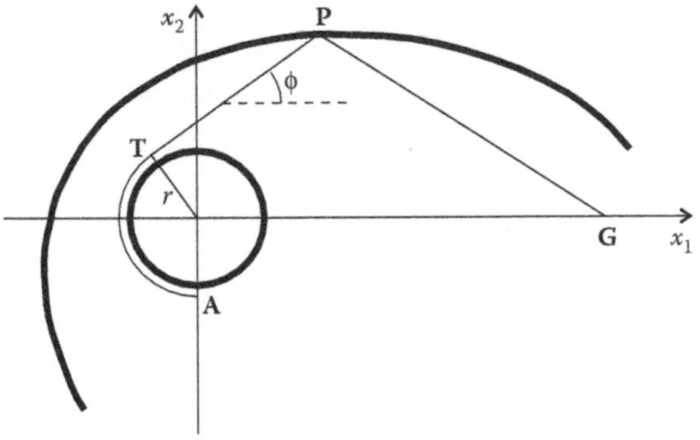

FIGURE 21.30
Unwinding macrofocal ellipse generated by a string **G-P-T-A** of constant length that unwinds around the macrofocus as angle ϕ increases.

where $K_U = K_W - 2\pi r$ and the unwinding macrofocal ellipse is given by

$$r(\cos(\phi + \pi/2), \sin(\phi + \pi/2)) + t(\phi)(\cos\phi, \sin\phi) \tag{21.74}$$

or

$$r(-\sin\phi, \cos\phi) + t(\phi)(\cos\phi, \sin\phi) \tag{21.75}$$

An unwinding macrofocal ellipse rotated by an angle α around the origin with the center of its macrofocus **F** displaced to a position $\mathbf{F} = (F_1, F_2)$ is given by

$$r(-\sin(\phi + \alpha), \cos(\phi + \alpha)) + t(\phi)(\cos(\phi + \alpha), \sin(\phi + \alpha)) + (F_1, F_2) \tag{21.76}$$

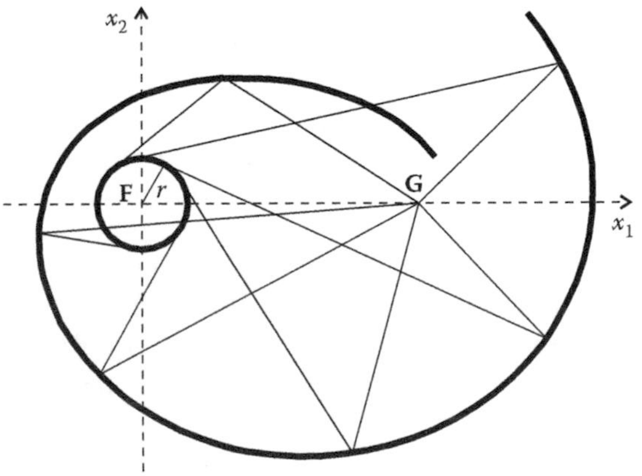

FIGURE 21.31
A macrofocal ellipse reflects rays emitted by a point source **G** tangent to the circular macrofocus.

Given a point **P** defined by t_P and ϕ_P, constant K_U can be obtained from expression (21.73) as

$$K_U = t_P - r\phi_P \pm \sqrt{f^2 + r^2 + t_P^2 - 2ft_P\cos\phi_P + 2fr\sin\phi_P} \tag{21.77}$$

Again we choose the solution with the positive sign

$$K_U = t_P - r\phi_P + \sqrt{f^2 + r^2 + t_P^2 - 2f(t_P\cos\phi_P - r\sin\phi_P)} \tag{21.78}$$

The values of t_P and ϕ_P can be calculated from the position of point **P** by using expressions (21.15) and (21.19).

21.11 Cartesian Oval for Parallel Rays

An optical refractive surface can redirect the light rays emitted by a point source **F** immersed in a medium of refractive index n_1, in such a way that they are parallel after refraction as they enter into a medium of refractive index n_2. This curve verifies

$$n_1 t + n_2 s = K^* \tag{21.79}$$

where K^* is a constant and t and s are defined as shown in Figure 21.32. Replacing

$$s = r - t\cos\phi \tag{21.80}$$

we get

$$t(\phi) = \frac{K^* - n_2 r}{n_1 - n_2\cos\phi} = \frac{K}{n_1 - n_2\cos\phi} \tag{21.81}$$

where K is another constant, since r is also a constant. Now for $\phi = 0$ we have $t = t_0$, and, therefore, we can write

$$t(\phi) = \frac{t_0(n_1 - n_2)}{n_1 - n_2\cos\phi} \tag{21.82}$$

Now making

$$C = 2t_0\frac{n_1}{n_1 + n_2}, \quad f = 2t_0\frac{n_2}{n_1 + n_2} \tag{21.83}$$

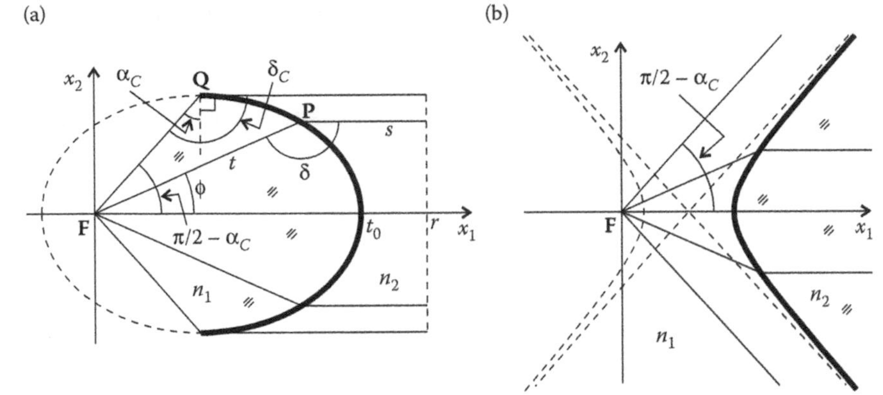

FIGURE 21.32
Cartesian oval curve that refracts light rays from a point source and makes them parallel. (a) The source is immersed in a medium with high refractive index and the curve is an ellipse. (b) The source is in a low-refractive index medium and the curve is a hyperbola.

We can also write

$$t(\phi) = \frac{C^2 - f^2}{2C - 2f\cos\phi} \tag{21.84}$$

This is the equation of an ellipse or a hyperbola, depending on whether $n_1 > n_2$ or $n_1 > n_2$, respectively. The variable f is the distance between the foci and C the distance between the vertices of the conic curves.

The limits of parameter ϕ are shown in Figure 21.32a, where α_C is the critical angle, given by $\alpha_C = \arcsin(\min(n_1, n_2)/\max(n_1, n_2))$, and where functions min and max give the minimum and maximum, respectively, between their two variables. If the incoming ray from **F** to a point **P** and the ray leaving **P** make an angle $\delta > \delta_C = \pi/2 + \alpha_C$, then refraction is possible, but for the portions of the ellipse for which $\delta < \delta_C = \pi/2 + \alpha_C$ refraction is not possible, thus, this portion of the curve cannot be used for a refractive optic. Something similar happens in the case of the hyperbola of Figure 21.32b. This condition translates to $-(\pi/2 - \alpha_C) \le \phi \le \pi/2 - \alpha_C$ for the parameter, in both the cases in Figure 21.32.

Now consider the general case of a Cartesian oval tilted by an angle α and with focus at a position $\mathbf{F} = (F_1, F_2)$ and that passes through a given point **P** defined by an angle ϕ_P and distance t_P to the focus, as shown in Figure 21.33.

In this case, the constant K can be obtained from expression (21.81) as $K = t_P(n_1 - n_2 \cos \phi_P)$, and the parameterization of the curve becomes

$$\frac{t_P(n_1 - n_2\cos\phi_P)}{n_1 - n_2\cos\phi}(\cos(\phi + \alpha), \sin(\phi + \alpha)) + (F_1, F_2) \tag{21.85}$$

for $-(\pi/2 - \alpha_C) \le \phi \le \pi/2$.

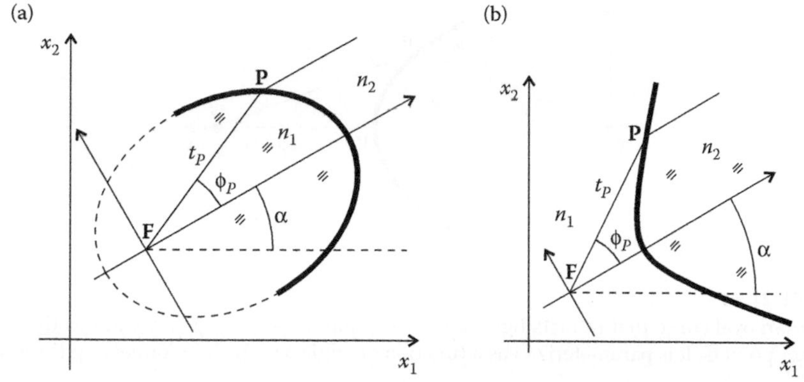

FIGURE 21.33
A general Cartesian oval that refracts the rays from a point source and makes them parallel can be defined by the position of the source **F**, the angle the parallel rays make to the horizontal, the refractive indices of the two media, and a point **P** on the curve. Cases in which $n_1 > n_2$ (a) or in which $n_1 < n_2$ (b).

Note that not every point **P** can be chosen for the Cartesian oval to pass through. Point **P** must fulfill $\delta \geq \delta_C = \pi/2 + \alpha_C$, where δ is defined in Figure 21.32. In the general case presented in Figure 21.33, this condition can be written as

$$\delta = \arccos\left(\frac{(\mathbf{F}-\mathbf{P})\cdot(\cos\alpha, \sin\alpha)}{\sqrt{(\mathbf{F}-\mathbf{P})\cdot(\mathbf{F}-\mathbf{P})}}\right) \geq \frac{\pi}{2} + \alpha_C \tag{21.86}$$

This curve will also converge parallel rays traveling in a medium of refractive index n_2 and being refracted into another medium of refractive index n_1 to point **F**.

21.12 Cartesian Oval for Converging or Diverging Rays

The light emitted from a point source **F** in a medium of refractive index n_1 can be concentrated onto a point **G** in a medium of refractive index n_2 by a refractive surface, as shown in Figure 21.34. In this case, the optical path length from **F** to **G** is a constant given by $n_1 t + n_2 s = K$.

Also, the light emitted from **F** immersed in a medium of refractive index n_1 can be refracted into a medium of refractive index n_2 in such a way that it appears to be diverging from another point **G**, as shown in Figure 21.35. In this case, if c is a circle with center **G** and radius r, the optical path length

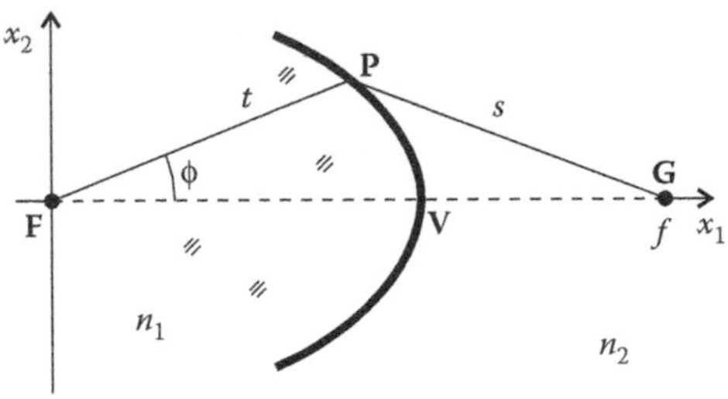

FIGURE 21.34
Cartesian oval curve that refracts light rays from a point source and concentrates them onto another point **G**. It is parameterized as a function of angle ϕ to the line connecting **F** and **G**.

$n_1t + n_2d = K^*$, where $K^* > 0$ is a constant. We can write $d = r - s$ and, therefore, $n_1t - n_2s = K^* - n_2r$, that is, $n_1t - n_2s = K$, where K is another constant.

Referring to Figure 21.34, consider that $\mathbf{F} = (0, 0)$ so that the point **P** is given by coordinates $\mathbf{P} = (P_1, P_2) = t(\cos\phi, \sin\phi)$. The points on the curve of Figure 21.35 can be obtained in a similar way. We also consider that $\mathbf{G} = (f, 0)$. The distance s from point **P** to **G** is given by

$$s = \sqrt{(\mathbf{G}-\mathbf{P})\cdot(\mathbf{G}-\mathbf{P})} = \sqrt{f^2 + t^2 - 2ft\cos\phi} \tag{21.87}$$

The points on these curves can be obtained from

$$n_1t \pm n_2s = K \Rightarrow s^2 = \left(\frac{K - n_1t}{n_2}\right)^2 \tag{21.88}$$

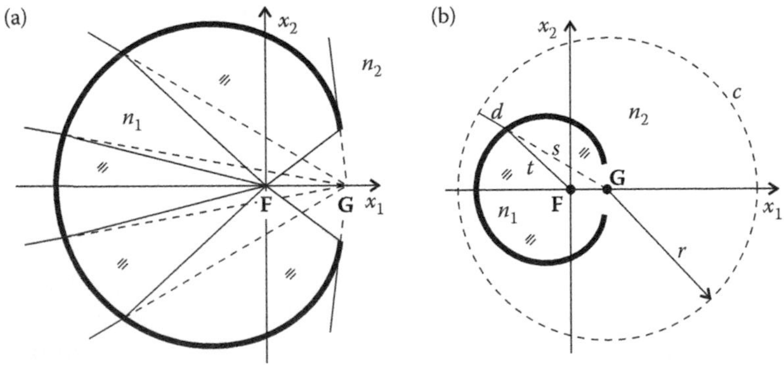

FIGURE 21.35
(a) Cartesian oval curve that refracts light rays from a point source and makes them diverge as if they appear to come from another point **G**. (b) It is defined by condition $n_1t + n_2d = K$ with K constant and is parameterized as a function of angle ϕ that line t makes with the line connecting **F** and **G**.

where K is a constant. We get

$$t_1(\phi) = \frac{Kn_1 - fn_2^2\cos\phi - n_2\sqrt{D}}{n_1^2 - n_2^2}$$
$$t_2(\phi) = \frac{Kn_1 - fn_2^2\cos\phi + n_2\sqrt{D}}{n_1^2 - n_2^2} \tag{21.89}$$

with

$$D = (fn_1 - K\cos\phi)^2 + (K^2 - f^2n_2^2)\sin^2\phi \tag{21.90}$$

The two possible plane curves c_1 and c_2 are then defined by

$$c_1 = t_1(\phi)(\cos\phi, \sin\phi)$$
$$c_2 = t_2(\phi)(\cos\phi, \sin\phi) \tag{21.91}$$

We will consider the case in which $n_1 > n_2$. Figure 21.36 shows an example of one of these curves.

Note that in the particular case where $K^2 - f^2n_2^2 = 0$, then $t_2(\phi)$ and $t_1(\phi)$ can be written as

$$t(\phi) = -\frac{fn_2^2 \pm Kn_2}{n_1^2 - n_2^2}\cos\phi + \frac{Kn_1 \pm fn_1n_2}{n_1^2 - n_2^2} \tag{21.92}$$

which has the form $t(\phi) = 2a\cos\phi + b$ as in the case of a Limaçon of Pascal.

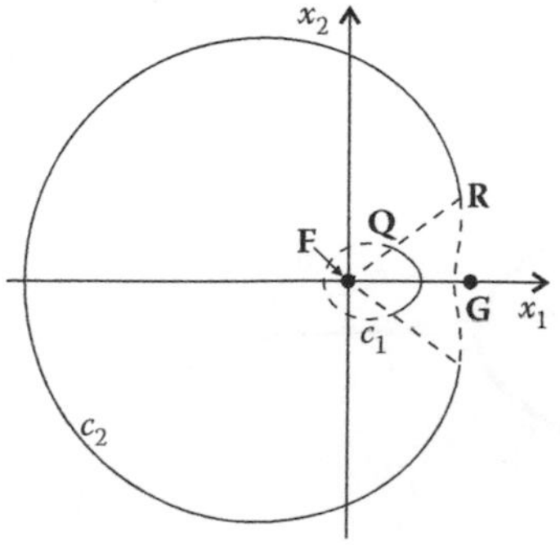

FIGURE 21.36
Cartesian ovals may completely surround the source, but it does not mean that the whole curve can be used as a refractor. Only those portions where refraction is possible can be used optically.

Although curves c_1 and c_2 defined by Equation 21.91 completely surround the point source **F**, only some portions of the curves can be used as refractors. The portions of the curves that can be used depend on the curves' parameters. The reason for this is that, as we move along the curves, we may reach a critical point, where the light emitted from **F** hits the surface at the critical angle. Beyond that point, light rays no longer converge, or appear to diverge from **G**. These critical points are **Q** for curve c_1 and **R** for curve c_2 in Figure 21.36. Figure 21.37 shows the geometry of the curves at those points.

Since the surfaces separate two media of refractive indices n_1 and n_2 with $n_1 > n_2$, the critical angle α_C fulfills $\sin \alpha_C = n_2/n_1$. We then have the following for points **Q** and **R**,

$$f \cos\phi_C = t \pm s \sin\alpha_C \Leftrightarrow \cos\phi_C = \frac{n_1 t \pm n_2 s}{f n_1} \tag{21.93}$$

and we can define

$$\begin{aligned} \phi_C &= \arccos\left(\frac{K}{f n_1}\right) && \text{if } |K| \le f n_1 \\ \phi_C &= 0 && \text{if } |K| > f n_1 \end{aligned} \tag{21.94}$$

where the positive and negative signs are for the cases of Figure 21.37a and b, respectively.

For the converging Cartesian oval shown in Figure 21.34, its vertex **V** must be contained between the foci **F** and **G**. Vertex **V** also verifies the condition $K = n_1[\mathbf{F},\mathbf{V}] + n_2[\mathbf{V},\mathbf{G}]$ that defines the curve. The maximum value of K is then obtained making $\mathbf{V} = \mathbf{G}$, so that $K = n_1[\mathbf{F},\mathbf{G}] = n_1 f$. The minimum value of K

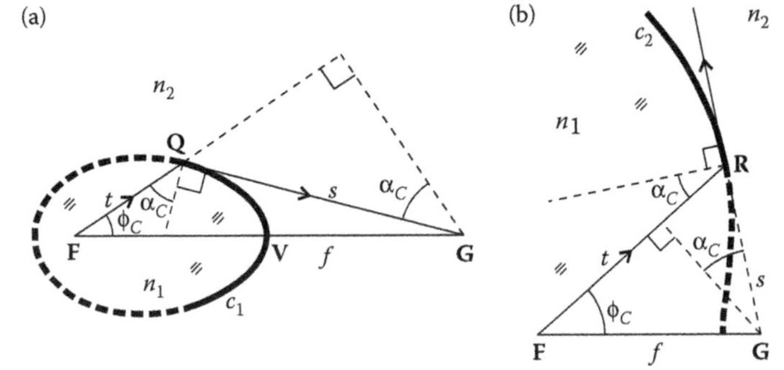

FIGURE 21.37
(a) Rays emitted from **F** converge at **G**. (b) Rays emitted from **F** appear to diverge from **G**. In both (a) and (b) the point **Q** or **R** on the curves for which the incidence angle reaches the critical angle limits the region of the Cartesian oval curves that can be used to refract the light.

is obtained when $\mathbf{V} = \mathbf{F}$ so that $K = n_2[\mathbf{F},\mathbf{G}] = n_2 f$. The value of K must then be contained between these two extreme values. A point $\mathbf{P}$ on the curve must fulfill

$$K = n_1[\mathbf{F},\mathbf{P}] + n_2[\mathbf{P},\mathbf{G}] \quad \text{with } n_2 f < K < n_1 f \tag{21.95}$$

In this case, the parameter range for angle ϕ is $-\phi_C \le \phi \le \phi_C$ for the portion of the curve where refraction is possible to guarantee that the point $\mathbf{P}$ is on the portion of the curve that can be used to refract the light, we must have $\phi_P = \theta \le \phi_C$, where θ is given by expression (21.4) with $\mathbf{v} = \mathbf{P} - \mathbf{F}$ and $\mathbf{u} = \mathbf{G} - \mathbf{F}$.

The limitations on K for the diverging solution of Figure 21.35 are different. When $K > fn_1$ then ϕ_C does not exist and the whole curve can be used as a refractive optic. In this case, the parameter range is $0 \le \phi \le 2\pi$, as shown in Figure 21.38a, but when $K = -fn_2$ then

$$t_2(\phi) = \frac{n_2\left[-fn_1 - fn_2\cos\phi + \sqrt{(fn_1 + fn_2\cos\phi)^2}\right]}{n_1^2 - n_2^2} \tag{21.96}$$

and, therefore, the curve c_2 described by $t_2(\phi)$ tends to a point at position $\mathbf{F}$ as $K \to -fn_2$. This case is shown in Figure 21.38b, where the value for K is close to $-fn_2$. This imposes a lower limit on K, so that $K > -fn_2$. For $-fn_2 < K < fn_1$, we can calculate the value of ϕ_C and the parameter range for curve c_2 is $\phi_C \le \phi \le 2\pi - \phi_C$ To guarantee that a point $\mathbf{P}$ is on the portion of the curve that can be used to refract the light, we must also have $\phi_P = \theta \ge \phi_C$, where θ is given by expression (21.4) with $\mathbf{v} = \mathbf{P} - \mathbf{F}$ and $\mathbf{u} = \mathbf{G} - \mathbf{F}$. These two parameter ranges can be written as $\phi_C \le \phi \le 2\pi - \phi_C$ for $K > -fn_2$, if ϕ_C is defined by expression (21.94).

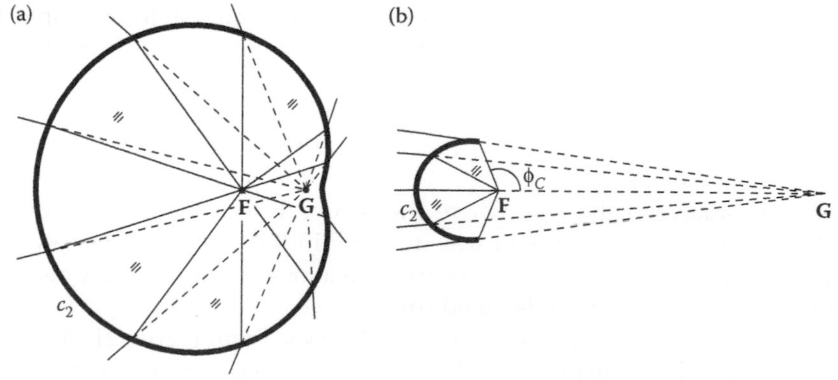

FIGURE 21.38
(a) For large values of K ($K > fn_1$), the Cartesian oval does not cross the line **FG** and the entire curve can be used as a refractor. (b) As the value of K diminishes, the size of the curve that can be used as a refractor gets smaller and smaller when K tends to $-fn_2$.

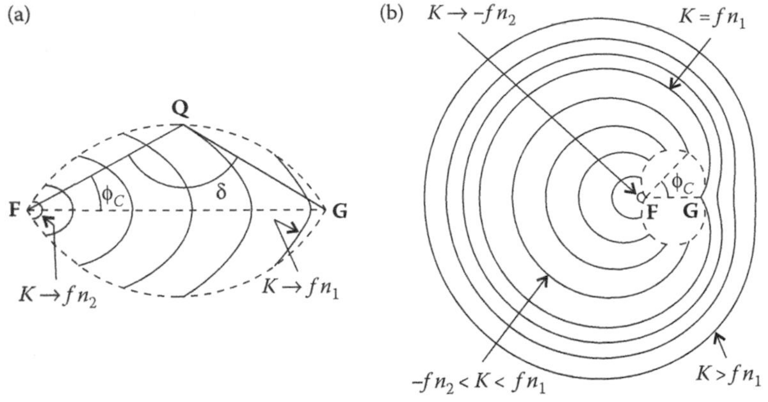

FIGURE 21.39
Regions of space where the Cartesian oval curves can be used as refractors. (a) The light emitted by point source **F** converges to point **G**. (b) The light emitted by **F** appears to diverge from **G**.

Figure 21.39a and b shows the shape of these curves for different values of K, for the converging (light converges to **G**) and diverging (light appears to diverge from **G**) Cartesian ovals, respectively.

The Cartesian ovals are limited to the region of space in which refraction is possible. In the case of the converging ovals of Figure 21.39a, this region is the interior of the curved dashed line. For the points **Q** on this line, the angle the incident and refracted rays make is $\delta = \pi/2 + \alpha_C$, where α_C is the critical angle, as shown in Figure 21.37a. For the points outside this curve, we would have $\delta < \pi/2 + \alpha_C$ and refraction would be impossible due to total internal reflection (TIR). For the points inside the dashed curved line angle $\delta > \pi/2 + \alpha_C$, refraction is possible, so we can design the Cartesian oval curves as shown. The shape of this dashed curve can be obtained, for example, by making $\phi = \phi_C$ in the expression for curve c_1. The result can now be plotted as a function of K for $n_2 f < K < n_1 f$.

Something similar happens in the case of the diverging Cartesian ovals of Figure 21.39b, but now these curves are limited to the space outside the curved dashed line shown.

We have analyzed the case where the source **F** was in a high-refractive index medium and the focus **G** was in a low-index medium. We now consider the opposite case where the source is in a low-refractive index medium and the focus is in a high-index medium.

The curves, however, are the same in both cases. From Figure 21.34 we can see that, if now **G** is considered as the point source, the refractive surface will concentrate the light emitted from **G** on to point **F**. Something similar happens for the diverging Cartesian oval of Figure 21.35.

Figure 21.40a shows point **F** immersed in a low-refractive index medium (e.g., air) of refractive index n_1. The surface separates this medium from a

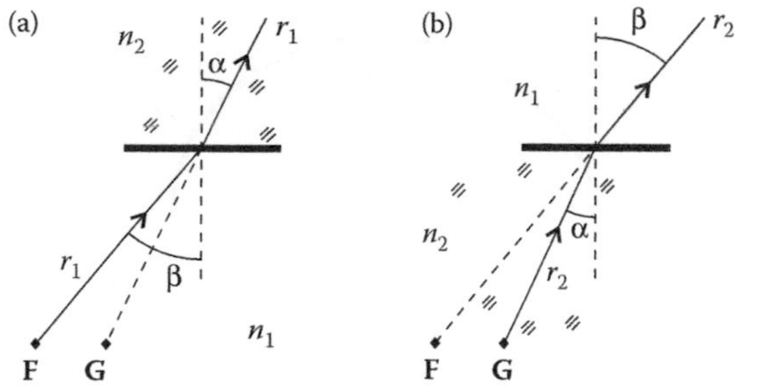

FIGURE 21.40
(a) A light ray emitted by **F** is refracted at a surface so that it appears to diverge from **G**. (b) A similar situation but now with the refractive indices interchanged. Now **G** is the source emitting a ray that is refracted as if it appears to come from **F**. In both cases, we have $n_2\sin\alpha = n_1\sin\beta$.

higher refractive index medium of refractive index n_2. A ray r_1 leaves **F**, making an angle β to the normal to the surface, it refracts and enters medium n_2 with a smaller angle α to the vertical, where $n_2 \sin \alpha = n_1 \sin \beta$. After refraction, this ray now appears to come from point **G**.

Figure 21.40b shows a similar situation, but now with points **G** immersed in a high refractive index medium of refractive index n_2. The surface now separates this medium from a lower refractive index medium of refractive index n_1 (e.g., air). A ray r_2 leaves **G**, making an angle α to the normal to the surface, it refracts and enters medium n_1 with a higher angle β to the vertical, where $n_2 \sin \alpha = n_1 \sin \beta$. After refraction, this ray now appears to come from point **F**.

So, if we have a surface that refracts rays emitted by **F** in such a way that they appear to be coming from **G**, that same surface will refract rays coming from **G** in a way that they appear to be coming from **F**, provided that the refractive indices this surface separates are interchanged.

Figure 21.41 shows the same curve as Figure 21.35 but now with the values of the refractive indices interchanged. Point **G** is now a source and **F** the focus from where the rays appear to diverge.

In any case, the refractive index n_1 associated with **F** is larger than the refractive index n_2 associated with **G**. The same curve then describes the situations in which the light of a point source **F** in a high-refractive index medium (n_1) is refracted as if it appears to come from a point **G** in a low-refractive index medium n_2 (Figure 21.35), as well as the case in which the light of a point source **G** in a low-refractive index medium (n_2) is refracted as if it appears to come from a point **F** in a high-refractive index medium (n_1) (Figure 21.41).

We now summarize the expressions given earlier, which enable us to calculate Cartesian oval curves for some particular cases. One of the foci, **F**, is at the origin **F** = (0,0) surrounded by a medium of refractive index n_1. The other focus, **G**, is at a position **G** = (f,0) and is surrounded by a medium of refractive index n_2 with $n_2 < n_1$. Now, we may give a point **P** to define the curve.

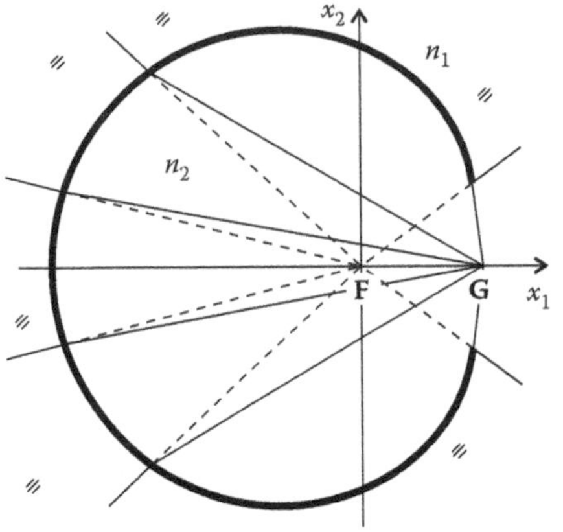

FIGURE 21.41
The same Cartesian oval curve of Figure 21.35 but with the refractive indices interchanged. We also interchange the roles of **F** and **G**, and we have a point source **G** that emits light, which is refracted at the Cartesian oval in such a way that it appears to come from a point **F**.

If we want to design a converging Cartesian oval, we may calculate $K = n_1[\mathbf{F},\mathbf{P}] + n_2[\mathbf{P},\mathbf{G}]$. If we have $n_2 f < K < n_1 f$ then the Cartesian oval exists and is given by $C_1(\phi) = t_1(\phi)(\cos\phi, \sin\phi)$, where $t_1(\phi)$ is defined by expression (21.89) and D by Equation 21.90. The parameter range for this curve is $-\phi_C \le \phi \le \phi_C$, where ϕ_C is given by expression (21.94). Point **P** must also fulfill $\phi_P = \theta \le \phi_C$, where θ is given by expression (21.4) with $\mathbf{v} = \mathbf{P} - \mathbf{F}$ and $\mathbf{u} = \mathbf{G} - \mathbf{F}$. If point **F** is the source, this curve focuses the light rays to point **G**. If point **G** is the light source, this curve focuses the light rays to point **F**.

If we need to design a diverging Cartesian oval, we must calculate $K = n_1[\mathbf{F},\mathbf{P}] - n_2[\mathbf{P},\mathbf{G}]$. If we have $K > fn_2$ then the Cartesian oval exists and is given by $C_2(\phi) = t_2(\phi)(\cos\phi, \sin\phi)$, where $t_2(\phi)$ is defined by expression (21.89) and D by Equation 21.90. The parameter range for the curve c_2 is $\phi_C \le \phi \le 2\pi - \phi_C$, where ϕ_C is given by expression (21.94). In this case, point **P** must also fulfill $\phi_P = \theta \ge \phi_C$, where θ is given by expression (21.4) with $\mathbf{v} = \mathbf{P} - \mathbf{F}$ and $\mathbf{u} = \mathbf{G} - \mathbf{F}$. If point **F** is the source, this curve causes the light rays to diverge from point **G**. If point **G** is the light source, this curve causes the light rays to diverge from point **F**.

Now consider the more general case of arbitrary positions for foci **F** and **G**. We start by calculating $f = [\mathbf{F},\mathbf{G}]$. Focus **F** is surrounded by a medium of refractive index n_1 and focus **G** is surrounded by a medium of refractive index n_2, with $n_2 < n_1$. Now, we may point **P** to define the curve.

If we need to design a converging Cartesian oval, we must calculate $K = n_1[\mathbf{F},\mathbf{P}] + n_2[\mathbf{P},\mathbf{G}]$. If we have $n_2 f < \mathrm{K} < n_1 f$ then the Cartesian oval exists and is given by $C_1(\phi) = \mathbf{F} + t_1(\phi)(\cos(\phi + \alpha), \sin(\phi + \alpha))$ where $t_1(\phi)$ is defined by expression (21.89), D by Equation 21.90, and α is the angle vector $\mathbf{G} - \mathbf{F}$ makes to the horizontal given by $\alpha = \mathrm{angh}(\mathbf{G} - \mathbf{F})$. Point **P** must also fulfill ang(**P** – **F**,**G** – **F**) ≤ ϕ_C. The parameter range for this curve is $-\phi_C \le \phi \le \phi_C$ where ϕ_C is given by expression (21.94). If point **F** is the source, this curve focuses the light rays to point **G**. If point **G** is the light source, this curve focuses the light rays to point **F**.

If we want to design a diverging Cartesian oval, we may calculate $K = n_1$[**F**,**P**] – n_2[**P**,**G**]. If we have $K > -fn_2$ then the Cartesian oval exists and is given by $C_2(\phi) = \mathbf{F} + t_2(\phi)(\cos(\phi + \alpha), \sin(\phi + \alpha))$, where $t_2(\phi)$ is defined by expression (21.89), D by Equation 21.90, and α is the angle vector **G** – **F** makes to the horizontal given by α = angh(**G** – **F**). The parameter range for the curve c_2 is $\phi_C \le \phi \le 2\pi - \phi_C$, where ϕ_C is given by expression (21.94). In this case, point **P** must also fulfill ang(**P** – **F**,**G** – **F**) ≥ ϕ_C. If point F is the source, this curve causes the light rays to diverge from point **G**. If point **G** is the light source, this curve causes the light rays to diverge from point **F**.

21.13 Cartesian Ovals Calculated Point by Point

The Cartesian ovals presented earlier can also be calculated point by point. Figure 21.42 presents the case in which we have a light source **F** immersed in a medium of refractive index n_1. There is also a focus **G**, immersed in a medium of refractive index n_2, onto which the light may be concentrated, or from where it may appear to diverge.

If t is the distance from **F** to **P** and s the distance from **P** to **G**, the optical path length S between **F** and **G** is $S = n_1 t + n_2 s$ in the case where the light from **F** is refracted and concentrated to **G** (converging case).

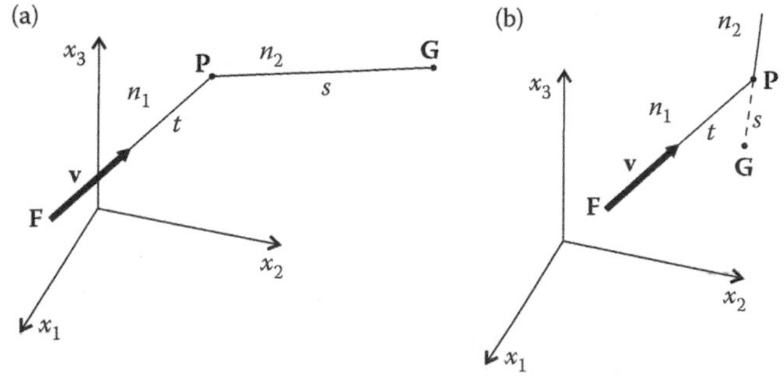

FIGURE 21.42
(a) A point source **F** in a medium of refractive index n_1 emits a light ray in a given direction. If the optical path length to another point **G** in a medium of refractive index n_2 is known, then the point **P** on the Cartesian oval curve that refracts light from **F** to **G** can be calculated by constant optical path length. (b) The same thing happens when the ray is refracted at point **P** as if it appears to come from point **G**.

When light appears to diverge from **G** after refraction (diverging case), we may also calculate an "optical path length" given by $S^* = n_1 t - n_2 s$. Note that quantity S^* may be negative, whereas an optical path length is always a positive quantity. S^* is the "optical path length" between **F** and a virtual "point wavefront" at **G**. However, to make the functions definitions (Section 21.15) simpler, we also use S for this case and make $n_1 t - n_2 s = S$.

Given this optical path length, it is possible to calculate the points on the Cartesian oval by constant optical path length. Points **P** on this curve must then satisfy the following condition:

$$n_1 t \pm n_2 s = S \tag{21.97}$$

Point **P** is defined by

$$\mathbf{P} = \mathbf{F} + t\mathbf{v} \tag{21.98}$$

The distance from **P** to **G** is

$$s = \sqrt{(\mathbf{F} + t\mathbf{v} - \mathbf{G}) \cdot (\mathbf{F} + t\mathbf{v} - \mathbf{G})} \tag{21.99}$$

replacing Equation 21.99 in expression (21.97), we can solve for t and we get the following for point **P**

$$\mathbf{P} = \mathbf{F} + \frac{C_1 + \delta\sqrt{C_2\left(n_2^2 - n_1^2\right) + C_1^2}}{n_1^2 - n_2^2}\mathbf{v} \tag{21.100}$$

with

$$\begin{aligned} C_1 &= n_1 S + n_2^2(\mathbf{F} - \mathbf{G}) \cdot \mathbf{v} \\ C_2 &= S^2 - n_2^2(\mathbf{F} - \mathbf{G}) \cdot (\mathbf{F} - \mathbf{G}) \\ \delta &= \pm 1 \end{aligned} \tag{21.101}$$

with $\delta = -1$ for $n_1 > n_2$ and $\delta = 1$ for $n_1 < n_2$ for the converging case and $\delta = 1$ for $n_1 > n_2$ and $\delta = -1$ for $n_1 < n_2$ for the diverging case. Varying the direction of vector **v** gives the points on the Cartesian oval that concentrates the light emitted by **F** onto **G**. In 2-G geometry, one of these curves is presented in Figure 21.34.

Expression (21.97) can be written as $(t - S/n)^2 = s^2$ for the case in which $n_1 = n_2 = n$ (the case of reflection) and we get

$$\mathbf{P} = \mathbf{F} + \frac{(S/n)^2 - (\mathbf{F} - \mathbf{G}) \cdot (\mathbf{F} - \mathbf{G})}{2(S/n + (\mathbf{F} - \mathbf{G}) \cdot \mathbf{v})}\mathbf{v} \tag{21.102}$$

for both the converging and diverging cases.

A similar procedure can be used to calculate the points of a Cartesian oval that collimates (makes parallel) the light rays emitted by a point source **F**. This situation is depicted in Figure 21.43, where a light source **F** is immersed in a medium of refractive index n_1 and we are calculating the points of a surface that refracts the light in a direction perpendicular to a plane wavefront μ defined by a point **Q** and a unit normal vector **n** pointing in the direction of propagation of the light.

If the optical path length *S* between **F** and the wavefront is given, points **P** in this curve must then satisfy the following condition,

$$n_1 t + n_2 s = S \tag{21.103}$$

where *t* is the distance from **F** to **P** and *s* the distance from **P** to the wavefront. Point **P** is defined by

$$\mathbf{P} = \mathbf{F} + t\mathbf{v} \tag{21.104}$$

The distance from **P** to the wavefront is

$$s = (\mathbf{Q} - \mathbf{P}) \cdot \mathbf{n} \tag{21.105}$$

substituting expression (21.105) in Equation 21.103, we can solve for *t* to get point **P** as

$$\mathbf{P} = \mathbf{F} + \frac{S - n_2(\mathbf{Q} - \mathbf{F}) \cdot \mathbf{n}}{n_1 - n_2 \mathbf{v} \cdot \mathbf{n}} \mathbf{v} \tag{21.106}$$

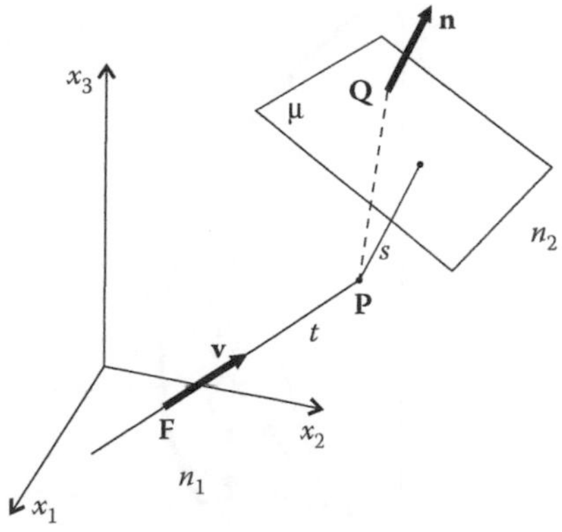

FIGURE 21.43
A point source **F** in a medium of refractive index n_1 emits a light ray in a given direction **v**. We want it to be refracted perpendicular to a plane wavefront μ in a medium of refractive index n_2 and defined by its normal vector **n** and a point **Q**. The point **P** on the corresponding Cartesian oval curve can be calculated by constant optical path length.

Varying the direction of vector **v** gives the points on the Cartesian oval that collimates the light emitted by **F** in the direction **n**. In 2-D geometry, the resulting curve is shown in Figure 21.33b.

21.14 Equiangular Spiral

The rays emitted by a point source F can be reflected by TIR at the critical angle by a curve separating two media of different refractive indices, as shown in Figure 21.44. This curve is called equiangular spiral, logarithmic spiral, or logistique.

The shape of this curve can be calculated from a differential equation. Figure 21.45a shows the geometry of an element of curve with an infinitesimal length.

The equiangular spiral can be parameterized by $t(\phi)(\cos\phi,\sin\phi)$. Function $t(\phi)$ is given by a differential equation, where α_C is the critical angle

$$\frac{dt}{t\,d\phi} = \tan\alpha_C \Leftrightarrow \ln t = \phi\tan\alpha_C + K \Leftrightarrow t = C\exp(\phi\tan\alpha_C) \qquad (21.107)$$

with $C = e^K$, where K and C are constants. If the curve must pass through a given point **P**, as shown in Figure 21.45b, the initial condition for this

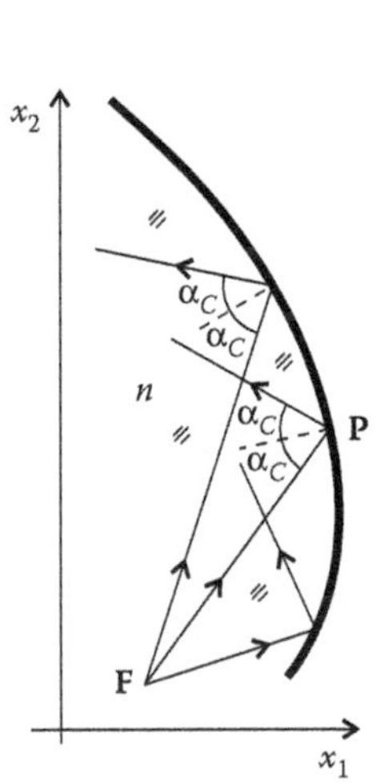

FIGURE 21.44
An equiangular spiral reflects (by TIR) the light coming from a point source at the critical angle for all its points.

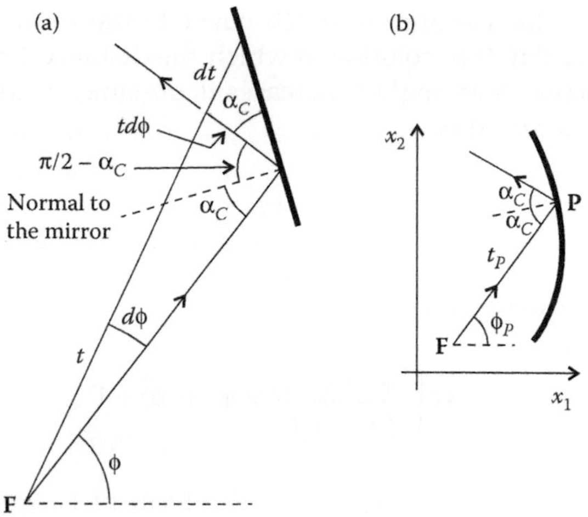

FIGURE 21.45
An equiangular spiral can be defined by a differential equation. (a) The geometry of an element of infinitesimal length on the curve. (b) Once its equation has been obtained, the position of a point **P** completely determines the shape of the curve.

equation is that, for $\phi = \phi_P$ we have $t = t_P$ where $\phi_P = \text{angh}(\mathbf{P} - \mathbf{F})$ and $t_P = [\mathbf{F},\mathbf{P}]$. We then get

$$C = \frac{t_P}{\exp(\phi_P \tan\alpha_C)} \quad \text{and} \quad t(\phi) = t_P \exp((\phi - \phi_P)\tan\alpha_C) \tag{21.108}$$

In the particular case where the surface separates a medium with refractive index n from air ($n = 1$), the critical angle α_C is related to the refractive index of the dielectric by $\sin\alpha_C = 1/n$, so that $\cos^2\alpha_C = 1 - 1/n^2$, giving

$$\tan\alpha_C = \frac{1}{\sqrt{n^2 - 1}} \tag{21.109}$$

and

$$t(\phi) = t_P \exp\left(\frac{\phi - \phi_P}{\sqrt{n^2 - 1}}\right) \tag{21.110}$$

If the focus of the spiral is at a point **F**, then its parameterization becomes

$$t_P \exp\left(\frac{\phi - \phi_P}{\sqrt{n^2 - 1}}\right)(\cos\phi, \sin\phi) + \mathbf{F} \tag{21.111}$$

In this solution, the distance from the curve to the source **F** increases as angle ϕ increases. Another solution in which the distance from the curve to the source **F** decreases as angle ϕ increases is possible. In this case, the differential equation would be

$$\frac{dt}{td\phi} = -\tan\alpha_C \tag{21.112}$$

and the curve is described by

$$t_P \exp\left(\frac{\phi_P - \phi}{\sqrt{n^2 - 1}}\right)(\cos\phi, \sin\phi) + \mathbf{F} \tag{21.113}$$

21.15 Functions Definitions

The functions and curve equations given earlier can be summarized in a list of functions that can then be used to calculate a variety of nonimaging optical devices.

1. The magnitude of a vector is given by

$$\|\mathbf{v}\| = \sqrt{\mathbf{v} \cdot \mathbf{v}} \tag{21.114}$$

2. A unit vector (magnitude 1) in direction **v** is given by

$$\text{nrm}(\mathbf{v}) = \frac{\mathbf{v}}{\|\mathbf{v}\|} = \frac{\mathbf{v}}{\sqrt{\mathbf{v} \cdot \mathbf{v}}} \tag{21.115}$$

3. The distance between two points **A** and **B** is given by

$$[\mathbf{A}, \mathbf{B}] = \|\mathbf{B} - \mathbf{A}\| \tag{21.116}$$

4. The angle between two vectors **u** and **v** in the range 0–π is given by

$$\text{ang}(\mathbf{v}, \mathbf{u}) = \arccos\left(\frac{\mathbf{v} \cdot \mathbf{u}}{\|\mathbf{v}\|\|\mathbf{u}\|}\right) = \arccos\left(\frac{\mathbf{v} \cdot \mathbf{u}}{\sqrt{\mathbf{v} \cdot \mathbf{v}}\sqrt{\mathbf{u} \cdot \mathbf{u}}}\right) \tag{21.117}$$

5. The angle between two vectors $\mathbf{u} = (u_1, u_2)$ and $\mathbf{v} = (v_1, v_2)$ in the plane measured in the positive direction from **u** to **v** (i.e., the angle of **v** relative to **u** measured in the positive direction) is given in the range 0–2π by (see Figure 21.1)

$$\begin{aligned} &\text{angp}(\mathbf{v},\mathbf{u}) = \text{ang}(\mathbf{v},\mathbf{u}) && \text{if } u_1 v_2 - u_2 v_1 \geq 0 \\ &\text{angp}(\mathbf{v},\mathbf{u}) = 2\pi - \text{ang}(\mathbf{v},\mathbf{u}) && \text{if } u_1 v_2 - u_2 v_1 < 0 \end{aligned} \tag{21.118}$$

or in the range $-\pi$—π by

$$\begin{aligned} &\text{angpn}(\mathbf{v},\mathbf{u}) = \text{ang}(\mathbf{v},\mathbf{u}) && \text{if } u_1 v_2 - u_2 v_1 \geq 0 \\ &\text{angpn}(\mathbf{v},\mathbf{u}) = -\text{ang}(\mathbf{v},\mathbf{u}) && \text{if } u_1 v_2 - u_2 v_1 < 0 \end{aligned} \tag{21.119}$$

6. The angle of a vector to the horizontal (axis x_1) is given (see Figure 21.1) by

$$\text{angh}(\mathbf{v}) = \text{angp}(\mathbf{v},(1,0)) \tag{21.120}$$

7. A rotation matrix $R(\alpha)$ is given (see Figure 21.2) by

$$R(\alpha) = \begin{pmatrix} \cos\alpha & -\sin\alpha \\ \sin\alpha & \cos\alpha \end{pmatrix} \tag{21.121}$$

8. A vector **v** can be rotated by an angle α if multiplied on the left by a rotation matrix as

$$R(\alpha)\cdot\mathbf{v} \tag{21.122}$$

9. The intersection between a straight line defined by a point **P** and vector **v** and a plane defined by a point **Q** and normal **n** is given (see Figure 21.3) by

$$\text{islp}(\mathbf{P},\mathbf{v},\mathbf{Q},\mathbf{n}) = \mathbf{P} + \frac{(\mathbf{Q}-\mathbf{P})\cdot\mathbf{n}}{\mathbf{v}\cdot\mathbf{n}}\mathbf{v} \tag{21.123}$$

The intersection between two straight lines, one of them defined by point **P** and vector **v** and the other defined by point **Q** and vector **u** is given (see Figure 21.3) by

$$\text{isl}(\mathbf{P},\mathbf{v},\mathbf{Q},\mathbf{u}) = \mathbf{P} + \frac{(\mathbf{Q}-\mathbf{P})\cdot\mathbf{n}}{\mathbf{v}\cdot\mathbf{n}}\mathbf{v} \tag{21.124}$$

where

$$\mathbf{n} = R(\pi/2)\cdot\mathbf{u} \tag{21.125}$$

10. A parabola tilted by an angle α to the horizontal, with focus at a point **F** and passing through a point **P** is given (see Figure 21.8) by

$$\text{par}(\alpha,\mathbf{F},\mathbf{P})=\frac{[\mathbf{P},\mathbf{F}]-(\mathbf{P}-\mathbf{F})\cdot(\cos\alpha,\sin\alpha)}{1-\cos\phi}(\cos(\phi+\alpha),\sin(\phi+\alpha))+\mathbf{F} \tag{21.126}$$

where ϕ is the parameter.

11. An ellipse with foci **F** and **G** and passing through a point **P** is given (see Figure 21.11) by

$$\text{eli}(\mathbf{F},\mathbf{G},\mathbf{P})=\frac{([\mathbf{F},\mathbf{P}]+[\mathbf{P},\mathbf{G}])^2-[\mathbf{F},\mathbf{G}]^2}{2([\mathbf{F},\mathbf{P}]+[\mathbf{P},\mathbf{G}])-2[\mathbf{F},\mathbf{G}]\cos\phi}(\cos(\phi+\alpha),\sin(\phi+\alpha))+\mathbf{F} \tag{21.127}$$

where

$$\alpha=\text{angh}(\mathbf{G}-\mathbf{F}) \tag{21.128}$$

and ϕ is the parameter.

12. A hyperbola with foci **F** and **G** and passing through a point **P** is given (see Figure 21.13) by

$$\text{hyp}(\mathbf{F},\mathbf{G},\mathbf{P})=\frac{([\mathbf{F},\mathbf{P}]-[\mathbf{P},\mathbf{G}])^2-[\mathbf{F},\mathbf{G}]^2}{2\left|[\mathbf{F},\mathbf{P}]-[\mathbf{P},\mathbf{G}]\right|-2[\mathbf{F},\mathbf{G}]\cos\phi}(\cos(\phi+\alpha),\sin(\phi+\alpha))+\mathbf{F} \tag{21.129}$$

where

$$\alpha=\text{angh}(\mathbf{G}-\mathbf{F}) \tag{21.130}$$

and ϕ is the parameter. Alternatively, if $U=2n\left|[\mathbf{F},\mathbf{P}]-[\mathbf{P},\mathbf{G}]\right|$, we can also write

$$\text{hyp}(\mathbf{F},\mathbf{G},U,n)=\frac{(U/2n)^2-[\mathbf{F},\mathbf{G}]^2}{U/n-2[\mathbf{F},\mathbf{G}]\cos\phi}(\cos(\phi+\alpha),\sin(\phi+\alpha))+\mathbf{F} \tag{21.131}$$

13. A winding involute passing through a point **P** and designed for a circle with center **F** and radius r is given (see Figure 21.15) by

$$\text{winv}(\mathbf{P},\mathbf{F},r)=r(\sin\phi,-\cos\phi)+(K-r\phi)(\cos\phi,\sin\phi)+\mathbf{F} \tag{21.132}$$

where

$$\begin{aligned} \phi_P &= \text{angh}(\mathbf{P} - \mathbf{F}) + \arcsin(r/[\mathbf{P},\mathbf{F}]) \\ K &= \sqrt{[\mathbf{P},\mathbf{F}]^2 - r^2} + r\phi_P \end{aligned} \tag{21.133}$$

14. An unwinding involute passing through a point **P** and designed for a circle with center **F** and radius r is given (see Figure 21.16) by

$$\text{uinv}(\mathbf{P},\mathbf{F},r) = r(-\sin\phi, \cos\phi) + (K + r\phi)(\cos\phi, \sin\phi) + \mathbf{F} \tag{21.134}$$

where

$$\begin{aligned} &\phi_P = \text{angh}(\mathbf{P} - \mathbf{F}) - \arcsin([r/\mathbf{P},\mathbf{F}]) \\ &\text{If } \phi_P < 0 \quad \text{then } \phi_P = 2\pi + \phi_P \\ &K = \sqrt{[\mathbf{P},\mathbf{F}]^2 - r^2} - r\phi_P \end{aligned} \tag{21.135}$$

15. A winding macrofocal parabola tilted by an angle α to the horizontal, with macrofocus having center **F** and radius r and passing through a point **P** is given (see Figure 21.20) by

$$\begin{aligned} \text{wmp}(\alpha,\mathbf{F},r,\mathbf{P}) = {} & r(\sin(\phi+\alpha), -\cos(\phi+\alpha)) \\ & + \frac{K + r(\sin\phi - 1 - \phi)}{1 - \cos\phi}(\cos(\phi+\alpha), \sin(\phi+\alpha)) + \mathbf{F} \end{aligned} \tag{21.136}$$

with

$$\begin{aligned} \phi_P &= \text{angp}(\mathbf{P} - \mathbf{F}, (\cos\alpha, \sin\alpha)) + \arcsin(r/[\mathbf{P},\mathbf{F}]) \\ K &= \sqrt{[\mathbf{P},\mathbf{F}]^2 - r^2(1 - \cos\phi_P)} + r(1 + \phi_P - \sin\phi_P) \end{aligned} \tag{21.137}$$

16. An unwinding macrofocal parabola tilted by an angle α to the horizontal, with macrofocus having center **F** and radius r and passing through a point **P** is given by

$$\begin{aligned} \text{ump}(\alpha,\mathbf{F},r,\mathbf{P}) = {} & r(-\sin(\phi+\alpha), \cos(\phi+\alpha) \\ & + \frac{K + r(\phi - 1 - \sin\phi)}{1 - \cos\phi}(\cos(\phi+\alpha), \sin(\phi+\alpha)) + \mathbf{F} \end{aligned} \tag{21.138}$$

with

$$\begin{aligned} &\phi_P = \text{angp}(\mathbf{P} - \mathbf{F}, (\cos\alpha, \sin\alpha)) - \arcsin(r/[\mathbf{P},\mathbf{F}]) \\ &\text{If } \phi_P < 0 \quad \text{then } \phi_P = 2\pi + \phi_P \\ &K = \sqrt{[\mathbf{P},\mathbf{F}]^2 - r^2(1 - \cos\phi_P)} + r(1 - \phi_P + \sin\phi_P) \end{aligned} \tag{21.139}$$

17. A winding macrofocal ellipse with macrofocus of center **F** and radius r, point focus **G** and passing through a point **P** is given (see Figure 21.29) by

$$\begin{aligned}\mathrm{wme}(\mathbf{F},r,\mathbf{G},\mathbf{P}) = {} & r(\sin(\phi+\alpha),-\cos(\phi+\alpha)) \\ & + \frac{(K-r\phi)^2 + 2fr\sin\phi - f^2 - r^2}{2(K-r\phi-f\cos\phi)} \\ & \times(\cos(\phi+\alpha),\sin(\phi+\alpha)) + \mathbf{F}\end{aligned} \tag{21.140}$$

with

$$\begin{aligned}\alpha &= \mathrm{angh}(\mathbf{G}-\mathbf{F}) \\ f &= [\mathbf{G},\mathbf{F}] \\ \phi_P &= \mathrm{angp}(\mathbf{P}-\mathbf{F},(\cos\alpha,\sin\alpha)) + \arcsin(r/[\mathbf{P},\mathbf{F}]) \\ t_P &= \sqrt{[\mathbf{P},\mathbf{F}]^2 - r^2} \\ K &= t_P + r\phi_P + \sqrt{f^2 + r^2 + t_P^2 - 2f(t_P\cos\phi_P + r\sin\phi_P)}\end{aligned} \tag{21.141}$$

18. An unwinding macrofocal ellipse with macrofocus of center **F** and radius r, point focus **G** and passing through a point **P** is given by

$$\begin{aligned}\mathrm{ume}(\mathbf{F},r,\mathbf{G},\mathbf{P}) = {} & r(-\sin(\phi+\alpha),\cos(\phi+\alpha)) \\ & + \frac{(K+r\phi)^2 - 2fr\sin\phi - f^2 - r^2}{2(K+r\phi-f\cos\phi)} \\ & \times(\cos(\phi+\alpha),\sin(\phi+\alpha)) + \mathbf{F}\end{aligned} \tag{21.142}$$

with

$$\begin{aligned}&\alpha = \mathrm{angh}(\mathbf{G}-\mathbf{F}) \\ &f = [\mathbf{G},\mathbf{F}] \\ &\phi_P = \mathrm{angp}(\mathbf{P}-\mathbf{F},(\cos\alpha,\sin\alpha)) - \arcsin\left(\frac{r}{[\mathbf{P},\mathbf{F}]}\right) \\ &\text{If } \phi_P < 0 \quad \text{then } \phi_P = 2\pi + \phi_P \\ &t_P = \sqrt{[\mathbf{P},\mathbf{F}]^2 - r^2} \\ &K = t_P - r\phi_P + \sqrt{f^2 + r^2 + t_P^2 - 2f(t_P\cos\phi_P - r\sin\phi_P)}\end{aligned} \tag{21.143}$$

19. A Cartesian oval that receives the rays emitted by a point source **F** immersed in a medium of refractive index n_1 and collimates them

(makes them parallel) into a medium of refractive index n_2 and that passes through a point **P** and the axis and the emitted parallel rays of which are tilted by an angle α to the horizontal is given (see Figure 21.33) by

$$\text{cop}(\mathbf{F},n_1,n_2,\mathbf{P},\alpha) = \frac{[\mathbf{F},\mathbf{P}](n_1 - n_2\cos\phi_P)}{n_1 - n_2\cos\phi}(\cos(\phi+\alpha),\sin(\phi+\alpha)) + \mathbf{F} \tag{21.144}$$

where

$$\phi_P = \text{ang}(\mathbf{P}-\mathbf{F},(\cos\alpha,\sin\alpha)) \tag{21.145}$$

20. A converging Cartesian oval with focus **F** immersed in a medium of refractive index n_1 and another focus **G** immersed in a medium of refractive index n_2, where $n_1 > n_2$ and passing through a point **P** is given by cco(**F**, n_1, **G**, n_2, **P**). See Figure 21.34 but note that points **F** and **G** are arbitrary and do not have to be on the x_1 axis.

 We first calculate

$$\begin{aligned} K &= n_1[\mathbf{F},\mathbf{P}] + n_2[\mathbf{P},\mathbf{G}] \\ f &= [\mathbf{F},\mathbf{G}] \\ \phi_C &= \begin{cases} \arccos(K/(fn_1)) & \text{if } |K| \le n_1 f \\ 0 & \text{if } |K| > n_1 f \end{cases} \end{aligned} \tag{21.146}$$

If $n_2 f < K < n_1 f$ and ang(**P** – **F**,**G** – **F**) $\le \phi_C$ the Cartesian oval is possible through point **P** and is given by

$$\text{cco}(\mathbf{F},n_1,\mathbf{G},n_2,\mathbf{P}) = \frac{Kn_1 - fn_2^2\cos\phi - n_2\sqrt{D}}{n_1^2 - n_2^2}(\cos(\phi+\alpha),\sin(\phi+\alpha)) + \mathbf{F} \tag{21.147}$$

where

$$\begin{aligned} D &= (fn_1 - K\cos\phi)^2 + (K^2 - f^2n_2^2)\sin^2\phi \\ \alpha &= \text{angh}(\mathbf{G}-\mathbf{F}) \end{aligned} \tag{21.148}$$

Note: If **F** is a point source in a high-refractive index medium of refractive index n_1, this curve refracts light to a point **G** in a low-refractive index medium of refractive index n_2. If **G** is a point source in a low-refractive index medium of refractive index n_2, this curve refracts light to a point **F** in a high-refractive index medium of refractive index n_1.

21. A diverging Cartesian oval with focus **F** immersed in a medium of refractive index n_1 and another focus **G** immersed in a medium of refractive index n_2, where $n_1 > n_2$ and passing through a point **P** is given by dco(**F**,n_1,**G**,n_2,**P**). See Figure 21.35 but note that points **F** and **G** are arbitrary and do not have to be on the x_1 axis.
We first calculate

$$\begin{aligned} K &= n_1[\mathbf{F},\mathbf{P}] - n_2[\mathbf{P},\mathbf{G}] \\ f &= [\mathbf{F},\mathbf{G}] \\ \phi_C &= \begin{cases} \arccos(K/(fn_1)) & \text{if } |K| \le n_1 f \\ 0 & \text{if } |K| > n_1 f \end{cases} \end{aligned} \tag{21.149}$$

If $K > -n_2 f$ and ang(**P** − **F**,**G** − **F**) $\ge \phi_C$ the Cartesian oval is possible through point **P** and is given by

$$\text{dco}(\mathbf{F},n_1,\mathbf{G},n_2,\mathbf{P}) = \frac{Kn_1 - fn_2^2\cos\phi + n_2\sqrt{D}}{n_1^2 - n_2^2}(\cos(\phi+\alpha),\sin(\phi+\alpha)) + \mathbf{F} \tag{21.150}$$

where

$$\begin{aligned} D &= (fn_1 - K\cos\phi)^2 + (K^2 - f^2 n_2^2)\sin^2\phi \\ \alpha &= \text{angh}(\mathbf{G} - \mathbf{F}) \end{aligned} \tag{21.151}$$

Note: If **F** is a point source in a high-refractive index medium of refractive index n_1, this curve refracts light as if it appears to come from a point **G** in a low-refractive index medium of refractive index n_2. If **G** is a point source in a low-refractive index medium of refractive index n_2, this curve refracts light as if it appears to come from a point **F** in a high refractive index medium of refractive index n_1.

22. A ray coming from a point **F** immersed in a medium of refractive index n_1 in a direction **v** is refracted at a point **P**
    a. Toward another point **G** immersed in a medium of refractive index n_2 (Figure 21.42a). The optical path length between **F** and **G** is $S = n_1[\mathbf{F},\mathbf{P}] + n_2[\mathbf{P},\mathbf{Q}]$
    b. And appears to diverge from a point **G** immersed in a medium of refractive index n_2 (Figure 21.42b). The "optical path length" between **F** and **G** is $S = n_1[\mathbf{F},\mathbf{P}] - n_2[\mathbf{P},\mathbf{Q}]$.

The point **P** (on a Cartesian oval) at which the refraction occurs is given by

$$\text{coptpt}(\mathbf{F}, n_1, \mathbf{v}, \mathbf{G}, n_2, S, \gamma) = \mathbf{F} + \frac{C_1 + \delta\sqrt{C_2\left(n_2^2 - n_1^2\right) + C_1^2}}{n_1^2 - n_2^2}\mathbf{v} \tag{21.152}$$

with

$$\begin{aligned} C_1 &= n_1 S + n_2^2(\mathbf{F} - \mathbf{G}) \cdot \mathbf{v} \\ C_2 &= S^2 - n_2^2(\mathbf{F} - \mathbf{G}) \cdot (\mathbf{F} - \mathbf{G}) \end{aligned} \tag{21.153}$$

and

$$\begin{aligned} \delta &= -\gamma \quad \text{for } n_1 > n_2 \\ \delta &= \gamma \quad \text{for } n_1 < n_2 \end{aligned} \tag{21.154}$$

In this function, parameter γ is

$$\begin{aligned} \gamma &= 1 \quad \text{when light converges to } \mathbf{F} \\ \gamma &= -1 \quad \text{when light appears to diverge from } \mathbf{G} \end{aligned} \tag{21.155}$$

We can also define

$$\text{ccoptpt}(\mathbf{F}, n_1, \mathbf{v}, \mathbf{G}, n_2, S) = \text{coptpt}(\mathbf{F}, n_1, \mathbf{v}, \mathbf{G}, n_2, S, 1) \tag{21.156}$$

for the converging case ($\gamma = 1$) and

$$\text{dcoptpt}(\mathbf{F}, n_1, \mathbf{v}, \mathbf{G}, n_2, S) = \text{coptpt}(\mathbf{F}, n_1, \mathbf{v}, \mathbf{G}, n_2, S, -1) \tag{21.157}$$

for the diverging case ($\gamma = -1$).

In the case where $n_1 = n_2 = n$ we have

$$\text{coptpt}(\mathbf{F}, \mathbf{v}, \mathbf{G}, n, S) = \mathbf{F} + \frac{(S/n)^2 - (\mathbf{F} - \mathbf{G}) \cdot (\mathbf{F} - \mathbf{G})}{2(S/n + (\mathbf{F} - \mathbf{G}).\mathbf{v})}\mathbf{v} \tag{21.158}$$

for both the converging and diverging cases.

23. A ray coming from a point **F** immersed in a medium of refractive index n_1 in a direction **v** is refracted into a direction perpendicular

to a plane wavefront defined by a point **Q** and a normal vector **n**. The optical path length between **F** and the plane wavefront is S. The point (on a Cartesian oval) at which refraction occurs is given (see Figure 21.43) by

$$\text{coptsl}(\mathbf{F}, n_1, \mathbf{v}, \mathbf{Q}, n_2, \mathbf{n}, S) = \mathbf{F} + \frac{S - n_2(\mathbf{Q} - \mathbf{F}) \cdot \mathbf{n}}{n_1 - n_2 \mathbf{v} \cdot \mathbf{n}} \mathbf{v} \tag{21.159}$$

24. An incident ray with direction **i** is reflected at a point on the surface with normal **n**. The reflected ray is given (see Chapter 16) by

$$\text{rfx}(\mathbf{i}, \mathbf{n}) = \mathbf{i} - 2(\mathbf{i} \cdot \mathbf{n})\mathbf{n} \tag{21.160}$$

where $\|\mathbf{i}\| = \|\mathbf{n}\| = 1$.

25. An incident ray with direction **i** is refracted at a point on the surface with normal $\mathbf{n}_S$ and that separates two media of refractive indices n_1 and n_2. The direction of the refracted ray is given by

$$\text{rfr}(\mathbf{i}, \mathbf{n}_S, n_1, n_2) = \begin{cases} \dfrac{n_1}{n_2}\mathbf{i} + \left(-(\mathbf{i} \cdot \mathbf{n})\dfrac{n_1}{n_2} + \sqrt{\Delta}\right)\mathbf{n} & \text{if } \Delta > 0 \\ \text{rfx}(\mathbf{i}, \mathbf{n}_S) & \text{if } \Delta \leq 0 \end{cases} \tag{21.161}$$

where the second case ($\Delta \leq 0$) refers to TIR and

$$\mathbf{n} = \begin{cases} \mathbf{n}_S & \text{if } \mathbf{i} \cdot \mathbf{n}_S \geq 0 \\ -\mathbf{n}_S & \text{if } \mathbf{i} \cdot \mathbf{n}_S < 0 \end{cases}$$
$$\Delta = 1 - \left(\frac{n_1}{n_2}\right)^2 [1 - (\mathbf{i} \cdot \mathbf{n})^2] \tag{21.162}$$

and $\|\mathbf{i}\| = \|\mathbf{n}_S\| = 1$.

26. Given an incident ray **i** and a refracted ray **r** of a surface that separates two media of refractive indices n_1 and n_2, the normal to the surface can be calculated as

$$\text{rfrnrm}(\mathbf{i}, \mathbf{r}, n_1, n_2) = \frac{n_1\mathbf{i} - n_2\mathbf{r}}{\| n_1\mathbf{i} - n_2\mathbf{r} \|} \tag{21.163}$$

where $\|\mathbf{i}\| = \|\mathbf{r}\| = 1$.

27. Given an incident ray **i** and a reflected ray **r** of a surface, the normal to the surface can be calculated as

$$\text{rfxnrm}(\mathbf{i},\mathbf{r}) = \text{rfrnrm}(\mathbf{i},\mathbf{r},1,1) = \frac{\mathbf{i}-\mathbf{r}}{\|\mathbf{i}-\mathbf{r}\|} \tag{21.164}$$

where $\|\mathbf{i}\| = \|\mathbf{r}\| = 1$.

References

1. Lawrence, J.D., *A Catalog of Special Plane Curves*, Dover Publications, New York, 1972.
2. Spencer, D.E., Montgomery, E.E., and Fitzgerald, J.F., Macrofocal conics as reflector contours, *J. Opt. Soc. Am.*, 55, 5, 1965.

Index

J

K

M

N

R

U

V

W

X

Printed and bound by CPI Group (UK) Ltd, Croydon, CR0 4YY
22/10/2024
01777613-0017